Innovative Water Resources Management – Understanding and Balancing Interactions Between Humankind and Nature

8th International Symposium on Integrated Water Resources Management (IWRM 2018)

Proceedings of the International Association of Hydrological Sciences (PIAHS) Volume 379, 2018

Beijing, China
13 – 15 June 2018

Editors:

Z. Xu
D. Peng
W. Sun
B. Pang
D. Zuo
A. Schumann
Y. Chen

ISBN: 978-1-5108-6372-9

Printed from e-media with permission by:

Curran Associates, Inc.
57 Morehouse Lane
Red Hook, NY 12571

Some format issues inherent in the e-media version may also appear in this print version.

Printed by Curran Associates, Inc. (2018)

International Association of Hydrological Sciences
Centre for Ecology and Hydrology
Wallingford, Oxfordshire OX10 8BB
United Kingdom

Phone: 441 491 692 442
Fax: 441 491 692 448

http://iahs.info/

Additional copies of this publication are available from:

Curran Associates, Inc.
57 Morehouse Lane
Red Hook, NY 12571 USA
Phone: 845-758-0400
Fax: 845-758-2633
Email: curran@proceedings.com
Web: www.proceedings.com

TABLE OF CONTENTS

https://doi.org/10.5194/piahs-379-463-2018

Preface: Innovative Water Resources Management in a Changing Environment – Understanding and Balancing Interactions between Humankind and Nature

Zongxue Xu[1,2], Dingzhi Peng[1,2], Wenchao Sun[1,2], Bo Pang[1,2], Depeng Zuo[1,2], Andreas Schumann[3], and Yangbo Chen[4]

[1]College of Water Sciences, Beijing Normal University, Beijing, China
[2]Beijing Key Laboratory of Urban Hydrological Cycle and Sponge City Technology, Beijing, China
[3]Institute for Hydrology, Water Management and Environmental Technology, Ruhr-University Bochum, 44780 Bochum, Germany
[4]Department of Water Resources and Environment, Sun Yat-Sen University, Guangzhou 510275, China

Correspondence: Zongxue Xu (zongxuexu@vip.sina.com)

Published: 5 June 2018

The 8th International Symposium on Integrated Water Resources Management (IWRM2018) brings together experts and scholars from nearly 30 countries and regions to present their researches, and discuss related issues of innovative water resources management. This symposium, part of the series of International Symposium on Integrated Water Resources Management (IWRM) organized by International Commission on Water Resources Systems (ICWRS) of IAHS (International Association of Hydrological Sciences), is focused on the innovative and integrated management of water resources.

During the past hundreds of years, various kinds of water resources systems have been developed to support efficient uses of water resources for humankinds worldwide. Such kinds of systems were mainly used to provide humankind with clean water for irrigation, industry and domestic uses. The history of water resources systems is part of the history of humankind itself. The development of water resources systems needs a careful analysis on water demands and environmental impact. Unfortunately, consideration on the water needs for nature is not enough for many water resources systems. This fact has resulted in the deterioration of water quality and even degradation of ecosystem in many river basins. Due to the rapid development of economics and society worldwide, present challenges for water resources systems are mostly unprecedented. A new effort to devise innovative technologies and solutions for present water problems should be paid much attention by research community including IAHS. These challenges for water resources systems highlight that the adaptation to the changing environment is an essential field of research, which is also involved in the Panta Rhei Scientific Decade (2013–2022) of IAHS. Efficient adaptation requires gaining a forward vision on future water demands and water availability. A deep understanding of the two-way interaction between nature and humankinds is needed to develop such vision.

This volume is the outcome of the 8th International Symposium on Water Resources Management of ICWRS, held at Beijing Normal University in Beijing China, on 13–15 June, 2018. Beijing Normal University, Beijing Key Laboratory of Urban Hydrological Cycle and Sponge City Technology, ICWRS of IAHS and Chinese National Committee for IAHS, jointly organized the symposium. More than 100 participants from Australia, Canada, China, France, Germany, Japan, USA, UK, and so on 30 countries and regions, assembled in Beijing. Fanghua Hao (Vice President of Beijing Normal University), Zhiyu Liu (Secretary-General, Chinese National Committee for IAHS) and Andreas Schumann (President of IAHS-ICWRS), delivered welcome speeches on behalf of Beijing Normal University, Chinese National Committee for IAHS and IAHS-ICWRS, respectively. Zongxue Xu, Vice President of IAHS-ICWRS, Chairman of Organizing Committee, chaired the opening ceremony.

Changming Liu, Kinzelbach Wolfgang, Kuniyoshi Takeuchi (former President of IAHS), Dan Rosbjerg, Jun Xia (former Vice-President of IAHS), Andreas Schumann

Published by Copernicus Publications on behalf of the International Association of Hydrological Sciences.

(President of IAHS-ICWRS), gave the Loucks lectures and keynote speeches on 13 June, respectively. In the afternoon of 13 June, on 14 June and in the morning of 15 June, there are three parallel sessions held in three meeting rooms, including the following themes:

- Water resources management under changing environment
- Socio-hydrology as the basis of water resources management
- Assessment of available water resources at regional and basin scales

During the preparation of the symposia, the secretariat received more than 140 abstracts. The Organizing Committee of IWRM 2018 reviewed all the submitted papers, and finally accepted 60 papers. The editors would like to thank all anonymous reviewers and Katrin Krüger at Copernicus for the production of this volume.

https://doi.org/10.5194/piahs-379-1-2018

Open Access
PIAHS

Development of an integrated model for the Campaspe catchment: a tool to help improve understanding of the interaction between society, policy, farming decision, ecology, hydrology and climate

Takuya Iwanaga[1], Fateme Zare[1], Barry Croke[1,2], Baihua Fu[1], Wendy Merritt[1], Daniel Partington[3], Jenifer Ticehurst[1], and Anthony Jakeman[1]

[1]Fenner School of Environment and Society, Australian National University, Canberra, Australia
[2]Mathematical Sciences Institute, Australian National University, Canberra, Australia
[3]National Centre for Groundwater Research and Training, Flinders University, Adelaide, Australia

Correspondence: Barry Croke (barry.croke@anu.edu.au)

Received: 20 December 2017 – Revised: 23 January 2018 – Accepted: 3 February 2018 – Published: 5 June 2018

Abstract. Management of water resources requires understanding of the hydrology and hydrogeology, as well as the policy and human drivers and their impacts. This understanding requires relevant inputs from a wide range of disciplines, which will vary depending on the specific case study. One approach to gain understanding of the impact of climate and society on water resources is through the use of an integrated modelling process that engages stakeholders and experts in specifics of problem framing, co-design of the underpinning conceptual model, and discussion of the ensuing results. In this study, we have developed such an integrated modelling process for the Campaspe basin in northern Victoria, Australia. The numerical model built has a number of components:

- Node/link based surface water hydrology module based on the IHACRES rainfall-streamflow model
- Distributed groundwater model for the lower catchment (MODFLOW)
- Farm decision optimisation module (to determine irrigation requirements)
- Policy module (setting conditions on availability of water based on existing rules)
- Ecology module (determining the impacts of available streamflow on platypus, fish and river red gum trees)

The integrated model is component based and has been developed in Python, with the MODFLOW and surface water hydrology model run in external programs, controlled by the master program (in Python). The integrated model has been calibrated using historical data, with the intention of exploring the impact of various scenarios (future climate scenarios, different policy options, water management options) on the water resources. The scenarios were selected based on workshops with, and a social survey of, stakeholders in the basin regarding what would be socially acceptable and physically plausible options for changes in management. An example of such a change is the introduction of a managed aquifer recharge system to capture dam overflows, and store at least a portion of this in the aquifer, thereby increasing the groundwater resource as well as reducing the impact of existing pumping levels.

1 Introduction

Effective, and holistic, water management is contingent on understanding the stressors that affect water resources. Such stressors may come from a variety of physical and social (anthropogenic) sources. In the field of water resource management physical influences typically include the hydrology, geology, ecology/biology, and climatic processes. Social systems that influence and affect water management include agricultural enterprises, water policies, and the social factors that influence the acceptability of water use and management practices. Due to the complexity and interconnected nature of this system of systems, water resource managers and researchers often turn to integrated models to assess potential management actions within the specific context.

Managing such system of systems requires the consideration of a wide range of factors across the interconnected physical and social domains. Integrated Assessment (IA) should not be conducted by individual disciplines in isolation as the issues faced do not fall neatly into traditional academic disciplines. Each represented system may have differing problem frames which influence and affect each other due to the interconnected nature of socio-enviro systems. Therefore, the development of Integrated Assessment Models (IAM) should not be conducted by model developers alone but in conjunction with experts and stakeholders so that the specifics of the problem frame(s) are accounted for. Collectively these problem frames make up a management context, a term used here to refer to the physical properties, the social elements and activities that have influence, and the management scope and objectives of concern.

The lower Campaspe basin in the north-central region of Victoria, Australia is an example of a highly interconnected socio-enviro system. We describe herein an iterative integrative process used to develop an Integrated Assessment Model (IAM) suited for the specific management context. Model development was continuously informed by stakeholder and expert knowledge throughout the process from initial conceptualization through to completion. As new information and knowledge became available and challenges encountered, stakeholders and experts were re-engaged to update the problem frames and model design; a beneficial co-design process. Incorporation of feedback at each iterative stage then helps to ensure that the model remains relevant for its given purpose and to the stakeholders themselves.

The principal aim of the model is to inform stakeholders of the impacts of a range of possible combinatory policy and on-farm water management decisions under a variety of climate conditions. These collectively represent a set of possible "futures". The model will be used in an exploratory manner through which a multitude of such possible "futures" are generated. The combination of factors that led to positive (or at least effective compromises) and negative future conditions can then be identified and communicated to stakeholders through this exploratory process. Because the geophysical, geographical, and social elements are found in a range of contexts, this iterative process is a generally applicable integrative water management approach.

2 Management context

Defining the management context through systems analysis with the aid of stakeholder knowledge is a crucial first step in an integrated assessment process, and a key aspect of Integrated Assessment Modelling (IAM, as in Jakeman and Letcher, 2003). Kraft et al. (2010) argues the importance of stakeholder involvement as incorporation of local domain knowledge ensures that key features of the management context are captured and subsequently represented in the model. Stakeholders further represent an important source of local knowledge which may in turn drive both information need and data accessibility, as well as playing an important role in validating model outputs (Krueger et al., 2012).

The involvement of stakeholders increases the transparency of the development process as it is exposed for critique and review by stakeholders. Through this stakeholder engagement process the scope and objectives of the model can be iteratively developed and refined so that the final model is suitable and relevant (and therefore useful) for the end purpose and users (Jakeman and Letcher, 2003). The process for gathering information and knowledge of the management context and the subsequent influences and implications on the model design and approach is described in later sections.

One motivation for this study was the adoption of the Murray-Darling Basin Plan developed under the Australian Government Water Act 2007. The Basin Plan defines environmental objectives which includes increasing water availability for the environment. To this end the Basin Plan sets Sustainable Diversion Limits, which will be applicable from 1 July 2019 for both ground and surface water (NCCMA, 2014a).

2.1 The Lower Campaspe

The Lower Campaspe catchment covers the northern portion of the Campaspe catchment in North-Central Victoria, an area that is approximately 150 km long and 25 km wide (NCCMA, 2014b), and is itself a part of the Murray-Darling Basin. The Campaspe River starts from the Great Dividing Range in the south, flowing in a northerly direction into Lake Eppalock from which the Lower Campaspe River begins. The Lower Campaspe River then continues northwards (downstream) into the Murray River which flows in a westerly direction. Population centres along the Lower Campaspe River include (from south to north) Axedale, Barnadown, Elmore, Rochester, and Echuca. The Lower Campaspe River itself is highly regulated by the operation of a dam at Lake Eppalock, the Campaspe Weir and Siphon located north of

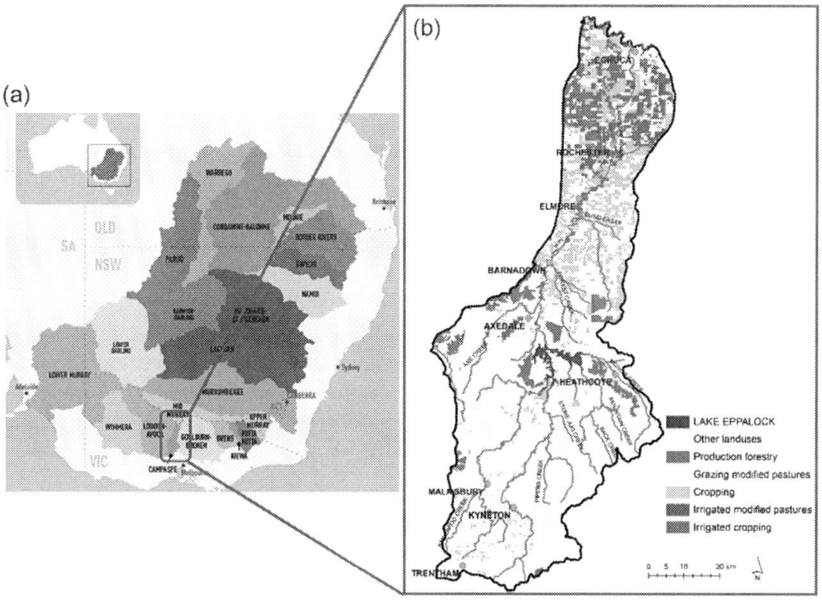

Figure 1. Panel **(a)** depicts the Campaspe catchment in the Murray-Darling Basin (adapted from MDBA, 2017) while **(b)** shows the Campaspe catchment proper with land use in the region.

Rochester (NCCMA, 2014b). A map indicating the catchment location is shown in Fig. 1.

Climate

The lower Campaspe is reported to be a dry semi-arid area which is evident in the historic rainfall records, with the median yearly rainfall being 434 mm (see Fig. 2). Two notable dry periods are identifiable in the historic rainfall records for the past 30 years which have influenced irrigators and water management. The first is a severe drought that occurred during 1982/1983 during which almost no rainfall occurred during the growing season resulting in severe (wheat) crop loss across eastern Australia (ABS, 1988; Arad and Evans, 1987; BoM, 2009). The second was the millennium drought, described as the worst drought on record for southeast Australia, defined as starting in 2001 with the drought eventually broken in 2009 (Van Dijk et al., 2013).

Climate projections for Northern Victoria, of which the Campaspe catchment is a part of, describe drier conditions with rainfall expected to decrease compared to the historic 20-year average. Decreases in mean rainfall of 12–13 % across south and east Australia compared to the 100-year average (1900–2000) have already been experienced within the first decade of the new millennium (2001 to 2009, Van Dijk et al., 2013).

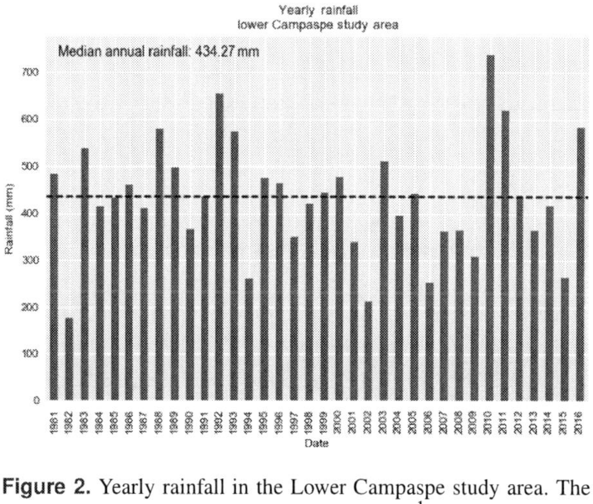

Figure 2. Yearly rainfall in the Lower Campaspe study area. The median amount was found to be 434 mm yr^{-1} (indicated by the dashed black line).

2.2 Hydrology

The area of the Campaspe basin is 4179 km^2, with a river length of 220 km, and a mean annual streamflow volume of 352 GL. The elevation in the southern part of the basin is around 600 m AHD (Australian Height Datum), with mean annual rainfall up to 1000 mm, and estimated mean annual pan evaporation of approximately 1300 mm. Near the catchment outlet (elevation 98 m AHD), the mean annual rainfall is approximately 430 mm, while the estimated mean annual pan evaporation is approximately 1700 mm. The main storage in the basin is Lake Eppalock, which has a catchment area of 2124 km^2, a storage capacity of 304 GL, and is located about 135 km from the catchment outlet at an elevation of about 160 m AHD. A further 3 large storages are located on the Coliban River (upstream of Lake Eppalock), with a total storage of 70 GL (BoM, 2017; GM-W, 2017a, b; MDBA, 2017).

2.3 Hydrogeology

The Campaspe region comprises the recent Coonambidgal Formation incised by the Campaspe River through the Shepparton formation, Parilla/Loxton sands, Newer Volcanic Basalts, with the primary productive aquifers of the region in the Calivil Formation and Renmark Group (collectively known as the Deep Lead)which overlay the Palaeozoic bedrock. The majority of the lower Campaspe consists of the Shepparton formation and the Deep Lead. The Deep Lead aquifers are the primary source of groundwater in the lower Campaspe irrigation areas, in which the Shepparton Formation has low permeability and is not very transmissive. Further details of the local hydrogeology may be found in Chiew et al. (1995).

2.4 Stakeholders

The local water corporation, Goulburn-Murray Water (GM Water), manages both surface and groundwater resources in the Lower Campaspe. Management includes the operation of the dam, water delivery infrastructure maintenance and investment, and the water accounts and licences in the region. GM Water is additionally responsible for determining the amount of water allocations – a percentage of water that an irrigator is entitled to – during each irrigation season.

The Department of Economic Development, Jobs, Transport and Resources (EcoDev) is a state level Government department that is interested in the water resource management and policy aspects, as well as providing advice and assistance to farmers regarding on-farm activities. The North Central Catchment Authority (NCCMA) and an expert from the Australian Platypus Conservancy were engaged for their input and feedback on the ecological system. Farmers themselves are an important stakeholder group to include as they will be impacted by any policy and climatic changes as well as being an important influencer of ecological and recreational water availability. Recreational users of the reservoir at Lake Eppalock were involved due to concerns that over-allocation of water for agricultural purposes have, and will exacerbate, negative impacts on recreational activities.

To gain further insights into the socio-agricultural system Ticehurst and Curtis (2016, 2017) conducted a survey of irrigators during 2016. The survey gathered responses from 254 participants (of 754 surveys sent out) that were later determined to be representative of irrigators in the region. The findings relevant to the model development process are repeated here, however readers are directed to Ticehurst and Curtis (2016, 2017) for further detail on the survey process.

These stakeholder groups were all engaged with through a series of workshops and discussions from late 2015 onwards. The latest workshop was run in October 2017, and another scheduled for March 2018. These stakeholder engagement activities aided in the selection of scenarios which describe plausible, and socially acceptable, options for changes in water management. Examples of such scenarios include the introduction of a managed aquifer recharge system to capture dam overflows for storage in the aquifer for use in times of water scarcity. Recharging the groundwater resource in this manner increases the availability of groundwater as well as reducing the impact of existing pumping levels.

3 Modelling process

The model development process followed a participatory process in which multi-disciplinary practitioners engaged with stakeholders. Through this process the model and its purpose was collaboratively defined and developed. Stakeholders play an additional important role in the development of scenarios of interest and validating model scope and behaviour (Krueger et al., 2012). Participatory engagement elicited a key set of management objectives including holistic management of water resources to improve crop yields, reliability of water availability, and beneficial improvements to environmental and socio-economic outcomes. Inclusion of stakeholders in the design and development process additionally fosters trust between stakeholders and modellers, and as a consequence model results (Franzén et al., 2011).

Another perspective is that of a software developer, as model implementations will largely be expressed in computer code. It is perhaps of interest to note that both software and model development best practices suggest an iterative process and arrived at these processes seemingly independently of each other. Sletholt et al. (2012) for example details and identifies software development practices that can be found in the model development process that have direct counterparts to software development practices.

A key point of interest is that iterative development is regarded as best practice in both model and software development paradigms. From a model developers' perspective continuous engagement with stakeholders has a hand in early detection and correction of faulty assumptions (Jakeman et al., 2006). Continuous exposure to the development process and incorporating feedback can drive stakeholder acceptance of the model by ensuring that the modelling process is transparent and relevant (Chan et al., 2010; Voinov and Gaddis, 2008). Jakeman et al. (2006) suggest an iterative approach to model development, where progress is reviewed at various steps, and part of the process repeated if issues are found. For software developers, iterative processes enable continu-

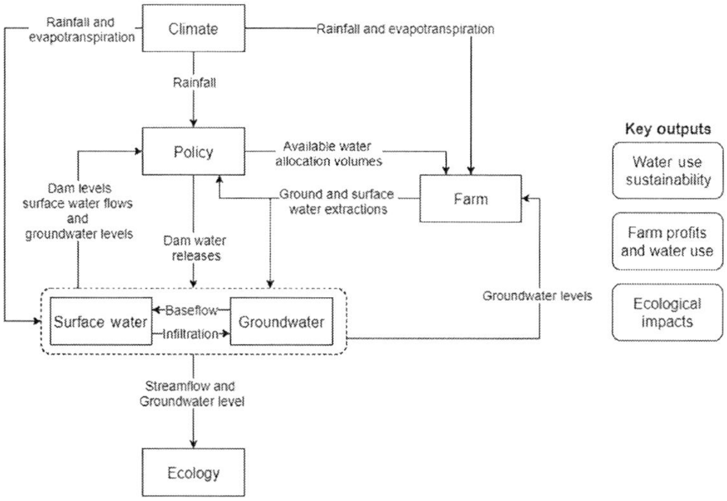

Figure 3. The interactions between component models and key model outputs. The dashed box around surface and groundwater models was inserted to simplify the model diagram and is not intended to indicate a separate coupled model.

ous validation of model implementation and the adjustments necessary to incorporate stakeholder feedback (Slethholt et al., 2012). To this end a component-based approach was applied for the implementation of the model, and so the development process could be described as applying a component-based participatory approach.

Component-based approaches compose a collection of compartmentalized models coupled loosely through a common framework. Loose coupling is achieved using specifically defined interfaces which handle data exchange and refers to the fact that the connections, and thus the feedbacks, between models are no longer "hardwired" to specific models. Malard et al. (2017) refers to the use of interfaces as a "wrapper approach" wherein the individual component models are "wrapped" and interactions channelled through the interfaces. Benefits of such an approach include the ability to reuse or "swap" a given component model for another (either new or pre-existing) as the need arises (de Kok et al., 2015). Changes within a model that do not affect the interface (i.e. the inputs and outputs) are safely abstracted and as such do not propagate and affect other component models. Incorporation of stakeholder feedback then becomes less problematic due to this model compartmentalization allowing model developers to focus on the modelling process instead of issues that may arise from direct coupling. Expected behaviour can then be verified through testing and comparisons against previous model outputs. As a consequence modellers are then able to progress through the iterative loops at a faster pace.

4 Model framing

To support these water reforms Federal and State Governments invested heavily in a modernization program in 2007 (State of Victoria, 2011), what is now known as the Connections Project and managed by GM Water. This infrastructure investment was described as the largest investment in irrigation infrastructure by the Australian Government (a total of AUD 1.1 Billion as reported in Bowler, 2015). A primary aim of the Connections Project was to improve the efficiency of water delivery and on-farm water use to meet sustainable water use goals as defined in the national Murray-Darling Basin Plan introduced in 2012 (Bowler, 2015).

Conjunctive use of water resources were identified by Ticehurst and Curtis (2017) as one method of improving water availability in the catchment. Here conjunctive water use was broadly defined as the multi-use of water sourced from both surface and groundwater for agricultural, recreational, and environmental purposes.

5 Model components

Each component represents a system of interest which collectively describes a system of systems. The developed integrated model represents a sociohydrological-environmental system including a farm model, surface water representing the lower Campaspe River and tributaries, groundwater hydrology, and a water management policy model. A climate component is also included which serves to provide the necessary rainfall and evapotranspiration data at the requisite spatial and temporal scales. The component models are coupled through a common framework developed in the Python programming language. Component models are not required to be developed in the same language as the framework as Python has robust language interoperability capabilities. The interactions between component models through their interfaces are depicted in Fig. 3.

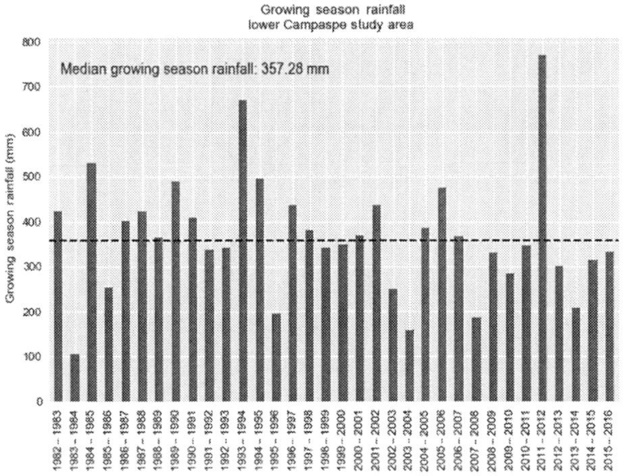

Figure 4. Growing season rainfall over the 1982–2015 growing seasons. Median growing season (dashed black line) was found to be 357.28 mm, below the usual growing season rainfall of 420 mm (see Fig. 5 below).

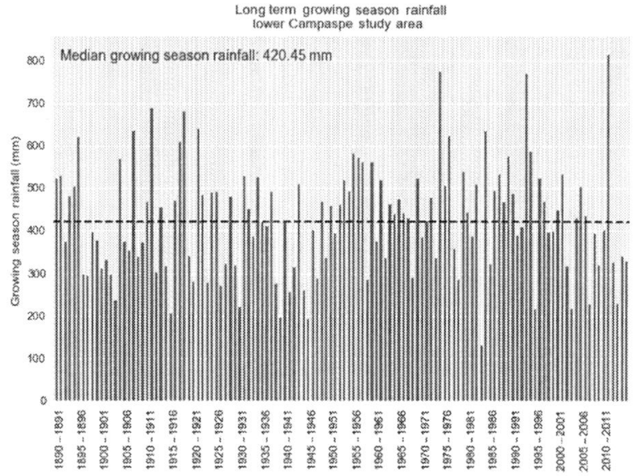

Figure 5. Long term growing season rainfall. Median in-season rainfall was found to be 420.45 mm (dashed black line).

In this section examples of the implications and influences from stakeholder feedback on each of the component models are given. The model continues to be developed in light of findings described herein, and as such is not made publicly available, although public release is intended. Additionally the data comes from various sources and so possible issues regarding intellectual property and data ownership will have to be cleared before public availability is possible. Model development utilizes version control which allows for the release of the model (in its current and future state) and requisite data at a later date.

5.1 Climate

Ticehurst and Curtis (2016) found that over 80 % of farmers surveyed believed that the impact of drought and changing rainfall patterns were important or very important. This finding in conjunction with the observed decrease in rainfall (see Sect. 2.1.1) shows that it is necessary to consider the impact of further climate variability. To this end (30 year) historic and future climate data were sourced via Climate Change in Australia (https://www.climatechangeinaustralia.gov.au, CSIRO, 2017). These datasets are described as being application ready. Long term ($\sim$ 100 years) climate records were developed through the use of interpolated historic rainfall and pan evapotranspiration data (see Vaze et al., 2011). The recent decrease in rainfall is evident within a typical growing season (defined as May to February). The median in-season rainfall during 1982 to 2016 was found to be 357 mm (see Fig. 4), compared to the reported usual growing season rainfall of 400 to 500 mm (EcoDev, 2015) and as indicated in the long term growing season rainfall records (see Fig. 5).

5.2 Surface water

The surface water module estimates the flows and water levels at selected nodes in the Campaspe Catchment. The nodes have been selected based on the location of gauges with suitable data, taking into consideration the needs of the integrated model (Fig. 6). As the focus of the integrated model is the lower Campaspe Catchment (below Lake Eppalock), the majority of the nodes are located in that region. To model the surface water flows, this means having information on releases and spills from Lake Eppalock, thereby requiring an estimate of the inflows to the reservoir. The resulting nodes are shown in Fig. 7. The surface water flows also depend on interaction with the groundwater, requiring a comparison of surface water levels with groundwater levels. This means that the surface water module needs to estimate the surface water levels at the nodes, and that this information is passed to the groundwater model in order to estimate the infiltration loss/baseflow contribution to surface water flow.

The surface water module has three components: a rainfall-streamflow model, a routing module and a rating curve module. Inputs required by the model are climate data (rainfall and potential evaporation), as well as estimates of the groundwater/surface water interactions (from the groundwater module), releases from Lake Eppalock reservoir (from the policy and the farm modules) and extractions from the surface water flows (from the farm model).

The rainfall-streamflow model used here is a variant of the IHACRES model, incorporating a non-linear loss module which converts rainfall into effective rainfall (rainfall that contributes to streamflow), and a unit hydrograph module that represents the dynamics of the water moving through the catchment (river network and landscape). The non-linear loss module used is based on the CMD version of the non-linear module (Croke and Jakeman, 2004), modified to produce two inputs to the unit hydrograph module (Croke et al., 2015): a

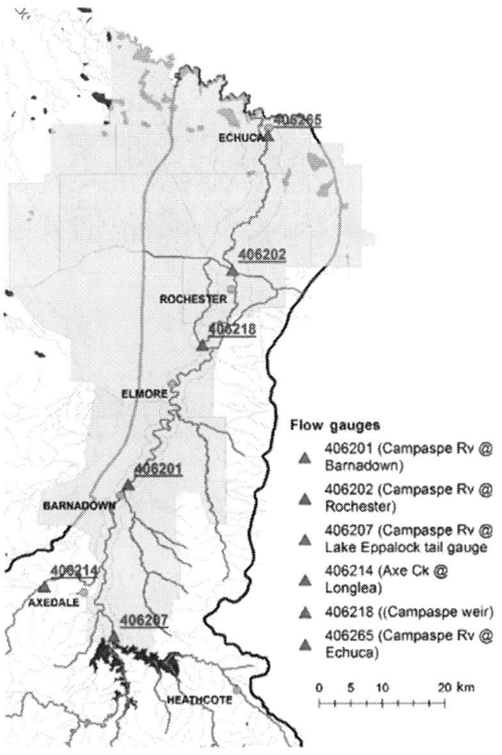

Figure 6. Flow gauges in the lower Campaspe.

contribution to the quick flow component (u_k) and a contribution to the slow flow component (r_k). This permits the model to partition effective rainfall between the two components based on the modelled catchment moisture status. The unit hydrograph module comprises two exponentially decaying stores arranged in parallel (a quick and a slow flow component), modified from the original to take the inputs to each store directly from the CMD module outputs.

An exponentially decaying store is also used to route the flows between nodes (a lag-route approach, Croke et al., 2006). In both the routing and the rainfall-streamflow models, the impact of losses from the river network are taken into consideration using the approach of Ivkovic et al. (2014). The rating module makes use of the rating curve data available at most gauge sites (the exception is gauge 406218, where only water level data is available).

5.3 Groundwater

The groundwater flow module is used to estimate the surface water-groundwater exchanges along the Campaspe River between flow gauges, and to provide information on the groundwater levels at specific locations as well as groundwater levels averaged over larger areas. The groundwater flow module interacts with the hydrology, farm, ecology and policy modules as detailed in Fig. 7.

The first stakeholder engagement workshop and subsequent communication with stakeholders (in particular, a local hydrogeologist) led to the definition of the groundwater model boundary, delineation of the hydrogeological units and provision of input data for groundwater pumping as well as observational head and chemistry data. Defining the model boundaries through this engagement ensured that the area covered as well as representation of hydrogeological units was consistent with their interpretation of the system and met the requirements for areas of interest. Furthermore the boundary conditions used came out of this initial consultation, in particular the consideration of the Campaspe River and smaller inflowing tributaries, the latter of which are not represented in the groundwater model due to their ephemeral nature and low flows.

The groundwater flow model of the Lower Campaspe Valley region is a finite difference representation. The model was constructed with Python scripts utilising Flopy (Bakker et al., 2016), and uses MODFLOW NWT (Niswonger et al., 2011). Representation of the hydrogeologic units (HGUs) is based on rasters (100 m resolution) from the Victorian Aquifer Framework (DSE, 2012). The model is made up of 7 layers, with a horizontal resolution of 1 km, and vertical spacing of the model grid informed by the HGU rasters. The 5 km resolution was chosen for computational speed to avoid the groundwater model becoming a computational bottleneck for the integrated model. Some HGUs span multiple layers where they are not overlain by other HGUs. There are 41 209 active cells within the model.

The groundwater model is driven by rainfall, river stage, groundwater extraction via pumping wells and groundwater head data via a series of boundary conditions shown in Fig. 7. Recharge in the model is implemented in the top layer of the model with the RCH package, and is calculated as a reduction of rainfall using a rainfall reduction parameter, and as such evapotranspiration is not directly modelled. River boundary conditions are implemented using the RIV package for the Campaspe River and Murray River. To allow outflow below the Murray River, through the subsurface in the north of the catchment, a general head boundary condition is implemented with the GHB package.

The model was calibrated using PEST (Doherty, 2016) to groundwater head data by modifying the HGU properties (i.e. hydraulic conductivity, specific yield and specific storage) and also a rainfall reduction parameter, applied statically from the period 1966–2015 and based on monthly stress periods. Initial conditions for the model were established by running the model in steady-state using long-term average rainfall and river stages.

As depicted in Fig. 7, the groundwater model is forced by:

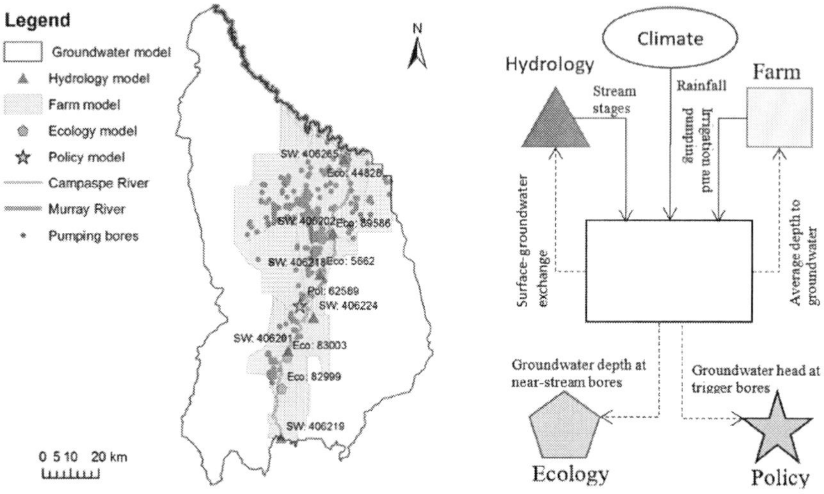

Figure 7. Groundwater model components and model area as well as points and interactions with other component models from the perspective of the groundwater model.

- Distributed rainfall (to be reduced through the rainfall reduction parameter) and irrigation water from the Farm model;
- Pumping volume from the Farm model (uniformly applied across pumps in the area);
- River stages from the Hydrology model
- After running each daily time-step, the model returns the:
- Surface water-groundwater exchange along reaches of the river consistent with the Hydrology model
- the average depth to groundwater for the Farm model
- depth to groundwater at key sites dictated by the Ecology model
- groundwater head at trigger level bores as dictated by the Policy model

Increases and decreases to pumping driven by the Farm model were applied to relevant wells within each farm zone. Surface water-groundwater responses lag behind the surface water forcing from the Hydrology model due to the use of a sequential coupling; it was assumed that a daily lag would not create significant differences in model behaviour. Outputs from the groundwater model, while not precise at the scale of local wells due to model resolution, were fit for purpose for indicative average groundwater levels at points of interest. In the case of the Ecology model, this is subject to the most variability as the levels are near-stream where the depth to groundwater table can change rapidly as it converges to the river. For the Policy model the trigger bores are chosen to be indicative of larger scale behaviour and hence the use of the average head in cells that correspond with the trigger bores is deemed adequate.

5.4 Policy

The current policy setting in the Campaspe is quite sophisticated reflecting extensive water reforms which introduced water trading, carryover, and environmental water provisions (Alston and Whittenbury, 2011; McKay, 2005; Wheeler and Cheeseman, 2013). The policy component of the integrated model provides a representation of policies determining the water allocation and carryover for entitlement holders (farm and environment). Use of these policies as a scenario supports further investigation of the implications and viability, as well as the opportunities, of the given policy condition(s) in the context of climate variability. The design of the policy component was such that it would allow scenarios that fit with current policies (e.g. increased groundwater use) but also the capability to explore alternate policy futures (e.g. conjunctive management of surface and groundwater, Managed Aquifer Recharge, and inter-catchment transfers). The latter include some of the conjunctive use opportunities explicitly identified with Campaspe stakeholders. For example, groundwater and surface water are managed separately in the current policy space. One option identified with Campaspe stakeholders was the temporary relaxation of groundwater restriction trigger levels during dry times when surface water allocations are low, with compensatory actions to increase recharge when climate conditions improve.

Within the current policy setting, groundwater use can be increased as most irrigators surveyed in the region are primarily reliant on surface water resources. Reported figures include 91 % of irrigators holding surface water licences compared to 22 % that additionally hold groundwater licences. Groundwater use historically reach a maximum of 60 % of allocated volumes, although this has increased to 80 % in recent times (2016 water usage, reported in Ticehurst and Curtis, 2017).

5.5 Farming

The farm component was developed with the aim of enabling investigation into the effect of water policy and water availability on farm financial performance under variable climate scenarios. Key attributes that are represented include the crop – currently one of wheat, barley or canola and tomato – irrigation system, pumping systems and soil types. The farming system is represented in a lumped manner with the study area divided into 12 zones configured to represent a mix of applicable surface and groundwater policy, water entitlements and usual cropping practices.

The principal agricultural enterprise in the Lower Campaspe is dairy farming with 55 % of land use devoted to annual and perennial pastures, 70 % (i.e. 38.5 % of reported farming area) of which is irrigated. Cereal cropping amounted to 35.8 % of land use, although the majority of this (68 %, i.e. $\sim$ 24 % of reported farming area) is dryland. Dairy farming is to be represented in the model through the use of an indicator crop to represent annual and perennial pasture crops and discussions with local experts are ongoing to determine how best to implement this.

Historically, the Campaspe region was an irrigation intensive area however most irrigators (90 %, concentrated in the middle of the study area) stopped irrigation practices in 2010 (NVIRP, 2010). This exit occurred during the millennium drought period (2001–2009) during which irrigators' water allocations were significantly reduced (NCCMA, 2014a). Irrigation is currently concentrated in the lower portion of the catchment the northern area surrounding Echuca, with dryland cropping in the mid and upper areas. A return to irrigation practices in the future remains a possibility due to the network of accessible irrigation infrastructure modernized under the Connections Project.

In the North Central region flood irrigation is the most common irrigation system in use accounting for 99 % of irrigated area (Ash, 2006). Flood irrigation is said to be 50–80 % water use efficient, meaning that 50–20 % of water applied to the field is lost (Clemmens, 2000; Finger and Morris, 2005; Tennakoon et al., 2013). Of those surveyed 77 % of respondents reported having undertaken additional improvements to flood irrigation such as laser grading and tail-water reuse, increasing the water use efficiency. Flood irrigation was then modelled as being 70 % water use efficiency based on this information. Other improvements can be achieved through the adoption of a piped system or investing in spray irrigation which is said to be 80 % water use efficient (Clemmens, 2000; Finger and Morris, 2005).

Outside of the survey, further information was gained through stakeholder engagement. It was highlighted, for example, that the choice to invest in a more water efficient irrigation system depends on the soil type. As such, generating a suitable representation of the soil textures in the modelled farming zones becomes a necessity and acts as a constraint to the choice of irrigation system adopted at each zone. It was also initially assumed that the vast majority of pumping systems in the area were diesel based rather than electric due to the substantial capital costs involved in developing the necessary infrastructure to operate electric pumps. This assumption was also corrected with local knowledge – electric pumping is in reality quite common and is used over the weekends due to off-peak electricity prices. The model will be modified to incorporate this elicited information.

5.6 Ecology

Management decisions that affect the lower Campaspe ecosystems flow on to the Murray River as the Campaspe flows into the Murray. The local ecology has historically been neglected due to over allocation of water resources for agricultural purposes. Decline in riverine health have been reported over the years including substantial decreases in biodiversity (MDBA, 2012; NCCMA, 2014b). The Murray-Darling Basin Plan includes provisions for increased environmental flows to support ecosystem maintenance and recovery (Bowler, 2015; GM-W, 2013; Hughes et al., 2015), and to meet these obligations up to 75 GL of water savings through infrastructure improvements through the GM Water Connections Project (formerly NVIRP) were intended for environmental purposes (NVIRP, 2011).

The local ecology along the Lower Campaspe River include communities of River Red Gum, a eucalyptus tree that is considered iconic (NCCMA, 2014b), platypus colonies, and two native fish populations: the Murray Cod and Golden Perch. Conceptualization and design of the model incorporated feedback from ecology experts from the NCCMA and the Australian Platypus Conservancy. Stakeholder feedback in combination with prior ecological studies and data availability resulted in the development of methods to generate indices that indicate the suitability of water flow for these flora and fauna. These consist of three indices for the River Red Gum which represent the suitability of groundwater availability and surface water flows for the maintenance and regeneration of the iconic tree. The fish indices capture key flow requirements as recommended in the Campaspe River Environmental Water Management Plan (NCCMA, 2014a). Indices developed for platypi indicate flow conditions that sustain food supply and movement, breeding cycles, and the avoidance of burrow flooding during the mating season.

6 Conclusion

Both model and software development best practices recommend working within an iterative cycle that moves the project towards a continually (re)defined goal, informed by stakeholders. Including stakeholders in the iterative development of integrated models was found to be useful in ensuring model validity, relevance, transparency and acceptability. Participatory engagement acts as a peer review process within each iteration of the model development cycle whilst also fostering trust between all participants, modellers and stakeholders alike.

Component-based development processes were found to be complementary to the participatory modelling approach. Throughout each iteration the implementation of the described component models were influenced by stakeholder knowledge and information. The compartmentalization of models that collectively represent a system of systems disentangles their implementation allowing specific and targeted modifications based on stakeholder feedback. Such changes then do not propagate throughout the model as a whole, allowing modellers to progress through each iteration quicker whilst simultaneously ensuring that the model development process is transparent.

Data availability. Climate time series data presented here is available from the Australian Bureau of Meteorology http://www.bom.gov.au/.

Competing interests. The authors declare that they have no conflict of interest.

Special issue statement. This article is part of the special issue "Innovative water resources management – understanding and balancing interactions between humankind and nature". It is a result of the 8th International Water Resources Management Conference of ICWRS, Beijing, China, 13–15 June 2018.

Acknowledgements. The primary author is supported through an Australian Government Research Training Program (RTP) Scholarship. The work was carried out as part of the Murray-Darling Basin Authority's partnership with the National Centre for Groundwater Research and Training (NCGRT). The authors would like to thank representatives from Goulburn-Murray Water and the Victorian Department of Economic Development, Jobs, Transport and Resources for their valuable insights and assistance over the course of this work. The authors would like to additionally thank the reviewers for their valuable comments and suggestions to improve the quality of this paper.

Edited by: Dingzhi Peng
Reviewed by: Qinglan Li and one anonymous referee

References

ABS: Drought in Australia, available at: http://www.abs.gov.au/ausstats/abs@.nsf/0/068F13BCCD03F27BCA2569DE001F1072?OpenDocument (last access: 26 February 2018), 1988.

Alston, M. and Whittenbury, K.: Climate change and water policy in Australia's irrigation areas: a lost opportunity for a partnership model of governance, Environ. Polit., 20, 899–917, https://doi.org/10.1080/09644016.2011.617175, 2011.

Arad, A. and Evans, R.: The hydrogeology, hydrochemistry and environmental isotopes of the Campaspe River aquifer system, North-Central Victoria, Australia, J. Hydrol., 95, 63–86, https://doi.org/10.1016/0022-1694(87)90116-8, 1987.

Ash, L.: Irrigation Farm Survey 2004/2005 Summary Extract from the final document 17/10/06, NSW Department of Primary Industries, Sydney, Australia, 2006.

Bakker, M., Post, V., Langevin, C. D., Hughes, J. D., White, J. T., Starn, J. J., and Fienen, M. N.: Scripting MODFLOW Model Development Using Python and FloPy, Groundwater, 54, 733–739, 2016.

BoM: Western Australia to the Rescue, avaiable at: http://pandora.nla.gov.au/pan/96122/20090317-1643/www.bom.gov.au/lam/climate/levelthree/c20thc/flip4.html (last access: 11 May 2016), 2009.

BoM: Climate and past weather information, available at: http://www.bom.gov.au/climate/ (last access: 18 December 2017), 2017.

Bowler, M.: Report for the Department of Agriculture and Water Resources – Goulburn-Murray Water Connections Project Stage 2 – Mid Term Review, GHD, Melbourne, Victoria, Australia, 2015.

Chan, T., Ross, H., Hoverman, S., and Powell, B.: Participatory development of a Bayesian network model for catchment-based water resource management, Water Resour. Res., 46, W07544, https://doi.org/10.1029/2009WR008848, 2010.

Chiew, F. H. S., McMahon, T. A., Dudding, M., and Brinkley, A. J.: Technical and economic evaluation of the conjunctive use of surface and groundwater in the Campaspe Valley, north-central Victoria, Australia, Water Resour. Manage., 9, 251–275, https://doi.org/10.1007/BF00872487, 1995.

Clemmens, A. J.: Measuring and improving irrigation system performance at the field level, in: Connellan, G. J., Irrigation Association of Australia 2000 National Conference and Exhibition: Water – Essential for Life. Proceedings. Melbourne Exhibition and Convention Centre, 23–25 May 2000, Irrigation Association of Australia, Melbourne, 190–199, 2000.

Croke, B. F. W. and Jakeman, A. J.: A Catchment Moisture Deficit module for the IHACRES rainfall-runoff model, Environ. Modell. Softw., 19, 1–5, 2004.

Croke, B. F. W., Letcher, R. A., and Jakeman, A. J.: Development of a flow network for the Namoi River Basin, Australia, J. Hydrol., 319, 51–71, https://doi.org/10.1016/j.jhydrol.2005.07.001, 2006.

Croke, B., Cornish, P., and Islam, A.: Modeling the Impact of Watershed Development on Water Resources in India, in: Integrated Assessment of Scale Impacts on Watershed Intervention: Assessing Hydrogeological and Bio-physical Influences on Livelihoods, edited by: Reddy, V. R. and Syme, G. J., Elsevier, 100–148, 2015.

CSIRO: Climate Change in Australia, available at: https://www.climatechangeinaustralia.gov.au/en/ (last access: 13 December 2017), 2017.

de Kok, J. L., Engelen, G., and Maes, J.: Reusability of model components for environmental simulation – Case studies for integrated coastal zone management, Environmental Modelling and Software, 68, 42–54, https://doi.org/10.1016/j.envsoft.2015.02.001, 2015.

Doherty, J. E.: PEST, SENSAN and Global Optimisers, in: PEST, Model-Independent Parameter Estimation User Manual Part I, Watermark Numerical Computing, Brisbane, Australia, available at: http://www.pesthomepage.org/getfiles.php?file=newpestman2.pdf&usg=AOvVaw0XsrKp9kwYUaX_DkwULET9 (last access: 26 February 2018), 2016.

DSE: Victorian Aquifer Framework: Updates for Seamless Mapping of Aquifer Surfaces, available at: http://www.vvg.org.au/cb_pages/files/207086.pdf (last access: 26 February 2018), 2012.

EcoDev: Victorian Winter Crop Summary 2015, Department of Economic Development, Jobs, Transport and Resources, Horsham, Victoria, Australia, available at: https://grdc.com.au/resources-and-publications/all-publications/publications/2015/03/nvt-victorian-winter-crop-summary-2015 (last access: 26 February 2018), 2015.

Finger, L. and Morris, M.: Water Management Options: Assisting irrigators with Stream Flow Management Plan implementation, Department of Primary Industries (DPI), State of Victoria, Tatura, Victoria, available at: http://vro.agriculture.vic.gov.au/dpi/vro/vrosite.nsf/0d08cd6930912d1e4a2567d2002579cb/adbda16966d0e530ca2574c8002ccbfa/$FILE/WaterManagementOptionsReport_2005.pdf (last access: 22 March 2018), 2005.

Franzén, F., Kinell, G., Walve, J., Elmgren, R., and Söderqvis, T.: Participatory social-ecological modeling in eutrophication management: The case of Himmerfjärden, Sweden, Ecol. Soc., 16, 27, https://doi.org/10.5751/ES-04394-160427, 2011.

GM-W: Lake Eppalock Land and On-Water Management Plan, Tatura, Victoria, 2013.

GM-W: Campaspe Basin – Goulburn Murray Water, available at: https://www.g-mwater.com.au/water-resources/catchments/gmwcatchments/campaspebasin (last access: 18 December 2017), 2017a.

GM-W: Lake Eppalock – Goulburn Murray Water, available at: https://www.g-mwater.com.au/water-resources/catchments/storages/campaspe/lakeeppalock (last access: 18 December 2017), 2017b.

Hughes, L., Steffen, W., Rice, M., and Pearce, A.: Feeding a hungry nation: Climate change, food and farming in Australia, Climate Council of Australia Limited, Sydney, Australia, available at: https://www.climatecouncil.org.au/uploads/7579c324216d1e76e8a50095aac45d66.pdf (last access: 26 February 2018), 2015.

Ivkovic, K. M., Croke, B. F. W., Kelly, R. A.: Overcoming the challenges of using a rainfall-runoff model to estimate the impacts of groundwater extraction on low flows in an ephemeral stream, Hydrol. Res., 45, 58–72, https://doi.org/10.2166/nh.2013.204, 2014.

Jakeman, A. J. and Letcher, R. A.: Integrated assessment and modelling: Features, principles and examples for catchment management, Environ. Modell. Softw., 18, 491–501, https://doi.org/10.1016/S1364-8152(03)00024-0, 2003.

Jakeman, A. J., Letcher, R. A., and Norton, J. P.: Ten iterative steps in development and evaluation of environmental models, Environ. Modell. Softw., 21, 602–614, https://doi.org/10.1016/j.envsoft.2006.01.004, 2006.

Kraft, P., Multsch, S., Vaché, K. B., Frede, H.-G., and Breuer, L.: Using Python as a coupling platform for integrated catchment models, Adv. Geosci., 27, 51–56, https://doi.org/10.5194/adgeo-27-51-2010, 2010.

Krueger, T., Page, T., Hubacek, K., Smith, L., and Hiscock, K.: The role of expert opinion in environmental modelling, Environ. Modell. Softw., 36, 4–18, https://doi.org/10.1016/j.envsoft.2012.01.011, 2012.

Malard, J. J., Inam, A., Hassanzadeh, E., Adamowski, J., Tuy, H. A., and Melgar-Quiñonez, H.: Development of a software tool for rapid, reproducible, and stakeholder-friendly dynamic coupling of system dynamics and physically-based models, Environ. Modell. Softw., 96, 410–420, https://doi.org/10.1016/J.ENVSOFT.2017.06.053, 2017.

McKay, J.: Water institutional reforms in Australia, Water Policy, 7, 35–52, https://doi.org/10.1080/02508069908692147, 2005.

MDBA: Sustainable Rivers Audit 2: The ecological health of rivers in the Murray–Darling Basin at the end of the Millennium Drought (2008–2010), Summary, available at: https://www.mdba.gov.au/sites/default/files/pubs/SRA2-SUMMARY-FINAL_0.pdf (last access: 26 February 2018), 2012.

MDBA: Campaspe/Murray-Darling Basin Authority, available at: https://www.mdba.gov.au/discover-basin/catchments/campaspe (last access: 18 December 2017), 2017.

NCCMA: The North Central CMA Region Environmental Water Management Plan for the Campaspe River System, available at: http://www.vewh.vic.gov.au/__data/assets/pdf_file/0004/368698/Campaspe-River-EWMP-FINAL.pdf (last access: 26 February 2018), 2014a.

NCCMA: Campaspe River resource Guide, available at: http://www.nccma.vic.gov.au/sites/default/files/publications/campaspe_river_resource_guide.pdf (last access: 26 February 2018), 2014b.

Niswonger, R. G., Panday, S., and Ibaraki, M.: MODFLOW-NWT, A Newton Formulation for MODFLOW-2005, in: Section A, Groundwater; Book 6 Modeling Techniques (p. 44), available at: https://pubs.usgs.gov/tm/tm6a37/pdf/tm6a37.pdf (last access: 26 February 2018), 2011.

NVIRP: Majority of Campaspe irrigators to exit permanently, available at: http://www.g-mwater.com.au/downloads/gmw/connections/Communication/Media/MediaReleases2010/Majority_of_Campaspe_irrigators_to_exit_permanently.pdf (last access: 26 February 2018), 2010.

NVIRP: NVIRP Annual Report 2010–2011, Northern Victoria Irrigation Renewal Project (NVIRP), Victoria, Australia, 2011.

Sletholt, M. T., Hannay, J. E., Pfahl, D., and Langtangen, H. P.: What do we know about scientific software development's agile practices?, 24–36, https://doi.org/10.4324/9780203894644, London, Routledge, 2012.

State of Victoria: NVIRP Stage 2 On Farm Project.pdf, State of Victoria, Northern Victoria Irrigation Renewal Project: On Farm Priority Project, available at: http://www.agriculture.gov.au/SiteCollectionDocuments/water/vic-farm-project.pdf (last access: 1 March 2018), 2011.

Tennakoon, S., Richards, D., Milroy, S., and Harris, G.: Water use efficiency in the Australian cotton industry, in: Waterpak: A Guide for Irrigation Management in Cotton and Grain Farming Systems, edited by: Wiggington, D., 22–27, Cotton Research and Development Corporation, 2013.

Ticehurst, J. and Curtis, A.: Improving on-farm water use efficiency amongst irrigators in the Campaspe River system and the Rochester Irrigation Area, Wagga Wagga, NSW, 2650, Australia, 2016.

Ticehurst, J. and Curtis, A.: Preliminary assessment of conjunctive use opportunities for in the Murray Darling Basin: a case study in the lower Campaspe catchment, Wagga Wagga, NSW, 2650, Charles Sturt University, 2017.

Van Dijk, A. I. J. M., Beck, H. E., Crosbie, R. S., De Jeu, R. A. M., Liu, Y. Y., Podger, G. M., Timbal, B., and Viney, N. R.: The Millennium Drought in southeast Australia (2001–2009): Natural and human causes and implications for water resources, ecosystems, economy, and society, Water Resour. Res., 49, 1040–1057, https://doi.org/10.1002/wrcr.20123, 2013.

Vaze, J., Teng, J., and Chiew, F. H. S.: Assessment of GCM simulations of annual and seasonal rainfall and daily rainfall distribution across south-east Australia, Hydrol. Process., 25, 1486–1497, https://doi.org/10.1002/hyp.7916, 2011.

Voinov, A. and Gaddis, E. J. B.: Lessons for successful participatory watershed modeling: A perspective from modeling practitioners, Ecol. Model., 216, 197–207, https://doi.org/10.1016/j.ecolmodel.2008.03.010, 2008.

Wheeler, S. A. and Cheeseman, J.: Key Findings from a Survey of Sellers to the Restoring the Balance Programme, Economic Papers, 32, 340–352, https://doi.org/10.1111/1759-3441.12038, 2013.

https://doi.org/10.5194/piahs-379-13-2018

Open Access
PIAHS

Hydrological regionalisation based on available hydrological information for runoff prediction at catchment scale

Qiaoling Li[1], Zhijia Li[1], Yuelong Zhu[2], Yuanqian Deng[1], Ke Zhang[1], and Cheng Yao[1]

[1]College of Hydrology and Water Resources, Hohai University, Nanjing 210098, China
[2]College of Computer and Information, Hohai University, Nanjing, 210098, China

Correspondence: Qiaoling Li (liqiaolinghhu@hhu.edu.cn)

Received: 3 January 2018 – Accepted: 12 February 2018 – Published: 5 June 2018

Abstract. Regionalisation provides a way of transferring hydrological information from gauged to ungauged catchments. The past few decades has seen several kinds of regionalisation approaches for catchment classification and runoff predictions. The underlying assumption is that catchments having similar catchment properties are hydrological similar. This requires the appropriate selection of catchment properties, particularly the inclusion of observed hydrological information, to explain the similarity of hydrological behaviour. We selected observable catchments properties and flow duration curves to reflect the hydrological behaviour, and to regionalize rainfall-runoff response for runoff prediction. As a case study, we investigated 15 catchments located in the Yangtze and Yellow River under multiple hydro-climatic conditions. A clustering scheme was developed to separate the catchments into 4 homogeneous regions by employing catchment properties including hydro-climatic attributes, topographic attributes and land cover etc. We utilized daily flow duration curves as the indicator of hydrological response and interpreted hydrological similarity by root mean square errors. The combined analysis of similarity in catchment properties and hydrological response suggested that catchments in the same homogenous region were hydrological similar. A further validation was conducted by establishing a rainfall-runoff coaxial correlation diagram for each catchment. A common coaxial correlation diagram was generated for each homogenous region. The performances of most coaxial correlation diagrams met the national standard. The coaxial correlation diagram can be transferred within the homogeneous region for runoff prediction in ungauged catchments at an hourly time scale.

1 Introduction

Flash floods are one of the most destructive hazards in the world, particularly in China. According to the statistics of the National Development and Reform Commission, China has 670 large rivers with the catchment areas over 3000 km^2, more than 8600 small and medium rivers with the catchment areas ranging between 200 and 3000 km^2, and much more mountainous gullies with the areas under 200 km^2. There has been no or few available streamflow data (ungauged) to derive reasonable rainfall runoff relationships for the small and medium rivers, despite the national initiative for hydrological monitoring system for small and medium rivers in China from 2011. The question is how to build an appropriate rainfall runoff relationship for accurate and timely runoff prediction at the ungauged catchments? Many hydrologists focus on the concept and method of regionalisation transferring of hydrological information/response from gauged to ungauged environments.

In the literature, numerous regionalisation approaches have been proposed for runoff predictions from a catchment scale to global scale and from a short-term flood event to long-term series (Li and Zhang, 2017). The most popular regionalisation approaches are generally based on (1) spatial proximity, (2) regression, (3) physical and climatic similarity, (4) hydrological signatures/indices, (5) integrated similarity such as the combination of spatial proximity and physical similarity.

Table 1. Information and characteristics of the study sites.

NO	Station	River System	Area (km^2)	Annual Rainfall (mm)	Flood events NO	Record period
1	Zhashui	Yangtze River	493	780.1	17	2004–2013
2	Qingniwan	Yangtze River	1377	780.1	15	2004–2013
3	Banqiao	Yangtze River	493	729	13	2000–2010
4	Daheba	Yangtze River	2198	843	5	2013–2014
5	Lianghekou	Yangtze River	2829	843	5	2013–2014
6	Youshuijie	Yangtze River	911	938	24	2000–2011
7	Changtancun	Yangtze River	237	837	17	2003–2013
8	Yingge	Yellow River	507	706	26	2000–2013
9	Manwan	Yellow River	122	731	17	2002–2013
10	Chenhe	Yellow River	1380	798	20	2003–2012
11	Laoyukou	Yellow River	347	768	11	2000–2007
12	Qinduzhen	Yellow River	566	833	27	2000–2014
13	Maduwang	Yellow River	1601	630.9	11	2000–2010
14	Xinghe	Yellow River	479	513	29	2000–2012
15	Zhidan	Yellow River	774	509.8	18	2000–2012

Hydrologists have tried to compare the performances of different approaches in ungauged catchments but results are not in agreement. Young (2006) compares two regionalisation approaches in 260 catchments in UK and finds that the regression-based approach is more accurate than the spatial proximity-based approach. Oudin et al. (2008) test three approaches in 913 catchments in Frances and conclude that the spatial proximity-based approach performs the best, and the regression-based approach perform the worst with the physical similarity approach in the middle. Zhang and Francis (2009) analyse several approaches in 210 catchments in Australia and show that the spatial proximity-based approach is slightly better than the physical similarity approach. Bao et al. (2012) investigate different regionalisation approaches for 55 catchments in China and find that the similarity based approaches are more accurate than the regression based approach, and become much more accurate from humid to arid regions. Parajka et al. (2013) conducts an integrated analysis on 34 studies including 3874 catchments and conclude that the performances of different approaches largely depend on the selected rainfall runoff models and research sites. The similarity-based approaches outperform other approaches in many catchments.

The main difficulty for the similarity approaches is to identify a suitable metric of similarity for the target catchments. Several recent studies have implemented data mining and machine learning techniques to identify similarity metrics for catchments in different climates. Further improvement is still necessary to increase regionalisation performance in catchment of the dryer regions. Only few regionalisation studies have been published with Chinese data and much less for prediction at short time scale for ungauged Chinese catchments (Ragettli et al., 2017). This could be attributed to strict Chinese data sharing policies that limit data availability.

The aim of this study is to regionalize rainfall-runoff response for short term runoff prediction in small and medium catchments of China. For this purpose, we performed a principal component analysis for each category of catchments properties and developed a clustering scheme to classify catchments. Daily flow duration curves were employed as an index of hydrological response. The integrated catchment properties and daily flow duration curves were investigated to assess hydrological similar catchment and determine the homogenous regions. We further evaluated the transferability of coaxial correlation diagrams within the homogenous region.

2 Study sites and data

Fifteen gauged catchments located in Shanxi Province were identified as the study area (Fig. 1) based on three criteria (1) data availability of daily/hourly rainfall and runoff (2) the catchments were unimpaired and unregulated (3) the catchments provide a range of catchment types and climatology. The catchments are classified by the Qin Mountain into "South" and "North" with climate changing from humid to semi-humid and semi-arid zones. The catchment areas range from 122 to 2829 km^2. Each catchment has 10–15 years of available daily continuous rainfall and runoff data with 11–29 flood events except Daheba and Lianghekou. Table 1 lists the period of record and the number of flood events used for the generation of flow duration curves and coaxial correlation diagrams for each catchment. These hydro-meteorological data were obtained under the collaboration with the Hydrology and Water Resources Bureau of Shanxi Province. Except for the hydro-meteorological data, the regionalisation methods require catchment attributes such as topography attributes extracted from a digital elevation model (http://www.csdb.cn/), fraction of land cover and land use derived

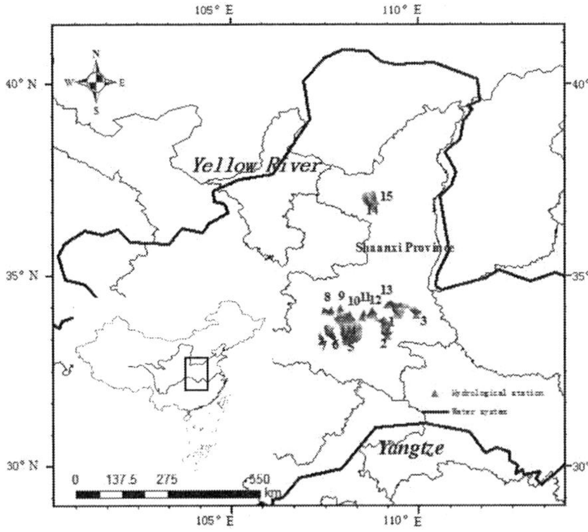

Figure 1. Locations of the study sites.

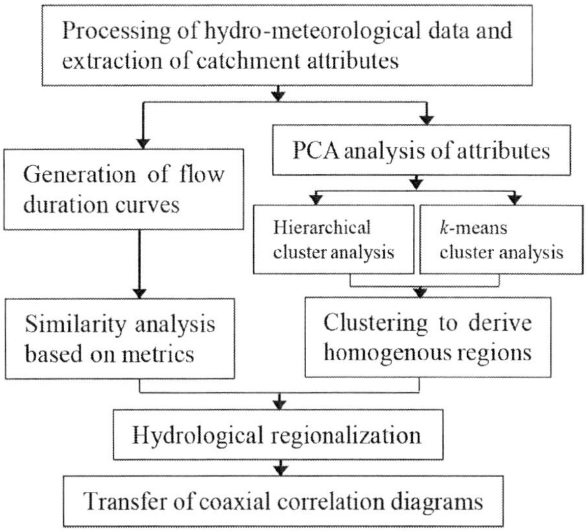

Figure 2. Schematic flow chart of study design.

from the Global Land Cover Database and Soil Map of China (Wu et al., 2003), to reflect the characteristics of rainfall runoff responses.

3 Methods

3.1 Study design

The hydrological regionalisation study is conducted by two ways (Fig. 2). One is the similarity analysis based on flow duration curves using root mean square errors:

$$\mathrm{RMSE} = \sqrt{\frac{\sum_{i=1}^{n} (Qx_i - Qy_i)^2}{n}} \quad (1)$$

Where n is the total number of percent exceedance, Qx_i and Qy_i are the discharges of any two catchments at the ith percent exceedance. The RMSE measures the difference between the discharges of each two catchments at the same percent exceedance.

The other is clustering analysis to derive homogenous regions. Firstly we performed a principal component analysis (PCA, Legendre and Legendre, 1998) for catchment characteristics including topography, land cover and land use, soil, and catchment structure such as area and channel length, to select uncorrelated metrics representing the catchment responses. Then k-means clustering analysis and hierarchical clustering analysis were selected to group the catchments into a number of classes (homogenous regions). Hydrological regionalisation was based on the results of both flow duration curves and homogenous regions. Finally the coaxial correlation diagram was generated for each catchment and transfer was performed within the homogenous regions.

3.2 Clustering analysis

In this study, clustering was defined as the grouping of similar catchments. Cluster analysis methods are broadly classified into hierarchical and partitional clustering methods.

Hierarchical clustering analysis grouped PCA results (in a transformed space of catchment attributes) step by step to build clusters. Inter-class distances were calculated using between-groups linkage, within-groups linkage, nearest neighbour, furthest neighbour, centroid clustering, median clustering, and ward's method based on a Euclidean distance measure respectively. Cluster analyses were carried out using these linkage methods for 4–7 clusters.

k-means clustering is one of the most used partitional cluster analysis methods (Parracho et al., 2016). The method was applied to the PCA results in SPSS and the number of clusters $k = 5$ was subjectively determined based on the number and geographical location of catchments and the PCA results. 5 random cluster centers were initialized and each attribute in the domain was assigned to the nearest center. Then the optimum centers were recalculated and the assignment of each attribute was repeated until the clustering did not change.

3.3 Coaxial correlation diagrams

A coaxial correlation diagram, commonly used tools for runoff prediction, particularly in China, integrates the main factors affecting rainfall-runoff response (Li et al., 2015). To make a coaxial correlation diagram, we need as many flood events as possible with areal average rainfall and the consequent runoff, as well as antecedent rainfall. The coaxial correlation diagram was established for each catchment using observed daily and hourly rainfall, and hourly runoff data. The performances of the coaxial correlation diagrams were

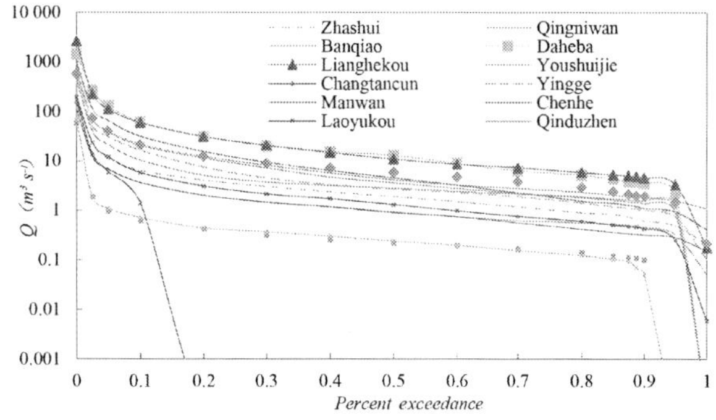

Figure 3. Flow duration curves for the fifteen catchments.

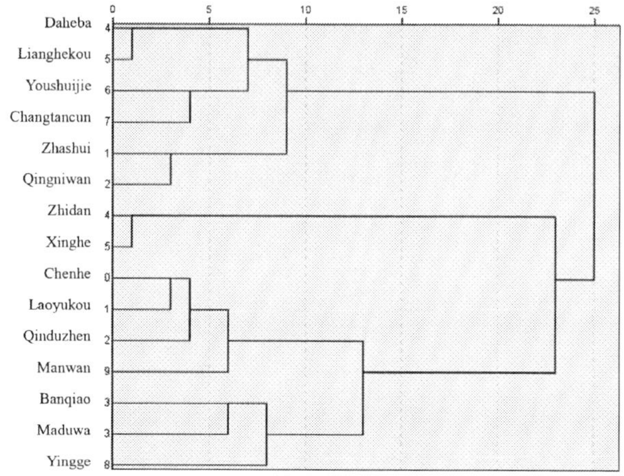

Figure 4. The dendrogram using Ward's method.

evaluated using a Qualified Rate which is computed by the Runoff Absolute Error and Allowed Error (Sun et al., 2008). The coaxial correlation diagrams were also established for homogenous zones and evaluation was conducted for each catchment.

4 Results

4.1 The Similarity based on flow duration curves

Fifteen flow duration curves were constructed from daily runoff data of the entire record period listed in Table 1. Figure 3 shows the flow duration curve for each catchment. Daily discharges changed slightly at the south catchments and greatly at the north catchments, particularly during the high flow period (with the percent exceedance between 0 and 10 %) and low flow period (with the percent exceedance between 90 and 100 %).

Table 2 summarises the root mean square errors between each pair of catchments. The smaller the value was, the more similar the two catchments were. Most of the values were smaller than 0.1 indicating that similarities lies in the corresponding catchments. The catchments of Qingniwan, Banqiao, Lianghekou, and Laoyukou had smaller root mean square errors with other catchments. The catchments of Yingge, Manwan, Qinduzhen, Xinghe, Zhidan, and particularly Changtancun showed larger root mean square error with other catchments.

4.2 Homogeneous regions based on cluster analyses

The fifteen catchments were classified into different groups depend on the number of clusters. 4, 5, 6, and 7 clusters were selected with seven linkage methods. Figure 4 showed the results of hierarchical clustering analyses combining ward's method and Euclidean distance using 4 clusters. The figure illustrates that the fifteen catchments were grouped into four regions: (1) Daheba, Lianghekou, Youshuijie, Changtancun, Zhashui, and Qingniwan; (2) Zhidan and Xinghe; (3) Chenhe, Laoyukou, Qinduzhen, and Manwan; (4) Banqiao, Maduwang, and Yingge. The results produced from other linkage methods and other number of clusters were different from this method and they were not shown in this paper.

Table 3 presents the grouping results using the k-means clustering with the number of clusters $k = 5$. The distances between the cases and the centers of their affiliated groups were shown in the fourth column.

4.3 Integrated hydrological regionalisation

An integrated analysis was conducted employing flow duration curves, the hierarchical clustering method with seven linkage methods, and k-means clustering method. It was found that the fifteen catchments could be classified into four homogeneous regions: (1) Daheba, Lianghekou, Youshuijie, Changtancun, Zhashui, and Qingniwan belong to the I zone; (2) Chenhe, Laoyukou, Qinduzhen, Manwan, and Yingge belong to the II zone; (3) Banqiao and Maduwang belong to the III zone; (4) Zhidan and Xinghe belong to the IV zone.

For a further evaluation, the coaxial correlation diagram was generated for each catchment based on visual judgment and computer aids design according to the distribution of flood events plots (Fig. 5). The performances of each coaxial correlation diagram were presented in Table 4. The results suggest that the performances of most coaxial correlation diagram met the national standard with the Qualified Rates no less than 60 % (Sun et al., 2008) except for Changtancun and Yingge, as the characteristics of flood events within these two catchments differed greatly from each other.

Four coaxial correlation diagrams were generated for the corresponding homogeneous zones (Fig. 6). The coaxial correlation diagrams of zone I and zone II were closer to the 45° line, indicating that catchments had a larger antecedent soil moisture. They were generally located in semi-humid and humid regions. In contrast to this, the coaxial correlation di-

Table 2. Root mean square errors between catchments.

NO.Catchments	1	2	3	4	5	6	7	8	9	10	11	12	13	14	15
1.Zhashui	0.00														
2.Qingniwan	0.09	0.00													
3.Banqiao	0.06	0.06	0.00												
4.Daheba	0.13	0.06	0.09	0.00											
5.Lianghekou	0.11	0.03	0.07	0.06	0.00										
6.Youshuijie	0.13	0.05	0.10	0.09	0.04	0.00									
7.Changtancun	0.54	0.50	0.51	0.45	0.50	0.53	0.00								
8.Yingge	0.21	0.14	0.17	0.16	0.12	0.10	0.57	0.00							
9.Manwan	0.10	0.09	0.09	0.13	0.09	0.09	0.58	0.12	0.00						
10.Chenhe	0.18	0.10	0.15	0.10	0.08	0.07	0.50	0.09	0.14	0.00					
11.Laoyukou	0.09	0.03	0.05	0.05	0.04	0.07	0.50	0.14	0.08	0.10	0.00				
12.Qinduzhen	0.10	0.11	0.06	0.10	0.12	0.15	0.47	0.21	0.14	0.18	0.09	0.00			
13.Maduwang	0.04	0.08	0.05	0.12	0.09	0.11	0.55	0.17	0.06	0.16	0.07	0.10	0.00		
14.Xinghe	0.10	0.16	0.11	0.18	0.18	0.21	0.52	0.28	0.18	0.26	0.15	0.11	0.12	0.00	
15.Zhidan	0.05	0.14	0.10	0.16	0.16	0.18	0.55	0.24	0.13	0.22	0.12	0.11	0.08	0.08	0.00

Table 3. k-means clustering results ($k = 5$).

Cases	1	2	3	4	5	6	7	8
Catchments	Zhashui	Qingniwan	Banqiao	Daheba	Lianghekou	Youshuijie	Changtancun	Yingge
Groups	1	1	3	5	5	5	5	1
Distances	1.506	1.781	1.816	0.959	1.703	1.281	2.111	2.511
Cases	**9**	**10**	**11**	**12**	**13**	**14**	**15**	
Catchments	Manwan	Chenhe	Laoyukou	Qinduzhen	Maduwang	Xinghe	Zhidan	
Groups	4	4	4	4	3	2	2	
Distances	2.171	1.596	1.462	1.354	1.816	0.481	0.481	

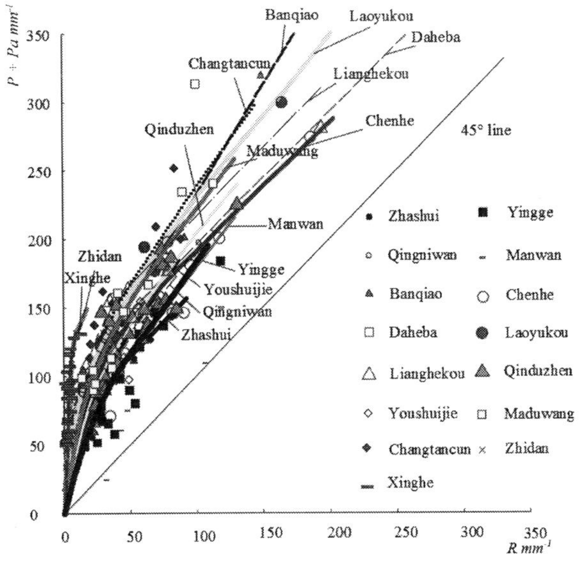

Figure 5. The coaxial correlation diagram for each catchment.

agrams of zone I and zone II were farther suggesting that the antecedent soil was drier in these regions. The performance of the common correlation diagram in a homogenous zone was a bit worse than the performance of the independent diagram for each catchment.

5 Discussion and conclusion

This paper explored hydrological regionalisation using flow duration curves and clustering analysis methods in order to provide a possible way of runoff prediction in ungauged catchments. The flow duration curves could give steady-state pictures of the entire hydrological regimes and represented similarities among them. It is worth noting that flow duration curves cannot keep steady-state under climate change and human activities. The root mean square errors of the fifteen flow duration curves between catchments suggested the similarities among them. The regionalisation was then performed employing hierarchical clustering methods with seven linkage method and k-means clustering method. The clustering results differed as the linkage methods and clusters varied for hierarchical clustering analyses. The number $k = 5$ was

Table 4. The performance of coaxial correlation diagrams.

NO.Catchments	1	2	3	4	5	6	7	8	9	10	11	12	13	14	15
Qualified rate (%)	65	80	62	80	80	70.8	52.9	53.8	64.7	70	90.9	74.1	72.7	93.1	100
Accuracy level	3	2	3	2	2	2	<3	<3	3	2	1	2	2	1	1

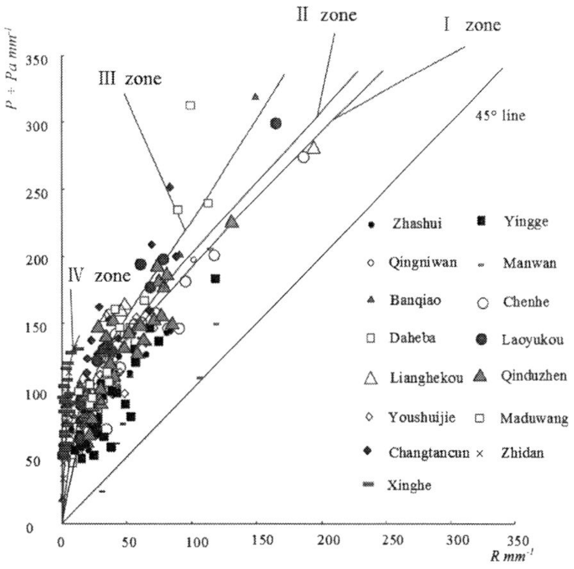

Figure 6. Regionalisation of coaxial correlation diagrams.

adopted for k-means clustering method and it is suggested that analyses with other numbers of clusters be carried out.

The fifteen catchments were grouped into four homogenous regions based on an integrated analysis. The coaxial correlation diagram was generated for each catchments and each homogenous zone with the performance of the latter a bit worse. It is argued that the common coaxial correlation diagram for the homogenous zone could be transferred within the catchments of the zone for runoff prediction. Further study is needed to combine several coaxial correlation diagrams according to the characteristics of flood events. In summary, this study presents a way of hydrological regionalisation using available hydrological information which could be useful for runoff prediction in ungauged catchments.

Data availability. The hydrological data used in this study was obtained under the collaboration with Hydrological Bureau of Shanxi Province of China, since we have a joint project (National Key R&D Program of China, 2016YFC0402705). The data cannot open to the public at the present time.

Competing interests. The authors declare that they have no conflict of interest.

Special issue statement. This article is part of the special issue "Innovative water resources management – understanding and balancing interactions between humankind and nature". It is a result of the 8th International Water Resources Management Conference of ICWRS, Beijing, China, 13–15 June 2018.

Acknowledgements. This work was supported by the Fundamental Research Funds for the Central Universities (2018B11014), the National Key R&D Program of China (2016YFC0402705), and Natural Science Foundation of China (41201028, 51679061, 41130639).

Edited by: Dingzhi Peng
Reviewed by: two anonymous referees

References

Bao, Z., Zhang, J., Liu, J., Fu, G., Wang, G., He, R., Yan, X., Jin, J., and Liu, H.: Comparison of regionalization approaches based on regression and similarity for predictions in ungauged catchments under multiple hydro-climatic conditions, J. Hydrol., 466–467, 37–46, https://doi.org/10.1016/j.jhydrol.2012.07.048, 2012.

Legendre, P. and Legendre, L.: Numerical Ecology, Elsevier, Amsterdam, 1998.

Li, H. and Zhang, Y.: Regionalising rainfall-runoff modelling for predicting daily runoff: Comparing gridded spatial proximity and gridded integrated similarity approaches against their lumped counterparts, J. Hydrol., 550, 279–293, https://doi.org/10.1016/j.jhydrol.2017.05.015, 2017.

Li, Q., Li, Z., Chen, L., and Yao, C.: Regionalization of coaxial correlation diagrams for the semi-humid and semi-arid catchments in Northern China, Proc. IAHS, 368, 317–322, https://doi.org/10.5194/piahs-368-317-2015, 2015.

Oudin, L., Andreassian, V., Perrin, C., Michel, C., and Le Moine, N.: Spatial proximity, physical similarity, regression and ungaged catchments: a comparison of regionalization approaches based on 913 French catchments, Water Resour. Res., 44, W03413, https://doi.org/10.1029/2007WR006240, 2008.

Parajka, J., Viglione, A., Rogger, M., Salinas, J. L., Sivapalan, M., and Blöschl, G.: Comparative assessment of predictions in ungauged basins – Part 1: Runoff-hydrograph studies, Hydrol. Earth Syst. Sci., 17, 1783–1795, https://doi.org/10.5194/hess-17-1783-2013, 2013.

Parracho, A. C., Melo-Goncalves, P., and Rocha, A.: Regionalisation of precipitation for the Iberian Peninsula and climate change, Phys. Chem. Earth, 94, 146–154, https://doi.org/10.1016/j.pce.2015.07.004, 2016.

Ragettli, S., Zhou, J., Wang, H., Liu, C., and Guo, L.: Modeling flash floods in ungauged mountain catchments of China: A decision tree learning approach for parameter regionalization, J. Hydrol., 555, 330–346, https://doi.org/10.1016/j.jhydrol.2017.10.031, 2017.

Sun, J., Zhang, J., Wang, J., Liang, J., Zhang, S., Chen, S., Le, J., Rui, X., Ge, S., Li, L., Liu, Z., Cheng, L., Sun, C., Yu, R., Tao, Y., Gao, X., Deng, C., Zhou, L., Li, Y., Chen, Z., and Wang, J.: Standard for hydrological information and hydrological forecasting, Beijing: Standards Press of China, GB/T 22482–2008, 2008.

Wu, B., Xu, W., Huang, H., and Yan, C.: The Land Cover Map for China in the Year 2000, GLC2000 database, European Commission Joint Research Centre, 2003.

Young, A. R.: Stream flow simulation within UK ungauged catchments using a daily rainfall-runoff model, J. Hydrol., 320, 155–172, https://doi.org/10.1016/j.jhydrol.2005.07.017, 2006.

Zhang, Y. and Francis, H.: Relative merits of different methods for runoff predictions in ungauged catchments, Water Resour. Res., 45, https://doi.org/10.1029/2008WR007504, 2009.

https://doi.org/10.5194/piahs-379-21-2018

Effect of reservoir zones and hedging factor dynamism on reservoir adaptive capacity for climate change impacts

Adebayo J. Adeloye[1] **and Bankaru-Swamy Soundharajan**[2]

[1]Institute for Infrastructure and Environment, Heriot-Watt University, Edinburgh EH14 4AS, UK
[2]Department of Civil Engineering, Amrita Vishwa Vidyapeetham, Coimbatore, India

Correspondence: Adebayo J. Adeloye (a.j.adeloye@hw.ac.uk)

Received: 19 December 2017 – Revised: 15 February 2018 – Accepted: 26 February 2018 – Published: 5 June 2018

Abstract. When based on the zones of available water in storage, hedging has traditionally used a single hedged zone and a constant rationing ratio for constraining supply during droughts. Given the usual seasonality of reservoir inflows, it is also possible that hedging could feature multiple hedged zones and temporally varying rationing ratios but very few studies addressing this have been reported especially in relation to adaptation to projected climate change. This study developed and tested Genetic Algorithms (GA) optimised zone-based operating policies of various configurations using data for the Pong reservoir, Himachal Pradesh, India. The results show that hedging does lessen vulnerability, which dropped from $\geq 60\,\%$ without hedging to below 25 % with the single stage hedging. More complex hedging policies, e.g. two stage and/or temporally varying rationing ratios only produced marginal improvements in performance. All this shows that water hedging policies do not have to be overly complex to effectively offset reservoir vulnerability caused by water shortage resulting from e.g. projected climate change.

1 Introduction

The realization that projected climate change will affect reservoir future inflow series and hence performance has led to the intensification of research efforts to assess these impacts as a precursor to the development of effective mitigation and adaptation strategies (see e.g. Nawaz and Adeloye, 2006; Fowler et al., 2003; Li et al., 2009). Most of these studies have reported deteriorating performance notably with regard to vulnerability unless improved operational practices are developed.

The use of rule curves for guiding reservoir operation (Yin et al., 2015) is schematically illustrated in Fig. 1a. Full satisfaction of demand is attempted whenever the water available (WA) is in the interval [LRC_m, URC_m], where LRC_m and URC_m are, respectively, the ordinates of the lower and upper rule curves for month m. The supply exceeds the demand when WA is above URC_m but no water is supplied if the WA is below LRC_m.

Although they are easy to use, rule curves often produce large water shortages or vulnerability (see e.g. Chiamsathit et al., 2014). Hedging, which is the deliberate cutting back of supplied water even when there is sufficient water, has been shown to moderate the vulnerability associated with rule curve operation (Eum et al., 2011).

Figure 1b illustrates a critical rule curve (CRC) delineated hedging zone integrated with the rule curve of Fig. 1a. Below the CRC, a maximum fraction "α" of the demand will be attempted, i.e. $D'_t \leq \alpha D_t$, where D'_t is the supply, D_t is the demand and α ($0 \leq \alpha \leq 1$) is the constant rationing ratio. Possibilities for enhancing the hedging policy exist, such as by incorporating multiple hedging zones (see Fig. 1c) and/or temporally varying (e.g. monthly, seasonally) rationing ratios. What is certain is that such enhancements will complicate the policy; however, the degree of performance improvement of such enhancements remains largely unknown.

The aim of this work is to evaluate the effectiveness of GA-optimised hedging policies in moderating reservoir vulnerability during projected climate change perturbations. The

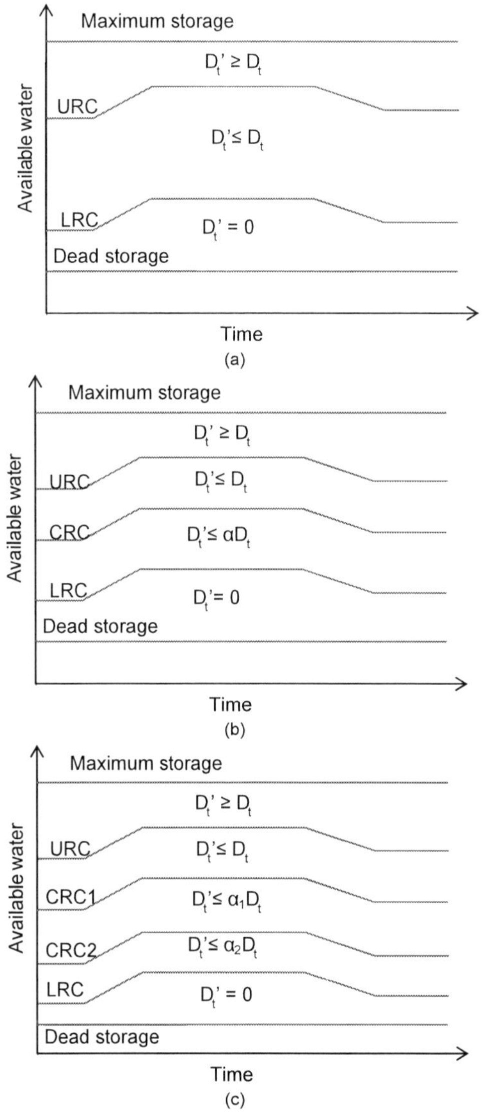

Figure 1. Schematic illustration of zone-based operating rules for **(a)** no-hedging; **(b)** single-stage hedging; and **(c)** two-stage hedging.

work used the Pong reservoir located on the Beas River in Himachal Pradesh, India and serves irrigation and hydropower purposes. In the following section, further details of the adopted methodology are given. This is then followed by the description of the case study, following which the results are presented and discussed. The final section contains the main conclusions and recommendations of the study.

2 Methodology

The methodological approach is illustrated in Fig. 2; brief details of the different aspects are provided in the sub-sections below. Additional information is available elsewhere (see Adeloye et al., 2016).

2.1 Genetic algorithm optimisation for hedging policy

Genetic Algorithms (GA) (see Michalewicz, 1992; Wardlaw and Sharif, 1999) were used for optimising the hedging policies because of its various advantages, including their potential to search the solution from population of points (not a single point), its use of objective function information itself (not any derivatives), and its use of probabilistic transition rules. For these reasons, GA have been widely used for solving complex optimisation problems in various branches of science including water resources. The objective function for the optimisation was:

$$\text{Minimise} \sum_{i=1}^{N} \left(D_t - D_t'\right)^2; \quad t \in N \tag{1}$$

subject to the reservoir mass balance equation constraint:

$$S_{t+1} = S_t + Q_t - D_t' - E_t, \tag{1a}$$

Other constraints determine the release decision and are operating-policy-specific. For example for single stage hedging:

Let total available water, $\mathrm{WA}_t = S_t + Q_t$

$$\text{if } \mathrm{WA}_t \geq \mathrm{URC}_m, D_t' = S_t + Q_t - E_t - \mathrm{URC}_m \\ \&\mathrm{ER}_t = D_t' - D_t \tag{1b}$$

$$\text{if } \mathrm{URC}_m \geq \mathrm{WA}_t > \mathrm{CRC}_m, D_t' \leq D_t \\ \&\mathrm{ER}_t = 0 \tag{1c}$$

$$\text{if } \mathrm{CRC}_m \geq \mathrm{WA}_t > \mathrm{LRC}_m, D_t' \leq \alpha_m D_t \\ \&\mathrm{ER}_t = 0 \tag{1d}$$

$$\text{if } \mathrm{WA}_t \leq \mathrm{LRC}_m, D_t' = 0 \\ 1 \geq \alpha_m > 0; \mathrm{URC}_m \geq \mathrm{CRC}_m \geq \mathrm{LRC}_m \tag{1e}$$

where S_t and S_{t+1} are reservoir storage at the beginning and the end of time period t; Q_t is the inflow during time period t; E_t is the net evaporation during time period t; ER_t is excess release during time period t; N is the total number of periods; URC_m and LRC_m are the ordinates of the upper and lower rule curves for the month m; CRC_m is the ordinate of the critical rule curve for the month m; and all other symbols are as defined previously.

For the single stage hedging, the decision variables for the optimisation are the CRC_m m $1, 2, \ldots 12$ ordinates for

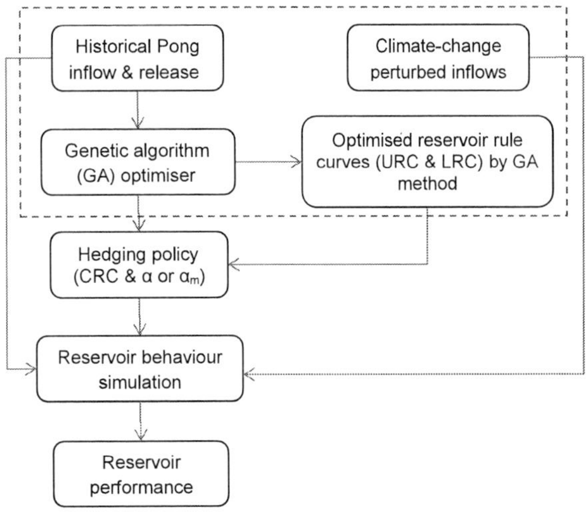

Figure 2. Methodology flow chart for optimising and performance evaluation of the hedging policy.

each month m of the year, and α_m; $m = 1, 2, .., 12$ are the corresponding monthly rationing ratios. Where hedging is static, the rationing ratio is a constant value α; for seasonal rationing, each of the four seasons will be represented by a constant rationing ratio α_s; $s = 1, 2..., 4$. Thus, the number of decision variables is 13 in the case of constant hedging, 24 in the case of monthly dynamic hedging and 16 for seasonal hedging. For two stage hedging, there will be two critical rule curves, making the associated number of decision variables to be 26 (constant hedging), 48 (monthly hedging) or 32 (seasonal hedging). The GA optimisation involved the usual selection, crossover and mutation operations. A real-value coding was used with the following parameters: crossover fraction = 0.8; mutation rate = 0.01; number of elite children = 2. The genetic operations were repeated for 500 generations.

2.2 Climate change Impacts on reservoir inflows

The assessment of the effects of projected climate change on reservoir inflows employed the delta perturbation approach (see Anandhi et al., 2011; Vicuna et al., 2012) in order to avoid the notorious uncertainties associated with GCM projections and their downscaling. The delta perturbations in temperature ΔT (°C) ranged from 0–5 °C, step of 1 °C, i.e. [0, 5; 1] while the corresponding perturbations in precipitation ΔP (%) were [−10, 10; 5]. These were applied to the Beas Basin historic temperature and precipitation data records respectively. The resulting series were then used to force a calibrated HYSIM catchment model (Manley and WRA, 2006) to derive the corresponding reservoir inflow series.

2.3 Reservoir behaviour simulation and performance indices

Reservoir simulation used the reservoir mass balance equation and associated constraints shown in Eqs. (1a–e). At the end of the simulation, reservoir key performance indicators were evaluated as follows (McMahon and Adeloye, 2005; Sandoval-Solis et al., 2011):

a. Reliability

Time-based: $R_t = N_s / N(2)$ (2)

Volume-based: $$R_v = \sum_{t=1}^{N} D_t{}' / \sum_{t=1}^{N} D_t \quad (3)$$

b. Resilience, ϕ

$$\phi = 1/((N - N_s)/f_s) \quad (4)$$

c. Vulnerability, η

$$\eta = \sum_{t=1}^{f_d} [(D_t - D_t{}')/D_t]/(N - N_s) \quad (5)$$

d. Sustainability, λ

$$\lambda = (R_t \phi (1 - \eta))^{1/3} \quad (6)$$

where N_s is the number of periods in which $D_t' \geq D_t$, f_s is the number of continuous failure sequences and f_d is the number of failure durations.

3 Case study and data

The Pong dam (and its reservoir, see Fig. 3) is located at longitude 76°05 E and latitude 32°05 01 N, drains a catchment area of 12 561 km^2, out of which the permanent snow catchment is 780 km^2 (Jain et al., 2007). The active storage capacity of the reservoir is 7051 Mm^3, which is primarily used for irrigating 1.6×10^6 ha total area dedicated for the cultivation of rice, wheat and cotton, although the released water first passes through turbines for hydropower generation. Consequently, the focus of the current study is the irrigation demand.

Monthly reservoir inflow and irrigation demands from January 2000 to December 2008 (9 years) were available for the study. The historic mean annual runoff (MAR) at dam site is 8485 Mm^3 (annual coefficient of variation is 0.225) and the seasonal distribution of the annual runoff is shown in Fig. 4.

The genetic algorithm (GA) optimised basic (i.e. without water hedging) rule curves were developed by Adeloye et al. (2016) for the reservoir as part of the wider study and are shown in Fig. 5a. These rule curves formed the bases for

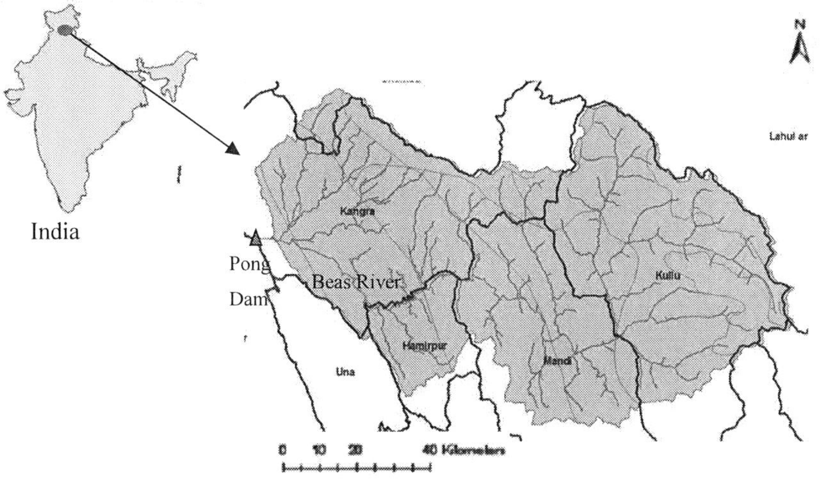

Figure 3. Map of India and the Beas river catchment.

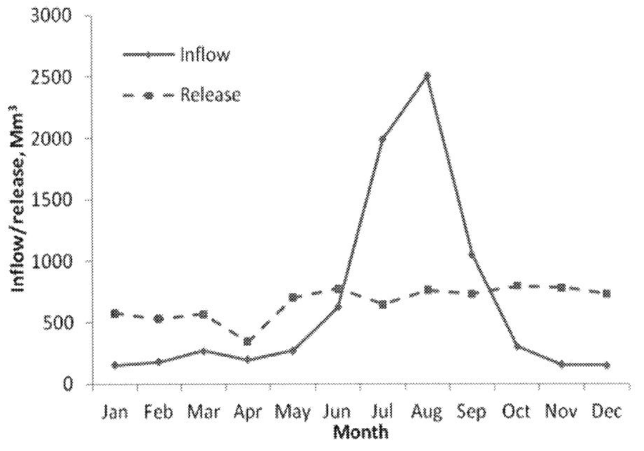

Figure 4. Average monthly inflow and irrigation demands from the Pong reservoir (data: 2000–2008).

the development of the hedging-integrated rule curves as outlined in Sect. 2.1.

4 Results and discussions

4.1 Climate change impacts on reservoir inflows

Table 1 summarises the effects of the projected climate change perturbations on the reservoir inflow. As expected, more inflows are recorded as the catchment precipitation increases. Table 1 also shows that the runoff increased for all the temperature increases even when the precipitation had remained unchanged. As noted earlier, there are significant glaciers and seasonal snow in the upper part of the Beas catchment. The simulated increase in the runoff at high temperatures is due to the additional runoff generated from the melting of the snow/glaciers at elevated temperatures, which appeared to have more than compensated for any evapotranspiration increases.

4.2 Optimised Rule Curves

The optimised hedging integrated rule curves are shown in Fig. 5b and c; as noted previously, Fig. 5a is the basic, no-hedging set of rule curves. While Fig. 5b is single stage, Fig. 5c has two stages; both policies, however, used constant hedging (or rationing ratios).

The CRC that triggers the water rationing in Fig. 5b lies everywhere between the URC and LRC as expected and allows attempting to supply the full demand over a very wide range of water availability in the system during the high flow monsoon (May–September) season. The range of water availability in which the full supply can be attempted is much narrower for the drier, post-monsoon periods. Water rationing will occur whenever the available water falls below the CRC. However, as shown in Fig. 5b when water rationing takes place, only 17 % of the full demand is cut back, leaving 83 % of the full demand being attempted. This deliberate cut-back is moderate and is thus not expected to cause undue hardship for water users.

In Fig. 5c with two hedging stages, the upper critical curve is everywhere lower than the single critical curve of Fig. 5a, implying that more water will be available for release in the upper hedging zone of the two stage hedging policy than from the single hedging zone of Fig. 5a. This has been confirmed by the value of the optimised rationing ratio for the upper hedging zone of Fig. 5c, which at 85 % is higher than the 83 % optimised for the single stage hedging policy of Fig. 5a. As expected, the lower critical curve (and hedging zone) has intensified the rationing in response to the dwindling water stock by allowing only 76 % of the full demand to be attempted whenever the available water falls into the lower

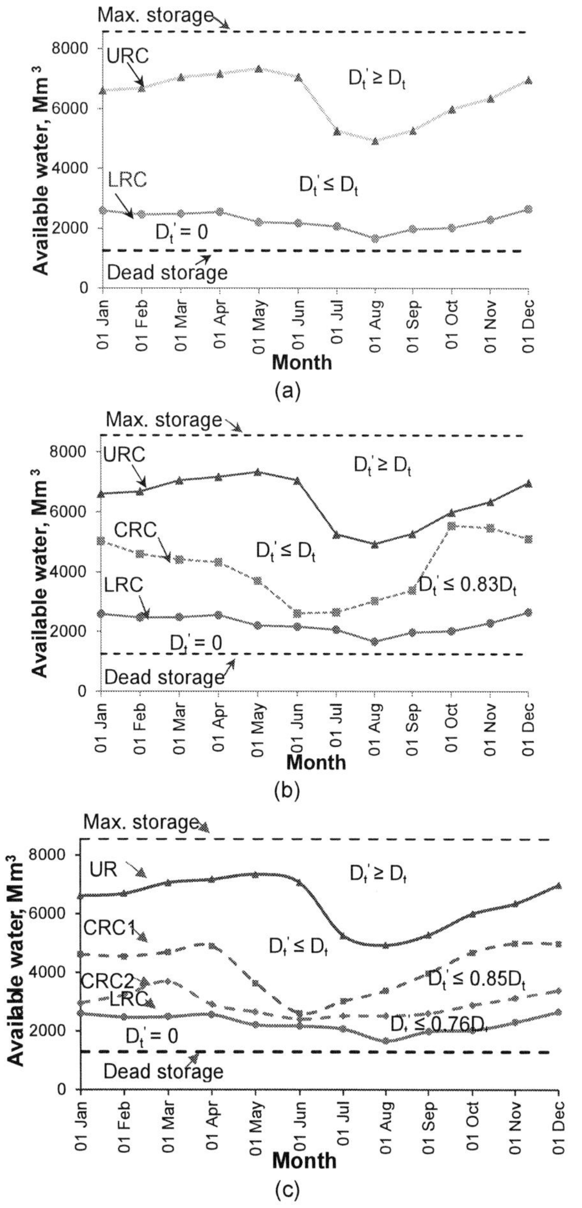

Figure 5. GA optimised rule curves **(a)** No hedging; **(b)** Single-stage constant hedging; **(c)** two-stage constant hedging.

Table 1. Change (%) in mean annual and seasonal runoff under climate change.

Temperature change, °C	Annual precipitation change, % −10	−5	0	+5	+10
Annual					
0	−12.11	−6.25	0.00	6.70	13.77
1	−7.08	−1.63	4.17	10.21	16.44
2	6.98	12.41	18.19	24.27	30.51
Season: Winter					
0	−13.25	−6.82	0.00	7.77	16.40
1	−18.32	−13.55	−8.19	−2.36	3.81
2	−5.90	−1.23	4.00	9.75	15.83
Season: Post-Monsoon					
0	−10.80	−5.54	0.00	5.83	11.93
1	−7.60	−2.80	2.30	7.54	12.81
2	6.88	11.63	16.71	21.99	27.29
Season: Monsoon					
0	−12.29	−6.35	0.00	6.76	13.85
1	−4.74	1.03	7.13	13.48	20.03
2	9.25	15.01	21.07	27.42	33.96
Season: Pre-Monsoon					
0	−15.09	−7.92	0.00	8.89	18.82
1	−17.76	−12.05	−5.72	1.09	8.29
2	−2.96	3.03	9.70	16.94	24.58

hedging zone. As noted previously, hedging is expected to temper reservoir vulnerability and any evidence of this will be reported in the next section when the reservoir performance is discussed. However, a re-assuring aspect of the hedging policies is that the amounts of water cut back represented by the optimised rationing ratios as obtained herein are modest and lesser than the 25 % tolerable shortage suggested by Fiering (1982).

The hedging policies in Fig. 5 relate to static, i.e. constant rationing ratios situation; however, a further aim of the work had been to investigate dynamically varying rationing ratios. Although both the monthly and seasonally varying options were investigated, only the results of the monthly are presented to save space. The CRC for the dynamically varying hedging policies are shown in Fig. 6a (single stage) and 6b (two-stage), respectively; their associated optimised supply limits (or rationing ratios) are shown in Table 2. Unlike the static situation for which the rationing ratio was constant, dynamic rationing has varied monthly, reflecting the relative water abundance in the various months.

Thus, as seen in Fig. 6a for example, the proportion of the demand supplied in the monsoon months was highest, almost approaching 100 %. As the available water reduces, e.g. during the winter and pre-monsoon seasons, the proportion of the demand supplied attained its least value of < 80 %. Another feature of the dynamic scheme is that the optimised critical storage curves that trigger hedging have also responded to the reservoir inflow situation in that during the low inflow winter season, the curves are below those for the static case ensuring that the water available for meeting the full demand is more and hence occasions when reductions will be needed will be few. On the other hand during the high inflow, monsoon seasons, the critical curves for the dynamic policies are higher than those of constant hedging,

Table 2. Rationing ratios for monthly varying hedging policies.

Month	Rationing Ratios (%)		
	Single stage (α_m)	Two-stage	
		($\alpha_{1,m}$)	($\alpha_{2,m}$)
January	76.0	78.2	69.8
February	78.9	77.9	68.2
March	75.0	77.2	68.8
April	75.0	77.0	69.0
May	84.7	86.9	78.5
June	95.0	97.0	89.0
July	94.3	96.5	88.1
August	94.9	96.9	88.7
September	87.7	94.9	86.5
October	90.0	92.2	86.0
November	85.3	87.3	79.1
December	83.8	86.0	77.6

meaning that rationing will occur more frequently albeit the cut back amounts would be very small since the associated rationing ratios are close to unity. For the 2-stage policy in Fig. 6b, the monthly rationing ratios presented in Table 2 behave as expected, with those for the lower zone being less than those for the upper zone.

The implication of this is that the dynamic policies will offer improvement in performance over the static policy but the question remains by how much? The attractiveness of the dynamic policies therefore would stem from their effect on the system performance: a very significant improvement in performance over the static policy case would be required to justify the preference for the former.

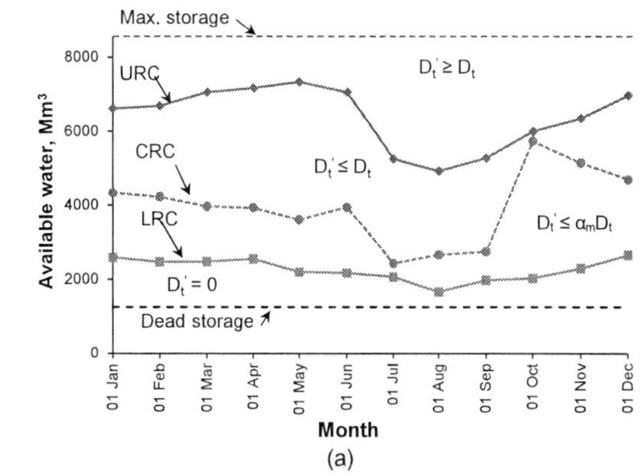

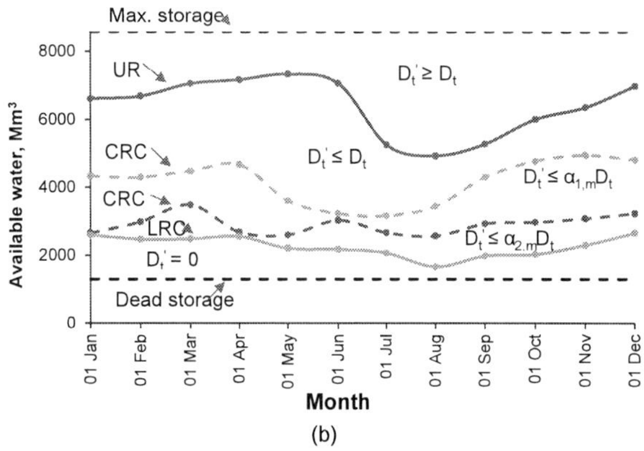

Figure 6. GA optimised rule curves **(a)** Single-stage monthly varying hedging; **(b)** two-stage monthly varying hedging.

4.3 Reservoir performance

To save space, discussion on the performance evaluation will be limited to the reliability (time- and volume-based) and the vulnerability. Figures 7–9 show the results of the performance evaluation using: time-based reliability (Fig. 7); volume-based reliability (Fig. 8) and the vulnerability (Fig. 9).

With no hedging, the time-based reliability is high under existing conditions but decreases when less water is projected by climate change, and increases when more inflow is projected. The effect of temperature rise on the inflow contributions from melting ice and snow has manifested in the time reliability, with the projected 2 °C rise in temperature in particular resulting in the highest reliability. The effect of hedging is to cause the time-reliability to deteriorate significantly. This is to be expected given that hedging deliberately introduces additional failures (i.e. occasions of non-supply of full demand). The greatest reduction in the time reliability occurred with the static, constant hedging policy where the time reliability reduced by almost 50 % relative to the no-hedging situation. When hedging is dynamically altered, the deterioration in the time reliability is reduced but only marginally. The 2-stage policy also provided marginal improvements in the R_t when compared to the single stage policy.

The volume reliability shown in Fig. 8 has confirmed that $R_v \geq R_t$ and that R_v is not as drastically affected by hedging as R_t. The R_v situation improved with both the dynamic and 2-stage policies but this improvement was rather marginal.

The vulnerability (or maximum single-period water shortage) is shown in Fig. 9. Also shown on the plots is the horizontal line for a vulnerability of 25 % which, as remarked previously, represents the suggested tolerable shortage limit for most water users. With no hedging, the vulnerability is high (approximately 60 %) under existing conditions and intensifies to about 65 % when the catchment becomes drier due to projected reduction in precipitation by climate change. The vulnerability is tempered for wetter conditions but even for the most benign of these, i.e. projected 10 % rise in precipitation, the recorded vulnerability was still above 47 %, much higher than the 25 % tolerable vulnerability threshold suggested by Fiering (1982).

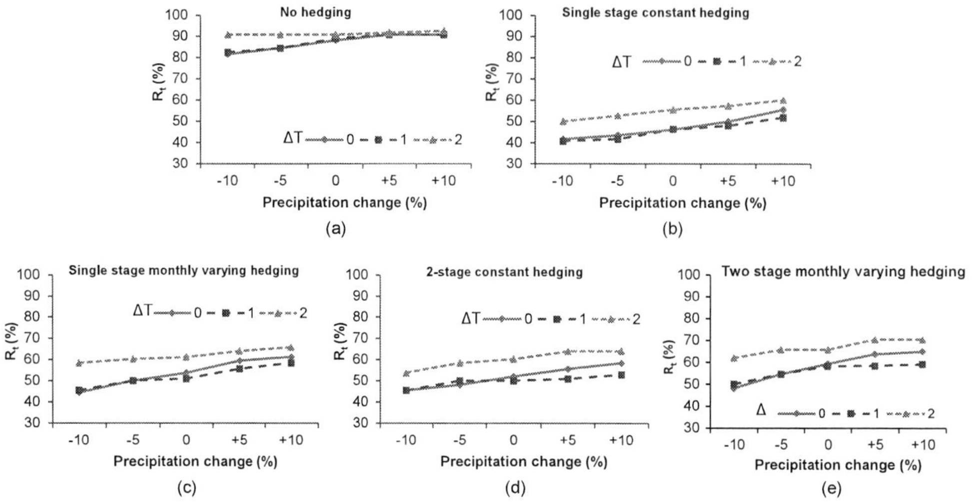

Figure 7. Hedging effect on reservoir performance (time-reliability).

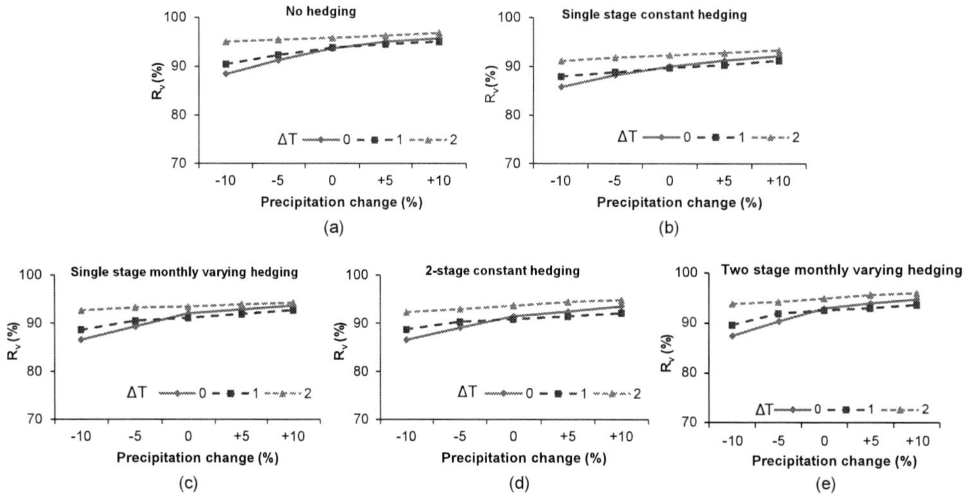

Figure 8. Hedging effect on reservoir performance (volume-reliability).

The dramatic effect of hedging on the vulnerability can be seen in the evaluations for both the static and the dynamic policies. For example for the static policy, hedging reduced the vulnerability to below the 25 % from the high of 65 %. It should be noted that only 17 % of the demand was hedged back when necessary by the single-stage, constant rationing ratio policy; yet what these results show is that such modest reductions that can be easily tolerated by most water users have almost eliminated the calamitous 65 % water shortage that could occur without hedging on some occasions. As was the case with the other performance indices, the vulnerability was slightly less relative to the single stage, constant rationing policy when the dynamic (monthly) and 2-stage policies were deployed. This reduction in vulnerability does not justify the extra efforts in developing and deploying the more complicated policies.

5 Conclusions

This study has developed optimised static and dynamic zone-based hedging policies for the Pong reservoir in India and compared its performance with that of a basic, zone-based policy that incorporates no hedging as a way of testing the usefulness of hedging in moderating the vulnerability of climate-change induced water shortages. GA was used to optimise the decision variables for the policies, including the rationing ratios and the target storage values that trigger hedging in each month of the year. The optimisation carried

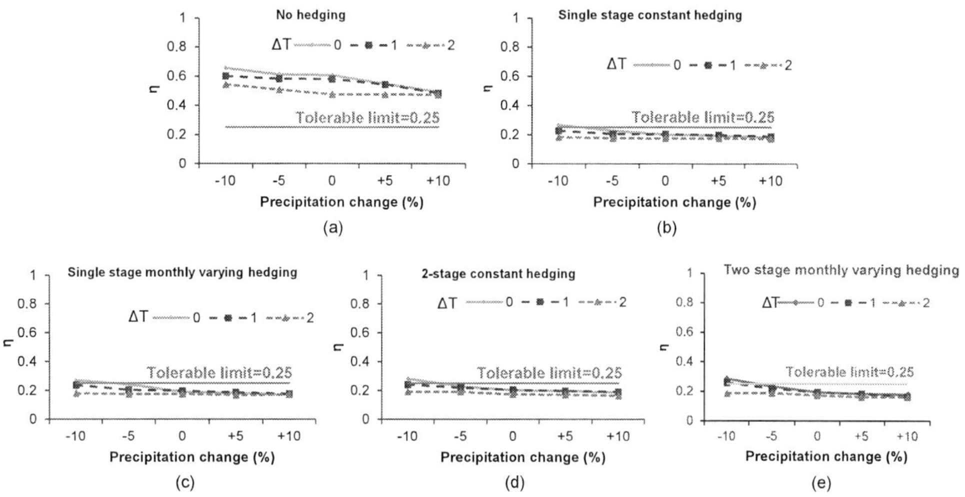

Figure 9. Hedging effect on reservoir performance (vulnerability).

out considered 1-stage and 2-stage hedging policies and rationing ratios that were either static (constant all year round) or dynamic (i.e. varying monthly or seasonally). A delta perturbation approach was used to develop alternative reservoir inflow responses to plausible changes in temperature and precipitation. Subsequent reservoir simulations to test the effectiveness of the various operating policies showed that without hedging, performance of the reservoir (reliability-time- and volume-based and vulnerability) deteriorated significantly when the reservoir inflow is projected to reduce due to climate change; the opposite occurred when the future is wetter.

The vulnerability was particularly high, reaching over 60 %. However, as hedging was introduced, the vulnerability reduced significantly because the modest, deliberate cut-backs during hedging prevented the occurrence of large, single-period water shortages. Indeed, for both the static and dynamic hedging policies, the vulnerability was reduced to below 25 % even for the worst (direst) climate change projections. However, because hedging deliberately fails to meet full demand on occasions, the occurrences of failures increased, which led to significant deterioration in the evaluated time-based reliability. However, since the amount of water shortages for most of these additional shortage periods was low-to-moderate, the overall volumetric reliability of the reservoir was practically unaffected. This is re-assuring since what should matter most in reservoir operation is not the number of failure occasions but the deficit sustained during such failures. All this confirms that water resources systems have inherent buffering capacity that if well-harnessed through improved operating practices such as the hedging policies developed in this work will offer effective and low-cost mitigation for climate change induced water shortages.

In terms of the overall system's performance, the dynamic hedging policies outperformed the constant hedging policy but only marginally. The same marginal improvement was recorded for the 2-stage policy when its performance was juxtaposed with that of the 1-stage policy. Given the complexity associated with the development and deployment of the dynamic and multi-stage policies, the marginal improvement recorded here is not sufficient reason for preferring the dynamic policy for reservoir operation.

Data availability. The data used in developing the results presented in this paper are openly available at https://doi.org/10.17861/5d02c025-f457-4511-a724-487407932495 (Adeloye, 2017).

Competing interests. The authors declare that they have no conflict of interest.

Special issue statement. This article is part of the special issue "Innovative water resources management – understanding and balancing interactions between humankind and nature". It is a result of the 8th International Water Resour. Manage. Conference of ICWRS, Beijing, China, 13–15 June 2018.

Acknowledgements. The work reported here was funded by the UK-NERC (Project NE/N016394/1) – "Sustaining Himalaya Water Resources in a Changing Climate (SusHi-Wat)" – as part of the UK-India Newton-Bhabha Sustainable Water Resources (SWR) thematic Programme.

Edited by: Wenchao Sun
Reviewed by: two anonymous referees

References

Adeloye, A. J., Soundharajan, B.-S., Ojha, C. S. P., and Remesan, R.: Effect of hedging-integrated rule curves on the performance of the Pong reservoir (India) during scenario-neutral climate change perturbations, Water Resour. Manage., 30, 445–470, https://doi.org/10.1007/s11269-015-1171-z, 2016.

Adeloye, A. J.: Simulated historic and climate-change-perturbed runoff inflow series data for Pong reservoir, available at: https://doi.org/10.17861/5d02c025-f457-4511-a724-487407932495 (last access: February 2018), 2017.

Anandhi, A., Frei, A., Pierson, D. C., Schneiderman, E. M., Zion, M. S., Lounsbury, D., and Matonse, A. H.: Examination of change factor methodologies for climate change impact assessment, Water Resour. Res., 47, W03501, https://doi.org/10.1029/2010WR009104, 2011.

Chiamsathit, C., Adeloye, A. J., and Soundharajan, B.: Assessing competing policies at Ubonratana reservoir, Thailand, Proceedings of the ICE-Water Management, 167, 551–560, 2014.

Eum, H., Kim, Y., and Palmer, R.: Optimal Drought Management Using Sampling Stochastic Dynamic Programming with a Hedging Rule, J. Water Resour. Plann. Manage., 137, 113–122, 2011.

Fiering, M. B.: Estimates of resilience indices by simulation, Water Resour. Res., 18, 41–50, 1982.

Fowler, H. J., Kilsby, C. G., and O'Connell, P. E.: Modeling the impacts of climatic change and variability on the reliability, resilience and vulnerability of a water resource system, Water Resour. Res., 39, 1222, https://doi.org/10.1029/2002WR001778, 2003.

Jain, S. K., Agarwal, P. K., and Singh, V. P.: Hydrology and water resources of India, Springer, the Netherlands, 2007.

Li, L., Xu, H., and Chen, X.: Streamflow forecast and reservoir operation performance assessment under climate change, Water Resour. Manage., 24, 83–104, 2009.

Manley, R. E. and Water Resources Associates (WRA): A guide to using HYSIM, R. E. Manley and water resources associates Ltd., 2006.

McMahon, T. A. and Adeloye, A. J.: Water Resources Yield, Water Resources Publications, Littleton, CO, USA, 2005.

Michalewicz, Z.: Genetic algorithms + data structures = evolution programs, Springer, New York, 1992.

Nawaz, N. R. and Adeloye, A. J.: Monte Carlo assessment of sampling uncertainty of climate change impacts on water resources yield in Yorkshire, England, Climatic Change, 78, 257–292, 2006.

Sandoval-Solis, S., Mckinney, D. C., and Loucks, D. P.: Sustainability index for water resources planning and management, J. Water Resour. Plann. Manage., 137, 381–389, 2011.

Vicuna, S., McPhee, J., and Garreaud, R. D.: Agriculture vulnerability to climate change in a snowmelt-driven basin in semiarid Chile, J. Water Resour. Plann. Manage. ASCE, 138, 431–441, 2012.

Wardlaw, R. and Sharif, M.: Evaluation of genetic algorithms for optimal reservoir operation, J. Water Resour. Plann. Manage. ASCE, 125, 25–33, 1999.

Yin, X.-A., Mao, X.-F., Pan, B.-Z., and Zhao, Y.-W.: Suitable range of reservoir storage capacities for environmental flow provision, Ecol. Eng., 76, 122–129, https://doi.org/10.1016/j.ecoleng.2014.04.002, 2015.

https://doi.org/10.5194/piahs-379-31-2018

Open Access
PIAHS

Stability and tilting of regional water cycle over Tarim Basin

Hongquan Zhang[1] **and Zhuguo Ma**[2]

[1]McDougall School of Petroleum Engineering, the University of Tulsa, Tulsa, OK 74104, USA
[2]Key Laboratory of Regional Climate-Environment Research for Temperate East Asia, Chinese Academy of Science, Beijing 100029, China

Correspondence: Hongquan Zhang (hong-quan-zhang@utulsa.edu)

Received: 13 December 2017 – Revised: 13 March 2018 – Accepted: 27 March 2018 – Published: 5 June 2018

Abstract. The Tarim Basin is located upwind of the Gobi Desert where individual deserts have expanded significantly during the last 50 years. In recent history, stable runoff in the Tarim Basin has been observed despite the Lop Nur dry up and dramatic water consumption shift from east to west. This regional water cycle stability is conceptually explained based on the relationship between precipitation and evapotranspiration. The water consumption imbalance is caused by human activities near the river sources, which tilts the humidity profile over the basin. As a result, more water vapour spills from the western part of the basin and causes precipitation to increase in adjacent areas. At the same time, the Westerlies carry the low humidity air mass out of the eastern part of the basin to make the downwind Gobi Desert and surrounding areas drier. Therefore, the observed wetting on the west and drying on the east of northwest China are coupled.

1 Introduction

During the last 50 years, deserts in northwest China have expanded significantly (Bai et al., 2002). Many residents had to relocate as a result of the ongoing desertification. Considerable precipitation decrease has been observed around the Gobi Desert area (Yang and Zhai, 2005; Guo et al., 2013). Running through the downwind area of the Gobi Desert, the Yellow River discharge rate has dropped more than 70 % (Ren et al., 2015). This regional environmental degradation has also negatively affected the surrounding provinces with frequent sand storms. These dramatic changes in such a short time are very unusual even considering the ongoing global warming process. Understanding the mechanism of this devastating process is crucial to finding solutions for the problem.

West of the Gobi area, the Tarim Basin (TB) is surrounded by high rise mountain ranges on its southern (Kunlun Mountains and Tibetan Plateau), western (Pamir Mountains) and northern (Tian Shan) sides (Fig. 1). On the eastern side are the much lower Beishan Mountain and the Hexi Corridor. Nine river basins and 114 tributaries in TB provide water for oases. Most tributaries connect to the Tarim River, the only mainstream in the basin flowing 1321 km from west to the Taitema Lake. Before AD 1960, the Tarim River and several other rivers flowed into the terminal lake Lop Nur, which is located at the east end of TB. About 2000 years ago, Lop Nur was described in the ancient Chinese history book "*Han Shu*" to be as "wide as 150 km, with still water and unnoticeable change through winter and summer". The estimated water surface area was 8426 km^2 (Yuan and Yuan, 1998). From then to 1950, Lop Nur gradually shrank to about 2000 km^2 with a basin population increase from 0.23 to 3.03 million. In 1962, it suddenly dried up due to upstream water withdrawals for agricultural irrigation with more than 200 newly constructed reservoirs (Fan et al., 2009).

After 1970, average precipitation has increased about 30 % in the Tian Shan Mountain region and 8 % in the adjacent areas on the Tibetan Plateau (Zhang et al., 2012; Huang et al., 2015). Many believed a "warm-dry" to "warm-wet" climate transition was happening (Shi et al., 2003; Li et al., 2003; Huang et al., 2011; Yao et al., 2013). There were even speculations that this wetting trend would consequentially extend to the eastern areas. Researchers have tried to relate the precipitation increase to large-scale atmospheric circulation and

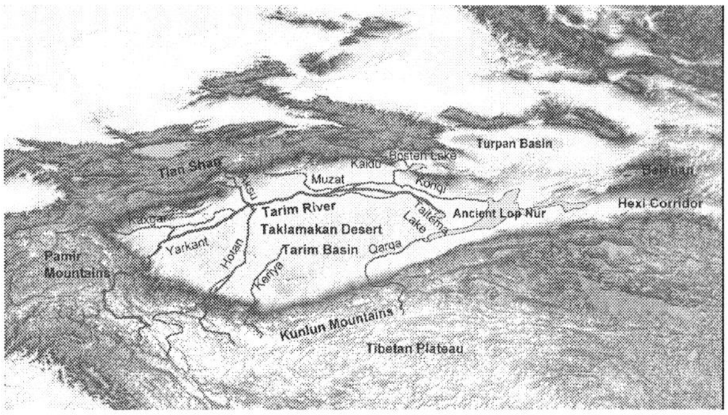

Figure 1. Rivers and ancient Lop Nur in Tarim Basin.

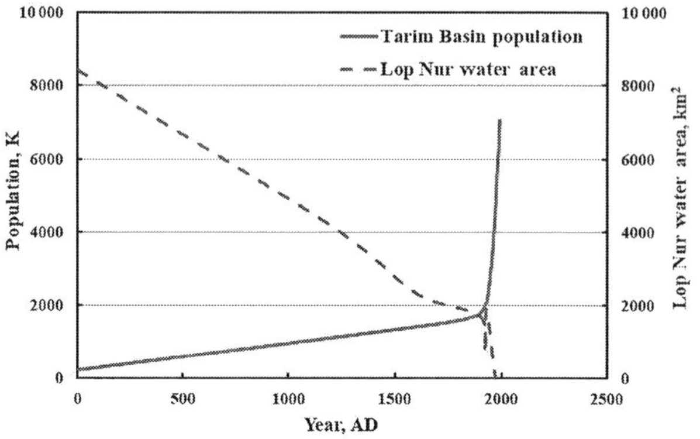

Figure 2. Tarim Basin population trend and correspondent Lop Nur water.

climate changes (Tao et al., 2016; Zuo et al., 2004). However, global warming is believed to result in dry areas becoming drier and wet areas becoming wetter (Trenberth, 2011). Studies show that the precipitation increase has not changed the area around west TB from dry to wet category (Ma et al., 2005; Ma and Fu, 2006). Therefore, the precipitation increase was not due to the area being wetter than normal. It is worth noting that the afore mentioned drying process around the Gobi Desert in the downwind area east of TB has been taking place simultaneously with this wetting trend around the west TB (Song and Zhang, 2003; Yang, 2014). To date, no study has made a connection between these two opposing changes.

2 Water cycle in Tarim Basin is stable

Evidently, the shrinkage of the Lop Nur water surface area was due to the population growth in TB over the last 2000 years (as shown in Fig. 2). Since the Lop Nur dry up, water entering the Tarim River has dropped to less than 10 % of the total annual runoff in the basin which is about 39.8 billion m^3 year^{-1} (Fan et al., 2009). Although the main water consumption has shifted to the western half of the basin, the overall runoff from the river sources in TB has not changed (Yang and He, 2003). Many studies have found that the water retreat from the east side was mainly caused by human activities rather than climate change (Chen and Xu, 2004; Hao et al., 2008).

On the opposite side of the globe, the Great Salt Lake in the Great Basin of western USA has shrunk about 50 % since 1847. A recent study reveals that the shrinkage of Great Salt Lake is solely due to consumptive water use by humans (Wurtsbaugh et al., 2016). Despite this shrinkage, the river flow rate ahead of diversions has been very stable over the past 150 years. The study demonstrates that the long-term trend in the lake level would have been flat if water had not been diverted for agriculture, industrial, urban and impounded wetland uses. Thus, water cycle stability appears to be a universal phenomenon occurring in all endorheic basins.

The water cycle stability in TB is governed by the relationship between precipitation and evapotranspiration. Most of the runoff originates from orographic precipitation spillover in the surrounding mountain ranges when atmospheric circulation passes over the TB area. However, most of the evapotranspiration occurs inside the basin during the course of surface water flow. Due to the distance and low humidity, the variation of evapotranspiration does not have a significant impact on precipitation. Assuming all the precipitation is taken away from the basin without any evapotranspiration, the surrounding mountain ranges would still provide a similar (but slightly lower) amount of precipitation (P_{AO}). Surface water evapotranspiration is a water vapour mass transfer process which determines the water concentration (WC) or humidity distribution across the atmospheric boundary layer (Katsaros, 2003). As shown in Fig. 3a, the WC profile from the basin surface to its top would be almost constant if there were no water vapour transfer from ground to atmosphere. At the same time, the average WC profile on the top of the basin would also remain flat because there is no net water vapour outflow from the basin top to the surrounding atmosphere (Bergman et al., 2011).

In TB, all the precipitation must be returned to the atmosphere through evapotranspiration so that there is no continuous water accumulation in the basin. Assuming evenly distributed evapotranspiration across the basin, WC will be higher on the surface and decrease with the height corresponding to the water vapour mass transfer from surface to the basin top (as shown by the middle red chart in Fig. 3b). On the basin top WC will be higher than that of the surrounding atmosphere in order to facilitate the water vapour outflow from the basin (as shown by the top purple chart in Fig. 3b). There will be a small precipitation increase (P_{AI}) due to the evapotranspiration inside the basin, compared to the assumed no-evaporation case. Therefore, at equilibrium the total average precipitation (P_A) and evapotranspiration (E_A) in the basin will be the same and equal to $E_{AE} = P_{AO} + P_{AI}$. The water vapour outflow from the basin remains the same (P_{AO}).

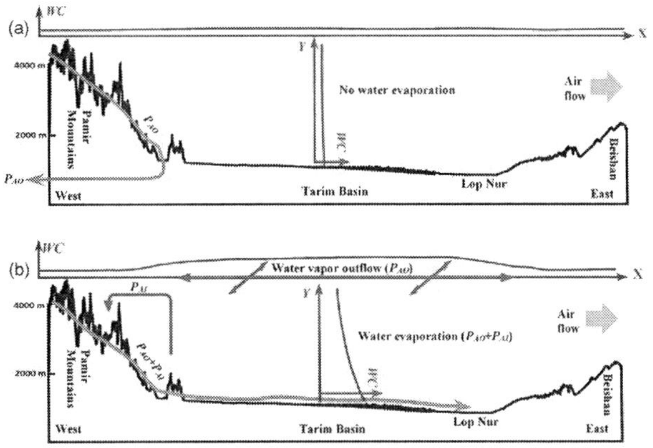

Figure 3. Precipitation in mountain ranges around Tarim Basin and evapotranspiration inside the basin. **(a)** Assuming all precipitation is taken away without evapotranspiration, humidity in the basin would be low and uniform. **(b)** In reality, all runoff must be returned to atmosphere as water vapour, and the humidity at the top of the basin is higher than that of the surrounding atmosphere.

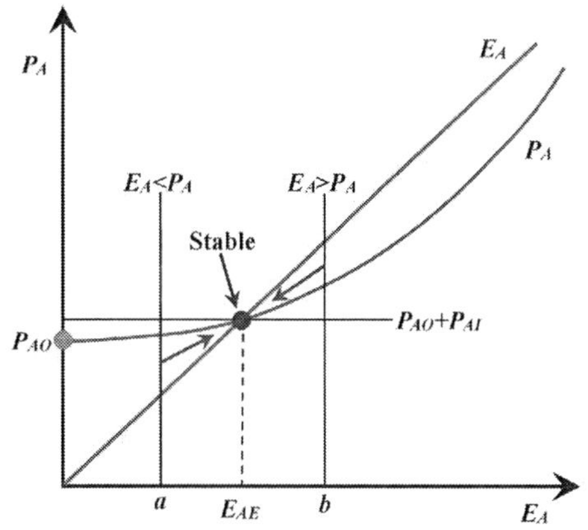

Figure 4. Non-linear relationship between precipitation (P_A) and evapotranspiration (E_A), and basin water cycle stabilization.

The relationship between P_A and E_A in the Tarim Basin is non-linear. If E_A is changed from low to high by output (when $E_A < E_{AE}$) and input (when $E_A > E_{AE}$) of surface water, P_A must follow the blue curve shown in Fig. 4. The green dot corresponds to the case displayed in Fig. 3a, assuming all precipitation is removed from the basin without any evapotranspiration. The blue dot corresponds to the case displayed in Fig. 3b, when all precipitation evaporates without any water input or output on the basin surface.

External weather systems may temporarily bring large amounts of water into the basin. When the surface evaporable water in the basin is above the equilibrium level, evapotranspiration will be higher due to the increased water evaporation surface. Precipitation will also be higher but its increase is much less than that of evapotranspiration. As a result, $E_A > P_A$, and the surface water amount in the system will decrease. Then, evapotranspiration will fall back to its equilibrium value (E_{AE}). If surface water in the basin falls below the equilibrium level due to drought, evapotranspiration will be lower due to the reduced water evaporation surface. Precipitation will also be lower but its decrease is much less than that of evapotranspiration. As a result, $P_A > E_A$, and the surface water amount in the basin will increase. Then, evapotranspiration will increase to E_{AE}. This water cycle stabilization mechanism must exist in all endorheic basins.

3 Humidity tilting over Tarim Basin

The water distribution shift in TB is a result of increased water consumption due to population expansion and other developments. When more water is consumed near the water heads by convenience, less water is left for downstream areas. After the Lop Nur dried up, the Tarim River flowed into the Taitema Lake about 200 km west. In 1974, the Taitema Lake also dried up. Since then, the dried-up length of the Tarim River increased continuously and reached 1200 km of the total river length of 1321 km in 2009 (Chen et al., 2009). The majority of the water runoff in TB was consumed in the northwest part of the basin where more oases and population were located (Fan et al., 2001). As a result, the local air humidity became much higher. Since the 1970's, the west TB and surrounding areas have seen precipitation increases (Hu et al., 2002; Shi et al., 2002). At the same time the dry area around Lop Nur in the east part of the basin has expanded significantly. The relative humidity in the Lop Nur area is only about 10 % compared to the basin's average relative humidity about 50 % (Luo et al., 2005; Li et al., 2012). The annual precipitation is about 20 mm compared to 90 mm in the west part of the basin (Shang et al., 2016). The Lop Nur became the driest area in China.

As illustrated in Fig. 5, corresponding to the current water consumption imbalance in TB, the water concentration (WC) profiles from basin surface to top are very different in the west and east. In the west, where most runoff evaporates, WC is much higher on the surface and it decreases with height corresponding to the water vapour mass transfer from ground to atmosphere. At the top of the basin west, WC is much higher than that corresponding to an even water distribution shown in Fig. 3b, to drive a greater mass transfer of water vapour from the basin to outside. On the basin east around the dried Lop Nur, there is almost no water evaporation. The WC profile from basin surface to top is low and flat. The humidity on top of the basin east is close to that of the surrounding atmosphere, and the water vapour outflow is small. A recent analysis shows that from 1979 to 2012, there was a large decline of moisture transport out of the basin across the eastern boundary (Ren et al., 2016), which agrees with the effect of humidity tilting after the dry-up of Lop Nur.

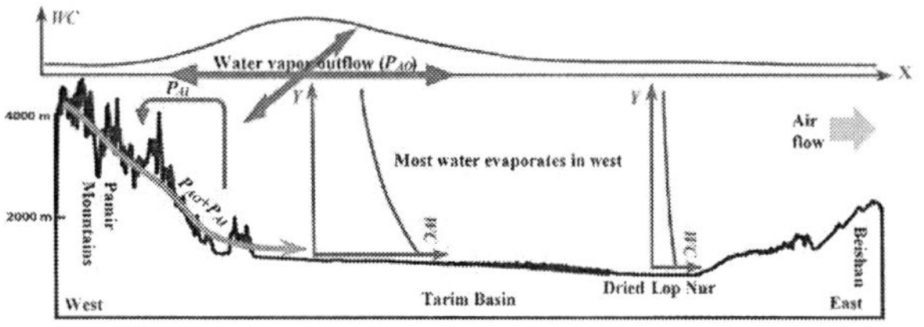

Figure 5. Currently most water evaporates in west Tarim Basin causing top humidity tilting.

4 Antiphase effects of humidity tilting

As shown in Fig. 6a, TB is surrounded by high rise mountain ranges on its south, west and north sides. Studies show a stable divergence area in west TB which helps diffuse the water vapour outward (Wu et al., 2016; Wang et al., 2006). The water vapour outflow from the basin west increases the precipitation in adjacent areas. Part of the increased precipitation falls in the Aksu River basin and flows back into TB through the Aksu River, increasing its runoff (Chen and Dai, 2009). The local northward wind direction and surface water concentration on the northwest side of TB explain why the precipitation increase is greater in the Tian Shan Mountains than on the Tibetan Plateau.

The shape of TB is like a dustpan with the opening facing the east. The Gobi Desert (encircled by the dashed red line in Fig. 6a) extends from TB to the east with a vague and ever changing boundary. The low humidity air mass from the Lop Nur area is carried out by the west wind to the Gobi Desert areas and makes the environment even drier. Rapid desert expansions occurred in this region following the desiccation of the Lop Nur. The Baidan Jaran desert (the largest on the Gobi Desert) expanded about 80 % from 1970's to 1990's (Bai et al., 2002). The second and third largest deserts in the area, the Tengger desert and Ulan Buh desert, expanded about 16 and 38 % respectively during the same time. These three previously separate deserts connected recently (Liu et al., 2011). Many naturally formed lakes on the Gobi Desert and surrounding areas disappeared or shrank, but lakes on the Tibetan Plateau expanded (Zhang et al., 2013, 2017). The significant air humidity decrease in the upwind Lop Nur area must be responsible for these rapid transitions.

Figure 6b shows the five-year average annual precipitation distribution around TB from January 1955 to December 1959, right before Lop Nur dried up. Figure 5c shows the five-year average annual precipitation distribution in the same area from January 2000 to December 2004, about 40 years after Lop Nur dried up. Clearly, there was a considerable shrinkage of the low precipitation area over TB, but the dry area on the Gobi Desert east of TB expanded significantly at the same time. The large-scale atmospheric circulation cannot provide explanations for the antiphase trends simultaneously occurring in two neighbouring areas, although

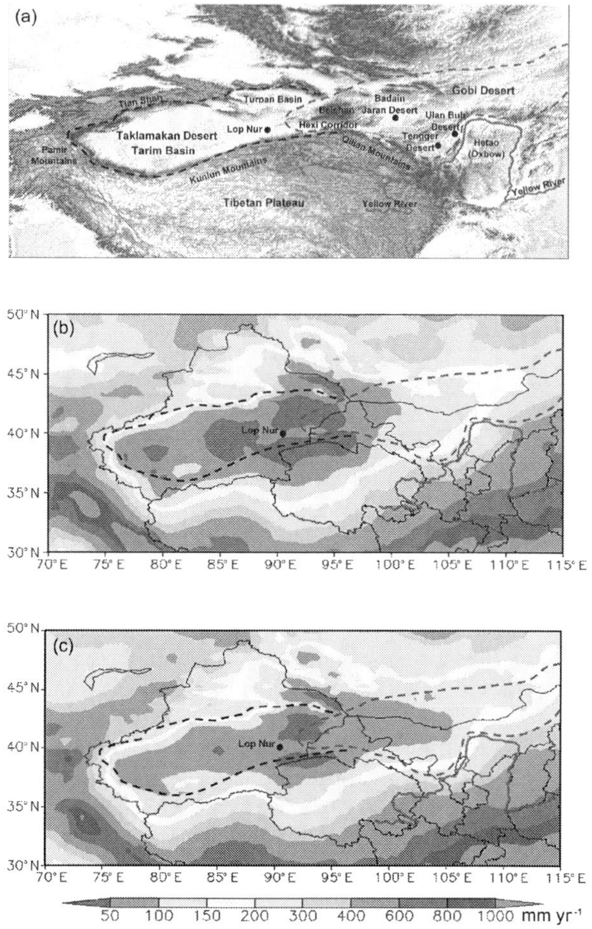

Figure 6. Change of precipitation distribution around TB with water consumption shift to west (precipitation data from Climate Research Unit (Harris et al., 2014) **(a)** Gobi Desert is downwind of Lop Nur. **(b)** Annual precipitation distribution around TB averaged over January 1955 to December 1959. **(c)** Annual precipitation distribution around TB averaged over January 2000 to December 2004.

individual draught and wetness extremes may be correlated to the long distance upstream anomalies (Tao et al., 2014; Borth et al., 2016). These two antiphase changes are obviously coupled. We believe that the root cause is local rather than remote, namely the humidity tilting over TB as a result of the dramatic water consumption shift to west TB is responsible for these opposite changes.

Shown in Fig. 6, the Yellow River Hetao (Oxbow) area serves as the front line of defence against desertification. However, precipitation in this area and adjacent regions dropped about 15 % from 1960 to 2010 (Wang et al., 2015; Zhong et al., 2006; Chen et al., 2008). A significant amount of water was drained from the Yellow River to meet the water deficit for irrigation and other consumptions. During the same time, the Yellow River discharge rate decreased about

47.5 % at the top of the Hetao area, and decreased about 68.9 % after passing the Hetao area (Li et al., 2014; Piao et al., 2010; Ma, 2005). This must have contributed to the Yellow River dry up in its downstream section which started in 1972. Therefore, the significant flow rate decrease and frequent dry-ups of the Yellow River are correlated to the dry-ups of Lop Nur and Tarim River in TB.

5 Concluding remarks

The regional water cycle in TB is stable because precipitation mostly originates from the high altitude surrounding mountain ranges. The evapotranspiration inside the basin has an insignificant effect on precipitation. Any water fluctuation in the basin will be drawn back to the balance point. This means that the available water in TB for consumption (evapotranspiration) is almost constant.

With the water consumption shift to the west of TB, the humidity profile over the basin has become significantly tilted. The area around TB on the west has become wetter while the Gobi Desert and its surrounding areas have become drier. These antiphase trends are coupled with the humidity tilting over TB. The precipitation increase around west TB is a moisture holdup which deprives humidity from the air flowing into the eastern Gobi Desert causing rapid expansion of this arid area. Therefore, the solution to the environmental degradation in northwest China is directly related to the water cycle in TB where water distribution must to be rebalanced.

Data availability. No data sets were used in this article.

Competing interests. The authors declare that they have no conflict of interest.

Special issue statement. This article is part of the special issue "Innovative water resources management – understanding and balancing interactions between humankind and nature". It is a result of the 8th International Water Resources Management Conference of ICWRS, Beijing, China, 13–15 June 2018.

Acknowledgements. We thank Liang Chen, Ziyan Zheng and Qing Yang for generating some of the figures.

Edited by: Andreas Schumann
Reviewed by: two anonymous referees

References

Bai, M., Shen, J., Pei, H., and Hao, R.: Assessment of climate change impact on desertification, Climatic and Environmental Research, 7, 457–464, 2002.

Bergman, T. L., Lavine, A. S., Incropera, F. P., and Dewitt, D. P.: Fundamentals of heat and mass transfer, USA, John Wiley & Sons, 2011.

Borth, H., Tao, H., Fraedrich, K., Schneidereit, A., and Zhu, X.: Hydrological extrems in the Aksu-Tarim River Basin: Mid-latitude dynamics, Clim. Dynam., 46, 2039–2050, 2016.

Chen, D. and Dai, Y.: Characteristics of northwest China rainfall intensity in recent 50 years, Chinese Journal of Atmospheric Sciences, 33, 923–935, 2009.

Chen, Y. and Xu, Z.: Possible effect of climate change on water resources in Tarim Basin, Science China D, 34, 1047–1053, 2004.

Chen, Y., Chen, N., Zheng, G., Mu, J., Ma, S., Na, L., and Shao, J.: Change of temperature, precipitation and NDVI in recent 45 years in Ningxia, Journal of Natural Resources, 23, 626–634, 2008.

Chen, Y., Ye, Z., Mao, X., Zhang, X., and Luo, J.: Dried-up trend of Tarim River and the countermeasures for mitigation, Arid land Geography, 32, 813–820, 2009.

Fan, Z., Ma, Y., Ji, F., and Wang, R.: Relations between exploitation-utilization of water resources and oasis evolution and ecological balance in Tarim Basin, Journal of Natural Resources, 16, 22–27, 2001.

Fan, Z., Kurban, A., Xu, H., Zhang, Q., and Abdumijiti: Changes of Tarim River and Evolution of Lop Nur, Quaternary Sci., 29, 233–240, 2009.

Guo, H., Li, D., Lin, S., Dong, Y., Sun, L., Huang, L., and Lin, J. : Temperal and spatial variation of precipitation over western China during 1954–2006, Journal of Glaciology and Geocryology, 35, 1165–1175, 2013.

Hao, X., Li, W., Chen, Y., and Li, C.: Indentification of human activity and climate change effects on Tarim River discharge, Prog. Nat. Sci., 18, 1409–1416, 2008.

Harris, I., Jones, P., Osborn, T., and Lister, D.: Updated high-resolution grids of monthly climatic observations – the CRU TS3.10 Dataset, Int. J. Climatol., 34, 623–642, 2014.

Hu, R., Jiang, F., Wang, Y., and Fan, Z.: A study on signals and effects of climatic pattern change from warm-dry to warm-wet in Xinjiang, Arid Land Geography, 25, 194–200, 2002.

Huang, X., Zhang, M., Jia, W. W. S., and Zhang, N.: Variations of surface humidity and its influential factors in Northwest China, Adv. Water Sci., 22, 151–159, 2011.

Huang, X., Li, Y., Feng, J., Wang, J., Wang, Z., Wang, S., and Zhang, Y.: Climate characteristics of precipitation and extreme drought events in Northwest China, Acta Ecologica Sinica, 35, 1359–1370, 2015.

Katsaros, K.: Encyclopedia of Atmospheric Sciences, USA, Elsevier, 2003.

Li, D., Wei, L., Cai, Y., Zhang, C., Feng, J., Yang, Q., Yuan, Y., and Dong, A.: The present facts and the future tendency of the climate change in Northwest China, Journal of Claciology and Geocryology, 25, 135–142, 2003.

Li, E., Mu, X., and Zhao, G.: Temporal changes in annual runoff and influential factors in the upper and middle reaches of Yellow

River from 1919 to 2010, Advances in Water Science, 25, 155–163, 2014.

Li, H., Mao, W., Zhao, Y., Wang, M., and Huo, W.: Trends and abrupt changes in surface vapor content over Tarim Basin during the last 50 years, J. Arid Land, 4, 260–270, 2012.

Liu, Y., Wang, X., Zhang, X., and Zhang, B.: Desertification process and its driving forces in the dune-activation region between Badain Jaran and Tegger deserts, Arid Zone Research, 28, 957–966, 2011.

Luo, C., Peng, Z., Liu, W., and Liu, G.: Research on the environmental evolution of Lop Nur in Xinjiang, China, Sciencepaper Online, 2005.

Ma, Z.: Historical regular patterns of the discharge in the Yellow River and the cause of their formation, Chinese J. Geophys.-Ch., 48, 1270–1275, 2005.

Ma, Z. and Fu, C.: Some evidence of drying trend over north China from 1951 to 2004, Chinese Sci. Bull., 51, 2931–2925, 2006.

Ma, Z., Huang, G., Gan, W., and Chen, M.: Multi-scale temporal characteristics of the dryness/wetness over north China during the last century, Chinese Journal of Atmospheric Sciences, 29, 671–681, 2005.

Piao, S., Ciais, P., Huang, Y., Shen, Z., Peng, S., Li, J., Zhou, L., Liu, H., Ma, Y., Ding, Y., Friedlingstein, P., Liu, C., Tan, K., Yu, Y., Zhang, T., and Fang, J.: The impacts of climate change on water resources and agriculture in China, Nature, 467, 43–51, 2010.

Ren, G., Yuan, Y., Liu, Y., Ren, Y., Wang, T., and Ren, X.: Changes in precipitation over northwest China, Arid Zone Research, 33, 1–19, 2016.

Ren, H., Li, G., Cui, L., and He, L.: Multi-scale variability of water discharge and sediment load into the Bohai Sea from 1950 to 2011, J. Geogr. Sci., 25, 85–100, 2015.

Shang, H., Tuohuti, M., Zhang, R., Zhang, T., Zhang, K., Wei, W., Fan, Z., and Chen F.: Annual precipitation reconstruction and analysis in the western edge of Tarim Basin during AD1715-2014, Desert and Oasis Meteorology, 10, 9–17, 2016.

Shi, Y., Shen, Y., and Hu, R.: Preliminary study on signal impact and foreground of climatic shift from warm-dry to warm-humid in Northwest China, Journal of Glaciology and Geocryology, 24, 219–226, 2002.

Shi, Y., Shen, Y., Li, D., Zhang, G., Ding, Y., Hu, R., and Kang, E.: Discussion on the present climate change from warm-dry to warm-wet in northwest China, Quaternary Sci., 23, 152–164, 2003.

Song, L. and Zhang, C.: Changing features of precipitation over northwest China during the 20th century, Journal of Glaciology and Geocryology, 25, 143–148, 2003.

Tao, H., Borth, H., Fraedrich, K., Su, B., and Zhu, X.: Drought and wetness variability in the Tarim River Basin and connection to large-scale atmospheric circulation, Int. J. Climatol., 34, 2678–2684, 2014.

Tao, H., Borth, H., Fraedrich, K., Schnerdereit, A., and Zhu, X.: Hydrological extremes in the Aksu-Tarim River Basin: Climatology and regime shift, Clim. Dynam., 46, 2029–2037, 2016.

Trenberth, K. E.: Changes in precipitation with climate change, Clim. Res., 47, 123–138, 2011.

Wang, K., Jiang, H., and Zhang, H.: Advection and convergence of water vapor transport over the northwest China, Advances in Water Science, 17, 164–169, 2006.

Wang, Y., Xiao, T., Na, L., and Zhu, X.: Variation characteristics of precipitation in Ningxia and its response to heating field over Tibet Plateau, Journal of Arid Meteorology, 33, 574–580, 2015.

Wu, Y., Feng, G., and Li, B.: Interactions of multiple atmospheric circulation drive the drought in Tarim River Basin, Sci. Rep.-UK, 6, 1–7, 2016.

Wurtsbaugh, W., Miller, C., Null, S., Wilcock, P., Hahnenberger, M., and Howe, F.: Impacts of water development on Great Salt Lake and the Wasatch front, s.l.: Watershed Science Faculty Publications, 2016.

Yang, Q. and He, Q.: Interrelationship of climate change, runoff and human activities in Tarim River Basin, Journal of Applied Meteorological Science, 14, 309–321, 2003.

Yang, X. and Zhai, P.: Changes in precipitation intensity frequency and total in Northwest China, Science and Technology Review, 23, 24–26, 2005.

Yang, Y.: Characteristics of precipitation anomalies in the last 50 years in eastern part of Northwest China, Journal of Arid Meteorology, 32, 701–711, 2014.

Yao, J., Yang, Q., Chen, Y., Hu, W., Liu, Z., and Zhao, L.: Climate change in arid areas of northwest China in past 50 years and its effects on the local ecological environment, Chinese Journal of Ecology, 32, 1283–1291, 2013.

Yuan, G. and Yuan, L.: An approach to the environmental changes in Lop Nur history, Acta Geographica Sinica, 53, 83–89, 1998.

Zhang, G., Yao, T., Piao, S., Bolch, T., Xie, H., Chen, D., Gao, Y., O'Reilley, C.M., Shum, C.K., Yang, K., Yi, S., Lei, Y., Wang, W., He, Y., Shang, K., Yang, X., and Zhang, H.: Extensive and drastically different alpine lake changes on Asia's high plateaus during the past four decades, Geophys. Res. Lett., 44, 252–260, 2017.

Zhang, Z., Liu, L., and Tang, X.: The regional difference and abrupt events of climatic change in Tianshan mountains during 1960–2010, Progress in Geography, 31, 1475–1484, 2012.

Zhang, Z., Wang, N., Wu, Y., Shen, S., Zhang, X., and Chang, J.: Remote sensing on spatial changes of lake area in Badain Jaran desert hinterland during 1979–2010, Journal of Lake Science, 25, 514–520, 2013.

Zhong, H., Li, D., and Chen, X.: Study on variation trend of the precipitation in the oxbow area in recent 40 years, Plateau Meteorology, October, 25, 900–905, 2006.

Zuo, H., Lv, S., and Hu, Y.: Variations trend of yearly mean air temperature and precipitation in China in the last 50 years, Plateau Meteorology, 23, 238–244, 2004.

https://doi.org/10.5194/piahs-379-37-2018

Open Access
PIAHS

Understanding the Impacts of Climate Change in the Tana River Basin, Kenya

Lal Muthuwatta[1], Aditya Sood[2], Matthew McCartney[3], Nishchitha Sandeepana Silva[4], and Alfred Opere[5]

[1]International Water Management Institute, Colombo, Sri Lanka
[2]The Nature Conservancy, New Delhi, India
[3]International Water Management Institute, Vientiane, Lao PDR
[4]Department of Estuarine and Ocean Sciences, University of Massachusetts Dartmouth, School for Marine Science and Technology, New Bedford, Massachusetts, USA
[5]African Collaboration Centre for Earth System Science, Department of Geology, University of Nairobi, Nairobi, Kenya

Correspondence: Lal Muthuwatta (l.mutuwatte@cgiar.org)

Received: 31 December 2017 – Revised: 10 April 2018 – Accepted: 12 April 2018 – Published: 5 June 2018

Abstract. In the Tana River Basin in Kenya, six Regional Circulation Models (RCMs) simulating two Representative Concentration Pathways (RCPs) (i.e., 4.5 and 8.5) were used as input to the Soil and Water Assessment Tool (SWAT) model to determine the possible implications for the hydrology and water resources of the basin. Four hydrological characteristics – water yield, groundwater recharge, base flow and flow regulation – were determined and mapped throughout the basin for three 30-year time periods: 2020–2049, 2040–2069 and 2070–2099. Results were compared with a baseline period, 1983–2011. All four hydrological characteristics show steady increases under both RCPs for the entire basin but with considerable spatial heterogeneity and greater increases under RCP 8.5 than RCP 4.5. The results have important implications for the way water resources in the basin are managed. It is imperative that water managers and policy makers take into account the additional challenges imposed by climate change in operating built infrastructure.

1 Introduction

The Tana River plays a prominent role in Kenya. It starts from the Aberdare Mountains, west of Nyeri, Kenya and flows into the Indian Ocean after covering about 1000 km (Fig. 1). The catchment area of the river (83 972 km^2) covers 21 % of the Kenya's land area and holds about 18 % of the country's population (Sood et al., 2017). The river basin supplies 80 % of Nairobi's domestic water supply via transfer from the Thika dam (Baker et al., 2015). Approximately 50 % of Kenya's total electricity demand is met by the series of five dams built along the river (Baker et al., 2015). Fisheries and agriculture in the basin provide a major source of food and employment for the people that live in the basin and in other parts of the country. One of the basin's important ecosystems is the Tana Delta at the coast: almost 50 % of the country's freshwater discharge into the Indian Ocean is made up of the flow from the Tana River. The delta was designated as a Ramsar site in 2012 and is home to several endangered species. The basin has two rainy seasons (i.e. bimodal distribution) with the mean annual rainfall of about 648 mm (1983–2011). Approximately 92 % of the rain falls during these wet seasons, which are from March–May, and October–January.

As part of the "Water Infrastructure Solutions from Ecosystem Services underpinning Climate Resilient Policies and Programmes" (WISE-UP) project, six Regional Circulation Models (RCMs), simulating two Representative Concentration Pathways (i.e. RCPs 4.5 and 8.5), were used as an input to a hydrological model. Representative Concentration Pathways (RCPs) are greenhouse gas concentration trajectories adapted by the IPCC for its fifth Assessment Report (AR5) in 2014. RCP 4.5 assumes that the GHG emissions will in-

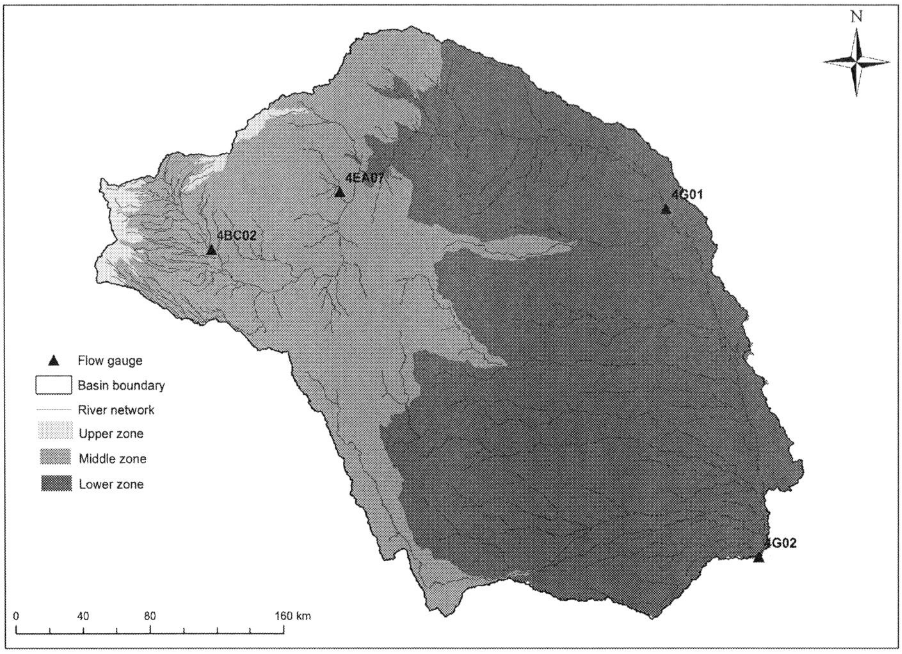

Figure 1. Tana River Basin showing three elevation zones, river network and flow gauging stations.

crease till about 2040 and then decline, whereas RCP 8.5 assumes that the emissions will continue to rise throughout the century and represents the worst-case scenario. RCP 4.5 and RCP 8.5 corresponds to 4.5 and 8.5 W m^{-2} increase in radiative forcing by the year 2100 as compared to pre-industrial values.

2 Methods and data

The Soil and Water Assessment Tool (SWAT) (Arnold et al., 1998) was used to simulate the hydrology of the basin for current and future scenarios. The model was configured and calibrated for the current situation and then used to determine the possible impacts of climate change on the hydrology of the basin.

SWAT is a publically available rainfall–runoff hydrology and water quality model. The model possesses adequate representation of physical processes governing hydrology and is particularly suitable for application in large river basins. In the SWAT model, a river basin is subdivided into multiple sub-catchments, each sub-catchment consisting of at least one representative stream. The sub-catchments are further divided into hydrologic response units (HRUs), which are lumped land areas within the catchment comprising unique land cover, soil, and slope combinations. For a detailed description of the SWAT model, reference is made to Muthuwatta et al. (2014). Various data sets were accumulated from global and local sources. The major data sets used in this study are listed in Table 1.

The ArcSWAT interface was used to pre-process the spatial data for the Tana River Basin. The SWAT Calibration and Uncertainty Program (SWATCUP) (Abbaspour, 2011) was used to calibrate the model. The model was calibrated and validated using monthly flow data available at the four locations indicated in Fig. 1. The performance indicators for the model calibration (i.e NSE and R^2) are presented in Table 2. The performance indicators, show that the model calibrated well for gauging stations 4EA07, 4G01 and 4G02. These stations are located downstream and represent the larger river basin. The most upstream gauging station (4BC02) did not perform well.

Once calibrated, the SWAT model was run from 1983–2011 to develop a baseline scenario. Multiple climate change scenarios were then run through the SWAT model to determine the impact of climate change on the hydrology of the Tana River basin. In this study, 6 RCMs of the Coordinated Regional Climate Downscaling Experiment (CORDEX: Giorgi and Gutowski, 2016) were used. Data for these RCMs and for each of two Representative Concentration Pathways (RCPs), i.e. RCP 4.5 and RCP 8.5, were used to create 12 CC "scenarios" (Table 3). The model was run for the period 2011 to 2099 and data were analysed for the three periods: 2020–2049, 2040–2069 and 2070–2099. Grid resolution of RCMs range from 25 to 50 km for the African region. Therefore, all six SWAT inputs were downscaled using the methodology described in El-Khoury et al. (2015). The centre of each grid was considered as an independent meteorological station and data were input into the SWAT

Table 1. Input data used in setting up the SWAT model for the Tana River Basin.

Category	Data	Data source	Spatial Resolution
Topography	Digital elevation model (DEM)	Shuttle Radar Topography Mission	90 m
Land use	Land use map	AFRICOVER (http://www.un-spider.org/links-and-resources/data-sources/land-cover-kenya-africover-fao, last access: 23 December 2017)	1 : 200 000
Soils	Digital map of the soils and soil properties	FAO Harmonized Soil Maps (http://www.fao.org/soils-portal/soil-survey/soil-maps-and-databases/harmonized-world-soil-database-v12/en/, last access: 23 December 2017)	30 arcsec
Climate	Rainfall, temperature, relative humidity, solar radiation, wind speed	The National Centers for Environmental Prediction (NCEP) Climate Forecast System Reanalysis (CFSR) (http://rda.ucar.edu/pub/cfsr.html, last access: 23 December 2017)	27 × 27 km
Hydrology	River discharge from 4 flow gauging stations	The Water Resource Management Agency (WRMA), Kenya	

Table 2. Model performance indicators for four locations in Tana River Basin.

Station	Station	Catchment	Coordinates	Calibration			Validation		
Id	Name	area (km^2)		Time Period	NSE	R^2	Time Period	NSE	R^2
4EA07	Mutonga	1835	−0.373275° N, 37.895803° E	Jan 1993–Dec 1998	0.79	0.79	Jan 1988–Dec 1992	0.38	0.43
4BC02	Sagana	520	−0.673711° N, 37.204516° E	Jan 2001–Jul 2004	0.34	0.63	Jan 1988–Dec 1992	0.39	0.47
4G01	Garissa	32551	−0.460717° N, 39.632683° E	Jan 2002–Sep 2006	0.58	0.61			
4G02	Garsen	83971	−2.281649° N, 40.122678° E	Feb 1988–Dec 1992	0.50	0.51			

model. The model assigns the station closest to the centroid of a sub-basin as a representative weather station for that sub-basin and utilizes the corresponding climate data.

For this study, the Tana Basin was divided into three elevation zones (Fig. 1). Sub-basins with average altitude greater than 2400 m were designated as the "Upper" zone, those with average altitude between 600 and 2400 m were grouped into the "Middle" zone and those below 600 m were grouped into the "Lower" zone. The area in the Upper, Middle and Lower zones are 1514, 29 890 and, 52 568 km^2 respectively. In February 2017, stakeholders from institutions with an interest in the basin, including the Kenya Electricity Generating Company, the Ministry of Water and Irrigation, the National Irrigation Board, the Water Resources Management Authority and the Tana and Athi Development Authority, were trained in the use of SWAT and its application to the Tana River Basin.

3 Results and discussion

3.1 Baseline simulation

Four hydrological characteristics – water yield, groundwater recharge, base flow and flow regulation were investigated. Water yield is the total water produced in a catchment that flows out of the catchment. Groundwater recharge is the water percolating beyond the root zone. Base flow is the groundwater contribution to the river flow. "Flow regulation" is the ratio of dry season flow to total annual flow in a stream. The spatial variation of the average annual values (1983–2011) for these hydrological variables and the rainfall are summarized in Table 4.

There is considerable heterogeneity in the response to rainfall throughout the basin, with large differences between the elevation zones. The average annual water yield in the lower zone is only 4 % of the annual rainfall. This compares to 39 % in the upper zone and 23 % in the middle zone. One major reason for higher water yield in the upper zone is topography; the greater slope in the upper zone results in higher runoff

Table 3. The climate change scenarios used for the Tana River Basin.

Institution	Driving GCM	RCM	RCP	Scenario Name
Canadian Centre for Climate Modelling and Analysis	CCCma-CanESM2	CCCma-CanRCM4	RCP4.5	TCM1-45
	CCCma-CanESM2	CCCma-CanRCM4	RCP8.5	TCM1-85
Climate Limited-area Modelling Community	CNRM-CERFACS-CNRM-CM5	CLMcom-CCLM4-8-17	RCP4.5	TCM2-45
	CNRM-CERFACS-CNRM-CM5	CLMcom-CCLM4-8-17	RCP8.5	TCM2-85
	ICHEC-EC-EARTH	CLMcom-CCLM4-8-17	RCP4.5	TCM3-45
	ICHEC-EC-EARTH	CLMcom-CCLM4-8-17	RCP8.5	TCM3-85
	MPI-M-MPI-ESM-LR	CLMcom-CCLM4-8-17	RCP4.5	TCM4-45
	MPI-M-MPI-ESM-LR	CLMcom-CCLM4-8-17	RCP8.5	TCM4-85
Royal Netherlands Meteorological Institute	ICHEC-EC-EARTH	KNMI-RACMO22T	RCP4.5	TCM5-45
	ICHEC-EC-EARTH	KNMI-RACMO22T	RCP8.5	TCM5-85
Swedish Meteorological and Hydrological Institute	IPSL-IPSL-CM5A-MR	SMHI-RCA4	RCP4.5	TCM6-45
	IPSL-IPSL-CM5A-MR	SMHI-RCA4	RCP8.5	TCM6-85

Table 4. Annual averages of hydrological outputs in the three zones of Tana River Basin for the current scenarios.

Tana Zone	Rainfall (mm)	Water yield (mm)	GW Recharge (mm)	Base Flow (mm)	Flow Regulation (%)
Upper	1289.1	498.2	257.0	81.3	31.2
Middle	896.2	207.5	171.9	41.2	35.1
Lower	475.7	17.6	30.1	3.5	33.4

than in the middle and lower zones. The annual groundwater recharge in the upper zone is about 20 % of the annual rainfall, it is 19 % for the middle zone and just 6 % in the lower zone. The base flow contribution to water yield is 16 % in the upper zone and 20 % in both the middle and lower zones.

3.2 Climate change – Rainfall trends

The results of the annual rainfall for the model simulations indicate that, compared to the baseline, both mean annual rainfall and inter-annual variability increase for both RCPs, with greater increases in RCP 8.5 for all three time periods. Figure 2 presents an inter-year comparison of annual rainfall for the period 2020–2040, two RCPs and six simulations (TCMs). Annual catchment average rainfall in the baseline period is 640 mm divided between two main rainy seasons. It is projected to increase to 787 and 912 mm, with earlier onset of both rainy seasons, by the end of the 21st century under RCP 4.5 and RCP 8.5, respectively.

Monthly rainfall, averaged across the ensemble of models (Fig. 3), separately for each RCP, indicate higher extreme values in March and April for RCP 8.5 compared to RCP 4.5 for the period 2020–2049. With the exception of RCP 8.5 for the period 2020–2049, all other periods show a reduction in average rainfall in April and May compared to the baseline. Rainfall in November and December also shows reductions other than the simulated values for the period 2070–2099 under RCP 8.5. These reductions are comparatively higher for RCP 4.5 than in RCP 8.5.

3.3 Climate change – Hydrological characteristics

Table 5 shows average values of rainfall and the four hydrological characteristics for all scenarios for RCP 4.5 and RCP 8.5.The flow generated in all CC scenarios is higher than the average flow generated in the baseline scenario. In general, compared to the baseline scenario, the average flow is higher in all the months in all three time periods with the largest increase in RCP 8.5. Furthermore, the extreme values clearly indicate that, with some exception in the 2020–2049 period, there is greater variability of flow under RCP 8.5 than RCP 4.5. The results also reflect the gradual shift to the earlier onset of rainfall with peak flows moving to earlier in the year.

Of all four variables considered in this study, water yield, groundwater recharge and base flow, increase significantly under both RCPs, across all three-elevation zones. However, flow regulation increases progressively in the middle and lower zones but not in the upper zone during the period 2070–2099. The increases in the hydrological variables are greater for the RCP 8.5 simulations than for RCP 4.5. For instance, for the whole basin, increase of water yield, groundwater recharge and base flow under RCP 4.5 for the period 2070–2099 compared to the baseline are 10.6, 94.8 and 227.9 % respectively. For the same period, respective in-

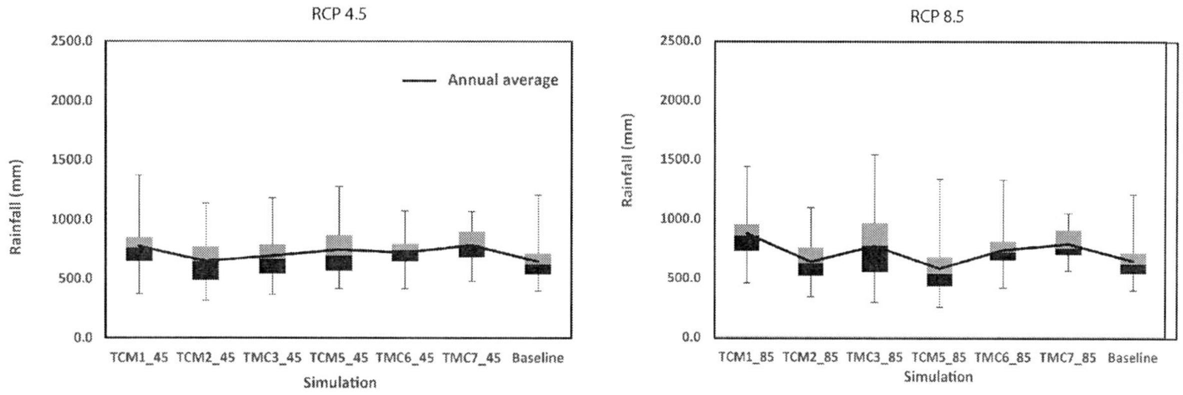

Figure 2. Rainfall as annual average for an inter-annual comparison for 2020–2049 and two RCPs.

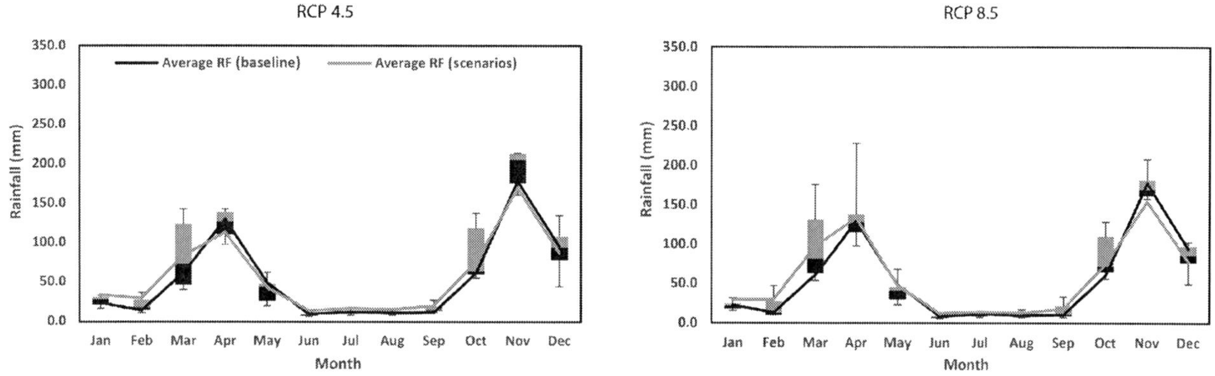

Figure 3. Simulated monthly rainfall (averaged across the model ensemble) for 2020–2049 and two RCPs.

Table 5. Impact of CC on rainfall, water yield, GW recharge, base flow and flow regulation **(a)** average values for all scenarios RCP 4.5; **(b)** average values for all scenarios RCP 8.5 in mm.

(a)	2020–2049					2040–2069					2070–2099				
	Rainfall	Water Yield	GW Recharge	Base Flow	Flow Reg	Rainfall	Water Yield	GW Recharge	Base Flow	Flow Reg	Rainfall	Water Yield	GW Recharge	Base Flow	Flow Reg
Upper	1400.7	643.3	339.9	116.9	29.8	1438.1	672.5	355.2	123.2	29.7	1500.7	719.1	383.3	134.7	32.2
Middle	1000.8	319.5	249.6	78.2	34.8	1029.1	344.0	269.0	89.3	34.5	1096.0	389.4	298.2	103.1	30.5
Lower	542.3	42.1	55.9	12.2	35.2	568.5	50.8	67.4	16.9	33.8	590.9	63.8	73.7	24.2	33.5
Average	721.0	151.7	130.0	37.6	35.0	748.1	166.4	144.3	44.6	34.0	787.1	191.5	159.2	54.3	32.4
(b)	2020–2049					2040–2069					2070–2099				
	Rainfall	Water Yield	GW Recharge	Base Flow	Flow Reg	Rainfall	Water Yield	GW Recharge	Base Flow	Flow Reg	Rainfall	Water Yield	GW Recharge	Base Flow	Flow Reg
Upper	1460.3	705.0	372.4	130.1	29.2	1472.8	700.4	368.3	127.5	30.0	1857.2	1021.5	532.0	194.2	35.1
Middle	1018.7	339.7	262.2	84.6	34.8	1047.1	352.6	273.7	89.3	35.1	1286.2	543.2	394.7	147.4	30.7
Lower	538.9	44.9	55.8	13.6	34.8	564.5	51.7	62.7	16.7	34.3	672.9	108.9	108.1	46.3	32.7
Average	726.3	161.7	134.9	40.9	34.7	752.6	170.5	143.3	44.5	34.5	912.5	279.9	217.7	84.9	32.0

creases in RCP 8.5 are 197.3, 160.2, and 391.9 %. The overall reductions of flow regulation for the basin are −17.6 and −18.5 % for RCP 4.5 and RCP 8.5 respectively.

4 Conclusions

The results confirm the findings of other studies that there will likely be a reversal of recent historic trends of declining rainfall. Rather, rainfall is projected to increase across the basin over the remainder of the twenty-first century, with greater increases (up to 43 % by the end of the century) under RCP 8.5 than RCP 4.5. The results also indicate an earlier onset of rainfall for both the long and the short rainy season under both RCPs. A consequence of the increased rainfall is disproportionate increase in flow from the basin.

The impact of climate change on water yield, groundwater recharge, base flow and flow regulation were estimated. The first three variables (water yield, groundwater recharge and base flow) progressively increase under both RCPs, but with greater increases under RCP 8.5 than 4.5. In contrast, natural flow regulation reduces across most of the basin in most future time periods, resulting in increased seasonal variability in flows and potentially larger floods. The reductions in flow regulation are comparable under both RCPs.

Increases in rainfall and flow broadly indicate an improved water resource situation in the future, with opportunities for increasing benefits from built infrastructure (i.e., hydropower generation as well as irrigation and water supply diversions). However, declining natural flow regulation, increased variability, and large increases in the frequency and magnitude of floods pose a significant risk that threatens to undermine development opportunities. Water resources management is likely to be more difficult than under historic climatic conditions. To build resilience, water resource managers need to adapt to changing conditions; increased flows but also greater variability.

Data availability. Sources of the climate, landuse and soils data used in this study are shown in Table 1. Stream flow data sets used here were obtained from The Water Resource Management Agency (WRMA), Kenya and are not available to the public as far as the authors know. Therefore, the authors cannot provide them. Further, the Tana SWAT model will soon be added to: http://waterdata.iwmi.org/pages/model_inventory.php (last access: 28 February 2018).

Author contributions. LM and AS developed the SWAT model, and NSS and AO have contributed to the data processing and analysis. MM overlooked this work as the project leader. LM prepared the manuscript with contributions from all co-authors.

Competing interests. The authors declare that they have no conflict of interest.

Special issue statement. This article is part of the special issue "Innovative water resources management – understanding and balancing interactions between humankind and nature". It is a result of the 8th International Water Resources Management Conference of ICWRS, Beijing, China, 13–15 June 2018.

Acknowledgements. This work was undertaken as part of the Water Infrastructure Solutions from Ecosystem Services Underpinning Climate Resilient Policies and Programmes (WISE-UP to Climate) project coordinated by the International Union for Conservation of Nature (IUCN). It was funded by the International Climate Initiative, Bundesministerium für Umwelt, Naturschutz, Bau und Reaktorsicherheit (BMUB) (Federal Ministry for the Environment, Nature Conservation, Building and Nuclear Safety), Germany, with additional support from the CGIAR Fund Donors and Research Program on Water, Land and Ecosystems (WLE).

Edited by: Zongxue Xu
Reviewed by: two anonymous referees

References

Abbaspour, K. C.: SWAT-CUP4: SWAT Calibration and Uncertainty Programs: A user manual, Department of Systems Analysis, Integrated Assessment Modelling (SIAM), Swiss Federal Institute of Aquatic Science and Technology (Eawag), Dübendorf, Switzerland, 2011.

Arnold, J. G., Srinivasan, R., Muttiah, R. S., and Williams, J. R.: Large area hydrologic modeling and assessment Part 1: Model development, J. Am. Water Resour. As., 34, 73–89, 1998.

Baker, T., Kiptala, J., Olaka, L., Oates, N., Hussain, A., and McCartney, M.: Baseline Review and Ecosystem Services Assessment of the Tana River Basin, Kenya. IWMI Working paper 165. International Water Management Institute, Colombo, Sri Lanka, 2015.

El-Khoury, A., Seidou, O., Lapen, D. R. L., Que, Z., Mohammadian, M., Sunohara, M., and Bahram, D.: Combined impacts of future climate and land use changes on discharge, and nitrogen and phosphorus loads for a Canadian river basin, J. Environ. Manage., 151, 76–86, 2015.

Giorgi, F. and Gutowski, W. J.: Coordinated experiments for projections of regional climate change, Current Climate Change Reports, 2, 202–210, 2016.

Muthuwatta, L. P., Sood, A., and Sharma, B.: Model to assess the impacts of external drivers on the hydrology of the Ganges River Basin, IAHS Publ., 364, 76–81, 2014.

Sood, A., Muthuwatta, L., Silva, N., and McCartney, M.: Understanding the hydrological impacts of climate change in the Tana River Basin, Kenya, IWMI Working paper 178, International Water Management Institute, Colombo, Sri Lanka, 2017.

https://doi.org/10.5194/piahs-379-43-2018

Centralized versus distributed reservoirs: an investigation of their implications on environmental flows and sustainable water resources management

Nishadi Eriyagama[1], **Vladimir Smakhtin**[2], **and Lakshika Udamulla**[3]

[1]International Water Management Institute, Battaramulla, 10120, Sri Lanka
[2]UNU Institute for Water Environment and Health, Hamilton, Ontario, Canada
[3]Department of Civil Engineering, The Open University of Sri Lanka, Nugegoda, 10250, Sri Lanka

Correspondence: Nishadi Eriyagama (n.eriyagama@cgiar.org)

Received: 31 December 2017 – Revised: 24 March 2018 – Accepted: 26 March 2018 – Published: 5 June 2018

Abstract. Storage of surface water is widely regarded as a form of insurance against rainfall variability. However, creation of surface storage often endanger the functions of natural ecosystems, and, in turn, ecosystem services that benefit humans. The issues of optimal size, placement and the number of reservoirs in a river basin – which maximizes sustainable benefits from storage – remain subjects for debate. This study examines the above issues through the analysis of a range of reservoir configurations in the Malwatu Oya river basin in the dry zone of Sri Lanka. The study produced multiple surface storage development pathways for the basin under different scenarios of environmental flow (EF) releases and reservoir network configurations. The EF scenarios ranged from "zero" to "very healthy" releases. It is shown that if the "middle ground" between the two extreme EF scenarios is considered, the theoretical maximum "safe" yield from surface storage is about 65–70 % of the mean annual runoff (MAR) of the basin. It is also identified that although distribution of reservoirs in the river network reduces the cumulative yield from the basin, this cumulative yield is maximized if the ratio among the storage capacities placed in each sub drainage basin is equivalent to the ratio among their MAR. The study suggests a framework to identify drainage regions having higher surface storage potential, to plan for the right distribution of storage capacity within a river basin, as well as to plan for EF allocations.

1 Introduction

Storage of water is widely regarded as a form of insurance against rainfall variability. Although the debate over the most appropriate forms of water storage continues, historically, the "natural" first choice among water planners and managers has been surface storage. This is evidenced by the ancient tank irrigation systems that existed in Sri Lanka from around 300 BC, as well as multiple large dams that were constructed all over the world, during the latter half of the twentieth century. However, creation of surface storage often endanger the functions of natural ecosystems, and, in turn, ecosystem services that benefit humans. The optimal size, placement, and the number of reservoirs in a river basin, which maximizes sustainable benefits, while minimizing negative impacts on ecosystems, remain subjects for debate. This study examines some of the above issues through the analysis of a range of reservoir configurations in the Malwatu Oya river basin in the dry zone of Sri Lanka. Optimal sustainable limits to surface storage development for water supply, and the differences between centralized large reservoirs and distributed small reservoirs, are examined, by studying the behaviour of "water supply yield" versus "environmental flow (EF) yield" with increasing storage under each configuration.

2 Study Area

The Malwatu Oya basin in the dry zone of Sri Lanka (Fig. 1) has a drainage area of 3246 km^2 and a mean annual runoff (MAR) of 0.79 km^3. It is dotted with *small tank cascade systems*, an ancient irrigation system consisting of a series of interconnected tanks, acting as storage and regulating reser-

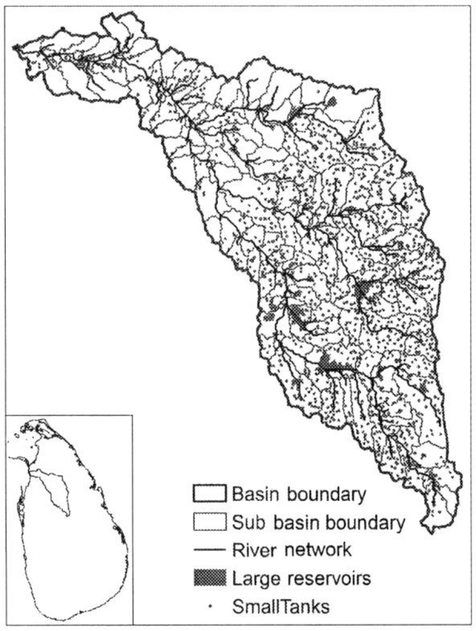

Figure 1. The Malwatu Oya basin and river network.

voirs (Panabokke et al., 2002; Jayatillake et al., 2003). Apart from these small tanks the basin also consists of a number of larger irrigation reservoirs, also built in ancient times, but subsequently modified to suit present day demands. The combination of small and large reservoirs satisfy irrigation and other needs of the basin particularly during the long dry period from May to September.

3 Methodology

3.1 Simulation of Natural River Flow

A Water Evaluation and Planning (WEAP) Model (Yates et al., 2005a, b) of the Malwatu Oya basin was developed in order to model its current hydrological processes, physical infrastructure and water abstractions. The larger basin was disaggregated into 147 sub watersheds (Fig. 1), for which climate input was provided with data acquired from six gauging stations in the vicinity. The sub watersheds were delineated in such a manner that they overlapped with the watersheds of the existing tank cascades as much as possible. The average area of a sub watershed was 22.7 km^2. Water inflows, outflows, demands and currently available storage capacity (large reservoirs and small tanks) within a sub watershed were lumped together. The model was calibrated and validated ($R^2 = 0.92$; Nash–Sutcliffe Coefficient = 0.84) for the period 1961–2013 against measured flow data, and simulated discharges output by a calibrated Soil and Water Assessment Tool (SWAT) model ($R^2 = 0.88$; Nash–Sutcliffe Coefficient = 0.86), set up by the International Water Management Institute, for the same basin. Once the model calibration was completed, natural river discharge of each branch of the river was simulated by removing all reservoirs and water abstractions. This natural river discharge served as the basis for analysing different storage and environmental flow scenarios.

3.2 Behaviour of Water Supply and EF Yields under a Centralized Large Reservoir

The maximum potential annual "Safe Yield" from reservoir storage in the basin was estimated under a range of reservoir sizes and network configurations while allowing different (with respect to magnitude and timing) EF releases. The "Safe Yield" is the target draft that can be supplied at 100 % reliability for a given flow record by a reservoir which starts full and refills at least once after the worst drought on record (Vogel et al., 2007). The Sequent Peak Algorithm (Thomas and Burden, 1963), implemented in WEAP, which is the automated equivalent of the mass curve approach developed by Rippl (1883) was used for this purpose.

3.2.1 Mean Annual Safe Water Supply Yield and EF Yield under a Centralized Large Reservoir

In the initial configuration, the cumulative runoff from the basin was routed through a hypothetical reservoir of increasing capacity while a varying monthly water demand, closely aligned with the actual demand distribution (which is mainly for irrigating rice) of the basin, was imposed on the reservoir. EF releases below the reservoir ranged from "zero" to "very healthy". Six EF scenarios (time series of flow releases) were generated using the method of Smakhtin and Anputhas (2006). In this approach, the natural monthly flow time series at the outlet of the Malwatu Oya basin is used as input to develop six other monthly flow time series, which differ in magnitude, but are similar in pattern (following the monthly pattern of the natural flows – the pattern considered as the ideal for EF releases). The flow time series correspond to six Environmental Management Classes (EMCs), named A through F, and represent the EF requirements at the downstream end of the basin for the river to be maintained in each of the EMCs. Class A represents the ecological status of a protected healthy river whereas class F represents that of a highly degraded river. Table 1 summarises the mean annual "EF yield" required to maintain the river in each of the EMCs.

The behaviour of the Safe Yield under the zero EF (no EF) and the six EMC scenarios (with equal priorities assigned to both water supply and EF yields) is shown in Fig. 2. The yields shown are gross yields with no evaporation or seepage losses, and, no flood control, buffer or inactive storage assumed for the reservoir. The reservoir storage was allowed to become zero if it was necessary to satisfy the demands, but was required to fill up at least once during the total simulation period (53 years) in order to ensure continuity of storage. The curves represent the theoretical upper thresholds for cumulative annual surface water withdrawals (in units of MAR) from reservoir storage in the basin under a range of ecologi-

Table 1. Mean annual EF requirements (yields) under each Environmental Management Class (EMC).

EMC	EF as % of MAR
Natural	100
A	58.3
B	35.5
C	23.2
D	16.6
E	12.6
F	9.7

cal protection categories. The curve for Class A provides the highest level of protection, while the curve with no EF requirements provides the lowest level of protection. The "middle ground" between the two extreme scenarios is shown as the "Mid" curve. The inverse of the curves correspond to the annual EF yields under the same set of scenarios, but are not shown on the figure for clarity.

3.2.2 Monthly Distribution of Safe Water Supply Yield and EF Yield under a Centralized Large Reservoir

The monthly distribution (quantity and timing) of both water supply and EF yields is as important as the mean annual quantity. The monthly water supply yield determines the "health" of the supply delivered, whereas the monthly EF yield determines the "health" of the river. Following the well documented *reliability, resilience vulnerability and sustainability* criteria (Hashimoto et al., 1982; Loucks, 1997) to evaluate the effectiveness of water storage, a Water Supply Sustainability (WSS) index and an Environmental Flow Sustainability (EFS) index (Eq. 1) was adopted to evaluate: how well a given demand is met by a given storage capacity and withdrawal scenario; and how well a prescribed EF regime is met by the in-stream flow releases below a reservoir, in terms of quantity and timing. Although the index of *sustainability* has been formulated in a number of ways by different authors, its formulation here (Eq. 1) follows that of McCartney et al. (2013). The optimum storage development pathway in a river basin should attempt to maximise both WSS and EFS.

$$\text{Sustainability} = \frac{\text{Reliability} + \text{Resilience} + (1 - \text{Relative Vulnerability})}{3} \quad (1)$$

The WSS versus EFS space was mapped out for the two reservoir sizes of 0.4 and 1.0 MAR assuming: (a) the "health of the water supply" is at its best (i.e. WSS = 1) when the water supply yield is at the level of the no EF scenario, and deteriorates when gradually increasing levels of EF is released, and, (b) the health of the river is at its best (i.e. EFS = 1) if the water supply yield is at the level of EMC scenario A, and deteriorates when gradually increasing levels of yields

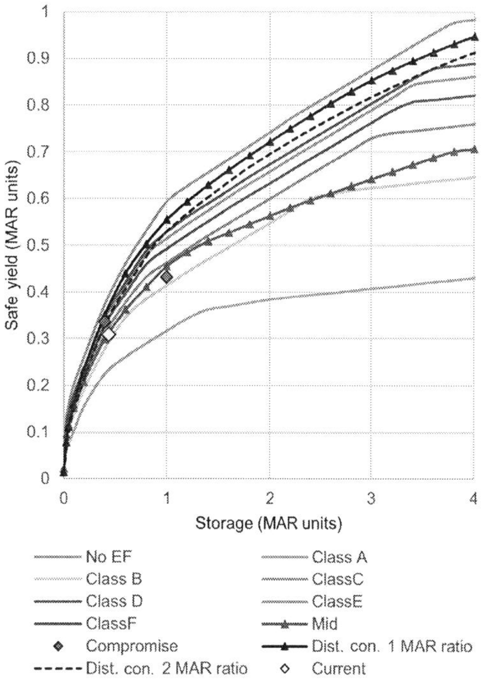

Figure 2. Behaviour of Safe Yield for a centralized hypothetical reservoir under different EF releases, compared with those for two distributed reservoir configurations under which no explicit EF releases are made.

are drawn out. The resulting graphs of WSS versus EFS for scenarios (a) and (b) above for a reservoir size of 1.0 MAR (with different assigned priorities to water supply (WS), EF releases and reservoir filling) are shown on Fig. 3. The Safe Yields at which a compromise is reached between scenarios (a) and (b) for reservoir sizes of 0.4 and 1.0 MAR (when equal priorities to water supply and EF yields, and a lower priority to reservoir filling are assigned) are shown on Fig. 2 as "Compromise" points.

3.3 Behaviour of Water Supply and EF Yields under Distributed Small Reservoirs

In the second reservoir configuration, hypothetical reservoirs were placed at the outlets of 14 major sub drainage basins (identified by the Department of Agrarian Development) of the Malwatu Oya basin. Maximum potential Safe Yield curves for each of these drainage basins under the no EF scenario were developed, considering only the runoff generated within each sub drainage basin. This exercise helped to identify drainage regions having higher surface storage potential. Next, the cumulative Safe Yield curve for the entire basin was estimated by aggregating the individual yields under two sub scenarios as follows: (a) reservoirs placed in each sub drainage basin are of equal capacity, and, (b) reservoirs placed in each sub drainage basin follow a capacity ra-

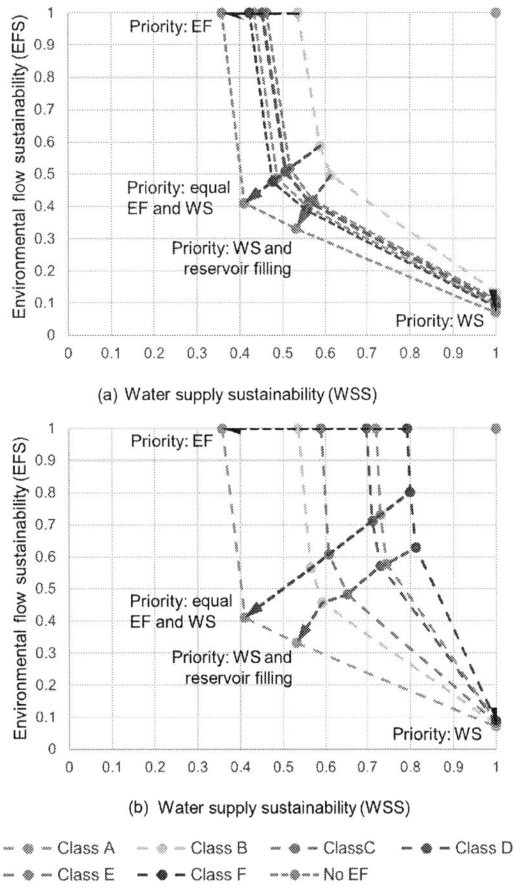

Figure 3. Variation of WSS and EFS when increasing levels of yields are extracted while trying to satisfy EF at EMC "A" **(a)**; when increasing levels of EF are released while trying to extract yields at "No EF" level **(b)** for a storage capacity of 1 MAR with different priorities assigned to water supply and EF demands.

tio equivalent to the MAR ratio among sub drainage basins. The above exercise was then repeated under a third configuration where hypothetical reservoirs were placed at the outlets of the 147 sub watersheds delineated in the initial WEAP model. These 147 sub watersheds represent a further sub division of the 14 major drainage basins considered in scenario 2. The resulting Safe Yield curves under both distributed reservoir configurations are also shown on Fig. 2 (Dist. Con.1 MAR ratio, Dist. Con.2 MAR ratio), along with the earlier curves for the centralized reservoir configuration.

4 Results and discussion

When attempting to compromise the mean annual water supply and EF yields for the entire basin (Mid curve of Fig. 2), for cumulative storage capacities below 1.5 MAR units, the EMC scenario C curve represents a reasonable "middle ground" storage development pathway to be adopted, unless there is a specific need for the river to be at a higher state of protection. However, as storage capacities become larger than 1.5 MAR units, adoption of the EMC scenario B curve is more appropriate. Considering the Mid curve, the theoretical maximum (gross) Safe Yield for the Malwatu Oya basin is about 65–70 % of the MAR.

For any given storage capacity, the performance of the two indicators, EFS and WSS (which assess the "health" of the actual pattern of EF releases and water supply yields withdrawn), depend on the magnitude of the yield withdrawn, and the order of priority assigned to meeting each demand and filling up of the reservoir (Fig. 3). Considering where a compromise is reached between WSS and EFS for 0.4 and 1.0 MAR capacities (Fig. 2), the EMC Scenario C curve represents a reasonable storage development pathway to be adopted for cumulative storage capacities below 1.0 MAR units. For storage capacities above 1.0 MAR units, adoption of the Class B scenario is more appropriate.

Figure 2 illustrates that the maximum achievable yield from a lumped reservoir of a certain capacity is higher than the cumulative yield obtained from a network of distributed reservoirs of the same aggregated capacity. The cumulative yield is maximized when the storage capacity in each sub drainage basin follows a ratio equivalent to the MAR ratio of each sub drainage basin (also observed by Pitman, 1995). The cumulative yield of the entire basin declines with subsequent levels of distribution of reservoirs. When the cumulative Safe Yield curves for the two distributed reservoir configurations were compared with the EMC scenario curves in Fig. 2, it was observed that they closely follow the curves for EMC scenarios F and D respectively up to a reservoir size of 0.5 MAR. This implies that more releases are available at the downstream of the basin as EF, even though no explicit EF releases were made. Therefore, planned distribution of reservoirs in the river network, and limiting abstractions from them to a fraction of the local runoff generated within their individual watersheds is a strategy that can be adopted (subject to other requirements in the basin) to ensure that sufficient environmental flow releases are maintained in the river. It is also possible to identify the appropriate distribution of storage capacity in individual watersheds, and threshold levels of withdrawals to ensure that the entire basin stays on an identified sustainable storage development pathway. The actual reservoir configuration in the Malwatu Oya basin is even more densely distributed than any of the ones examined here. The position of the current cumulative storage capacity and current annual yield (approximately 0.43 and 0.31 MAR units respectively) shown on Fig. 2 also agrees with the finding that higher levels of distribution lowers cumulative yields. The current position is almost coincident with the "Mid" curve (Fig. 2). Although this implies that there is a reasonable quantity of annual EF releases at the downstream end, their temporal and spatial distribution (especially during the dry season) needs to be further investigated.

5 Conclusions

The objective of this research was to analyse a series of surface storage scenarios, consisting of hypothetical reservoirs of increasing storage capacity, placed at three levels of distribution across the river network, to investigate how an optimum reservoir arrangement (which maximizes water supply yields) under a given EF release scenario may be formulated. It provides tools and approaches to identify "sustainable" storage development pathways and theoretical maximum "safe" yields for whole river basins. Developing storage yield relationships for sub drainage regions (sub watersheds) helps to identify: regions in a river basin having higher surface storage potential; and, the right distribution of storage capacity (and allowable withdrawals) in individual drainage regions, to stay on an identified storage development pathway, which maximizes water supply yields while accommodating targeted EF yields.

When modelling current river flow, all water transfers and available storage capacity within a sub watershed was lumped together in order to reduce model complexity and processing time. Although the model may have not captured the heterogeneity within a sub watershed, it was still able to simulate river flows at sub watershed outlets (the hydrological resolution of our interest) reasonably well. However, use of simulated flow data in subsequent analyses introduces uncertainty into the end results, no matter how well the hydrological models have performed. Therefore, the use of measured flow time series, wherever possible, is recommended in order to reduce this uncertainty. In addition, further research should also investigate the impact of river fragmentation when reservoirs are distributed across the river network.

Data availability. The daily climate data and monthly flow data used in this study were obtained from the recording gauges of the Meteorological and Irrigation Departments of Sri Lanka respectively. They are currently not publicly available.

Competing interests. The authors declare that they have no conflict of interest.

Special issue statement. This article is part of the special issue "Innovative water resources management – understanding and balancing interactions between humankind and nature". It is a result of the 8th International Water Resources Management Conference of ICWRS, Beijing, China, 13–15 June 2018.

Acknowledgements. This research was supported by the International Water Management Institute (IWMI) as part of the CGIAR Research Programs on Climate Change Agriculture and Food Security (CCAFS) and Water Land and Ecosystems (WLE). The authors would like to acknowledge the support provided by Madusanka Thilakarathne for processing the data required by the WEAP model.

Edited by: Zongxue Xu
Reviewed by: two anonymous referees

References

Hashimoto, T., Stedinger, J. R., and Loucks, D. P.: Reliability, resiliency, and vulnerability criteria for water resource system performance evaluation, Water Resour. Res., 18, 14–20, 1982.

Jayatillake, C. J., Sakthivadivel, R., Shinogi, Y., Makin, I., and Witharana, P.: A simple water balance modelling approach for determining water availability in an irrigation tank cascade system, J. Hydrol., 273, 81–102, 2003.

Loucks, D. P.: Quantifying trends in system sustainability, Hydrolog. Sci. J., 42, 513–530, 1997.

McCartney, M., Rebelo, L.-M., Xenarios, S., and Smakhtin, V.: Agricultural water storage in an era of climate change: assessing needs and effectiveness in Africa, IWMI Research Report 152, International Water Management Institute (IWMI), Colombo, Sri Lanka, 2013.

Panabokke, C. R., Sakthivadivel, R., and Weerasinghe, A. D.: Small tanks in Sri Lanka: evolution, present status and issues, International Water Management Institute (IWMI), Colombo, Sri Lanka, 2002.

Pitman, W. V.: Towards Improved Utilization of South Africa's Water Resources, Proceedings of the Seventh South African National Hydrology Symposium, 4–6 September 1995, Grahamstown, South Africa, 1995.

Rippl, W.: The capacity of storage-reservoirs for water supply, P. I. Civil Eng., 71, 270–278. 1883.

Smakhtin, V. U. and Anputhas, M.: An Assessment of Environmental Flow Requirements of Indian River Basins, IWMI Research Report 107, International Water Management Institute (IWMI), Colombo, Sri Lanka, 2006.

Thomas Jr., H. A. and Burden, R. P.: Operations research in water quality management, Division of Engineering and Applied Physics, Harvard University, Cambridge, MA, USA, 1963.

Vogel, R., Seiber, J., Archfield, S., Smith, M., Apse, C., and Huber-Lee, A.: Relations among storage, yield, and instream flow, Water Resour. Res., 44, W05403, https://doi.org/10.1029/2006WR005226, 2007.

Yates, D., Sieber, J., Purkey, D., and Huber-Lee, A.: WEAP21: A demand, priority, and preference-driven water planning model: 1. Model characteristics, Water Int., 30, 487–500, 2005a.

Yates, D., Purkey, D., Sieber, J., Huber-Lee, A., and Galbraith, H, WEAP21: A demand, priority, and preference-driven water planning model: 2. Aiding freshwater ecosystem service evaluation, Water Int., 30, 501–512, 2005b.

https://doi.org/10.5194/piahs-379-49-2018

Application of remote sensing data for measuring freshwater ecosystems changes below the Zeya dam in the Russian Far East

Oxana I. Nikitina[1,2], **Kirill Y. Bazarov**[3,4], **and Evgeny G. Egidarev**[1,3]

[1]World Wide Fund for Nature (WWF-Russia), Moscow, 109240, Russia
[2]Water Problems Institute of the Russian Academy of Sciences, Moscow, 119333, Russia
[3]Pacific Geographical Institute of the Far Eastern Branch of the Russian Academy of Sciences, Vladivostok, 690041, Russia
[4]Far Eastern Federal University, Vladivostok, 690090, Russia

Correspondence: Oxana I. Nikitina (onikitina@wwf.ru)

Received: 30 December 2017 – Revised: 28 February 2018 – Accepted: 2 March 2018 – Published: 5 June 2018

Abstract. The large Zeya hydropower dam is located on the Zeya River, the largest left-bank tributary of the Amur-Heilong River in Russia. The dam had been constructed by 1980 and its operation has significantly transformed the flow regime of the Zeya River. The flow regulation has reduced the magnitude of periodic flooding of the floodplain areas located downstream from the Zeya dam and disrupted habitats of flora and fauna. An estimation of the transformation of the freshwater ecosystems is required to develop measures necessary either to maintain or restore disrupted ecosystems. Application of remote sensing methods allows measuring characteristics of the ecosystem's components. Two sections of a floodplain below the Zeya dam were considered for analysis in order to detect changes in objects at each site during the comparison of remote data from 1969/1971 and 2016.

1 Introduction

The Amur-Heilong River, one of the ten largest rivers on Earth, flows in Northeast Asia through Mongolia, China and Russia. The large-scale Zeya hydropower dam is located on the Zeya River which is the largest left-bank tributary of the Amur-Heilong in the Russian Far East. The location of the basin and the dam are shown in Fig. 1. The construction of the Zeya dam began in 1964, the reservoir created by the dam was filled between 1974 and 1980.

The environmental consequences of large dams include changes of hydrological regime and sediment runoff, fragmentation of the basin, flooding of lands and transformation of the natural ecosystem dynamics in the areas located upstream and downstream of dams (Nilsson et al., 2005; Dams and Development, 2000). River damming provides a considerable impact on downstream freshwater ecosystems in many cases extending for hundreds of kilometres below a dam (Richter and Thomas, 2007; Podolsky et al., 2006). The territory affected by the Zeya dam comprises biologically diverse areas which depend upon the hydrological regime of the floodplain (Simonov et al., 2016; Nikitina et al., 2015). Sections of the Zeya's floodplain are important stopovers for migratory birds along their routes (Simonov and Dahmer, 2008). Wetlands of the Muraviovka Park for Sustainable Land Use are included in the List of Wetlands of International Importance (the Ramsar List); they are affected by the flow regimes of the Amur and Zeya rivers (Smirenski, 2016).

Periodic flooding of the floodplain is an essential ecohydrological indicator of both the flow regime and the quality and productivity of floodplain areas (Fashchevsky, 2007). The flow regulation has reduced the magnitude of periodic flooding of the floodplain areas located downstream from the Zeya dam and has also disrupted habitats of flora and fauna. As shown in Fig. 2, under natural conditions, small floods used to occur almost annually at the Belogorye stream gauge located 617 km downstream. These used to last anything from 19 to 58 days. However, under flow regulation,

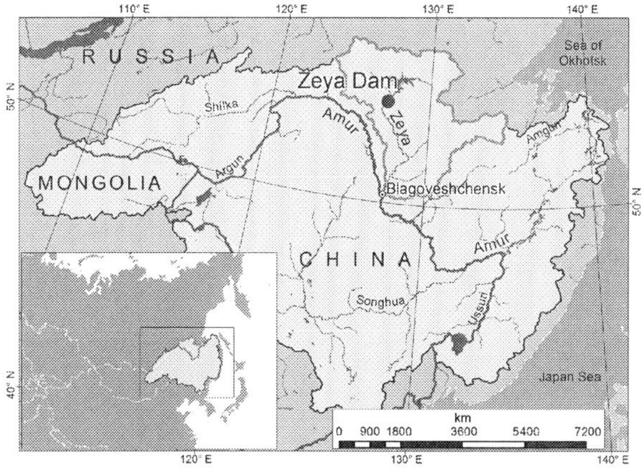

Figure 1. Map of the Amur River basin showing the location of the Zeya River basin and the Zeya dam.

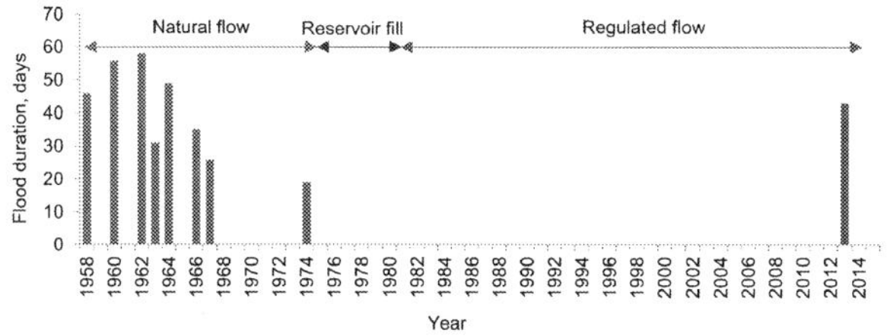

Figure 2. Changes in flood duration at the Belogorye stream gauge located 617 km downstream from the Zeya dam.

floods have now practically disappeared. Reduction of maximum flow magnitude and duration leads not only to the degradation of freshwater habitats and floodplain's ecosystems but also to changes in the composition of vegetation. The extent of changes depends on the degree of river flow regulation, specifically on the level, time and duration of natural floods and released water from reservoirs (Fashchevsky, 2007).

An estimation of the transformation of the freshwater ecosystems is required to develop measures necessary either to maintain or restore disrupted ecosystems. The application of remote sensing methods allows us to estimate the characteristics of the components of the ecosystem (Dorber et al., 2018; Kravtsova, 1995). These characteristics can serve as a basis for an assessment of changes in the freshwater ecosystems.

The research aims to identify and analyse the trend and intensity of changes in the components of the floodplain located below the Zeya dam when they are exposed to its impact. The initial assumption is that flow regulation by the dam has had a decisive effect on the changes in the ecosystems. Other possible and additional factors affecting transformation of the ecosystems include climate change, development of agriculture and fires.

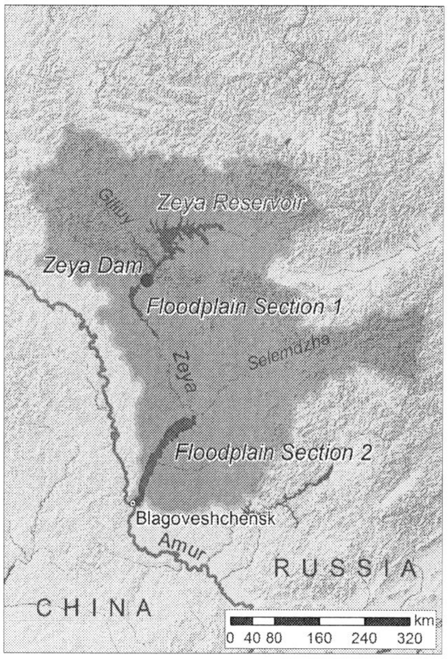

Figure 3. Location of the wide floodplain sections described in this article.

2 Materials and methods

The methodology involves estimation of spatiotemporal changes of components located within the wide floodplain of the Zeya River downstream from the dam. Due to its size, the wide floodplain serves as a habitat for numerous species of birds, fish and other animals; it is crucial to the biodiversity of the freshwater ecosystem.

Expert interpretation using 1 : 100 000 topographic maps revealed the borders of the floodplain where flooding rarely occurs (approximately once in fifty years) (Egidarev, 2012; Farr et al., 2000). Two sections of a wide floodplain located below the Zeya dam were considered during the analysis as shown in Fig. 3. The first section is located in the immediate vicinity of the Zeya dam; its length is 140 km with a maximum width of 17 km. The second section begins 350 km away from the dam of the Zeya hydropower station below the confluence of the Selemdzha River (the Zeya's biggest tributary) and ends in the estuary of the Zeya; the length of the section is 200 km, its maximum width being 23 km.

With the use of ArcGIS 10.5 software and remote imageries, two thematic layers that are demonstrated in Fig. 4 have been created. These layers represent both archive (1969/1971) and contemporary (2016) pictures with a scale of 1 : 50 000, reflecting the structure and changes of the land cover in two sections of the wide floodplain located downstream from the Zeya dam.

The retrospective state is represented by archival imageries from missions flown by the CORONA, ARGON and LANYARD satellite systems. Mosaics are collected from im-

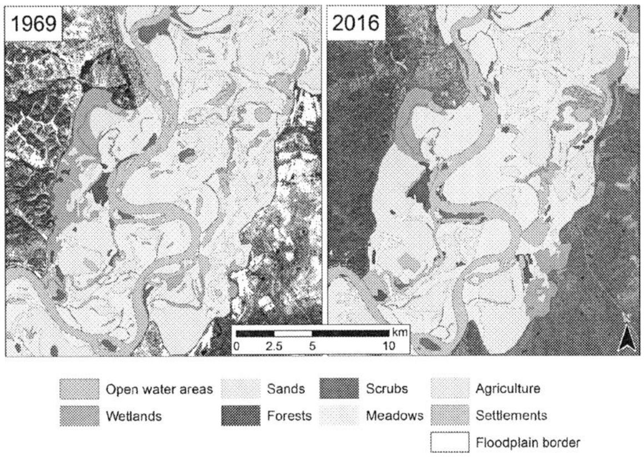

Figure 4. Land cover of the floodplain sections identified for the analysis.

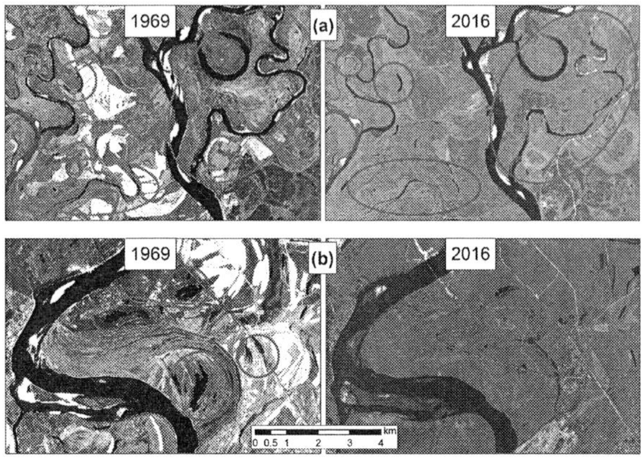

Figure 5. Examples of the changes in the floodplain objects. Red circles mark the differences detected for water bodies: **(a)** a river channel that has become a chain of oxbow lakes; the surface area of the lakes has been reduced in the Floodplain Section 1; **(b)** the surface area of the lakes has been reduced in the Floodplain Section 2.

ageries dated 23 September 1969 and 14 September 1971. The imageries of 1969 and 1971 reflect the state of the area and are typical of the Zeya River's natural (little disrupted) flow regime. The contemporary picture is presented by images captured by the Sentinel-2 satellite. The mosaics of the images, obtained on 11 September 2016, were created for the study area.

Two basic maps reflecting land usage during the 2002 and 2016 vegetation periods were created on the basis of the ArcGIS 10.5, and remote data collected by the Landsat satellite were analyzed.

Indicators of Hydrologic Alteration software were used to calculate the characteristics of both the natural and altered flow regimes shown on Fig. 2. Sets of observations for the Belogorye stream gauge for the period 1957–2013 were used to detect hydrologic alteration. Two distinct time periods of daily water discharge were compared to analyse how the flow regime was affected: (1) 1957–1975 – pre-impact regime; (2) 1980–2013 – post-impact regime.

3 Preliminary results

The interpretation of both archive and contemporary images revealed different water conditions: 2016 was a wet year compared with 1969 and 1971, with increased river inflow into the Zeya below the dam; the difference is pronounced in the second section of the floodplain in the lower reaches of the Zeya.

Table 1 presents data on the total area of the objects and the changes which occurred to the sections of a wide floodplain in 1969–2016. The area of the objects under the natural flow regime (in 1969/1971) was taken as 100 % when the relative figures were calculated. The examples of revealed differences between floodplain objects of 1969/1971 and 2016 are demonstrated in Fig. 5.

3.1 Change of objects in the Floodplain Section 1

The surface area of the floodplain and oxbow lakes has reduced from 950 to 617 hectares (35 %). Supposedly, the main reason for this reduction is the absence of periodic flooding in the floodplain due to river flow regulation by the Zeya dam combined with the reduced level and duration of natural floods. Another additional factor could be climate change. A comparison of regional trend estimates for the two periods 1976–2006 and 1976–2012 has revealed a trend of temperature rise (+0.44 °C/10 years) in the Amur region; however, this trend is statistically insignificant at a level of 5 %. The average amounts of precipitation for the periods of 1936–2010 and 1976–2010 demonstrate that the areas with trends of decreasing annual and summer precipitation predominate in the Amur region but the percentage of their decrease is insignificantly small (Katsov et al., 2014). These remarks are consistent with the assumption that climate change has not been the principal reason for changes in the floodplain objects.

Water discharges at the lower reaches of the Zeya dam in wet season of 2016 were lower than the water discharges observed under the natural (little disrupted) flow regime in 1969. This is typical of regulated flow regime of Zeya and explains the reduced area of the Zeya River's main channel in 2016. Small tributaries and river branches within the floodplain were also reduced by 2016 with some of them changing from a continuous stream into a chain of oxbow lakes.

The meadows area has increased from 33.7 to 38.8 thousand hectares. This is mainly due to previously farmed but currently abandoned agricultural fields being turned into meadows. The forest area of the floodplain has seen a 70 % growth from 3.2 to 5.5 thousand hectares. It confirms the thesis of Martynov (2013) that changes in flood regime, or

Table 1. Land cover in the study area for the time period 1969–2016.

Object types	Floodplain Section 1				Floodplain Section 2			
	Area in 1969/1971 (ha)	Area in 2016 (ha)	Area changes in 1969–2016 (ha)	Area changes in 1969–2016 (%)	Area in 1969/1971 (ha)	Area in 2016 (ha)	Area changes in 1969–2016 (ha)	Area changes in 1969–2016 (%)
Lakes	950	617	−332	−35.0	5109	2932	−2177	−42.6
Wetlands	–	–	–	–	30 883	19 588	−11 296	−36.6
Rivers	8343	7537	−805	−9.7	28 532	30 593	2061	7.2
Sands	929	937	8	0.9	7346	839	−6508	−88.6
Meadows	33 666	38 841	5175	15.4	127 747	137 892	10 144	7.9
Forests	3244	5521	2276	70.2	1396	5325	3929	281.5
Scrubs and trees	4005	3105	−900	−22.5	5259	10 292	5033	95.7
Settlements	1951	2172	220	11.3	3785	5608	1823	48.2
Agriculture	14 628	8999	−5629	−38.5	48 984	45 732	−3252	−6.6
Total area	67 717	67 730	13	0.02	259 041	258 800	−241	−0.1

disappearance of floods lead to an acceleration of vegetation fluctuations in the floodplain, ultimately resulting in the emergence of forests or steppe meadows. The scrubs and trees area has been reduced from 4 to 3.1 thousand hectares; some of these objects have turned into forests.

The settlements area has grown from 1.9 to 2.1 thousand hectares. A comparison between contemporary and archive images has revealed a significant reduction of farmed lands area from 14.6 to 9.0 thousand hectares. This is due to the active development of virgin lands in 1954–1965 that was undertaken to increase grain production when large areas of such lands were ploughed in the various states of the former Soviet Union. This included the Russian Far East, the Zeya's floodplain being amongst them. Floodplains typically boast fertile land with high crop capacity which makes them highly suitable for efficient agriculture. The expansion of farmed land area for soya growth continued throughout the period under consideration (i.e. 1969/1971). The crisis which began in the 1980s has led to a reduction of the farmlands revealed during the analysis.

3.2 Change of objects in the Floodplain Section 2

2016 was a wet year compared to 1969: the tributaries flowing into the Zeya were abundant forming wetlands in some of the mouth sections. This explains the 7.2 % increase in the rivers' water surface: even given the regulation of the Zeya flow higher upstream, high water discharges were noted compared to the unaltered flow regime of 1969. Obviously in its lower reaches, the Zeya is less dependent on the hydropower station's operating mode due to the effect of the large free-flowing tributary – the Selemdzha which flows into the Zeya at the beginning of the floodplain section. At the same time, the smaller floodplain streams have reduced which indicates changing overall flooding conditions in the floodplain.

The water surface of the lakes has seen a 43 % drop from 5.1 to 2.9 thousand hectares; the wetland area – a 36 % drop from 30.8 to 19.6 thousand hectares. Despite the significant impact of the large tributary Selemdzha, the Zeya floodplain's flooding regime has changed in the river's lower reaches as well.

The meadows area has grown from 127.7 to 137.9 thousand hectares (7.9 %) due to abandoned farmlands being turned into meadows. The forest area has almost quadrupled (from 1.4 to 5.3 thousand hectares). The scrubs and trees area has nearly doubled in size, from 5.3 to 10.3 thousand hectares; such changes could happen due to frequent fires. According to the global fire monitoring program and the archive data (Giglio et al., 2016), about 4.5 thousand sites of fire were registered in the lower reaches of the river Zeya throughout the previous decade only (in 2006–2016). Due to this permanent pyrogenic factor, the dominant vegetation type is changing from forest to arboreal and shrubby; the species composition of flora and fauna alike is becoming less diverse.

The settlements area has grown from 3.8 to 5.6 thousand hectares. The farmlands area has reduced insignificantly (a 6.6 % reduction from 49.0 to 45.7 thousand hectares): development of virgin lands and the ploughing of large land areas were major activities in 1969/1971. However, since the early 2000s, the farmed land area in the lower reaches of the Zeya River has grown again. Calculations based on the remote data collected by the Landsat satellite revealed that arable land area had doubled between 2002 and 2016, from 14.5 to 29.8 thousand hectares. The area of abandoned agricultural lands, including the so-called fallow lands (ploughed and weeds-free but not sown) has grown too, from 8 thousand hectares in 2002 to 13.7 thousand in 2016. The analysis confirms the thesis that flow regulation and flood control lead to the transformation of natural ecosystems caused by the subsequent active development of agriculture (Brykina, 2017; Shalikovsky and Kurganovich, 2011).

4 Conclusions

Large dams affect freshwater ecosystems and cause extensive environmental impacts downstream from the dam. The alteration of the freshwater ecosystems affects the floodplain habitats of flora and fauna and causes changes in land use; this can lead to biodiversity loss. Remote sensing data is useful for monitoring surface area of the objects and their changes. As for the study area, the surface areas of lakes in both floodplain sections of the Zeya River have decreased in size which likely indicates that the landscape is changing due to the changed flow regime. Small streams of the floodplain have also reduced, with some of them turning from a continuous stream into a chain of oxbow lakes. The wetland area has significantly decreased in the second floodplain section. The impact of the dam in the lower reaches of the Zeya is less pronounced due to the long distance from the dam and the increased lateral inflow. The forest area has grown in both sections; a possible explanation for this being the accelerated vegetational fluctuations in the floodplain due to the changed flow regime. The abandoned agricultural lands have mostly become meadows and it explains the growth of meadow territories in both sections. The conducted study can serve as a basis to assess the impact of the Zeya dam operation on the biodiversity of the freshwater ecosystems of the Amur-Heilong basin.

Data availability. The majority of the data presented in the article is from an ongoing project. The complete data set will be made public on the Amur information center, a web portal hosted by Amur Branch of WWF-Russia. Further information can be obtained from Oxana I. Nikitina.

Competing interests. The authors declare that they have no conflict of interest.

Special issue statement. This article is part of the special issue "Innovative water resources management – understanding and balancing interactions between humankind and nature". It is a result of the 8th International Water Resources Management Conference of ICWRS, Beijing, China, 13–15 June 2018.

Acknowledgements. We would like to express our gratitude to Mikhail V. Bolgov, Water Problems Institute of the Russian Academy of Sciences, for his advice on this research.

The work was funded by World Wide Fund for Nature (WWF-Russia).

Edited by: Dingzhi Peng
Reviewed by: two anonymous referees

References

Brykina, I. G.: Agricultural use of floodplains under river regulation, Agricultural Sciences for Agriculture, Publishing House of Altay State Agrarian University, 406–406, 2017.

Dams and Development: A New Framework For Decision-Making, The Report Of The World Commission On Dams, Earthscan Publications Ltd, London and Sterling, VA, 2000.

Dorber, M., May, R., and Verones, F.: Modeling Net Land Occupation of Hydropower Reservoirs in Norway for Use in Life Cycle Assessment, Environ. Sci. Technol., 52, 2375–2384, https://doi.org/10.1021/acs.est.7b05125, 2018.

Egidarev, E. G.: Mapping and estimation of floodplain complexes in the Amur River valley, Vestnik of the Far East Branch of the Russian Academy of Sciences, 2, 9–16, 2012.

Farr, T. G., Hensley, S., Rodriguez, E., Martin, J., and Kobrick, M.: The shuttle radar topography mission, CEOS SAR Workshop, Toulouse, 26–29 October 1999, Noordwijk, 361–363, 2000.

Fashchevsky, B. V.: Ecological value of floodplains in riverine ecosystems, Russian State Hydrometeorological University Proceedings, 5, 118–129, 2007.

Giglio, L., Schroeder, W., and Justice, C.O.: The collection 6 MODIS active fire detection algorithm and fire products, Remote Sens. Environ., 178, 31–41, 2016.

Katsov, V. M., Semenov, S. M., Alekseev, G. V., Ananicheva, M. D., et al.: Second Roshydromet Assessment Report on Climate Change and its Consequences in the Russian Federation, Moscow, Roshydromet, 2014.

Kravtsova, V. I.: Space mapping techniques, Publishing House of Moscow State University, Moscow, 1995.

Martynov, A. V.: Soil cover structure of bottomland of major rivers in the Amur Region (exemplified by the Zeya and Selemdja rivers), Bulletin of the North-East Scientific Center, Russian Academy of Sciences, Far East Branch, 2, 108–116, 2013.

Nikitina, O. I., Simonov, E. A., and Egidarev, E. G.: River flood adaptation in the Amur basin and nature conservation, Use and Protection of Natural Resources of Russia, 3, 15–24, 2015.

Nilsson, C., Reidy, C. A., Dynesius, M., and Revenga, C.: Fragmentation and flow regulation of the world's large river systems, Science, 308, 405–408, 2005.

Podolsky, S. A., Simonov, E. A., and Darman, Yu. A.: Where does the Amur flow?, WWF-Russia, 2006.

Richter, B. D. and Thomas, G. A.: Restoring environmental flows by modifying dam operations, Ecol. Soc., 12, 12, http://www.ecologyandsociety.org/vol12/iss1/art12/, 2007.

Shalikovsky, A. V. and Kurganovich, K. A.: Hazard and risk assessment for commercial use of the Upper and Middle Amur basin floodplains, Transbaikal State University Journal, 11, 119–124, 2011.

Simonov, E. A. and Dahmer, T. D.: Amur-Heilong River Basin Reader, Hong Kong, Ecosystems Ltd, 2008.

Simonov, E. A., Nikitina, O. I., Osipov, P. O., Egidarev, E. G., and Shalikovsky, A. V.: We and the Amur floods: Lesson (Un)Learned?, WWF-Russia, 2016.

Smirenski, S. M. (Ed.): Giltchin river watershed, History. Wetlands. Water resources, Dalnauka, Vladivostok, 2016.

https://doi.org/10.5194/piahs-379-55-2018

Impact of urbanization on flood of Shigu creek in Dongguan city

Luying Pan, Yangbo Chen, and Tao Zhang

Department of Water Resources and Environment, Sun Yat-sen University, Guangzhou, 510275, China

Correspondence: Yangbo Chen (eescyb@mail.sysu.edu.cn)

Received: 22 November 2017 – Revised: 11 January 2018 – Accepted: 19 January 2018 – Published: 5 June 2018

Abstract. Shigu creek is a highly urbanized small watershed in Dongguan City. Due to rapid urbanization, quick flood response has been observed, which posted great threat to the flood security of Dongguan City. To evaluate the impact of urbanization on the flood changes of Shigu creek is very important for the flood mitigation of Shigu creek, which will provide insight for flood planners and managers for if to build a larger flood mitigation system. In this paper, the Land cover/use changes of Shigu creek from 1987–2015 induced by urbanization was first extracted from a local database, then, the Liuxihe model, a physically based distributed hydrological model, is employed to simulate the flood processes impacted by urbanization. Precipitation of 3 storms was used for flood processes simulation. The results show that the runoff coefficient and peak flow have increased sharply.

1 Introduction

Flooding is the natural disaster that has caused the most damages to the world. The Pearl River Delta Area (PRDA) in southern China is in the monsoon area, where heavy storms in the summer are observed very often, and flooding is the most frequent natural disaster threatening the safety of the local society. For the past decades, PRDA has experienced rapid urbanization, and caused significant changes of land use/cover (LUC). Urban land area increased sharply and has become the major or even the dominant land use type, thus speeding up the flood responses, i.e., the increased flood peak flow, and the decreased runoff routing time. This, in one side, strengthens the flood quantities, while on the other side, as more and more people move and live and work in the urban area, flood damages increased largely with urbanization. Dongguan City is one of the newly developed international metropolitans in PRDA in the past three decades, but the flood mitigation measurements were not built to catch up with the flood changes due to rapid urbanization. Purpose of this study is to evaluate the impact of urbanization (LUC change) on flood peak flow and flood runoff coefficient on the typical watershed, the Shigu creek, so to provide scientific base for smart flood measurement planning and building in Dongguan City.

Watershed hydrological model is the tool for watershed flood process simulation in evaluating the impact of LUC change on flooding. Early hydrological model used for flood simulation/prediction is the so called lumped model, such as the SCS model (USDA, 1972), and the Xinanjiang model (Zhao, 1977). Lumped model calibrates model parameters with the observed hydrological data series, for this reason it could not be used in data spare watershed, i. e., in the watersheds with rapid urbanization. Distributed hydrological model is the further development of hydrological model under the big data era. With the development of satellite remote sensing, large quantities of the watershed terrain data was prepared and could be used for hydrological model setting up, thus providing opportunity for distributed hydrological model. Many distributed hydrological models have been proposed, such as the SHE model (Abbott et al., 1986), and the VIC model (Liang et al., 1994).

In this study, the Liuxihe model (Chen et al., 2010), a physically based distributed hydrological model is employed to simulate the flood processes of Shigu creek in Dongguan City, which is a highly urbanized small watershed that is also a very typical watershed in Dongguan City.

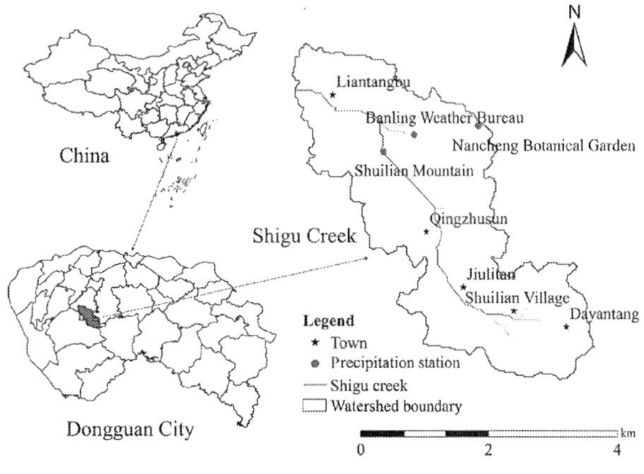

Figure 1. Sketch map of Shigu creek.

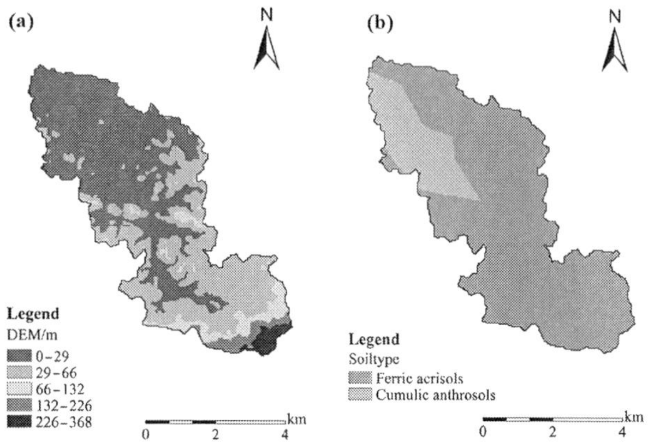

Figure 2. (a) DEM. (b) Soil type. DEM and soil type of Shigu creek.

2 Shigu creek and data

2.1 Shigu creek

Shigu creek locates in the south of Dongguan City center, it originates from the south peak of Shuilian mountain, and is the first order branch of Dongjiang River. Shigu creek is a small watershed in Dongguan City, previously it is a watershed in the suburb area of Dongguan City center, but with the rapid urbanization of Dongguan City in the past three decades, Shigu creek becomes a part of Dongguan City. The drainage area of Shigu creek is 19.29 km^2, and the length is 11.87 km. Climate in the creek is belong to subtropical marine climate with an annual average temperature of 23.6°, and an annual average precipitation of 1769.5 mm. Heavy storms in the summer monsoon season usually induces serious flood which is a great threat to the security of the downtown area of Dongguan City. Rapid urbanization in the creek makes flooding worse. Figure 1 is a sketch map of Shigu creek, and shows its location in China.

2.2 Terrain property data

Terrain property data used for Liuxihe model structure setting up and parameter deriving are DEM, soil type and land use/cover (LUC) type. In this study, the DEM was downloaded from the Shuttle Radar Topography Mission database website at http://srtm.csi.cgiar.org, its spatial resolution is at 30 m × 30 m, as shown in Fig. 2a. From the result it has been found that the highest elevation in the watershed is 368 m, the lowest elevation is 0 m, and the average elevation is 40.94 m.

The soil type was downloaded from FAO soil type dataset at http://www.isric.org, the spatial resolution is 1000 m × 1000 m, as shown in Fig. 2b. There are only two soil types, including Ferric acrisols and Cululic anthrosols, accounting for 83.23, 16.77 % of the total watershed area respectively.

Due to the rapid urbanization, the LUC type in Shigu creek is in constant changing. Chen et al. (2017a) prepared a LUC dataset for the whole Dongguan City at 12 date from 1987 to 2015 by using the Landsat series satellite imageries, and the SVM algorithm is employed to estimate the LUCs. Spatial resolution of this dataset is at 30 m × 30 m. Figure 3 shows the LUCs of Shigu creek extracted from this dataset from 1987 to 2015 respectively. From the results of Fig. 3, it could be found that the LUC types have not changed. There are 6 LUC types, including forestry land, grass land, farm land, urban land, water body and bare land. But the rates of the LUCs have changed drastically from 1987 to 2015, particularly the urban land area increased rapidly, while the grass land decreased significantly.

2.3 Hydrological data

In the Shigu creek, there is no river discharge observation, but there are three rain gauges installed in recent years, and only one rain gauge data is available, so it is a hydrological data spare watershed. For this reason, model parameters could not be optimized, and can only be derived directly from the terrain properties. In this study, precipitation of 3 storms was collected at half hour interval, which will be used as the input to the hydrological model for flood simulation.

3 Liuxihe Model and parameter

3.1 Introduction of Liuxihe model

Liuxihe Model (Chen, 2009; Chen et al., 2010) is a physically based distributed hydrological model proposed for watershed flood forecasting. In Liuxihe model, the studied watershed is divided into grids horizontally by using the DEM, and are treated as a uniform watershed. Every grid has its own elevation, LUC type and soil type, and is categorized as one of the 3 types, namely hill slope cell, river channel cell

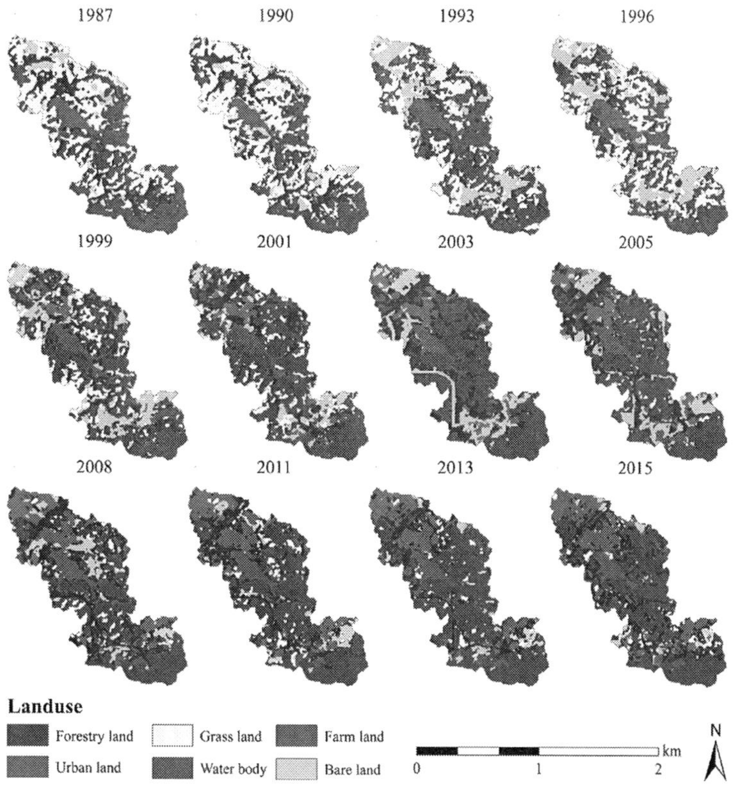

Figure 3. Land use/cover from 1987 to 2015 of Shigu creek.

and reservoir cell. The grid is divided into canopy layer, soil layer and underground layer vertically. Precipitation in every cell is unique and different from other cells, and the model parameters are also considered different in different cells. Evapotranspiration is calculated at cell scale, so does the runoff production. Runoff is categorized as surface runoff, subsurface runoff and underground runoff. Surface runoff produced on the hill slope first, and is routed through the hill slope to its adjacent river cell using the kinematical wave approximation method. Runoff routing in the river channel is treated as river flow routing, and calculated by using the one dimensional diffusive wave approximation method.

The parameters of Liuxihe model are categorized as climate-related parameters, topography-related parameters, soil-related parameters and LUC-related parameters. All parameters are derived directly from their related terrain properties, this makes it be used in hydrological data spare watershed. If there is hydrological data observation, Liuxihe model can optimize model parameters by using PSO algorithm automatically (Chen et al., 2016).

Standard software has been developed for Liuxihe model set up, parameter optimization and flood forecasting (Chen and Dong, 2013), which has several independent software components, including Evapotranspiration Modular, Runoff Production Modular, Runoff Routing Modular and Parameter Deriving Modular. A public computer infrastructure for implementing Liuxihe model has been built which utilizes big data, parallel computing and supercomputing technologies.

Liuxihe model has been widely used in China for small and medium sized watershed flood forecasting (Liao et al., 2012; Chen et al.,2013), reservoir inflow flood forecasting (Chen et al., 2010; Huang et al., 2017), and large scale watershed flood forecasting (Chen et al., 2017b; Li et al., 2017), and experiences have been accumulated in the past application.

3.2 Structure set up of Liuxihe model

The watershed is divided into cells with the collected DEM in this study, the model spatial resolution is set to be 30 m × 30 m. There are 21 434 cells in total, which is further categorized as 21106 hill slope cells and 328 river channel cells, no reservoir cell is included. The river channel system is divided into 2 orders, and 13 virtual sections have been prepared after setting 8 nodes. The cross-section sizes of the virtual sections have been estimated by referencing to satellite remote sensing imageries. The Liuxihe model structure is shown in Fig. 4.

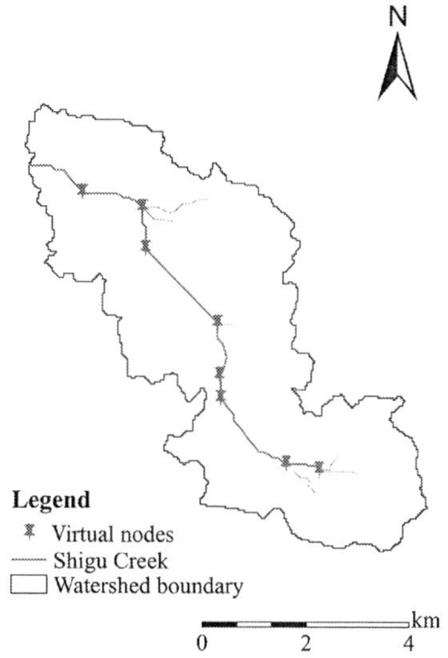

Figure 4. Liuxihe model structure in Shigu creek.

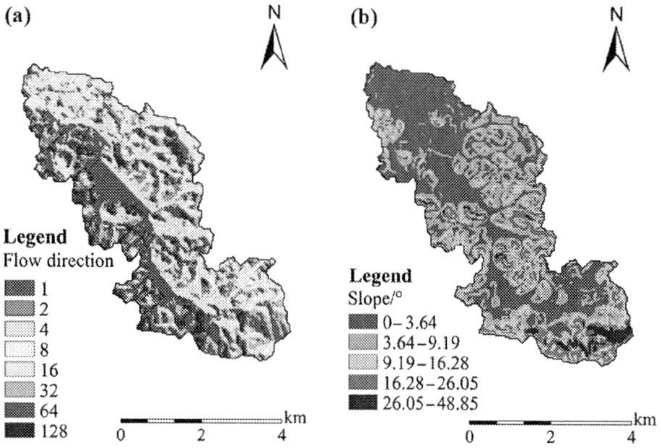

Figure 5. **(a)** Flow direction. **(b)** Slope. Topography-related model parameters.

3.3 Parameterization of Liuxihe model

Topography-related model parameters are flow direction and slope, these two parameters are derived from the topography data and are not adjusted. In this study, flow direction and slope are derived by using D8 method (Jenson and Dominggue, 1988; Strahler, 1957) based on the DEM. The derived flow direction and slope of Shigu creek are shown in Fig. 5.

Climate-related parameter is evaporation capacity, it is estimated as 5 mm day^{-1} for every cells based on local experiences and Liuxihe model parameterization practice. LUC-related parameters are evaporation coefficient and roughness. According to the past experiences of Liuxihe model parameterization (Chen et al., 2010), evaporation coefficients for all cells are set to 0.7, and the roughness of forestry land, grass land, farm land, urban land, water body and bare land are proposed as 0.55, 0.18, 0.36, 0.01, 0.06 and 0.12.

Table 1. Runoff coefficients of the 3 storms. The 3 storms were collected on 7 May 2010, 26 July 2010 and 28 May 2016 respectively. In this paper, they were written as 20100507, 20100726 and 20160528 respectively.

Time	Urbanization Rate (%)	20160528	20100507	20100726
1987	1.0	0.461	0.625	0.595
1990	1.8	0.458	0.622	0.594
1993	2.6	0.489	0.645	0.619
1996	3.4	0.490	0.647	0.620
1999	5.7	0.501	0.657	0.633
2001	8.7	0.518	0.670	0.645
2003	10.6	0.536	0.684	0.660
2005	10.8	0.544	0.691	0.668
2008	15.9	0.602	0.734	0.715
2011	17.4	0.613	0.744	0.724
2013	16.8	0.595	0.731	0.712
2015	20.2	0.616	0.745	0.725

Table 2. Peak Flows of the 3 storms ($m^3\ s^{-1}$).

Time	Urbanization Rate (%)	20160528	20100507	20100726
1987	1.0	23.99	68.07	82.35
1990	1.8	23.34	67.82	80.39
1993	2.6	28.33	72.22	89.97
1996	3.4	31.00	73.33	96.23
1999	5.7	34.48	73.79	92.34
2001	8.7	29.89	72.47	90.60
2003	10.6	30.61	76.47	91.33
2005	10.8	32.79	78.74	99.79
2008	15.9	42.03	87.30	106.37
2011	17.4	40.92	86.82	105.60
2013	16.8	37.65	83.97	99.47
2015	20.2	41.08	86.13	105.61

The soil-related parameters are soil water content at saturated condition, field condition and wilting condition, the soil layer thickness, the soil hydraulic conductivity at saturation and the soil characteristics coefficient. According to local experiences, the soil characteristics coefficient takes 2.5, and the soil layer thickness of Ferric acrisols and Cululic anthrosols are set to 850 and 1000 mm respectively. The soil water content at saturation condition and field condition, and the hydraulic conductivity at saturation condition are determined by the Soil Water Characteristics Hydraulic Properties Calculator (Arya and Paris, 1981), and the soil water content at wilting condition is estimated as 30 % of the soil water content at saturated condition.

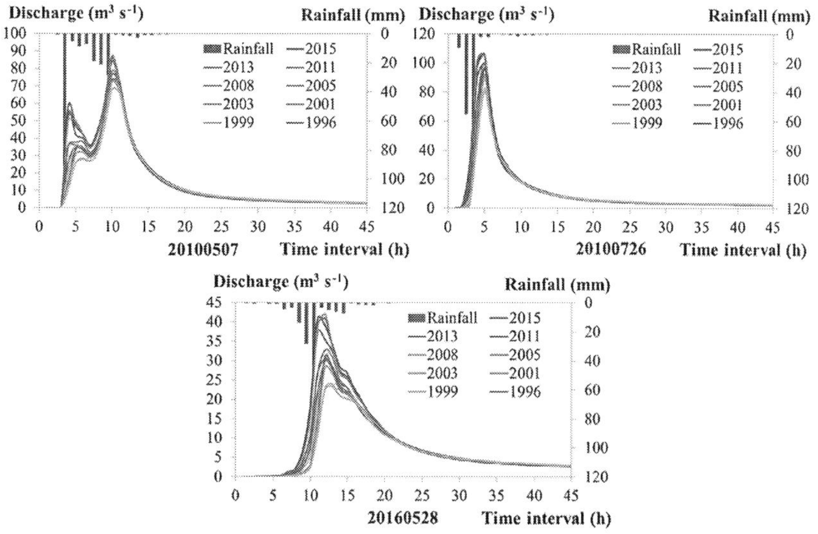

Figure 6. 3 of the simulated flood hydrographs.

4 Results and discussion

The flood events were simulated by using the model set up in this paper, the input is the precipitation collected in this paper. 3 floods have been simulated, and the simulated flood hydrographs are shown in Fig. 6. Runoff coefficients and peak flows for each storms were calculated as shown in Tables 1 and 2. The maximum runoff coefficient is 0.745, the minimum runoff coefficient is 0.461, and the averaged runoff coefficient of all the 3 storms is 0.626. The urbanization has a significant influence on the flood characteristics of the Shigu creek. Compared with 1987, the average peak flow of 2015 increased by 19.47 $m^3\,s^{-1}$ (42 %), while the average runoff coefficient increased by 0.135 (24.9 %). With the increasing of the urbanization rate, the runoff coefficient and the peak flow have been increasing sharply.

It can be seen that the results are in line with our expectations. The impact of urbanization on flood response is significant. In the process of urbanization, the LUCs have changed greatly. Therefore, the LUC-related parameters and soil-related parameters in the model should be different during the study period. However, for lacking of observed discharge, the parameters cannot be optimized. As a result, in the course of the follow-up study, it is necessary to give comprehensive analysis of parameter optimization in urban area without observed discharge.

5 Conclusion

Results in this study shows that the urbanization of Shigu creek is very rapid, and this change has a significant impact on the flood characteristics of the Shigu creek. Compared with 1987, the average peak flow of 2015 increased by 19.47 $m^3\,s^{-1}$ (42 %), while the average runoff coefficient increased by 0.135 (24.9 %). This will provide sound input for flood planners and managers in planning and managing the flood of Shigu creek.

Data availability. In this study, the DEM was downloaded from the Shuttle Radar Topography Mission database website (http://srtm.csi.cgiar.org), and the soil type was downloaded from FAO soil type dataset (http://www.isric.org). The LUC types were extracted from the whole Dongguan City LUC dataset that was prepared by Chen et al. (2017a). For which, the raw data is not put in an open access dataset, but available by writing to the contacting author by email at eescyb@mail.sysu.edu.cn. The hydrological data was provided by the department of water resources of Dongguan City inclusively for this study, and is not available for the public.

Competing interests. The authors declare that they have no conflict of interest.

Special issue statement. This article is part of the special issue "Innovative water resources management – understanding and balancing interactions between humankind and nature". It is a result of the 8th International Water Resources Management Conference of ICWRS, Beijing, China, 13–15 June 2018.

Acknowledgements. This study is supported by the National Science & Technology Pillar Program during the Twelfth Five-year Plan Period (funding no. 2015BAK11B02), the Science and Technology Program of Guangdong Province (funding no. 2014A050503031), and the National Natural Science Foundation of China (NSFC) (funding no. 51379222).

Edited by: Wenchao Sun
Reviewed by: two anonymous referees

References

Abbott, M. B., Bathurst, J. C., Cunge, J. A., O'Connell, P. E., and Rasmussen, J.: An introduction to the European Hydrological System – Systeme Hydrologique Europeen, "SHE", 1: History and philosophy of a physically-based, distributed modelling system, J. Hydrol., 87, 45–59, 1986.

Arya, L. M. and Paris, J. F.: A physioempirical model to predict the soil moisture characteristic from particle-size distribution and bulk density data, Soil Sci. Soc. Am. J., 45, 1023–1030, 1981.

Chen, Y.: Liuxihe Model, Beijing, Science Press, 198 pp., 2009.

Chen, Y. and Dong, Y.: Design and development of Liuxihe model cloud, Proceeding of the 2013 annual meeting of the Chinese Society of Hydraulic Engineering, 1231–1237, 2013.

Chen, Y., Ren, Q., Huang, F., Xu, H., and Cluckie, I.: Liuxihe Model and Its Modeling to River Basin Flood, J. Hydrol. Eng., 16, 33–50, 2010.

Chen, Y., Dong, Y., and Zhang, P.: Study on the method of flood forecasting of small and medium sized catchment, Proceeding of the 2013 annual meeting of the Chinese Society of Hydraulic Engineering, 1001–1008, 2013.

Chen, Y., Li, J., and Xu, H.: Improving flood forecasting capability of physically based distributed hydrological models by parameter optimization, Hydrol. Earth Syst. Sci., 20, 375–392, https://doi.org/10.5194/hess-20-375-2016, 2016.

Chen, Y., Zhang, T., Dou, P., Dong, L., and Chen, H.: Error Sources and Post Processing Method for Land Use/cover Change Estimation of Dongguan City – based on Landsat Remote Sensing Imagery with SVM, Remote Sensing Technology and Application, 32, 893–903, 2017a.

Chen, Y., Li, J., Wang, H., Qin, J., and Dong, L.: Large-watershed flood forecasting with high-resolution distributed hydrological model, Hydrol. Earth Syst. Sci., 21, 735–749, https://doi.org/10.5194/hess-21-735-2017, 2017b.

Huang, J., Dong, L., Chen, Y., Wang, H., and Chen, L.: Liuxihe Model-based study on inflow flood forecasting model of Lechangxia Reservoir, Water Resources and Hydropower Engineering, 48, 1–7, 2017.

Jenson, K. and Domingue, O.: Extracting Topographic Structure from Digital Elevation Data for Geographic Information System Analysis, Photogramm. Eng. Rem. S., 54, 1593–1600, 1988.

Li, J., Chen, Y., Wang, H., Qin, J., Li, J., and Chiao, S.: Extending flood forecasting lead time in a large watershed by coupling WRF QPF with a distributed hydrological model, Hydrol. Earth Syst. Sci., 21, 1279–1294, https://doi.org/10.5194/hess-21-1279-2017, 2017.

Liang, X., Lettenmaier, D. P., Wood, E. F., and Burges, S. J.: A simple hydrologically based model of land surface water and energy fluxes for general circulation models, J. Geophys. Res., 99, 14415–14428, 1994.

Liao, Z., Chen, Y., Xu, H., and He, J.: Study of Liuxihe Model for flood and rainfall forecast of Tiantoushui Watershed, Yangtze River, 43, 12–16, 2012.

Strahler, A. N.: Quantitative Analysis of watershed Geomorphology, T. Am. Geophys. Un., 8, 913–920, 1957.

USDA: National engineering handbook, section 4, hydrology, USDA, Soil Conservation Service, US Government Printing Office, Washington, DC, 1972.

Zhao, R. J.: Flood Forecasting Method for Humid Regions of China, East China College of Hydraulic Engineering, Nanjing, China, 1977.

https://doi.org/10.5194/piahs-379-61-2018

Understanding the potential sources and environmental impacts of dissolved and suspended organic carbon in the diversified Ramganga River, Ganges Basin, India

Mohd Yawar Ali Khan and Fuqiang Tian

Department of Hydraulic Engineering, Tsinghua University, Beijing 100084, China

Correspondence: Mohd Yawar Ali Khan (yawar@mail.tsinghua.edu.cn)

Received: 16 December 2017 – Revised: 2 March 2018 – Accepted: 6 March 2018 – Published: 5 June 2018

Abstract. The river network is one of the important transporters of nutrients from the environment and land masses to the oceans and regularly provides storage for several compounds. The variations in suspended and dissolved discharge of the river are more substantial than the changes in water discharge. (Suspended and dissolved) organic carbons (SOC and DOC) are imperative segments in the carbon cycle and fill in as essential food sources for amphibian sustenance networks. In the present study, 26 samples of water were collected from different locations over the 642 km stretch of the Ramganga River and its adjoining tributaries to observe the spatial variation of DOC, dissolved inorganic carbon (DIC), SOC and suspended inorganic carbon (SIC) in river water. The DOC and DIC values of Ramganga River goes between 1.49 to 4.65 and 9.61 to 36.6 mg L^{-1} with an average convergence of 2.5 and 20 mg L^{-1}, individually, while in case of tributaries, these values extends between 0.09 to 4.52 and 4.61 to 42.36 mg L^{-1} with an average convergence of 2.13 and 21.1 mg L^{-1}, separately. The estimations of SOC and SIC in the Ramganga River extend between 1.31 to 22.15 and 1.27 to 10.14 g kg^{-1} with an average convergence of 6.29 and 4.24 g kg^{-1}, individually, though in tributaries, these values run between 0.80 to 47.23 and 0.31 to 22.94 g kg^{-1} with an average convergence of 9.25 and 5.14 g kg^{-1}, separately. The results also show the higher values of DOC as compared with SOC and these values shows an increasing pattern with a decrease in elevation.

1 Introduction

The imbalance in the carbon cycle is an outcome of large scale human activities that enhance the atmospheric CO_2 concentration. This enhanced amount of atmospheric CO_2 has led to global warming. For the purpose of studying the role played by nature in global warming, quantification of each source and sink of the carbon cycle is immensely important. Rivers are immensely important components of the ecosystem as they relate soil with groundwater, oceans as well as the atmosphere (Khan et al., 2017). ~2 Petagram (Pg) of terrigenous organic carbon is channelled, transformed and stored by fluvial networks transport annually (Battin, 2008). The role of fluvial network in the transportation, transformation as well as outgassing of organic carbon (OC) to the atmosphere has been highlighted by Battin (2008), on the other hand, hypothesis referred by several researchers, for example, Raymo and Ruddiman (1992) supported by Berner and Berner (1996); Gaillardet et al. (1999); Galy and France-Lanord (2001) emphasise that occurrence of high rate of physical and chemical weathering of the Himalayan terrain during the Cenozoic period, burial of huge masses of organic carbon as well as its segregation has caused a cooling trend on the climate globally. Over the past 15 Myr, the river system of Ganga-Brahmaputra has deposited 0.6×10^{12} mol organic carbons in the Bay of Bengal; this accounts for 15 % of the total burial flux (Galy et al., 2007). The current study has been conducted in the Ramganga River, a tributary of River Ganges to determine the concentrations of total dissolved carbon (TDC) comprising DOC and dissolved inorganic carbon (DIC) and total suspended carbon (TSC) comprising SOC and suspended inorganic carbon (SIC) on spatial scale. The objectives of this study also include under-

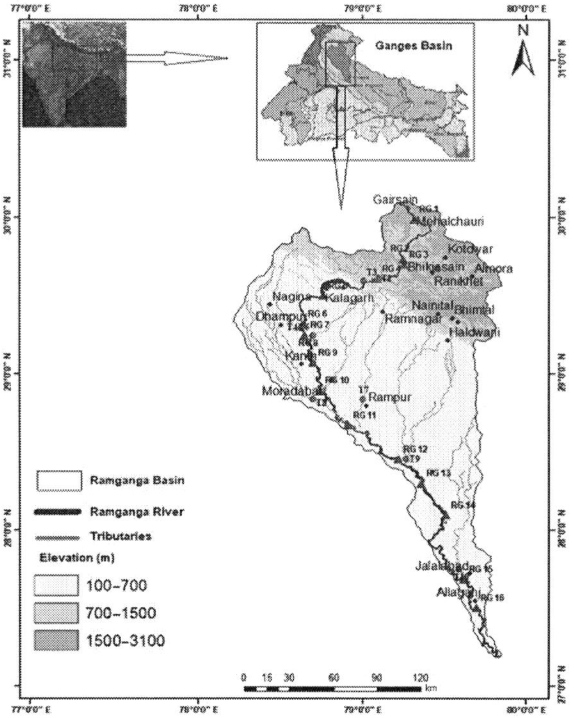

Figure 1. The Ramganga River basin with the sampling locations for the present study.

standing the variation of TDC and TSC in the upstream and downstream of the Ramganga Basin.

2 Study area

Ramganga River originates from the Dudhotali Mountains in the Chamoli district of Uttarakhand, and flows through the Kumaun Himalayas and the Ganga Flood Plains (GFP) before joining Ganges River (Khan and Chakrapani, 2016) (Fig. 1).

It has a catchment area of about 22 685 km^2 (Khan and Chakrapani, 2016; Khan, 2018) of which mostly covered by forests and vegetation in the Uttarakhand state and agriculture in the Uttar Pradesh state (Table 1). Ramganga River flows through a length of 642 km (Daityari and Khan, 2017; Khan et al., 2016b). The river flows through the Kumaun Himalayas for about 158 km, covering the Jim Corbett National Park. Thereafter, it enters the GFP at Kalagarh, where the Ramganga Dam has been constructed. The river then flows through the highly industrialized and agricultural land of densely populated districts of Uttar Pradesh viz., Bijnor, Moradabad, Rampur, Bareilly, Shahjahanpur, Hardoi and Farrukhabad (Khan et al., 2016a).

Figure 2 shows the average annual temperature, precipitation and water discharge at three gauging site on the Ramganga River. Geologically, the entire basin has been divided

Table 1. Land Use and Land Cover area of Ramganga Basin.

Class	Area (km^2)	Area (%)
Agriculture	15 109.4	66.6
Built-up	497.5	2.2
Deciduous Forest	128.3	0.6
Evergreen Broadleaved Forest	529.1	2.3
Evergreen Needle leaved Forest	1439.2	6.3
Grassland	119.7	0.5
Mixed Forest	2585.4	11.4
Plantation	505.0	2.2
Scrubland	709.3	3.1
Wasteland	341.5	1.5
Water Body	721.3	3.2
Total	22 685.6	100.0

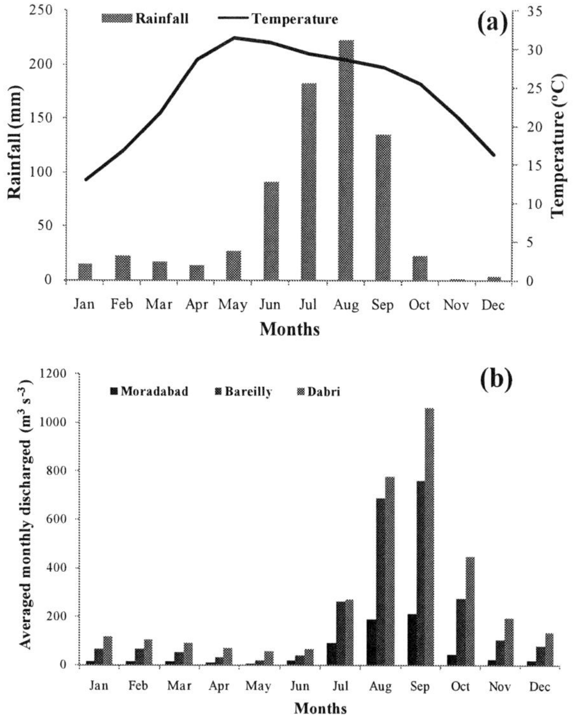

Figure 2. (a) Averaged monthly rainfall (mm) and temperature (°C) (b) Averaged monthly water discharge ($m^3 s^{-1}$) at three gauging sites of the study area.

into three zones, namely Lesser Himalayas, Sub Himalayas and GFP (Khan et al., 2016c).

3 Material and methods

10 L of water samples were collected from 26 different locations from the three stretches of the Ramganga River and its tributaries viz., upper (RG1–RG6 and T1–T5), middle

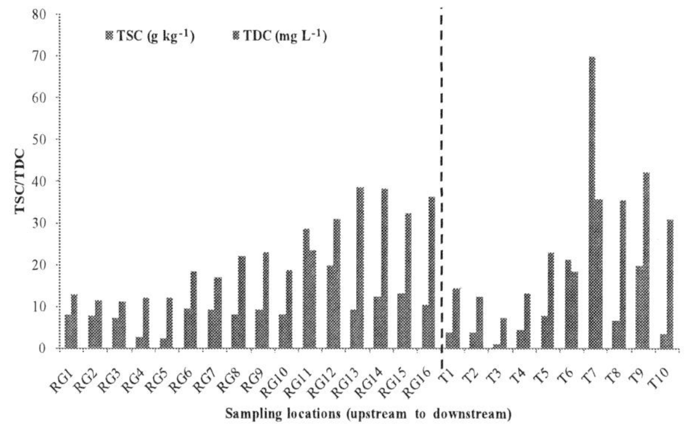

Figure 3. Spatial variation of TSC and TDC in Ramganga River and its tributaries.

(RG7–RG11 and T6–T8) and lower (RG12–RG16 and T9–T10) (Fig. 1). All the sampling has been done in the month March and April 2014. The suspended sediment concentration (SSC) was obtained by filtering the water sample using cellulose nitrate membrane filter paper of 0.45 µm and the material retained was weighed after dried. The collected water samples were placed in freezer at 4 °C and the process of filtering was done within 2 days from the collection of samples to avoid microbial growth (Chakrapani and Saini, 2009).

The estimation of TDC, DOC and DIC in the river water was done by liquid module of Analytik Jena TC Analyser Multi N/C 3100 instrument. TSC, SOC and SIC were estimated in the solid module of the same instrument by using calcium carbonate as a standard.

4 Results and discussion

TDC shows very clear increasing pattern from upstream to downstream in case of Ramganga River and its tributaries which indicate the impact of human influence on the river; whereas TSC did not show any consistency in its concentration from upstream to downstream (Fig. 3).

4.1 Spatial variations in dissolved organic and inorganic carbon

DOC and DIC are the two important forms in which dissolved carbon is present in the river water. There is huge spatial variation has been shown in the concentration of DOC and DIC in the Ramganga basin. In Ramganga River and its tributaries, DOC varied from 1.49–4.65 and 0.097–4.52 mg L^{-1}, respectively, whereas, the concentration of DIC ranges between 9.6–36.6 and 4.6–42.36 mg L^{-1}, respectively (Fig. 4). The average DOC concentration in the Ramganga Basin in this study was 2.35 mg L^{-1}, which is around half of the global average of 5.75 mg L^{-1}, Indus River (4.6 mg L^{-1}), Ganges (16.1 mg L^{-1}) and Brahmaputra (3.2 mg L^{-1}) DOC concentrations stated by Meybeck (1993) and Ramesh et

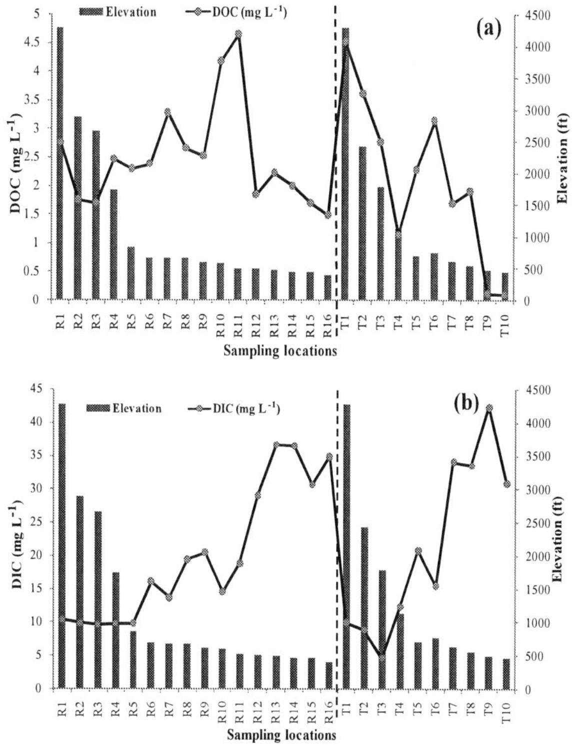

Figure 4. Spatial variation in the DOC and DIC concentration of the Ramganga River and its tributaries.

al. (1995). The average concentration of DOC in Ramganga basin is two times higher from closely associated Alaknanda Basin, which was around 1.2 mg L^{-1} (Panwar et al., 2017). From the Fig. 4, it is evident that the value of DOC decreases in the upper and the lower reaches of the Ramganga River and its tributaries whereas increases in the middle reach of the river. On the other hand, a value of DIC decreases in the upper and increases in the middle and lower reaches of the Ramganga River and its tributaries.

According to Sun et al. (2011), for most rivers, bicarbonate is the dominant form of DIC, primarily from chemical weathering of silicates and carbonates in the drainage basin. Soil erosion is the essentially derived source for DOC and SOC in the river which shows higher turbidity values, however, commitment of aquatic biomass may possibly enhance considerably in low turbid rivers (Ludwig et al., 1996).

Through temporary mechanism, for instance primary production and respiration, microbial consumption, organic and inorganic carbon in the river can transform to each other (Sun et al., 2011). The upper reach of the Ramganga River and its tributaries flowing in Himalayas shows low concentrations of DOC and DIC as compared to middle and lower reaches and the value of turbidity is also low in this stretch (Fig. 5). The DIC concentrations in this reach are mainly attributed to the dilution of carbonate and silicate rocks by

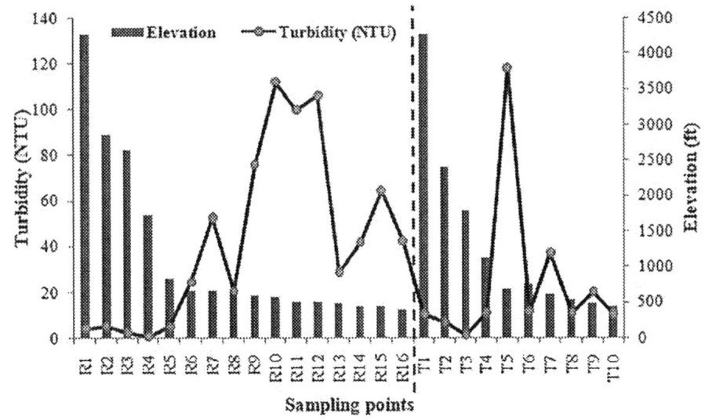

Figure 5. Spatial variation in turbidity values of the Ramganga River and its tributaries.

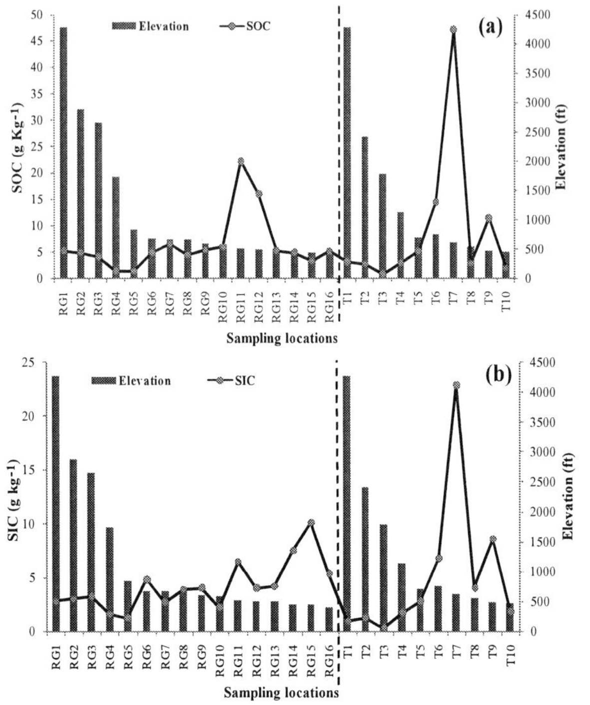

Figure 6. Spatial variation in the SOC and SIC concentration of the Ramganga River and its tributaries.

chemical weathering. Forest covers which occupied > 75 % of the upper region of the Ramganga basin might be a possible factor for the DOC concentration. Evans et al. (2007) and Butman et al. (2015) observed that DOC has been regularly of recent source acclimatized by plants and old carbon (fossilized) includes just 3–9 % of the total DOC conveyed by the global rivers. The $\delta^{13}C_{org}$ investigations by Aucour et al. (2006) in the Bengal fan sediments revealed that that Himalayan Rivers conveyed 70–85 % of late organic carbon. Vegetations comprised of both C_3 and C_4 were found in the Upper Ganges Basin (Chakrapani and Veizer, 2005). The dissimilarity amongst C_3 and C_4 plants is that C_4 plants influence a four-carbon sugar amid the Calvin cycle rather than two three-carbon sugars as in C_3 plants.

The huge elevation in the DOC and DIC concentration (Fig. 4) is shown in the middle and lower reaches of the Ramganga River and its tributaries, which is situated in the GFP. This area has largely influenced by dense population and highly industrial and agricultural activities and covers around 66 % area of the total basin. The higher values of DOC in sampling sites RG6 to RG11 and T5 to T8 in Fig. 4a may be due to the non-point sources of pollution from semi-urban and rural areas, human interaction with river such as direct discharge of waste material, mass bathing during festivals, religious funerals and cremation on the bank of river, cattle bathing etc. Whereas, the higher values of DIC from RG6 to RG11 and T5 to T8 in Fig. 4b is mainly attributed to the presence of silicate and carbonate rocks in the upper catchment area, soil erosion, agricultural activities and runoff, sand mining and river meandering etc.

4.2 Spatial variations in suspended organic and inorganic carbon

SOC and SIC are the two forms of suspended carbon or particulate carbon (SC or PC) are present in river systems. High concentrations of suspended organic and inorganic matters are a relentless component of numerous substantial streams (Smith et al., 1987). As similar to the dissolved carbon, huge spatial variation has also been shown in the concentration of SOC, whereas SIC does not show the similar pattern in the Ramganga basin (Fig. 6).

In Ramganga River and its tributaries, SOC varied from 1.31–22.15 and 0.80–47.23 $g\,kg^{-1}$, respectively, whereas, the concentration of SIC ranges between 1.27–10.14 and 0.31–22.94 $g\,kg^{-1}$, respectively (Fig. 6). It is evident from the figures that the value of SOC decreases in the upper and the lower reaches of the Ramganga River and its tributaries while increases in the middle reach of the river and its tributaries. Dissimilar to the case of DIC, SIC does not show any spatial variation from upstream and downstream of the river.

Panwar et al. (2017), stated that due to variations in topography and rates of chemical weathering, the organic carbon transported by River Ganges mainly in the form DOC in its upstream and mostly as SOC down the Himalayan foothills. The possible factors behind the SIC concentrations in the upper reaches of the basin, which flows through Kumaon Himalayas is may be due to the dilution of carbonate and silicate rocks by chemical weathering. Forest covers which occupied > 75 % of the upper reaches of the Ramganga basin might be a possible factor for the SOC concentration. SC mostly shows positive correlation with SSC (Panwar et al., 2017). In case of Ramganga River and its tributaries, a positive correlation between SSC with SIC and SOC is shown in Fig. 7.

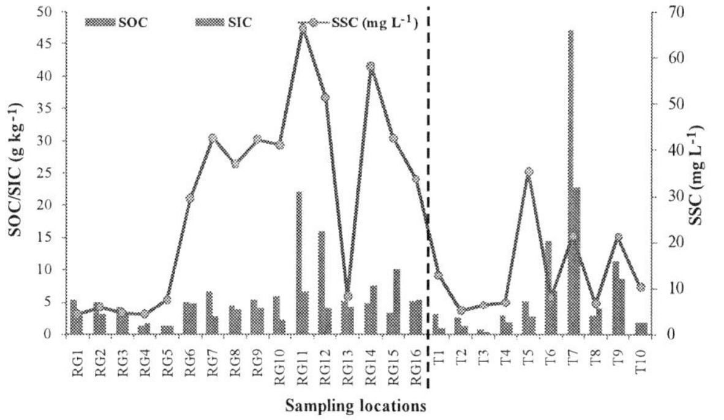

Figure 7. Correlation between SOC and SIC with SSC.

A regular downstream increase pattern in SSC, 4.4 to 66.36 $mg\,L^{-1}$, SOC, 0.80–47.23 $g\,kg^{-1}$ and SIC, 0.31–22.94 $g\,kg^{-1}$ was observed during sampling movement, mainly marked below 304.8 m above sea level, demonstrating that most suspended matter transported towards the River Ganges instigated from the middle and lower regions of the Ramganga River, which is situated in GFP and also low altitude zones rather than from upper region or headwater regions. This indicates that the Ramganga dam, which has been constructed in Kalagarh town between the sampling stations RG5 and RG6 act as tremendously efficient sediment trap. It is concluded that the influence of anthropogenic activity in the GFP due to dense populations and highly industrial setup in one of the major reason of SOC concentration in the Ramganga river, on the other hand, soil erosion, agricultural activities, and sand mining etc., are the possible factors for SIC concentration.

5 Conclusions

The significant conclusions of this investigation are as per the following:

- The Ramganga river and its tributaries show abundance amount of TSC (SOC and SIC) and TDC (DOC and DIC) both in the upstream and downstream.
- TDC accounts more in river concentration as compared to TSC, especially in the downstream of the river.
- Urbanization has driven the expansion of fluvial DOC in the Ramganga basin, which is likely connected to the expansion in sewage discharge because of expanded population.
- Agricultural activities and sand mining have also plays significant role in triggering the river TSC in Ramganga River.

This study only shows the concentration of TSC and TDC on spatial scales, but in order to get the role Ramganga Basin's in transporting TSC and TDC to the River Ganges, this study can be further extended to the temporal scale in order to obtain total carbon flux and role of meteorological factors on regulating this phenomenon.

Data availability. Data are available upon request by the corresponding author.

Competing interests. The authors declare that they have no conflict of interest.

Special issue statement. This article is part of the special issue "Innovative water resources management – understanding and balancing interactions between humankind and nature". It is a result of the 8th International Water Resources Management Conference of ICWRS, Beijing, China, 13–15 June 2018.

Acknowledgements. Mohd Yawar Ali Khan thanks the Indian Institute of Technology, Roorkee, India. The authors gratefully acknowledge the comments of the reviewers and the editor, which enormously improved the presentation of the final manuscript.

Edited by: Wenchao Sun
Reviewed by: two anonymous referees

References

Aucour, A. M., France-Lanord, C., Pedoja, K., Pierson-Wickmann, A. C., and Sheppard, S. M.: Fluxes and sources of particulate organic carbon in the Ganga-Brahmaputra river system, Global Biogeochem Cy., 20, GB2006, https://doi.org/10.1029/2004GB002324/, 2006.

Battin, T. J.: Biophysical controls on organic carbon fluxes in fluvial networks, Nat. Geosci., 1, 95–100, 2008.

Berner, R. A. and Berner, E. K.: Global environment: water, air, and geochemical Cycles, Prentice-Hall, Englewood Cliffs, NJ, p. 376, 1996.

Butman, D. E., Wilson, H. F., Barnes, R. T., Xenopoulos, M. A., and Raymond, P. A.: Increased mobilization of aged carbon to rivers by human disturbance, Nat. Geosci., 8, 112–116, 2015.

Chakrapani, G. J. and Saini, R. K.: Temporal and spatial variations in water discharge and sediment load in the Alaknanda and Bhagirathi rivers in Himalaya, India, J. Asian Earth Sci., 35, 545–553, 2009.

Chakrapani, G. J. and Veizer, J.: Dissolved inorganic carbon isotopic compositions in the upstream Ganga River in the Himalayas, Curr. Sci., 89, 553–556, 2005.

Daityari, S. and Khan, M. Y. A.: Temporal and spatial variations in the engineering properties of the sediments in Ramganga River, Ganga Basin, India, Arab. J. Geosci., 10, 134, https://doi.org/10.1007/s12517-017-2915-2, 2017.

Evans, C. D., Freeman, C., Cork, L. G., Thomas, D. N., Reynold, B., Billett, M. F., Garnett, M. H., and Norris, D.: Evidence against recent climateinduced destabilisation of soil carbon from 14C analysis of riverine dissolved organic matter, Geophys. Res. Lett., 34, L07407, https://doi.org/10.1029/2007GL029431, 2007.

Gaillardet, J., Dupre, B., Louvat, P., and Allegre, C. J.: Global silicate weathering and CO_2 consumption rates deduced from the chemistry of large rivers, Chem. Geol., 159, 3–30, 1999.

Galy, A. and France-Lanord, C.: Higher erosion rates in the Himalaya: geochemical constraints on riverine fluxes, Geology, 29, 23–26, 2001.

Galy, V., France-Lanord, C., Beyssac, O., Faure, P., Kudrass, H., and Palhol, F.: Efficient organic carbon burial in the Bengal fan sustained by the Himalayan erosional system, Nature, 450, 407–410, 2007.

Khan, M. Y. A.: Spatial Variation in the Grain Size Characteristics of Sediments in Ramganga River, Ganga Basin, India, Handbook of Environmental Materials Management, 1–11, 2018.

Khan, M. Y. A. and Chakrapani, G. J.: Particle size characteristics of Ramganga catchment area of Ganga River, in: Geostatistical and Geospatial Approaches for the Characterization of Natural Resources in the Environment (pp. 307–312), Springer, Cham, 2016.

Khan, M. Y. A., Hasan, F., Panwar, S., and Chakrapani, G. J.: Neural network model for discharge and water-level prediction for Ramganga River catchment of Ganga Basin, India, Hydrol. Sci. J., 61, 2084–2095, 2016a.

Khan, M. Y. A., Gani, K. M., and Chakrapani, G. J.: Assessment of surface water quality and its spatial variation. A case study of Ramganga River, Ganga Basin, India, Arab. J. Geosci., 9, 28, https://doi.org/10.1007/s12517-015-2134-7, 2016b.

Khan, M. Y. A., Daityari, S., and Chakrapani, G. J.: Factors responsible for temporal and spatial variations in water and sediment discharge in Ramganga River, Ganga Basin, India, Environ. Earth Sci., 75, 283, https://doi.org/10.1007/s12665-015-5148-2, 2016c.

Khan, M. Y. A., Gani, K. M., and Chakrapani, G. J.: Spatial and temporal variations of physicochemical and heavy metal pollution in Ramganga River – a tributary of River Ganges, India, Environ. Earth Sci., 76, 231, https://doi.org/10.1007/s12665-017-6547-3, 2017.

Ludwig, W., Probst, J. L., and Kempe, S.: Predicting the oceanic input of organic carbon by continental erosion, Global Biogeochem Cy., 10, 23–41, 1996.

Meybeck, M.: Riverine transport of atmospheric carbon: sources, global typology and budget, Water Air Soil Pollut., 70, 443–463, 1993.

Panwar, S., Gaur, D., and Chakrapani, G. J.: Total organic carbon transport by the Alaknanda River, Garhwal Himalayas, India, Arab. J. Geosci., 10, 207, https://doi.org/10.1007/s12517-017-3003-3, 2017.

Ramesh, R., Purvaja, G. R., and Subramanian, V.: Carbon and phosphorus transport by the major Indian rivers, J. Biogeogr., 22, 409–415, 1995.

Raymo, M. E. and Ruddiman, W. F.: Tectonic forcing of late Cenozoic climate, Nature, 359, 117–122, 1992.

Smith, R. A., Alexander, R. B., and Wolman, M. G.: Water-quality trends in the nation's rivers, Science, 235, 1607–1616, 1987.

Sun, H. G., Han, J. T., Zhang, S. R., and Lu, X. X.: Transformation of dissolved inorganic carbon (DIC) into particulate organic carbon (POC) in the lower Xijiang River, SE China: an isotopic approach, Biogeosciences Discuss., 8, 9471–9501, https://doi.org/10.5194/bgd-8-9471-2011, 2011.

https://doi.org/10.5194/piahs-379-67-2018

Assessment of freshwater ecosystem services in the Beas River Basin, Himalayas region, India

Sikhululekile Ncube, Lindsay Beevers, Adebayo J. Adeloye, and Annie Visser

Institute for Infrastructure and Environment, School of Energy, GeoScience, Infrastructure and Society, Heriot-Watt University, Edinburgh, EH14 4AS, UK

Correspondence: Sikhululekile Ncube (s.ncube@hw.ac.uk, snmancue@gmail.com)

Received: 30 December 2017 – Revised: 4 April 2018 – Accepted: 5 April 2018 – Published: 5 June 2018

Abstract. River systems provide a diverse range of ecosystem services, examples include: flood regulation (regulating), fish (provisioning), nutrient cycling (supporting) and recreation (cultural). Developing water resources through the construction of dams (hydropower or irrigation) can enhance the delivery of provisioning ecosystem services. However, these hydrologic alterations result in reductions in less tangible regulating, cultural and supporting ecosystem services. This study seeks to understand how multiple impoundments, abstractions and transfers within the upper Beas River Basin, Western Himalayas, India, are affecting the delivery of supporting ecosystem services. Whilst approaches for assessing supporting ecosystem services are under development, the immediate aim of this paper is to set out a framework for their quantification, using the macroinvertebrate index Lotic-Invertebrate Index for Flow Evaluation (LIFE). LIFE is a weighted measure of the flow velocity preferences of the macroinvertebrate community. Flow records from multiple gauging stations within the basin were used to investigate flow variability at seasonal, inter-annual and decadal time scales. The findings show that both mean monthly and seasonal cumulative flows have decreased over time in the Beas River Basin. A positive hydroecological relationship between LIFE and flow was also identified, indicative of macroinvertebrate response to seasonal changes in the flow regime. For example, high LIFE scores (7.7–9.3) in the winter and summer seasons indicate an abundance of macroinvertebrates with a preference for high flows; this represents a high potential for instream supporting ecosystem services delivery. However, further analysis is required to understand these hydroecological interactions in the study basin and the impact on instream supporting ecosystem services delivery.

1 Introduction

River systems provide an array of ecosystem services, ranging from tangible benefits such as freshwater supply (provisioning) to the less tangible where examples include religious enrichment (cultural), climate regulation (regulating) and nutrient cycling (supporting) (Gilvear et al., 2017). Some riverine ecosystem services may be enhanced through the development of water resources such as the construction of dams for hydropower and/or irrigation to meet prioritised human needs. However, there are often trade-offs between the ecosystem service categories, with downstream ecosystems and the ecosystem services they provide, paying the price (Millennium Ecosystem Assessment, 2005). More often, such trade-offs result from developments which increase the delivery of provisioning ecosystem services, but negatively impact less tangible cultural, regulating and supporting ecosystem services.

In river basins such as the Beas, Western Himalayas, India, increasing water demands for domestic, industrial and irrigation purposes and hydropower generation have led to rapid development. Additionally, changing climatic conditions have led to increases in temperature, decreased snow fall and glacier shrinkage (Jain et al., 2010). These drivers of change have, and continue, to impact on the delivery of ecosystem services, particularly those which are less tangible.

Supporting ecosystem services underpin the delivery of all other ecosystem services. Therefore, to recognise the im-

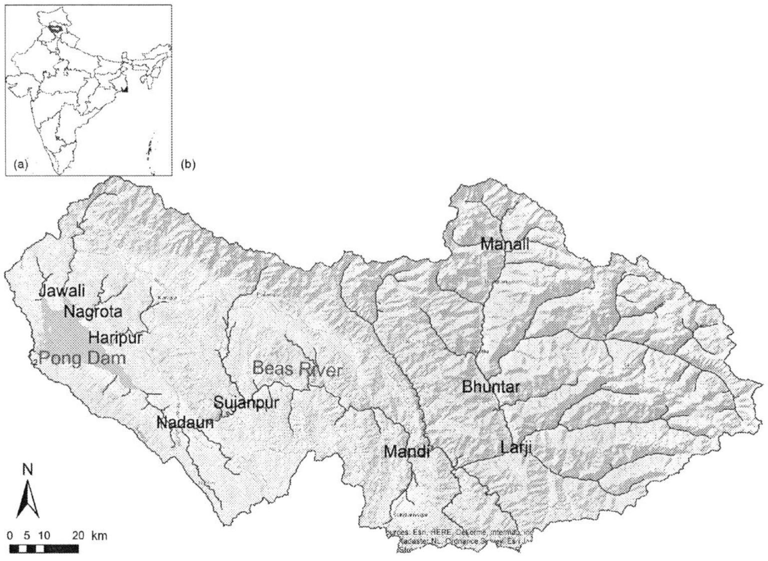

Figure 1. Map of the Beas River Basin and its situation (inset).

pact of developments on ecosystem service delivery, process understanding of the formation and regulation of these supporting ecosystem services is essential. Due to their sensitivity to change (Chaube et al., 2008; Chauhan and Verma, 2016), macroinvertebrates have widely been used as ecological indicators of the impact of flow regulation and abstraction in river systems (Ghosh and Biswas, 2015; Monk et al., 2008). Macroinvertebrates play a significant role in the delivery of supporting ecosystem services in river systems (Thorp et al., 2010). As primary consumers, detritivores, predators and pollinators, macroinvertebrates contribute to supporting ecosystem services such as nutrient cycles, primary productivity, and decomposition. Additionally, in freshwaters, macroinvertebrates operate at intermediate trophic levels (approximately the middle of the food web) and are generally the main in-stream food source of fish (Wallace and Webster, 1996).

To assess the ecological response of macroinvertebrates to flow regime the Lotic-invertebrate Index for Flow Evaluation (LIFE) (Extence et al., 1999) is commonly used (Monk et al., 2006). LIFE is abundance-weighted, based on macroinvertebrate flow velocity preferences. Knowledge on the applicability of LIFE on snow fed river systems, such as the Beas River, is limited. The results presented in this paper are part of ongoing work. The overall aim of this study is to understand the impact of the altered flow regime on supporting ecosystem services delivery potential in the case study basin, using macroinvertebrates as an indicator.

2 Materials and Methods

2.1 Study Area

The source of the Beas River is located in the Himalayas, central Himachal Pradesh, India; the river merges with the Satluj River at Harike in Punjab. The Beas River is one of the main tributaries of the Indus River system, which is critical for food, energy and economic development in India. Figure 1 shows the location of the Beas River Basin in North India. The Pong Dam reservoir is situated downstream in the basin; the dam is mainly used for hydropower generation and irrigation water supply.

2.2 Daily stream flow discharge data

Historic daily mean stream flow discharge data for the study area was made available by the Bhakra Beas Management Board (BBMB) for two gauging stations: Nadaun (from 1980 to 2017) and Thalout (from 1967 to 2007) (Fig. 2).

Time varying seasonal and decadal summary flow statistics were calculated to understand the long-term flow variability which may influence macroinvertebrates, and hence, the provision of supporting ecosystem services. These included mean monthly flows, the cumulative seasonal flow exceeded 10 % of the time (Q_{10}) and the cumulative seasonal flow exceeded 90 % of the time (Q_{90}). While Q_{10} is a measure of high flows, the Q_{90} is a measure of low flow. Seasons were defined as the following hydrological periods: Summer (pre monsoon) (March–May), Monsoon (June–September), Autumn (Post monsoon) (October–November) and Winter (December–February).

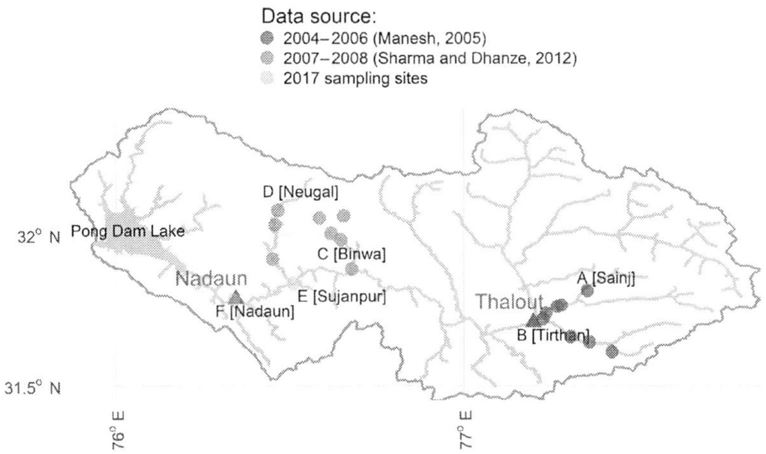

Figure 2. Historic and recent (2017) sampling sites in the Beas River Basin. Sampling reaches are labelled A to F.

2.3 Macroinvertebrate data

The macroinvertebrate sampling locations (historic and from 2017) are detailed in Fig. 2. The historic sampling sites are located on the tributaries of the Beas River whilst the 2017 sampling sites are located on the main stem.

2.3.1 Historic macroinvertebrate data

Two previous studies were used as sources of historic macroinvertebrate data. Historic macroinvertebrate data, sampled monthly, was made available from: Manesh (2005) in 2005–2006 at A and B (Fig. 2, Sainj River and Tirthan River) and Sharma and Dhanze (2012) in 2007–2008 at C and D (Fig. 2; Binwa River and Neugal River). Data from these historic studies was used to calculate seasonal LIFE scores. This seasonality was considered with reference to seasonal changes in the Beas River flow regime.

2.3.2 Current macroinvertebrate sampling

Additional macroinvertebrate sampling was carried out as part of this study at E and F (Fig. 2; Sujanpur and Nadaun on the main stem). Sampling was carried out during the post-monsoon period in November 2017. The kick-sampling method (Barbour et al., 1999) was used to collect the macroinvertebrates from both sampling sites. A handnet (1 mm mesh size) was used to collect two samples at each site to form a single composite sample per site. Kick-sampling required disturbing the river bed for a duration of 3 min whilst the handnet is held vertically to the river bed, with the frame of the net at right angles to river flow (downstream of one's feet). Dislodged macroinvertebrates within an area close to the net were gathered and subsequently stored in containers and preserved with 90 % ethanol for identification in the laboratory. The samples were sorted and taxonomically identified to Family level following Subramanian and Sivaramakrishnan (2007). Family LIFE scores were calculated via the method described by Extence et al. (1999) (refer to Supplement 1).

3 Results and Discussion

3.1 Flow variation over time

Figure 3 depicts the decomposed time series of the mean monthly flows at both gauging stations. There are strong similarities in flow seasonality, with high flows during the monsoon season and low flows during the winter season. Overall, the mean monthly flows show a decreasing trend after the 1990s. The observed facet (Fig. 3) also shows trend change points in the 1990s, thereafter the flows appear to decrease.

Mean monthly flows from Nadaun gauging station are more variable compared to Thalout upstream; this is also reflected by irregularities in flows at Nadaun during the change point in the 1990s.

Similarly, analysis of seasonal cumulative flows shows a shift over time in both the Q_{90} and Q_{10} as revealed in Fig. 4.

As illustrated in Fig. 4, both gauging stations depict similar trends of decreasing cumulative flows over the four seasons, especially the monsoon flows. The Nadaun station recorded a noticeable decrease in Q_{10} and Q_{90} monsoon flows from the 1980s to 2017. On the other hand, the Thalout station shows a noticeable downward shift in Q_{10} and Q_{90} during the winter season from the 1960s to the 2000s.

These decreases in both mean monthly flows and seasonal cumulative flows in the Beas River could be the result of climatic change and human activity impacts linked to increases in abstraction activities and hydropower generation in the basin (Singhvi and Krishnan, 2014).

3.2 LIFE scores

Figure 5 provides a comparison of the LIFE scores.

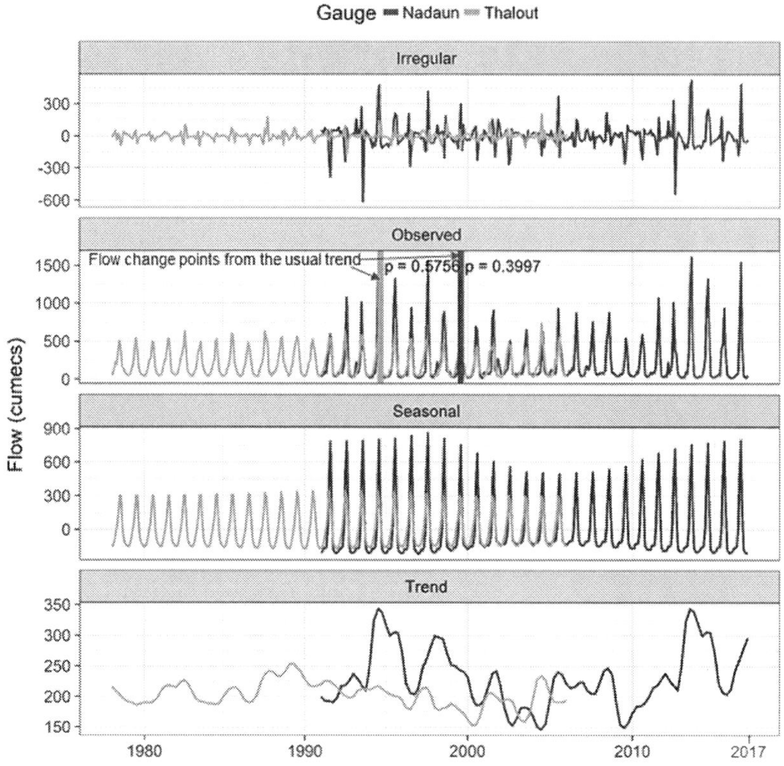

Figure 3. Decomposed time series of mean monthly flows at the two gauging stations.

Despite differences in the spatial location and timing of sampling, similar patterns in macroinvertebrate diversity and abundance were observed; this is represented by the seasonal variations in LIFE scores. Previous studies e.g. Joshi et al. (2007) on macroinvertebrate assemblage in the Himalayas region also observed maximum macroinvertebrate densities during winter and lower densities during the monsoon. Overall, LIFE scores were generally high (7.7–9.3) during the winter and summer seasons and low (minimum of 4.5) during the monsoon period. See Supplement 2 for the LIFE scores per site. The lower diversities and LIFE scores could be the result of the high monsoon flows washing the macroinvertebrates away.

Macroinvertebrate community structure changes across the seasons. High LIFE scores (winter and summer seasons) indicate an abundance of macroinvertebrates with a preference for high flows; this represents a high potential for instream supporting ecosystem services delivery. During the monsoon season, changes in the flow impacts on the macroinvertebrate structure resulting in low LIFE scores. Although macroinvertebrate response to these seasonal changes in river flows could be a natural process, human induced changes in the flow regime could negatively impact on the macro invertebrate community structure and the potential for supporting ecosystem services delivery.

3.3 Influence of river flow changes on LIFE and supporting ecosystem services

The observed seasonal variation in LIFE scores suggests that the macroinvertebrate community is responding to changes in river flow. This reflects the importance of the flow regime in the ecological health of rivers and their capacity to deliver ecosystem services (Vörösmarty et al., 2010). This implies that the observed decrease in temporal and seasonal river flows in the study basin could have a detrimental impact on macroinvertebrates and, consequently, the instream supporting ecosystem services. However, a greater understanding of the hydroecological relationship in the Beas River is required. To this end, the next steps in this study focus on hydroecological modelling, where numerical models are developed to explore the relationship.

Due to data limitations, long-term historic understanding of the macroinvertebrate community and LIFE in the basin are limited. Consequently, this narrows the scope within which the changes in LIFE scores could be assessed. Generally, research on macroinvertebrates in river systems in India has been limited (Mishra et al., 2013). Despite this, the successful application of the LIFE methodology and subsequent findings, helps to confirm the applicability of the methodology in snow fed river systems in India. Additionally, this

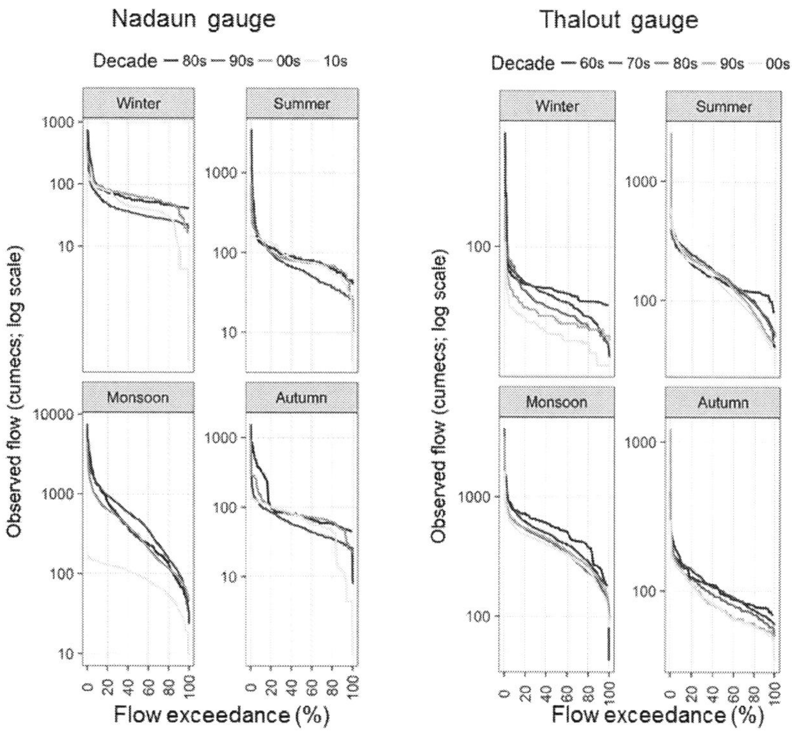

Figure 4. Flow distribution curves (by season and decade) at the two gauging stations.

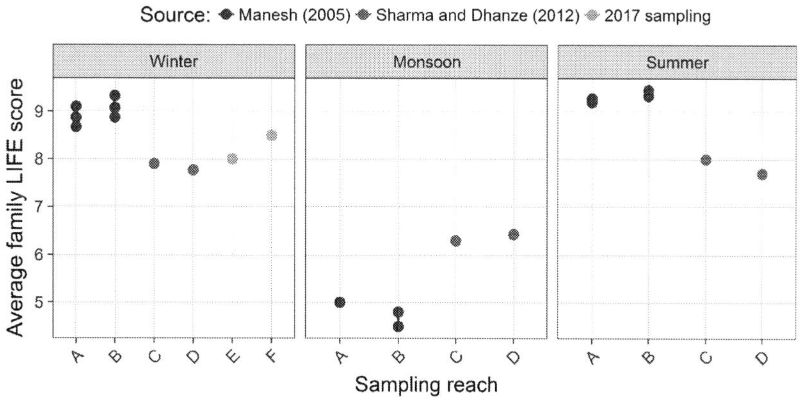

Figure 5. Seasonal variations in LIFE score.

study highlights the need for further work in the form of hydroecological modelling.

4 Conclusions

As part of an ongoing study, current findings presented in this paper, show that both mean monthly and seasonal cumulative flows have decreased overtime in the Beas Basin. Linked to this were seasonal variations in LIFE scores, indicative of macroinvertebrate response to flow changes and their potential to deliver supporting ecosystem services. However, further analysis is required to understand these hydroecological interactions and their impact on instream supporting ecosystem services. Given that more hydropower dams and abstraction activities and changing climatic conditions are set to continue, such an understanding is important in future water resources management planning in the Beas River Basin.

Data availability. All macroinvertebrate data is provided in Manesh (2005), Sharma and Dhanze (2012) and in the Supplement of this article. All hydrological data is not publicly accessible and is provided for use in the project only from the Bhakra-Beas Management Board.

The Supplement related to this article is available online at https://doi.org/10.5194/piahs-379-67-2018-supplement.

Competing interests. The authors declare that they have no conflict of interest.

Special issue statement. This article is part of the special issue "Innovative water resources management – understanding and balancing interactions between humankind and nature". It is a result of the 8th International Water Resources Management Conference of ICWRS, Beijing, China, 13–15 June 2018.

Acknowledgements. This work is part of an ongoing Sustaining Himalayan Water Resources in a changing climate (SusHi-Wat) research project and we acknowledge funding support from the UK Natural Environment Research Council (NERC; Project NE/N016394/1) and the Indian Ministry of Earth Sciences (MoES) under the aegis of the Newton-Bhabha scheme. We are also grateful to our project partners i.e. IITRoorkee, NIT-Hamirpur, NIH-Roorkee and IISc-Bangalore, Cranfield University, the British Antarctic Survey and the Bhakra-Beas Management Board.

Edited by: Zongxue Xu
Reviewed by: two anonymous referees

References

Barbour, M. T., Gerritsen, J., Snyder, B. D., and Stribling, J. B.: Rapid Bioassessment Protocols for Use in Streams and Wadeable Rivers: Periphyton, Benthic Macroinvertebrates and Fish, 2nd Edn., Environmental Protection Agency, Washington, DC, 1999.

Chaube, U. C., Sharma, S., Sharma, P., and Kumar, P.: Water quality assessment of River Satluj using benthic macroinvertebrates, in: Proceedings of the Scientific Conference on Rivers in the Hindu Kush-Himalaya – Ecology and Environmental Assessment, Nepal, 3–7 March 2008, 77–83, 2008.

Chauhan, A. and Verma, S. C.: Distribution and Diversity of Aquatic Insects in Himachal Pradesh, India: A Review, Int. J. Curr. Microbiol. Appl. Sci., 5, 273–281, 2016.

Extence, C. A., Balbi, D. M., and Chadd, R. P.: River flow indexing using British benthic macroinvertebrates: a framework for setting hydroecological objectives, Regul. Rivers, 15, 545–574, 1999.

Ghosh, D. and Biswas, J.: Macroinvertebrate diversity indices: A quantitative bioassessment of ecological health status of an oxbow lake in Eastern India, J. Adv. Environ. Health Res., 3, 78–90, 2015.

Gilvear, D., Beevers, L., O'keeffe, J., and Acreman, M.: Environmental Water regimes and natural capital – Free Flowing Ecosystem Services, in: Water for the Environment: From Policy and Science to Implementation and Management, 1st Edn., edited by: Horne, A., Webb, A., Stewardson, M., Richter, B., and Acreman, M., Academic Press, Elsevier, 151–173, 2017.

Jain, S. K., Goswami, A., and Saraf, A. K.: Assessment of Snowmelt Runoff Using Remote Sensing and Effect of Climate Change on Runoff, Water Resour. Manag., 24, 1763–1777, 2010.

Joshi, P. C., Negi, R. K., and Negi, T.: Seasonal variation in benthic macro-invertebrates and their correlation with environmental variables in fresh water stream in Garhwal region (India), Life Sci., 4, 227–234, 2007.

Manesh, K.: Ecology of some selected hillstreams of the river Beas in the vicinity of the upper Himalayas of Himachal Pradesh, PhD thesis, Department of Zoology, Panjab University, India, http://hdl.handle.net/10603/82916 (last access: 25 April 2018), 2005.

Millennium Ecosystem Assessment: Ecosystems and Human Well-being: Synthesis, Washington, DC, 2005.

Mishra, A. S., Nautiyal, P., and Semwal, V. P.: Distributional Patterns of Benthic Macro-Invertebrate Fauna in the Glacier Fed Rivers of Indian Himalaya, Our Nature, 11, 36–44, 2013.

Monk, W. A., Wood, P. J., Hannah, D. M., Wilson, D. A., Extence, C. A., and Chadd, R. P.: Flow variability and macroinvertebrate community response within riverine systems, River Res. Appl., 22, 595–615, 2006.

Monk, W. A., Wood, P. J., Hannah, D. M., and Wilson, D. A.: Macroinvertebrate community response to inter-annual and regional river flow regime dynamics, River Res. Appl., 24, 988–1001, 2008.

Sharma, I. and Dhanze, R.: Evaluation of macrobenthic fauna in hill stream environment of Western Himalaya, India, JoTT, 4, 2875–2882, http://threatenedtaxa.org/index.php/JoTT/article/view/790/1415 (last access: 25 April 2018), 2012.

Singhvi, A. K. and Krishnan, R.: Past and the Present Climate of India, in: Landscapes and Landforms of India, edited by: Kale, V. S., Dordrecht, 2014.

Subramanian, K. A. and Sivaramakrishnan, K. G.: Aquatic Insects of India-A Field Guide, Ashoka Trust for Ecology and Environment (ATREE), Bangalore, India, 2007.

Thorp, J. H., Flotemersch, J. E., Delong, M. D., Casper, A. F., Thoms, M. C., Ballantyne, F., Williams, B. S., O'Neill, B. J., and Haase, C. S.: Linking Ecosystem Services, Rehabilitation, and River Hydrogeomorphology, BioScience, 60, 67–74, 2010.

Vörösmarty, C. J., McIntyre, P. B., Gessner, M. O., Dudgeon, D., Prusevich, A., Green, P., Glidden, S., Bunn, S. E., Sullivan, C. A., Liermann, C. R., and Davies, P. M.: Global threats to human water security and river biodiversity, Nature, 467, 555-561, 2010.

Wallace, J. B. and Webster, J. R.: The Role Of Macroinvertebrates In Stream Ecosystem Function, Annu. Rev. Entomol., 41, 115–139, 1996.

https://doi.org/10.5194/piahs-379-73-2018

Spatiotemporal variability and assessment of drought in the Wei River basin of China

Siyang Cai[1,2], **Depeng Zuo**[1,2], **Zongxue Xu**[1,2], **Xianming Han**[1,2], **and Xiaoxi Gao**[1,2]

[1]College of Water Sciences, Beijing Normal University, Beijing 100875, China

[2]Beijing Key Laboratory of Urban Hydrological Cycle and Sponge City Technology, Beijing 100875, China

Correspondence: Depeng Zuo (dpzuo@bnu.edu.cn)

Received: 31 December 2017 – Revised: 14 March 2018 – Accepted: 27 March 2018 – Published: 5 June 2018

Abstract. The temporal and spatial variations of drought in the Wei River basin (WRB) were investigated by calculating the meteorological drought Index (Standardized Precipitation Index, SPI) and the agricultural drought index (Vegetation Health Index, VHI). Monthly precipitation and air temperature were from 22 meteorological stations over the region from 1960 to 2015. Monthly Normalized Difference Vegetation Index (NDVI) and 8-days Land Surface Temperature (LST) were provided from the National Aeronautics and Space Administration (NASA) for the period 2000–2015 were also adopted. The results showed that the drought initially increased and then decreased, reaching at the maximum value in 1990s. The spatial pattern of meteorological drought showed that the drought in northern WRB was heavier than that in southern WRB before 1990s, after that, the situation had the opposite. By comparing the agricultural drought index (VHI) with crop yield, it was proved that VHI was applicable in the WRB and could well reflect the fluctuation of agricultural drought. The WRB suffered from serious agricultural drought in 2000, 2001, 2007 and 2008. Through analysis of the historical precipitation and temperature data, it was found that precipitation had a greater contribution to creating agricultural drought conditions than temperature in the Wei River basin.

1 Introduction

Drought is a natural and recurring feature of climate; occurring in virtually all climatic regimes (Mishra and Singh, 2010) and frequent drought have been concerned for many years (Lei et al., 2016). On a global scale, the frequency, duration and severity of droughts have increased substantially in recent decades (Dai, 2011), especially in the regions of arid and semi-arid (Solomon, 2007). Furthermore, drought directly or indirectly affect social and economic sustainability (Alam et al., 2012). Generally divided drought into four categories, meteorological drought, hydrological drought, agricultural drought, as well as socio-economic drought (Wilhite and Glantz, 1985).

Drought severity or magnitude can be illustrated by drought indices (Steinemann, 2003). Numerous indices have been developed for identifying the characteristic of drought over the past few decades. Standardized Precipitation Index (McKee et al., 1993) is a commonly used drought monitoring index, which is calculated based on precipitation data and could have multiple time scales. The SPI has been widely used for drought assessment all over the world (Karavitis et al., 2011; Moreira et al., 2015; Zarch et al., 2015; Oloruntade et al., 2017). Palmer Drought Severity Index (PDSI) is a drought index for drought assessment, which based on precipitation and air temperature to calculating moisture supply and demand by a two-layer soil moisture model (Palmer, 1965). Rayne and Forest (2016) found that there was a substantial increase in drought variability at short time scales based on the PDSI in the United States since 1895. Liu et al. (2013) employed the PDSI for analyzing the spatiotemporal characteristics of dryness conditions across Qinghai Province, Northwest China. The Standardized Precipitation Evapotranspiration Index (SPEI) (Vicente-Serrano et al., 2010) is also based on the supply and demand concept of the water balance equation. Zuo et al. (2016) showed that the applicability of the SPEI in Shandong Province were verified by comparing the SPEI, crop yield and drought-affected crop area.

In addition, remote sensing technology has made it possible to monitor the condition of vegetation across large areas (Lei et al., 2015; Zhang et al., 2017). Thus, remote sensing, which can be used to extract meteorological or biophysical characteristics of terrestrial surfaces, has gained more attention for drought monitoring (Rhee et al., 2010). The widely used vegetation indices include VCI (Vegetation Condition Index) (Kogan, 1995), Temperature Condition Index (TCI) (Kogan, 1995), Soil Moisture Condition Index (SMCI) and Precipitation Condition Index (PCI) (Zhang and Jia, 2013). Furthermore, Martinez-Fernandez et al. (2016) proved that SMOS Soil Water Deficit Index (SWDI) well reproduces the soil water balance dynamics and was able to appropriately track agricultural drought. Unganai and Kogan (1998) clearly showed that temporal and spatial characteristics of drought in Africa could be detected, tracked and mapped from AVHRR data based on Vegetation Health Index (VHI). The Normalized Vegetation Supply Water Index (NVSWI) registered correlation coefficients with the AMSR-E soil moisture data at the values of 0.53 in September and indicated its applicability in monitoring drought in Northeast China (Cong et al., 2017).

China is one of the major crop production countries around the world, which accounted for 20 % of global cereal production in 2011 (FAO, 2014). However, Yu et al. (2014) also revealed that dry areas were reported to increase by 3.72 % per decade in the past five decades. Since the late 1990s, droughts have become more frequent and severe across China (Chen and Sun, 2015). Both the frequency and severity of droughts over the Yun–Gui Plateau in Southwest China were intensified during the recent decades from total water storage anomalies (TWSA) generated using GRACE satellite data and ANN models (Long et al., 2014). The Wei River is of great importance in social and economic development of Shan'xi, Gausu and Ningxia Provinces in China. Zou et al. (2017) showed that the whole Wei River basin exhibited a dry trend, with more significant trends in the northern, southeastern and western WRB than the remaining regions based on PDSI_SWAT. Zhai and Feng (2009) showed that short time drought had affected the Yellow River Basin and Yangtze River Basin more than the Inland River Basin. However, there is no comprehensive research on spatiotemporal variability of drought based on remote sensing dataset in the Wei River basin. Hence, the main process of the study are as follows: (1) to analysis temporal trends of annual total precipitation and average air temperature during the period 1960–2015; (2) to calculate frequency occurrence of different degree of drought at various time scales and the variation of drought intensity for the recent six decades based on the SPI; (3) to identify spatial variations of drought and drought rating based on the SPI; (4) to estimate the spatiotemporal variability of agricultural drought based on the VHI; (5) to evaluate the capability of the VHI; (6) to analysis the relationship between the VHI and meteorological factors. The results gained in this study could provide useful information for drought assessment in the Wei River basin and other regions in China.

2 Study Area and Data Description

2.1 Study Area

Wei River is the largest tributary of the Yellow River in North China, with a drainage area of $1348 \times 10^5\,\text{km}^2$ and originate from the Gansu Province and flows into the Yellow River at the Tongguan port with its length of approximately 818 km (Fig. 1). It lies between latitude 33.5 to 37.5° N and 103.5 to 110.5° E. The mean annual temperature in the WRB is between 7.8 and 13.5°, mean annual precipitation is 400–800 mm^{-1} and potential evapotranspiration is 800–1000 mm yr^{-1}. In winter the climate is cold, dry and rainless, controlled by the Mongolia high pressure system, and it is hot and rainy in summer due to the effects of the West Pacific subtropical high pressure system.

2.2 Data Description

The monthly precipitation and average air temperature data at 22 national meteorological stations analyzed in this paper collected from the Chinese Meteorological Data Sharing Service System, China Meteorological Administration over the Wei River Basin from 1960 to 2015, twelve of them were in Shanxi Province, seven of them were in Gansu Province and others in Ningxia Province. The spatial distribution and detail information of the 22 gauging stations are shown in Fig. 1 and Table 1.

All the products of the remote sensing dataset are provided by the Moderate Resolution Imaging Spectroradiometer (MODIS), which are acquired from the National Aeronautics and Space Administration (NASA). Monthly Normalized Difference Vegetation Index (NDVI) information are obtained from MODIS13A3 NDVI product and eight-days Land Surface Temperature (LST) information are provided by MOD11A2 during 2000–2015. All the MODIS datasets have a spatial resolution of 1 km × 1 km. To remove noise from the remote sensing data caused by clouds or weather, the maximum value composite (MVC) method is used for data processing. The selected MODIS data products are preprocessed using the MODIS Reprojection Tools (MRT) (Yang et al., 2012).

3 Methodology

3.1 Calculation of the SPI

Computation of the SPI involves fitting a gamma probability density function to a given time series of precipitation. This is performed separately for each month (or any other temporal basis of the raw precipitation time series) and for

Table 1. The meteorological gauging stations in the Wei River Basin.

Station Name	Longitude (° E)	Latitude (° N)	Altitude (m)	Precipitation Average (mm)	Precipitation Trend (mm yr^{-1})	Air temperature Average (mm)	Air temperature Trend (mm yr^{-1})
wuqi	108.17	36.92	1331.40	468.33	−1.226	7.93	0.027
huanxian	107.30	36.58	1255.60	431.83	−0.734	8.89	0.039
xiji	105.72	35.97	1916.50	409.19	−1.064	5.65	0.029
Pingliang	106.67	35.55	1346.60	502.98	−0.847	9.06	0.034
xifeng	107.63	35.73	1421.00	549.40	−0.923	8.94	0.044
changwu	107.80	35.20	1206.50	584.21	−0.035	9.31	0.016
luochuan	109.50	35.82	1159.80	610.63	−0.690	9.68	0.030
tongchuan	109.07	35.08	978.90	584.92	−0.494	10.75	0.015
tianshui	105.75	34.58	1141.70	517.90	−0.594	11.09	0.028
baoji	107.13	34.35	612.40	659.11	−1.046	13.27	0.025
wugong	108.22	34.25	447.80	599.71	−0.863	13.31	0.022
xian	108.93	34.30	397.50	561.48	−0.408	13.84	0.028
huashan	110.08	34.48	2064.90	810.33	−3.699	6.30	0.026
yanan	109.50	36.60	958.50	538.22	−1.011	9.99	0.035
foping	107.98	33.52	827.20	920.36	−0.172	11.96	0.037
shangzhou	109.97	33.87	742.20	690.71	−0.312	12.89	0.005
zhenan	109.15	33.43	693.70	787.21	−1.351	13.15	0.044
lintao	103.85	35.35	1893.80	538.93	−0.860	7.31	0.020
huajialing	105.00	35.38	2450.60	484.22	−1.618	3.77	0.023
minxian	104.02	34.43	2315.00	577.42	−1.209	5.97	0.027
yanchi	107.38	37.80	1349.30	292.00	−0.148	8.27	0.029
guyuan	106.27	36.00	1753.00	451.44	−1.071	6.74	0.038

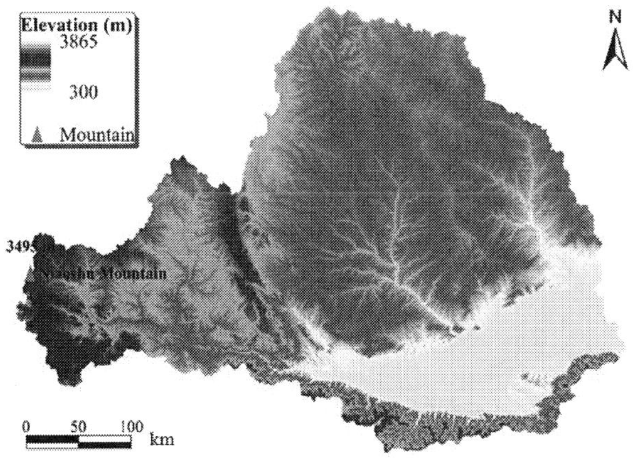

Figure 1. Distribution of meteorological stations in the Wei River Basin, China.

each location in space (McKee et al., 1993). The SPI can be calculated as the following:

It is firstly fitted with an incomplete gamma probability density function to a given frequency distribution of precipitation at a station.

$$p(x) = \frac{1}{\beta^{\alpha}\Gamma(\alpha)} x^{\alpha-1} e^{-x/\beta}, \quad x > 0 \tag{1}$$

where x is the precipitation amount, $\alpha$is the shape parameter, β is the scale parameter, and the gamma function is expressed as:

$$\Gamma(\alpha) = \int_0^{\infty} x^{\alpha-1} e^{-x} \mathrm{d}x \tag{2}$$

The maximum likelihood method is used to estimate the optimal values of $\alpha$and β (Thom, 1966).

$$\hat{\alpha} = \frac{1}{4A}\left(1 + \sqrt{1 + \frac{4A}{3}}\right) \tag{3}$$

$$\hat{\beta} = \frac{\overline{x}}{\hat{\alpha}} \tag{4}$$

$$A = \ln(\overline{x}) - \frac{\sum \ln(x)}{n} \tag{5}$$

where n is the number of precipitation records.

The cumulative probability for a given month then can be obtained by the following equation:

$$G(x) = \int_0^x g(x)\mathrm{d}x = \frac{1}{\hat{\beta}^{\hat{\alpha}}\Gamma(\hat{\alpha})} \int_0^x x^{\hat{\alpha}-1} e^{-x/\hat{\beta}} \mathrm{d}x \tag{6}$$

Setting $t = x/\hat{\beta}$, the incomplete gamma function can be obtained:

$$G(x) = \frac{1}{\Gamma(\hat{\alpha})} \int_0^x t^{\hat{\alpha}-1} e^{-t} \mathrm{d}t \tag{7}$$

Parameters of the gamma distribution are estimated for every individual station in each month of the year. Considering the probability for a precipitation amount of zero ($x = 0$) which is not defined in the gamma function, the cumulative probability becomes:

$$H(x) = q + (1-q) G(x) \tag{8}$$

where q is the probability of a zero and is estimated by m/n, m is the number of zeros in precipitation records, and n is the length of precipitation records.

The cumulative probability of the distribution for each value of precipitation is then transformed using equal probability to a normal distribution with a mean of zero and standard deviation of one, which is the value of the SPI. For a given cumulative probability $H(x)$, the corresponding normal distribution probability can be estimated by using following equation (Abramowitz and Stegun, 1964):

$$\mathrm{SPI} = \begin{cases} k - \dfrac{(c_2k + c_1)k + c_0}{((d_3k + d_2)k + d_1)k + 1.0}, \\ k = \sqrt{\ln\left(1/(1-H(x))^2\right)},\ 0.5 < H(x) \le 1 \\ -\left(k - \dfrac{(c_2k + c_1)k + c_0}{((d_3k + d_2)k + d_1)k + 1.0}\right), \\ k = \sqrt{\ln\left(1/H(x)^2\right)},\ 0 < H(x) \le 0.5 \end{cases} \tag{9}$$

where $c_0 = 2.515517$, $c_1 = 0.802853$, $c_2 = 0.010328$, $d_1 = 1.432788$, $d_2 = 0.189269$, $d_3 = 0.001308$. Table 2 shows the categorization of drought and wet classified based on the SPI.

3.2 Calculation of the VHI

The Vegetation Index was defined as combination of VCI and TCI. It has been widely used to detect drought in many places, such as Africa, China, America, etc. The VHI was defined by the following formulae:

$$\mathrm{VCI} = \frac{\mathrm{NDVI} - \mathrm{NDVI}_{\min}}{\mathrm{NDVI}_{\max} - \mathrm{NDVI}_{\min}} \times 100 \tag{10}$$

$$\mathrm{TCI} = \frac{T_{\max} - T}{T_{\max} - T_{\min}} \times 100 \tag{11}$$

$$\mathrm{VHI} = \alpha * \mathrm{VCI} + (1-\alpha) * \mathrm{TCI} \tag{12}$$

where $\mathrm{NDVI}_{\max}$ and $\mathrm{NDVI}_{\min}$ are the multi-year absolute maximum and minimum of NDVI, respectively; $T_{\max}$ and $T_{\min}$ are the multi-year absolute maximum and minimum of T, respectively; and α is the weight to measure the contribution of VCI and TCI for assessing the status of drought. Generally, α is set as 0.5 because it is difficult distinguish the contribution of crop surface temperature and NDVI when measuring drought stress. VHI was proposed to address the limitations of VCI in areas with excessive soil wetness and/or long periods of cloudiness (Kogan, 1995), under which conditions NDVI is normally very depressed and VCI has low values, which can be interpreted as drought in error. TCI is expected to correct low values because of the variation in response of vegetation to temperature as compared with NDVI. Table 3 shows the categorization of drought based on the VHI.

Table 2. Classification of drought calculated by the SPI.

Categories of drought	SPI
Extreme drought	SPI ≤ -2.0
Severe drought	$-2.0 <$ SPI ≤ -1.5
Moderate drought	$-1.5 <$ SPI ≤ -1.0
Near normal	$-1.0 <$ SPI < 1.0
Moderate wet	$1.0 \le$ SPI < 1.5
Severe wet	$1.5 \le$ SPI < 2.0
Extreme wet	SPI ≥ 2.00

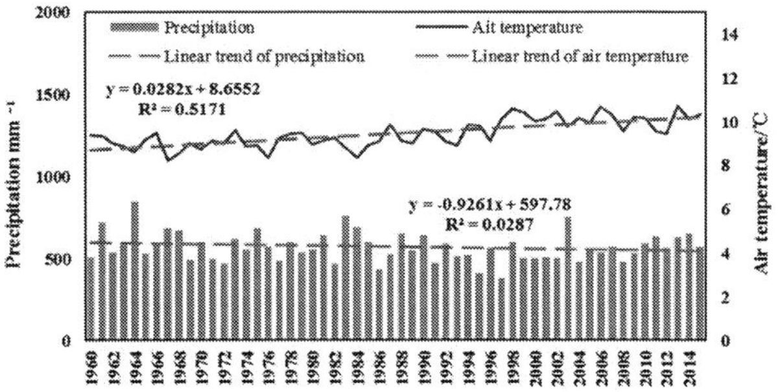

Figure 2. Temporal variations of annual average air temperature and total precipitation in the Wei River Basin.

4 Results and Discussion

4.1 Trend analysis of precipitation and air temperature

The temporal trends of annual total precipitation and average air temperature over the Wei River Basin WRB at 22 meteorological stations during the period 1960–2015 are shown in Table 1. The results shows that all stations showed a decreasing trend in annual total precipitation, ranging from -3.7 to $-0.15\ \mathrm{mm\,yr^{-1}}$. The largest decreasing trend ($-3.7\ \mathrm{mm\,yr^{-1}}$) was identified at Huashan station, which is located in the southeastern area.

For annual average air temperature, an increasing trend was found at all the 22 stations, ranging from 0.005 to $0.044\ ^{\circ}\mathrm{C\,yr^{-1}}$. The highest increase was observed at Zhen'an station, located in the south area, while the lowest increase was detected at Shangzhou station, located in the southeastern area.

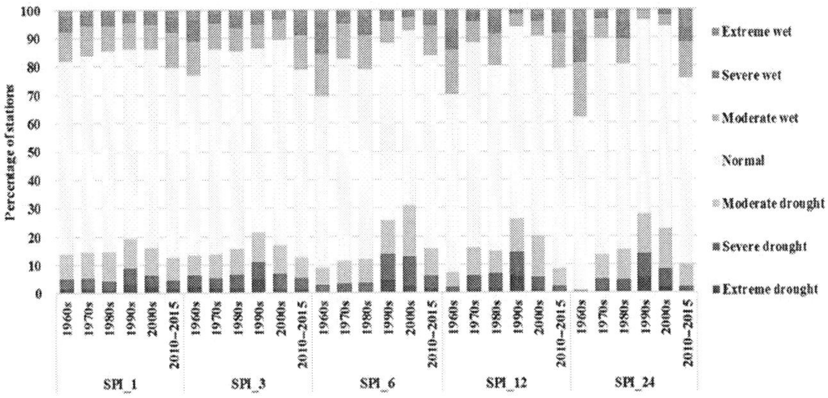

Figure 3. Drought frequency in different decades at different time scales in the Wei River Basin.

Table 3. Classification of drought calculated by the VHI.

Categories of drought	VHI
Extreme drought	VHI ≤ 10
Severe drought	10 < VHI ≤ 20
Moderate drought	−20 < VHI ≤ 30
Light drought	30 < VHI ≤ 40
No drought	40 < VHI

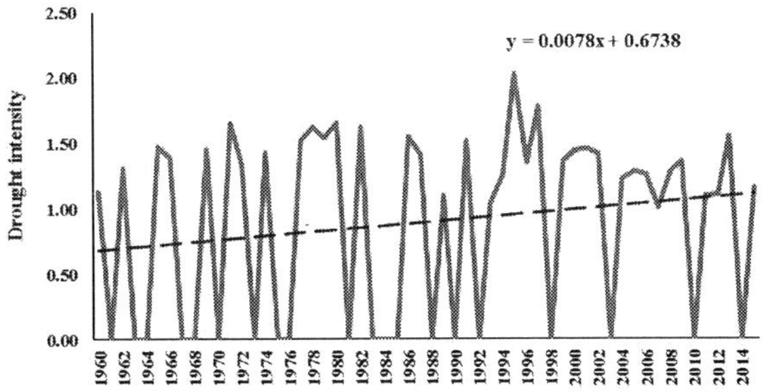

Figure 4. The variation of drought intensity in the Wei River Basin during 1960–2015.

The annual total precipitation and average air temperature series over the WRB are shown in Fig. 2. The annual total precipitation series in Fig. 2 shows that the driest year was 1997 (380 mm), while the year 1964 was the wettest year (845 mm) from 1960 to 2015. The hottest year was 2013 (10.7 °C), while the coldest year was 1967 (8.2 °C). The results showed that the annual total precipitation and average air temperature generally exhibited negative and positive trends across the WRB, respectively, which is prone to occur drought events.

4.2 Spatiotemporal characteristics of the SPI at various time scales

4.2.1 Temporal characteristics of the SPI at various time scales

The frequency of drought and wet occurrences at different time scales during the six decades over the Wei River Basin based on SPI (Fig. 3). In the 1960s, the percentages of extreme drought and severe drought increased first and then decreased as the increased SPI time scales, all the maximum values reached at 3 month. While the percentages of moderate drought increased as the increased SPI time scales. In the 1970s, the percentages of extreme drought ranged from 0.53 to 1.74 %; the percentages of severe drought almost increased first and then decreased as the increased SPI time scales, the maximum values was reached at 12 month (4.81 %); the percentage of moderate drought was about 9 %. In the 1980s, the percentages of extreme drought and severe drought fluctuated with the increased SPI time scale; the frequency of moderate drought decreased first and then increased as the increased SPI time scales, the minimum value reached at 12 month (7.92 %), the maximum value reached at 24 month (10.64 %). In the 1990s, the percentages of extreme drought and moderate drought increased with the increased SPI time scales; the percentages of severe drought increased first and then decreased with the increased SPI time scales, the maximum value reached at 6 month (9.32 %). In the 2000s, the percentages of extreme drought ranged from 1.21 to 2.54 %; the percentages of severe drought and moderate drought reached minimum values at 1 month (3.79 and 9.96 %) and reached maximum values at 6 month (10.45 and 18.11 %). During the period 2010–2015, the percentages of extreme drought decreased as the increased SPI time scales; the percentages of severe drought and moderate drought increased first and then decreased as the increased SPI time scales, the maximum values reached at 6 month (4.23 and 0.79 %).

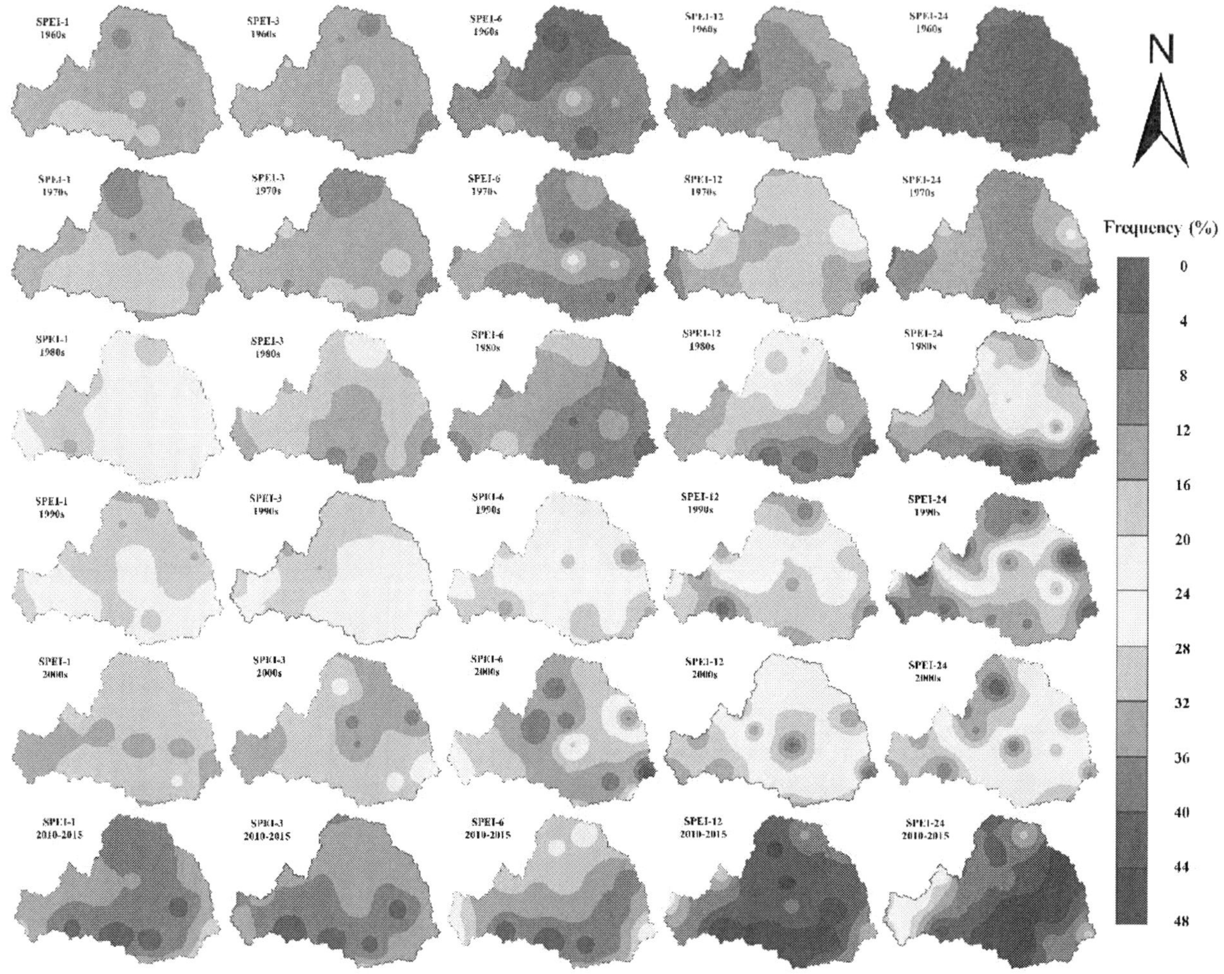

Figure 5. Spatial distributions of drought frequency at various time scales during the six decades in the Wei River Basin.

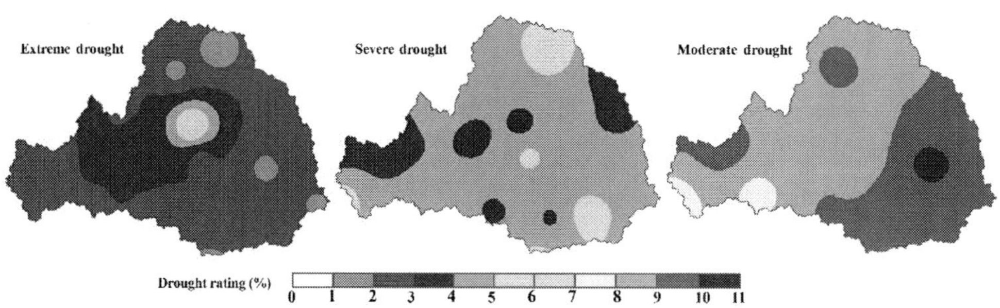

Figure 6. Drought rating of extreme drought, severe drought and moderate drought based on SPI-3 in the Wei River Basin during the period 1960–2015.

The annual drought intensity time series of Wei River Basin was constructed by the absolute value of the sum of each site's annual SPI (≤ -1) based on SPI-12. Due to the difference of the number of stations in each region, this method calculated the drought intensity of each region was not comparable, so the value divided by the number of stations (Fig. 4). The increases in value of drought intensity was 0.078 per 10-years for Wei River Basin during 1960–2015. The drought intensity was greater, but the duration was short before 1976; a continuous severe drought occurred during

1977–1980. After 1990, drought intensity increased and long duration. A continuous severe drought also occurred from 1993 to 1997 and the value of drought intensity reached the maximum value (2.03). A continuous moderate drought happened in both 1999–2002 and 2004–2009.

4.2.2 Spatial patterns of the SPI at various time scales

Frequency distributions of extreme drought, severe drought and moderate drought at 1, 3, 6, 12 and 24 months during the six decades in the Wei River Basin were shown in Fig. 5. The frequency of drought decreased as the time scale of the SPI increased during 1960s. In 1970s, the frequency of drought increased in varying degree at different time scales. In 1980s, the values was relatively small in south of Wei River Basin; the value was found in the north and west for SPI-1 at about 18 %; the value was found in the north for SPI-12 and SPI-24 approximately 30 %, which were longer time scales. The frequency of drought reached the maximum value at SPI-3, SPI-12 and SPI -24 during 1990s. In 2000s, the frequency of drought continuously increased and the extent of drought area expanded for SPI-6; the value decreased at other time scales. The frequency of drought was basically below 20 % during 2010–2015.

The drought rating of extreme drought, severe drought and moderate drought based on SPI-3 during the period 1960–2015 were shown in Fig. 6. The moderate drought rating of southeast was greater than the northwest and the maximum value was in Tongchuan. The drought rating of severe drought ranged from 2.98 to 6.25 % and the value of median was smaller than the south and the maximum value was in Wuqi area in the northern part of the basin. The drought rating of extreme drought relatively smaller and the value of southeast was greater than the northwest and the maximum value was in Xifeng.

4.3 Assessment of drought in the Wei River basin based on the VHI

4.3.1 Spatiotemporal characteristics of the VHI in the Wei River Basin

In order to investigate the percentage of drought occurrence during the period of 2000–2015 in the Wei River Basin, the frequencies of the VHI values in five categories (Table 3) were calculated at 22 stations (Fig. 7). From 2000 to 2015, the percentages of extreme drought was almost zero, and that of severe drought ranged from 0 to 1.37 % The percentages of moderate drought were relatively higher than those of extreme drought and severe drought, which were between 0.24 to 15.32 %. The percentages of light drought ranged from 4 to 25.97 %. Furthermore, 59.33 to 95.7 % of the VHI belongs to the no drought.

In additional, the values of mean, maximum and minimum of the VHI during 2000–2015 were shown in Fig. 8. The mean of VHI ranged from 46.71 to 61.59 and an increasing

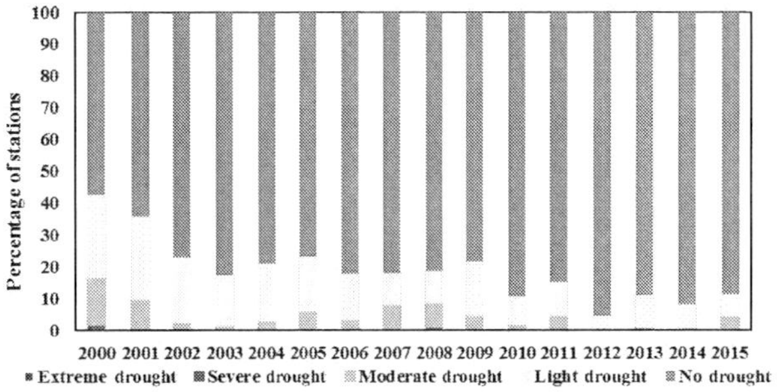

Figure 7. Frequency of occurrence of drought in six categories in the Wei River Basin during the period of 2000–2015.

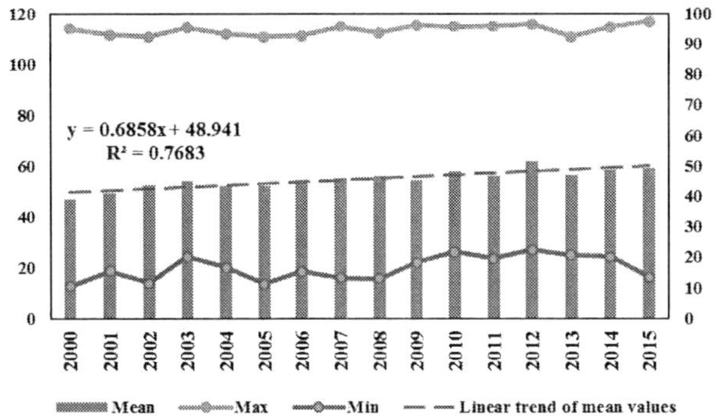

Figure 8. The value of maximum, minimum and mean of the VHI during 2000–2015.

trend had been detected ($R = 0.88$). The mean of maximum value were 95 and there was no significant fluctuation found. The minimum value of VHI ranged from 11.45 to 22.32 and the minimum of that was in 2000 (10.45), the maximum of that was in 2012 (22.32).

The spatial characteristics of the VHI drought index, the value of drought calculated from 2000 to 2015 in the Wei River Basin were shown in Fig. 9. According to the spatial patterns showed that crops in the WRB were affected by different levels of drought. Overall, there were few extreme drought occurred from 2000–2015. Severe drought occurred in 2000, 2005, 2007, 2008 and 2015, and all occurred in the northern part of the WRB, which occurred in the west of the WRB in 2000 and 2002. Different degrees of moderate drought was happened every year and all happened in the northern part of the WRB. For light drought, in addition to the northern regions, there were also some areas in the west had a light drought.

4.3.2 Impact of drought on agriculture in the Wei River Basin

Correlationship between crop yield and the VHI during 2000–2015 in the Wei River Basin are shown in Fig. 10. The relationship between the VHI and crop yield was detected

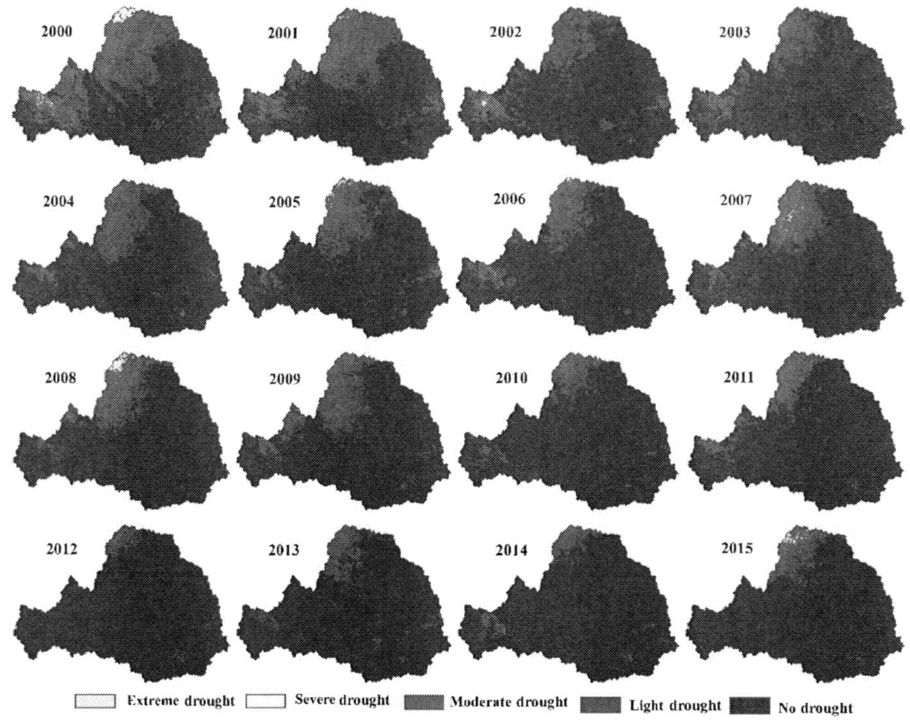

Figure 9. Spatial distributions of drought from 2000 to 2015 based on the VHI in the Wei River Basin.

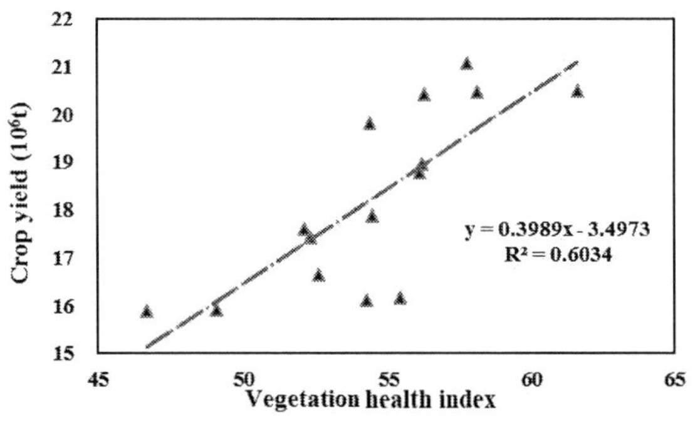

Figure 10. Correlationship between crop yield and the VHI during 2000–2015 in the Wei River Basin.

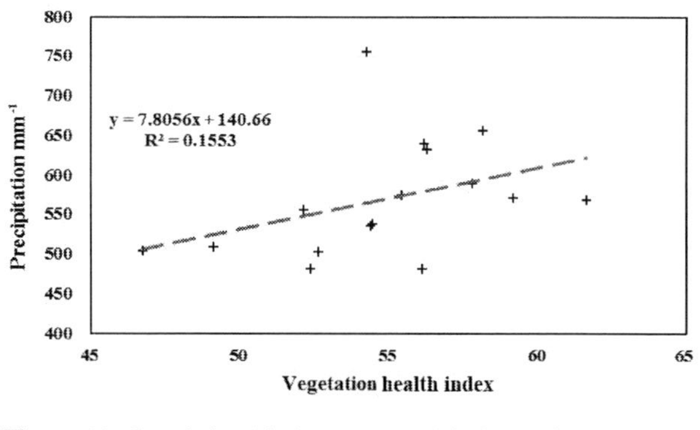

Figure 11. Correlationship between precipitation and the VHI during 2000–2015 in the Wei River Basin.

with R value of 0.78. The result showed that the VHI had a promising potential of quantifying the agricultural yield reduction caused by agricultural drought in the WRB.

4.3.3 Comparison with the meteorological factors

The correlationship between VHI and precipitation and average air temperature are shown in Figs. 11 and 12. The VHI showed a relationship with the precipitation and the R value of the linear regression equation reached at $R = 0.4$ during 2000–2015. While the correlation between VHI and average air temperature showed a polynomial correlation with the $R = 0.37$. In a word, the effect of precipitation on agricultural drought was greater than that of air temperature.

5 Conclusion

This study detected the temporal and spatial pattern of drought in the Wei Rvier Basin from 1960–2015. The SPI drought index at different time scales were calculating using monthly precipitation and air temperature. The VHI was calculating based on monthly NDVI and 8-days LST. A regres-

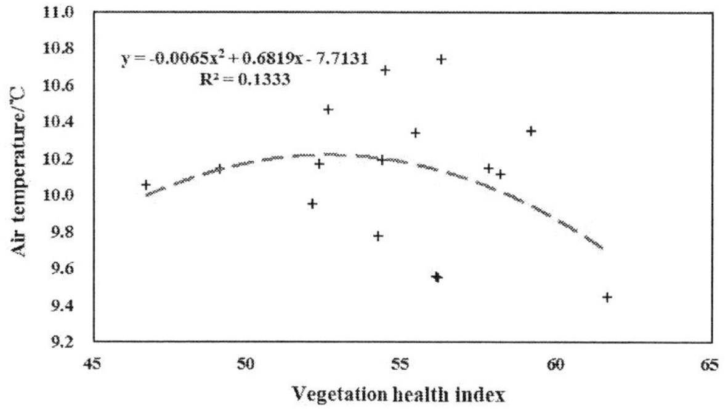

Figure 12. Correlationship between air temperature and the VHI during 2000–2015 in the Wei River Basin.

sion was employed to identify the capability of VHI to evaluating the drought events. Major conclusions can be summarized as follows:

1. The total precipitation had a decreasing trend with 0.9261 mm yr^{-1} and average air temperature had an increasing trend with 0.0282° yr^{-1} and the situation was prone to occur drought events,

2. The percentages of severe drought initially increased and then decreased with decades in 1970s and 1990s while those of moderate drought increased with decades in 1960s and 1990s. Based on drought intensity, a continuous moderate drought occurred in both 1992–2002 and 2004–2009. The drought rating of moderate drought was greater than that of severe drought and extreme drought. Meanwhile, the drought rating was greater in southern than that in northern. Before 1990s, the degree of drought in southern was heavier than that in northern, however, the degree of drought was the opposite.

3. The average had showed an increasing trend with $R = 0.88$. Based on the VHI, the degree of northern was heavier than that in southern, the degree of western was heavier than in eastern. The heaviest drought was occurred in 2000, 2007 and 2008.

4. Compared to crop yield, VHI had a well capability to evaluating agricultural drought in the Wei River Basin. And precipitation had greater effect on agricultural drought than temperature

Data availability. Meoteological data were obtained from China Meteorological Data Sharing Services System Network (http://data.cma.cn/; CMDC, 2018). Modis datasets (MODIS 13A3 and MODIS 11A2) were obtained from https://modis.gsfc.nasa.gov/ (last access: 30 October 2017). Crop yield datasets were obtained from National Bureau of Statistics of China (http://data.stats.gov.cn, last access: 10 January 2018).

Competing interests. The authors declare that they have no conflict of interest.

Special issue statement. This article is part of the special issue "Innovative water resources management – understanding and balancing interactions between humankind and nature". It is a result of the 8th International Water Resources Management Conference of ICWRS, Beijing, China, 13–15 June 2018

Acknowledgements. This study is jointly supported by the National Key Research and Development Program of China (Grant No. 2017YFC1502703), the National Natural Science Foundation of China (Grant No. 91647202), and the Major Science and Technology Program for Water Pollution Control and Treatment (Grant No. 2017ZX07302-04).

Edited by: Andreas Schumann
Reviewed by: two anonymous referees

References

Abramowitz, M. and Stegun, A. I.: Handbook of mathematical functions with formulas, graphs, and mathematical tables, Courier Dover Publications, New York, USA, 1964.

Alam, M. M., Siwar, C., Toriman, M. E., and Molla, R. I.: Climate change induced adaptation by paddy farmers in Malaysia, Mitig. Adapt. Strat. Gl., 17, 173–186, 2012.

Chen, H. and Sun, J.: Changes in Drought Characteristics over China Using the Standardized Precipitation Evapotranspiration Index, J. Climate, 28, 5430–5447, https://doi.org/10.1175/JCLI-D-14-00707.1, 2015.

CMDC (China Meteorological Data Service Center): Dataset of daily climate data from China Ground Climate Data Monthly Dataset, available at: http://data.cma.cn/, last access: 1 May 2018.

Cong, D., Zhao, S., Chen, C., and Duan, Z.: Characterization of droughts during 2001-2014 based on remote sensing: A case study of Northeast China, Ecol. Inform., 39, 56–67, https://doi.org/10.1016/j.ecoinf.2017.03.005, 2017.

Dai, A.: Drought under global warming: a review, Wires Clim. Change, 2, 45–65, https://doi.org/10.1002/wcc.81, 2011.

Food and Agriculture Organization of the United Nations: FAOSTAT, The Statistics Division of FAO, Rome, Italy, 2014.

Karavitis, C. A., Alexandris, S., Tsesmelis, D. E., and Athanasopoulos, G.: Application of the Standardized Precipitation Index (SPI) in Greece, WATER, 3, 87–805, https://doi.org/10.3390/w3030787, 2011.

Kogan, F. N.: Droughts of the late 1980s in the United States as derived from NOAA polar-orbiting satellite data, B. Am. Meteorol. Soc., 76, 655–668, 1995.

Lei, T., Wu, J., Li, X., Geng, G., Shao, C., Zhou, K., Wang, Q., and Liu, L.: A new framework for evaluating the impacts of drought on net primary productivity of grassland, Sci. Total Environ., 536, 161–172, https://doi.org/10.1016/j.scitotenv.2015.06.138, 2015.

Lei, Y., Zhang, H., Chen, F., and Zhang, L.: How rural land use management facilitates drought risk adaptation in a changing climate – A case study in arid northern China, Sci. Total Environ., 550, 192–199, https://doi.org/10.1016/j.scitotenv.2016.01.098, 2016.

Liu, Z., Zhou, P., Zhang, F., Liu, X., and Chen, G.: Spatiotemporal characteristics of dryness/wetness conditions across Qinghai Province, Northwest China, Agr. Forest Meteorol., 182–183, 101–108, https://doi.org/10.1016/j.agrformet.2013.05.013, 2013.

Long, D., Shen, Y., Sun, A., Hong, Y., Longuevergne, L., Yang, Y., Li, B., and Chen, L.: Drought and flood monitoring for a large karst plateau in Southwest China using extended GRACE data, Remote Sens. Environ., 155, 145–160, https://doi.org/10.1016/j.rse.2014.08.006, 2014.

Martinez-Fernandez, J., Gonzalez-Zamora, A., Sanchez, N., Gumuzzio, A., and Herrero-Jimenez, C. M.: Satellite soil moisture for agricultural drought monitoring: Assessment of the SMOS derived Soil Water Deficit Index, Remote Sens. Environ., 177, 277–286, https://doi.org/10.1016/j.rse.2016.02.064, 2016.

McKee, T. B., Doedken, N. J., and Kleist, J.: The relationship of drought frequency and duration to time scales, Proceeding of the 8th Conference on Applied Climatology, Anaheim, Calif., USA, 17–22 January 1993, Am. Meteorol. Soc., 17, 179–182, 1993.

Mishra, A. K. and Singh, V. P.: A review of drought concepts, J. Hydrol., 391, 202–216, 2010.

Moreira, E. E., Martins, D. S., and Pereira, L. S.: Assessing drought cycles in SPI time series using a Fourier analysis, Nat. Hazards Earth Syst. Sci., 15, 571–585, https://doi.org/10.5194/nhess-15-571-2015, 2015.

Oloruntade, A. J., Mohammad, T. A., Ghazali, A. H., and Wayayok, A.: Analysis of meteorological and hydrological droughts in the Niger-South Basin, Nigeria, Glob. Planet. Change, 155, 225–233, https://doi.org/10.1016/j.gloplacha.2017.05.002, 2017.

Palmer, W. C.: Meteorological drought, US Department of Commerce, Weather Bureau, Washington, DC, USA, 1965.

Rayne, S. and Forest, K.: Evidence for increasingly variable Palmer Drought Severity Index in the United States since 1895, Sci. Total Environ., 544, 792–796, https://doi.org/10.1016/j.scitotenv.2015.11.167, 2016.

Rhee, J., Im, J., and Carbone, G. J.: Monitoring agricultural drought for arid and humid regions using multi-sensor remote sensing data, Remote Sens. Environ., 114, 2875–2887, 2010.

Solomon, S.: The Physical Science Basis. Working Group I Contribution to the Fourth Assessment Report of the IPCC[J], edited by: Chang, I. C., 17 November 2007, Comp. Geom., 18, 95–123, 2007.

Steinemann, A.: Drought indicators and triggers: A stochastic approach to evaluation, J. Am. Water Resour. As., 39, 1217–1233, https://doi.org/10.1111/j.1752-1688.2003.tb03704.x, 2003.

Thom, H. C. S.: Some methods of climatological analysis, World Meteorological Organization (WMO), Technical Note no. 82, Geneva, Switzerland, 20–22, 1966.

Unganai, L. S. and Kogan, F. N.: Drought monitoring and corn yield estimation in Southern Africa from AVHRR data, Remote Sens. Environ., 63, 219–232, https://doi.org/10.1016/S0034-4257(97)00132-6, 1998.

Vicente-Serrano, S. M., Begueria, S., and Lopez-Moreno J. I.: A Multi-scalar Drought Index Sensitive to Global Warming: The Standardized Precipitation Evapotranspiration Index, J. Climate, 23, 1696–1718, https://doi.org/10.1175/2009JCLI2909.1, 2010.

Wilhite, D. A. and Glantz, M. H.: Understanding: the Drought Phenomenon: The Role of Definitions, Water Int., 10, 111–120, https://doi.org/10.1080/02508068508686328, 1985.

Yang, B., Ma, S., Li, J., Liao, Y., Zhao, B., and Claudia, K.: Agriculture drought monitoring in Dongting lake basin by MODIS data, 1st International Conference on Agro-Geoinformatics, 2–4 August 2012, Shanghai, China, 2012.

Yu, M., Li, Q., Hayes, M. J., Svoboda, M. D., and Heim, R. R.: Are droughts becoming more frequent or severe in China based on the Standardized Precipitation Evapotranspiration Index: 1951–2010?, Int. J. Climatol., 34, 545–558, https://doi.org/10.1002/joc.3701, 2014.

Zarch, M. A. A., Sivakumar, B., and Sharma, A.: Droughts in a warming climate: A global assessment of Standardized precipitation index (SPI) and Reconnaissance drought index (RDI), J. Hydrol., 526, 183–195, https://doi.org/10.1016/j.jhydrol.2014.09.071, 2015.

Zhai, L. and Feng, Q.: Spatial and temporal pattern of precipitation and drought in Gansu Province, Northwest China, Nat. Hazards, 49, 1–24, https://doi.org/10.1007/s11069-008-9274-y, 2009.

Zhang, A. and Jia, G.: Monitoring meteorological drought in semiarid regions using multi-sensor microwave remote sensing data, Remote Sens. Environ., 134, 12–23, 2013.

Zhang, L., Jiao, W., Zhang, H., Huang, C., and Tong, Q.: Studying drought phenomena in the Continental United States in 2011 and 2012 using various drought indices, Remote Sens. Environ., 190, 96–106, https://doi.org/10.1016/j.rse.2016.12.010, 2017.

Zuo, D., Cai, S., Xu, Z., Yang, X., and Li, F.: Spatiotemporal patterns of drought at various time scales in Shandong Province of Eastern China, Theor. Appl. Climatol., 2016, 1–14, https://doi.org/10.1007/s00704-016-1969-5, 2016.

Zou, L., Xia, J., and She, D.: Drought Characteristic Analysis Based on an Improved PDSI in the Wei River Basin of China, WATER, 9, 1783, https://doi.org/10.3390/w9030178, 2017.

https://doi.org/10.5194/piahs-379-83-2018

Methodology to explore emergent behaviours of the interactions between water resources and ecosystem under a pluralistic approach

Glenda García-Santos[1], Mariana Madruga de Brito[2], Britta Höllermann[2], Linda Taft[2], Adrian Almoradie[2], and Mariele Evers[2]

[1]Department of Geography and Regional Studies, Alpen-Adria-University Klagenfurt, Klagenfurt, 9020, Austria
[2]Department of Geography, University of Bonn, Bonn, 53115, Germany

Correspondence: Glenda García-Santos (glenda.garciasantos@aau.at)

Received: 28 December 2017 – Accepted: 21 January 2018 – Published: 5 June 2018

Abstract. Understanding the interactions between water resources and its social dimensions is crucial for an effective and sustainable water management. The identification of sensitive control variables and feedback loops of a specific human-hydro-scape can enhance the knowledge about the potential factors and/or agents leading to the current water resources and ecosystems situation, which in turn supports the decision-making process of desirable futures. Our study presents the utility of a system dynamics modeling approach for water management and decision-making for the case of a forest ecosystem under risk of wildfires. We use the pluralistic water research concept to explore different scenarios and simulate the emergent behaviour of water interception and net precipitation after a wildfire in a forest ecosystem. Through a case study, we illustrate the applicability of this new methodology.

1 Introduction

The management of water systems is influenced both by physical and social aspects such as hydrology, climate, technological advances, economy, and politics. Nevertheless, while the understanding of the dynamics of the vegetation, soils, and hydrologic processes has greatly improved over the past few decades (Roobavannan et al., 2017; Vose et al., 2011), the complex interactions between social and environmental systems are often ignored (Evers et al., 2017). Part of this complexity arises from the fact that the linkages between human and water systems have often proved difficult to quantify (Patterson et al., 2004), plus existing frameworks for dealing with water management problems are vague, which make them difficult to apply in real case studies (Biswas, 2004).

Despite these challenges, Evers et al. (2017) point out that understanding the co-evolution or interaction of human-water systems and its social dimensions is essential to effectively tackle the shortcomings in sustainable water management. Thus, the authors propose the use of a Pluralistic Water Research (PWR) concept, in which key system elements, sensitive control variables, and feedback loops are identified to enhance the knowledge about the potential set of factors that lead to the current water resources and ecosystems situation. This in turn supports the decision-making process of desirable futures.

In forest ecosystems, fires influence the composition and structure of forests, changing the amount of intercepted rainfall and consequently the net precipitation, recharge of aquifers, aesthetics, water and air quality, which increase the costs for society. Hence, knowing how intercepted rainfall and net precipitation rates vary over time and space after such perturbations is required for assessing their social and ecological implications. This is especially relevant in areas with scarce water resources, where a good understanding of the water dynamics is essential for managing water resources in often missing long-term planning strategies and tools (Maddock, 2016).

Available models of the canopy interception process like Gash (1979) and van Dijk and Bruijnzeel (2001) are designed to predict accumulated net precipitation rates over time. However, how net precipitation rates influence other parameters of the water cycle remained for further research. First attempts to dynamically model rainfall interception are found in Keim and Skaugset (2004) and Keshta et al. (2009). In those models, the re-growth of leaves following a fire and information on the post-fire recovery of the leaf area index (LAI) are not integrated, most presumably because they are rarely empirically studied. Other advances in the original Gash model include the use of measured parameters of forest structure from satellite data (Cui and Jia, 2014). Interestingly, we observe that social aspects are systematically neglected.

To tackle this, approaches that consider the interactions between human and water systems such as the enhanced Driving force-State-Impact-Response (e-DPSIR) (Niemeijer and de Groot, 2008) could be used. To represent processes dynamically the system dynamics (SD) approach is recommended. The SD methodology is based on the understanding of the complex relationships that exist among the different elements within a system (Keshta et al., 2009). The main goal of SD is to comprehend the system and its boundaries by identifying key building blocks, and the proper representation of the physical processes through mathematical relationships (Elshorbagy et al., 2007). Besides its capability for incorporating interdependences between variables, SD-based simulation approaches allow for increased speed of model development, ease of model structure modification, ability to perform sensitivity analysis and effective communication of model results.

This paper aims to model the dynamics of rainfall interception and net precipitation after a human-induced fire under the pluralistic water research (PWR) framework (Evers et al., 2017). We highlight the utility of the SD approach to model the social-physical interlinkages existing in the e-DPSIR framework. We explore different scenarios to simulate the emergent behaviour of interception and net precipitation after a wildfire. A case study in La Gomera, Spain, is used to illustrate the applicability of the proposed model.

2 Methods

2.1 Case study: La Gomera island

The study area comprises La Gomera and its National Park, Garajonay, Spain. The regional climate is characterized by low mean annual precipitation of 660 ± 247 mm (García-Santos, 2007, 2012), high relative humidity, mild temperatures (5–7 °C in winter and 25 °C in summer) and low insolation due to the frequent presence of clouds. The humid conditions provided by fog enable the occurrence of Laurisilva woodlands on the island. The forest is protected as a national park since 1981 due to its high biodiversity and water resources. Around the park the land use is observed as semi-abandoned areas.

The local culture influences land use, e.g. where quality water is available, consumption as drinking water, leisure, industrial infrastructures and agricultural purposes is ensured, whereas places with a lack of water, risk being abandoned and have a high risk of natural fires such as the wildfire of 2012. La Gomera was selected as the case study because a central human-water research question can be identified after that hazard, i.e. how the conditions of the Laurisilva cloud forest influence water availability for entire ecosystems and human consumption?

2.2 Modelling approach

Based on the PWR framework (Fig. 1a) and e-DPSIR approach (Fig. 1b), we identified key elements within the physical human-hydro-scape and human and physical boundaries to understand the effects wildfires on water availability, quality and land use.

We built a dynamic model in STELLA (Systems Thinking, Experimental Learning Laboratory with Animation) software v.9 to understand the dynamics of net precipitation to model the vertical connectivity between atmosphere-plant-soil surface by implementing the Gash model (1979). Furthermore, to consider changes in canopy LAI, we modelled the rainfall interception by vegetation using the framework developed by van Dijk and Bruijnzeel (2001). This model considers that the canopy storage capacity is linearly related to the LAI rather than to the canopy cover. Hence, it allows understanding how intercepted rainfall varies over time.

The initial values used for the modelling parameters (the free throughfall coefficient, the stemflow coefficient, trunk storage capacity (mm), mean rainfall intensity falling onto a saturated canopy (mm h^{-1}), mean wet canopy evaporation rate (mm h^{-1}), specific leaf storage (mm), leaf area index (m^2 m^2)) were based on the work of García-Santos (2007). For the extinction coefficient, which depends on leaf distribution and inclination angle, we used the values found by Ross (1975), 0.70. The specific leaf storage was optimized. Furthermore, we assumed that the energy exchange coefficient between the canopy and atmosphere is equal to the extinction coefficient. To simulate the conditions after a fire, the LAI values measured by Bond-Lamberty et al. (2002) after fires were used. Nevertheless, it is important to highlight that the values adopted were calculated for a Boreal Black Forest. Hence, when available, data for Laurisilva forest should be used instead.

Changes in evaporation were not considered due to lack of data but will be addressed in the future. In addition, other elements and connections to water quality and the social scape, i.e. water management, activities and limitations in the National Park (dashed lines in Fig. 1b), will be included in the dynamic model in the future. Special attention will be given to model fog interception.

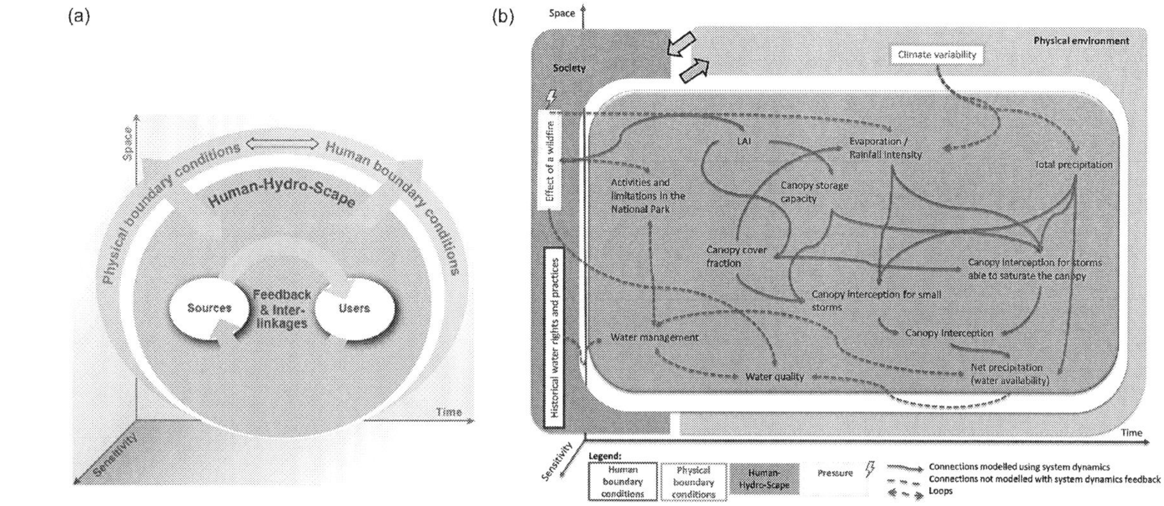

Figure 1. **(a)** The pluralistic water research (PWR) framework (Evers et al., 2017). **(b)** Representation of the key elements within the physical human-hydro-scape and human and physical boundaries to understand the effects of human-induced wildfires on rainfall interception, water availability, quality and land use based on the e-DPSIR approach for the case of La Gomera.

2.3 Model calibration and validation

Calibration of the two rainfall interception models was conducted daily to obtain the parameters using data measured in 2003 (García-Santos, 2007). For validation, data during 2004 were used (February–September) (García-Santos, 2007). The goodness of fit of measured data was studied through the r^2. A r^2 close to 1 expresses a good agreement between measured and predicted data. The Nash-Sutcliffe model efficiency coefficient was calculated. An efficiency of 1 corresponds to a perfect match of simulated to the observed data.

2.4 Scenarios

We studied two scenarios. Scenario 1 represents the baseline scenario for those days with precipitation only and modelled canopy interception and net precipitation, whereas scenario 2 represents those variables after a wildfire which sets canopy water interception to zero in the first year and as a function of LAI for the rest of the years. Precipitation inputs were kept as in year 2.

3 Results and discussion

3.1 Canopy interception and net precipitation dynamics under a PWR framework

Through the new pluralistic water research framework coupled with an e-DPSIR causal change approach, we were able to identify the key elements within the physical human-hydro-scape and human and physical boundaries due to the effect of an induced wildfire (Fig. 1b).

3.1.1 Key control variables

After the sensitivity analysis of the model parameters, evaporation was the most sensitive parameter followed by LAI. Interestingly, rainfall intensity had a moderate influence. This highlights the importance of changing evaporative conditions such as solar radiation, air temperature, relative humidity above the canopy and canopy structure.

3.1.2 Emergent behaviors

As expected in our still linear model representation, the interlinkages between the continuum atmosphere-plant-soil are evident (Fig. 2). In the year before the fire, the net precipitation was 132 mm and after the fire it was 153 mm, having the same precipitation inputs due to the interaction between water resources balance and ecosystem variation (LAI). Thus, higher net precipitation is possible because low overstory LAI during the first 3 years allows higher stand drainage. LAI strongly influences energy, water and carbon dioxide exchange between terrestrial ecosystems and the atmosphere (Campbell and Norman, 1998) and at the same time, it is strongly influenced by species, climate, soil fertility, water availability, tree density, evapotranspiration and photosynthesis (in Bond-Lamberty et al., 2002). Changes in net precipitation might be more visible under slightly higher evaporative conditions (most sensitive parameter). Empirical evaporation data after the fire needs to be obtained and incorporated into the dynamic model as well as non-linear relationships.

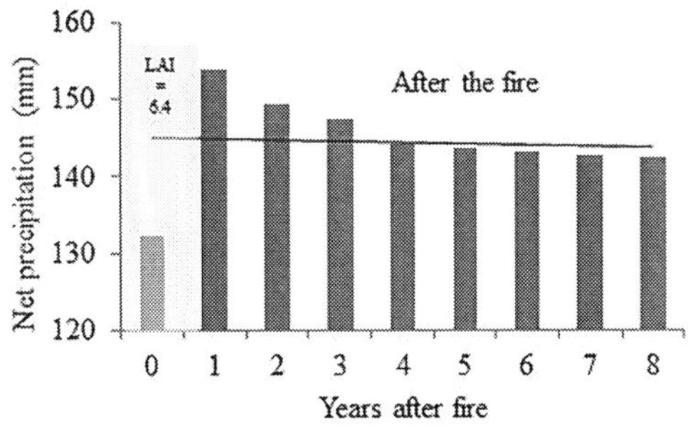

Figure 2. Modeled annual net precipitation before and 8 years after a forest fire after van Dijk and Bruijnzeel (2001) interception model. Precipitation conditions are those measured during 2004. LAI = 0.7497 ln(x) + 1.6512 (Bond-Lamberty et al., 2002).

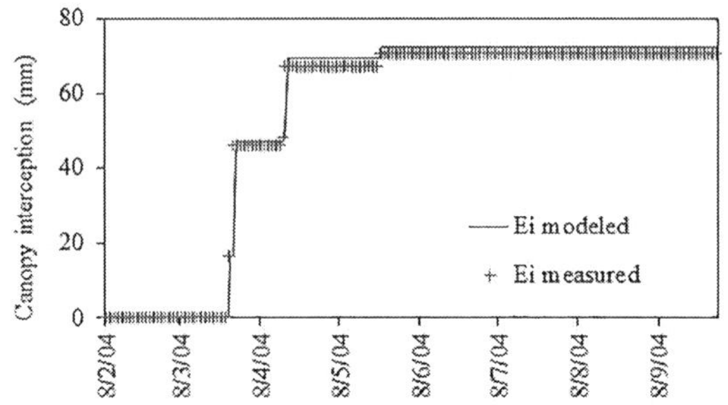

Figure 3. Cumulative measured versus modeled canopy interception data after van Dijk and Bruijnzeel (2001).

3.1.3 Feedbacks

Feedbacks were found between net precipitation and water management, between water quality and wildfire, between water management and activities and limitation in the park and between those and wildfires (Fig. 1b). Though those were not included in the dynamic model due to lack of empirical data, further work will make possible to capture that less water storage in the canopy increases net precipitation in a short-term but decreases net precipitation in a long-term and then decreases again due to the recovery of the woody vegetation.

3.1.4 Delays

Delays to growth understory and woody vegetation were modelled and affected canopy interception after the fire following the LAI function (Fig. 2).

3.1.5 Recovery mechanisms

Recovery mechanisms of Laurisilva forest, like the fast growth of under story and slow woody vegetation, will have a decisive impact on net precipitation rates. Further work should investigate actual changes of LAI in Laurisilva forest. Opportunities after a fire such as invasive species, changes in physiology should be also investigated in order to incorporate real conditions. The causal changes between water availability, water quality and the human boundary conditions were identified but not modeled. Thus, effects on water quality and land use remain still under work until empirical functions can be derived.

3.2 Modelled net precipitation

Cumulative totals of measured and modelled net precipitation are shown in Fig. 3. Modelled interception explained 86 % of the variance of measured interception, with an overall overestimation of 3 %. The Nash-Sutcliffe model efficiency coefficient was 0.99 for Gash (1979) and 0.97 for van Dijk and Bruijnzeel (2001). Therefore, both models performed fairly well. The later model was used in the dynamic model. Thus, we were able to model and assess the main dynamics of net precipitation after the wildfire. However, further work needs to include changes in evaporation conditions after a fire, as well as identified feedback loops and water inputs like fog and mixed precipitation (rainfall and fog) in the dynamic model (García-Santos and Bruijnzeel, 2011). This is essential to understand cloud forest dynamics. The causal changes between water availability, water quality and the human boundary conditions were identified but not modeled and therefore some effects remain unknown until empirical functions can be derived.

4 Conclusions

This study shows, for the first time, an application of the new pluralistic water research framework coupled with an e-DPSIR causal change approach with most of the parameters modeled analytically using SD modeling to explore different scenarios and simulate the emergent behaviour of interception and net precipitation after a wildfire. We studied a baseline scenario for those days with precipitation only and scenario 2 representing the conditions after a wildfire. Frist, we identified the key elements within the physical human-hydro-scape and human and physical boundaries due to the effect of an induced wildfire. Then, after a sensitivity analysis of the analytical model parameters, we founded that evaporation was the most sensitive parameter followed by LAI. Four feedbacks were found between net precipitation and water management, between water quality and wildfire, between water management and activities and limitation in the National park and between those and wildfires. Using van Dijk and Bruijnzeel (2001), we were able to model and assess the main dynamics of net precipitation after the wildfire. However, actual delays to growth understory and woody vegetation which affect canopy interception after the fire in Laurisilva forest need to be further validated. Thus, so fully understand cloud forest dynamics, the SD model will need to include changes in evaporation conditions after a fire, feedback loops and water inputs like fog and mixed precipitation. In the future the causal changes between water availability, water quality and the human boundary conditions will be dynamically modeled and therefore we will gain understanding on the effects of wildfires in the human-hydro-scape.

Data availability. Data used in this study are available on request from the corresponding author, Glenda García-Santos (glenda.garciasantos@aau.at). Data are part of the PhD thesis of the first author, which is available at http://dare.ubvu.vu.nl/handle/1871/12697.

Competing interests. The authors declare that they have no conflict of interest.

Special issue statement. This article is part of the special issue "Innovative water resources management – understanding and balancing interactions between humankind and nature". It is a result of the 8th International Water Resources Management Conference of ICWRS, Beijing, China, 13–15 June 2018.

Acknowledgements. The fieldwork phase to obtain precipitation and throughfall data was carried out by the first author, sponsored by a pre-doctoral scholarship to the first author from the Instituto Nacional de Investigación y Tecnología Agraria y Alimentaria (INIA) of the Ministry of Science and Technology of Spain. The scholarship was to be partly at the Instituto Canario de Investigaciones Agrarias (ICIA) in Tenerife (Canary Islands) within the framework of project RTA01-97. These data are part of the PhD thesis, which was carried out at VU University, Amsterdam. The authors gratefully acknowledge the support received by the Director from the National Park of Garajonay, Ángel Fernández and the staff from the National Park.

Edited by: Dingzhi Peng
Reviewed by: two anonymous referees

References

Biswas, A. K.: Integrated Water Resources Management: A Reassessment, Water Int., 29, 248–256, 2004.

Bond-Lamberty, B., Wang, C., Gower, S. T., and Norman, J.: Leaf area dynamics of a boreal black spruce fire chronosequence, Tree Physiol., 22, 993–1001, 2002.

Campbell, G. S., Norman, J. M.: An introduction to environmental biophysics, 2nd Edn., Springer-Verlag Inc., New York, 286 pp., 1998.

Cui, Y. and Jia, L.: A modified gash model for estimating rainfall interception loss of forest using remote sensing observations at regional scale, Water, 6, 993–1012, 2014.

Elshorbagy, A., Jutla, A., and Kells, J.: Simulation of the hydrological processes on reconstructed watersheds using system dynamics, Hydrolog. Sci. J., 52, 538–562, 2007.

Evers, M., Höllermann, B., Almoradie, A., García-Santos, G., and Taft, L.: The Pluralistic Water Research Concept: A New Human-Water System Research Approach, Water, 9, https://doi.org/10.3390/w9120933, 2017.

García-Santos, G.: An ecohydrological and soils study in a montane cloud forest in the National Park of Garajonay, La Gomera (Canary Islands, Spain), PhD Thesis, VU Amsterdam, The Netherlands, available at: https://research.vu.nl/en/publications/an-ecohydrological-and-soils-study-in-a-montane-cloud-forest-in-t (lsat access: 14 February 2018), 2007.

García-Santos, G.: Transpiration in a sub-tropical ridge-top cloud forest, J. Hydrol., 462–463, 42–52, 2012.

García-Santos, G. and Bruijnzeel, L. A.: Rainfall, fog and throughfall dynamics in a subtropical ridge top cloud forest, National Park of Garajonay (La Gomera, Canary Islands, Spain), Hydrol. Process., 25, 411–417, 2011.

Gash, J. H. C.: An analytical model of rainfall interception by forests, Q. J. Roy. Meteor. Soc., 105, 43–55, 1979.

Keim, R. F. and Skaugset, A. E.: A linear system model of dynamic throughfall rates beneath forest canopies, Water Resour. Res., 40, 1–12, 2004.

Keshta, N., Elshorbagy, A., and Carey, S.: A generic system dynamics model for simulating and evaluating the hydrological performance of reconstructed watersheds, Hydrol. Earth Syst. Sci., 13, 865–881, https://doi.org/10.5194/hess-13-865-2009, 2009.

Maddock, N.: Reducing hazards impacts with smarter spending, Fire Aust., 14–16, available at: https://www.bnhcrc.com.au/news/2016/reducing-hazard-impacts-smarter-spending (last access: 14 February 2018), 2016.

Niemeijer, D. and de Groot, R. S.: Framing environmental indicators: moving from causal chains to causal networks, Environ. Dev. Sustain., 10, 89–106, 2008.

Patterson, T., Gulden, T., Cousins, K., and Kraev, E.: Integrating environmental, social and economic systems: A dynamic model of tourism in Dominica, Ecol. Model., 175, 121–136, 2004.

Roobavannan, M., van Emmerik, T. H. M., Elshafei, Y., Kandasamy, J., Sanderson, M., Vigneswaran, S., Pande, S., and Sivapalan, M.: Norms and values in socio-hydrological models, Hydrol. Earth Syst. Sci. Discuss., https://doi.org/10.5194/hess-2017-432, in review, 2017.

Ross, J.: Radiative transfer in plant communities, in: Vegetation and the Atmosphere, edited by: Monteith, J. L., Academic Press, London, 13–15, 1975.

van Dijk, A. I. J. and Bruijnzeel, L.: Modelling rainfall interception by vegetation of variable density using an adapted analytical model. Part 1. Model description, J. Hydrol., 247, 230–238, 2001.

Vose, J. M., Sun, G., Ford, C. R., Bredemeier, M., Otsuki, K., Wei, X., Zhang, Z., and Zhang, L.: Forest ecohydrological research in the 21st century: what are the critical needs?, Ecohydrology, 4, 146–158, 2011.

https://doi.org/10.5194/piahs-379-89-2018

Village-level supply reliability of surface water irrigation in rural China: effects of climate change

Yanrong Li[1] **and Jinxia Wang**[2]

[1]Beijing Research Institute of Commercial Machinery, All China Federation of Supply and Marketing Cooperatives, Beijing, 100070, China
[2]China Center for Agricultural Policy, School of Advanced Agricultural Sciences, Peking University, 100871 Beijing, China

Correspondence: Jinxia Wang (jxwang.ccap@pku.edu.cn)

Received: 29 December 2017 – Accepted: 9 March 2018 – Published: 5 June 2018

Abstract. Surface water, as the largest part of water resources, plays an important role on China's agricultural production and food security. And surface water is vulnerable to climate change. This paper aims to examine the status of the supply reliability of surface water irrigation, and discusses how it is affected by climate change in rural China. The field data we used in this study was collected from a nine-province field survey during 2012 and 2013. Climate data are offered by China's National Meteorological Information Center which contains temperature and precipitation in the past 30 years. A Tobit model (or censored regression model) was used to estimate the influence of climate change on supply reliability of surface water irrigation.

Descriptive results showed that, surface water supply reliability was 74 % in the past 3 years. Econometric results revealed that climate variables significantly influenced the supply reliability of surface water irrigation. Specifically, temperature is negatively related with the supply reliability of surface water irrigation; but precipitation positively influences the supply reliability of surface water irrigation. Besides, climate influence differs by seasons. In a word, this paper improves our understanding of the impact of climate change on agriculture irrigation and water supply reliability in the micro scale, and provides a scientific basis for relevant policy making.

1 Introduction

Surface water, as the largest part of water resources, plays an important role on China's agricultural production and food security. China's surface water is $2.63 \times 10^{12}\,m^3$, containing almost 96.7 % of the total water volume. There are 48.4 % provinces whose surface water supply contains more than 90 % of the total water supply. And the share of provinces that surface water supply contains more than 70 % of the total water supply is even 67.7 %. Only 12.9 % provinces that surface water supply is less than 50 % of the total water supply (NBSC, 2015).

Surface water is vulnerable to climate change. Human activities and climate change first influence moisture and heat transfer, and then influence precipitation and temperature variations and change patterns (Trenberth et al., 2003; Hao et al., 2007; Piao et al., 2010; IPCC, 2014). These changes, through impacts on evaporation and surface runoff, could greatly alter land surface water balance in turn (Hao et al., 2007; Prowse and Belataos, 2006; IPCC, 2014). During recent 60 years (1961–2011), China's surface runoff decreased remarkably. For instance, river runoff in six large river basins presented a declining trend, they were Hai, Yellow, Liao, and Songhua in northern China, and Yangtze and Pearl in southern China (Wang et al., 2017). The onset of daily drying-up occurs more frequently and earlier than before and the period of zero flow also lengthens (Long et al., 2006). Even some river basins (e.g., Hai and Yellow) have changed from open to closed ones in some years, which negatively influenced the ecological environment (such as causing biodiversity degradation, atrophy and sea water intrusion) (Wang et al., 2016). Besides researches about the past reality, there are also many prediction studies on relationship of climate change and surface water resources. For instance, Zhang and Wang (2007) predicted an overall increase of river runoff over China (7.5 % in B1 scenario by 2100, and 9.7 % in A1 scenario). And in the Yellow River basin, annual runoff will increase by 11 % in A2 scenario and by 5 % in B2 scenario. In Shanxi Province, by 2030, with temperature gradually increasing, there will be less surface water resources than in 2010, but the overall trend is stable. By 2050, the influence of the temperature will increase more than that of precipitation, except for the regions of Taiyuan and Yangquan, in other parts of Shanxi, surface water resources will generally reduce (Hao et al., 2007). Supply and demand balance of water resources affected greatly by climate change will further affect agricultural production. In Haihe River in China, by 2030, changes of water resources balance will cause wheat irrigated area decreasing by 6.9 %, wheat raining area increasing by 19.9 %, and wheat yield decreasing 1.2 % (Wang et al., 2013). But all the prediction studies on relationship of water resources and climate change are greatly sensitive to biases of climate projections used in the selected models (Piao et al., 2010; Hao et al., 2007; Zhang and Wang, 2007; Wang et al., 2013).

Not only the quantity but also the quality of surface water is influenced by climate change (Xia et al., 2010, 2012; Zhang et al., 2009). Through affecting process of biochemical reaction rate, the ecological effect, the rise of temperature or migration type of water pollutants, the changes of wind speed, light duration and radiation enhancement can directly or indirectly impact on water environment (Dalla et al., 2007; Delpla et al., 2009; Wilby et al., 2006). Such as temperature rising affects the eutrophication of lake sediment and water pollutants in secondary interpretation (Lamon et al., 2009). Vanvliet (2008) quantified the influence of drought event on water quality of Meuse River in Western Europe in 1976 and 2003. Results showed that, compared with normal years, the concentration of NH_4^+, NO_2^- and Chlorophyll A respectively increased 1.9, 1.3 and 1.2 times. Precipitation and natural runoff decreasing, temperature increasing caused by climate change were helpful to increase the concentration of Ca^{2+} and Mg^{2+} in the Yellow River (Xia et al., 2012).

Besides climate change, irrigation infrastructure also plays an important role on the use of surface water for irrigation. Poor construction of rural irrigation infrastructure was one of the mayor reasons for increasing serious drought in agricultural production in China (Wang et al., 2013; Chen et al., 2012; Yang et al., 2012, Jia, 2016). Investment on village-level irrigation facilities is helpful for increasing farmers' income (Ma, 2008). Thankfully, construction of irrigation and water conservancy facilities has been paid great attention in recent years. By 2010, there were 87 873 reservoirs and 5795 irrigation districts that are more than 10 000 μ (MWR, 2011). Investment on irrigation infrastructure has increased by 27 times in the past 60 years (from CNY 700 million in 1958 to CNY 19.5 billion in 2009) (MWR, 2011). Compared with other water conservancy facilities investment, the irrigation infrastructure investment is still far not enough (Ma, 2008). Even in 2009, the share of irrigation infrastructure investment on the total water resources investment is only 10 % (MWR, 2011).

Despite more and more scholars begin to study the impacts of climate change or irrigation infrastructure on surface water supply, empirical analysis on this aspect is still very limit. In terms of the supply reliability of surface water irrigation, there are few works that address status of it, especially with data from large scale village survey and associating it with climate change. As for irrigation infrastructure, researches mainly focus on supply or requirement desire of irrigation infrastructure (Wang, 2015; Zhu and Wang, 2014; He and Guo, 2010). Few studies combine irrigation infrastructure and supply reliability of surface water irrigation, let alone under changing climate condition.

So with research limitation of this filed, our study focused on the following questions: how much reliable of surface water irrigation supply in rural area in China? Whether climate change influences the supply reliability of surface water irrigation? If yes, how can we quantify this influence? What is the relationship between the supply reliability of surface water irrigation and irrigation infrastructure?

In order to find answers of the questions above, the rest of this paper is organized as follows. Section 2 introduces the data used in this study. Section 3 displays local climate situation, and describes rural irrigation infrastructures and supply reliability of surface water irrigation. Econometric analyses in Sect. 4, including the influence of climate variables, rural irrigation infrastructures and other factors on the supply reliability of surface water irrigation. The final section concludes.

2 Sampling approach and data

2.1 Field survey data

The data we used in this study was collected from a nine-province field survey during 2012 and 2013. The nine provinces contain six Northern provinces (Hebei, Henan, Shandong, Jiangsu, Anhui and Jilin) and three southern provinces (Jiangxi, Guangdong and Yunnan). Based on local water resources, climate condition and economic development level, the nine provinces are firstly collected from the whole nation, and then the following strategies were taken to select samples.

First, in each province except for Jiangxi (10 counties) and Guangdong (6 counties), randomly selected three counties. After selecting the counties, we invited local experienced water conservancy workers to stratify all townships of the three counties into three groups according to their rural water infrastructure: above average, average, or below average. And then we randomly selected one township from each of the three groups and randomly chose three villages from each of the selected townships. Finally, 330 villages in 37 counties in 9 provinces were selected. Since we only analyze the supply reliability of surface water irrigation in this study, we just kept villages that use surface water for irrigation. So, the samples used in the study are 256 villages in 94 towns, 37 counties and 9 provinces. The sample distribution of this study is summarized in Table A1 in the Appendix.

The survey contains a three-year (2010, 2011 and 2012) and three- level (town, village and household) data collection.[1] In terms of this study, we mainly use village level questionnaires. A most important point was the supply reliability of surface water irrigation, meaning that whether farmers who needed irrigate their crops using surface water could get enough water and in a timely fashion. If the response was "yes", we define surface water supply for irrigation to be reliable; otherwise, it was seen as being unreliable. Specifically, in the survey, we asked village leaders that, in each of the past 3 years (2010–2012), what percentage of irrigated crop-sown areas could not get surface water when needed.[2]

In addition to the information on the supply reliability of surface water irrigation, following information was collected: (i) Village socio-economic characteristics: such as total cultivated land and number of population in the village. (ii) Village physical characteristics: such as topography (plain or mountain) and soil type (sand, clay or loam). (iii) Rural irrigation infrastructures: such as whether irrigating by the following kinds of water sources: large reservoir, middle reservoir, small reservoir, tubewell, pool, river or pump.

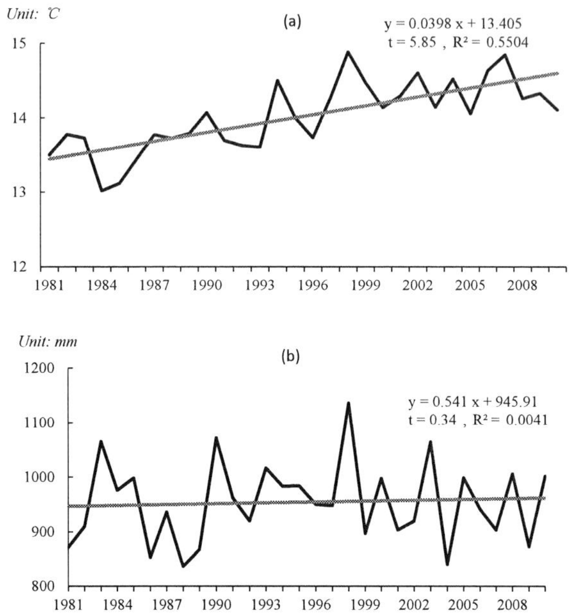

Figure 1. Change of annual temperature and precipitation in the last 30 years (1981–2010): **(a)** Change of annual temperature; **(b)** Change of annual precipitation. Data sources: National Meteorological Information Center in China.

2.2 Secondary climate data

In the study, climate data are offered by China's National Meteorological Information Center. We assume that both temperature and precipitation were homogenous within the county, so we use county level climate data. However, in the 37 sampled counties, only 23 counties contain national meteorological stations. In order to obtain relative climate data for other counties, we use the spatial interpolation method proposed by Thornton et al. (1997). More details about secondary climate data introduction can be seen in another paper of our team (Li, 2017).

All variables' descriptive statistics are displayed in Table A3.

3 Climate change in the past 30 years and supply reliability of surface water irrigation

3.1 Climate change in the past 30 years

What is the characteristic of climate change for sample area in long run? Figure 1a reflects the trend of average temperature[3] changes of nine provinces related in this research in the past 30 years. Remarkably, there is an increasing trend and it is statistically significant (the method of significance test is t-test). This in fact is highly consistent with many other scholar's study results, for instance, Ding et al. (2006) pointed that temperature has a most remarkable increase in recent 50 years. In our sample area, the annual average temperature increases about 0.6° in the past 30 years. What' more, fluctuations of annual average temperature are apparent, which reflects that the inter-annual temperature changes are very intensive.

In regard to annual total precipitation, changes are not obvious. As shown in Fig. 1b, the average annual total precip-

[1]Not all villages used surface water for irrigation in all 3 years (2010–2012), although we have 256 villages totally, some villages did not irrigate by surface water during 2010–2012 and we have not included these villages into our analysis; therefor, there are 256 villages in the analysis for 3 years and the total sample is 762, but not $256 \cdot 3 = 768$.

[2]We ask village leaders the following question: In your village, when crops need surface water irrigation, how much percentage of irrigated crop sown areas that could not get sufficient and timely surface water supply during 2010–2012? Based on this information, we can get indicator of the supply unreliability of surface water irrigation (share of irrigated crop-sown areas having access to an unreliable surface water supply), then using 1 to minus this number and we can get the indicator of the supply reliability of surface water irrigation.

[3]The method for calculating the spatial average value is as follows: first, calculate the average value of each month's temperature of all samples; second, calculate the annual average value, which is 12-month average.

itation changes appear slightly upward trend and it is statistically insignificant. But the fluctuation of it is much more remarkable, which implies that the inter-annual variability of precipitation may be obvious for a long time.

3.2 Rural irrigation infrastructures of surface water irrigation

As shown in Table 1, distributions of rural water infrastructures are different by province. Totally, there are 34.2 % villages using one water source, 50.3 % using two water sources, 14.9 % using three water sources and only 0.5 % using four water sources. Specifically, in Jilin Province, all sample villages are using two water sources for irrigation. In Henan Province, there are 90.6 % villages that using two water sources and others are all using one water source. Situation in Hebei and Shandong is similar. While in Jiangsu, villages that using one water source (82.2 %) are much more than that using two water sources (17.8 %). For Guangdong and Yunnan, numbers of water sources farmers use are 1 to 3. For Jiangxi and Anhui, numbers of water sources farmers use are 1 to 4, although percentages of villages using four water sources are only 1.1 and 1.8 % respectively. More details about distributions of water sources in sample regions of this study are displayed by Table A4.

3.3 Supply reliability of surface water irrigation

Through Table 2 we can see that, on average the total reliability of surface water is 74 %. Jilin, Henan and Jiangxi own relatively higher reliability with 91, 85 and 83 %. The worst appears in Yunnan, with 48 % only; and next is in Hebei, with 50 %. For other four provinces (Shandong, Jiangsu, Anhui, Guangdong), the supply reliability of surface water irrigation is among 72–76 %. Besides, through the past 3 years, data shows that in total sample, there exists a little fluctuation of the supply reliability of surface water irrigation with changes from 75 to 71 % , and then from 71 to 76 %. In terms of each province, Jiangxi appears a remarkable fluctuation with the supply reliability of surface water irrigation changing from 85 to 76 %, and then from 76 to 87 % during the past 3 years. The reason for this remarkable fluctuation is that, in 2011, in almost 70 % area in Jiangxi it happened drought damage. So the supply reliability of surface water irrigation in 2011 decreased greatly. Similar but slighter fluctuation also appears in Anhui and Henan. What's more, there appears a remarkable rising trend in Yunnan, with the supply reliability of surface water irrigation changing from 42 % in 2010 to 54 % in 2012. But in Jiangsu, a slightly decreasing trend can be seen with the supply reliability of surface water irrigation changing from 80 % in 2010 to 73 % in 2012. For other four provinces (Hebei, Shandong, Jilin and Guangdong), the supply reliability of surface water irrigation appears no obvious inter-annual changes.

4 Econometric model and estimation results

4.1 Specification of econometric model

In this study, we specify the following four econometric models to better identify and quantify the impacts of different factors on the supply reliability of surface water irrigation:

$$W_{ij} = \alpha_{ij} + \beta_1 C_{ij} + \beta_2 I_{ij} + \beta_3 V_{1ij} + \beta_4 P_{ij} + \varepsilon_{ij} \quad (1)$$

$$W_{ij} = \omega_{ij} + \partial_1 C_{ij} + \partial_2 I_{ij} + \partial_3 V_{2ij} + \partial_4 P_{ij} + \pi_{ij} \quad (2)$$

$$W_{ij} = \theta_{ij} + \lambda_1 C_{ij} + \lambda_2 I_{ij} + \lambda_3 V_{3ij} + \lambda_4 P_{ij} + \zeta_{ij} \quad (3)$$

$$W_{ij} = \delta_{ij} + \varphi_1 C_{ij} + \varphi_2 I_{ij} + \varphi_3 V_{4ij} + \varphi_4 P_{ij} + \mu_{ij} \quad (4)$$

In the four models, i and j indicate the ith village in the jth county, and the dependent variable W_{ij} are all the same. Here W_{ij} measures the share of irrigated crop-sown areas having reliable surface water supply.

We included similar independent variables on the right side of the four models. (i) C_{ij} is a set of climate variables. For Model 1 and Model 2 we use annual climate variables (average annual temperature, average annual temperature squared, annual total precipitation and annual total precipitation squared), and for Model 3 and Model 4, we use seasonal climate variables (Spring/Summer/Fall/Winter temperature and precipitation, and their squared term). Considering the possible relationship between temperature and precipitation, this study also included interactive variables into Model 2 and Model 4; (ii) treating the variable whether irrigating by more than two kinds of water sources as baseline, I_{ij} contains the following 0–1 irrigation variables: irrigating by two kinds of water sources, irrigating only by large reservoir/middle reservoir/small reservoir/tubewell/pool/river/pump; and (iii) V_{1-4ij} is a set of five variables that measure village characteristics: whether the soil type is sand (1 = yes; 0 = no) and whether the soil type is clay (1 = yes; 0 = no)(the baseline is loam); landform (1 if the village is located in plain, 0 otherwise) (we predict that plain is a better geographical condition than mountain for getting surface water), arable area per person (ha) and population (we assume that the two variables are related to water demand and then influences water reliability).

The models also include a set of dummy variables P_i to control for the impact of regional characteristics that do not change over time but might affect water reliability. Leaving one of them, Yunnan province, as the default category, each model has eight province dummy variables. Besides, in the four models, $\beta_1 - \beta_4$, $\partial_1 - \partial_4$, $\lambda_1 - \lambda_4$, and $\varphi_1 - \varphi_4$ are the parameters to be estimated; α_{ij}, ω_{ij}, θ_{ij} and δ_{ij} are the constant. ε_{ij}, Π_{ij}, ζ_{ij} and μ_{ij} are random error terms, and are all assumed to be subjected to independent identical distribution.

Tobit model (or censored regression model) are used to estimate the determinants of the supply reliability of surface water irrigation in this study. More introductions about Tobit model can be found in another paper of our team (Li, 2017).

Table 1. Percentage of villages using various irrigation sources in the sample regions (%).

	Number of water sources			
	One water source	Two water sources	Three water sources	Four water sources
Total samples	34.2	50.3	14.9	0.5
Hebei	33.3	66.7	0	0
Henan	9.4	90.6	0	0
Shandong	23.4	76.6	0	0
Jilin	0	100.0	0	0
Anhui	5.3	64.9	28.1	1.8
Jiangsu	82.2	17.8	0	0
Jiangxi	33.7	47.0	18.1	1.1
Guangdong	46.5	39.0	14.5	0
Yunnan	11.1	52.8	36.1	0

Data sources: Authors' survey.

Table 2. Supply reliability of surface water irrigation over 3 years by province in China.

	Supply reliability of surface water irrigation (Share of crop sown area accessing to reliable surface water supply when not enough surface water) (%)			
	Average	2010	2011	2012
Total sample	74	75	71	76
Hebei	50	50	50	50
Henan	85	86	85	84
Shandong	70	70	71	69
Jilin	91	91	91	92
Anhui	72	70	69	77
Jiangsu	76	80	74	73
Jiangxi	83	85	76	87
Guangdong	73	73	73	72
Yunnan	48	42	47	54

Data sources: Authors' survey.

4.2 Estimation results

All four models above perform well. The pseudo R^2 values of the four Tobit models are among 0.1361–0.2067, which are reasonably high values for multivariate analysis based on cross-sectional data. Moreover, among all the models, the signs of most climate variables, rural irrigation infrastructures and other control variables were statistically significant and also highly consistent with our expectation.

4.2.1 Influence of annual temperature and precipitation on the supply reliability of surface water irrigation

Estimation results show that, average annual temperature was statistically insignificant at the 10 % level (Table 3). But for annul total precipitation, the first degree term is positive and significant, and the quadratic term is negative and significant in the model without interactive variable. And the turning point is 1667 mm which means that as to the result of the model without interactive variable, when the quantum of rainfall is less than 1667 mm, the supply reliability of surface water irrigation will be increased with increasing precipitation, and otherwise, the supply reliability of surface water irrigation will be decreased with increasing precipitation. However, with interactive variable, the first degree term is positive and significant, but the quadratic term is negative and insignificant, and the coefficient of the climate interactive variable is negative and significant. This indicated that after considering the correlation between temperature and precipitation, if both temperature and precipitation increase simultaneously, the supply reliability of surface water irrigation will be significantly reduced. Or in other words, even having higher precipitation, if the temperature is also higher, the supply reliability of surface water irrigation will be lower.

What's more, keeping the constant of other factors, if we consider the interactive and non-interactive variables together, we can get the separate impacts of annual temperature or precipitation on supply reliability of surface water irrigation. As shown in Table 5, increasing temperature by 0.5° will reduce the supply reliability of surface water irrigation by 1.1 % (reducing from 62.4 to 61.3 %). If the temperature continues to increase, the supply reliability will be further reduced. In the future, if the temperature increases by 2°, the supply reliability will be reduced by 3.8 % (reducing from 62.4 to 58.6 %). Besides, increasing precipitation will result in increase of supply reliability. For example, when precipitation increases by 10 %, the supply reliability can be increased

Table 3. Marginal effects of the determinants of the supply reliability of surface water irrigation from regression results of annual climate variables model.

	Supply reliability of surface water irrigation: share of crop sown area not affected as not enough surface water	
	Without interactive variable	With interactive variable
	Model 1	Model 2
Climate variables		
Average annual temperature (°)	0.024	−0.011
	(0.81)	(0.30)
Average annual temperature squared	−0.002	0.003
	(1.76)	(0.92)
Annual total precipitation (mm)	0.002***	0.003***
	(5.22)	(4.71)
Annual total precipitation squared	−0.0000006***	−0.0000001
	(4.43)	(0.41)
Interactive variable		−0.0001**
Average annual temperature* Annual total precipitation		(1.57)
	0.024	−0.011
Irrigation variables		
Whether irrigating by two kinds of water sources (1 = yes; 0 = no)	−0.123***	−0.116***
	(4.19)	(3.85)
Whether irrigating only by large reservoir (1 = yes; 0 = no)	−0.243**	−0.242**
	(2.49)	(2.47)
Whether irrigating only by middle reservoir (1 = yes; 0 = no)	0.059	0.070
	(0.83)	(0.98)
Whether irrigating only by small reservoir (1 = yes; 0 = no)	0.061	0.066
	(1.04)	(1.11)
Whether irrigating only by pool (1 = yes; 0 = no)	−0.054	−0.043
	(0.97)	(0.76)
Whether irrigating only by river (1 = yes; 0 = no)	−0.105*	−0.094
	(1.84)	(1.63)
Whether irrigating only by pump (1 = yes; 0 = no)	−0.097**	−0.081*
	(2.27)	(1.86)
Village characteristics		
Whether the soil type is sand (1 = yes; 0 = no)	−0.039	−0.041
	(1.43)	(1.50)
Whether the soil type is clay (1 = yes; 0 = no)	−0.011	−0.008
	(0.41)	(0.30)
Landform (1 = Plain; 0 = Mountain)	−0.034	−0.038
	(1.36)	(1.51)
Arable area per person (ha)	0.561**	0.589**
	(1.99)	(2.08)
Density of population	−0.002*	−0.002*
	(1.84)	(1.77)

by 4.6 % (increasing from 62.4 to 67.0 %), when precipitation increases by 25 %, the supply reliability can be increased by 10.7 % (increasing from 62.4 to 73.1 %), when precipitation increases by 50 %, the supply reliability can be increased by 17.0 % (increasing from 62.4 to 79.4 %). However, reducing precipitation by 10 % will result in reduction of supply reliability by 5.3 % (reducing from 62.4 to 57.1 %), reducing precipitation by 25 % will result in reduction of supply reliability by 14.7 % (reducing from 62.4 to 47.7 %), and reducing precipitation by 50 % will result in reduction of supply

Table 3. Continued.

	Supply reliability of surface water irrigation: share of crop sown area not affected as not enough surface water	
	Without interactive variable	With interactive variable
	Model 1	Model 2
Province dummy (versus Yunnan province)		
Hebei	0.755***	0.808***
	(4.74)	(4.95)
Henan	0.841***	0.785***
	(7.47)	(6.46)
Shandong	0.661***	0.640***
	(6.25)	(6.01)
Jilin	0.922***	1.028***
	(5.30)	(5.51)
Anhui	0.372***	0.348***
	(5.99)	(5.43)
Jiangsu	0.386***	0.375***
	(6.42)	(6.18)
Jiangxi	−0.147**	−0.246***
	(−1.78)	(2.37)
Guangdong	0.014	−0.041
	(0.11)	(0.31)
Constant	−2.778***	−3.261***
	(3.67)	(3.98)
Observations	762	762
Fit statistics of Tobit		
LR chi^2	158.15	160.66
Log likelihood	−522.3194	−521.0652
Prob > chi^2	0.0000	0.0000
Pseudo R^2	0.1315	0.1336

Absolute t statistics in parentheses. * $p < 0.10$, ** $p < 0.05$, *** $p < 0.01$.

reliability by 28.6 % (reducing from 62.4 to 33.8 %). In summary, on the average, temperature is negatively related with the supply reliability of surface water irrigation; but precipitation positively influences the supply reliability of surface water irrigation.

4.2.2 Influence of seasonal temperature and precipitation on surface water supply reliability

As to seasonal climate variables, let's first discuss temperature variables. For spring temperature, the first degree term is negative and significant, the quadratic term is positive and significant (Table 4). So, the impact of spring temperature on the supply reliability of surface water irrigation presents a u-shape. And the turning point is 14.4°/11.7° in the model without/with interactive terms. This means that the change trend of marginal effect is first downward (when temperature is below 14.4° in the model without interactive terms or 11.7° in the model with interactive terms) and then becomes upward (when temperature is beyond 14.4° in the model without interactive terms or 11.7° in the model with interactive terms). Similar analysis can be done in terms of winter temperature. Specifically, the impact of winter temperature on the supply reliability of surface water irrigation presents an inverted u-shape. And the turning point is 7.9°/0° in the model without/with interactive terms. Maybe this is the reason why when we treat average annual temperature as independent variables in annual climate models (Table 3), it is statistically insignificant. As temperature in spring and winter has opposite influence on the supply reliability of surface water irrigation, they offset each other.

In terms of seasonal precipitation variables, expect summer precipitation, other three seasonal precipitation variables are all statistically significant, not only the first de-

Table 4. Marginal effects of the determinants of the supply reliability of surface water irrigation from regression results of seasonal climate variables model.

	Supply reliability of surface water irrigation: share of crop sown area not affected as not enough surface water	
	Without interactive variable	With interactive variable
	Model 3	Model 4
Climate variables		
Spring temperature (°)	−1.409***	−1.127*
	(2.44)	(1.68)
Spring temperature squared	0.049***	0.048***
	(3.54)	(2.74)
Summer temperature (°)	0.304	1.022
	(0.28)	(0.94)
Summer temperature squared	−0.010	−0.031
	(0.43)	(1.33)
Fall temperature (°)	−0.109	−0.618
	(0.16)	(0.91)
Fall temperature squared	0.002	0.027
	(0.09)	(1.01)
Winter temperature (°)	0.333*	0.183
	(1.88)	(1.09)
Winter temperature squared	−0.021*	−0.030***
	(1.85)	(2.58)
Spring precipitation (mm)	−0.013**	−0.019***
	(2.41)	(3.26)
Spring precipitation squared	0.000008*	0.00001**
	(1.85)	(2.24)
Summer precipitation (mm)	−0.0006	−0.003
	(0.27)	1.16)
Summer precipitation squared	0.000002	0.000002
	(1.50)	(1.41)
Fall precipitation (mm)	−0.039***	−0.034***
	(5.08)	(4.60)
Fall precipitation squared	0.00005***	0.00006***
	(4.75)	(5.25)
Winter precipitation (mm)	0.050***	0.058***
	(4.86)	(5.51)
Winter precipitation squared	−0.0001***	−0.0001***
	(4.48)	(4.66)
Interactive variable		
Spring temperature* Spring precipitation		0.0002
		(1.55)
Summer temperature* Summer precipitation		0.00008
		(1.44)
Fall temperature* Fall precipitation		−0.0003**
		(2.34)
Winter temperature* Winter precipitation		−0.0007
		(1.49)

Table 4. Continued.

	Supply reliability of surface water irrigation: share of crop sown area not affected as not enough surface water	
	Without interactive variable	With interactive variable
	Model 3	Model 4
Irrigation variables		
Whether irrigating by two kinds of water sources (1 = yes; 0 = no)	−0.119***	−0.122***
	(4.14)	(4.29)
Whether irrigating only by large reservoir (1 = yes; 0 = no)	−0.257***	−0.178**
	(2.66)	(1.81)
Whether irrigating only by middle reservoir (1 = yes; 0 = no)	0.064	0.070
	(0.92)	(1.02)
Whether irrigating only by small reservoir (1 = yes; 0 = no)	0.036	0.052
	(0.60)	(0.87)
Whether irrigating only by pool (1 = yes; 0 = no)	−0.049	−0.064
	(0.88)	(1.17)
Whether irrigating only by river (1 = yes; 0 = no)	−0.095**	−0.104*
	(1.67)	(1.84)
Whether irrigating only by pump (1 = yes; 0 = no)	−0.092**	−0.077*
	(2.18)	(1.84)
Village characteristics		
Whether the soil type is sand (1 = yes; 0 = no)	−0.052*	−0.051*
	(1.94)	(1.55)
Whether the soil type is clay (1 = yes; 0 = no)	−0.040	−0.044*
	(1.56)	(1.69)
Landform (1 = Plain; 0 = Mountain)	−0.036	−0.038
	(1.46)	(1.55)
Arable area per person (ha)	0.322	0.434
	(1.11)	(1.48)
Density of population	−0.002*	−0.002
	(1.84)	(1.26)
Province dummy (versus Yunnan province)		
Hebei	1.153	0.478
	(0.78)	(0.31)
Henan	0.612	0.160
	(0.45)	(0.11)
Shandong	0.469	−0.025
	(0.34)	(0.02)
Jilin	4.143**	3.858**
	(2.39)	(2.17)
Anhui	0.070	−0.380
	(0.05)	(0.25)
Jiangsu	0.032	−0.429
	(0.02)	(0.28)
Jiangxi	0.562	0.201
	(0.51)	(0.17)
Guangdong	0.007	−0.007
	(0.01)	(0.01)
Constant	27.626	16.020
	(1.68)	(1.06)
Observations	762	762
Fit statistics of Tobit		
LR chi^2	226.61	239.31
Log likelihood	−4880	−481.7394
Prob > chi^2	0.0000	0.0000
Pseudo R^2	0.1884	0.1990

Absolute t statistics in parentheses. * $p<0.10$, ** $p<0.05$, *** $p<0.01$.

Table 5. Change in supply reliability of surface water irrigation due to climate change in the future.

	The percentage of irrigated crop-sown areas having reliable surface water supply (%)
baseline	62.4
Temperature change	
0.5° (+)	61.3
1° (+)	60.3
1.5° (+)	59.4
2° (+)	58.6
Precipitation change	
10 % (+)	67.0
10 % (-)	57.1
25 % (+)	73.1
50 % (+)	79.4
25 % (−)	47.7
50 % (−)	33.8

Data sources: Authors' survey.

gree term but also quadratic term. Specifically, for spring and fall precipitation, the first degree terms are negative and quadratic terms are positive, which means that the relationship of the two displays a U-shape. For winter precipitation, the first degree term is positive and quadratic term is negative, which means that there was an inverse U-shaped relationship between the two. Specifically, the turning point of spring precipitation is 812.5 mm in the model without interactive terms or 783.8 mm in the model with interactive terms, higher than 393.6 mm (the average spring precipitation in Table A3), so the marginal effect of spring precipitation is negative. In other words, there was a negative correlation between spring precipitation and supply reliability of surface water irrigation. Similarly, the turning point of winter precipitation is 250.0 mm in the model without interactive terms or 267.9 mm in the model with interactive terms, higher than 131 mm (the average winter precipitation in Table A3), so there was a positive correlation between spring precipitation and supply reliability of surface water irrigation. As to fall precipitation, the turning point is 390 mm in the model without interactive terms or 327.8 mm in the model with interactive terms, and the average fall precipitation is 211. 9 mm (in Table A3), so, when fall precipitation is lower than 211.9 mm, the relationship of fall precipitation and supply reliability of surface water irrigation is positive, otherwise, the relationship of fall precipitation and supply reliability of surface water irrigation is negative.

4.2.3 Influence of rural irrigation infrastructure and other control variables on the supply reliability of surface water irrigation

Results of four models show that, based on irrigating by more than two kinds of water sources, the coefficients of most other rural irrigation infrastructure are statistically significant with negative sign. This means that, most rural irrigation infrastructures are playing statistically significant roles in the supply reliability of surface water irrigation. And among all of them, the role of the water source, more than two kinds of water sources, is a most obvious one. Specifically, the coefficient of two kinds of water resources is statistically significant in each model. And the coefficients are −0.123, −0.116, −0.119 and −0.122 respectively in four models (Tables 3 and 4). This means that, in the case of all other conditions remaining unchanged, compared with irrigating by more than two kinds of water sources, if farmers irrigate by two kinds of water sources, the share of irrigated crop-sown areas having reliable surface water supply will decrease by 11.6–12.3 %. Similarly, if farmers irrigate by large reservoir/river/pump, the share of irrigated crop-sown areas having reliable surface water supply will decrease by 8.1–11.1/10.5–12.4/24.2–25.2 %. Besides, different from all water infrastructures discussed above, the roles of middle reservoir, small reservoir and pool in the supply reliability of surface water irrigation are statistically insignificant. The coefficients of middle and small reservoirs are positive, and the coefficient of pool is negative. This probably means that, compared with more than two water sources, middle reservoir and small reservoir are playing more significant roles and pool is playing a less significant role in the supply reliability of surface water irrigation, but these results are statistically insignificant. Or this maybe indicates that, keeping all other conditions unchanged, compared with irrigating by more than two kinds of water sources, irrigating by middle reservoir, small reservoir or pool, farmers' share of irrigated crop-sown areas with reliable surface water supply has no significant difference.

What's more, some physical and socio-economic conditions of villages are also closely related to the supply reliability of surface water irrigation. For instance, in terms of population, its coefficients in all models are all negative and statistically significant with small numerical value. Specifically, with all other conditions unchanged, if increase one person in one hectare, the supply reliability of surface water irrigation will decrease by about 0.2 %. Besides, in annual climate models, the coefficients of arable area per person are both positive and statistically significant at the 10 % level. Maybe this is because that, surface water usually can concentrate to supply with large quantity, so, when we assume other conditions unchanged, if arable area per person increase, the share of irrigated crop-sown areas having reliable surface water supply will not decrease but increase. What's more, in models with seasonal climate variables, the coefficients of whether the soil type is sand are both negative and statistical significant at the 10 % level. With other conditions unchanged, comparing to villages with loam, the supply reliability of surface water irrigation of villages with sand, will decrease about 5.1 % in the model without interactive terms or 5.2 % in the model with interactive terms (Table 4). And the supply reliability of surface water irrigation of villages with clay (comparing to villages with loam) will decrease 4.0 % in the model without interactive terms or 4.4 % in the model with interactive terms (Table 4).

Finally, the supply reliability of surface water irrigation also differs by province, especially in models with annual climate variables. As shown in Table 3, leaving Yunnan Province as baseline, the coefficient of each other province is statistically significant, except Guangdong province. But for models with seasonal climate variables, except Jilin Province, the coefficient of each other province is statistically insignificant (Table 4).

5 Conclusion

Based on nine-province large-scale field surveys in China, this study mainly examines the status of current surface water supply reliability and identifies how climate change and rural water infrastructures influence it.

Firstly, through descriptive analysis, we conclude: on average the total reliability of surface water is 74 %. Jilin, Henan and Jiangxi own relatively higher reliability with 91, 85 and 83 %. The worst appears in Yunnan, with 48 % only. And there exists a little fluctuation of the supply reliability of surface water irrigation in the past 3 years.

Secondly, through estimation results of econometric models, we conclude: (a) Average annual temperature was statistically insignificant, as temperature in spring and winter has opposite influence on the supply reliability of surface water irrigation, they offset each other. But as for annul total precipitation, supply reliability of surface water irrigation will be increased with increasing precipitation. However, after considering the correlation between temperature and precipitation, if both temperature and precipitation increase simultaneously, the supply reliability of surface water irrigation will be significantly reduced. (b) In the future, keeping the constant of other factors, increasing temperature by 0.5° will reduce the supply reliability of surface water irrigation by 1.1 %. At the same time, when precipitation increases by 10 %, the supply reliability can be increased by 4.6 %, when precipitation increases by 25 %, the supply reliability can be increased by 10.7 %. Reducing precipitation by 10 % will result in reduction of supply reliability by 5.3 %, reducing precipitation by 25 % will result in reduction of supply reliability by 14.7 %. (c) In terms of seasonal precipitation variables, expect summer precipitation, other three seasonal precipitation variables are all statistically significant, not only the first degree term but also quadratic term. Specifically, there was a negative correlation between spring precipitation and supply reliability of surface water irrigation. There was a positive correlation between spring precipitation and supply reliability of surface water irrigation. When fall precipitation is lower than 211.9 mm, the relationship of fall precipitation and supply reliability of surface water irrigation is positive, otherwise, the relationship of fall precipitation and supply reliability of surface water irrigation is negative.

Thirdly, compared with irrigating by more than two kinds of water sources, if farmers irrigate by two kinds of water sources, the supply reliability of surface water irrigation will decrease by 11.6–12.3 %. Similarly, if farmers irrigate by large reservoir/river/pump, the share of irrigated crop-sown areas having reliable surface water supply will decrease by 8.1–11.1/10.5–12.4/24.2–25.2 %.

Finally, some physical and socio-economic conditions of villages are also closely related to the supply reliability of surface water irrigation. And the supply reliability of surface water irrigation also differs by province, especially in annul climate models.

Data availability. The underlying research data in the paper contains field data and secondary climate data. The field data in this study was collected by our research team from a nine-province field survey during 2012 and 2013. The nine provinces contain six Northern provinces (Hebei, Henan, Shandong, Jiangsu, Anhui and Jilin) and three southern provinces (Jiangxi, Guangdong and Yunnan). Based on local water resources, climate condition and economic development level, the nine provinces are firstly collected from the whole nation, and then the following strategies were taken to select samples. Climate data are offered by China's National Meteorological Information Center. Both the two kinds of data are confidential to the public.

Appendix A

Table A1. Sample distribution of this study.

Province	County	Township	Village
Hebei	3	4	7
Henan	3	5	11
Shandong	3	8	22
Jilin	3	3	5
Anhui	3	8	19
Jiangsu	3	9	25
Jiangxi	10	30	90
Guangdong	6	18	53
Yunnan	3	9	24
Total	37	94	256

Data sources: Authors' survey.

Table A2. Sample distribution based on types of water source.

Types of water sources	Province	County	Townshihp	Village	Samples
One kind of water sources	7	25	45	79	230
Two kinds of water sources	9	36	78	138	400
More than two kinds of water sources	4	21	33	46	132

Data sources: Authors' survey.

Table A3. Statistical descriptions of major variables.

	Mean	Min	Max	SD
Dependent variable				
Share of crop sown area not affected as not enough surface water	0.74	0.00	1.00	0.33
Climate variables				
Average annual temperature (°)	16.69	2.67	22.34	3.51
Average annual temperature squared	290.82	7.11	498.88	111.87
Annual total precipitation (mm)	1341.07	399.56	2126.33	505.07
Annual total precipitation squared	2 053 737	159 646	4 521 271	1 294 184
Average annual temperature* Annual total precipitation	23 851.43	1065.569	47 078.66	12 216.82
Spring temperature (°)	16.62	4.02	22.24	3.13
Spring temperature squared	286.10	16.18	494.75	102.12
Summer temperature (°)	26.03	17.45	28.13	2.50
Summer temperature squared	683.88	304.33	791.22	117.87
Fall temperature (°)	17.79	2.88	23.77	3.92
Fall temperature squared	331.77	8.32	565.11	133.29
Winter temperature (°)	6.31	−13.68	15.50	5.88
Winter temperature squared	74.29	0.21	240.31	82.35
Spring precipitation (mm)	393.60	52.70	697.08	231.35
Spring precipitation squared	208 375.80	2777.63	485 920.50	175 342.10
Summer precipitation (mm)	604.48	265.38	1040.02	180.74
Summer precipitation squared	398 020.80	70 427.92	1 081 648.00	245 034.80
Fall precipitation (mm)	211.89	51.88	386.59	70.50
Fall precipitation squared	49 862.01	2691.54	149 453.10	31 638.58
Winter precipitation (mm)	131.10	5.69	257.43	86.38
Winter precipitation squared	24 637.92	32.38	66 272.62	24 681.82
Spring temperature* Spring precipitation	6983.04	164.70	13 238.57	4488.35
Summer temperature* Summer precipitation	15 977.25	4259.04	29 056.33	5522.83
Fall temperature* Fall precipitation	4015.55	128.19	9198.73	1994.22
Winter temperature* Winter precipitation	1012.23	−162.22	1977.91	779.29
Irrigation variables				
Whether irrigating only by one kind of water sources (1 = yes; 0 = no)	0.30	0.00	1.00	0.46
Whether irrigating by two kinds of water sources (1 = yes; no)	0.52	0.00	1.00	0.50
Whether irrigating by more than two kinds of water sources (1 = yes; no)	0.17	0.00	1.00	0.38
Whether irrigating only by large reservoir (1 = yes; 0 = no)	0.01	0.00	1.00	0.10
Whether irrigating only by middle reservoir (1 = yes; 0 = no)	0.03	0.00	1.00	0.16
Whether irrigating only by small reservoir (1 = yes; 0 = no)	0.04	0.00	1.00	0.20
Whether irrigating only by tubewell (1 = yes; 0 = no)	0.0013	0.00	1.00	0.04
Whether irrigating only by pool (1 = yes; 0 = no)	0.04	0.00	1.00	0.20
Whether irrigating only by river (1 = yes; 0 = no)	0.04	0.00	1.00	0.20
Whether irrigating only by pump (1 = yes; 0 = no)	0.13	0.00	1.00	0.34
Village characteristics				
Whether the soil type is sand (1 = yes; 0 = no)	0.30	0.00	1.00	0.46
Whether the soil type is clay (1 = yes; 0 = no)	0.42	0.00	1.00	0.49
Landform (1 = Plain; 0 = Mountain)	0.28	0.00	1.00	0.45
Arable area per person (ha)	0.09	0.00	0.46	0.06
Density of population	8.24	0.50	110.00	9.13
Province dummy				
Hebei	0.03	0.00	1.00	0.16
Henan	0.04	0.00	1.00	0.20
Shandong	0.08	0.00	1.00	0.28
Jilin	0.02	0.00	1.00	0.14
Anhui	0.07	0.00	1.00	0.26
Jiangsu	0.10	0.00	1.00	0.29
Jiangxi	0.35	0.00	1.00	0.48
Guangdong	0.21	0.00	1.00	0.41
Yunnan	0.09	0.00	1.00	0.29

Note: Observation number is 762.

Table A4. Percentage of villages using kinds of water sources in sample regions (%).

Water sources	Total samples	Henan	Jiangxi	Guangdong	Hebei	Shandong	Jiangsu	Jilin	Anhui	Yunnan
Only one kind of water source										
Tubewell	1.3	9.4	0	0.6	4.8	0	0	0	6	2.2
River	16.4	0	20.3	8.3	0	0	69.9	0	0	2.2
Reservoir	10.1	0	8.9	26.9	14.3	0	0	0	0	8.7
Pond	2.7	0	3.7	3.8	0	4.7	0	0	0	2.2
Irrigation District	5.1	0	3.7	3.8	0	18.8	12.3	0	0	0
Dam	0.1	0	0.4	0	0	0	0	0	0	0
Waterjam	0.1	0	0	0	0	0	0	0	0	2.2
Others	1.3	0	0	3.8	14.3	0	0	0	0	0
Two kinds of water sources										
Tubewell, River	10.1	12.5	8.1	1.9	38.1	9.4	17.8	20	28	0
Tubewell, Reservoir	2.6	0	1.2	3.8	28.6	0	0	0	0	6.5
Tubewell, Pond	3.3	9.4	0	1.9	0	0	0	0	34	0
Tubewell, Irrigation district	12.7	68.8	2.4	1.9	0	67.2	0	80	0	6.5
Tubewell, Dam	0.4	0	0	1.9	0	0	0	0	0	0
Tubewell, Waterjam	0.4	0	0	0	0	0	0	0	0	6.5
Tubewell, Others	1.1	0	0	1.3	0	0	0	0	12	0
River, Reservoir	7.5	0	14.2	11.5	0	0	0	0	0	0
River, Pond	2.3	0	5.3	1.9	0	0	0	0	0	0
River, Irrigation district	0.4	0	0	1.9	0	0	0	0	0	0
River, Dam	0.9	0	1.2	0	0	0	0	0	0	6.5
River, Waterjam	0.6	0	0	0	0	0	0	0	0	8.7
River, Others	1	0	1.2	1.3	0	0	0	0	0	4.3
Reservoir, Pond	2.8	0	6.5	2.6	0	0	0	0	0	0
Reservoir, Irrigation district	0.9	0	0	3.8	0	0	0	0	0	0
Reservoir, Dam	2.6	0	4.1	3.8	0	0	0	0	0	4.3
Reservoir, Others	0.6	0	0.8	0	0	0	0	0	0	4.3
Pond, Irrigation district	0.4	0	1.2	0	0	0	0	0	0	0
Pond, Dam	1.6	0	4.5	0	0	0	0	0	0	0
Pond, Waterjam	0.9	0	0	0	0	0	0	0	0	13
Pond, Others	0.3	0	0	0	0	0	0	0	0	4.3
Dam, Waterjam	0.9	0	0	0	0	0	0	0	0	13
Dam, Others	0.3	0	0.8	0	0	0	0	0	0	0
Waterjam, Others	0.3	0	0	0	0	0	0	0	0	4.3
More than two kinds of water sources										
Tubewell, River, Reservoir	0.9	0	1.2	1.9	0	0	0	0	0	0
Tubewell, River, Pond	1.8	0	0	1.9	0	0	0	0	20	0
River, Reservoir, Pond	2.6	0	4.9	3.8	0	0	0	0	0	0
River, Reservoir, Dam	1.3	0	3.7	0	0	0	0	0	0	0
River, Reservoir, Others	0.1	0	0.4	0	0	0	0	0	0	0
Reservoir, Pond, Irrigation district	0.3	0	0	1.3	0	0	0	0	0	0
Reservoir, Pond, Dam	0.3	0	0	1.3	0	0	0	0	0	0
Reservoir, Pond, Waterjam	0.3	0	0	1.3	0	0	0	0	0	0
Reservoir, Pond, Others	0.3	0	0	1.3	0	0	0	0	0	0
River, Reservoir, Pond, Dam	0.4	0	1.2	0	0	0	0	0	0	0
Total	100	100	100	100	100	100	100	100	100	100

Note: others contain rain gutter, fountain and sewage disposal. Data sources: Authors' survey.

Competing interests. The authors declare that they have no conflict of interest.

Special issue statement. This article is part of the special issue "Innovative water resources management – understanding and balancing interactions between humankind and nature". It is a result of the 8th International Water Resources Management Conference of ICWRS, Beijing, China, 13–15 June 2018.

Acknowledgements. We acknowledge the financial support from National Natural Sciences Foundation in China (71161140351, 91325302 and 71333013) and Chinese Academy of Sciences (Y02015004 and KSZD-EW-Z-021-1).

Edited by: Bo Pang
Reviewed by: two anonymous referees

References

Chen, H., Wang, J., and Huang, J.: Evaluation of Drought Resisting Effects of Rural Irrigation Infrastructure: Based on Empirical Study in Seven Provinces in China, Journal of Natural Resources, 27, 1656–1365, 2012 (in Chinese).

Dalla, V. M., Marcomini, A., and Codato, E.: Climate change influence on POPs distribution and fate: A case study, Chemosphere, 67, 1287–1295, 2007.

Delpla, I., Jung, A. V., Baures, E., Clement, M., and Thomas, O.: Impact of climate change on surface water quality in relation to drinking water production, Environ. Int., 35, 1225–1233, 2009.

Ding, Y., Ren, G., and Shi, G.: National assessment of climate change (I): The history and future trends of China's climate change, Progress of Climate Change Research, 2, 3–8, 2006 (in Chinese).

Hao, Z., Li, L., Wang, J., Wang, Z., and Shi, X.: Impact of Climate Change on Surface Water Resources, Earth Science-Journal of China University of Geosciences, 32, 425–433, 2007.

He, X. and Guo, L.: The stakeholders of farmland water conservancy and their cost-benefit analysis – based on the survey of farmland water conservancy in Shayang County, Hubei Province, Management World (Monthly), Phase 7, 86–99, 2010.

IPCC: Climate Change 2014, Impacts, Adaptation, and Vulnerability. Part A: Global and Sectoral Aspects. Contribution of Working Group II to the Fifth Assessment Report of the Intergovernmental Panel on Climate Change, Cambridge, Cambridge University Press, 2014.

Jia, X.: Effectiveness of irrigation supply from the perspective differential farms' income, Dissertation for Doctor Degree, Northwest A&F University, 73–74, 2016.

Lamon, L., Valle, M. D., Critto, A., and Marcomini, A.: Introducing an integrated climate change perspective in POPs modeling, monitoring and regulation, Environ. Pollu., 157, 1971–1980, 2009.

Li, Y.: Impacts of Climate Change on Reliability of Irrigation Wate Supply and Responses of Cropping Pattern-Empirical Research Based on 9 Provinces of China, Dissertation for PHD, The university of Chinese Academy of Science, 2017 (in Chinese).

Long, H., Mao, L. Q., Che, Z. X., and Yang, G. W.: Impact on water resources in Yellow River with climate change, J. Water Resour, 17, 74–77, 2006 (in Chinese).

Ma, L.: Performance Analysis of China's Rural Water Conservancy Irrigation Facilities Investment – the Influence of the Farmers' Income per acre, for example, Chinese Rural Economy, 4, 55–62, 2008.

MWR (Ministry of Water Resources): China Water Resources Yearbook, Ministry of Water Resources, Beijing, China, 2011.

NBSC (National Bureau of Statistics in China): China Statistical Yearbook, China Statistics Press, Beijing, China, 2015.

Prowse, T. D. and Belataos, S.: Climate change, flow regulation and land-use effects on the hydrology of the peace-athabasca-slave system; findings from the north rivers ecosystem initiative, Environ. Monit. Assess., 113, 167–197, 2006.

Piao, S., Ciais, P., Huang, Y., Shen, Z., Peng, S., Li, J., Zhou, L., Liu, H., Ma, Y., Ding, Y., Friedlingstein, P., Liu, C., Tan, K., Yu, Y., Zhang, T., and Fang, J.: The impacts of climate change on water resources and agriculture in China, Nature, 467, 43–51, 2010.

Thornton, P. E., Running, S. W., and White, M. A.: Generating surfaces of daily meteorological variables over large regions of complex terrain, J. Hydrol., 190, 214–251, 1997.

Trenberth, K. E., Dai, A., Rasmussen, R. M., and Parsons, D. B.: The changing character of precipitation, B. Am. Meteorol. Soc., 84, 1205–1217, 2003.

Vanvliet, M. T. H.: Impact of summer droughts on the water quality of the Meuse river, J. Hydrol., 353, 1–7, 2008.

Wang, J., Huang, J., and Yan, T.: Impacts of Climate Change on Water and Agricultural Production in Ten Large River Basins in China, J. Integr. Agr., 12, 1267–1278, 2013.

Wang, J., Huang, Q., Huang, J., and Rozelle, S.: Managing Water on China's Farms for a Sustainable Future: Institutions, Policies and the Transformation of Irrigation under Scarcity, Elsevier, 14–16, 2016.

Wang, J., Yanrong, L., Jikun, H., Tingting, Y., and Tianhe, S.: Growing water scarcity, food security and government responses in China, Globle Food Security, 1, 9–17, 2017.

Wang, X.: Supply and management of Public Irrigation Infrastructure, Journal of Shanxi Agricultural University (Social Science Edition), 14, 936–941, 2015.

Wilby, L., Whitehead, P. G., Wade, A. J., Butterfield, D , Davis, R. J., and Watts, G.: Integrated modeling of climate change impacts on water resources and quality in a lowland catchment: River Kennet, UK, J. Hydrol., 330, 204–220, 2006.

Xia, J., Cheng, S., Hao, X., Xia, R., and Liu, X.: Potential impacts and challenges of climate change on water quality and ecosystem: Case studies in representative rivers in China, Journal of Resources and Ecology, 11, 31–35, 2010.

Xia, X., Wu, Q., and Mou, X.: Advances in impacts of climate change on surface water quality, Advances in Water Science, 23, 124–133, 2012 (in Chinese).

Yang, Y., Wang, J., and Chen, H.: "Assessment on water supply reliability of different irrigation water sources. China Population", Resources and Environment, 22, 1–4, 2012.

Zhu, Y. and Wang, L.: Analysis on different income level' farmers' demand willingness to farmland water conservancy facilities – based on Field Data from Shaanxi and Henan, Rural Economy in China, 76–87, 2014 (in Chinese).

Zhang, J. and Wang, G.: The Impacts of Climate Change on Hydrology and Water Resources, Science Press, 138–181, 2007 (in Chinese).

Zhang, J., Wang, G., He, R., and Liu, C.: Variation trends of runoffs in the Middle Yellow River basin and its response to climate change, Advances in Water Science, 20, 153–158, 2009 (in Chinese).

https://doi.org/10.5194/piahs-379-105-2018

Open Access
PIAHS

Analysis of vegetation condition and its relationship with meteorological variables in the Yarlung Zangbo River Basin of China

Xianming Han[1,2], Depeng Zuo[1,2], Zongxue Xu[1,2], Siyang Cai[1,2], and Xiaoxi Gao[1,2]

[1]College of Water Sciences, Beijing Normal University, Beijing 100875, China
[2]Beijing Key Laboratory of Urban Hydrological Cycle and Sponge City Technology, Beijing 100875, China

Correspondence: Depeng Zuo (dpzuo@bnu.edu.cn)

Received: 31 December 2017 – Accepted: 12 January 2018 – Published: 5 June 2018

Abstract. The Yarlung Zangbo River Basin is located in the southwest border of China, which is of great significance to the socioeconomic development and ecological environment of Southwest China. Normalized Difference Vegetation Index (NDVI) is an important index for investigating the change of vegetation cover, which is widely used as the representation value of vegetation cover. In this study, the NDVI is adopted to explore the vegetation condition in the Yarlung Zangbo River Basin during the recent 17 years, and the relationship between NDVI and meteorological variables has also been discussed. The results show that the annual maximum value of NDVI usually appears from July to September, in which August occupies a large proportion. The minimum value of NDVI appears from January to March, in which February takes up most of the percentage. The higher values of NDVI are generally located in the lower elevation area. When the altitude is higher than 3250 m, NDVI began to decline gradually, and the NDVI became gradual stabilization as the elevation is up to 6000 m. The correlation coefficient between NDVI and precipitation in the Yarlung Zangbo River Basin is greater than that with temperature. The Hurst index of the whole basin is 0.51, indicating that the NDVI of the Yarlung Zangbo River Basin shows a weak sustainability.

1 Introduction

As an important natural resource, vegetation is an indispensable part of the living environment of mankind. Vegetation coverage is closely related with meteorological factors. By means of the transfer of water vapor, climate change is adapted by vegetation and characterized to some extent. At the same time, the global water cycle, biochemical cycle and energy balance are affected by vegetation, which have had a profound impact on ecosystem changes. Meteorology and hydrology are connected by vegetation as well. Vegetation index plays an important role in evaluating vegetation status information such as vegetation coverage, growth vigor and biomass. Normalized Difference Vegetation Index (NDVI) is an indicator of vegetation growth status and vegetation spatial density (Nouri et al., 2017), which can well characterize the dynamic changes of vegetation. NDVI is widely used in the fields of meteorology and hydrology. Many studies have explored the relationship between NDVI and meteorological elements in different regions (Chen et al., 2016; Guo et al., 2017; Karkauskaite et al., 2017; Lumbierres et al., 2017; Riihimäki et al., 2017). Bing et al. (2014) found that the vegetation grown has been positively influenced by the rainfall and precipitation both in moving tracks of gravity center and time-lag effect especially for the growing season. The NDVI lower areas was mainly located in the lower altitude regions (Yu et al., 2017).

As the largest outflowing river in southwest China, the Yarlung Zangbo River Basin is located in the hinterland of the Qinghai-Tibet Plateau. Its elevation is very large, its vegetation and ecological environment are very fragile and is an important geographical barrier in China. Therefore, the stations in the Yarlung Zangbo River Basin are sparse, so the remote sensing is particularly important for the study. Salameh et al. (2017) analyzed the change of surface water storage of the Ganges-Brahmaputra river basin over the period of 1993–

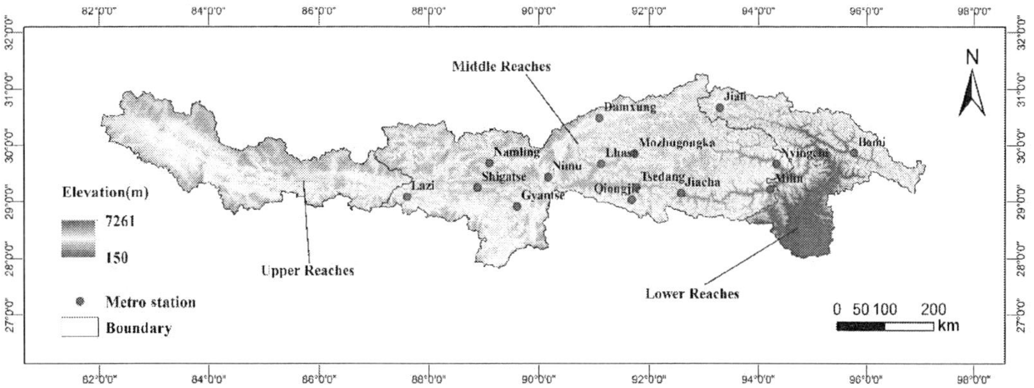

Figure 1. Overview of the study area and meteorological stations.

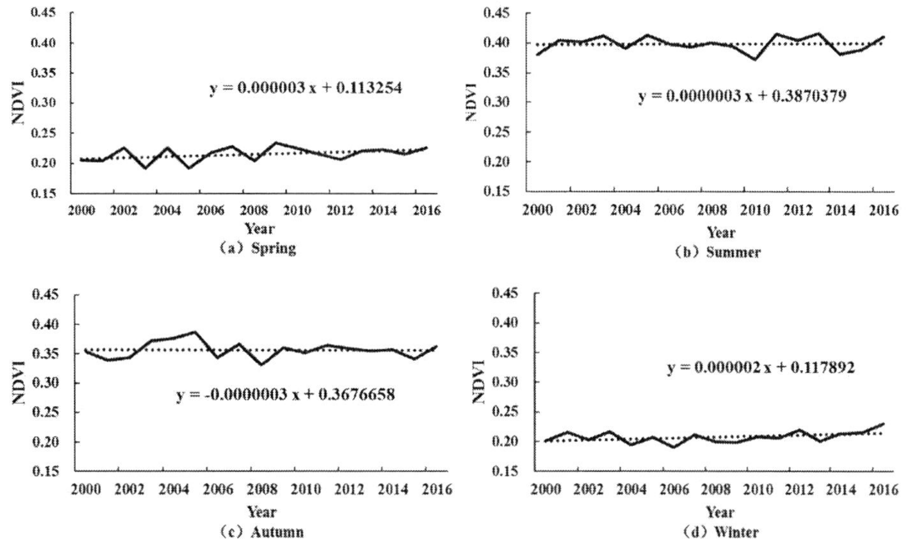

Figure 2. Seasonal variation of NDVI in the Yarlung Zangbo River Basin.

2007 through remote sensing, and found that the dataset of SWS is valuable for hydrological and climate modeling studies. Cai et al. (2017) build a model of TRMM (tropical rainfall measuring mission) precipitation spatial-temporal downscaling for higher temporal-spatial resolution rainfall data production in the data scarce Yalung Zangbo River basin. Yang et al. (2011) found that there is a significant positive correlation between NDVI and precipitation in the Yarlung Zangbo River Basin.

The analysis of vegetation condition and its relationship with meteorological variables in the Yarlung Zangbo River Basin of China is of great significance to the development of water energy and the protection of nature in the transboundary rivers (Li et al., 2015; Lv et al., 2014; Peng et al., 2014; Zhang et al., 2015). At the same time, the research on vegetation status of the Yarlung Zangbo River Basin also helps to further protect the ecology of the study area. However, there is not a comprehensive and detailed assessment of the spatial and temporal changes of vegetation in the Yarlung Zangbo River basin during the past 17 years. The aim of this paper is to discuss this issue. The main objectives of this study are to: (1) make a thorough inquiry about vegetation condition changed with temporal and spatial variation; (2) discuss the trend of vegetation change in the future Yarlung Zangbo River Basin; (3) quantify the effect of relationships with meteorological variables in the Yarlung Zangbo River Basin.

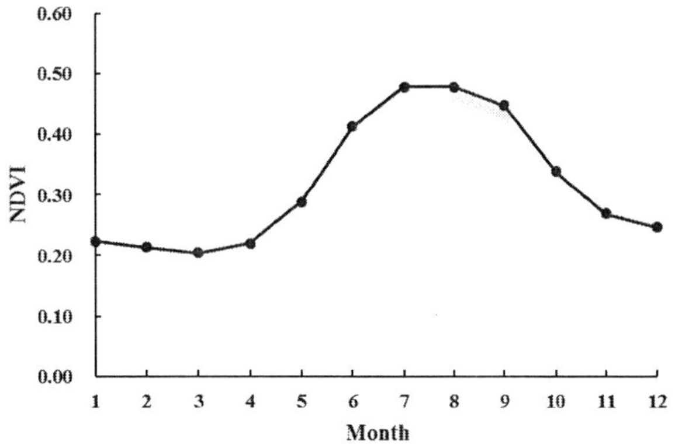

Figure 3. Inner-annual variation of average NDVI in the Yarlung Zangbo River Basin.

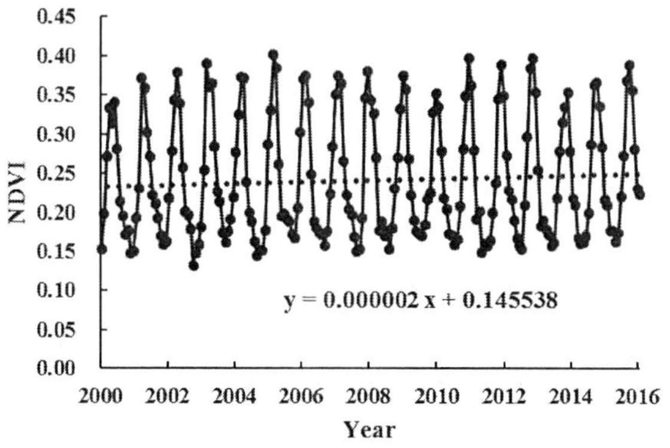

Figure 4. Monthly variation of NDVI in the Yarlung Zangbo River Basin.

Table 1. Persistence of monthly NDVI in the Yarlung Zangbo River Basin.

Hurst index	Persistence level	Percentage	Area (km^2)
Hurst ≤ 0.4	Strong anti-continuity	7.69 %	18 498
0.4 < Hurst ≤ 0.5	Weak anti-continuity	35.69 %	85 826
0.5 < Hurst ≤ 0.6	Weak continuity	43.93 %	105 633
Hurst > 0.6	Strong continuity	12.69 %	30 526

2 Study area and date description

2.1 Study area

The Yarlung Zangbo River Basin (Guo et al., 2017), extending from longitude 82°0′ to 97°1′E and latitude 27°8′ to 31°2′N, is located in the southern part of Tibet Autonomous Region, China (Fig. 1). The basin originates from the Gyimah Yangzon Glacier in the north of the Himalayas, known as the Maquan River. Most of the watershed is above 3000 m in altitude. The basin consists of arid areas, semi-dry and semi-humid regions and humid areas, from the Gyimah Yangzon Glacier in the upper stream to the valley in the downstream. The basin is long and narrow in the east-west direction, short in the north-south direction, whose south is the Himalayas and north is the Gangdise Mountains and the Nyainqentanglha Mountains. A big bend is located in the downstream area (Huang et al., 2013). The average annual precipitation in the basin is from 300 to 500 mm, which is very different in spatial distribution and uneven during the year. Hydrological and meteorological stations are mainly located in the midstream and downstream, whose distribution is inhomogeneous, so the basin is a lack of data area.

2.2 Data description

The quality-controlled NDVI remote sensing product (MOD13A3) is selected in this study, obtained from the observation of MODIS (Moderate Resolution Image Spectroradiometer) data provided by NASA, spanning 17 years (February 2000 to December 2016). MOD13A3 is the third-level product, based on the secondary product, corrected the edge distortion (Bowtie effect) produced by the sensor imaging process. The spatial resolution of the product is 1 km × 1 km while the time resolution is monthly. The larger the value, the better the vegetation coverage. The data was processed into GEO TIFF format by MRT software and processed by ArcGIS projection splicing. using maximal synthesis method. The DEM (Digital Elevation Model) of the Basin is provided by SRTM (Shuttle Radar Topography Mission), and the projection is longitude and latitude one, which is transferred to ALBERS projection by ArcGIS. The spatial resolution of the data DEM is 90 m. The meteorological data used to calculate the relationship between meteorological elements and NDVI in this paper include monthly precipitation and average temperature, both from Climatic Data Center, National Meteorological Information Center, China Meteorological Administration (http://data.cma.cn/).

3 The disposal method of data

3.1 Correlation analysis

Partial correlation coefficient (Linn et al., 1969) is a research method adopted when examining the correlation between multivariate variables, a measure of the direct linear correlation between two variables under the control of other variables. The definition of the first-order partial correlation coefficient can be estimated as:

$$r_{12-3} = \frac{r_{12} - r_{13}r_{23}}{\sqrt{\left(1 - r_{13}^2\right)\left(1 - r_{23}^2\right)}} \quad (1)$$

where r_{12-3} is the relationship between the first variable and the second variable in the case of the third variable; r_{12} is the correlation between the first variable and the second; r_{13} is

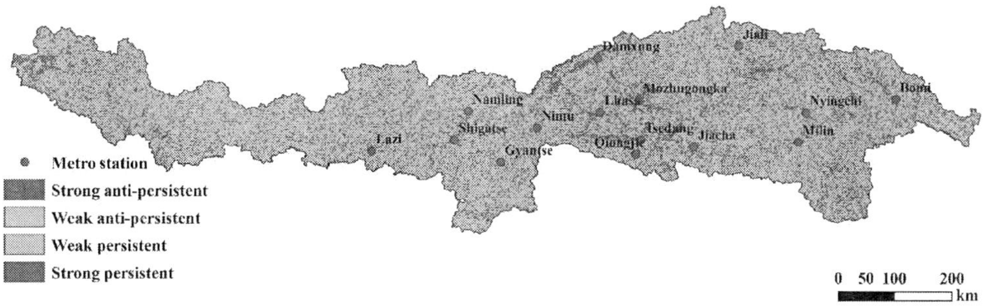

Figure 5. The spatial distribution of Hurst index of monthly NDVI in the Yarlung Zangbo River Basin.

Figure 6. Spatial distribution of NDVI during the growing season of 2002 in the Yarlung Zangbo River Basin.

the correlation between the first variable and the third variable; r_{23} is the correlation between the second variable and the third variable; the closer r_{12-3} to 1, the higher the partial correlation coefficient.

3.2 Trend analysis

The Hurst exponent (Feng et al., 2009) based on R/S analysis is an effective method to quantitatively describe the long-term dependence of time-series information. It was first proposed by Hurst, a British hydrologist, whose basic principle is:

Given a time series $\{\xi(t)\}, t = 1, 2, \ldots$, for any positive integer $\tau = 1, 2, \ldots$, define the mean sequence:

$$\langle \xi_\tau \rangle = \frac{1}{\tau} \sum_{t=1}^{\tau} \xi(t) \quad \tau = 1, 2, \ldots \tag{2}$$

Accumulated deviation:

$$X(t, \tau) = \sum_{u=1}^{t} (\xi(u) - \xi_\tau) \quad 1 \le t \le \tau \tag{3}$$

Max Difference:

$$R(\tau) = X(t, \tau) X(t, \tau) \quad \tau = 1, 2, \ldots \tag{4}$$

Standard deviation:

$$S(\tau) = \left[\frac{1}{\tau} \sum_{t=1}^{\tau} \left(\xi(t) - \overline{\xi}_\tau \right)^2 \right]^{\frac{1}{2}} \quad \tau = 1, 2, \ldots \tag{5}$$

If $R/S \propto \tau^H$ could established in the time series, $t = 1, 2, \ldots$, the Hurst phenomenon exist, Hurst index (H) can be obtained in the fitted log-log plot ($\ln\tau$, $\ln R/S$) by the least square method. When $0.5 < H < 1$, it indicates that the time series is a continuous sequence, which means the trend of the future is consistent with the past trend, and the closer to 1, the stronger the continuity; When $H = 0.5$, it shows that the time series is a random sequence, that is, the trend of the future has nothing to do with the past trend; When $0 < H < 0.5$, it shows that the time series has the anti-persistence. And the closer H is to 0, the stronger the anti-sustainability.

4 Results and discussion

4.1 Seasonal variation of NDVI

In order to show NDVI in different seasons, seasonal variation of NDVI in the Yarlung Zangbo River Basin is shown in Fig. 2 according to the season of spring (March–May), summer (June–August), autumn (September–November) and winter (December–February).

As can be seen from Fig. 2, the NDVI value reached the maximum in summer, with an average of 0.399, while the minimum occurred in winter, with an average of 0.208. In summer, it reached a maximum of 0.416 in 2013 and reached a minimum of 0.372 in 2010. Followed by the autumn, with an average NDVI of 0.357. The NDVI showed a slow upward trend between 2000 and 2005, reaching a maximum of 0.387 in 2005 and then fluctuating down to reach a minimum of 0.330 in 2008, NDVI value remained at about 0.350 after 2009. The NDVI values in spring and winter were relatively small, with an average of about 0.21. The NDVI values in autumn showed a downward trend, which was −0.0003 %/10a. NDVI values in spring, summer and winter increased with 0.003 %/10a, 0.0003 %/10a and 0.002 %/10a, respectively.

4.2 Inner-annual variation of average NDVI

To reveal the law of inner-annual variation of average NDVI, The NDVI values during the study period (2000–2016) were transformed to 12 months. It is clear that NDVI values change greatly with time (Fig. 3).

From Fig. 3, the monthly NDVI value increased first and then decreased with the change of the month. From January to March, the NDVI value basically maintains at around 0.215 and reaches the minimum value of 0.205 in March, then steadily increased until reaching its maximum in July. Although the NDVI values dropped in August and September, they still maintained a relatively large NDVI value of 0.479 and 0.449 respectively. After September, the NDVI value declined more obviously. The average annual NDVI was 0.320.

4.3 Monthly variation of NDVI

Before further trend analysis, the relationship of NDVI in a long time series should be researched. The statistical analysis of NDVI values in monthly and annual scale during the study period (2000–2016) in the Yarlung Zangbo River Basin was carried out. The results are shown in Fig. 4.

As shown in Fig. 4, NDVI generally shows an upward trend, with an increase rate of 0.002 %/10a ($p < 0.001$), indicating that the vegetation coverage in the Yarlung Zangbo River Basin has been slowly improving over the past 16 years. The maximum annual NDVI occurred from July to September, of which more appeared in August. The minimum appeared in January to March, of which more occurred in February. From 2000 to 2005, the overall trend from 2010 to 2013 showed an upward trend. From 2006 to 2009, the trend from 2014 to 2015 showed a downward trend.

4.4 The spatial distribution of Hurst index of monthly NDVI

The above analysis shows that the NDVI changes in the Yarlung Zangbo River Basin have obvious periodicity. Therefore, it is necessary to study the persistent of the NDVI changes in the Yarlung Zangbo River Basin. The Hurst index is calculated by the monthly grid point of NDVI. The spatial distribution is shown in Fig. 5, and the statistical results are shown in Table 1.

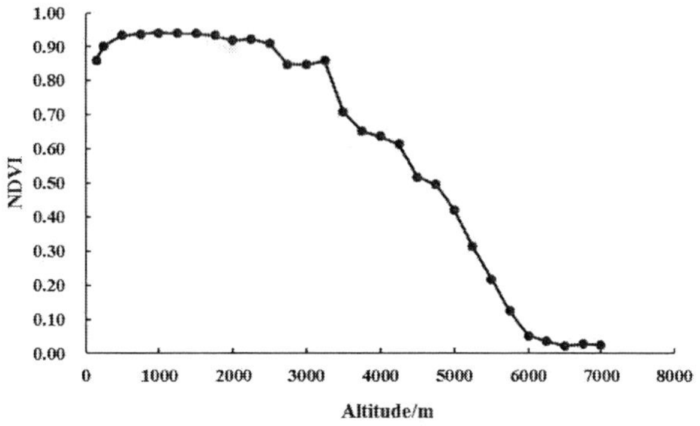

Figure 7. Distribution of NDVI along the elevation of the Yarlung Zangbo River Basin.

Figure 5 illustrating that the distribution of Hurst index has some spatial differences. There is a strong continuity in Lhasa River basin and the upper reaches of the basin. The Hurst index of the whole basin is 0.51, which shows a weak persistence in the Yarlung Zangbo River Basin. As shown in Table 1, the ratio of weak persistence is the largest, reaching 43.93 %, and the drainage area is 105 633 km^2. The second is weak anti-persistence, with an area of 85 826 km^2 accounting for 35.69 %. The strong persistent area and strong anti-persistent area account for the smallest proportion, accounting for 12.69 and 7.69 % respectively.

4.5 NDVI spatial distribution characteristics

From 2000 to 2016, the average NDVI of the growing season in the Yarlung Zangbo River was 0.407, and the NDVI in the growing season of 2002 was 0.407. Therefore, the spatial distribution of NDVI in the growing season was selected as a typical year, as shown in Fig. 6.

From Fig. 6, the NDVI values downstream NDVI values of Millin and Bomi stations are higher than 0.6, and the NDVI values in the river basin upstream of Lazi during the growing season are still less than 0.4, indicating that the NDVI high value areas are mainly distributed in the middle and lower reaches of the Yarlung Zangbo River Basin, especially in the lower reaches. However, due to the high altitude in the upper reaches, snow and ice don't thaw and the rare vegetation grows through a whole year in many areas.

4.6 Distribution of NDVI along the elevation

To obtain the distribution of NDVI along elevation, the relationship between the elevation of the Yarlung Zangbo River Basin and the NDVI value was statistically analyzed with an index of 250 m. The result is shown in Fig.7.

As can be seen from Fig. 7, a distinct feature is shown that higher the altitude, smaller the NDVI value. The middle and low altitude areas are the high value areas of NDVI. At the altitude of 1000–2500 m, the NDVI values are basically

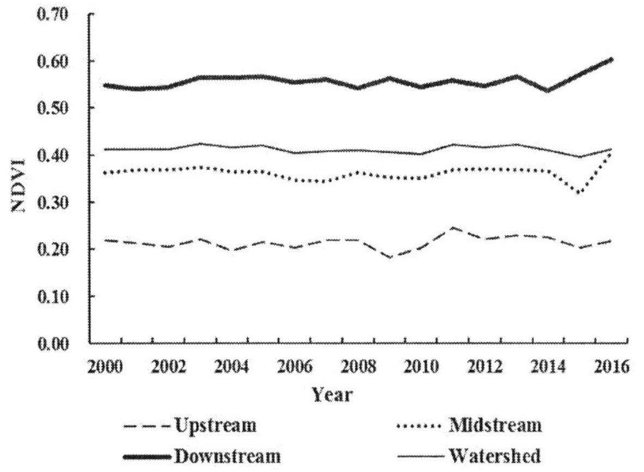

Figure 8. Distribution of NDVI in sub-basins of the Yarlung Zangbo River Basin.

Table 2. Correlation analysis of NDVI and meteorological factors in the Yarlung Zangbo River Basin.

Station	Partial correlation between NDVI and precipitation	Partial correlation between NDVI and temperature
Nyingchi	−0.041	0.352
Tsedang	0.543	0.436
Shigatse	0.568	0.266
Qiongjie	0.574	0.414
Nimu	0.568	0.339
Namling	0.422	0.380
Mozhugongka	0.437	0.521
Milin	−0.058	0.548
Lhasa	0.385	0.519
Lazi	0.241	0.462
Gyantse	0.503	0.415
Jiali	−0.026	0.402
Jiacha	0.238	0.664
Damxung	0.503	0.487
Bomi	−0.281	0.853

the same, about 0.913. Above 2500 m, the NDVI value began to decline, showing a sharp decline after more than 3000 m. When the altitude is above 6000 m, the NDVI value is only about 0.029, and the NDVI values remain unchanged as the altitude increases.

4.7 Distribution of NDVI in sub-basins

Through consultation with experts and access to the relevant papers (Chen et al., 2015), the Yarlung Zangbo River Basin can be divided into upper, middle and lower reaches. The upstream is the wide valley area above the Lazi prefecture in Shigatse. The midstream is valley while the downstream is located in the Pai town of Milin district. The mean NDVI values in sub-basins are calculated respectively. The results are shown in Fig. 8.

From Fig. 8, the annual value of NDVI shows the order of lower, middle and upper reaches. It is obvious that this is related to the elevation difference of the sub-basins. Vegetation is rarely grown above the snow line and at the edge of the valley. Due to uneven terrain and elevation distribution, the vegetation distribution in the Yarlung Zangbo River Basin is uneven, the average value of NDVI in the Yarlung Zangbo River Basin is 0.41. The average NDVI values in the upper, middle and lower reaches are 0.21, 0.36 and 0.55 respectively.

4.8 The relationship between meteorological elements and NDVI

In order to explore the relationship between regional NDVI and meteorological elements in the study area of the Yarlung Zangbo River Basin, this paper uses the ArcGIS to extract the average NDVI of the region centered around each meteorological station with a radius of 1.5 km. The trend of data of NDVI, temperature and precipitation were removed, then the partial correlation coefficients between NDVI and temperature and precipitation were calculated respectively. The results are shown in Table 2.

Table 2 shows that the correlation between NDVI and air temperature is about 0.465 (0.266–0.853), while the partial correlation coefficient between NDVI and precipitation is 0.281 (−0.281–0.574), which has passed the test of 5 % significant level. Therefore, the partial correlation coefficient between NDVI and temperature in the Yarlung Zangbo River Basin is greater than the correlation coefficient with precipitation, indicating that vegetation growth in the Yarlung Zangbo River Basin is more affected by temperature than precipitation.

5 Conclusions

This study investigated the distribution of vegetation in the Yarlung Zangbo River Basin in the Tibet Autonomous Region of China during the period from 2000 to 2016. The spatial and temporal distribution pattern of vegetation was analyzed by Hurst exponent method using the MODIS remote sensing data and monthly precipitation and temperature data. The correlation with NDVI between precipitation and temperature was analyzed using partial correlation coefficient method. Major conclusions can be summarized as follows:

1. the NDVI value decreases with increasing altitude. At the altitude of 1000–2500 and 6000–7000 m, the NDVI values are basically remain unchanged. Reasons may be further discussed in combination with botany.

2. The annual value of NDVI shows the order of downstream, whole basin, middle reaches and upstream, with the average NDVI of 0.41 in the Yarlung Zangbo River

Basin and 0.21, 0.36 and 0.55 of the upper and lower reaches respectively. It can be inferred that this is related to the characteristics of underlying surface in sub-basins

3. The correlation analysis between vegetation and meteorological elements in the Yarlung Zangbo River Basin shows vegetation growth is more affected by temperature than precipitation.

4. The mean Hurst index of the whole basin is 0.51, which shows that the NDVI in the Yarlung Zangbo River Valley shows a weak sustainability. The ratio of weak persistent to weak anti-persistent is the largest.

The results obtained in this study would provide more evidence and useful reference for study about vegetation in the Yarlung Zangbo River Basin.

Data availability. The raw data required to reproduce these findings are available to download from http://data.cma.cn/ and https://modis.gsfc.nasa.gov/. The processed data required to reproduce these findings cannot be shared at this time as the data also forms part of an ongoing study.

Competing interests. The authors declare that they have no conflict of interest.

Special issue statement. This article is part of the special issue "Innovative water resources management – understanding and balancing interactions between humankind and nature". It is a result of the 8th International Water Resources Management Conference of ICWRS, Beijing, China, 13–15 June 2018.

Acknowledgements. This study is supported by the Natural Science Foundation of China (Grant No. 91647202).

Edited by: Dingzhi Peng
Reviewed by: Liqun Sun and one anonymous referee

References

Cai, M., Lv, Y., Yang, S., and Zhou, Q.: TRMM precipitation downscaling in the data scarce Yarlung Zangbo River Basin, Journal of Beijing Normal University, Natural Science, 1, 111–119, 2017.

Chen, B., Li, H., Cao, X., Shen, W., and Jin, H.: Vegetation Pattern and Spatial Distribution of NDVI in the Yarlung Zangbo River Basin of China, Journal of Desert Research, 1, 120–128, 2015.

Chen, B., Li, H., Cao, X., and Tang, H.: Dynamic Changes in Vegetation Coverage in the Yarlung Zangbo River Basin Based on SPOT – VGT NDVI, Mountain Research, 2, 249–256, 2016.

Feng, X., Luo, L., and Feng, Z.: Hurst index experiment on precipitation change trend and mutation of China in the near 50 years, Arid Land Geography, 6, 859–866, 2009.

Guo, B., Zhou, Y., Wang, S., and Tao, H.: The relationship between normalized difference vegetation index (NDVI) and climate factors in the semiarid region: A case study in Yalu Tsangpo River basin of Qinghai-Tibet Plateau, J. Mt. Sci., 11, 926–940, https://doi.org/10.1007/s11629-013-2902-3, 2014.

Guo, B., Jiang, L., Ge, D., and Shang, M.: Driving Mechanism of Vegetation Coverage Change in the Yarlung Zangbo River Basin under the Stress of Global Warming, Journal of Tropical and Subtropical Botany, 3, 209–217, 2017.

Guo, J., Hu, Y., Xiong, Z., Yan, X., Li, C., and Bu, R.: Variations in Growing-Season NDVI and Its Response to Permafrost Degradation in Northeast China, Sustainability, 9, 551, https://doi.org/10.3390/su9040551, 2017.

Huang, W., Wang, G., Wang, A., Cao, S., and Cao, K.: The morphology of the Yalung Zangbo River in the Great Canyon Region and its implications, Geological Bulletin of China, 1, 130–140, 2013.

Karkauskaite, P., Tagesson, T., and Fensholt, R.: Evaluation of the Plant Phenology Index (PPI), NDVI and EVI for Start-of-Season Trend Analysis of the Northern Hemisphere Boreal Zone, Remote Sens., 9, 485, https://doi.org/10.3390/rs9050485, 2017.

Li, H., Li, Y., Shen, W., Li, Y., Lin, J., Lu, X., Xu, X., and Jiang, J.: Elevation-Dependent Vegetation Greening of the Yarlung Zangbo River Basin in the Southern Tibetan Plateau, 1999–2013, Remote Sens., 7, 16672–16687, https://doi.org/10.3390/rs71215844, 2015.

Linn, R. L. and Werts, C. E.: Assumptions in making causal inferences from part correlations, partial correlations, and partial regression coefficients, Psychol. Bull., 72, 307–310, https://doi.org/10.1037/h0028107, 1969.

Lumbierres, M., Méndez, P., Bustamante, J., Soriguer, R., and Santamaría, L: Modeling Biomass Production in Seasonal Wetlands Using MODIS NDVI Land Surface Phenology, Remote Sens., 9, 392, https://doi.org/10.3390/rs9040392, 2017.

Lv, Y., Dong, G., Yang, S., Zhou, Q., and Cai, M.: Spatio-Temporal Variation in NDVI in the Yarlung Zangbo River Basin and Its Relationship with Precipitation and Elevation, Resources Science, 3, 603–611, 2014.

Nouri, H., Anderson, S., Sutton, P., Beecham, S., Nagler, P., Jarchow, C., and Roberts, D.: NDVI, scale invariance and the modifiable areal unit problem: An assessment of vegetation in the Adelaide Parklands, Sci. Total Environ., 584–585, 11–18, https://doi.org/10.1016/j.scitotenv.2017.01.130, 2017.

Peng, D. and Du, Y.: Comparative analysis of several Lhasa River basin flood forecast models in Yarlung Zangbo River, International Conference on Bioinformatics and Biomedical Engineering, 1–4, 18–20 June 2010, Chengdu, China, 2010.

Riihimäki, H., Heiskanen, J., and Luoto, M.: The effect of topography on arctic-alpine aboveground biomass and NDVI patterns, Int. J. Appl. Earth Obs., 56, 44–53, https://doi.org/10.1016/j.jag.2016.11.005, 2017.

Salameh, E., Frappart, F., Papa, F., Güntner, A., Venugopal, V., Getirana, A., Prigent, C., Aires, F., Labat, D., and Laignel, B.: Fifteen Years (1993–2007) of Surface Freshwater Storage Variability in the Ganges-Brahmaputra River Basin Using Multi-Satellite Observations, Water, 9, 15–35, https://doi.org/10.3390/w9040245, 2017.

Yang, L., Cangjue, Z., Ji, T., Yang, M., Zhou, S., Li, J., and Li, L.: A Primary Investigation of Bird Resource in Drainage Area of

Yarlung Zangbo River and its Two Branches from Tibet, China, Sichuan Journal of Zoology, 3, 475–480, 2011.

Yu, L., Yao, K., Liu, H., Pu, S., and Jiang, Q.: NDVI variation at different elevation and its relationship with climatic factors, Computing Techniques for Geophysical and Geochemical Exploration, 2, 296–300, 2017.

Zhang, J. and Ren, Z.: Responses of Vegetation Changes in Growing Season to Precipitation in Yarlung Zangbo River Basin, Research of Soil and Water Conservation, 2, 209–212, 2015.

https://doi.org/10.5194/piahs-379-113-2018

Sources and behavior of perchlorate ions (ClO_4^-) in chalk aquifer of Champagne-Ardenne, France: preliminary results

Feifei Cao[1], Jessy Jaunat[1], Patrick Ollivier[2], Benjamin Cancès[1], Xavier Morvan[1], Daniel Hubé[2], Alain Devos[1], Nicolas Devau[2], Vincent Barbin[1], and Pierre Pannet[3]

[1]Université de Reims Champagne-Ardenne – GEGENAA – EA 3795, 2 esplanade Roland Garros, 51100 Reims, France

[2]BRGM, 3 av. C. Guillemin, BP 36009, 45060 Orléans Cedex 2, France

[3]BRGM, Pôle technologique Henri Farman, 12 rue Clément Ader, 51100 Reims, France

Correspondence: Feifei Cao (feifei.cao@etudiant.univ-reims.fr)

Received: 30 December 2017 – Revised: 16 April 2018 – Accepted: 20 April 2018 – Published: 5 June 2018

Abstract. Perchlorate (ClO_4^-) is an environmental contaminant of growing concern due to its potential human health effects and widespread occurrence in surface water and groundwater. Analyses carried out in France have highlighted the presence of ClO_4^- in drinking water of Champagne-Ardenne (NW of France), with two potential sources suspected: a military source related to the First World War and an agricultural source related to the past use of Chilean nitrates. To determine the sources of ClO_4^- in groundwater, major and trace elements, ^2H and ^{18}O, ClO_3^- and ClO_4^- ions and a list of 39 explosives were analyzed from 35 surface water and groundwater sampling points in the east of the city of Reims. ClO_4^- ions were found in almost all sampling points (32 out of 35) with a max value of $33\,\mu g\,L^{-1}$. ClO_4^- concentrations were highest in groundwater ranging from 0.7 to $33\,\mu g\,L^{-1}$ (average value of about $6.2\,\mu g\,L^{-1}$) against from < 0.5 to $10.2\,\mu g\,L^{-1}$ in surface water (average value of about $2.7\,\mu g\,L^{-1}$). Most of the water samples showing high ClO_4^- levels ($> 4\,\mu g\,L^{-1}$) were collected near a military camp, where huge quantities of ammunitions have been used, stored and destroyed during and after the First World War.

1 Introduction

Perchlorate (ClO_4^-) is an inorganic anion and a powerful oxidizer with high solubility and mobility in water. Relatively stable under ambient conditions, perchlorate may persist for many years, possibly decades, under typical surface water and groundwater conditions (Sturchio et al., 2014). Perchlorate salts are widely used as oxidizer in solid rockets fuel and as component of fireworks, pyrotechnics, flares and explosives (Urbansky, 1998). In addition, natural sources such as sodium nitrate deposits in Chile (Urbansky et al., 2001) and formation in the atmosphere have also been reported (Dasgupta et al., 2005). Perchlorate is an environmental contaminant of concern in water because of its potential ability to inhibit iodide uptake and to impact neurodevelopment, especially for foetuses and infants (Leung et al., 2010). Since 1997, perchlorate has been listed as a contaminant in drinking water monitoring programs in the USA (USEPA, 1998). The current health advisory level for ClO_4^- is set at $15\,\mu g\,L^{-1}$ based on the reference dose recommended by the US EPA. However, in many countries, drinking water standard for ClO_4^- is yet to be defined. Many studies have underlined perchlorate contamination in water from numerous countries such as the USA, Canada (Backus et al., 2005), China (Wu et al., 2010), Japan (Kosaka et al., 2007) and India (Kannan et al., 2009).

In France, the interest of research on ClO_4^- has increased since 2011 because of the discovery of ClO_4^- contamination in several drinking water resources in South-Western France, in the Paris Basin and in North-Eastern France (Lopez et al., 2015). In the Champagne-Ardenne region (NE of France), the measurement campaigns carried out by the

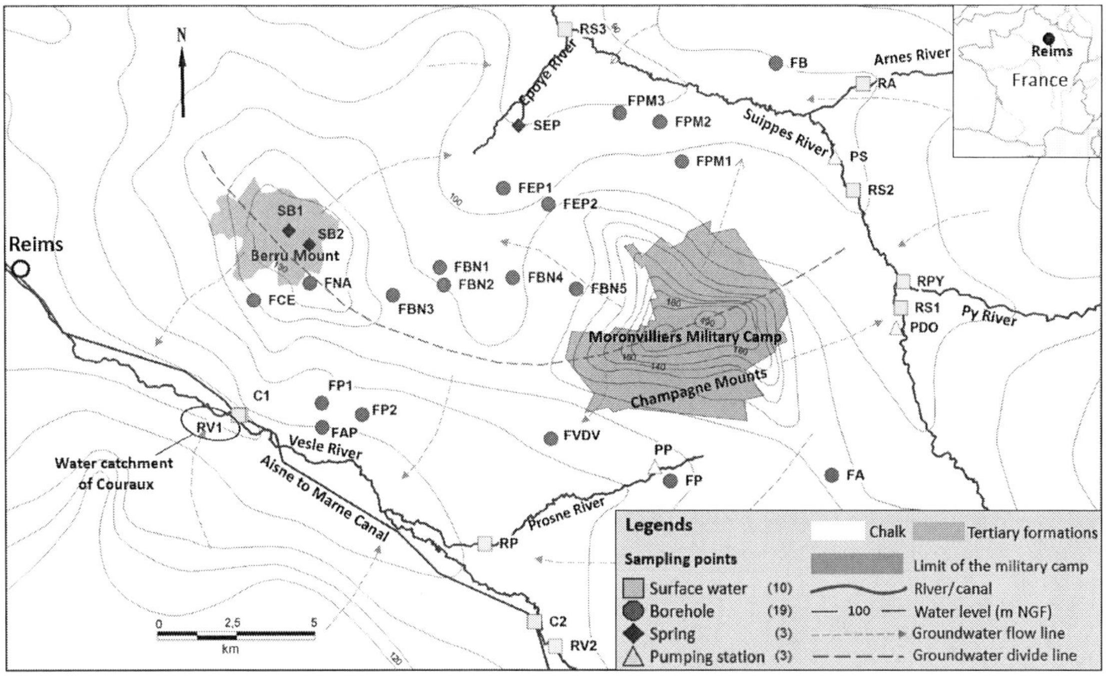

Figure 1. Location of the study area, main rivers and sampling points (water level data from Rouxel-David et al., 2002).

Regional Health Agency (ARS) in 2014 have showed high levels of ClO_4^- ions ($> 15\,\mu g\,L^{-1}$) in raw water resources intended for human consumption. Two potential sources of ClO_4^- in groundwater are suspected: military and agricultural. Indeed, the Champagne-Ardenne region was highly marked by the events of the First World War that may have emitted large quantities of perchlorates and other explosives in the environment, and France has imported, between 1875 and 1920, large quantities of natural nitrates as fertilizers from Chile, which are particularly rich in ClO_4^- ions. The study site in an intensive agricultural area, thereby it is concerned with the use of these Chilean nitrates.

The objectives of this study are (1) to identify the sources of ClO_4^- and other explosives in groundwater and to establish links between their occurrence and the type of human activities (military and/or agricultural), (2) to understand the transfer mechanisms of ClO_4^- in the Champagne chalk aquifer, with studies on flow and mineralization processes in the aquifer. A study area has been selected in the Champagne-Ardenne region for a continuous monitoring of water chemistry to observe a temporal and spatial evolution of the concentrations of the molecules concerned. In this paper, we will focus on the results of the 2 screening measurement campaigns carried out in 2017, with the purpose to obtain a water chemical map of the study area and to proceed a further monthly monitoring over the next 2 years.

2 Materials and methods

2.1 Study area

2.1.1 Location

The study area is located in the east of Reims (NE of France; Fig. 1) and covers approximately $600\,km^2$ between the Suippe River (as the northern and western boundary) and the Vesle River (as the southern boundary), where some water catchments intended for drinking water supply are significantly impacted by the presence of ClO_4^- ions.

It corresponds to an agricultural zone (traditional crops of wheat and beet, for which Chilean nitrates were largely used to increase the production until the 1920s) located on the former battlefield of the First World War (military camp of Moronvilliers and the surrounding trench areas). In addition to some forests covered at the Berru Mount and the Champagne Mounts, the whole zone is used as agricultural lands.

2.1.2 Hydrogeological context

The aquifer of the study area is the unconfined Champagne chalk aquifer, which is a crucial water resource of the region. In some areas, the chalk is partially covered with Tertiary (the Berru Mount; Fig. 1) or Quaternary formations (the valleys of the Vesle River, the Suippes River and their tributaries).

The total porosity of the Chalk is about 40 %, with only 1 to 5 % related to the effective porosity (Rouxel-David et al., 2002; Vachier et al., 1987). The Champagne Mounts constitute the major reliefs where the transmissivity ranges from 10^{-6} to $10^{-5}\,m^2\,s^{-1}$. In the valleys, these values are much higher, ranging from 10^{-2} to $0.3\,m^2\,s^{-1}$. The Darcy velocity of the water flow in the saturated zone varies between $0.16\,m\,yr^{-1}$ at the reliefs and $315\,m\,yr^{-1}$ in valleys (Rouxel-David et al., 2002).

The study area is divided into two parts by the groundwater divide line across the summit of Berru Mount and the Champagne Mounts, which delimits the Vesle River watershed in the south and the Suippe River watershed in the north. According to water levels (Rouxel-David et al., 2002), the chalk aquifer of the study area is generally drained by rivers except the water catchment of Couraux (Fig. 1) which supplies the drinking water for the Reims area. Due to the drawdown of groundwater level caused by pumping, the catchment is partially fed by river water, especially during low water period. The precipitation constitutes the only recharge of the Champagne chalk aquifer, mainly between September and April.

2.2 Sampling

Two chemical screening campaigns have been carried out in June and September 2017. In June, 25 groundwater (22 boreholes and 3 springs) were collected. All the selected sampling points are situated on areas both agricultural and concerned by military activities. Ten points of surface waters, distributed in the Aisne to Marne Canal, Vesle River, Suippes River and their tributaries (Table 1 and Fig. 1) were also sampled to take into account the groundwater/surface water exchanges in the diffusion of the molecules. In September, a total of 25 water samples were collected (Table 1). The points with non-detectable or very low ClO_4^- levels were not sampled and a new one has been added in the Suippes River. The physicochemical parameters (pH, conductivity, temperature, alkalinity, dissolved oxygen and redox potential) were measured at the sampling points.

2.3 Instrumental analysis

The quantification of ClO_4^- was performed using a two-dimension ion chromatography on an ion chromatography (DIONEX ICS 2000) at BRGM (Orléans, France), with a detection limit of $0.5\,\mu g\,L^{-1}$. Quantitative analysis of cations (Ca^{2+}, Mg^{2+}, Na^+ and K^+) and anions (NO_3^-, SO_4^{2-}, Cl^-, F^-) was done by ICP Optical Emission Spectrometry (ICAP 6300, THERMO) and Ion Chromatography system (DIONEX ICS 2000) respectively at GEGENAA (Reims, France). The analysis of 39 explosives (e.g. TNT, trinitrophenol, trinitronaphthalene) was performed by the company Envilytix GmbH (Wiesbaden, Germany).

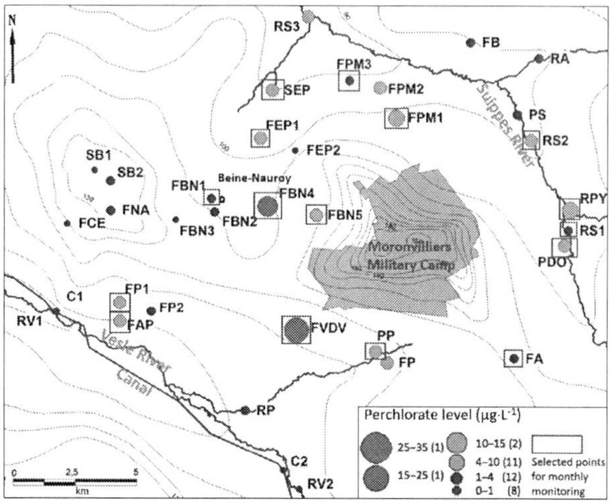

Figure 2. Perchlorate levels in waters in June 2017 and selected sampling points for the monthly monitoring.

3 Results and discussion

3.1 Major ions

Calcium and bicarbonate are the predominant ions in water, which is consistent for groundwater in Chalk aquifer. However, spatially heterogeneities are present not only on Ca^{2+} and HCO_3^- but also on Na^+, Cl^-, SO_4^{2-} and NO_3^- contents. This may be due to human activities and/or mineralization conditions of the aquifer, assumptions to be confirmed.

3.2 Perchlorate and explosives

The two screening campaigns show almost similar results for ClO_4^- concentrations (Table 1). Perchlorate ions were found in almost all the samples, and the mean ClO_4^- concentrations in groundwater and surface water samples were 6.2 and $2.7\,\mu g\,L^{-1}$, respectively. 15 among the 35 water points (43 %) present ClO_4^- levels higher than $4\,\mu g\,L^{-1}$ (concentration beyond which infants should not consume water according to French health authorities advices) and 1 sampling point exhibits ClO_4^- concentrations higher than $15\,\mu g\,L^{-1}$ (US EPA reference dose). Twenty sampling points (57 %) show ClO_4^- levels lower than $4\,\mu g\,L^{-1}$.

Most of the sampling points showing ClO_4^- concentrations above $4\,\mu g\,L^{-1}$ are found around the military camp of Moronvilliers, where large quantities of ammunition were used, stored and destroyed during and after the Great War. The maximum ClO_4^- level is $33\,\mu g\,L^{-1}$ (FVDV, south of military camp of Moronvilliers; Fig. 2). According to historical data, this sampling point is located at a former military airfield used during the Great War. In addition, the military camp of Moronvilliers is located upstream of this point, representing a potential source of ClO_4^-. Samples from FBN4, FVDV and

Table 1. Information on sampling locations and perchlorate concentrations. The values in bold are the maximum concentrations of perchlorate in June and September. The symbol "X" represents the points selected for monthly monitoring.

Point ID	Sample type	Latitude (° N)	Longitude (° E)	Altitude (m NGF)	Depth (m)	ClO_4^-_June ($\mu g L^{-1}$)	ClO_4^-_Sept ($\mu g L^{-1}$)	Selected for monthly monitoring
FCE	Exploited borehole	49.24205	4.12656	175	85	0.5	N.S.	
FNA	Exploited borehole	49.24662	4.15105	165	47	1.3	N.S.	
FBN3	Borehole	49.24280	4.18759	124.9	56	0.9	N.S.	
FBN1	Borehole	49.25048	4.20809	134	47.6	3.9	< 0.5	X
FBN2	Borehole	49.24554	4.20989	122	32	2.9	< 0.5	
FPM1	Exploited borehole	49.27875	4.31364	107	24	10.5	9.6	X
FBN4	Borehole	49.24725	4.23997	111	28	17.5	17	X
FBN5	Borehole	49.24383	4.26732	140	46.75	7.8	2.7	X
FEP1	Borehole	49.27210	4.23658	107	25	9.2	10.2	X
FEP2	Borehole	49.26752	4.25589	107	23	0.7	3.1	
FPM3	Borehole	49.29255	4.28730	103	21	2.6	2.4	X
FPM2	Borehole	49.28989	4.30454	121	34.3	5.5	5.8	
FP1	Borehole	49.21308	4.15558	95	18.9	6	3.8	X
FAP	Borehole	49.20634	4.15561	89.5	15	5.5	2.8	X
FP2	Borehole	49.20966	4.17316	97	20.75	1.3	1.8	
PP	Exploited borehole	49.19371	4.29981	110	80	6.6	6.3	X
FVDV	Borehole	49.20202	4.25525	107	21.65	**33**	**37.7**	X
FP	Borehole	49.18969	4.30624	116	23.6	4.8	6.6	
FB	Borehole	49.30563	4.35599	108	22.5	2.2	N.S.	
PS	Exploited borehole	49.27923	4.38166	98.5	16	2.7	3.3	
PDO	Exploited borehole	49.23141	4.40722	111	25	9	8.7	X
FA	Borehole	49.19046	4.37777	123	33.9	2	2.4	X
SB2	Spring	49.25732	4.15116	216		2.1	N.S.	
SB1	Spring	49.26133	4.14238	225		< 0.5	N.S.	
SEP	Spring	49.28945	4.24359	92		4.7	1.8	X
RV1	River	49.21076	4.11505	84		0.8	N.S.	
C1	Canal	49.20874	4.12446	85		< 0.5	N.S.	
RP	River	49.17311	4.22585	93		3.6	N.S.	
RV2	River	49.14414	4.25526	99		0.6	N.S.	
C2	Canal	49.15137	4.24642	96		< 0.5	N.S.	
RS1	River	49.23697	4.40956	107		1.8	4.3	X
RPY	River	49.24423	4.41038	110		10.2	12.3	X
RS2	River	49.27000	4.38921	110		4.5	6.7	X
RA	River	49.30098	4.40443	105		1.1	< 0.5	
RS3	River	49.31591	4.26468	110		4.2	4.6	
RS4	River	49.31420	4.27165	92		N.S.	5	

N.S.: Non sampled points.

FP show also chlorate (ClO_3^-) levels slightly higher than the other samples.

The sampling points showing ClO_4^- levels lower than $4\,\mu g L^{-1}$ are mainly located on the Berru Mount, in the Vesle River and the Aisne to Marne Canal. ClO_4^- concentrations in the canal are less than $0.5\,\mu g L^{-1}$. Since water in the canal is not (or little) related to the chalk aquifer, it verified the fact that ClO_4^- in river waters come from the exchanges with groundwater. ClO_4^- concentrations in rivers are partly diluted by upstream water and drainage of low-level parts of the aquifer so that generally lower levels of ClO_4^- than those in groundwater are measured.

However, some surface waters show significantly higher levels in ClO_4^- ($> 10\,\mu g L^{-1}$), such as the RPY (the Py River) and RS2 (downstream of the Py in the Suippe River; Fig. 2 and Table 1). ClO_4^- sources exist thus in the Py catchment area. As the point RS1 (upstream of the Py, in the Suippes river) has a low level of ClO_4^- ($< 4\,\mu g L^{-1}$), the high level on RS2 is most likely from the contribution of the Py river. Indeed, the Py River is originated from downstream of the military camp of Suippes, potentially representing a source of ClO_4^-.

Note that explosives have not been detected in surface and groundwater samples, which may be due to their low persistence in soil and water.

Sixteen sampling points were selected based on the results of the 2 sampling campaigns for a monthly monitoring over the next 2 years (Fig. 2). Sampling points showing high ClO_4^- concentrations were selected. Although the ClO_4^- levels on FBN1 and FA were low, these two sampling points will also be collected to ensure a consistent distribution of sampling points in the west and the southeast of the study area.

4 Conclusion and perspective

The preliminary results of the 2 sampling campaigns allowed us to obtain the chemical facies of groundwater, the levels and distribution of perchlorate ions in the study area. Most of sampling points showing high levels of ClO_4^- are located around the military camp of Moronvilliers, representing a potential source of perchlorate. That must be confirmed with further studies.

Over the next 2 years, measurements will be performed on 16 sampling points in the study area to determine the contents of major and trace elements, ClO_3^- and ClO_4^- ions in waters and stable isotopes of water molecule (^2H and ^{18}O). In parallel, the measurement of the isotopic signature of oxygen and chlorine in ClO_4^- will make it possible to precise the sources of perchlorates (agriculture and/or military). A better understanding of the chalk aquifer properties and flow mechanisms will be reached by the measurement of the residence time of groundwater and the monthly monitoring of the chemical composition of waters. Finally, this multi-tracer approach will allow us to predict the spatial and temporal evolution of ClO_4^- concentrations, and thus offer innovative information to make appropriate recommendations in terms of water management.

Data availability. The data supporting the findings of this study came from the site study results and are available within the article.

Competing interests. The authors declare that they have no conflict of interest.

Special issue statement. This article is part of the special issue "Innovative water resources management – understanding and balancing interactions between humankind and nature". It is a result of the 8th International Water Resources Management Conference of ICWRS, Beijing, China, 13–15 June 2018.

Acknowledgements. This study is financially supported by the BRGM, the Agence de l'Eau Seine-Normandie, the Region Grand-Est, the Grand-Reims Metropolis and the ARS Grand-Est.

Edited by: Depeng Zuo
Reviewed by: two anonymous referees

References

Backus, S. M., Klawuun, P., Brown, S., D'sa, I., Sharp, S., Surette, C., and Williams, D. J.: Determination of perchlorate in selected surface waters in the Great Lakes Basin by HPLC/MS/MS, Chemosphere, 61, 834–43, 2005.

Dasgupta, P. K., Martinelango, P. K., Jackson, W. A., Anderson, T. A., Tian, K., Tock, R. W., Tock, R. W., and Rajagopalan, S.:: The origin of naturally occurring perchlorate: the role of atmospheric processes, Environ. Sci. Technol., 39, 1569–1575, 2005.

Kannan, K., Praamsma, M. L., Oldi, J. F., Kunisue, T., and Sinha, R. K.: Occurrence of perchlorate in drinking water, ground water, surface water and human saliva from India, Chemosphere, 76, 22–6, 2009.

Kosaka, K., Asami, M., Matsuoka, Y., Kamoshita, M., and Kunikane, S.: Occurrence of perchlorate in drinking water sources of metropolitan area in Japan, Water Res., 41, 3474–82, 2007.

Leung, A. M., Pearce, E. N., and Braverman, L. W.: Perchlorate, iodine and the thyroid, Best Pract. Res. Cl. En., 24, 133–141, 2010.

Lopez, B., Vernoux, J. F., Neveux, A., Barrez, F., and Brugeron, A.: Recherche des origines possibles de la pollution en perchlorates impactant des captages en eaux souterraines du territoire Nemours-Bourron, Final report, BRGM/RP-64840-FR, 2015.

Rouxel-Davis, E., Cordonnier, G., and Dachy, S.: Synthèse des études menées sur le bassin versant de Couraux (Marne), BRGM, 2002.

Sturchio, N. C., Beloso Jr., A., Heraty, L. J., Wheatcraft, S., and Schumer, R.: Isotopic tracing of perchlorate sources in groundwater from Pomona, California, Appl. Geochem., 43, 80–87, 2014.

Urbansky, E. T.: Perchlorate chemistry: implications for analysis and remediation, Bioremediat. J., 2, 81–95, 1998.

Urbansky, E. T., Brown, S. K., Magnuson, M. L., and Kelty, C. A.: Perchlorate levels in samples of sodium nitrate fertilizer derived from Chilean caliche, Environ. Pollut., 112, 299–302, 2001.

USEPA: Drinking water contaminant list, EPA document No. 815-F-98-002, GPO, Washington, DC, 1998.

Vachier, P., Dever, L., and Fontes, J. C.: Mouvements de l'eau dans la zone non satureìe et alimentation de la nappe de la craie de champagne (France): Isotope Techniques in Water Resources Development, IAEA Conf., Vienna, 30 March–3 April 1987, 367–379, 1987.

Wu, Q., Zhang, T., Sun, H., and Kannan, K.: Perchlorate in tap water, ground water, surface waters and bottled water from China and its association with other inorganic anions and with disinfection byproducts, Arch. Environ. Con. Tox., 58, 543–50, 2010.

https://doi.org/10.5194/piahs-379-119-2018

Study on reservoir time-varying design flood of inflow based on Poisson process with time-dependent parameters

Jiqing Li, Jing Huang, and Jianchang Li

North China Electric Power University, Beijing, 102206, China

Correspondence: Jiqing Li (jqli6688@163.com)

Received: 29 December 2017 – Accepted: 25 January 2018 – Published: 5 June 2018

Abstract. The time-varying design flood can make full use of the measured data, which can provide the reservoir with the basis of both flood control and operation scheduling. This paper adopts peak over threshold method for flood sampling in unit periods and Poisson process with time-dependent parameters model for simulation of reservoirs time-varying design flood. Considering the relationship between the model parameters and hypothesis, this paper presents the over-threshold intensity, the fitting degree of Poisson distribution and the design flood parameters are the time-varying design flood unit period and threshold discriminant basis, deduced Longyangxia reservoir time-varying design flood process at 9 kinds of design frequencies. The time-varying design flood of inflow is closer to the reservoir actual inflow conditions, which can be used to adjust the operating water level in flood season and make plans for resource utilization of flood in the basin.

1 Introduction

Design floods are the key basis for reservoir designing and operating. Whether it can fully reflect the law of flood evolution in flood season is of great significance to the flood control and development of reservoirs. During the process of reservoir operation, it is necessary to determine the design floods which closes to the reservoir actual water conditions. The time-varying design flood is a dynamic design flood, which considers the shortage of water resources and followed the law of flood evolution in reservoir operation. The time-varying design flood is close to the actual runoff process of reservoir. Compared with the annual maximum design flood, it can not only dynamic reflect changes of inflow, but also do not require artificial smoothing. The method of deriving the time-varying design flood is a combination of the peak over threshold method (Wang, 1999) and the model of Poisson process with time-dependent parameters (Wang, 2000). In hydrology, the key of limiting the accuracy of design flood calculation is too little data available for the sample (Li et al., 2016). On this basis, this paper adopts the method of peak over threshold for flood sampling and takes fixed days as a calculation period. Compared with the traditional application of peak over threshold method, this way can well expand the capacity of sample data, dynamically reflect the actual flood characteristics of reservoir and improve the accuracy of time-varying design flood. This method is a good way to make up for the lack of research on time-varying design flood at home and abroad.

2 Theoretical model

Optimizing design floods and hydrological forecasting are two most effective ways to optimize flood control scheduling and improve profitability, attracting more and more scholars' attention. This paper improves the method of designing the flood process line from the aspects of sampling method and frequency curve, making it have the characteristic of time-varying and meet the requirements of time-varying design flood.

2.1 Peaks over threshold

In the process of deducing the design flood, the traditional annual maximum method only considers the distribution of extreme values in the measured data. It cannot reflect the actual characteristics of the flood and is not suitable for the period of reservoir operation. The peak over threshold method often takes one year as a calculation period. The traditional way of using the method of peak over threshold not only increased sample size but also enhanced the physical correlation among sample individuals (Eastoe, 2012). In order to meet the assumptions of independence and the series are drawn from the same distribution (Lang et al., 1999), it need to control the over-threshold intensity in ∼2–3 times/year (Zhou et al., 2011). Thus, the design flood process can only indicate the general trend of the inflow process. Applying the method of peak over threshold in a calculation period of fixed days can significantly enhance its own advantages.

2.2 Probability model

The flood over-threshold series is a series of the part of all individuals (flood peak, a fixed period floods) which exceed the threshold. It can be expressed as:

$$Y = X - y_0 \tag{1}$$

Where Y is the over-threshold; X is a random variable in the flood; y_0 is the threshold, which is selected based on the design frequency and the calculation period.

2.3 Threshold selection

When performing the over-threshold series frequency analysis, the samples must satisfy the assumptions of independence and the series are drawn from the same distribution. The assumptions can be achieved by choosing a reasonable threshold (y_0). At present, the criterion of threshold selection is mainly the American Water Resources Association Standard (Zhou and Huang, 2014). But, it is only suitable for the calculation period of one year. Considering the threshold selection criteria and the model assumptions, this paper proposes a way to solve problems of the threshold selection and the rationality of the model established. The way is to detect whether the intensity of over-threshold occurred at any given time always obeys Poisson distribution at a certain set threshold. Considering the selection of the threshold value is related to the design frequency and the calculation period of flood, this paper proposes a parallel judgment that the design parameters should be combined with two or more flood parameters (the peak flow, the maximum flood volume of fixed period) of reservoirs in comparative analysis.

2.4 Poisson process with time-dependent parameters

Compared to the traditional method of designing flood process, the advantage of the Poisson process with time-dependent parameters model lies in the two time-dependent parameters (over-threshold average b_t and over-threshold intensity r_t). The dependent parameter is the embodiment of the time-varying characteristics of the model. It makes the model can fully reflect the actual hydrological characteristics of the reservoir, and without the need to smooth.

The total duration of the study is divided into t periods ($t = 1, 2, \ldots, n$). The number of occurrences per unit time K is random. K is assumed to follow the Poisson distribution (Önöz and Bayazit, 2001). The probability of occurrence of k-times over-threshold per unit time can be expressed as:

$$P_{r_t}\{K = k\} = \frac{r_t^k}{k!} \times e^{r_t} \ (k = 1, 2, 3, \ldots) \tag{2}$$

Where: r_t is a parameter, which means the intensity of the over-threshold series occurrences over the period t.

Further, assuming the over-threshold series follows an exponential distribution (Wang, 2000), the frequency curves can be expressed as:

$$\mathrm{P}(y) = P_{r_t}\left\{Y \geq y \vdots y \geq y_0\right\} = e^{\frac{-y}{b_t}} \tag{3}$$

Where: b_t is a parameter, which means the average number of the over-threshold series over the period t; y is the value of over-threshold. Because of the over-threshold intensity conforms to the Poisson distribution, its recurrence period $E(T)$ can be expressed as:

$$E(T) = \frac{1}{r_t \times P(y)} \tag{4}$$

Based on the above calculation steps, the formula of time-varying design flood can be directly given here:

$$y = b_t \times \ln[r_t \times E(T)] \tag{5}$$

Each symbol has the same meaning as above.

3 Case study

3.1 Introduction of Longyangxia reservoir

Longyangxia, a large reservoir with multi-year regulation capacity in the upper reaches of the Yellow River, is characterized by a good correlation relationship between flood peak and flood volume (Li et al., 2007). The flood season of Longyangxia reservoir is from July to September. Most of the flood peak appeared in July or September, and August was relatively low. Usually, the peak of flood appears in July is sharp and skinny. On the contrary, the peak of flood appears in September is short and fat. The duration of each flood is about 40–50 days (China Hydropower Consulting Group, Northwest Institute of survey and design, 2005). Combining the peak over threshold method, which applied in a calculation period of fixed days, with the Poisson process with time-dependent parameters model, this paper will show everyone the process of deducing the time-varying design floods of Longyangxia reservoir.

Table 1. Threshold of peak flow and 3 day flood volume.

Characteristic parameters	Design frequency (%)	Period (day)	Threshold ($m^3 s^{-1}$)
Peak discharge	1	1	905
	0.5	1	905
	0.33	1	905
	0.2	1	905
	0.1	3	975
	0.05	4	1051
	0.02	5	1131
	0.01	7	1175
	PMF	8	1201
The maximum 3-day flood volume (million m^3)		1	248

3.2 Threshold selecting

According to the flood control tasks of the Longyangxia reservoir, this paper selects 9 kinds of design frequencies peak flows and flood volume of 3 days (this paper only takes flood volume of 3-day as an example to explain the time-varying design flood volume estimation method) by using peak over threshold method for the runoff data of 1956–1997 (42 years) from 15 June to 15 October. In order to judge the reasonableness of each threshold, the K-S test was used to test the fitting degree of over-threshold intensity with Poisson distribution. However, there are still a lot of threshold values after screening by K-S tests. This paper further proposes a screening method based on the comparative analysis of the peak flow, the maximum flood volume of 15 days and the maximum flood volume of 45 days of each design frequency for Longyangxia reservoir. The final results are shown in Table 1.

As the design frequency decreases, the length of the calculation period and the threshold gradually increase (Table 1). The fitting degree between the intensity of over-threshold and the Poisson distribution determines the threshold value and the length of calculation period at a certain design frequency. The larger the design frequency, the smaller the calculation period, which is in line with the laws of nature. When the design frequency is large, the model can provide detailed daily changes of flood characteristics to users. On the contrary, when the design frequency is small, the model can provide users with a reasonable flood trend line. However, for the calculation of the flood volume of fixed period, setting one day as a calculation period can help the user visually judge the information of flood. In this way, the model can not only ensure the rationality of the result but also improve the design flood accuracy.

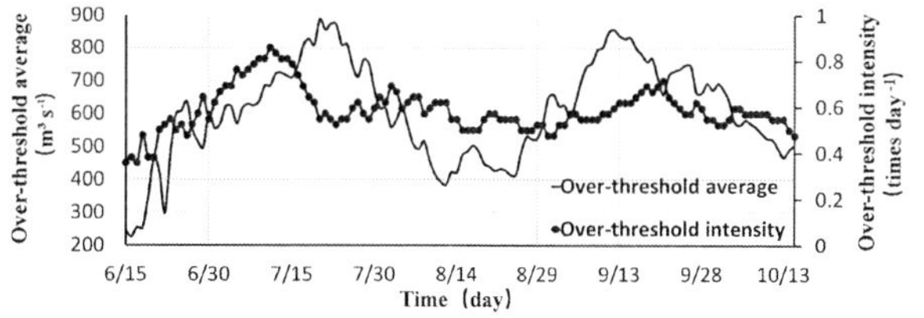

Figure 1. The change process of time-dependent parameters in design frequency of 1 %.

3.3 The time-varying design flood of Longyangxia reservoir

Take the 100 year time-varying design flood estimation process as an example, the data can be divided into 123 periods ($t = 1, 2, 3, \ldots, 123$) by the calculation period of 1 day. According to the peak flow data, this paper sets the threshold as $905\ m^3 s^{-1}$, and uses programming to filter out 3035 over-thresholds. The process of time-dependent parameters (over-threshold average b_t and over-threshold intensity r_t) changes with time, as shown in Fig. 1.

Two time-vary parameters are stochastic (Fig. 1). The peak appears in July is sharp and skinny. On the contrary, the peak appears in September is short and fat (shown by over-threshold average b_t). The trend of time-dependent parameters are in line with the characteristics of flood in the Yellow River basin. So, the time-varying parameters can characterize the actual flood characteristics of Longyangxia reservoir. After data test proved that the 9 kinds of design frequencies of time-dependent parameters are all able to reflect the actual characteristics of the flood. This conclusion is not repeated here.

The average duration of each flood is 45 days. The 9 kinds of design frequencies time-varying flood of Longyangxia reservoir shown in Fig. 2.

Considering the information given in Table 1 and Fig. 2, as the design frequency decreases from 1 % to PMF, the calculation period of time-varying design floods increases from 1 day to 8 days. The threshold has been increased from 905 to $1201\ m^3 s^{-1}$. Maybe someone will have questions. When the design frequency is small, the model does not give a detailed flood change process line. Will this affect the accuracy of flood control calculation? After simulating the reservoir flood control test, the result given by the model will not affect the flood control process. Compared with the traditional annual maximum design flood, the flood peak of time-varying design flood appears in the second half. This is a huge challenge for reservoir flood control. However, when using time-varying design flood as the basis for reservoir operation planning, it can not only ensure dam safety but also increase reservoir efficiency.

The 9 kinds of time-varying design frequencies flood volume of the 3 day line of Longyangxia, shown in Fig. 3.

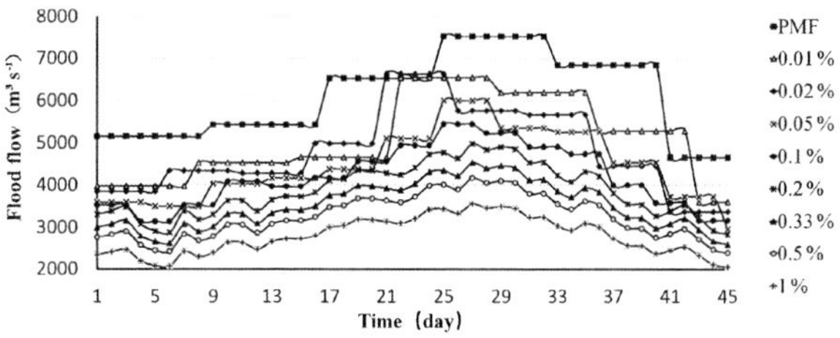

Figure 2. The change process of time-varying design flood at 9 kinds of design frequencies.

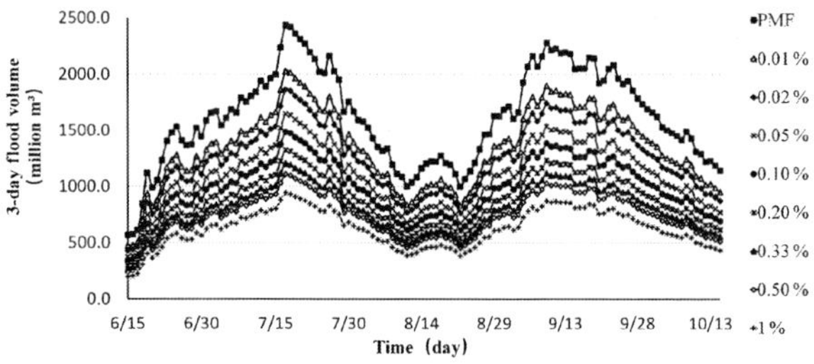

Figure 3. The change process of 3 day time-varying design flood volume at 9 kinds of design frequencies.

The time-varying design flood volume of the 3 days is in accordance with the actual hydrological characteristics of the Yellow River Basin (Fig. 3). When the date and the fixed period is known, the fixed period of time-varying design flood volume process can be directly used to judge the magnitude of flood. For example, on 1 July, when the 3 day flood volume of the reservoir is between 974 and 1079 million m^3, the flood can be regarded as reached the 1000 year design flood standards. So, the schedule of reservoir flood control can be prepared according the 1000 year design flood. In order to improve the accuracy, users can combine the time-varying design flood curve of peak flow with the time-varying design flood curve of flood volume in fixed periods, to make a comprehensive judgment of the magnitude of inflow.

4 Conclusion

This paper sets fixed days as the calculation period of the peak over threshold method, which greatly enhances its own advantages. Such as, this way makes full use of the measured data, greatly increases the sample size, and provides a basis for describing the flood features of the river basin. For the new application way of the peak over threshold method, this paper puts forward the principle of threshold selection. Compared with the traditional design flood, the time-varying design flood of inflow has the advantage of deduced flood process closer to the actual situation, and does not require manual smoothing.

Under the background of the global water resources shortage, it is more reasonable to design the reservoir operation plan based on the time-varying design flood. Because it can fully reflect the reservoir flood situation. In addition, the time-varying design flood can also be used for the research on the utilization of flood resources, such as the hydrological forecast, the adjustment of reservoir operation levels in flood season and so on.

Data availability. The data are not publicly accessible because they related to the reservoir distribution of benefits and other issues.

Competing interests. The authors declare that they have no conflict of interest.

Special issue statement. This article is part of the special issue "Innovative water resources management – understanding and balancing interactions between humankind and nature". It is a result of the 8th International Water Resources Management Conference of ICWRS, Beijing, China, 13–15 June 2018.

Acknowledgements. This study was financially supported by National Key Projects of China "Water resources efficient development and utilization" (2017YFC0405900, 2016YFC0402208, 2016YFC0401903) and National Natural Science Foundation of China (No. 51641901). The author would like to give special thanks to anonymous reviewers.

Edited by: Dingzhi Peng
Reviewed by: two anonymous referees

References

China Hydropower Consulting Group, Northwest Institute of survey and design: Study on the Recent Dispatch and Utilization of Longyangxia-Liujiaxia Reservoir, R, 2005.

Eastoe, E. F.: Modelling the distribution of the cluster maxima of exceedances of subasymptotic thresholds, J. Biometrika, 99, 43–55, https://doi.org/10.1093/biomet/asr078, 2012.

Lang, M., Ouarda, T. B. M. J., and Bobée, B.: Towards operational guidelines for over-threshold modelling, J. Hydrol., 255, 103–117, https://doi.org/10.1016/S0022-1694(99)00167-5, 1999.

Li, D., Guo, S. L., and Yin, J. B.: PDS/GP model with variable parameters and its application for flood frequency analysis, J. Hydraul. Eng., 47, 1269–1276, https://doi.org/10.13243/j.cnki.slxb.20151322, 2016.

Li, J. Q., Yao, Z. Z., Jia, H. S., and Ji, C. M.: Feasibility argumentation on the dynamical flood control limited water level of Liujiaxia reservoir in flood season, J. Hydraul. Eng., 5, 1–6, https://doi.org/10.3969/j.issn.1003-1243.2007.05.001, 2007.

Önöz, B. and Bayazit, M.: Effect of the occurrence process of the peaks over threshold on the flood estimates, J. Hydrol., 244, 86–96, https://doi.org/10.1016/S0022-1694(01)00330-4, 2001.

Wang, S. X.: Flood frequency analysis of PDS, Yangtze River, 30, 23–15, https://doi.org/10.16232/j.cnki.1001-4179.1999.08.010, 1999.

Wang, S. X.: Poisson process with time-dependent parameters model, J. Hydropower and New Energy, 39, 18–21, https://doi.org/10.13622/j.cnki.cn42-1800/tv.2000.04.006, 2000.

Zhou, C. and Huang, Y. J.: Comparison of Annual Maximum Sampling Method and Peaks over Threshold Method in Design Flood Analysis, J. Water Resources Power, 32, 172–175, 2014.

Zhou, C., Chen, Y. F., and Wei, L.: Application of fitting curve method to frequency analysis of POT floods series, J. Water Resources and Power, 29, 48–50, https://doi.org/10.3969/j.issn.1000-7709.2011.03.014, 2011.

https://doi.org/10.5194/piahs-379-125-2018

Open Access PIAHS

Reservoirs operation and water resources utilization coordination in Hongshuihe basin

Chonghao Li[1], **Kaige Chi**[2], **Bo Pang**[2], **and Hongbin Tang**[1]

[1]China Southern Power Grid Dispatching and Control Center, Guangzhou, 510520, China
[2]College of Water Sciences, Beijing Normal University, Beijing, 100875, China

Correspondence: Chonghao Li (lich@csg.cn)

Received: 28 December 2017 – Accepted: 14 January 2018 – Published: 5 June 2018

Abstract. In the recent decade, the demand for water resources has been increasing with the economic development. The reservoirs of cascade hydropower stations in Hongshuihe basin, which are constructed with a main purpose of power generation, are facing more integrated water resources utilization problem. The conflict between power generation of cascade reservoirs and flood control, shipping, environmental protection and water supply has become increasingly prominent. This paper introduces the general situation and integrated water demand of cascade reservoirs in Hongshuihe basin, and it analyses the impact of various types of integrated water demand on power generation and supply. It establishes mathematic models, constrained by various types of integrated water demand, to guide the operation and water resources utilization management of cascade reservoirs in Hongshuihe basin. Integrated water coordination mechanism of Hongshuihe basin is also introduced. It provides a technical and management guide and demonstration for cascade reservoirs operation and integrated water management at home and abroad.

1 Overview

Hongshuihe basin is located in the upper reach of Xijiang River, the mainstream of the Pearl River system. The basin, with an area of 190 000 km^2, locates in the subtropical area. The climate is mild and the rainfall is abundant. The annual rainfall is 1200 mm in its upstream and reached 1500 to 1600 mm in reaches in Guangxi province. From Tianshengqiao of Nanpanjiang in the upper reach to Datengxia of Qianjiang, the total length is 1050 km and the gap reaches 760 m. The average annual water volume reaches 133 billion m^3 at Datengxia dam site.

Hongshuihe Cascade Hydropower Base is one of the 13 major hydropower bases in China and also an important energy base in South China. It plays an important role in power supply in Guangdong and Guangxi. At present, the constructed cascade hydropower stations in Hongshuihe basin include Lubuge, Tianshengqiao I, Tianshengqiao II, Pingban, Longtan, Yantan, Dahua, Bailongtan, Letan, Qiaogong and Changzhou. Together with Datengxia Hydropower Station which is under construction, the total cascade generating capacity is 13 645 MW, and the average annual electricity generation is about 53.3 billion kWh. The sketch map of cascade hydropower stations in Hongshuihe basin is shown in Fig. 1.

In the recent decade, with the economic development the demand for water resources has been increasing rapidly. The reservoirs of cascade hydropower stations in Hongshuihe basin, which are constructed with a main purpose of power generation, are facing more and more serious integrated water use problems. The conflict between power generation of cascade reservoirs and flood control, shipping, environmental protection and water supply has become increasingly prominent. Power dispatching agencies and water resources management departments are facing new challenges on reservoirs operation. Usually multi-purposes operation method is studied and applied in the practical reservoir operations under the conflicting circumstances (Feng et al., 2014; Yang et al., 2010).

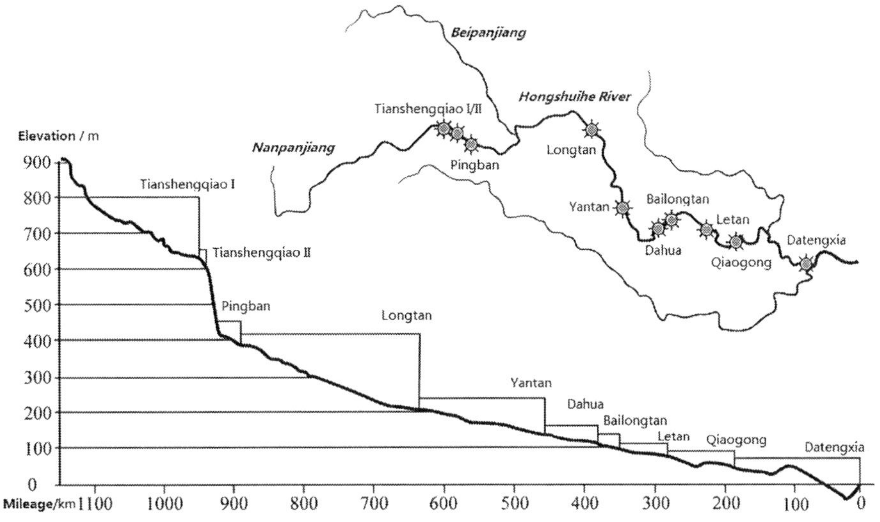

Figure 1. Sketch map of cascade hydropower stations in Hongshuihe basin.

2 Water utilization demands and their impacts on power dispatching

Water use for power generation and other purposes are all utilized through hydraulic engineering facilities such as reservoirs, tunnels and locks. Due to different purposes, conflicts often arise in the way of water use (Zhang, 1983; Feng, 1990). Take flood control for example, the flood control departments require the main reservoirs to reduce water storage and reserve enough flood control capacity to ensure flood control safety in flood season. However, the power dispatching departments need the main reservoirs to increase water storage to ensure the power supply in dry season. There are often contradictions in the way and time of water storing or releasing.

The cascade hydropower stations in Hongshuihe basin, which are constructed with a main purpose of power generation, are an important energy base of south China. Its total installed capacity accounts for about 65 % of the maximum load in Guangxi in 2017, playing an important role in the operation of the Southern Power Grid such as peak load regulation, frequency regulation, voltage control and accident backup in the power supply of Guangdong and Guangxi provinces. The impacts of various types of integrated water demands on power dispatching of cascade hydropower stations is analysed from the perspective of hydropower generation and power supply as below.

2.1 Flood control

The specific requirement of flood control is that water level of reservoir shall not exceed the flood limit water level, or outflow shall not be greater than the specified value. The impact on hydropower generation and power supply include:

1. The reservoir will lose water regulation ability if the flood limit water level is reached or exceeded, and the peak regulation and frequency modulation capacity of the power station will decrease. Water should be discharged to regulate the peak.
2. If water storage is not enough in flood season, the water capacity and power supply will be limited in dry season.

2.2 Shipping

The specific requirement of shipping is that reservoir shall discharge water in the determined flow at a specified time period to ensure the water level of the downstream channel is stable. In case of emergency, it may be necessary to increase or decrease the discharge flow significantly to meet the rescue requirements of ship accidents. The impacts on hydropower generation and power supply include:

1. There is water head loss or even water volume loss in the power station.
2. Power station can only generate power according to the determined output power, losing the capacity of peak regulation, frequency modulation and accident backup.
3. Power stations in upstream and downstream shall cooperate with each other. The cascade power stations will be limited as a whole. The operation mode of thermal power plants and other power plants need to be adjusted and the power supply reliability declined.
4. In case of emergency, a large number of power plants may need to be involved in adjustment, and the grid operation mode will be changed.

2.3 Water supply

The specific requirement of water supply is that reservoir shall discharge water in the determined flow at a specified time period, as required by shipping; or the reservoir water level and the downstream water level should not be less than the specified value in order to meet the water demand of relevant facilities. The impacts on hydropower generation and power supply include:

1. There is water head efficiency loss in the power station. The adjustable output of power stations units will be limited by the extra water supply in dry season.
2. Hydropower station may be forced to drain water during equipment maintenance period in dry season.
3. Power stations in upstream and downstream shall cooperate with each other. The cascade power stations will be limited as a whole. The operation mode of thermal power plants and other power plants needs to be adjusted.
4. Power supply reliability will be declined.

2.4 Environmental or ecological water

The specific requirement of environmental or ecological water is that reservoir shall control the outflow of reservoirs by not less than the specified lower limit so as to maintain the environmental and ecological water in the downstream. In case of emergency, it may be necessary to increase or decrease the discharge flow significantly. The impact on hydropower generation and power supply include:

1. Load peak regulation capacity of power stations and grid will be influenced in dry season.
2. In case of emergency, a large number of power plants may need to be involved in adjustment, the grid operation mode will be changed, and the power supply reliability will be declined.

What's more, power market reform is implementing in the power industry in China at present. The operation of hydropower stations depends not only on water supply, water use and power system operation requirements, but also on the results of the transactions in power market. Unplanned water use, especially water needs in emergency like water pollution accident or shipwreck rescue, may result in market breaches and economic penalties for hydropower owner as a market member. Market trading agencies may have to deal with other power plants temporarily. This may trigger a short-term sharp price increase in the power market.

3 Mathematical models

The contradiction between the integrated utilization of reservoirs and the power generation operation of cascade hydropower stations has become increasingly prominent. In order to guide the operation of cascade reservoirs in Hongshuihe basin and to provide decision support for the daily operation of hydropower stations, this paper presents mathematical models for the integrated utilization of cascade reservoirs. The models use various types of integrated water demand as constraints, including a long-term model and a short-term model according to the different operation periods.

The long-term dispatch model is mainly used for monthly and annual operation plan making of cascade reservoirs in the Hongshuihe basin. The model maximizes the total power generation of cascade, taking the operation requirements of reservoirs, various types of integrated water demand (like outflow and water level control) and power demands of each period of the power system as constraints.

The objective function:

$$E = \mathrm{Max} \sum_{i=1}^{N} \sum_{j=1}^{T} K_i \cdot Q_{i,j} \cdot H_{i,j} \cdot \Delta t \tag{1}$$

Constraints:

$$\begin{cases} \text{Water Level}: & Z_{i,\min} \le Z_{i,j} \le Z_{i,\max} \\ \text{Generating Flow}: & Q_{i,\min} \le Q_{i,j} \le Q_{i,\max} \\ \text{Power Capacity}: & P_{i,\min} \le P_{i,j} \le P_{i,\max} \end{cases} \tag{2}$$

In addition, it should obey the water balance equation:

$$V_{i,j+1} = V_{i,j} + (I_{i,j} Q_{i,j} q_{i,j}) \cdot \Delta t \tag{3}$$

Where E is the total power generation of the cascade reservoirs in the basin; K_i is the output coefficient of the hydropower station i; and $Q_{i,j}$, $H_{i,j}$, $Z_{i,j}$, $P_{i,j}$, $I_{i,j}$, $q_{i,j}$ and $V_{i,j}$ respectively represent for the generating water flow, water head, reservoir water level, power output, inflow, discharge and reservoir water storage of the hydropower station i during time period j.

The short-term dispatch model is mainly used for daily operation schedule of cascade reservoirs in the basin. It aims to minimize the total generating water consumption of the cascade hydropower stations, considering the determined intraday hourly power demand. It takes the operation requirements of reservoirs, various types of integrated water demand (like outflow and water level control) and hourly power grid demand as constraints.

The objective function:

$$W = \mathrm{Min} \sum_{i=1}^{N} \sum_{j=1}^{T} Q_{i,j} \cdot \Delta t \tag{4}$$

The constraints are basically the same with those in the long-term model. While the time period is different.

In addition, the water flow equation among cascade reservoirs, which used to calculate the inflow of reservoir from

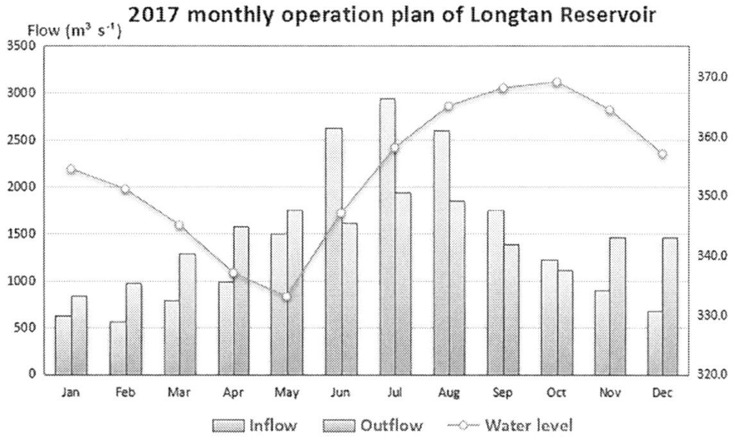

Figure 2. 2017 Monthly operation plan of Longtan reservoir.

its upstream reservoir outflow considering the flow lag time, should be obeyed:

$$I_{i,j} = Q_{i-1,j-dT} + R_{i,j} \tag{5}$$

Where W is the total generating water consumption of the cascade hydropower stations in the basin; $I_{i,j}$, $Q_{i,j}$ and $R_{i,j}$ are the average inflow, quantity of generating water flow and cascade flows between hydropower stations of the hydropower station i during period j respectively; and dT is the water flow lag time from the hydropower station $i-1$ to the hydropower station i.

Reservoirs operation is a typical multi-objective dynamic programming. Usually large system decomposition and coordination methods or dynamic programming algorithms are used in the optimizing solutions (Wang and Zhao, 2003; Shao et al., 1998; Mei et al., 2007; Huang et al., 2014).

Based on the mathematical models above, software of cascade hydropower dispatching system in Hongshuihe basin has been developed using Java and Oracle database. It provides good technical support for reservoirs operation decision-making and daily, monthly or annual plan making of hydropower operation. Figure 2 shows the 2017 long-term optimizing calculation results of Longtan reservoir from the dispatching system.

4 Water utilization coordination management

The dispatch management of basin cascade reservoirs involves many industries and departments like water, power, shipping and environmental protection. In fact, it requires not only reservoir dispatch optimization techniques, but also needs a sound and comprehensive water coordination mechanism to strengthen the dispatch and management of reservoirs in Hongshuihe basin. In order to further improve the integrated utilization of water resources and the overall benefits of the whole basin, a water integrated utilization working group, involving many industries and departments like power dispatching agency, flood control agency, waterway management bureau, environmental protection departments, industrial and agricultural water use organizations and hydropower stations, was established. A constant communication and coordination mechanism has formed to share the basin water information timely and resolve the contradictions in water use promptly, which effectively promote the security and efficiency of integrated water utilization. In addition, measures like the tracing and monitoring of integrated water use in the basin, water utilization benefit assessment afterwards furtherly improve the scientific management of water resources utilization in the basin.

Since 2010, the power supply and water demand of all parties have been well coordinated in a series of major integrated water using events through active coordination and scientific operation, such as basin flood control, water transfer in winter and spring, shipping solution of Changzhou navigation lock in dry season and dispose of water pollution accidents. Good results and significant social benefits have been achieved.

5 Summary

This paper introduces the integrated water demand of the cascade reservoirs in Hongshuihe basin, analysing their impact on power generation and supply, and sets up mathematic models, which are constrained by integrated water demand, for cascade reservoirs integrated operation in Hongshuihe basin. It also introduces the basin integrated water coordination mechanism in use and its active role in improving the water resources utilization efficiency in the basin, providing a technical and management demonstration for cascade reservoirs operation and integrated water management at home and abroad.

Data availability. The data are not publicly accessible as the reservoir operation case in this paper is production and operating relevant activity, and the data is under commercial privacy protection.

Competing interests. The authors declare that they have no conflict of interest.

Special issue statement. This article is part of the special issue "Innovative water resources management – understanding and balancing interactions between humankind and nature". It is a result of the 8th International Water Resources Management Conference of ICWRS, Beijing, China, 13–15 June 2018.

Acknowledgements. This research is supported by National Key Research and Development Program (during the 13th Five-year Plan) under Grant No.2016YFC0401309, Ministry of Science and Technology, PRC.

Edited by: Dingzhi Peng
Reviewed by: two anonymous referees

References

Feng, S.: Multi-objective Decision Theory and Its Application, Human: Huazhong University of Science and Technology Press, 1990.

Feng, Y., Tian, Y., Han, J., and Cui, Y.: Reservoir Ecological Power Optimal Operation and Its Applied Research, China Rural Water and Hydropower, 0, 137–141, 2014.

Huang, C., Wang, Z., Li, S., and Chen, S.: A Multi-reservoir Operation Optimization Model and Application in the Upper Yangtze River Basin I. Principle and Solution of the Model, J. Hydraul. Eng., 45, 1009–1018, 2014.

Mei, Y., Xiong, Y., and Chen, L.: A dynamic programming method for the multi-purpose operation of cascade reservoirs, Journal of Hydroelectric Engineering, 26, 1–4, 2007.

Shao, D., Xia, J., and Sun, Z.: Study on Real Optimal Operation Model for the Reservoir of Water Resources Multi-Objective Integrated Utilization, International Journal Hydroelectric Energy, 16, 7–11, 1998.

Wang, X. and Zhao, R.: Theory of Optimal Reservoir Dispatching with Multi-objective and Its Application, J. Hydraul. Eng., 104–109, 2003.

Yang, F., Zhang, X., and Tan, G.: Elementary Study of Multiobjective Operation Model Considering Ecological Operation, Engineering Journal of Wuhan University, 43, 433–437, 2010.

Zhang, Y.: Hydropower Station Optimal Operation, China Water & Power Press, Beijing, 1983.

https://doi.org/10.5194/piahs-379-131-2018

Framework for quantifying flow and sediment yield to diagnose and solve the aggradation problem of an ungauged catchment

Sagar Kumar Tamang[1], **Wenjun Song**[2], **Xing Fang**[2], **Jose Vasconcelos**[2], **and J. Brian Anderson**[2]

[1]Saint Anthony Falls Laboratory, University of Minnesota, Minneapolis, MN 55414, USA
[2]Department of Civil Engineering, Auburn University, Auburn, AL 36849, USA

Correspondence: Xing Fang (xing.fang@auburn.edu)

Received: 30 December 2017 – Accepted: 21 January 2018 – Published: 5 June 2018

Abstract. Estimating sediment deposition in a stream, a standard procedure for dealing with aggradation problem is complicated in an ungauged catchment due to the absence of necessary flow data. A serious aggradation problem within an ungauged catchment in Alabama, USA, blocked the conveyance of a bridge, reducing the clearance under the bridge from several feet to a couple of inches. A study of historical aerial imageries showed deforestation in the catchment by a significant amount over a period consistent with the first identification of the problem. To further diagnose the aggradation problem, due to the lack of any gauging stations, local rainfall, flow, and sediment measurements were attempted. However, due to the difficulty of installing an area-velocity sensor in an actively aggrading stream, the parameter transfer process for a hydrologic model was adopted to understand/estimate streamflow. Simulated discharge combined with erosion parameters of MUSLE (modified universal soil loss equation) helped in the estimation of sediment yield of the catchment. Sediment yield for the catchment showed a significant increase in recent years. A two-dimensional hydraulic model was developed at the bridge site to examine potential engineering strategies to wash sediments off and mitigate further aggradation. This study is to quantify the increase of sediment yield in an ungauged catchment due to land cover changes and other contributing factors and develop strategies and recommendations for preventing future aggradation in the vicinity of the bridge.

1 Introduction

Soil erosion is the process of degradation of the top layer of soils by mechanical forces of wind or water. About USD 30–40 billion is lost in the US alone due to on and off-site effects such as loss in agricultural productivity, blockage of conveyance of irrigation channel, etc. (Morgan, 2009). One of the most important dataset for modeling soil erosion and quantifying the sediment yield is the streamflow. Streamflow data can be obtained from gauges installed in a stream or be simulated/projected using a hydrologic model. Even in the case of a hydrologic model, streamflow data is necessary as the model's resemblance to reality can be increased through calibration with the existing gauged data (Sivapalan, 2003). However, gauged data is not available in all streams due to financial constraints and installation difficulties. Runoff response prediction in an ungauged catchment remains a complex problem. Considering the scope and importance of the prediction in the ungauged basin (PUB), the International Association of Hydrological Sciences put forward PUB as an initiative for the decade of 2003–2012.

2 Study Area and Input Data

Soapstone Branch, a tributary of the Little Choctawhatchee River located in Dale County, Alabama (AL) has been experiencing a serious aggradation problem (Fig. 1). This problem was first identified in 2013 and aggravated over time reducing the conveyance of the Dean Road bridge from 2.44 m (8 ft.) to a couple of inches by 2014. A detailed study of historical aerial imageries for the Soapstone branch catchment revealed significant land cover changes over a period of sev-

Figure 1. Location of Dean Road Bridge and Aggradation Problem.

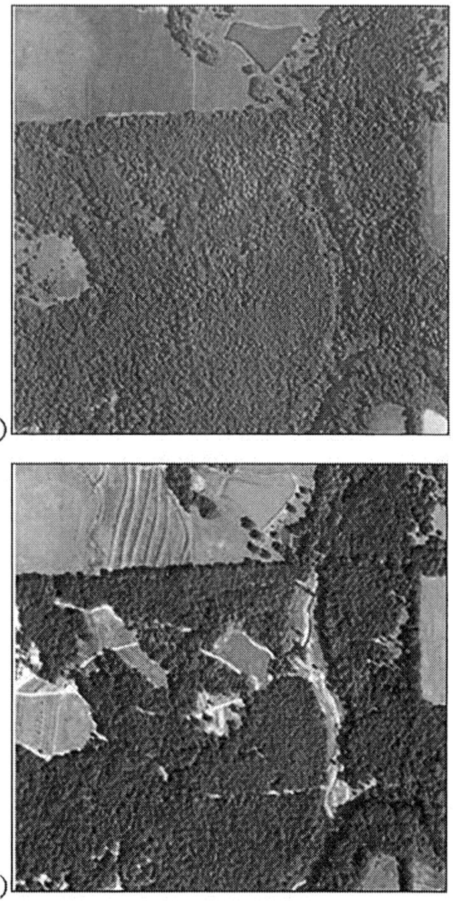

Figure 2. Aerial Imagery of a Portion of Soapstone Branch in **(a)** 2011 and **(b)** 2015.

eral years. In the period from 2011 to 2015, change in land cover due to clear cutting of the trees in the vicinity of the stream channel is clearly visible (Fig. 2). For understanding these effects on the process of aggradation and for quantifying the amount of sediments, a hydrological model together with a sediment model was necessary.

Installation attempts of the area-velocity sensor to record the streamflow data as required in the calibration of the hydrological model failed due to severe aggradation occurring in the stream. The sensor was buried after each storm event and was unable to capture flow dynamics. Therefore, a need for parameter transfer of a hydrological model from a nearby catchment was felt.

It has been well established that among different options for selecting donor catchment for parameter transfer process, spatial proximity performs best. Choctawhatchee river catchment draining near Newton, AL and covering an area of 1776.7 sq. km. (686 sq. miles) was selected as donor catchment for parameter transfer process to Soapstone Branch catchment (7 sq. km. (2.7 sq. miles)). Also, for verifying the parameter transfer process of Soil Moisture accounting (SMA) model, Double Bridges Creek catchment draining near Enterprise, AL and covering an area of 54.4 sq. km. (21 sq. miles) was also selected. Both of the catchments were selected based on their spatial proximity and gauged data availability.

Different types of data viz. digital elevation model (DEM from AlabamaView), land cover data (National Land Cover Database–NLCD), soil data (Soil Survey Geographic Database–SSURGO), streamflow data (U.S. Geological Survey), daily evapotranspiration data (from National Oceanic and Atmospheric Administration–NOAA), and precipitation data (from U.S. Climate Reference Network's quality controlled dataset; Auburn University Mesonet; Local Climatological data from NOAA) were obtained for donor and receiver catchments. Three rainfall stations viz. Troy, Union Springs, and Dothan were used for donor catchment whereas, for receiver catchments, rainfall data from Dothan was used due to data availability and spatial proximity (Tamang, 2017).

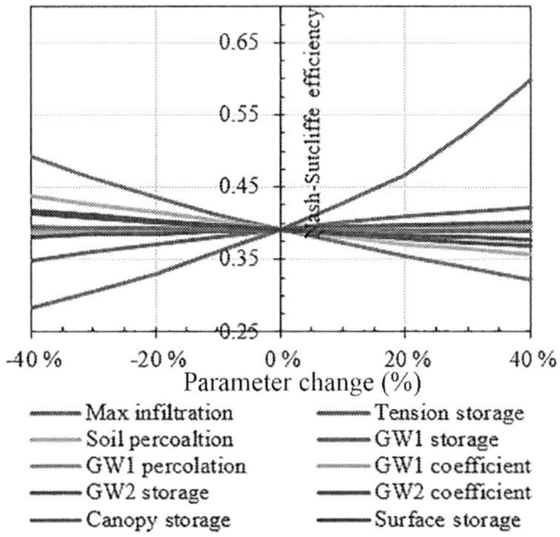

Figure 3. Nash-Sutcliffe Efficiency of the Model Output due to Percent Change in Parameter Values.

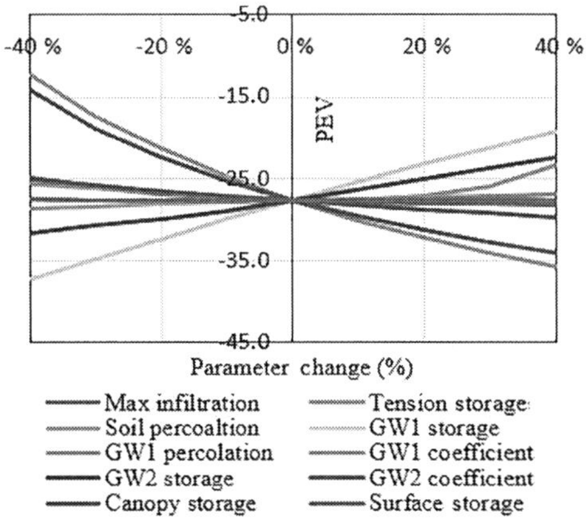

Figure 4. Percent Error in Volume of the Model Output due to Percent Change in Parameter Values.

3 Model Overview

3.1 Hydrologic Model

Hydrologic Engineering Center's Hydrologic Modeling System (HEC-HMS, Feldman, 2000) was developed by US Army Corps of Engineers. HEC-HMS consists of four different models to represent each component of the runoff process viz. models to compute runoff volume, direct runoff, baseflow, and channel routing. HEC-HMS is capable of performing both event and continuous hydrologic simulations. The Soil Moisture Accounting (SMA) algorithm is a continuous, semi-distributed and empirical loss method available within HEC-HMS. It consists of series of different layers for the movement of water within the land-based components (Bennett, 1998).

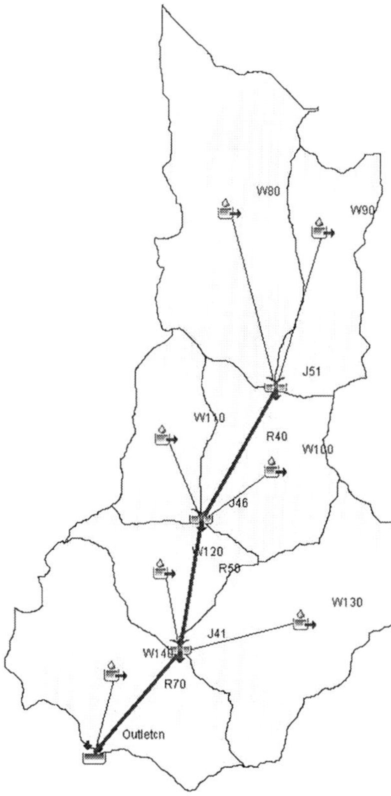

Figure 5. HEC-HMS Model Setup for Soapstone Branch Catchment.

3.2 Modified Universal Soil Loss Equation

One of the most widely adopted methods for estimating soil erosion worldwide is Universal Soil Loss Equation (USLE). Modified Universal Soil Loss Equation (MUSLE) is an advancement over USLE, developed by replacing the rainfall erosivity factor with the runoff energy factor (Williams, 1975). Unlike USLE for annual sediment application, MUSLE is an event-based soil loss model which considers the effect of runoff energy on generating sediment. The mathematical expression for MUSLE is given by:

$$S = 95 \times (Q \times q_p)^{0.56} \times K \times LS \times C \times P \quad (1)$$

Where S is the sediment yield in tons, Q is the runoff volume in acre-ft, q_p is the peak discharge ($ft^3\,s^{-1}$ or cfs), K is the soil erodibility factor, LS is the topographic factor (ft), C is the cover management factor, and P is the support practice factor.

3.3 Hydraulic Model

HEC's River Analysis System (HEC-RAS) version 5.0.3 (Brunner, 2016) was used in the present study. The model enables the simulation of the river using the two-dimensional

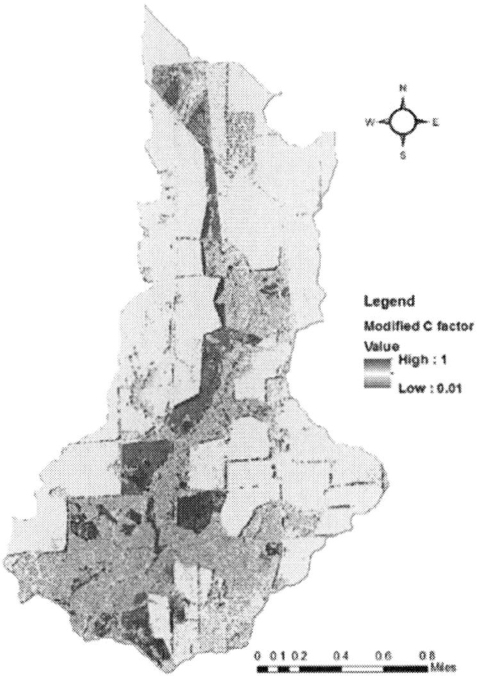

Figure 6. The C Factor Raster Grid of the Soapstone Branch Catchment in 2015.

(2-D) flow equations, also referred to as the shallow water equations. Inflows can be admitted through boundaries at the edge of the solution domain or even through direct rainfall. Use of 2-D solution is particularly adequate to consider effects of river meandering, proposed alternatives for stream modification and changes of velocity magnitude across the Dean-Road bridge cross-section.

4 Methodology

4.1 Catchment Delineation

Catchment delineation for the study area was performed using HEC's Geospatial Hydrologic Modeling Extension: HEC-GeoHMS (Doan, 2000). A stream definition of 0.4 sq. km was selected by using a trial and error method to match the generated streams with the natural streams. This procedure divided the donor catchment into 15 subcatchments and seven subcatchments for Soapstone Branch watershed.

4.2 Land Cover Map Generation

Unsupervised classification using the Iterative Self-Organizing Data Analysis Technique (ISODATA, Nellis et al., 1998) with 40 classes of the similar spectral signature was applied to 2011 and 2015 National Agriculture Imagery Program (NAIP) dataset using ERDAS IMAGINE 2016

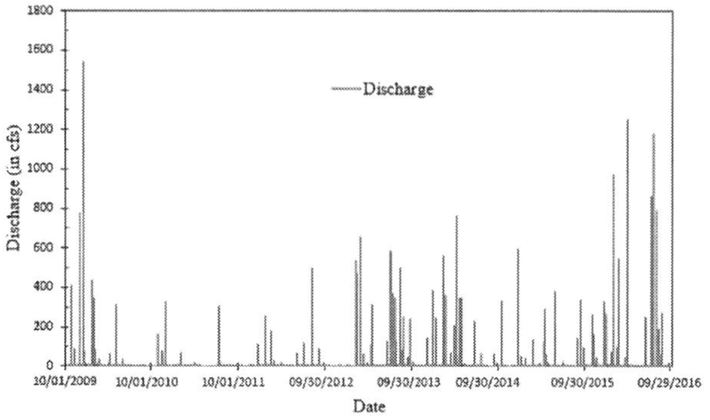

Figure 7. SMA Model Result of Soapstone Branch Catchment (October 2009–September 2016).

software. Using multispectral NAIP imagery, these 40 classes were categorized into 4 different land use types viz. forest (34.4 % in 2011; 28.7 % in 2015), agricultural land (56.3 % in 2011; 58.1 % in 2015), rangeland (8.7 % in 2011; 12.6 % in 2015), and water (0.6 % in 2011 and 2015). For improving the accuracy of land cover classification, the cluster busting technique was applied (Civco et al., 2002).

4.3 Parameter Transfer from Donor Catchment

Sensitivity analysis is an important tool for decision makers to identify sensitive or important variables (Pannel, 1997). Therefore, a local sensitivity analysis was performed by varying the values of parameters $\pm 10\,\%$ from initial estimates and their effect on Nash-Sutcliffe model efficiency (NSE) and percent error in volume (PEV) is shown in Figs. 3 and 4. On the basis of their sensitivities, parameters were ranked and then highly sensitive thus important parameters together with calibration parameters were then transferred to the receiver catchment. Finally, areal average values of highly sensitive and calibration parameters from the donor catchment were then applied to each of the seven subbasins (Fig. 5) of the Soapstone Branch catchment.

4.4 HEC-HMS Model Setup

HEC-GeoHMS was used for background map development and creating the distributed-basin schematic model file for each of three study catchments. It was also used in checking of errors in catchment model development and connectivity of streams. HEC-HMS model setup for Soapstone branch catchment is shown in Fig. 5 as an example.

4.5 MUSLE Model Development

K factor was obtained from soil data available from SSURGO using the online USDA soil data viewer (https://www.nrcs.usda.gov/wps/portal/nrcs/detail/soils/home/?cid=nrcs142p2_053620). LS factor was calculated

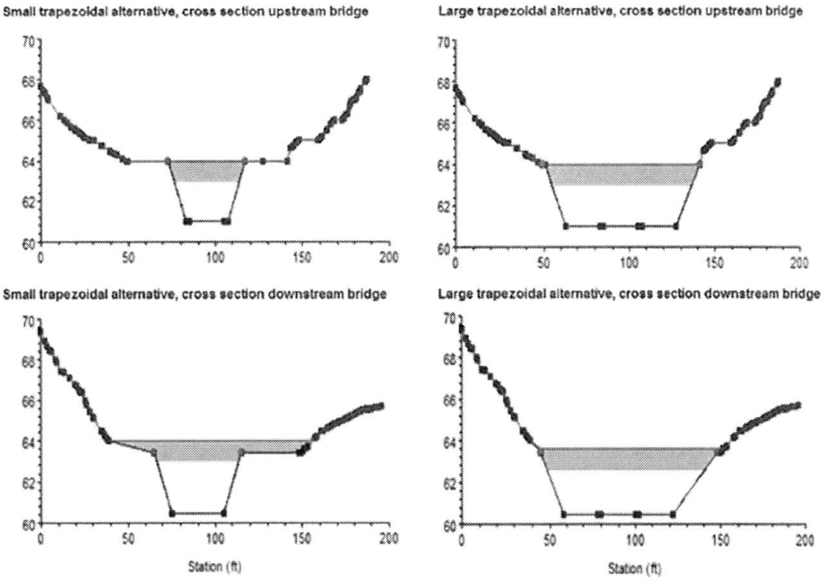

Figure 8. Geometric alternatives of stream modification considered in the HEC-RAS simulation.

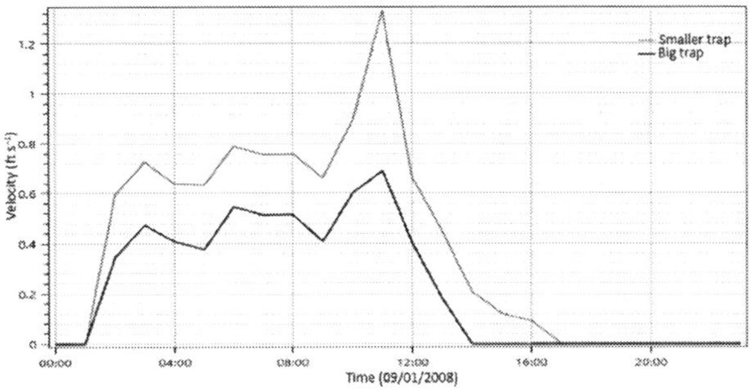

Figure 9. Effects of Stream Modification in the Water Velocity at the Dean-Road Bridge Cross Section Calculated by HEC-RAS.

from slope length and slope gradient values obtained from 3.28 m (1 ft.) resolution DEM of the catchment. The C factor was developed from land cover maps provided by NLCD. However, the most recent available NLCD data is for 2011 and unable to account for land cover changes that occurred during the year 2011–2015. NAIP provides multispectral imagery at a spatial resolution of 1 m every 2 years making it suitable for our study. The C factor for the catchment was developed using normalized difference vegetation index (Gitas et al., 2009) for two different years 2011 and 2015 (Fig. 6) to account for the land cover change in the catchment. Since NAIP imagery was collected during the agricultural growing season (August–September) every two years in the continental US, the computed C factor values are conservative due to the absence of specific information on agricultural practices in the catchment, a conservative value of 1 was selected for P factor.

4.6 HEC-RAS Model Setup

The DEM for Dale County, AL served as a base for the needed elevations for the HEC-RAS 5 model. Mesh sizes were selected so that the typical 2-D cell was around 1.5 to 2 m wide and there would be at least five cells across the bridge cross section. Recorded stage levels were used for model calibration and assessment. Through modification of DEMs using an algorithm implemented in Excel VBA, various alternatives of stream bed elevation near the bridge were considered, emulating the possible strategies for stream modification

5 Results and Discussions

5.1 Streamflow Simulation

A calibration period of three years from October 2009–September 2012 and a validation period of three years from October 2012–September 2015 were adopted for the donor catchment. NSE values of 0.73 during calibration and 0.63 during validation period were obtained which are rated as good and satisfactory performance by a continuous hydrologic model (Moriasi et al., 2007), respectively. The receiver catchment (Double Bridges Creek) is a gauged catchment, however to test the efficiency of the model parameter transfer process, it was assumed as an ungauged catchment for the model parameter transfer. The discharge was then simulated

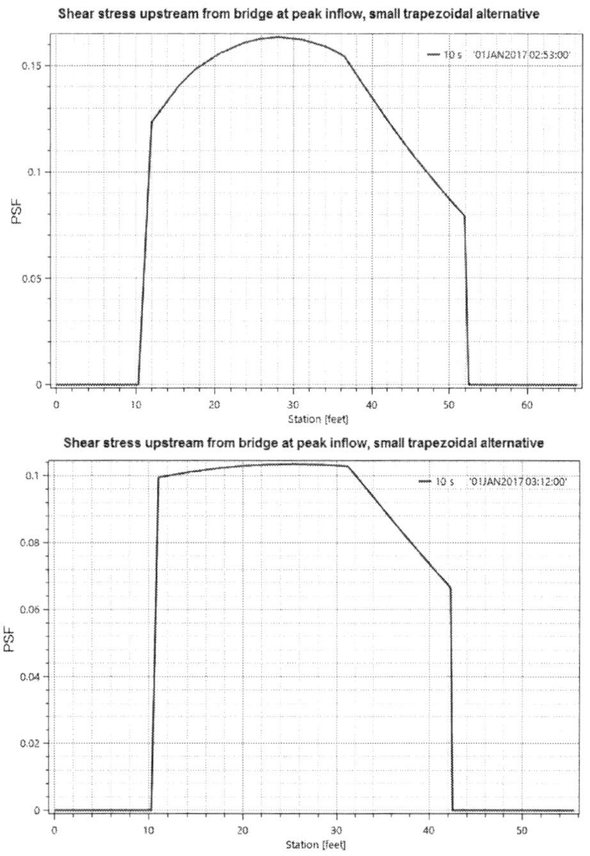

Figure 10. Comparison between Shear Stresses in the Channel Obtained with the Large Trapezoidal and Small Trapezoidal Cross Section Alternatives.

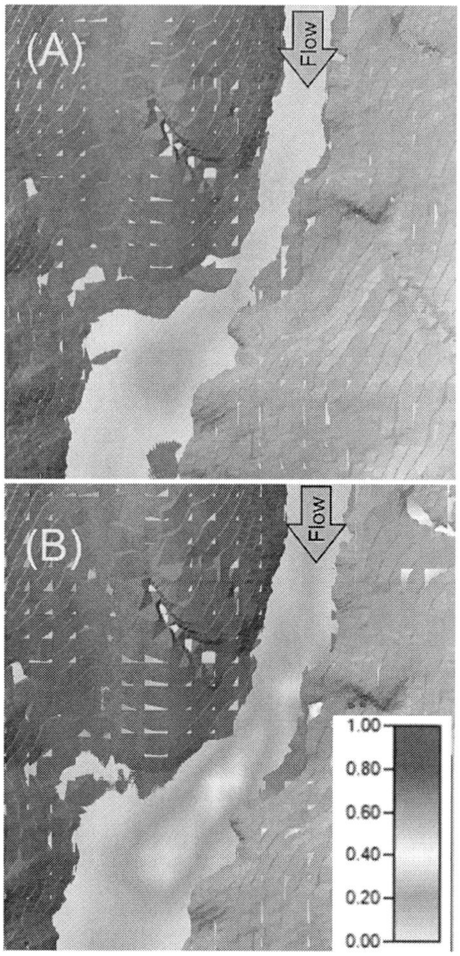

Figure 11. Comparison between Shear Stresses in the Channel Obtained with the Large Trapezoidal **(a)** and Small Trapezoidal **(b)** Cross Section Alternatives.

for the receiver catchment during a transfer validation period of three years from October 2009–September 2012. NSE value of 0.64 was obtained, which is rated as satisfactory performance for the continuous hydrological model (Moriasi et al., 2007). In 2010, some parts of Alabama experienced severe to extreme drought. Cumulative annual rainfall during this year varied from 508–1778 mm (20–70 in.). throughout Alabama (Tamang, 2017). Due to the fewer number of rainfall stations in both donor and receiver catchment and the discrepancy during this year introduced by the coarser spatial resolution of precipitation data have reduced the NSE values. The semi-distributed HEC-HMS model for Soapstone Branch catchment was then run from October 2009–September 2016 (Fig. 7) after transferring parameters from the donor catchment. During the study period, the annual average precipitation was 1361.4 mm (53.6 in.) with a standard deviation of 365.8 mm (14.4 in.) and range of 922 mm (36.3 in.). An initial warmup period of 9 months was selected to minimize the effects of initial estimated moisture value on the simulation. As seen from Fig. 7, streamflow was low and had fewer storm events in 2010 and 2011 whereas, the remaining years experienced higher streamflow and frequent storm events. The highest streamflow of 43.9 cumecs (1550 cfs) during the study period occurred on 14 December 2009 due to a storm event of 134.1 mm (5.28 in.).

5.2 Effects of stream modification in water velocity

Two alternative cross-sections modifications near the Dean Road bridge site were simulated thus far using HEC-RAS 5. The geometric characteristics of the stream modification are presented in Fig. 8. The small trapezoidal alternative creates more significant blockage to the flow, which reflects on the higher velocities across the bridge, as shown in Fig. 9.

The small trapezoidal stream modification alternative also creates an increase in backwater effect, and both these have an impact on the resulting shear stress under the bridge. The large trapezoidal peak shear stress is in the range

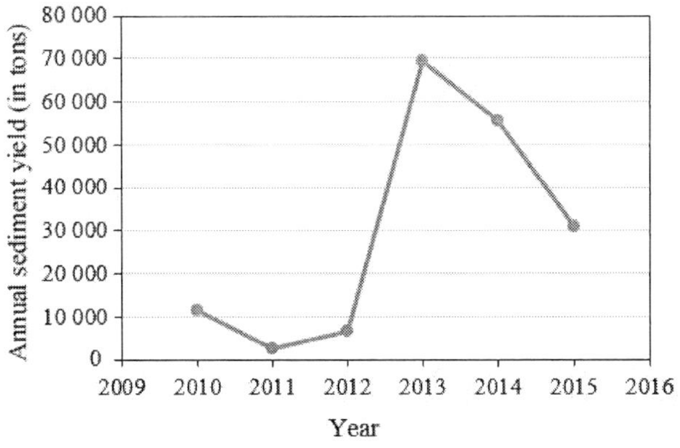

Figure 12. Simulated Annual Sediment Yield for Soapstone Branch Catchment (2010–2015).

of $0.49\,\mathrm{kg\,m^{-2}}$ ($0.10\,\mathrm{lb\,ft^{-2}}$), which is 70 % smaller than the corresponding value for the small trapezoidal solution (Fig. 10). Also, as it may be noticed, the region with larger shear stresses downstream from the bridge site are much wider in the small trapezoidal alternative (Fig. 11).

5.3 Annual Sediment Yield

In order to apply event-based MUSLE to continuous streamflow simulation, a threshold of 0.28 cumecs (10 cfs) was selected for sediment generation. SMA model discharge was applied to MUSLE equation to calculate event sediment yield and each event output for a year were summed up to obtain annual sediment yield from 2010–2015 (Fig. 12). In the calculation of sediment yield by MUSLE, 2011 C factor value of 0.347 was determined and applied from 2011–2012 whereas, 2015 C factor value of 0.648 was applied from 2013–2015. During the study period, 2013 had the highest number of storm events i.e. 24 generating sediment whereas 2011 had the lowest number of storm events that generated sediment i.e. 4. The event on 22 February 2013 produced the maximum sediment yield of $\sim$9400 tons. It was also found that the maximum annual sediment yield of 69 439 tons in the catchment was in 2013 which is consistent with the first identification of the aggradation problem. Also, the recent land-cover changes in the catchment have accounted for 34 % increase in sediment yield. The uncertainty of the proposed model for sediment yield computation is dependent on two different parameter sets viz. parameters from hydrologic model and parameters forming the remaining MUSLE equation. Since the discharge is computed for the Soapstone branch only after validating the parameter transfer to an assumed ungauged receiver catchment, the uncertainty associated with the discharge parameters is low and can be considered within the range of satisfactory performance. Also, the attempts were made to reduce the uncertainty of remaining MUSLE parameters by obtaining the data of highest spatial and temporal resolution available, and conservative values viz. 1 m resolution DEM for LS factor, conservative values of C and P factor.

6 Conclusions

Significant land cover changes in the vicinity of stream negatively altered the discharge and sediment output of the Soapstone Branch catchment and reduced the conveyance of the bridge in the downstream. In the absence of discharge data, parameter transfer approach with an empirical sediment yield method was used to simulate discharge and compute sediment yield. Different cross section modifications were simulated, and it was found that narrowing the stream works best in increasing velocity and washing the sediment off to downstream. An overall goal of the study to quantify the flow and sediment yield in an ungauged catchment was achieved and a recommendation strategy of narrowing the stream thus creating a small depth bank was suggested.

Data availability. The study is from an on-going project and data are not currently available.

Competing interests. The authors declare that they have no conflict of interest.

Special issue statement. This article is part of the special issue "Innovative water resources management – understanding and balancing interactions between humankind and nature". It is a result of the 8th International Water Resources Management Conference of ICWRS, Beijing, China, 13–15 June 2018.

Acknowledgements. The authors would like to thank two anonymous reviewers whose valuable suggestions and comments have greatly improved the quality of the paper. Also, the authors would like to thank Alabama Department of Transportation (ALDOT) for funding the project 930-925 (Grant number G00009876) "Analysis and potential solutions to sediment deposition in Dean Road Bridge watershed, Midland City, Alabama".

Edited by: Dingzhi Peng
Reviewed by: two anonymous referees

References

Bennett, T. H.: Development and application of a continuous soil moisture accounting algorithm for the Hydrologic Engineering Center Hydrologic Modeling System (HEC-HMS), University of California, Davis, 1998.

Brunner, G. W.: HEC-RAS River Analysis System User Manual, version 5.0, US Army Corps of Engineers Report No. CPD 68, 2016.

Civco, D. L., Hurd, J. D., Wilson, S. M., and Zhang, Z.: A comparison of land use and land cover change detection methods, in: ASPRS annual convention proceedings (on CD-ROM), Washington, DC, 2002.

Doan, J. H.: Geospatial hydrologic modeling extension HEC-GeoHMS user's manual, U.S. Army Corps of Engineers Hydrologic Engineering Center, Davis, Calif., 2000.

Feldman, A. D.: Hydrologic modeling system HEC-HMS technical reference manual, U.S. Army Corps of Engineers Hydrologic Engineering Center, Davis, Calif., 2000.

Gitas, I. Z., Douros, K., Minakou, C., Silleos, G. N., and Karydas, C. G.: Multi-temporal soil erosion risk assessment in N. Chalkidiki using a modified USLE raster model, EARSeL eProceedings, 8, 40–52, 2009.

Moriasi, D. N., Arnold, J. G., Van Liew, M. W., Bingner, R. L., Harmel, R. D., and Veith, T. L.: Model evaluation guidelines for systematic quantification of accuracy in watershed simulations, T. ASABE, 50, 885–900, 2007.

Morgan, R. P. C.: Soil erosion and conservation, John Wiley & Sons, 2009.

Nellis, M. D., Harrington, J. A., and Wu, J.: Remote sensing of temporal and spatial variations in pool size, suspended sediment, turbidity, and Secchi depth in Tuttle Creek Reservoir, Kansas, 1993, Geomorphology, 21, 281–293, 1998.

Pannel, D. J.: Sensitivity analysis: strategies, methods, concepts, examples, Agr. Econ., 16, 139–152, 1997.

Sivapalan, M.: Prediction in ungauged basins: a grand challenge for theoretical hydrology, Hydrol. Process., 17, 3163–3170, 2003.

Tamang, S. K.: Quantifying flow and sediment yield of an ungauged catchment using a combination of continuous soil moisture accounting and even-based curve number method, Department of Civil Engineering, Auburn University, Auburn, AL, 2017.

Williams, J. R.: Sediment-Yield Prediction with Universal Equation Using Runoff Energy Factor, Present and Prospective Technology for Predicting Sediment Yield and Sources ARS-S-40, 244-52, 1975.

https://doi.org/10.5194/piahs-379-139-2018

Climate change impact on streamflow in large-scale river basins: projections and their uncertainties sourced from GCMs and RCP scenarios

Olga N. Nasonova, Yeugeniy M. Gusev, Evgeny E. Kovalev, and Georgy V. Ayzel

Institute of Water Problems, Russian Academy of Sciences, Moscow, 119333, Russian Federation

Correspondence: Olga N. Nasonova (olniknas@yandex.ru)

Received: 29 December 2017 – Revised: 2 April 2018 – Accepted: 4 April 2018 – Published: 5 June 2018

Abstract. Climate change impact on river runoff was investigated within the framework of the second phase of the Inter-Sectoral Impact Model Intercomparison Project (ISI-MIP2) using a physically-based land surface model Soil Water – Atmosphere – Plants (SWAP) (developed in the Institute of Water Problems of the Russian Academy of Sciences) and meteorological projections (for 2006–2099) simulated by five General Circulation Models (GCMs) (including GFDL-ESM2M, HadGEM2-ES, IPSL-CM5A-LR, MIROC-ESM-CHEM, and NorESM1-M) for each of four Representative Concentration Pathway (RCP) scenarios (RCP2.6, RCP4.5, RCP6.0, and RCP8.5). Eleven large-scale river basins were used in this study. First of all, SWAP was calibrated and validated against monthly values of measured river runoff with making use of forcing data from the WATCH data set and all GCMs' projections were bias-corrected to the WATCH. Then, for each basin, 20 projections of possible changes in river runoff during the 21st century were simulated by SWAP. Analysis of the obtained hydrological projections allowed us to estimate their uncertainties resulted from application of different GCMs and RCP scenarios. On the average, the contribution of different GCMs to the uncertainty of the projected river runoff is nearly twice larger than the contribution of RCP scenarios. At the same time the contribution of GCMs slightly decreases with time.

1 Introduction

The most common approach for assessing the impact of climate change on river runoff is to run regional hydrological models (RHMs) or global hydrological models (GHMs) driven by meteorological projections from General Circulation Models (GCMs). In so doing, the obtained hydrological projections inevitably contain uncertainties. Many modeling studies have been carried out to explore uncertainties in projected hydrological variables (e.g., Wilby and Harris, 2006; Gosling et al., 2011; Vetter et al., 2017; Nasonova et al., 2018), according to which the following sources of uncertainties can be distinguished with respect to climatic runoff: scenarios of greenhouse-gas emissions to the atmosphere; application of different GCMs for simulating meteorological projections; downscaling and post-processing bias-correction techniques applied for GCMs' outputs; hydrological models applied for hydrological projections. In the latter case, uncertainties can be associated with differences in model structure, complexity, representation of hydrological processes, spatial and temporal resolution, estimation and calibration of model parameters. Increased attention to assessment of different types of uncertainties can be explained by the need to search a predictive range of hydrological models.

As it was shown in Vetter et al. (2017) the results of many studies have indicated that GCMs structure represent the main source of uncertainty in projected river runoff, however, in some regions and seasons other sources of uncertainties can also play an important role. It was concluded that studying the contribution of different sources of uncertainties to the overall uncertainty of hydrological projections should be continued with involvement of large amount of river basins located in different parts of the Earth under a wide variety of climatic conditions. This was done within the framework of the second phase of the Inter-Sectoral Impact Model Intercomparison Project (ISI-MIP2) (Warszawski et al., 2013; Krysanova and Hattermann, 2017).

The present work was carried out when participating in the aforementioned project. The main tool used in the given study for streamflow simulations is the land surface model (LSM) SWAP (Soil Water – Atmosphere – Plants) in contrast to hydrological models participated in the ISI-MIP2 and listed in Krysanova and Hattermann (2017).

The aim of the present paper is to investigate the contribution of different emission scenarios and different GCMs to the uncertainty range of the projected changes in runoff from 11 river basins (covering different climatic conditions on five continents) due to possible climate change during the 21st century.

2 Study basins and data

Eleven river basins provided within the ISI-MIP2 were used in this study. The basins are located in different regions of the globe under different natural conditions: the Rhine and Tagus in Europe; the Ganges, Lena, Upper Yellow and Upper Yangtze in Asia, the Niger in Africa; the Mackenzie and Upper Mississippi in North America, the Upper Amazon in South America and the Darling in Australia. The main characteristics of the selected river basins are given in Table 1.

For model simulations the basins were schematized as a set of regular grid cells with a spatial resolution $0.5° \times 0.5°$ in latitude and longitude connected by a river network (see Gusev et al., 2017, for details). The number of calculational grid cells for each river basin is also presented in Table 1.

Input data for the SWAP model includes the land surface parameters and meteorological forcing data (incoming longwave and shortwave radiation, air temperature and humidity, precipitation, wind speed, and air pressure) for each calculational grid cell.

For model calibration and validation, daily values of forcing data were taken from the WATCH data set (Weedon et al., 2011) for the period of 1970–2001.

For hydrological projections, daily values of meteorological outputs simulated by five GCMs (GFDL-ESM2M, HadGEM2-ES, IPSL-CM5A-LR, MIROC-ESM-CHEM, and NorESM1-M) for four Representative Concentration Pathway scenarios (RCP: RCP2.6, RCP4.5, RCP6.0, and RCP8.5) for 2006–2099 were provided by ISI-MIP2 organisers (Krysanova and Hattermann, 2017). All these meteorological projections were subject to a post-processing bias correction to the WATCH data performed within the framework of the ISI-MIP project. A detailed description of the bias-correction technique can be found in Hempel et al. (2013).

A priori model parameters were derived on the basis of information from the global ECOCLIMAP data set.

More details on data used can be found in this issue (Gusev et al., 2018).

3 Methodology

3.1 The SWAP model

River runoff was simulated by the land surface model SWAP. It is a spatially distributed physically-based model describing heat and water exchange at the land surface – atmosphere interface. Different versions of SWAP were developed for unfrozen soil, seasonally frozen soils, and permafrost. SWAP operates at different scales: from local to global. Model description and the results of its successful validation were detailed in a number of publications (e.g., Gusev and Nasonova, 2004, 2010; Gusev et al., 2008).

3.2 Calibration and validation

Here, SWAP was calibrated and validated against monthly values of measured river runoff from each basin with making use of forcing data from the WATCH data set for 1970–2001. Calibration was performed for eight years (different for different rivers in dependence of available measured runoff); the rest years were used for model validation. The details of the calibration (including a calibration technique, objective function and list of calibrated parameters) can be found in Gusev et al. (2017).

The agreement between simulated (with the WATCH data set) and observed streamflow for each river basin was estimated at monthly basis using several goodness-of-fit statistics: systematic error Bias (equalled to the difference between the simulated and observed mean values of river runoff):

$$\text{Bias} = \frac{\sum\limits_{\Omega}(x_{\text{cal}} - x_{\text{obs}})}{\sum\limits_{\Omega} x_{\text{obs}}} \cdot 100\,\% \tag{1}$$

and the Nash and Sutcliffe coefficient of efficiency NS:

$$\text{NS} = 1 - \frac{\sum\limits_{\Omega}(x_{\text{cal}} - x_{\text{obs}})^2}{\sum\limits_{\Omega}(x_{\text{obs}} - \overline{x_{\text{obs}}})^2}, \tag{2}$$

where x_{cal} and x_{obs} are calculated and observed values of a variable x (here, monthly river runoff), Ω is a discrete sample set of the variable x.

The optimal values of model parameters obtained for each river basin were then used for simulating river runoff by the SWAP model driven by meteorological forcing data simulated by five GCMs for the historical period from 1971 to 2005. The simulated runoff was validated against measured one on an annual basis.

3.3 Projecting changes in river runoff and their uncertainties

For the future period (2006–2099), river runoff was simulated by the SWAP model forced by each of 20 meteoro-

Table 1. River basins with their area and the number of calculational grid cells; gauge stations; averaged over 1971–2000 air temperature T, precipitation P, river runoff R and runoff ratio R / P. The values of T, P and R are taken from Krysanova and Hattermann (2017).

River	Streamflow gauge station	Area, km^2	Number of grid cells	T, °C	P, $mm\,yr^{-1}$	R, $mm\,yr^{-1}$	R / P
Lena	Stolb	2 460 000	1668	−10.2	384	201	0.52
U. Amazon	Sao Paulo de Olivenca	990 781	330	21.7	2122	1459	0.69
MacKenzie	Arctic Red River	1 660 000	1128	−4.3	435	171	0.39
U. Yangtze	Cuntan	804 859	325	6.8	768	389	0.51
Ganges	Farakka	835 000	340	21.1	1173	471	0.4
U. Yellow (Huanghe)	Tangnaihai	121 000	51	−2	506	169	0.33
Niger	Lokoja	2 074 171	678	27.7	625	77	0.12
Rhine	Lobith	160 800	83	8.7	1038	457	0.44
U. Mississippi	Alton	444 185	198	7.3	967	257	0.27
Tagus	Almourol	67 490	35	14	671	152	0.23
Darling	Louth	489 300	180	19.2	590	8	0.01

logical projections (5 GCMs × 4 RCP scenarios). The prognostic period was divided into three parts: P1 (2006–2036), P2 (2037–2067) and P3 (2068–2099). The projected annual runoff was averaged over each period to obtain the climatic runoff. Changes in the climatic runoff $\Delta R_{\mathrm{GCM,RCP,Pi}}$ obtained for each GCM, RCP scenario, and prognostic period Pi ($i = 1, 2, 3$) were calculated as the difference between the projected runoff $R_{\mathrm{GCM,RCP,Pi}}$ and historical runoff $R_{\mathrm{GCM,H}}$ simulated by appropriate GCM and averaged over 1971–2005:

$$\Delta R_{\mathrm{GCM,RCP,Pi}} = R_{\mathrm{GCM,RCP,Pi}} - R_{\mathrm{GCM,H}}. \quad (3)$$

Relative changes were calculated as follows:

$$\Delta R_{\mathrm{GCM,RCP,Pi}}\,\% = \frac{\Delta R_{\mathrm{GCM,RCP,Pi}}}{R_{\mathrm{GCM,H}}} \times 100\,\%. \quad (4)$$

So, 20 values of $\Delta R_{\mathrm{GCM,RCP,Pi}}\,\%$ were obtained for each prognostic period. They were used for estimation mean value M and standard deviation SD. As it was shown in Gelfan et al. (2017), the interval ($M \pm 1.96$ SD) can be treated as the index of hydrological projection uncertainty caused by both the climate model structural uncertainty and the climate scenario variability.

3.4 Contribution of GCM and RCP scenario uncertainties into hydrological projection uncertainty

The obtained relative changes in river runoff $\Delta R_{\mathrm{GCM,RCP,Pi}}\,\%$ allowed us to estimate their uncertainties resulted from application of various GCMs and RCP scenarios. For this purpose, for each prognostic period and each river, the range of variability of $\Delta R\,\%$ as a difference between the largest and the smallest values was estimated:

$$\mathrm{Range} = \Delta R\,\%_{\max} - \Delta R\,\%_{\min}. \quad (5)$$

If we use all 20 values of $\Delta R_{\mathrm{GCM,RCP,Pi}}\,\%$ for estimating the range ($\mathrm{Range}_{\mathrm{GCM,RCP}}$) we obtain uncertainty caused by both GCMs and climatic scenarios. Besides, we can calculate the variation ranges caused (1) only by a scatter among GCM's projections $\mathrm{Range}_{\mathrm{GCM}}$ and (2) by a scatter due to different RCP scenarios $\mathrm{Range}_{\mathrm{RCP}}$. The former indicates the contribution of the climate model structural uncertainty into the hydrological uncertainty, while the latter is associated with the contribution of the climate scenario differences.

4 Results

4.1 Historical simulations of river runoff

Analysis of the results of simulation of river runoff from 11 basins by the SWAP model with a priori land surface parameters, derived from the ECOCLIMAP, and meteorological forcing data from the WATCH for the period of 1970–2001 has shown a poor agreement between measured and simulated monthly runoff: for most basins NS < 0, Bias ranged from 3 % (for the Upper Mississippi) to 340 % (for the Darling).

Calibration of the most influencing parameters against measured monthly river runoff resulted in significant improvement of SWAP performance with respect to goodness-of-fit statistics and the shape of hydrograph. For the calibration period, |Bias| did not exceed 5 % for all rivers (mean value: 2.4 %), monthly NS varied from 0.68 to 0.95 (mean value: 0.85). For the validation period, |Bias| did not exceed 13.8 % (mean value: 4.4 %), monthly NS varied from 0.58 to 0.90 (mean value: 0.78). Comparison of annual values of simulated and measured climatic runoff is given in Fig. 1 (compare "Observation" and "WATCH").

The above results were considered as good and the obtained optimal values of model parameters were used for simulating river runoff by the SWAP model forced by meteorological data from five GCMs (bias-corrected to WATCH)

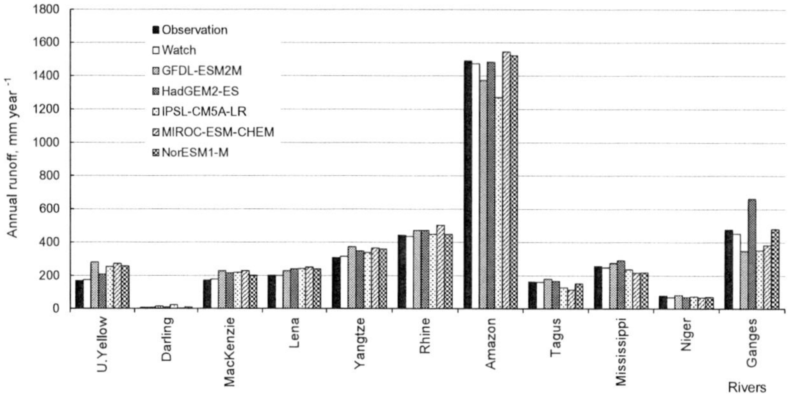

Figure 1. Comparison of annual values of climatic river runoff measured and simulated by the SWAP model driven by forcing data from the WATCH data set and five GCMs.

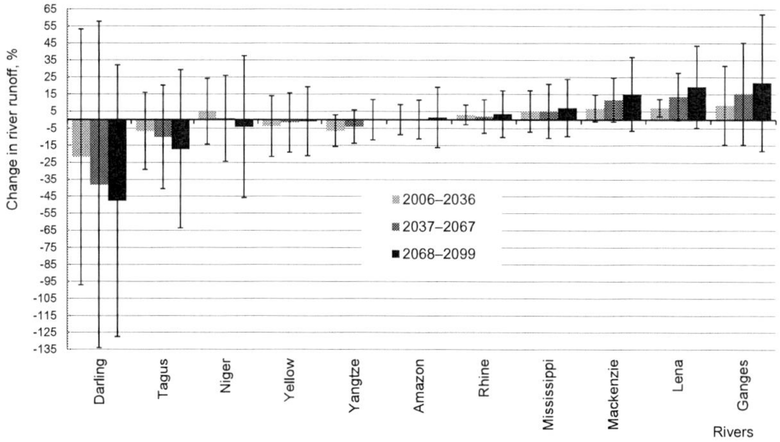

Figure 2. The projected mean relative changes in annual values of climatic river runoff and their uncertainties caused by both the GCMs structural uncertainty and the RCP scenario differences. Bars are the intervals ($M \pm 1.96\,\mathrm{SD}$) treated as hydrological uncertainty.

for the period from 1971 to 2099. Historical simulations (for 1971–2005) were compared with measured annual river runoff and showed a satisfactory agreement (Fig. 1).

4.2 River runoff projections

Figure 2 depicts the projected relative changes in annual values of climatic river runoff $\Delta R_{\mathrm{GCM,RCP,Pi}}$ % averaged over 5 GCMs and 4 RCP scenarios (corresponding to mean value M in Sect. 3.3). They are shown for three prognostic periods. As can be seen in Fig. 2, changes in river runoff are different (both in magnitude and sign) for different rivers. For most rivers, the changes in runoff increase by the end of the 21st century (with the exception of the Niger, Yellow, and Yangtze). For the Niger, the projected climatic precipitation increases during the first prognostic period and decreases during the last period. As a result, in the former case, climatic runoff has positive increment, while in the latter case the increment is negative (see Fig. 2). As to the Yellow and Yangtze rivers, the projected increase in climatic precipitation does not lead to runoff growth, because of increase in incoming radiation that results in that additional precipitation raises evapotranspiration rather than river runoff.

For some rivers (the Amazon, Rhine, Mississippi, Mackenzie, Lena, and Ganges), runoff will increase by 1.4–22 % by the end of the 21st century, while for the Niger, Tagus, and Darling runoff will decrease by 4–48 % (Fig. 2). This is in a good agreement with projected changes in climatic precipitation.

Bars in Fig. 2 show the intervals of uncertainty in the projected relative changes of river runoff caused by both the GCMs structural uncertainty and the climate scenario differences. The largest uncertainty was found for the Darling, then for the Tagus, Ganges and Niger. This can be explained by the largest variability of the projected changes in climatic

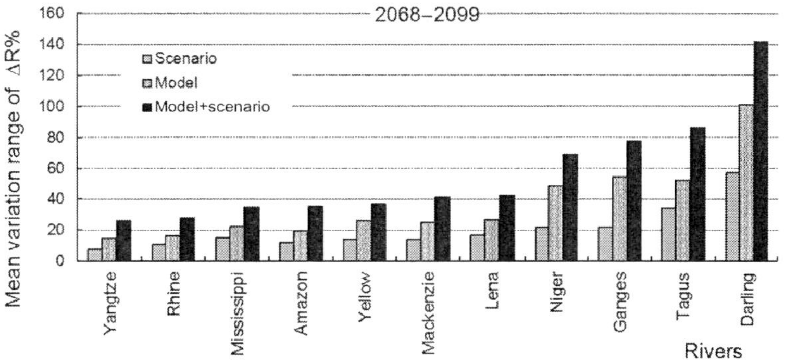

Figure 3. Mean variation range of projected relative changes in climatic runoff from 11 river basins due to differences in RCP scenarios ("Scenario"), GCMs ("Model"), as well as both scenarios and models ("Model + Scenario").

precipitation for these rivers. As it was found, the coefficient of correlation between SD of relative changes in projected climatic precipitation (averaged over all GCMs and RCP scenarios) and corresponding changes in climatic river runoff is 0.84 for the last climatic period (2068–2099).

4.3 Contribution of GCMs and RCP scenarios into hydrological uncertainty

Figure 3 presents mean variation ranges of the projected relative changes in annual values of climatic runoff from 11 river basins due to differences in RCP scenarios ("Scenario"), GCMs ("Model"), as well as both scenarios and models ("Model + Scenario") for the last prognostic period, which were calculated as $Range_{RCP}$, $Range_{GCM}$, and $Range_{GCM,RCP}$, respectively. The values of $Range_{RCP}$, $Range_{GCM}$, and $Range_{GCM,RCP}$ represent uncertainties in the projected relative changes of annual values of climatic river runoff caused, respectively, by differences in RCP scenarios, by structural differences in GCMs and by joint influence of scenarios and GCMs.

As can be seen in Fig. 3, uncertainties sourced from joint influence of RCP and GCMs ranges from 26 to 142 % for different rivers. The contribution of RCP scenarios into the total uncertainty is smaller than that of the GCMs: $Range_{RCP}$ varies from 8 to 57 %, while $Range_{GCM}$ is within the interval from 15 to 101 %. The ratio of $Range_{GCM}$ to $Range_{RCP}$ varies ranges from 1.4 to 2.5 for different rivers.

Figure 4 shows the same variation ranges averaged over 11 rivers for the three prognostic periods. All ranges, i.e. all types of uncertainties, grow with time. As to the ratio of $Range_{GCM}$ to $Range_{RCP}$, it varies from 2.1 for 2006–2036 to 1.8 for 2068–2099, i.e. decreases slightly with time.

5 Conclusions

Twenty projections of the dynamics of river runoff from 11 large-scale basins during the 21st century were calculated

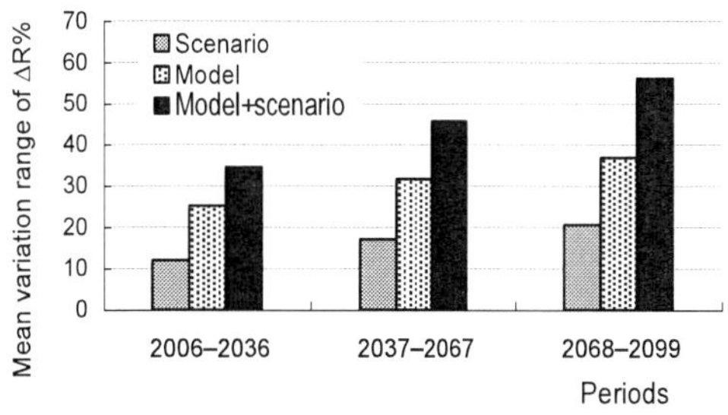

Figure 4. Mean variation range of projected relative changes in climatic river runoff (averaged over 11 river basins) due to differences in RCP scenarios ("Scenario"), GCMs ("Model"), as well as both scenarios and models ("Model + Scenario").

with the help of the SWAP model driven by meteorological projections simulated by 5 GCMs for 4 climatic scenarios of the RCP family. They allowed us to project relative changes in climatic river runoff for three prognostic periods and estimate their uncertainties.

Analysis of the uncertainties of the projected changes in river runoff has shown that: (1) the largest uncertainties resulted from application of different GCMs and different climatic scenarios were obtained for the Darling, Tagus, Niger and Ganges rivers; (2) the uncertainties increase by the end of the 21st century; (3) for all 11 rivers and for the three prognostic periods, the contribution of GCMs into the uncertainty of the projected relative changes in climatic river runoff is larger than that of the RCP scenarios; (4) on the average, the contribution of GCMs into hydrological uncertainty is nearly twice larger than that of RCP scenarios; (5) on the average, the contribution of GCMs into the total hydrological uncertainty slightly decreases with time, as compared with the contribution of RCP scenarios.

Data availability. Data were provided by ISI-MIP regional-scale water sector organizers, and by GRDC (the Global Runoff Data Centre, 56068 Koblenz, Germany).

Competing interests. The authors declare that they have no conflict of interest.

Special issue statement. This article is part of the special issue "Innovative water resources management – understanding and balancing interactions between humankind and nature". It is a result of the 8th International Water Resources Management Conference of ICWRS, Beijing, China, 13–15 June 2018.

Acknowledgements. The study was supported by the Russian Science Foundation (grant no. 16-17-10039). We are grateful to ISI-MIP regional-scale water sector organizers for providing with data and materials, and to GRDC for providing discharge data.

Edited by: Zongxue Xu
Reviewed by: two anonymous referees

References

Gelfan, A., Gustafsson, D., Motovilov, Yu., Arheimer, B., Kalugin, A., Krylenko, I., and Lavrenov, A.: Climate change impact on water regime of two great Arctic rivers: modeling and uncertainty issues, Climatic Change, 141, 499–515, 2017.

Gosling, S. N., Taylor, R. G., Arnell, N. W., and Todd, M. C.: A comparative analysis of projected impacts of climate change on river runoff from global and catchment-scale hydrological models, Hydrol. Earth Syst. Sci., 15, 279–294, https://doi.org/10.5194/hess-15-279-2011, 2011.

Gusev, E. M. and Nasonova O. N.: Simulation of heat and water exchange at the land–atmosphere interface on a local scale for permafrost territories, Eurasian Soil Sci., 37, 1077–1092, 2004.

Gusev, E. M. and Nasonova, O. N.: Modelling heat and water exchange between the land surface and the atmosphere, Nauka, Moscow, 2010.

Gusev, E. M., Nasonova, O. N., Dzhogan, L. Ya., and Kovalev, E. E.: The application of the land surface model for calculating river runoff in high latitudes, Water Resources, 35, 171–184, 2008.

Gusev, Ye. M., Nasonova, O. N., Kovalev, E. E., and Aizel, G. V.: Mdelling river runoff and estimating its weather-related uncertainty for 11 large-scale rivers located in different regions of the globe, Hydrol. Res., https://doi.org/10.2166/nh.2017.015, 2017.

Gusev, Ye. M., Nasonova, O. N., Kovalev, E. E., and Aizel, G. V.: Impact of possible climate changes on river runoff under different natural conditions, Proc. IAHS, in press, 2018.

Hempel, S., Frieler, K., Warszawski, L., Schewe, J., and Piontek, F.: A trend-preserving bias correction – the ISI-MIP approach, Earth Syst. Dynam., 4, 219–236, https://doi.org/10.5194/esd-4-219-2013, 2013.

Krysanova, V. and Hattermann, F. F.: Intercomparison of climate change impacts in 12 large river basins: overview of methods and summary of results, Climatic Change, 141, 363–379, 2017.

Nasonova, O. N., Gusev, Ye. M., Volodin, E. M., and Kovalev, Ye. E.: Application of the land surface model SWAP and global climate model INMCM4.0 for projecting runoff of northern Russian rivers. 2. Projections and their uncertainties, Water Resources, in press, 2018.

Vetter, T., Reinhardt, J., Flörke, M., van Griensven, A., Hattermann, F., Huang, S., Koch, H., Pechlivanidis, I. G., Plötner, S., Seidou, O., Su, B., Vervoort, R. W., and Krysanova, V.: Evaluation of sources of uncertainty in projected hydrological changes under climate change in 12 large-scale river basins, Climatic Change, 141, 419–433, 2017.

Warszawski, L., Frieler, K., Huber, V., Piontek, F., Serdeczny, O., and Schewe, J.: The inter-sectoral impact model intercomparison project (ISI–MIP): project framework, P. Natl. Acad. Sci. USA, 111, 3228–3232, 2013.

Weedon, G. P., Gomes, S., Viterbo, P., Shuttleworth, W. J., Blyth, E., Oesterle, H., Adam, J. C., Bellouin, N., Boucher, O., and Best, M.: Creation of the WATCH Forcing Data and its uses to assess global and regional reference crop evaporation over land during the twentieth century, J. Hydrometeorol., 12, 823–848, 2011.

Wilby, R. L. and Harris, I.: A framework for assessing uncertainties in climate change impacts: low-flow scenarios for the River Thames, UK, Water Resour. Res., 42, W02419, https://doi.org/10.1029/2005WR004065, 2006.

https://doi.org/10.5194/piahs-379-145-2018

Open Access
PIAHS

An assessment of temporal effect on extreme rainfall estimates

Samiran Das[1], **Dehua Zhu**[1], **and Cheng Chi-Han**[2]

[1]School of Hydrology and Water Resources, Nanjing University of Information Science and Technology, Nanjing, China

[2]Taiwan Research Institute on Water Resources and Agriculture, Taiwan (R.O.C.)

Correspondence: Samiran Das (samirandas@gmail.com)

Received: 31 December 2017 – Accepted: 17 January 2018 – Published: 5 June 2018

Abstract. This study assesses the temporal behaviour in terms of inter-decadal variability of extreme daily rainfall of stated return period relevant for hydrologic risk analysis using a novel regional parametric approach. The assessment is carried out based on annual maximum daily rainfall series of 180 meteorological stations of Yangtze River Basin over a 50-year period (1961–2010). The outcomes of the analysis reveal that while there were effects present indicating higher quantile values when estimated from data of the 1990s, it is found not to be noteworthy to exclude the data of any decade from the extreme rainfall estimation process for hydrologic risk analysis.

1 Introduction

The temporal assessment of rainfall extremes is necessary to understand its temporal behavior which will lead to better assess the frequency of extreme rainfall under climate change, and subsequently the flood risk. Since the climate change (e.g. IPCC, 1995) phenomena emerges, a large number of studies appear to find its impact on rainfall. Among them, a substantial research was carried out on rainfall extremes (Bülow et al., 2015; Das et al., 2013; Feng et al., 2007; Ntegeka and Willems, 2008; Su et al., 2008; Tabari et al., 2014; Zhang et al., 2008) as they impact the society most. The outcome is not one dimensional- a significant variability is observed in some regions whereas in some cases, a number of regions do not show much variation that prove to be significant (e.g. Alexander et al., 2006; Damberg and AghaKouchak, 2014; Feng et al., 2007). The data length and type of diagnostic tools can also influence the outcome (Ntegeka and Willems, 2008). A regional behavior is also recognized in those studies (Alexander et al., 2006).

The Mann Kendall and the Spearman rho tests are mostly used approaches to assess the temporal variability. In a more recent time, the quantile perturbation method (Ntegeka and Willems, 2008; Tabari et al., 2014) is used to assess the decadal variability in terms of identify the changes in the extremes for particular return periods. The peak-over-threshold (POT) data was used so that a longer data series can be gathered in a decade. All the above approaches are non parametric.

This study introduces a new parametric approach to assess the temporal variability of rainfall extremes for the Yangtze River Basin, a major development area of China. A number of studies were carried out to investigate observed long-term trends and patterns in extreme precipitation events in china (e.g. Fu et al., 2013) including the Yangtze Basin (e.g. Su et al., 2008) in order to assess the regional response to climate change. A summary of the outcomes was reported, among others, by Chen et al. (2012) who also recognized that only a few studies were concentrated on to assess decadal variability of rainfall extremes. This study aims to investigate the issue further in terms of inter-decadal variability of extreme daily rainfall of stated return period relevant for hydrologic risk analysis. In particular, the study aims to address the following research objectives:

- whether climate change has a significant role in decadal quantile variability of daily rainfall extremes in Yangtze Basin context

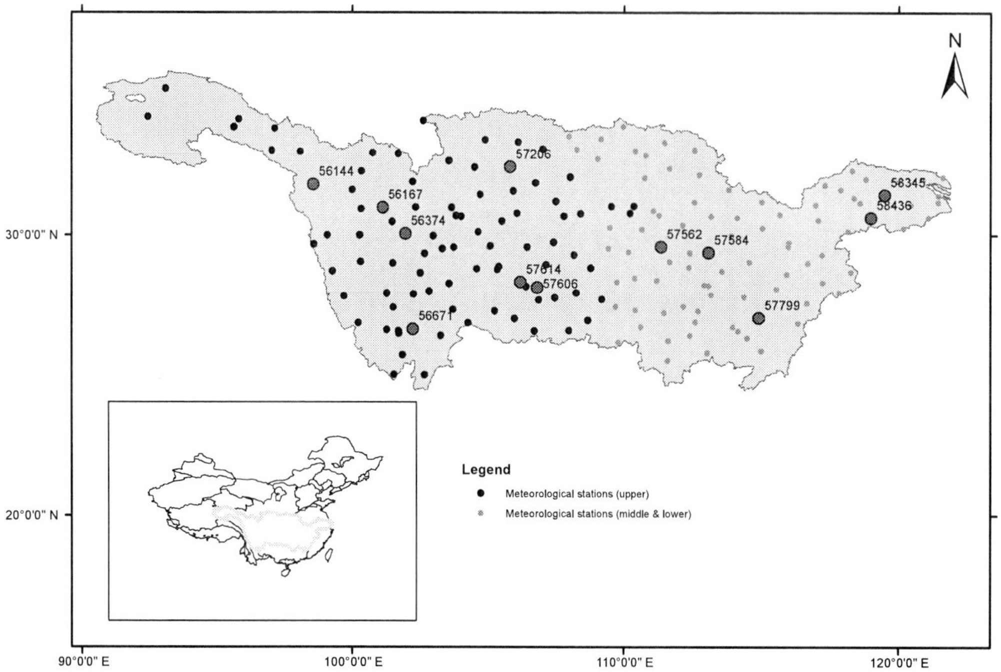

Figure 1. Locations of meteorological stations on the map of Yangtze River Basin.

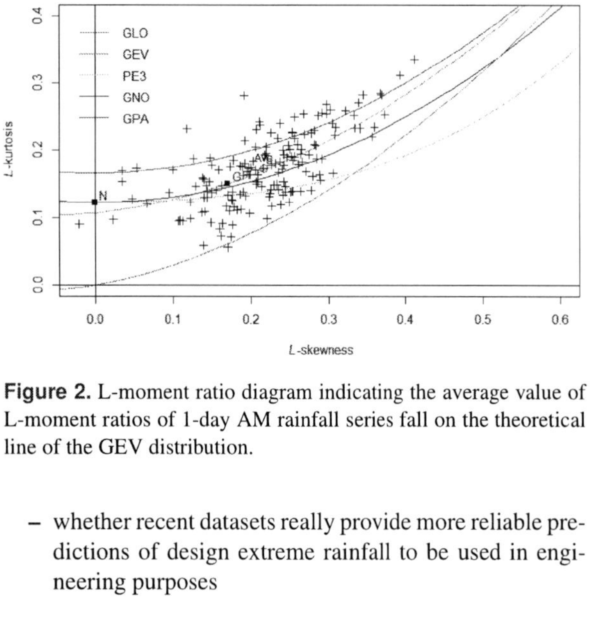

Figure 2. L-moment ratio diagram indicating the average value of L-moment ratios of 1-day AM rainfall series fall on the theoretical line of the GEV distribution.

- whether recent datasets really provide more reliable predictions of design extreme rainfall to be used in engineering purposes

2 Methodology

The approach examines whether rainfall quantiles estimated for a decade as well as the most recent changes in extreme quantiles can be qualified as statistically significant in comparison with the natural temporal variability. A major drawback of performing decadal variability in a parametric setup is that a limited number of data set is available in a decade to fit with the selected probability model. In this study, a regional approach in region-of-influence (ROI) form (Burn, 1990; Das and Cunnane, 2012; Institute of Hydrology, 1999) is used to study the decadal quantile behavior. The ROI method is an objective way of forming a homogeneous pooling group/region for a site with limited data aiming to perform a frequency analysis. In this approach, annual maximum rainfall series are pooled from other homogeneous stations and perform frequency analysis based on index-flood method (Dalrymple, 1960). The estimation of P_T, T-year return period rainfall depth, based on this approach involves derivation of a growth curve X_T which shows the relation between X_T and the return period T where $X_T = P_T/P_I$ and P_I is the index measurement (e.g. mean or median of the at-site annual maximum rainfall series) at the site of interest (Das, 2017). A suitable probability distribution function is required to estimate quantiles. A detailed explanation of the method including the criteria that form homogeneous groups can be obtained from Das (2017). This study uses geographical proximity to group stations which give promising results in many ROI based extreme rainfall study (Kyselý et al., 2011; Reed et al., 1999). The $5T$ rule (Institute of Hydrology, 1999) (the total number of station years of data to be included when estimating the T year event) is applied to delineate the size of the group for a subject location. Finally the homogeneity is assessed by the Hosking-Wallis heterogeneity measure H1 (Hosking and Wallis, 1997). The ROI

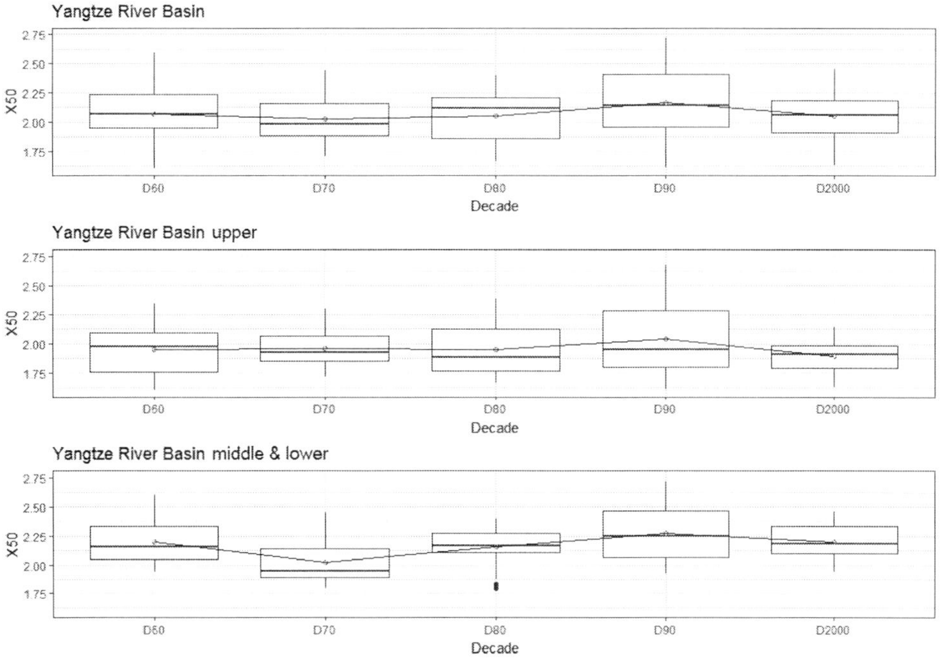

Figure 3. Variation of X50 with decades from each individual stations' pooling group.

regional approach is appealing because it reduces the variability of the estimates and reflects the changes in a best possible manner.

With this approach, a homogeneous pooling/regional group for a particular decade is selected from gauged stations that are available in that particular decade and a frequency analysis is performed for a target site. A decade consists of 10 annual maximum (AM) data points but with this approach a large number of AM data is available for that particular decade which will lead to achieve a reliable estimate.

In order to test whether the variability is statistically significant, the study outlines confidence intervals based on a parametric Monte Carlo method. In this test, a baseline AM rainfall series is first fitted to a distribution after which random samples are generated from the distribution, and confidence intervals are estimated. In the present study, the baseline series is chosen as the long-term historical datasets from which the base quantile is estimated while the other one is from the group of series taken from a particular decade of interest. After assessing the decadal variability, the confidence intervals are also estimated and overlaid on the same chart. It is therefore graphically possible to identify periods that demonstrate significant departures under the hypothesis of no trend of rainfall extremes.

3 Study area and data

This study selects Yangtze River Basin to analyze decadal behavior in annual maximum (AM) daily rainfall series of over a 50-year period (1961–2010). The basin is analyzed with two major distinct climatic conditions: upper and middle-lower. A total number of 180 stations are available in the study area. They are displayed on the map in Fig. 1. The datasets were obtained from the National Meteorological Information Center of China Meteorological Administration (CMA). Data of these stations have different record length and have different starting and ending year.

Table 1. Decadal datasets.

Decade	Period of record	No. of Stations available
1960	1961–1970	163
1970	1971–1980	165
1980	1981–1990	161
1990	1991–2000	157
2000	2001–2010	151

Table 1 lists the number of stations that are available for each decade. Only stations which have a minimum of 9 years of data available are included for that particular decade. There are 144 stations for which AM data are available for 5 decades.

4 Result and discussion

In a parametric approach, a suitable probability distribution is needed to describe the data. The L-moment ratio diagram

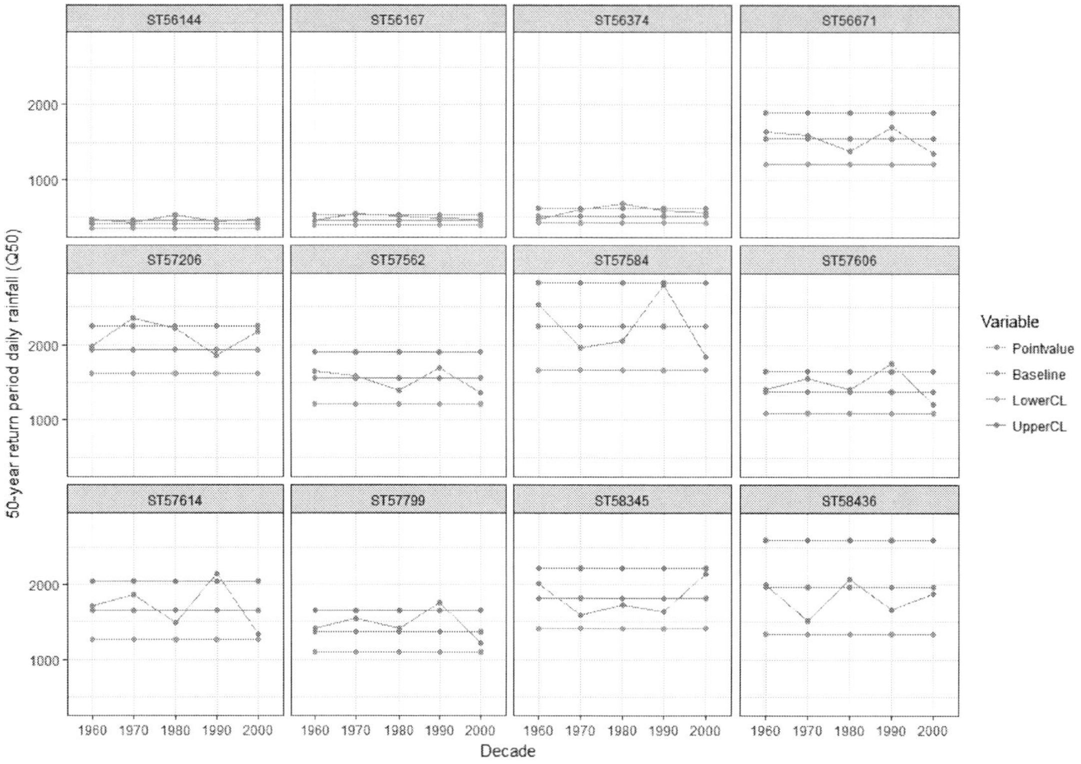

Figure 4. P_{50} (mm) for selected decades together with 95 % confidence intervals to define statistically significant changes.

(Das, 2016; Hosking and Wallis, 1997) is used to identify an appropriate distribution. The LMRD diagram shown in Fig. 2 identifies the generalized extreme value (GEV) as the most suitable distribution for the Yangtze Basin. The formula for estimating growth factors and the associated quantiles based on the GEV can be obtained from Das and Cunnane (2011).

Growth factors, X_T (e.g. standardized values) estimated for each decade of 144 stations are analyzed first to show the general temporal behavior. The study analyzes extremes in terms of 50 year return period values (X_{50}). They are estimated based on pooling groups of size 250 station-year (i.e. 25 stations) that satisfy the $5T$ rule. For each of these 144 stations, the pooling group for particular decade was selected from stations that are available in that particular group. The values of X_{50} for each decade for each station are displayed in Box-Plot form in Fig. 3. The largest average X_{50} value occurs in decade 90 for both the Upper and the Middle-Lower reaches. The average smallest value occurs in decade 70. The upward trend of average X_{50} in decade 90 is the main contribution to the increase in extreme rainfall in the Yangtze Basin, which can partly explain the intensification of flooding in the 1990s which was also noticed by Su et al. (2008). But the trend decreased in the following decade indicating that it is not worthwhile to take only recent data for frequency analysis.

This method also permits to identify whether the decadal variability occurred in an individual station is significant or not. A number of stations shown bold in Fig. 1 are chosen to represent for that particular region. They can be classified as Upper (Upper-West stations 56144, 56167, 56374; Upper-South station 56671; Upper-North station 57206; Upper-East stations 57614, 57606), Middle (stations 57562, 57584, 57799) and Lower (stations 58345, 58436). The decadal quantile, P_{50} of rainfall extremes are shown in Fig. 4. The confidence intervals in Fig. 4 represent the 95 % bounds of random variation under the null hypothesis of no trends.

The different decade's values (anomalies) came out significant for different location of Yangtze basin. In the Upper-West basin, the 1980 decade value of P_{50} is larger than those of the other decades with some cases significant values were also observed for that decade. However, the quantiles are significantly lower than those of the other parts of the Yangtze Basin. The 1970 decade value of P_{50} came out significant for station 57206. The decadal values of 1990s provide largest values for the Upper-East and Middle region. In many cases they are found to be significant which is in line what is observed by Su et al. (2008) that partly explain the frequent occurrence of floods in the 1990s. The decadal quantile values for the decade 2000 came out as the largest for Lower region but they are not proved to be significant which is similar what is found by Zhang et al. (2008).

5 Conclusion

This study examines the decadal variability in daily AM rainfall series of Yangtze River Basin over a 50-year period (1961–2010) with a novel regional parametric approach. Based on the dimensionless standardized growth factor of 50 year return period value, the average largest value occurred in the 1990s but a dip occurred in the following decade. There are some case studies conducted to see the nature of the decadal variability in several stations of the Yangtze Basin with significant test. In general, the 1980s decadal quantile value came out as the largest in the far Upper, the 1990s value came out as the largest in the Middle with some being proved significant, and the 2000s value came out as the largest for the lower basin. So far the outcome of the analysis reveal that while there were small effects present indicating higher quantile values when estimated from data of decade 1990, it is found not to be worthwhile to exclude the data of any decade from the extreme rainfall estimation process for hydrologic risk analysis.

Data availability. The observed daily rainfall data series of selected stations were obtained from the National Meteorological Information Center of China Meteorological Administration (CMA). CMA owns the datasets and a formal request is required to obtain the required datasets.

Competing interests. The authors declare that they have no conflict of interest.

Special issue statement. This article is part of the special issue "Innovative water resources management – understanding and balancing interactions between humankind and nature". It is a result of the 8th International Water Resources Management Conference of ICWRS, Beijing, China, 13–15 June 2018.

Acknowledgements. The research is funded by Nanjing University of Information Science and Technology in the form of a grant (Grant no. 2243141501015) of the first author.

Edited by: Dingzhi Peng
Reviewed by: Qinglan Li and one anonymous referee

References

Alexander, L. V., Zhang, X., Peterson, T. C., Caesar, J., Gleason, B., Klein Tank, A. M. G., Haylock, M., Collins, D., Trewin, B., Rahimzadeh, F., Tagipour, A., Rupa Kumar, K., Revadekar, J., Griffiths, G., Vincent, L., Stephenson, D. B., Burn, J., Aguilar, E., Brunet, M., Taylor, M., New, M., Zhai, P., Rusticucci, M., and Vazquez-Aguirre, J. L.: Global observed changes in daily climate extremes of temperature and precipitation, J. Geophys. Res.-Atmos., 111, 1–22, https://doi.org/10.1029/2005JD006290, 2006.

Bülow, I., Henrik, G., and Dan, M.: Long term variations of extreme rainfall in Denmark and southern Sweden, Clim. Dynam., 44, 3155–3169, https://doi.org/10.1007/s00382-014-2276-4, 2015.

Burn, D. H.: Evaluation of regional flood frequency analysis with a region of influence approach, Water Resour. Res., 26, 2257–2265, 1990.

Chen, H., Sun, J., and Fan, K.: Decadal features of heavy rainfall events in eastern China, Acta Meteorol. Sin., 26, 289–303, https://doi.org/10.1007/s13351-012-0303-0, 2012.

Dalrymple, T.: Flood frequency methods, U.S. Geol. Surv. 1543-A, 11–51, 1960.

Damberg, L. and AghaKouchak, A.: Global trends and patterns of drought from space, Theor. Appl. Climatol., 117, 441–448, https://doi.org/10.1007/s00704-013-1019-5, 2014.

Das, S.: An assessment of using subsampling method in selection of a flood frequency distribution, Stoch. Env. Res. Risk A., 31, 1–13, https://doi.org/10.1007/s00477-016-1318-3, 2016.

Das, S.: Performance of region-of-influence approach of frequency analysis of extreme rainfall in monsoon climate conditions, Int. J. Climatol., 37, 612–623, https://doi.org/10.1002/joc.5025, 2017.

Das, S. and Cunnane, C.: Examination of homogeneity of selected Irish pooling groups, Hydrol. Earth Syst. Sci., 15, 819–830, https://doi.org/10.5194/hess-15-819-2011, 2011.

Das, S. and Cunnane, C.: Performance of flood frequency pooling analysis in a low CV context, Hydrolog. Sci. J., 57, 433–444, https://doi.org/10.1080/02626667.2012.666635, 2012.

Das, S., Millington, N., and Simonovic, S. P.: Distribution choice for the assessment of design rainfall for the city of London (Ontario, Canada) under climate change, Can. J. Civil. Eng., 40, 121–129, https://doi.org/10.1139/cjce-2011-0548, 2013.

Feng, S., Nadarajah, S., and Hu, Q.: Modeling Annual Extreme Precipitation in China Using the Generalized Extreme Value Distribution, J. Meteorol. Soc. Jpn., 85, 599–613, https://doi.org/10.2151/jmsj.85.599, 2007.

Fu, G., Yu, J., Yu, X., Ouyang, R., Zhang, Y., Wang, P., Liu, W., and Min, L.: Temporal variation of extreme rainfall events in China, 1961–2009, J. Hydrol., 487, 48–59, https://doi.org/10.1016/j.jhydrol.2013.02.021, 2013.

Hosking, J. R. M. and Wallis, J. R.: Regional frequency analysis: an approach based on L-moments, Cambridge University Press, 1997.

Institute of Hydrology: Flood Estimation Handbook, Vol. 1–5, Institute of Hydrology, Wallingford, UK, 1999.

IPCC: Climate Changes 1995, Contribution of Working Group I to the Second Report of the Intergovernmental Panel on Climate Change, 141–193, 1995.

Kyselý, J., Gaál, L., and Picek, J.: Comparison of regional and at-site approaches to modelling probabilities of heavy precipitation, Int. J. Climatol., 31, 1457–1472, https://doi.org/10.1002/joc.2182, 2011.

Ntegeka, V. and Willems, P.: Trends and multidecadal oscillations in rainfall extremes, based on a more than 100-year time series of 10 min rainfall intensities at Uccle, Belgium, Water Resour. Res., 44, https://doi.org/10.1029/2007WR006471, 2008.

Reed, D. W., Faulkner, D. S., and Stewart, E. J.: The FORGEX method of rainfall growth estimation II: Description, Hydrol.

Earth Syst. Sci., 3, 197–203, https://doi.org/10.5194/hess-3-197-1999, 1999.

Su, B., Gemmer, M., and Jiang, T.: Spatial and temporal variation of extreme precipitation over the Yangtze River Basin, Quaternary Int., 186, 22–31, https://doi.org/10.1016/j.quaint.2007.09.001, 2008.

Tabari, H., AghaKouchak, A., and Willems, P.: A perturbation approach for assessing trends in precipitation extremes across Iran, J. Hydrol., 519, 1420–1427, https://doi.org/10.1016/j.jhydrol.2014.09.019, 2014.

Zhang, Q., Xu, C. Y., Zhang, Z., Chen, Y. D., Liu, C. L., and Lin, H.: Spatial and temporal variability of precipitation maxima during 1960–2005 in the Yangtze River basin and possible association with large-scale circulation, J. Hydrol., 353, 215–227, https://doi.org/10.1016/j.jhydrol.2007.11.023, 2008.

https://doi.org/10.5194/piahs-379-151-2018

Open Access
PIAHS

Coupling physically based and data-driven models for assessing freshwater inflow into the Small Aral Sea

Georgy Ayzel[1,2] and Alexander Izhitskiy[3]

[1]Institute of Earth and Environmental Science, University of Potsdam, 14476 Potsdam, Germany
[2]Institute of Water Problems, Russian Academy of Sciences, 119333 Moscow, Russia
[3]Shirshov Institute of Oceanology, Russian Academy of Science, 117997 Moscow, Russia

Correspondence: Georgy Ayzel (ayzel@uni-potsdam.de)

Received: 18 November 2017 – Revised: 18 January 2018 – Accepted: 25 January 2018 – Published: 5 June 2018

Abstract. The Aral Sea desiccation and related changes in hydroclimatic conditions on a regional level is a hot topic for past decades. The key problem of scientific research projects devoted to an investigation of modern Aral Sea basin hydrological regime is its discontinuous nature – the only limited amount of papers takes into account the complex runoff formation system entirely. Addressing this challenge we have developed a continuous prediction system for assessing freshwater inflow into the Small Aral Sea based on coupling stack of hydrological and data-driven models. Results show a good prediction skill and approve the possibility to develop a valuable water assessment tool which utilizes the power of classical physically based and modern machine learning models both for territories with complex water management system and strong water-related data scarcity. The source code and data of the proposed system is available on a Github page (https://github.com/SMASHIproject/IWRM2018).

1 Introduction

The Aral Sea and its basin are among the highly recognizable examples of significant environmental changes which took place in the Central Asia during the last decades (Izhitskiy et al., 2016; Micklin, 2007; Raskin et al., 1992; Zavialov et al., 2003). Induced by river runoff exploitation across huge irrigation systems the Aral Sea level has significantly decreased and run irreversible ecosystem and water balance shifts (Zmijewski and Becker, 2014). Nowadays the Small Aral Sea has a limited (by the Kokaral Dam) hydrological connection with dying southern sea basins and tends to stay a separate part under current social and political situation in the region. It is extremely important to devote scientific attention to this region as a real live example of the human-induced impact on water balance and its response (Immerzeel and Bierkens, 2012).

The main volume of the freshwater inflow into the Small Aral Sea is formed on the Syr Darya river basin which is among the largest and highly vulnerable river basins in the Central Asia. There are thirteen large reservoirs and much local water management related installations on the Syr Darya river and its tributaries which utilize full freshwater potential for irrigational, industrial, recreational, and social needs. This complex structure of water management system coupled with the total absence of data describes its functioning is a challenge for any approach directed to the accurate assessment of the Small Aral Sea freshwater budget formation and evolution across the basin (Lutz et al., 2012a; Raskin et al., 1992; Sorg et al., 2014).

There are three main categories of scientific literature devoted to the identification of modern water balance shifts in the Aral basin. The first group accumulates research directed to large-scale heat and water flux changes assessment based on remote sensing, climate modeling and reanalysis data for a whole basin area (López et al., 2017; Shi et al., 2014; Zmijewski and Becker, 2014). These investigations help us to identify patterns and key factors affect long-term hydrological changes and its trends (geographical approach), but cannot be easily scaled for providing quantitative predictions. The second group focuses attention mostly on the upstream Aral Sea basin area (mountainous zone and Ferghana Valley) by the reason of high-altitude glaciers and the biggest reservoirs presence here. Pereira-Cardenal et al. (2011), Siegfried et al. (2012),

Hagg et al. (2006, 2007), Gan et al. (2015) and Lutz et al. (2012b) modeled runoff in glacierized catchments and its contribution into underlying reservoirs inflow using conceptual and physically based models (NAM, HBV-ETH, OEZ, SWAT, AralMountain). Apel et al. (2017) evaluated the skill of simple statistical models for seasonal runoff forecast in this region. Radchenko et al. (2017) examined historical runoff for 18 river basins in Ferghana Valley using HBV-light model and estimated projected changes in streamflow characteristics according to the A1B climatic scenario for these basins. For the extensive review of hydrological modeling studies in glacierized catchments of Central Asia please refer to Chen et al. (2017). The third (and the least) group of papers being conducted on developing end-to-end hydrological modeling system for the whole Small Aral Sea basin. A simplified approach for assessing annual freshwater inflow based on hypothetical and general circulation model based scenarios of future climate temperature and precipitation has been applied in Shibuo et al. (2007) and Jarsjö et al. (2012) using Porflow model without any parameters calibration. The most comprehensible routine for model-based assessment of water balance components of Syr Darya river basin was proposed in Lutz et al. (2012a) and utilizes coupling of conceptual runoff formation model AralMountain (Lutz et al., 2012b) with Water Evaluation And Planning model (WEAP) which has been already implemented for the former Aral Sea basin in 1989 (Raskin et al., 1992).

In presented work we have tried to combine best practices in an existing scientific literature and modern advances in the field of machine learning to develop continuous hybrid hydrological model for investigating both runoff generation processes using physically based models and runoff transformation through one of the most complex water management systems in the world using machine learning algorithms and models. With our research, we want to fill a modern gap in developing a continuous runoff prediction system for the entire Syr Darya river basin domain using a combination of state-of-the-art modeling techniques. Our research does not pretend to cover the problem of freshwater inflow predictions in the Small Aral Sea in high details, but it is an attempt to map the efficiency level of the runoff prediction system which has been built only on open data sources.

2 Materials

2.1 Study area

The main part of the Small Aral Sea basin (Fig. 1) is occupied by the Syr Darya river and its tributaries which contribute around 40 km^3 of freshwater inflow annually (Radchenko et al., 2017). About 70 % of the runoff of the Syr Darya river basin originates in the Kyrgyzstan mountain ranges and the main contribution of this volume corresponds to Ferghana Valley river basins (Belyaev, 1995; Radchenko et al., 2017). In our research, we have selected 24 basins which run to the Ferghana Valley as the main source of hydrological insights and information about runoff generation in the freshwater formation zone of the Small Aral Sea (Fig. 1). These basins are highly contrasting in geographical and hydroclimatic conditions, and cover a range of areas from 150 to 24 000 km^2. For a detailed geographical description of Ferghana Valley river basins please refer to Radchenko et al. (2017).

2.2 Runoff and meteorological forcing data

Observed runoff data for selected basins were provided by the Global Runoff Data Centre (GRDC; http://www.bafg.de/GRDC/). Only for 2 basins of 24 there were daily observed runoff time series, therefore in our work we used only monthly observations for holding methodological consistency. Runoff data availability is the main limit for developing and validation of our methodology by the reason of the majority of available observations lie in the interval from 1975 to 1985. For the modern studies related to contemporary water resources assessment on vast territories, it is essential to use global gridded data products as the only spatial and temporal continuous source. For this reason, all models were driven by precipitation and temperature data from the ERA-40 reanalysis (1957–2002, 0.5° spatial resolution, http://apps.ecmwf.int/datasets/, Uppala et al., 2005). Potential evapotranspiration is another required forcing variable for all models and it was derived based on Oudin et al. temperature-based equation (Oudin et al., 2005).

3 Methods and software

3.1 Hydrological models

The HBV (Hydrologiska Byråns Vattenbalansavdelning, in Lindström et al., 1997), the GR4J (modele du Genie Rural a 4 parametres Journalier in Perrin et al., 2003), and the SIMHYD (in Chiew et al., 2009) models were used in this study according to its wide implementation for different hydrological applications, flexibility, proven effectiveness for runoff predictions in different geographical conditions, and numerous successful applying for prediction in ungauged basins related studies (Beck et al., 2016; Oudin et al., 2008; Reichl et al., 2009). All listed models have a typical conceptual, bucket type with lumped parameters representation of runoff formation processes at basin scale with daily timestep. GR4J and SIMHYD models have been updated by adding Cema-Neige snow module (Valéry et al., 2014). Models' source code is freely available as a component of LHMP tool (Lumped Hydrological Models Playground, http://github.com/hydrogo/LHMP, Ayzel, 2016). Models' parameters were automatically calibrated by maximizing Nash-Sutcliffe criteria (NSE, Nash and Sutcliffe, 1970) for a whole period of observations using differential evolution algorithm

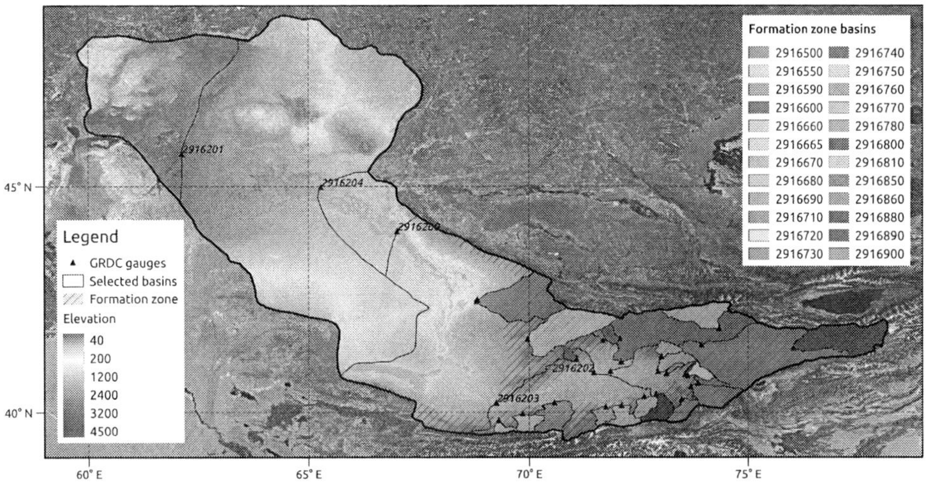

Figure 1. The Small Aral Sea basin and selected river basins.

which finds the global minimum of a multivariate function (http://docs.scipy.org, Storn and Price, 1997).

3.2 Machine learning models

For runoff modeling, we have used different machine learning models starting from simple MLP (Multiple Linear Regression) and wide-based decision tree ensembles of ETR (Extra Trees Regression) to the most complicated depth-based tree ensembles of LGB (Light Gradient Boosting machine) and XGB (eXtreme Gradient Boosting machine). MLP and RFR were implemented by using the Scikit-learn package (https://github.com/scikit-learn/scikit-learn, Pedregosa et al., 2011), LGB was taken from the LightGBM package (https://github.com/Microsoft/LightGBM, Zhang et al., 2017), LGB was taken from XGboost package (https://github.com/dmlc/xgboost, Chen and Guestrin, 2016). Machine learning model parameters tuning requires a lot of expertise and experimentation and cannot be resolved automatically because of high computational complexity (Snoek et al., 2012), therefore we had calibrated required parameters manually. For deriving predictions in ensemble manner and approaching realism in the model setting we have used leave-one-out cross-validation technique for machine learning model performance assessment (Ayzel, 2017; Hastie et al., 2001) – as a result, we evaluated model performance on every observational point independently and produced ensemble realization according to the amount of runoff observation we use. This setting provides us most comprehensive evaluation protocol for machine learning models and uncertainties related to models' structures.

3.3 Feature engineering

Feature engineering is an essential part of any routine of machine learning model developing. The general idea of feature engineering is to map already existed features of data to the new representation (dimension). Two (among others) classical implementations of these techniques are extending data with adding some features shifted in time (further referred as LAGS) and shrinking data dimensionality with principal component analysis (PCA) orthogonal transformation algorithm (Hastie et al., 2001). We have tested performances of our machine learning models with default input features, using LAGS and PCA separately and in a coupled setting, and then select the best combination in term of runoff predictions accuracy.

3.4 Workflow representation

The main idea of the presented work was to extract the value using all freely available hydrological information available for the Small Aral Sea basin. On the first stage of our research workflow (Fig. 2) we have calibrated parameters of three hydrological models for 24 rivers run to Ferghana Valley. During the calibration stage, every model had been running at daily temporal resolution then predicted runoff was aggregated at monthly scale for consistency with observational data for loss function calculation. On the second stage, we have implemented common spatial proximity based model parameters regionalization technique (Oudin et al., 2008) for transferring optimal sets of model parameters to meteorological forcing grid cells centroids. On the third stage we have run our models in a grid cell wise mode – for computing runoff in every grid cell in previously delineated formation zone (Fig. 1). As a result we have developed daily gridded multi-model runoff database for the Small Aral Sea

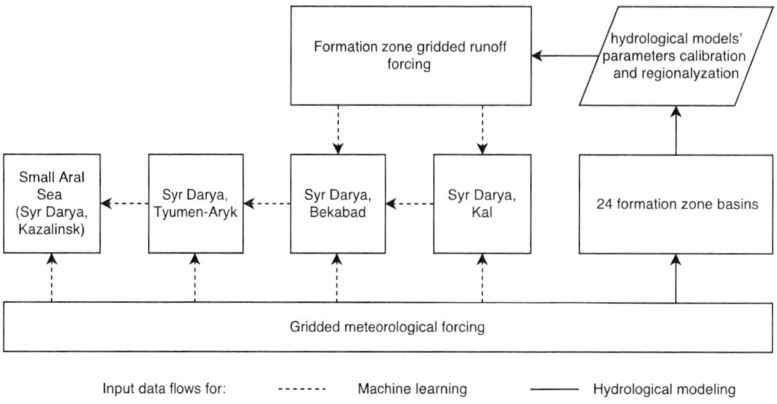

Figure 2. Runoff modeling workflow.

runoff formation zone which serves us as additional input data source for runoff modeling using machine learning models: for the first gauge in our cascade on the Syr Darya river – Kal – we have used both gridded meteorological and formation zone runoff forcing as input, the same for the next gauge in a cascade – Bekabad – but with added mean ensemble modeled runoff realization from Kal. For the remaining two gauges in a cascade (Tyumen Aryk and Kazalinsk), we used only meteorological forcing and mean ensemble modeled runoff realization from overlying gauge in a cascade.

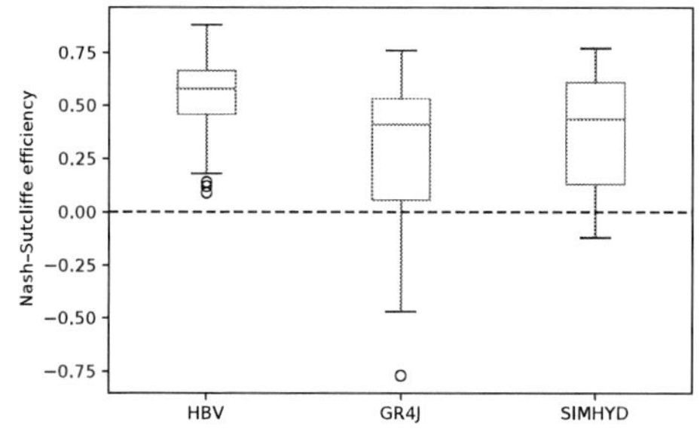

Figure 3. Boxplot of NSE for formation zone basins.

4 Results and discussion

Model calibration results differ from model to model and from different complexity of optimization algorithm. Only one setting with HBV model and the most computationally expensive realization of differential evolution algorithm (number of iteration equals 25) provides positive values of NSE for every single basin (Fig. 3) and we have decided to use only this set-up for further investigations. Only five of selected basins have an NSE less than 0.45 – all of them (GRDC ids: 2916590, 2916660, 2916665, 2916670, 2916810) are located on a north exposition of Alay range. These low efficiencies can be explained by errors in GRDC observational runoff data or errors in basins' metadata (wrong coordinates of outlets, basin areas) which are quite hard to detect and check in the open literature and web sources. Inter-comparison of obtained modeling results with different studies (Lutz et al., 2012b; Pereira-Cardenal et al., 2011; Radchenko et al., 2017; Siegfried et al., 2012) shows high consistency among different approaches for modeling water balance in the upper and mountainous part of the Syr Darya river. Therefore we showed positive value of using freely available data sources for water balance modeling in the study area and tried to transfer this value to the gridded runoff database of the Small Aral Sea (runoff) formation zone using the most robust way of model parameters regionalization (Ayzel et al., 2017). Using developed gridded runoff dataset for extracting runoff realization for 24 selected basins in a semi-distributed manner shows good consistency with realizations were produced by lumped model setting with optimal parameters.

There is no silver bullet in machine learning field in taking a priori decision about the best data preprocessing routine, the best model, the best validation technique, and the best measure of failure (or success) of the proposed approach. In our research, we have tried to investigate the most widespread solutions for tackling regression problems in machine learning using different state-of-the-art techniques. Results (Table 1) show a good efficiency of different machine learning models to predict monthly runoff alongside a cascade of gauges on the Syr Darya river. The high variance between models' efficiencies from gauge to gauge is explained by the various complexity of water management infrastructure and runoff formation complexes located between those gauges. The worst results for Kal and Kazalinsk gauges and the best efficiency for Tyumen Aryk gauge highly correspond with a complexity of runoff formation/transformation processes we want to map with our input data. The only inflow from the upper Bekabad gauge is enough for robust mapping to a runoff in Tyumen Aryk using simple linear regression

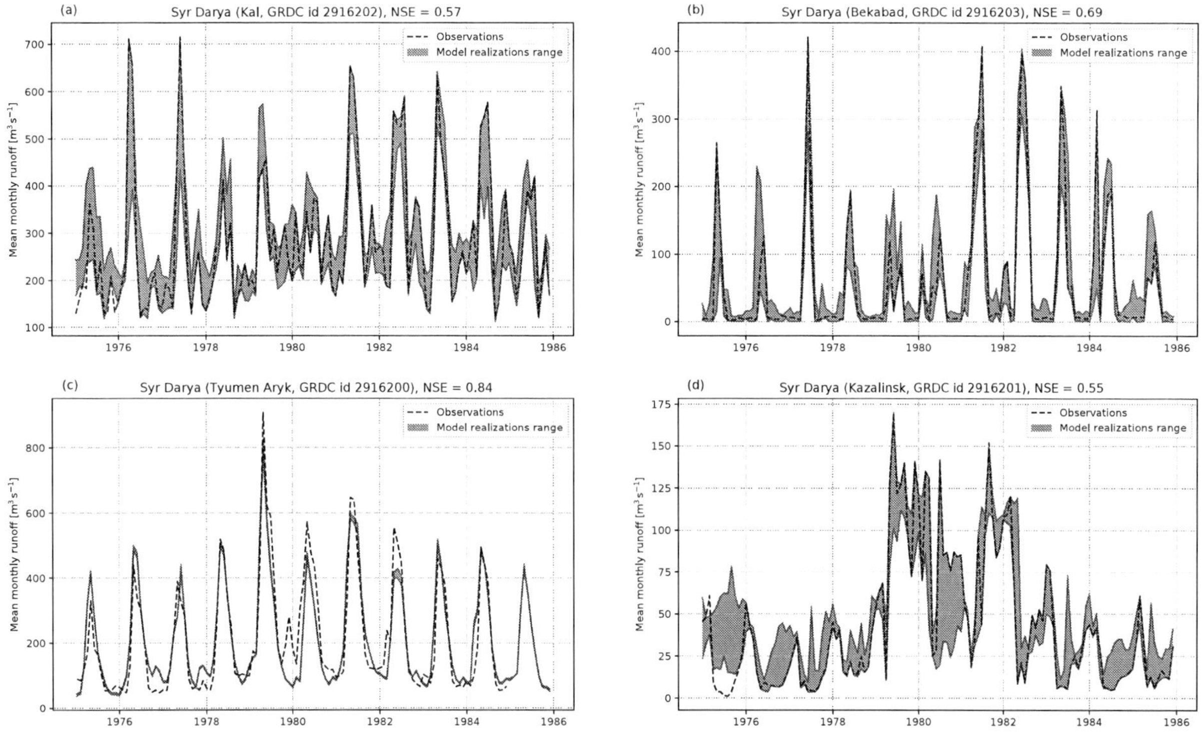

Figure 4. Predictions of machine learning model ensembles.

Table 1. Runoff modeling results.

Gauge, GRDC id	Machine learning models efficiency (NSE)				Feature engineering
	MLR	ETR	LGB	XGB	
Kal (2916202)		0.56	0.56	0.58	LAGS, PCA
Bekabad (2916203)		0.67	0.66	0.73	
Tyumen Aryk (2916200)	0.84				
Kazalinsk (2916201)		0.56		0.54	LAGS

model because of little factors of water redistribution in this local region. But we need more (in terms of quality, quantity, and diversity) data to map more complicated relationships for Kal and Kazalinsk because of high load of water management system on runoff formation/transformation processes. There is no clear pattern in models' efficiency for specific gauges – models rank differently in different settings. Nevertheless, we consider that the best solution in our case is to use both linear MLR and non-linear XGB with different feature engineering techniques for maximizing the spread of possible solutions. Obtained results are in the upper part of NSE range as in Gudmundsson and Seneviratne (2015) who provided monthly runoff predictions for a set of Europe river basins using Random Forest model and Watch Forcing Data (http://www.eu-watch.org/data_availability) as input forcing – this result underlines the crucial value of added gridded runoff information to machine learning model inputs which allows comparable model performance with European basins.

Ensemble runoff predictions produced by machine learning models (Fig. 4) depict significant rate of model-related uncertainties which highly correlates with model complexity. This highlights a statement "the simpler – the better" regarding scientific model robustness issues, but we have to mention that high prediction uncertainties are fair pay for the ability of complex model map input features to relevant output. It is also clear that obtained efficiency correlates with overall complexity of observed system – for Kal gauge station (Fig. 4a) wide range of overlying tributaries and water management rules on them significantly contribute to complexity of processes we have to consider, the same is relevant for the gauge station in Kazalinsk which affected by many, often fuzzy and unclear water management practices (Fig. 4d). This result is also confirmed by the complexity of preprocessing routine – for the simplest cases (Bekabad and Tyumen Aryk gauge stations) we do not need to implement either PCA or LAGS for mapping features for a different dimension.

There is about 100 km from Kazalinsk gauge station to the actual Syr Darya delta, and there are a lot of channels, ponds, and other water management infrastructure units (e.g. the Aklak water regulation station) which can affect total freshwater inflow in the Small Aral Sea basin, but for consistency with previous studies (Lutz et al., 2012a; Raskin et al., 1992) we consider the equality between observed runoff in Kazalinsk and freshwater inflow to the sea. Only a brief look on the observed runoff time series in Kazalinsk (Fig. 4d) gives any researcher a clear representation of high complexity of runoff formation system behavior here – we can only detect simple seasonal pattern with maximum water availability during winter, but it is impossible to generalize remaining runoff amplitude according to natural reasons. Nevertheless, XGB and ETR models utilize this complexity well due to their native algorithmic structures based on simple binarized decision rules which try to mimic decision-making process takes place in many real-life situations. Despite the clear attraction of observations to lower and upper boundaries of our prediction interval which may identify unstable system behavior, there is an obvious correlation between observed and modeled runoff.

Despite the limited observed runoff data availability for this region (mainly for 1975–1985) which was the main constraint in implementing comprehensive routines for proposed methodology validation, obtained machine learning model-based ensemble realizations of freshwater inflow into the Small Aral Sea for the period of 1958–2002 (alongside the forcing data availability) could form the basis for further "Soviet-driven water management" scenario predictions which help us better understand modern shifts in water resources distribution in post-Soviet time.

5 Conclusions

The complex structure of the Small Aral Sea basin water management system coupled with the total absence of data describes its functioning is a challenge for any approach directed to the accurate assessment of the freshwater budget formation and evolution across the basin. Our work shows the possibility to tackle these challenges by coupling hydrological models with the state-of-the-art machine learning techniques. In detail, we have evaluated the significant value of using physically based models for runoff predictions in ungauged upper part of the Syr Darya river for developing gridded runoff database which can be used as an additional feature for machine learning model in a coupled setting. Results show a positive skill and a high flexibility of the proposed methodology, and in our perspective, it can be used widely as a baseline approach for water balance research studies in arid, ungauged areas, with complex water management system and strong water-related data scarcity.

We understand that an equality between freshwater inflow into the Small Aral Sea and observed runoff in Kazalinsk is quite a rough assumption, and in the further studies, we will try to assess real inflow by coupling simple seawater balance model to our existing modeling system.

The code and data we have developed are totally open and freely accessible. We hope that this supports reproducibility of our research and provides easy access to the community to test, criticize, or apply our findings.

Code and data availability. Raw data were downloaded from ECMWF Public Datasets web interface (http://apps.ecmwf.int/datasets/) and GRDC archive (http://www.bafg.de/GRDC/, via request). For using raw data you have to agree with corresponding data policies from ECMWF and GRDC. You can find all code and data (under MIT license) on our project Github page (https://github.com/SMASHIproject/IWRM2018; Ayzel and Izhitskiy, 2017). There are no restrictions on use or distribution of our software code and data.

Competing interests. The authors declare that they have no conflict of interest.

Special issue statement. This article is part of the special issue "Innovative water resources management – understanding and balancing interactions between humankind and nature". It is a result of the 8th International Water Resources Management Conference of ICWRS, Beijing, China, 13–15 June 2018.

Acknowledgements. This work was undertaken within the frame of the SMASHI project (http://smashiproject.github.io) and was funded by the Russian Foundation for Basic Research (RFBR), project 17-05-01175 A. The part of presented study related to the developing, adapting and implementing of conceptual hydrological model was financially supported by the Russian Science Foundation (grant number 16-17-10039). The Global Runoff Data Centre (GRDC) is gratefully acknowledged for providing observed runoff data.

Edited by: Wenchao Sun
Reviewed by: two anonymous referees

References

Apel, H., Abdykerimova, Z., Agalhanova, M., Baimaganbetov, A., Gavrilenko, N., Gerlitz, L., Kalashnikova, O., Unger-Shayesteh, K., Vorogushyn, S., and Gafurov, A.: Statistical forecast of seasonal discharge in Central Asia for water resources management: development of a generic linear modelling tool for operational use, Hydrol. Earth Syst. Sci. Discuss., https://doi.org/10.5194/hess-2017-340, in review, 2017.

Ayzel, G.: LHMP: First major release, https://doi.org/10.5281/zenodo.59680, 2016.

Ayzel, G.: Use of machine learning techniques for modeling of snow depth, Ice and Snow, 34–44, https://doi.org/10.15356/2076-6734-2017-1-34-44, 2017.

Ayzel, G. and Izhitskiy, A.: Data, code, and results for the paper "Coupling physically based and data-driven models for assessing freshwater inflow into the Small Aral Sea", https://doi.org/10.5281/zenodo.1161906, 2017.

Ayzel, G. V., Gusev, E. M., and Nasonova, O. N.: River runoff evaluation for ungauged watersheds by SWAP model. 2. Application of methods of physiographic similarity and spatial geostatistics, Water Resour., 44, 547–558, https://doi.org/10.1134/S0097807817040029, 2017.

Beck, H. E., van Dijk, A. I. J. M., de Roo, A., Miralles, D. G., McVicar, T. R., Schellekens, J., and Bruijnzeel, L. A.: Global-scale regionalization of hydrologic model parameters, Water Resour. Res., 52, 3599–3622, https://doi.org/10.1002/2015WR018247, 2016.

Belyaev, A. V.: Water Balance and Water Resources of the Aral Sea Basin and Its Man-Induced Changes, GeoJournal, 35, 17–21, 1995.

Chen, T. and Guestrin, C.: XGBoost: Reliable Large-scale Tree Boosting System, available at: https://arxiv.org/abs/1603.02754 (last access: 13 February 2018), 2016.

Chen, Y., Li, W., Fang, G., and Li, Z.: Review article: Hydrological modeling in glacierized catchments of central Asia – status and challenges, Hydrol. Earth Syst. Sci., 21, 669–684, https://doi.org/10.5194/hess-21-669-2017, 2017.

Chiew, F. H. S., Teng, J., Vaze, J., Post, D. A., Perraud, J. M., Kirono, D. G. C., and Viney, N. R.: Estimating climate change impact on runoff across southeast Australia: Method, results, and implications of the modeling method, Water Resour. Res., 45, W10414, https://doi.org/10.1029/2008WR007338, 2009.

Gan, R., Luo, Y., Zuo, Q., and Sun, L.: Effects of projected climate change on the glacier and runoff generation in the Naryn River Basin, Central Asia, J. Hydrol., 523, 240–251, https://doi.org/10.1016/j.jhydrol.2015.01.057, 2015.

Gudmundsson, L. and Seneviratne, S. I.: Towards observation-based gridded runoff estimates for Europe, Hydrol. Earth Syst. Sci., 19, 2859–2879, https://doi.org/10.5194/hess-19-2859-2015, 2015.

Hagg, W., Braun, L. N., Weber, M., and Becht, M.: Runoff modelling in glacierized Central Asian catchments for present-day and future climate, Nordic Hydrology, 37, 93–105, https://doi.org/10.2166/nh.2006.001, 2006.

Hagg, W., Braun, L. N., Kuhn, M., and Nesgaard, T. I.: Modelling of hydrological response to climate change in glacierized Central Asian catchments, J. Hydrol., 332, 40–53, https://doi.org/10.1016/j.jhydrol.2006.06.021, 2007.

Hastie, T., Tibshirani, R., and Friedman, J.: The Elements of Statistical Learning, Springer Series in Statistics, Springer New York Inc., https://doi.org/10.1007/978-0-387-84858-7, 2001.

Immerzeel, W. W. and Bierkens, M. F. P.: Asia's water balance, Nat. Geosci., 5, 841–842, https://doi.org/10.1038/ngeo1643, 2012.

Izhitskiy, A. S., Zavialov, P. O., Sapozhnikov, P. V., Kirillin, G. B., Grossart, H. P., Kalinina, O. Y., Zalota, A. K., Goncharenko, I. V., and Kurbaniyazov, A. K.: Present state of the Aral Sea: diverging physical and biological characteristics of the residual basins, Scientific Reports, 6, 23906, https://doi.org/10.1038/srep23906, 2016.

Jarsjö, J., Asokan, S. M., Prieto, C., Bring, A., and Destouni, G.: Hydrological responses to climate change conditioned by historic alterations of land-use and water-use, Hydrol. Earth Syst. Sci., 16, 1335–1347, https://doi.org/10.5194/hess-16-1335-2012, 2012.

Lindström, G., Johansson, B., Persson, M., Gardelin, M., and Bergström, S.: Development and test of the distributed HBV-96 hydrological model, J. Hydrol., 201, 272–288, https://doi.org/10.1016/S0022-1694(97)00041-3, 1997.

López, O., Houborg, R., and McCabe, M. F.: Evaluating the hydrological consistency of evaporation products using satellite-based gravity and rainfall data, Hydrol. Earth Syst. Sci., 21, 323–343, https://doi.org/10.5194/hess-21-323-2017, 2017.

Lutz, A., Droogers, P., and Immerzeel, W.: Climate Change Impact and Adaptation on the Water Resources in the Amu Darya and Syr Darya River Basins. FutureWater Report 110, Tech. rep., FutureWater, available at: http://www.futurewater.eu/wp-content/uploads/2013/01/CC_Downstream_Report_V51.pdf (last access: 13 February 2018), 2012a.

Lutz, A., Immerzeel, W., and Droogers, P.: Climate Change Impacts on the Upstream Water Resources of the Amu and Syr Darya River Basins. FutureWater Report 107, Tech. rep., FutureWater, available at: http://www.futurewater.eu/wp-content/uploads/2012/03/Upstream_Report_FW_web.pdf (last access: 13 February 2018), 2012b.

Micklin, P.: The Aral Sea Disaster, Annu. Rev. Earth Planet. Sc., 35, 47–72, https://doi.org/10.1146/annurev.earth.35.031306.140120, 2007.

Nash, J. E. and Sutcliffe, J. V.: River flow forecasting through conceptual models part I – A discussion of principles, J. Hydrol., 10, 282–290, https://doi.org/10.1016/0022-1694(70)90255-6, 1970.

Oudin, L., Hervieu, F., Michel, C., Perrin, C., Andréassian, V., Anctil, F., and Loumagne, C.: Which potential evapotranspiration input for a lumped rainfall-runoff model? Part 2 – Towards a simple and efficient potential evapotranspiration model for rainfall-runoff modelling, J. Hydrol., 303, 290–306, https://doi.org/10.1016/j.jhydrol.2004.08.026, 2005.

Oudin, L., Andréassian, V., Perrin, C., Michel, C., and Le Moine, N.: Spatial proximity, physical similarity, regression and ungaged catchments: A comparison of regionalization approaches based on 913 French catchments, Water Resour. Res., 44, W03413, https://doi.org/10.1029/2007WR006240, 2008.

Pedregosa, F., Varoquaux, G., Gramfort, A., Michel, V., Thirion, B., Grisel, O., Blondel, M., Prettenhofer, P., Weiss, R., Dubourg, V., Vanderplas, J., Passos, A., Cournapeau, D., Brucher, M., Perrot, M., and Duchesnay, E.: Scikit-learn: Machine Learning in Python, J. Mach. Learn. Res., 12, 2825–2830, 2011.

Pereira-Cardenal, S. J., Riegels, N. D., Berry, P. A. M., Smith, R. G., Yakovlev, A., Siegfried, T. U., and Bauer-Gottwein, P.: Real-time remote sensing driven river basin modeling using radar altimetry, Hydrol. Earth Syst. Sci., 15, 241–254, https://doi.org/10.5194/hess-15-241-2011, 2011.

Perrin, C., Michel, C., and Andréassian, V.: Improvement of a parsimonious model for streamflow simulation, J. Hydrol., 279, 275–289, 2003.

Radchenko, I., Dernedde, Y., Mannig, B., Frede, H.-G., and Breuer, L.: Climate change impacts on runoff in the Ferghana Valley (Central Asia), Water Resour., 44, 707–730, https://doi.org/10.1134/S0097807817050098, 2017.

Raskin, P., Hansen, E., Zhu, Z., and Stavisky, D.: Simulation of Water Supply and Demand in the Aral Sea Region, Water Int., 17, 55–67, https://doi.org/10.1080/02508069208686127, 1992.

Reichl, J. P. C., Western, A. W., McIntyre, N. R., and Chiew, F. H. S.: Optimization of a similarity measure for estimating ungauged streamflow, Water Resour. Res., 45, W10423, https://doi.org/10.1029/2008WR007248, 2009.

Shi, W., Wang, M., and Guo, W.: Long-term hydrological changes of the Aral Sea observed by satellites, J. Geophys. Res.-Oceans, 119, 3313–3326, https://doi.org/10.1002/2014JC009988, 2014.

Shibuo, Y., Jarsjö, J., and Destouni, G.: Hydrological responses to climate change and irrigation in the Aral Sea drainage basin, Geophys. Res. Lett., 34, l21406, https://doi.org/10.1029/2007GL031465, 2007.

Siegfried, T., Bernauer, T., Guiennet, R., Sellars, S., Robertson, A. W., Mankin, J., Bauer-Gottwein, P., and Yakovlev, A.: Will climate change exacerbate water stress in Central Asia?, Climatic Change, 112, 881–899, https://doi.org/10.1007/s10584-011-0253-z, 2012.

Snoek, J., Larochelle, H., and Adams, R. P.: Practical Bayesian Optimization of Machine Learning Algorithms, in: Advances in Neural Information Processing Systems 25, edited by: Pereira, F., Burges, C. J. C., Bottou, L., and Weinberger, K. Q., 2951–2959, Curran Associates, Inc., 2012.

Sorg, A., Mosello, B., Shalpykova, G., Allan, A., Hill Clarvis, M., and Stoffel, M.: Coping with changing water resources: The case of the Syr Darya river basin in Central Asia, Environ. Sci. Policy, 43, 68–77, https://doi.org/10.1016/j.envsci.2013.11.003, 2014.

Storn, R. and Price, K.: Differential Evolution – A Simple and Efficient Heuristic for global Optimization over Continuous Spaces, J. Global Optim., 11, 341–359, https://doi.org/10.1023/A:1008202821328, 1997.

Uppala, S. M., Kållberg, P. W., Simmons, A. J., Andrae, U., Bechtold, V. D. C., Fiorino, M., Gibson, J. K., Haseler, J., Hernandez, A., Kelly, G. A., Li, X., Onogi, K., Saarinen, S., Sokka, N., Allan, R. P., Andersson, E., Arpe, K., Balmaseda, M. A., Beljaars, A. C. M., Berg, L. V. D., Bidlot, J., Bormann, N., Caires, S., Chevallier, F., Dethof, A., Dragosavac, M., Fisher, M., Fuentes, M., Hagemann, S., Hólm, E., Hoskins, B. J., Isaksen, L., Janssen, P. A. E. M., Jenne, R., Mcnally, A. P., Mahfouf, J.-F., Morcrette, J.-J., Rayner, N. A., Saunders, R. W., Simon, P., Sterl, A., Trenberth, K. E., Untch, A., Vasiljevic, D., Viterbo, P., and Woollen, J.: The ERA-40 re-analysis, Q. J. Roy. Meteor. Soc., 131, 2961–3012, https://doi.org/10.1256/qj.04.176, 2005.

Valéry, A., Andréassian, V., and Perrin, C.: "As simple as possible but not simpler": What is useful in a temperature-based snow-accounting routine? Part 1 – Comparison of six snow accounting routines on 380 catchments, J. Hydrol., 517, 1166–1175, https://doi.org/10.1016/j.jhydrol.2014.04.059, 2014.

Zavialov, P. O., Kostianoy, A. G., Emelianov, S. V., Ni, A. A., Ishniyazov, D., Khan, V. M., and Kudyshkin, T. V.: Hydrographic survey in the dying Aral Sea, Geophys. Res. Lett., 30, 1659, https://doi.org/10.1029/2003GL017427, 1659, 2003.

hang, H., Si, S., and Hsieh, C.-J.: GPU-acceleration for Large-scale Tree Boosting, available at: http://arxiv.org/abs/1706.08359 (last access: 13 February 2018), 2017.

Zmijewski, K. and Becker, R.: Estimating the Effects of Anthropogenic Modification on Water Balance in the Aral Sea Watershed Using GRACE: 2003–12, Earth Interact., 18, 1–16, https://doi.org/10.1175/2013EI000537.1, 2014.

https://doi.org/10.5194/piahs-379-159-2018

Open Access
PIAHS

Evaluation of blue and green water resources in the upper Yellow River basin of China

Xiaoxi Gao[1], Depeng Zuo[1,2], Zongxue Xu[1,2], Siyang Cai[1], and Han Xianming[1]

[1]College of Water Sciences, Beijing Normal University, Beijing 100875, China
[2]Beijing Key Laboratory of Urban Hydrological Cycle and Sponge City Technology, Beijing 100875, China

Correspondence: Depeng Zuo (dpzuo@bnu.edu.cn)

Received: 31 December 2017 – Revised: 3 March 2018 – Accepted: 6 March 2018 – Published: 5 June 2018

Abstract. The total amount of water resources severely affects socioeconomic development of a region or watershed, which means that accurate quantification of the total amount of water resources is vital for the area, especially for the arid and semi-arid regions. Traditional evaluation of water resources only focused on the qualification of blue water, while the importance of green water was not fully considered. As the second largest river in China, the Yellow River plays an important role in socioeconomic development of the Yellow River basin. Therefore, the blue and green water resources in the upper Yellow River basin (UYRB) were evaluated by the SWAT model in this study. The results show that the average annual total amount of water resources in the UYRB was 140.5 billion m^3, in which the blue water resources is 37.8 billion m^3, and green water resources is 107.7 billion m^3. The intra-annual variability of the blue water and green water is relatively similar during the same period. The higher temperature, the greater difference between the blue and green water. The inter-annual variability of the blue and green water shows that the trends in precipitation, blue and green water have a relatively similar characteristic. The spatial distribution of the blue and green water is characteristic with gradually decreasing from the northwest to the southeast, and the blue water around the main stream is greater than that in the other areas.

1 Introduction

Water is one of the essential resources for human lives because it is needed for municipal, industrial and agricultural water use, hydropower generation and recreational activities (Lee and Bae, 2015). However, many people lack awareness ofwater conservation. In addition, the improper utilization of water resources and the increasing demand of water for agriculture and industry, the shortage of water resources has restricted the development of human society. Therefore, the importance of water resources should be fully understood in a sustainable development way and the amount of water resources should be scientifically assessed, which are key factors to guarantee the human society healthy and ordered development. However, the traditional evaluation of water resources only considers the surface water and groundwater in hydrologic cycle, namely "blue water" (Shikiomanov, 1997, 2000) and often have ignored the "green water" in hydrologic cycle. Green water flow to the atmosphere through evapotranspiration of grassland, forest and farmland accounts for 65 % of the global precipitation, while traditional blue water resources only occupy 35 % of the global precipitation (Ringersmas et al., 2003). In arid and semi-arid regions, green water plays an essential role in crop production and the provision of ecosystem services (Falkenmark, 1995). Therefore, many studies begin integrating the two types of water into one concept or theoretical framework (Falkenmark and Rockström, 2006; Rockström et al., 2009), it may provide a scientific basis for the planning and management of water resources and optimal utilization of water resources in the water scarce region especially in arid and semi-arid areas.

The concepts of blue and green water were proposed firstly by Falkenmark (1995) at the conference held by the Food and Agriculture Organization of the United Nations (FAO). He thought that the amount of water on the continent is composed of blue water and green water. Blue water is surface runoff formed by precipitation and deep aquifer recharge formed by infiltration, mainly including visible liquid water stored in rivers, reservoirs, lakes, wetlands and shallow aquifers. Green water is the water that infiltrates into the unsaturated soils from precipitation to plant growth and the invisible water flow released to the atmosphere. Afterwards the concept of green and blue water was developed by many other researches (Falkenmark and Rockström, 2006; Schuol et al., 2008a, b; Faramarzi et al., 2009; Liu and Yang, 2010). By definition, blue water is consisted of surface water and groundwater. Green water consisted of green water flow and green water storage. Green water flow is actually evapotranspiration including evaporation from soil and water bodies and transpiration from vegetation. Green water storage is the amount of water in the soil profile at the end of a time period.

The method of estimating blue water is well established, but the blue water is needed to develop. There are many studies about the estimation method of green water (Cheng and Zhao, 2006). The green water resources are estimated based on the water consumption of dry matter produced by the ecosystem and the water demand. For example, Postel et al. (1996) estimated the global non-irrigation vegetation evapotranspiration as the amount of green water resources by using the net primary productivity data; estimating the green water flow of biotic formation based on the factors affecting evapotranspiration, for example, Rockström and Gordon (2001) estimated the green water flow based on the vegetation cover multiply by the evapotranspiration and considering the factor affecting the evapotranspiration; estimating the green water flow by hydrological model, such as Faramarzi et al. (2009) estimated the blue-green water resources of Iranian by using ArcSWAT model, taking into account the impact of reservoir operation and different irrigation conditions on wheat crop yield; coupling hydrological models with vegetation dynamics, biogeography and so on to estimate the green water flow. For example, Y. J. Wang et al. (2009) constructed eco-hydrological models to quantitatively simulate the ecological water consumption of Sanmenxia vegetation and analyzed the water consumption efficiency of different vegetation cover.

As the largest source of water supply in northwest and north China, the runoff of the Yellow River accounts for 2 % of the whole China, but it is responsible for the irrigation of 15 % cultivated land and the water use of 12 % population of China (Lan, 2005). Since the 20th century, the water resources problem in the Yellow River Basin have become more and more serious, including the prominent contradiction between the supply and demand of water resources, the sharp drop in water volume and the frequent interruptions in the lower reaches and so on. It has seriously affected the socioeconomic development of the Yellow River Basin. In order to ensure the sustainable development of water resources in the Yellow River Basin and to ameliorate the local situation of water use, more attention should be pay to the ecological environment in the upper Yellow River Basin. Since Two thirds of the water resources of the Yellow River come from the upper reaches, the water resources issue of the Yellow River should be focused on the upstream. During the past half century, the changes of drought innorthern and even all over Chinasignificantly aggravated, there also occurred rare and serious droughts in the upper reaches region of the Yellow River. Coupled with the adverse effects of human activities, it makes the extremely fragile ecology environment more worse.

Therefore, this paper intends to take advantages of the SWAT model that can directly output the blue and green water to scientifically and reasonably assess the amount of blue and green water resources and its spatiotemporal change in the upper reaches region of the Yellow River. It could make us better understand conversion process of blue and green water and consequently provide useful information for water resource planning and utilizing in other arid and semi-arid regions.

2 Study area and data description

2.1 Study area

The Yellow River Basin is located between 96–119° E and 32–42° N. The Yellow River rises in Bayan Har Mountan of Qinghai province and goes through the provinces of Qinghai, Sichuan, Gansu, Ningxia, Inner Mongolia, Shaanxi, Shanxi, Henan and Shandong and flows into Sea in Dongying City of Shandong Province with an overall length of 5464km and the drainage area of 795 000 km^2. The basin crosses the northeast of the loess plateau and northeast of the Tibetan Plateau,and between 95°52′–103°50′ E and 32°09′–38°21′ N. The mainstream length of the studied area is 21 199 km, accounting for 38.9 % of the total length of the Yellow River and 61.0 % of the total length of the upper Yellow River basin. The drainage area cover 22.31 km^2, accounting for 29.7 % of the total area of theYellow River basin (Fig. 1). The mean annual depth of runoff and precipitation are 147.1 and 426.2 mm respectively (G. Wang et al., 2009).

2.2 Data description

The databases used in this study include Digital Elevation Model (DEM), land-use, soil-type, meteorological and hydrological observed data. DEM with a spatial resolution of 90 m was collected from Geospatial Data Cloud. Based on the DEM extracted by SWAT model according to the location of the Basin export section, the digital river network was generated, and the sub-basins were divided.

A land use map at a scale of 1 : 250 000 was provided by the Data Center for Resources and Environmental Sciences, Chinese Academy of Sciences (RESDC). The original land use was reclassified into 12 different types by merging similar type of land use to match the SWAT land use classification. The land use mainly include forest-evergreen (FRSE), forest-deciduous (FRSD), forest-mixed (FRST), range-brush (RNGB), pasture (PAST) and so on.

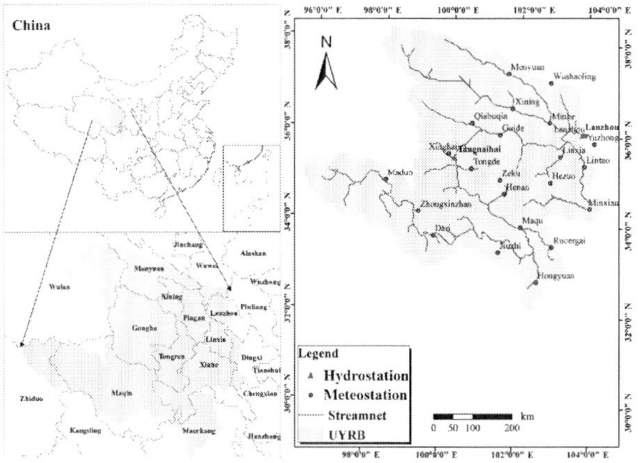

Figure 1. Location of the study area and the meteorological and hydrological stations selected.

A soil property map (1 : 1 000 000) was also obtained from the RESDC. Because the soil property database of SWAT model is based on US information and there was a little difference in grain size partition of soil between US and China, a local soil properties database for the UYRB was established combined with the Soil Plant Atmosphere Water (SPAW) software (Saxton and Rawls, 2006).

Weather input data (daily precipitation, maximum and minimum temperature, relative humidity, wind speed and sunshine hours) were obtained from China Meteorological Data Sharing Services System Network for 24 synoptic stations from 1968 to 1997. The daily solar radiation was calculated by daily sunshine hours with an empirical expression. River discharge data series needed for calibration and validation were obtained from the Yellow River Conservancy Commission (YRCC) for two gauging stations (Tangnaihai and Lanzhou) which have same period with weather data.

3 Methodology

3.1 SWAT model

The Soil and Water Assessment Tool (SWAT) is a basin hydrology model developed by the U.S. Department of Agriculture-Agricultural Research Service (USDA-ARS) in the early 1990s to assess the impact of land management practices on water, sediment, and agriculture chemical yield in large, complex watersheds with varying soils, land use, and management conditions over long periods (Zuo et al., 2016).

A watershed is first divided into sub-basin on the basis of the DEM and river and then into multiple hydrological response units (HRUs) based on the unique soils and land use types within the subbasins (Arnold et al., 1998). Due to the variability of spatial-temporal distribution for underlying surface and climatic elements, this SWAT model can improves the simulation accuracy to a certain extent. Hydrological processes can be divided into two phases. The former controls the amount of water, sediment and nutrients that will flow into the main channel at each sub-basin, while the latter controls transport of water, sediment and other substances in channels (Xu et al., 2009).

The preprocessing of the SWAT model input (e.g., watershed delineation, manipulation of the spatial and tabular data) was performed by ESRI ArcGIS10.1 using the ArcSWAT interface (Winchell et al., 2013).

3.2 SUFI-2 algorithm

The sequential uncertainty fitting algorithm (SUFI-2) procedure (Abbaspour, 2011) was used for parameter optimization. In this procedure, all uncertainty sources (parameters, models, measured data, and so on) are displayed onto parameter ranges, which are calibrated to bracket most of the measured data in the 95 % prediction uncertainty (95PPU) (Abbaspour et al., 2007). Overall uncertainty in the output is quantified by the 95PPU calculated at 2.5 % (L95PPU) and 97.5 % (U95PPU) levels of the cumulative distribution of an output variable obtained through Latin hypercube sampling (Zuo et al., 2015). And the P-factor and R-factor are used to measure the goodness of calibration/uncertainty. The P-factor indicates the percentage of data bracketed by the 95PPU band and the R-factor is the average width of the band divided by the standard deviation of the corresponding measured variable (Abbaspour et al., 2007; Yang et al., 2008). Theoretically, the calibration result is perfect when the P-factor is close to 1 and the R-factor is close to 0, so a better balance is needed to find to make both best.

The operation of SUFI-2 algorithm mainly includes:

1. Determine the objective function.

2. Select the parameters that are sensitive to the variables and give the limit of variation.

3. The model performs Latin Hypercube sampling according to the parameter range.

4. Simulating each group of parameters obtained by Latin Hypercube sampling.

5. Evaluate the objective function and calculate the final results.

3.3 Objective functions for model calibration and validation

There are nine types of objective function involved in SWAT. In this paper, two evaluation indexes was used including linear regression coefficient (R^2) and Nash efficiency coefficient (E_{ns}) to be as evaluation criteria for parameter opti-

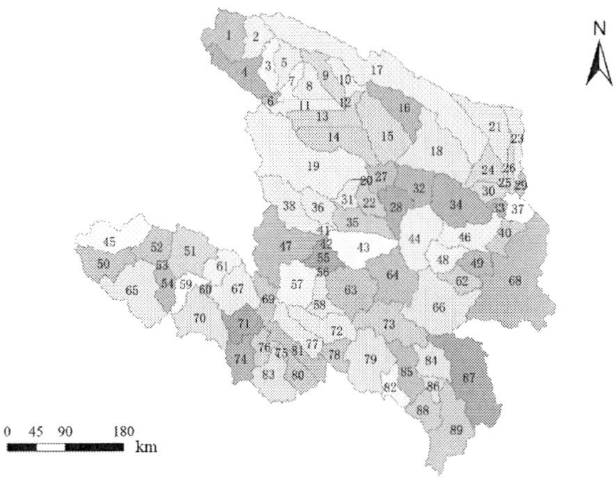

Figure 2. The number of every subbasin.

mization.

$$R^2 = \frac{\sum_{i=1}^{n}\left(Q_{\mathrm{m},i} - Q_{\mathrm{m,avg}}\right)\cdot\left(Q_{\mathrm{s},i} - Q_{\mathrm{s,avg}}\right)}{\sum_{i=1}^{n}\left(Q_{\mathrm{m},i} - Q_{\mathrm{m,avg}}\right)^2 \cdot \sum_{i=1}^{m}\left(Q_{\mathrm{s},i} - Q_{\mathrm{s,avg}}\right)} \tag{1}$$

$$E_{\mathrm{ns}} = 1 - \frac{\sum_{i=1}^{n}\left(Q_{\mathrm{m},i} - Q_{\mathrm{s},i}\right)}{\sum_{i=1}^{n}\left(Q_{\mathrm{m},i} - \overline{Q}_{\mathrm{s},i}\right)^2} \tag{2}$$

Nash and Sutcliffe (1970), Where $Q_{\mathrm{m},i}$ is measured value, $\mathrm{m^3\,s^{-1}}$; $Q_{\mathrm{s},i}$ is simulation value, $\mathrm{m^3\,s^{-1}}$; $Q_{\mathrm{m,avg}}$ is the average of measured value, $\mathrm{m^3\,s^{-1}}$; $Q_{\mathrm{s,avg}}$ for the average value of the simulation, $\mathrm{m^3\,s^{-1}}$; n is the length of the database.

The R^2 represent the consistency of trend for simulated and measured value, the larger the value of R^2 indicates the simulation agree better with the measurement. E_{ns}indicates the rate of deviation between the measured value and the simulated value, E_{ns} represents the degree of deviation for simulated and measured value, the value of E_{ns} are closer to 1which indicates the smaller deviation between the measured value and the simulated value when $E_{\mathrm{ns}} \leq 0.36$, the simulation is considered bad, when $0.36 \leq E_{\mathrm{ns}} < 0.75$, the simulation is considered satisfied, when $E_{\mathrm{ns}} \geq 0.75$, the simulation is thought good (Geza and Mccray, 2008).

4 Results and discussion

4.1 The model calibration and validation

According to the DEM of upper reaches of the Yellow River, the catchment area is divided into 89 sub-basins by the SWAT model (Fig. 2).

After completing the model specification, the model parameters need to be calibrated and validated to ensure the accuracy of the final results that estimating the blue and green water. The data from 1966 to 1967 were taken as warm-up period. The data from 1968–1982 were used for calibration and from 1983 to 1997 was used for verification purpose. Due to the large number of parameters contained in the SWAT model, 10 parameters that are sensitive to runoff are selected for the calibration and verification based on runoff data. The R^2 and E_{ns} are used for evaluating goodness of fit between the simulated and measured values.

The SUFI-2 algorithm is used to calibrate the parameters of model through two main hydrological stations (Tangnaihai and Lanzhou station) at the basin, the hydrographs for the calibration and validation periods are showed in the Figs. 3 and 4 respectively.

According to Figs. 3 and 4, it can be seen that the trend of the measured value and the simulated value of two main gauging stations is good at the calibration and validation periods, but there is an obvious deflection in the place where appear extremum. And there are some obvious deviation between measured value and simulated value at base flow part in Tangnaihai verification period, which may be related to the parameter of ALPHA_BF in validation period, or because of the self-defect of SWAT model for extremum simulation of runoff.

The calculation of snowmelt runoff in the SWAT model mainly considers the effect of air temperature. The degree-day factor method is suitable for the plain areas with abundant rainfall and flat terrain. However, portions of the study area where the altitude is high, the climate is cold and rainfall is scarce, the snowmelt runoff is not only affected by the temperature, but the topography, climate change and solar radiation also have a great impact on snowmelt runoff. The accuracy of snowmelt runoff simulated by the degree-day factor method of SWAT model is relatively low, which actually gives some errors to the assessment results of blue and green water resources. However, the relative error of the simulation period was −2.35 %, the correlation coefficient was 0.63, and the Nash efficiency coefficient was 0.74, which met the accuracy requirements of model simulation.

On the whole, the simulation accuracy of the model for the validation period is satisfactory that proving it is feasible to use the SWAT model to analyze the blue and green water in the upper Yellow River basin of China.

4.2 Assessment of blue and green water resources

According to output file of the model, the amount of blue and green water (1968–1997) in the upper Yellow River basin of China can be obtained (Table 1). The sum of WYLD and DA_RCHG can be considered as the blue water and the green water is equal to the sum of green water flow and green water storage.

According to Table 1, the average annual precipitation in the upper Yellow River basin of China was 140.48 billion $\mathrm{m^3}$, the total amount of blue and green water resources was

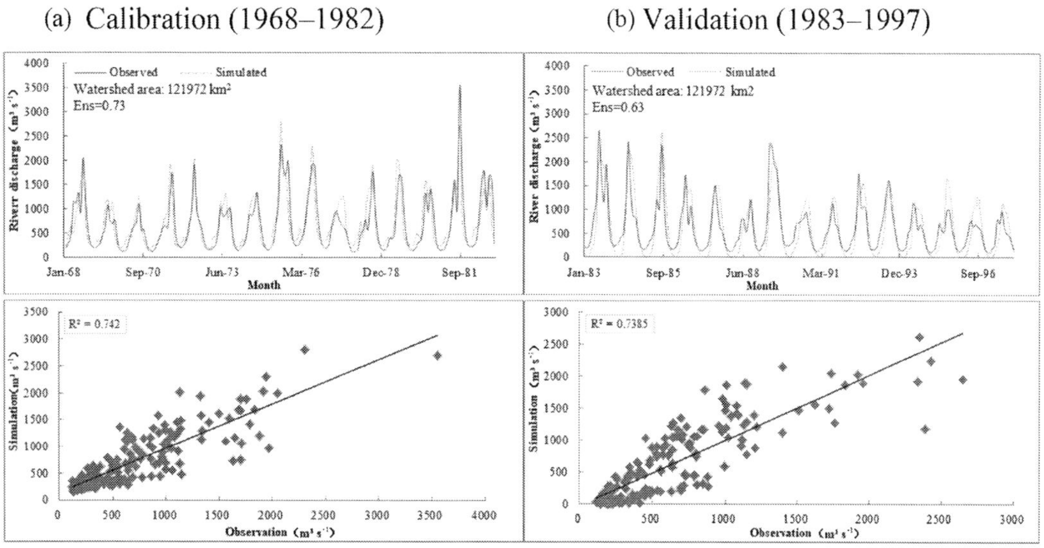

Figure 3. Comparison between the measured and the simulated discharges for Tangnaihai hydrometric station. Calibration **(a)** and Validation **(b)** are showed.

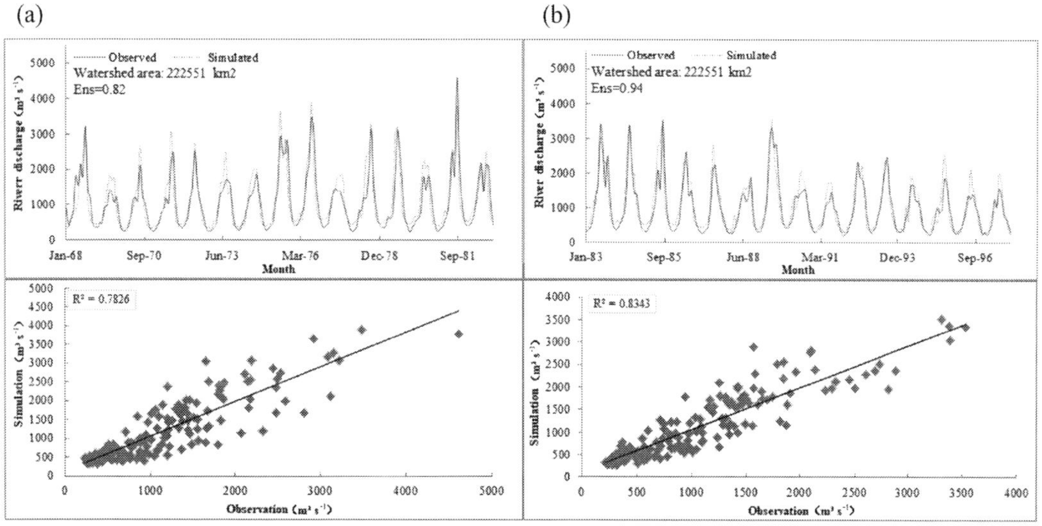

Figure 4. Comparison between the measured and the simulated discharges for Lanzhou hydrometric station. Calibration **(a)** and Validation **(b)** are showed.

145.42 billion m^3, of which 37.76 billion m^3 was blue water and 107.66 billion m^3 was green water. The amount of green water was about 2.85 times the amount of blue water. As a result, the upper Yellow River basin of China has abundant water resources. However, most of the precipitation in the basin is converted into green water because of the influence of regional climate and underlying surface, which cannot be directly used by human beings. As global temperatures rise making the situation worsen in the future. The continuous reduction of blue water resources will inevitably lead to the contradiction among supply and demand of water resources and ecological environment in the basin.

4.3 The temperal variations of blue and green water resources

The water quantity of blue and green water can be obtained from the SWAT model output file and the average value of blue and green water resources in each month from 1968 to 1997 had been shown in Table 1. The intra-annual variability of blue and green water was shown in Table 2.

Table 1. The amount of the precipitation and the blue and green water from 1968 to1997

Year	Precipitation (10^9 m^3)	Water resources (10^9 m^3)	Blue water (10^9 m^3)	Green water (10^9 m^3)
1968	143.20	144.59	39.33	105.26
1969	130.09	141.19	32.46	108.73
1970	136.07	144.17	32.54	111.64
1971	145.05	141.70	35.24	106.47
1972	130.09	141.98	33.99	107.99
1973	145.18	150.23	35.87	114.36
1974	143.23	144.25	34.68	109.57
1975	163.20	162.41	48.29	114.12
1976	153.14	161.06	46.30	114.76
1977	129.08	136.23	37.34	98.88
1978	151.60	154.39	42.62	111.77
1979	137.45	145.95	40.46	105.48
1980	131.57	138.84	36.83	102.01
1981	159.07	152.35	45.56	106.79
1982	136.99	146.10	38.74	107.36
1983	159.67	160.88	46.56	114.32
1984	142.90	150.64	44.53	106.12
1985	158.84	159.82	47.11	112.70
1986	129.55	141.04	39.29	101.76
1987	132.11	138.39	35.98	102.41
1988	138.22	140.38	33.26	107.12
1989	166.28	158.66	45.32	113.34
1990	127.64	143.42	36.20	107.22
1991	121.64	131.76	30.21	101.54
1992	145.13	142.75	36.04	106.70
1993	139.44	145.58	35.13	110.45
1994	132.78	141.19	31.87	109.31
1995	130.98	134.87	32.85	102.02
1996	126.24	136.42	29.76	106.66
1997	128.00	131.50	28.54	102.96
Mean	140.48	145.42	37.76	107.66

Table 2. The amount of blue and green water within the year.

Month	Precipitation (mm)	Green water (mm)	Blue water (mm)
1	3.50	2.71	4.70
2	5.25	5.74	4.51
3	13.12	7.21	4.92
4	27.05	227.86	8.33
5	64.18	52.80	11.19
6	90.76	66.74	17.42
7	110.69	76.06	24.28
8	98.58	70.29	22.91
9	77.73	49.71	20.58
10	31.86	26.49	11.19
11	4.92	8.28	5.53
12	1.17	1.55	3.39

As can be seen from Fig. 5, the amount of blue and green water varies irregularly during the year, and the trend was basically accord with precipitation. The precipitation reached the highest in the flood season (6–8) and the lowest in winter (12–2), showing a process of gradual advance and decline.

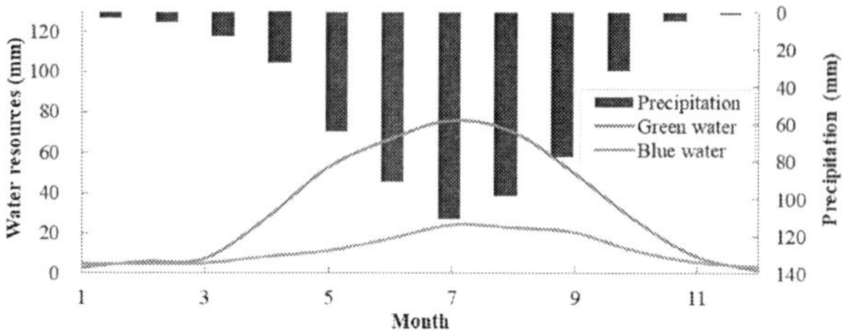

Figure 5. The intra-annual variability of the blue water and green water during the year.

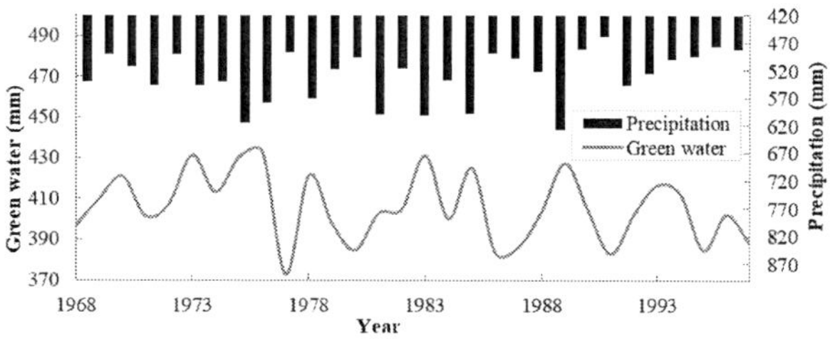

Figure 6. The inter-annual variability of the precipitation and green water.

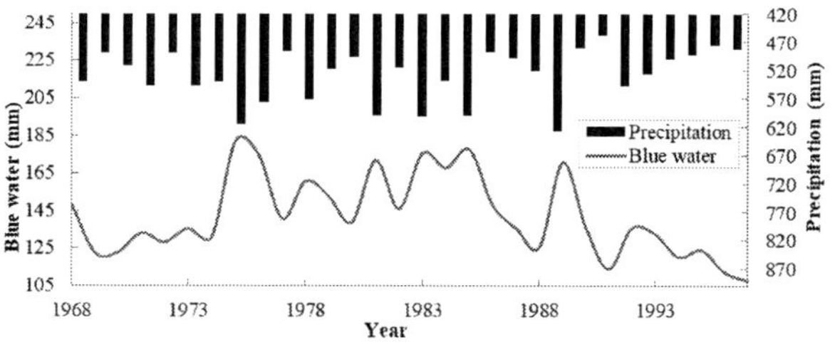

Figure 7. The inter-annual variability of the precipitation and blue water.

The blue and green water resource was low in winter, not only connected with less precipitation, but the temperature was also an important factor. With the temperature rising, the difference between the blue and green water became wider and wider with the highest in July. The reason was that the higher temperature, the greater evapotranspiration. Therefore the summer precipitation was mainly consumed by evapotranspiration.

By calculating the annual precipitation and blue and green water resources from 1968 to 1997, the intra-annual variation of precipitation and blue water and green water were got and shown in Figs. 6 and 7.

As can be seen from Figs. 6 and 7, the inter-annual variation of precipitation, blue water and green water were basically the same. Where the years with large variation of precipitation, the change of the blue and green water are also large, forming trough and crest. There is a typical phenomenon that more precipitation during 1975 and 1989 and more blue and green water during the period, indicating the

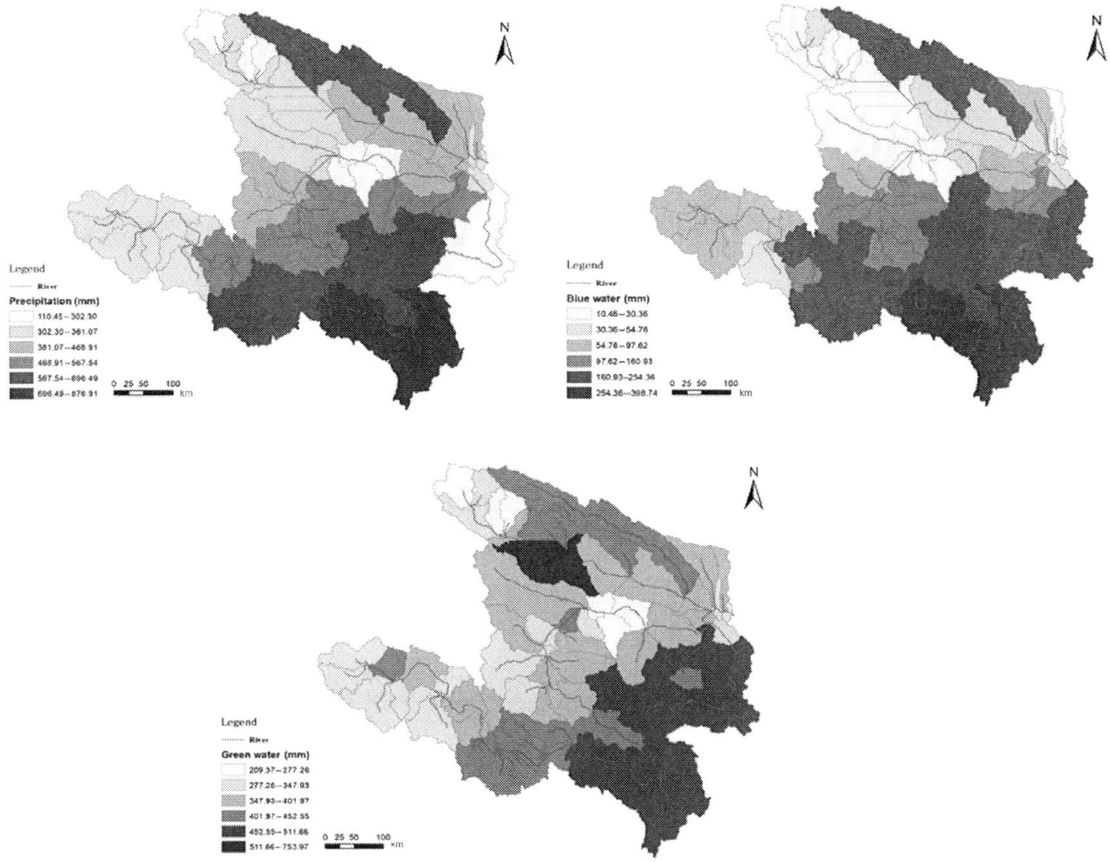

Figure 8. The spatial distribution of the precipitation and the blue and green water.

change of blue and green water were closely related to precipitation. In addition, the change of blue water in some years lagged behind that precipitation. One cause may be that the evaporation is relatively small during each precipitation in a year, cumulative processes are the major factors behind the phenomenon.

4.4 The spatial patterns of blue and green water resources

Through neatening the output files of SWAT model, the average annual precipitation (1968–1997) of precipitation, blue and green water at 89 subbasins can be obtained. The ArcGIS10.1 was used to express the spatial distribution of precipitation, blue and green water (Fig. 8).

The spatial distribution of precipitation, blue and green water were shown in Fig. 8. The Fig. 8 illustrates that mean annual precipitation was about 300 and 700 mm in the northwest and southeast of study area respectively, and showing a decreasing trend from southeast to northwest. Most of the green water comes from the actual evapotranspiration, so the green water is closely related to the distribution of precipitation in the basin. The spatial distribution of green water and precipitation are basically the same. Because of the existence of Qinghai Lake in the northwest of Qinghai Province, large quantity of green water evaporates from water surface. The spatial distribution of blue water is consistent with the precipitation and green water. The reason for the result is closely related to the underlying surface conditions in the basin. The southeastern of basin is distributed mainly over Hongyuan County, Zoige County and Aba Prefecture in Sichuan Province. There are much swamps and wetlands and the temperature is higher compared with other regions at the same latitude, resulting in a large amount of precipitation and evaporation in the region. At the same time, the upper Yellow River basin is located in inland basin of northwest China where far away from the oceans, making it hard to be affected by warm wet air current of Pacific Ocean, coupled with the impact of global warming, making increase of growing season for vegetation, in addition the vegetation coverage increases from the southeast to the northwest, which is another reasons for the decrease of green water from southeast to northwest.

5 Conclusions

The assessment on spatial and temporal distribution of blue and green water was evaluated in the upper Yellow River basin of China, the major conclusions obtained in this study are as follows:

1. The mean annual (1968–1997) precipitation of the upper basin of the Yellow River was 140.481 billion m^3, and the sum multi-year mean of blue and green water was 145.425 billion m^3, which was 3.85 times of the blue water that traditional evaluation of water resources. The annual average of blue and green water were 37.764 and 107.661 billion m^3 respectively, and green water accounts for 77 % of the total water resources, indicating that green water resources are the main part of the total water resources, while the upper basin of the Yellow River are located in arid and semi-arid area, so it is essential to evaluate the blue water and green water together for mitigating contradiction of water resource supply and demands. At the same time, water shortage in the Yellow River Basin can be improved by implementing green water management in the basin, reducing ineffective evaporation and surface runoff, and increasing downstream water yield.

2. The variation within the year for blue and green water and precipitation were similar, showing the highest in the flood season (6–8) and the lowest in winter (12–2). The dissipation of green water increases in pace with the temperature rise, and the gap between green water and blue water is also gradually increasing with the temperature rise. The green water in summer is the main component of the total green water throughout the year, while in winter, the green water is low because of low temperature.

3. The inter-annual variation (1968–1997) of blue and green water and precipitation showed a similar tendency, and the years with large precipitation, the blue and green water are also generally larger, demonstrating the precipitation is the dominant factor for blue and green water.

4. The spatial distribution of green and blue water and precipitation are basically the same in the whole upper reaches of the Yellow River which present decreasing trend from southeast to northwest. The amount of blue water in the basin is between 10.46–398.74 mm, and the blue water around the main stream is greater than that in the other areas. In addition, the sub-basins with large precipitation also have abundant blue water resources.

Data availability. Digital elevation Model (DEM) was collected from Geospatial Data Cloud (http://www.gscloud.cn/; International Scientific & Technical Data Mirror Site, 2003), and the weather input data were obtained from China Meteorological Data Sharing Services System Network (http://data.cma.cn/; CMDC, 2018). Land use map and soil property map were obtained from the Data Center for Resources and Environmental Sciences, Chinese Academy of Sciences (RESDC), and runoff data was collected from Yellow River Institute of Hydraulic Research, but these data required to reproduce these findings cannot be shared at this time due to legal or ethical reasons.

Competing interests. The authors declare that they have no conflict of interest.

Special issue statement. This article is part of the special issue "Innovative water resources management – understanding and balancing interactions between humankind and nature". It is a result of the 8th International Water Resources Management Conference of ICWRS, Beijing, China, 13–15 June 2018.

Acknowledgements. This study is jointly supported by the National Natural Science Foundation of China (Grant No. 91647202), the National Key Research and Development Program of China (Grant No. 2017YFC1502703), and the Major Science and Technology Program for Water Pollution Control and Treatment (Grant No. 2017ZX07302-04).

Edited by: Yangbo Chen
Reviewed by: two anonymous referees

References

Abbaspour, K. C.: SWAT-CUP, SWAT Calibration and Uncertainty Programs, Swiss Federal Institute of Aquatic Science and Technology, Eawag, Duebendort, Swizerland, p. 103, 2011.

Abbaspour, K. C., Yang, J., Maximov, I., Siber, R., Bogner, K., Mieleitner, J., Zobrist, J., and Srinivasan, R.: Modelling hydrology and water quality in the pre-alpine/alpine Thur watershed using SWAT, J. Hydrol., 333, 413–430, https://doi.org/10.1016/j.jhydrol.2006.09.014, 2007.

Arnold, J. G., Srinivasan, R., Muttiah, R. S., and Williams, J. R.: Large area hydrologic modeling and assessment part I: Model development, J. Am. Water Resour. Assoc., 34, 73–89, https://doi.org/10.1111/j.1752-1688.1998.tb05961.x, 1998.

Cheng, G. D. and Zhao, W. Z.: Green water and its research progresses, Adv. Earth Sci., 21, 221–227, 2006.

CMDC (China Meteorological Data Service Center): Dataset of daily climate data from Chinese surface stations ffor global exchange (V3.0), available at: http://data.cma.cn/ (last access: 13 April 2017), 2018.

Falkenmark, M.: Coping with water scarcity under rapid population growth, Conference of SADC Ministers, Pretoria, 1995.

Falkenmark, M. and Rockström, J.: The New Blue and Green Water Paradigm: Breaking New Ground for Water Resources Planning and Management, J. Water Res. Pl., 132, 129–132, 2006.

Faramarzi, M., Abbaspour, K. C., Schulin, R., and Yang, H.: Modelling blue and green water resources availability in Iran, Hydrol. Process., 23, 486–501, https://doi.org/10.1002/hyp.7160, 2009.

Geza, M. and Mccray, J. E.: Effects of soil data resolution on SWAT model stream flow and water quality predictions, J. Environ. Manage., 88, 393–406, 2008.

International Scientific & Technical Data Mirror Site: Computer Network Information Center, Chinese Academy of Sciences, SRTMDEM 90M, available at: http://www.gscloud.cn (last access: 1 March 2017), 2003.

Lan, Y. C.: Review on Impact of Climate Change on Water Resources System in the Upper Reaches of Yellow River, Advances in Climate Change Research, 64, 310–313, 2005.

Lee, M. H. and Bae, D. H.: Climate Change Impact Assessment on Green and Blue Water over Asian Monsoon Region, Water Resour. Manag., 29, 2407–2427, 2015.

Liu, J. and Hong, Y.: Spatially explicit assessment of global consumptive water uses in cropland: Green and blue water, J. Hydrol., 384, 187–197, 2010.

Nash, J. E. and Sutcliffe, J. V.: River flow forecasting through conceptual models part I – A discussion of principles, J. Hydrol., 10, 282–290, 1970.

Postel, S. L., Daily, G. C., and Ehrlich, P. R.: Human Appropriation of Renewable Fresh Water, Science, 271, 785–788, 1996.

Ringersmas, J., Batjes, N. H., and Dent, D. L.: Green Water: Definitions and Data for Assessment, Wagening, ISRIC-World Soil Information, 2003.

Rockström, J. and Gordon, L.: Assessment of green water flows to sustain major biomes of the world: Implications for future ecohydrological landscape management, Physics & Chemistry of the Earth Part B Hydrology Oceans & Atmosphere, 26, 843–851, 2001.

Rockström, J., Falkenmark, M., Karlberg, L., Hoff, H., Rost, S., and Gerten, D.: Future water availability for global food production: the potential of green water for increasing resilience to global change, Water Resour. Res., 45, 142–143, 2009.

Saxton, K. E. and Rawls, W. J.: Soil Water Characteristic Estimates by Texture and Organic Matter for Hydrologic Solutions, Soil Sci. Soc. Am. J., 70, 1569–1578, 2006.

Schuol, J., Abbaspour, K. C., Srinivasan, R., and Yang, H.: Estimation of freshwater availability in the West African sub-continent using the SWAT hydrologic model, J. Hydrol., 352, 30–49, https://doi.org/10.1016/j.jhydrol.2007.12.025, 2008a.

Schuol, J., Abbaspour, K. C., Yang, H., Srinivasan, R., and Zehnder, A. J. B.: Modeling blue and green water availability in Africa, Water Resour. Res., 44, 212–221, 2008b.

Shikiomanov, I. A.: Comprehensive Assessment of the Freshwater Resources of the World: Assessment of the Water Resources and Water Availability in the World, Stockholm, Sweden, World Meteorological Organization, 1997.

Shikiomanov, I. A.: Comprehensive Assessment of World Water Resources, Water Int., 25, 11–32, 2000.

Wang, G., Zhang, J., He, R., and Wang, Y.: Variation and Trends Detection of Precipitation and Temperature in the Upper Reaches of Yellow River, J. Arid Environ., 23, 77–81, 2009.

Wang, Y. J., Yang, S. T., Liu, C. M., Dai, D., Zheng, D. H., and Zeng, H. J.: The analysis on the pattern of eco-water use of vegetation and the availability of green water in Sanmenxia area of Yellow River, Geogr. Res., 10, 503–510, 2009.

Winchell, M., Srinivasan, R., Di Luzio, M., and Arnold, J. G.: ArcSWAT interface for SWAT2012-User's guide, Blackland Research Center, Texas Agricultural Experiment Station and Grassland, Soil and Water Research Laboratory, USDA Agricultural Research Service, Temple, Tex, 2013.

Xu, Z., Wang, G., Deng, J., Bao, W., Wu, C., Ren, L., Liu, Z., Liu, B., Xu, W., Yang, Y., Yang, J., Li, Z., Li, H., Li, F., Cheng, J., Chen, X., Zheng, B., Hu, T., Yao, W., Mei, Y., Xie, Z., and Cai, S.: Hydrological Model, Beijing, Science Press, 2009 (in Chinese).

Yang, J., Reichert, P., Abbaspour, K. C., Xia, J., and Yang, H.: Comparing uncertainty analysis techniques for a SWAT application to the Chaohe Basin in China, J. Hydrol., 358, 1–23, https://doi.org/10.1016/j.jhydrol.2008.05.012, 2008.

Zuo, D., Xu, Z., Peng, D., Song, J., Cheng, L., Wei, S., Abbaspour, K. C., and Yang, H.: Simulating spatiotemporal variability of blue and green water resources availability with uncertainty analysis, Hydrol. Process., 29, 1942–1955, 2015.

Zuo, D., Xu, Z., Yao, W., Jin, S., Xiao, P., and Ran, D.: Assessing the effects of changes in land use and climate on runoff and sediment yields from a watershed in the Loess Plateau of China, 544, 238–250, https://doi.org/10.1016/j.scitotenv.2015.11.060, 2015.

https://doi.org/10.5194/piahs-379-169-2018

Open Access
PIAHS

Detecting trend on ecological river status – how to deal with short incomplete bioindicator time series? Methodological and operational issues

Flavie Cernesson[1], Marie-George Tournoud[2], and Nathalie Lalande[3]

[1]TETIS, AgroParisTech, Cirad, CNRS, IRSTEA, Université de Montpellier, Montpellier, France
[2]HSM, Université de Montpellier, CNRS, IRD, Montpellier, France
[3]TETIS, IRSTEA, AgroParisTech, Cirad, CNRS, Université de Montpellier, Montpellier, France

Correspondence: Flavie Cernesson (flavie.cernesson@agroparistech.fr)

Received: 31 December 2017 – Revised: 16 April 2018 – Accepted: 20 April 2018 – Published: 5 June 2018

Abstract. Among the various parameters monitored in river monitoring networks, bioindicators provide very informative data. Analysing time variations in bioindicator data is tricky for water managers because the data sets are often short, irregular, and non-normally distributed. It is then a challenging methodological issue for scientists, as it is in Saône basin (30 000 km^2, France) where, between 1998 and 2010, among 812 IBGN (French macroinvertebrate bioindicator) monitoring stations, only 71 time series have got more than 10 data values and were studied here. Combining various analytical tools (three parametric and non-parametric statistical tests plus a graphical analysis), 45 IBGN time series were classified as stationary and 26 as non-stationary (only one of which showing a degradation). Series from sampling stations located within the same hydroecoregion showed similar trends, while river size classes seemed to be non-significant to explain temporal trends. So, from a methodological point of view, combining statistical tests and graphical analysis is a relevant option when striving to improve trend detection. Moreover, it was possible to propose a way to summarise series in order to analyse links between ecological river quality indicators and land use stressors.

1 Introduction

The improvement or degradation of the ecological status of a river often results from the combined effect of natural environmental conditions and anthropogenic actions. Detecting the effectiveness of national policies or actions taken by local authorities is then a major challenge for policy-makers and water managers (Lalande et al., 2014; Diamantini et al., 2018). Therefore, there is a great need for tools to quantify the evolution of the ecological river status over time what we call here: trend. The trend could evolve positively, negatively, or be stable, as the ecological river status is improved, is degraded or is constant.

Bioindicators are recognized to be good tools to monitor ecological river status, as they indicate both long-term disturbances as well as short-term perturbations that are sufficiently intense to cause the immediate death of the living organisms (Karr and Chu, 2000). Unfortunately these indicators are quite difficult to elaborate because sampling and analysing protocols are often time consuming and expensive. They are often modified by new regulation or new scientific knowledge. Hence, bioindicator series are often short and irregularly sampled data sets: stations are usually sampled at most twice a year but sometimes not at all for several consecutive years. Moreover the data sets may be non-normally distributed. So, it is not easy to analyse such series using statistical techniques (McLeod et al., 1991).

The objective of this paper is to propose a framework to qualify trends in short and irregular bioindicator time series with a focus on the French macroinvertebrate bioindicator time series: IBGN (Indice Biologique Global Normalisé). Series gathered in the Saône basin are studied here. The positive, negative or stable trends are discussed and attempt to link them to the overall characteristics of the stations is presented.

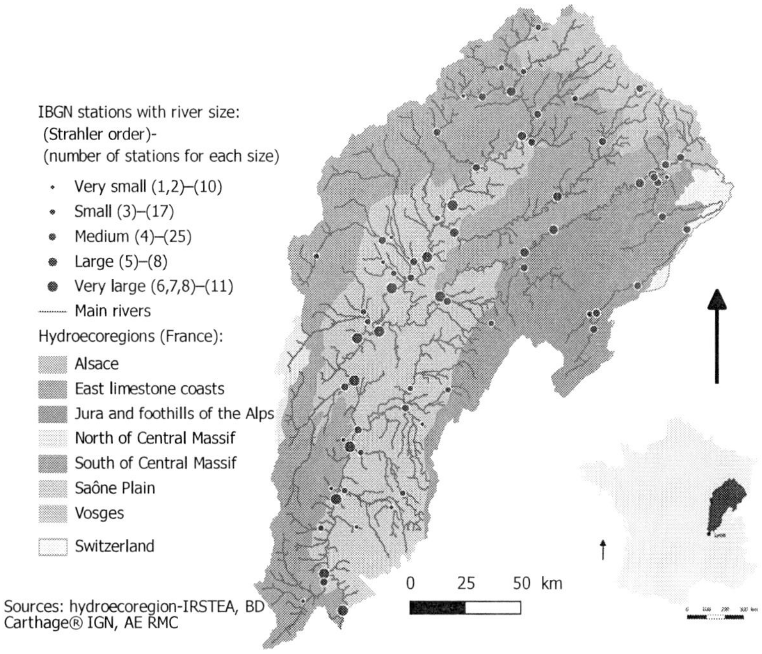

Figure 1. Saône basin: hydroecoregions, main rivers and IBGN monitoring stations network. Each station is localised by a blue point and ranked by the river size: the more the point is big, the more the river is large. For each river size, Strahler orders and IBGN stations count are mentioned into the first and second brackets respectively.

2 Material and methods

2.1 Study site: Saône basin

The Saône River, located in the East of France, is the main tributary of the Rhône River. It has a catchment area of 30 000 km^2 and a river network of 9000 km. With a total of 2.6 million inhabitants, the basin is relatively sparsely urbanized: urban areas represent 5 % of the basin. Industrial activities are located near the urban areas. Semi-natural forest areas cover about 35 % of the basin. About 30 % of the catchment is covered with crops and another 30 % with grasslands: livestock dominates in the upper basin, the left bank and the lower valley are mainly devoted to grain farming and market gardening, while the right bank is occupied by vineyards.

2.2 Ecological river quality data: IBGN data set

The IBGN macroinvertebrate index is based on the abundance of river benthic invertebrates and their selective sensitivity to environmental stressors (flow, substrate, dissolved substances, temperature, light, pH, turbidity, etc.). The IBGN index is mainly used to monitor organic pollution, but it could also indicate the presence of chemical or toxic substances, or local habitat deteriorations. In France it is the most commonly used standard index since 1985 (AFNOR, NF T90-350) and is therefore one of the most studied and most abundant in terms of available data. IBGN scores range from 1 (very bad status) to 20 (very good status) as the combination of two notes related to: (i) the taxon richness which is based on the total number of taxa families harvested during sampling; (ii) the observed benthic macroinvertebrate group which is the most pollution-sensitive.

Among the 812 stations monitored by the French national Rhone-Mediterranean and Corsica water agency in the Saône River basin, 71 IBGN data series that contained at least 10 values between 1988 and 2010 are studied here. These data series represent all kinds of rivers even if they are mainly located in the Saône main watercourse and in the Doubs River (in the mid-eastern part of the watershed) as shown Fig. 1.

Two parameters are used to characterize the monitoring stations: the hydroecoregion and the river size. The Saône basin consists of seven hydroecoregions (Fig. 1). Three main hydroecoregions are represented by 64 stations: "Saône plain", "East limestone coasts" and "Jura and foothills of the Alps" respectively include 28, 21 and 15 stations. "Vosges", "Alsace" and the "South of the Central Massif" are respectively represented by 3, 2 and 2 stations. Finally there is no station in the "North of the Central Massif" region. The river size is divided into five classes based on Strahler order (classes can include two or three Strahler orders). The stations are mainly monitoring medium and small size rivers, with respectively 25 and 17 stations (Fig. 1).

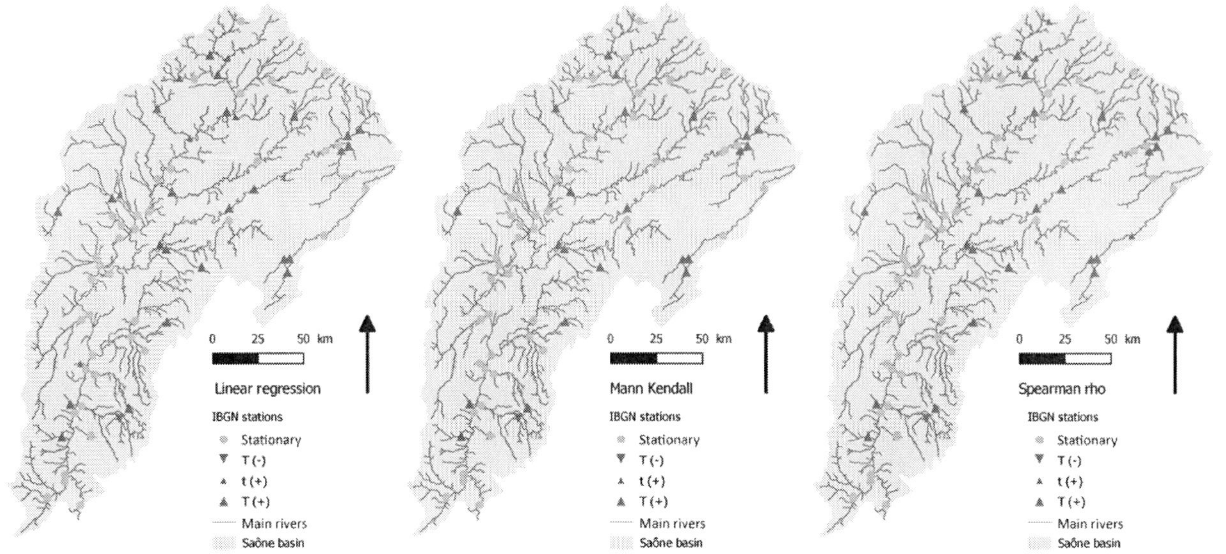

Figure 2. Detected trends for the 71 IBGN series (statistical tests and FDR correction). Stationary, t(+), t(−) stands for: stable, positive and negative trend respectively. The use of the capital letter "T" instead of the "t" means the test results remains significant after the FDR correction.

2.3 Method

Statistical tests are powerful tools to evaluate if a time- serie is stationary or not. For small and incomplete data sets, three statistical tests are particularly well-suited and commonly used: a parametric test – the linear regression (LR) – and two non-parametric tests – the Mann–Kendall (MK) and Spearman rho (SR) tests (Hirsch et al., 1991; Diamantini et al., 2018). In this study, the outputs of these three tests were analysed and compared simultaneously. So, an adjustment for multiple testing that reduces the risk of wrongly detecting significant effects was necessary. Among the existing methods, the False Discovery Rate adjustment (FDR), which is considered as one of the most robust was chosen here (Bar-Hen et al., 2005). A corrected p value was calculated for each test as follows: corrected_p $= \min(p \cdot \text{nbp}/j, 1)$, where p is the initial p value, nbp is the total number of p values calculated (here, nbp $= 3$) and j is the rank of the p value when the p values are ranked in ascending order. After this correction, the series were classified as stationary series (respectively non-stationary), if the three tests do not reject (respectively reject) the null hypothesis (i.e. if the three corrected_p were lower (respectively greater) to 0.05). If the three tests did not produce the same conclusion, a plot analysis was conducted to visually identify if the main characteristics (i.e. shape of plots) of the series are (or not) responsible for these discrepancies. Indeed, when the series are too short, statistical tests are sometimes inefficient while graphical methods offer a non-formal way of detecting trends in series. Unlike McLeod et al. (1991), we did not use the graphical method as explanatory tool but as an additional one, to enhance the reliability of the statistical analysis. To assist in reading the time plots we added a locally weighted scatterplot smoothing (LOESS) which is a robust regression smoothing, as the smoothed points are not distorted by extreme values or other kinds of deviant points.

The detection of trend in the IBGN time-series was then linked to the environmental conditions and characteristics of the monitoring stations. A Chi test was carried out (using a permutation procedure to consider small sample size) to study the effect of (i) the size of the river at the station, based on the Strahler order and (ii) the environmental context defined by the hydroecoregion. Indeed, Strahler river order is one of the most important factors influencing the structure and the function of river (Crunkilton and Duchrow, 1991; Hughes et al., 2011). Moreover, rivers located in the same hydroecoregion share the same natural background defined by geological, climate and relief parameters. These parameters are important factors that significantly influence river functioning and river fauna and flora composition (Cohen et al., 1998).

3 Results

3.1 Trends in IBGN data set

The outputs of the three tests are showed in Fig. 2. Respectively 32, 29, 24 IBGN series were qualified as non-stationary by LR, SR, MK statistical tests.

The three tests gave the same results for 60 series: 24 series were classified as non-stationary (23 with a positive trend and one with a negative one) and 36 as stationary. Graphical

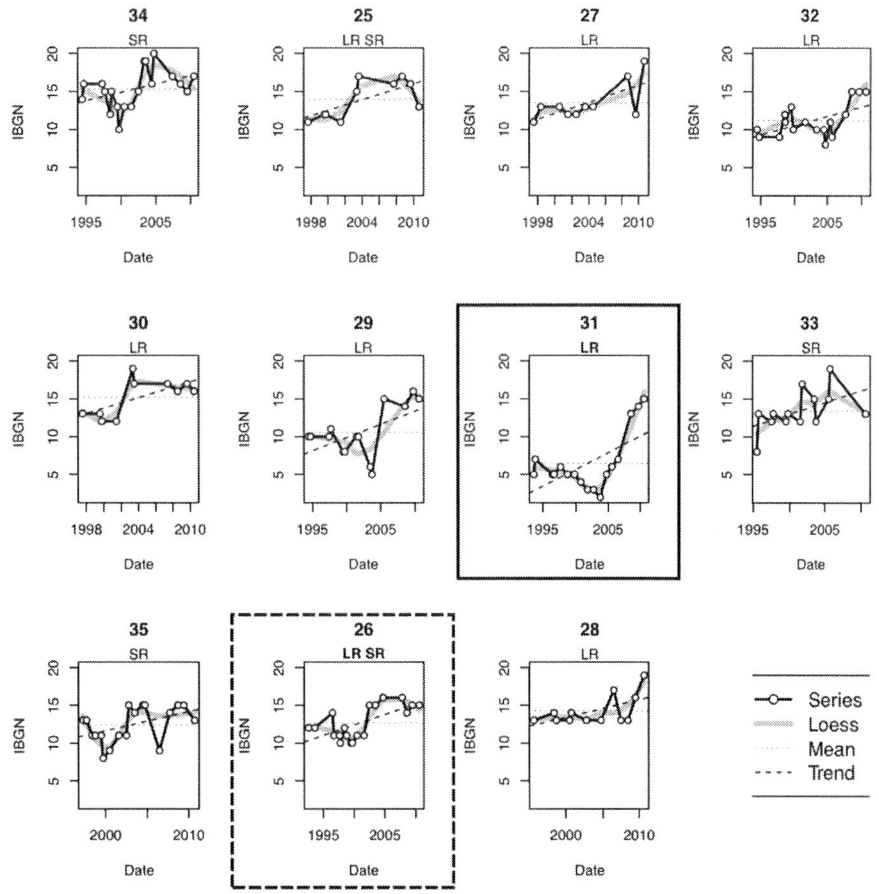

Figure 3. Plots of the eleven series for which the three tests, taken individually, gave different outputs. The two box- plots refer to unclassified series after the FDR correction. Black line with white dots are the observed IBGN values, light dotted line is the mean of the series, bold dashed line is the linear regression, bold light grey line is the LOESS fitting of the series.

analysis confirmed that the 24 series that were directly classified by the three tests as non-stationary showed a strong temporal evolution of IBGN over time. FDR correction did not modify these results.

The three tests disagreed on 11 series: the 11 series were all qualified as stationary by MK test while 6 of them were stood as non-stationary by LR test, 3 of them by SR test and the remaining 2 by both LR and SR tests. After FDR correction, the three tests MK, LR and SR agreed to qualify 9 out of 11 series as stationary series. There were still 2 unclassified series for which the test results were not modified by FDR correction: series 31 and 26 as shown on Fig. 3. A positive trend was identified in serie 31 by the LR test because IBGN scores increased by 13 points during the last 7 years. The non-stationarity of serie 26 was still significant for the LR and SR tests, because fluctuations were quite small and did not hide the significant positive jump in IBGN scores (plus 4 points) observed since 2003. Series 26 and 31 were then visually reclassified, as non-stationary series, with a positive trend.

The graphical analysis of these 11 cases showed very different shapes (Fig. 3): it indicates that trends in IBGN time series were sometimes not well-defined, due to (i) extremes values at the beginning or at the end of the time serie, (ii) fluctuations between two levels of IBGN values plotted as a two-blocks partition. Some series showed score fluctuations greater than 5 points, which makes very difficult graphical interpretation, and can introduce a statistical bias in tests and distort their results. Moreover outliers that significantly affected the shape of the graph of small series may have the same effect. Most of these outliers were due to the effect of the 2003 drought.

To conclude, out of the 71 series, our methodology classified 26 series as non-stationary (positive trend for 25 out of 26) and 45 as stationary. So, ecological river status was either stable (mainly for the Saône River) or improving (mainly for

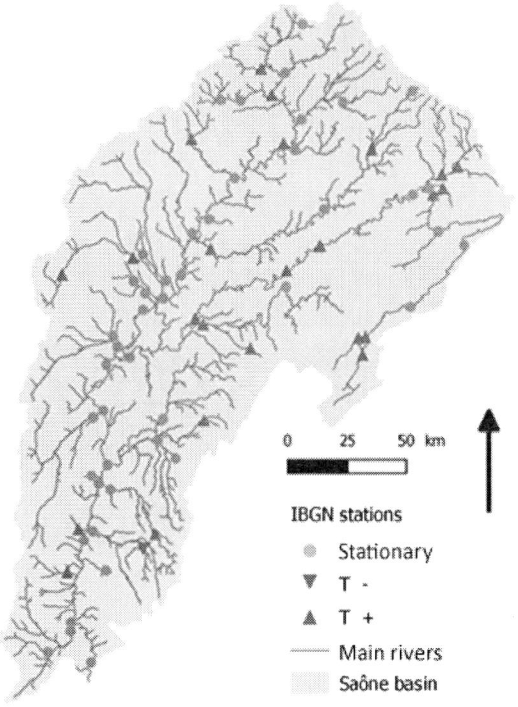

Figure 4. Detected trends for the 71 IBGN series (statistical tests plus FDR correction and graphical analysis). Stationary, T(+), T(−) stands for: stable, positive and negative trend respectively.

the Doubs River and the other Saône River tributaries) according to the IBGN index (Fig. 4).

3.2 Trend versus environmental characteristics

The Chi test applied to the data set showed that trends in ecological river status were significantly linked to the natural environmental background of each river defined by the hydroecoregions ($\chi 2 = 13.48$, $p = 0.01$), but were not related to the river size at the station ($\chi 2 = 9.01$, $p = 0.06$). Stationary and non-stationary stations are observed in all classes of river size. For very large river (i.e. Strahler order greater than 6), 10 stations among 11 showed stable trends in their ecological river status and only 1 has a positive trend. For all the other cases, both stable and non-stable trends were observed in equal parts. So, trends in the ecological river status were not really linked to river sizes in the Saône River. This result confirms the opinion issued by Hughes et al. (2011) that summarizing river size only by the Strahler order is an oversimplification of the issue. A significant link between the trend classification and hydroecoregions was found. Rivers located on the same hydroecoregion shared the same natural background. Hydroecoregion classification is a key-factor for river functioning and river fauna and flora composition. However, it is worth to remind that the number of monitoring stations in each hydroecoregion varies from 2 ("South of the Central Massif" region) to 28 ("Saône plain" region). This may have a strong influence on our findings.

Due to the very coarse scale of the two environmental factors we used here, the interpretation and explanation of observed patterns was difficult. More interesting patterns would probably emerge from analyses based on the type of waterbodies that combines both river size and natural background.

The average IBGN score of stationary series (mean 14.25 over 20–45 series) and non-stationary series (12.47 over 20–26 series) was significantly different ($F = 10.46$, $p = 0.001$). This could be explained by a saturation effect: as only one negative trend is observed, and as the IBGN score is limited to 20, positive trends can only be observed in sites where an IBGN increase is still possible (those with a moderate or low initial IBGN value).

4 Conclusion

Trend in ecological river status is difficult to qualify because the monitoring of bioindicators is not performed on regular basis (sampling started, is interrupted and then restarted, etc.). This leads to small and irregular time series. So a suitable methodology is required, that imposes the use of several analytical tools. It is important to apply a combination of statistical tools (e.g., parametric and not parametric methods) and less conventional ones, such as graphical analysis. The graphical analysis helped to confirm the type of trends and to underline the strengths and weaknesses of the various statistical procedures. The methodological framework proposed here combines three statistical tests (Mann–Kendall, Spearman's rho test, and the linear regression) and an a posteriori graphical visualization assisted by the LOESS procedure. On the example of the Saône River IBGN data, the combined use of different statistical and graphical tools was shown to be a relevant alternative when trying to improve trend detection, which is a key-issue in ecological river status assessment. This combined approach provides a powerful tool to improve the accuracy of the environmental diagnosis of water managers and the significance of the reporting of policy makers.

Data availability. All the data we used are open data produced by french public agencies or french research institute. Hydroecoregions data are the results of Wasson et al. (2002). and can be downloaded https://www.data.gouv.fr/fr/datasets/hydroecoregions-de-niveau-1-her-1/. The main rivers, extracted from Carthage® IGN data base, can be downloaded https://www.data.gouv.fr/fr/datasets/cours-deau-metropole-2014-bd-carthage/. River size informations can be downloaded https://www.data.gouv.fr/fr/datasets/masses-deau-cours-deau-metropole-version-rapportage-2010/. IBGN monitoring stations informations are produced by the French national Rhone-Mediterranean and Corsica Agency. Localisation of monitoring stations can be downloaded http://carmen.carmencarto.fr/74/sierm.map. IBGN data can be downloaded http://sierm.eaurmc.fr/surveillance/eaux-superficielles/index.php.

Competing interests. The authors declare that they have no conflict of interest.

Special issue statement. This article is part of the special issue "Innovative water resources management – understanding and balancing interactions between humankind and nature". It is a result of the 8th International Water Resources Management Conference of ICWRS, Beijing, China, 13–15 June 2018.

Acknowledgements. This work was funded by the French National Research Agency (ANR), as part of the ANR11_MONU14 FresQueau project and by Agence de l'Eau Rhône-Méditerranée et Corse (agreement no. 2010 0190). The authors thank the reviewers for their very useful comments.

Edited by: Depeng Zuo
Reviewed by: Feifei Cao and one anonymous referee

References

Bar-Hen, A., Daudin, J.-J., and Robin, S.: Comparaisons multiples pour les microarrays, Journal de la société française de statistique, 146, 45–62, 2005.

Cohen, P., Andriamahefa, H., and Wasson, J. G.: Towards a regionalization of aquatic habitat: Distribution of mesohabitats at the scale of a large basin, Regul. River., 14, 391–404, 1998.

Crunkilton, R. L. and Duchrow, R. M.: Use of stream order and biological indexes to assess water-quality in the osage and black river basins of Missouri, Hydrobiologia, 224, 155–166, 1991.

Diamantini, E., Lutz, S. R., Mallucci, S., Majone, B., Merz, R., and Bellin, A.: Driver detection of water quality trends in three large European river basins, Sci. Total Environ., 612, 49–62, 2018.

Hirsch, R. M., Alexander, R. B., and Smith, R. A.: Selection of Methods for the Detection and Estimation of Trends in Water-Quality, Water Resour. Res., 27, 803–813, 1991.

Hughes, R. M., Kaufmann, P. R., and Weber, M. H.: National and regional comparisons between Strahler order and stream size, J. N. Am. Benthol. Soc., 30, 103–121, 2011.

Karr, J. R. and Chu, E. W.: Sustaining living rivers, Hydrobiologia, 422, 1–14, 2000.

Lalande, N., Cernesson, F., Decherf, A., and Tournoud, M. G.: Implementing the DPSIR framework to link water quality of rivers to land use: methodological issues and preliminary field test, International Journal of River Basin Management, 12, 201–217, 2014.

McLeod, A. I., Hipel, K. W., and Bodo, B. A.: Trend analysis methodology for water-quality time-series, Environmetrics, 2, 169–200, 1991.

Wasson, G., Chandesris, A., Pella, H., and Blanc, L.: Les hydro-écorégions de France métropolitaine – approche régionale de la typologie des eaux courantes et éléments pour la définition des peuplements de référence d'invertébrés, IRSTEA-ex-Cemagref-MEDDE, 190 pp., 2002.

https://doi.org/10.5194/piahs-379-175-2018

The socio-economics dynamics of Dam on Rural Communities: A case study of Oyan Dam, Nigeria

Amidu Ayeni and Lawrence Ojifo

Department of Geography, University of Lagos, Nigeria

Correspondence: Amidu Ayeni (aayeni@unilag.edu.ng)

Received: 9 December 2017 – Revised: 23 February 2018 – Accepted: 27 February 2018 – Published: 5 June 2018

Abstract. Dams construction and operations have many benefits, nevertheless, they have also led to lots of negative social, health and human impacts. It is based on this that this study assesses the potential and socio-economics dynamics of Oyan dam between 1980 and 2016. The data used for this study include water level and discharge records of the dam between 2007 and 2016, Landsat imageries of 1984 and 2016 and socio-economic datasets for the period. Analysis of the dam potentials (water supply, agriculture and hydropower) and socio-economic impacts of the dam were carried out using basic statistical tools, land use change anaysis and field survey using questionnaire, structured interview with major stakeholders and personal observation. The results revealed that the water level and storage of the Oyan dam had a relative reduction of about 2 % as well as non-stationarity pattern of water abstraction and production for the period. The landuse classes show all classes decreased in extent except the cultivated landuse that acrued an increased of 19.9 % between 1984 and 2016. Furthermore, commercial water supply varied significantly between 2010 and 2016 while irrigation scheme is grossly under-utilized from the inception in 1983 to 2016. Finally, the result of socio-economic impacts revealed that majority of the selected communities' members are actually not benefiting from the dam and their livelihoods are not from the dam.

1 Introduction

The occurrence and impact of the 1972–1975 Sahelian drought aggravated the food security in the country which prompted various arms of government to embark on a rigorous policy to increase food production (Taviano and Marcelo, 2010). To accomplish the policy, river basins impoundment was seen as inevitable to provide sufficient water for year-round irrigation which led to the construction of over 246 dams (Imevbore et al, 1986). Of the 323 dams in Nigeria in 2002, 106 of them are large dams (dams with walls higher than 15 m or 10–15 m high with a crest length of over 500 m or having a reservoir capacity of 1 million m^3), 27 are medium sized dams (walls are 8–10 m high) and 192 are small dams (walls less than 8 m) (Ofoezie, 2002; Ogbeide and Uyigue, 2004).

The construction and operation of these dams have yielded many benefits, nevertheless, they have also have led to lots of negative social, health and human impacts (Webbe, 1981; Williams and Wolman, 1984; McCartney and King, 2011). The environmental changes and social disruption resulting from dams and associated infrastructure (e.g. irrigation schemes) have had adverse effect on health of the local population and downstream communities (Williams and Wolman, 1984; Collier et al., 1996; Richter et al., 2010; Sivongxay et al., 2017). Therefore, access to drinking water, health services and ability to cope with new social and physical environment determine socio-health conditions of the displaced population and/or communities (WHO, 1985; World Commission on Dams, 2000; Richter and Thomas, 2010). It is based on the aforementioned that this study tends to provide answers to these questions (i) What are the designed functions for the establishment of the dam? (ii) what is the present serving condition and/or capacity of the dam functionalities? (iii) What are the socio-economic impacts of the Oyan dam? (iv) what is the spatio-temporal dynamics of the Oyan Basin over the last 31 years?

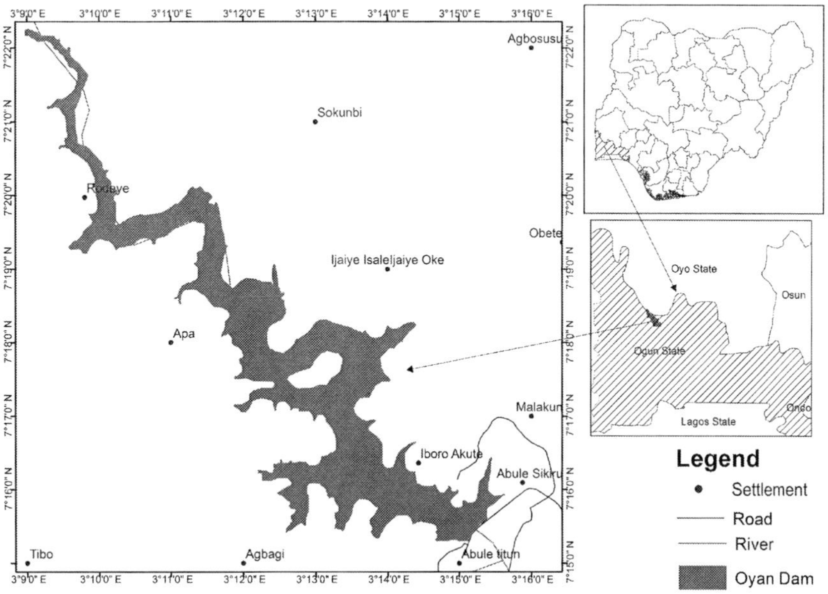

Figure 1. Oyan Dam (Insert – top right: Nigeria; Insert – bottom right: Ogun State).

2 Study Area

Oyan dam is located between latitude 7°15′ and 7°25′ N, and longitude 3°08′ and 3°16′ E near Abeokuta, the capital city of Ogun State, Nigeria (Fig. 1). The dam which is owned and operated by the Ogun-Osun River Basin Development Authority (OORBDA) was commissioned in 1984. Lying adjacent to the dam are three resettlement communities named Abule Sikiru, Abule Titun, and Ibaro. The dam is about 30 m high with a reservoir (gross storage) capacity of 270 million m^3 and a dead storage capacity of 16 million m^3 and 9-megawatt power generation capacity. It has a surface area of 40 km^2 and catchment area is approx. 9000 km^2 with embankment crest length of 1044 m. The dam design free board at 10 000 years flood is 2.10 m high with four spillway gates (each with 15 m wide and 7 m high), and three outlet valves (each with 1.8 m in diameter). It supplies water for irrigation and for domestic uses to the cities of Abeokuta and Lagos.

3 Methodology

The data used for this research work include hydrological, remotely sensed, and socio-economic datasets. Each of these datasets were used to achieve the objectives of the study. Hydrological dataset consists of the daily hydrological records for the last 10 years (i.e. 2007–2016). This data includes river water level, water discharge, and supply. These data were sourced from the Ogun-Oshun River Basin Development Authority (OORBDA). The collected daily hydrological datasets were analysed as a function of the monthly data using simple arithmetic mean equation.

The second dataset obtained through field survey using questionnaire and interview was socio-economic data. This data addresses community livelihood profiles, health status of the population, and social impact of the dam on the communities. Socio-economic impact assessment of the dam was carried out in three selected communities around the Oyan dam vis-à-vis Abule Sikiru, Abule Titun, and Ibaro. Structured questionnaire was used to extract information about the socio-economic impact of the dam. The sample size of the communities was based on OORBDC estimated population size of the communities, therefore, about 50, 60, and 50 questionnaires were randomly distributed in Abule Sikiru, Abule Titun, and Ibaro, respectively. The questions examined includes household characteristics, community livelihood profile, socio-economic benefits, and resource potential of the dam. The socio-economic datasets were analysed using basic descriptive statistics method of data analysis.

The last dataset for spatio-temporal dynamics assessment of the dam Basin was based on Landsat imageries of Thematic Mapper (TM) and Enhanced Thematic Mapper (ETM) for 1984 and 2016 respectively. The two periods dataset were obtained from Glovis portal. Six (6) landuse/landcover classes were extracted from each of the imageries using the standard procedures for LULC classification anaysis (de Vries et al., 2007; Chander et al., 2009; Ayeni et al., 2016). In view of this, a certain number of training dataset was randomly sampled from the spectral signature of each of the classes to define their respective landuse/landcover type.

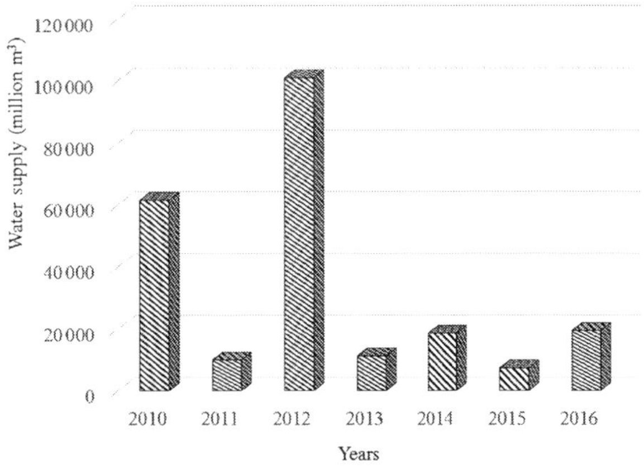

Figure 2. Water Supply of Oyan dam between 2010 and 2016.

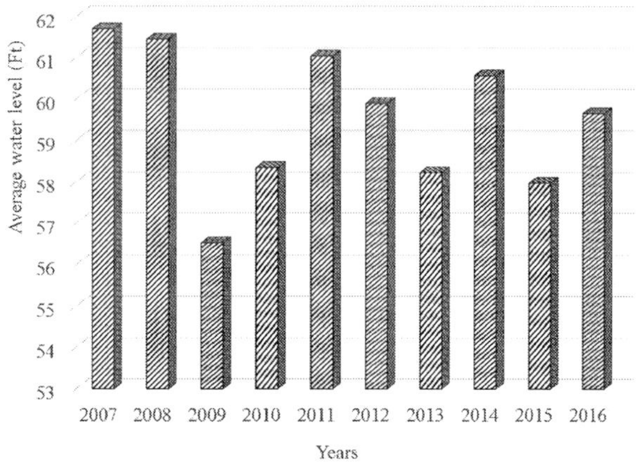

Figure 3. Average Yearly Water Level of Oyan dam Source: OORBDA, 2017.

Subsequently, the results of the datasets collected were presented in maps, tables and charts.

4 Results and discussions

4.1 The primary functions for the establishment of the dam

Oyan dam was primarily designed to: (i) generate about 9 megawatts hydro-electric power to support the existing national grid; (ii) supply raw water to Lagos and Abeokuta which are the immediate downstream urban cities, and, (iii) finally to support Lower Ogun basin 3000 ha Irrigation Project. The dam is used for breeding fish. Oyan dam was also designed to function as flood control flood through safe release of water in the downstream. Over the years, cases of flooding in the downstream had been attributed to excess water release from the Oyan dam. In addition, other operational problems have also led to conflicts over the adequacy of water released for downstream uses, irrigation, fisheries and raw water supply for water treatment in Abeokuta city and Lagos State.

Oyan dam capacity of generating over 9 mw of hydroelectric power has been locked since the commissioning in 1984. This implies that the dam currently generates zero megawatt of electricity making it a 100 % decrease in the power generation function. As stated earlier, Oyan dam was designed to supply raw water to the immediate downstream urban cities – Lagos and Abeokuta, and to support the 3000 ha Lower Ogun Irrigation Project. Three turbines of 3 megawatts each were installed in 1983 but they have not been in operation since the inception.

According to OORBDO (2017), the dam has a potential irrigation capacity of 32 000 ha of land around the dam and the lower basin irrigation project at Mokoloki. However, the dam presently supports only 8000 and 3000 ha around the dam and at the both upstream/downstream lower irrigation agricultural project respectively. The water supply pattern of Oyan dam revealed that water was only supply on demand from Ogun and Lagos Water Corporation. The study further revealed that water is supply mostly during the dry season. Thus, certain months received a high demand than the others. Figure 2 shows that 2012 has a highest water supply of about 101 041.2 million m^3 in the history of the dam. Information revealed that the dam has a capacity to supply more than this volume since there were no set limit for supply from inceptions. Furthermore, based on this data, it can be inferred that Oyan dam has a larger carrying capacity to supply more water than what is currently supplying.

Figure 3 shows the annual water level storage of Oyan dam between 2007 and 2016. The annual average water level storage ranging between 17.23 m (lowest) and 18.80 m (highest) in 2009 and 2007 respectively. The figure shows that there is significant high amount of water production in 2007 which account for 10.36 %. This explains the percentage of the water abstraction in each of the years, and a level of consistency in water level and storage capacity of dam over a 10 years period.

The monthly water discharge of Oyan dam releases a significant amount of water for domestic and industrial uses. Though, there was data paucity in all years due to certain factors except 2007 which its records show that water was released throughout the year with some months exhibit exceptional discharge rate than the others. It shows that from May to November, the amount of water that was discharged rose significantly (Fig. 4). The figure revealed that January to April record low amount of water discharge ranging from 112 to 192 million m^3 while May discharge increased significantly to about 1720 million m^3.

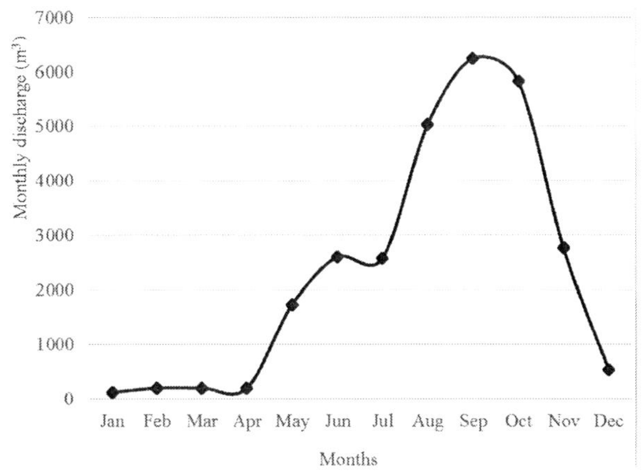

Figure 4. Monthly Water Discharge of Oyan Dam in 2007.

Table 1. Oyan Dam Landuse Dynamics between 1984 and 2016.

	1984		2016	
	Area (ha)	%	Area (ha)	%
Built-up	121.3	0.1	81.7	0.08
Cultivated	70 860.4	72.4	90 320.4	92.3
Floodplain	14 847.6	15.2	2191.1	2.2
Forest	6624.7	6.8	2613.7	2.7
Rock outcrop	2861.8	2.9	737.5	0.8
Waterbody	2524.9	2.6	1896.3	1.9
Total	97 840.7	100	97 840.7	100

4.2 Spatio-Temporal Dynamics of the Oyan Downstream Basin

Table 1 and Fig. 5 show the six major landuse/landcover classes identified around the Oyan dam for the two scenarios (1984 and 2016).

The classes of landuse/landcover identified include built-up, forest, water body, cultivated land, floodplain and rock outcrop/bare surface. The results show that built-up decreased from 121.3 ha (0.1 %) in 1984 to about 81.7 ha (0.08 %) in 2016 (Table 1 and Fig. 5). The changes signify the effect of dam development on small communities which were either relocated or lost to the development. On the other hand, forest decreased from 6624.7 ha (6.8 %) in 1984 to 2613.7 ha (2.7 %) in 2016.

However, water body relatively shows a stable slight change. It decreases from with about 2524.9 ha (2.6 %) in 1984 to about 1896.3 ha (1.9 %) in 2016. Thus, the dynamic nature of this landuse/landcover observed in this basin can be associated with the variation in the water abstraction and production of the dam where the most dominant landuse (Cultivated) demand more water.

Table 2. Purpose of the Water in the Dam.

	Frequency	Percent
Domestic purpose	98	62.4
Industrial purpose	13	8.3
Commercial purpose	36	22.9
Cultural purpose	5	3.2
Total	153	97.5
	157	100

Table 3. Effect of Dam on Livelihood/Standard of Living of the respondent.

	Frequency	Percent
Below average	14	8.9
Average	19	12.1
A little above average	109	69.4
very high	9	5.7
Total	151	96.2
	157	100

4.3 Socio-Economic Impacts of the Dam

The socio-economic impacts of the dam on the neighboring communities was evaluated on the basis of the community livelihood profiles, health status of the population, and social impact of the dam on the communities.

Table 2 shows the socio-economic benefits of the dam as argued by the respondents. The result shows that more than 86 % of the respondents were fully aware of the dam in their neighborhood. This signifies that the communities' member understands the potentials and socio-economic benefits they can derive directly or indirectly from the dam. It also indicates that more than 62.4, 8.3, and 22.9 % of the respondents use this dam for domestic, industrial and commercial purpose purposes respectively (Table 2).

The results on sources of livelihood derivable from Oyan dam show that about 70.1 % of the respondents are not deriving their livelihood from the dam while only 1.3 % of the respondents claimed they are deriving their livelihood from the dam. Table 3 reveals the effect of the dam on livelihood and/or standard of living of the respondents. It was observed that about 18.9 % of the respondents are living "below average", 12.1 % are on the "average", 63.3 % are above average' while a fraction of about 5.7 % of the respondents had "high" standard of living.

This study reveals that the health status of majority of the population is good as argued by 51.6 % respondents and the presence of dam has nothing to do with their state of health. Displacement and environmental degradation have been identified as the challenges experienced from time to time. Thus, communities expect more from this dam in terms

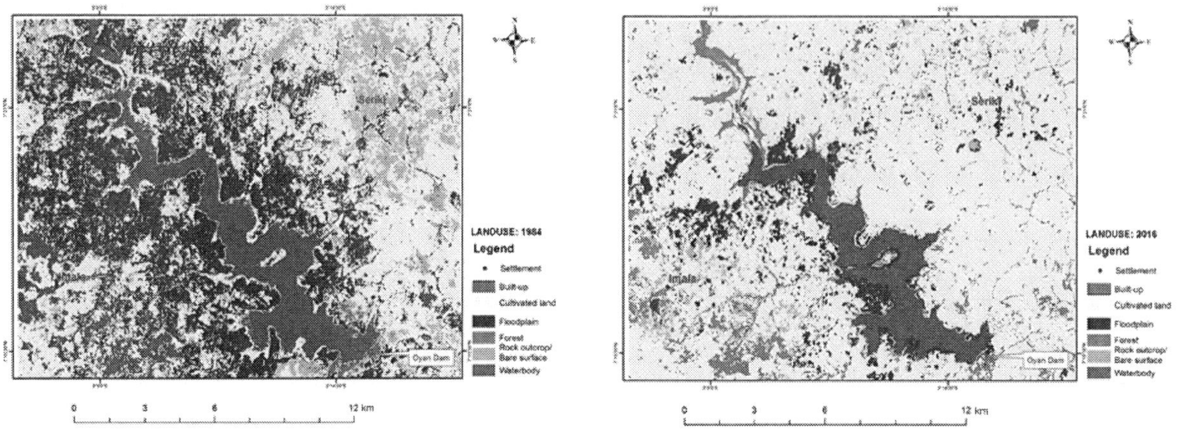

Figure 5. Landuse Pattern of Oyan Dam in 1984 and 2016.

of employment of the local communities, compensation as well as relocation of the displaced communities, and capacity building as pointed out by 36.3, 54.8, and 3.8 % respondents respectively.

These results revealed that the spatio-temporal changes (reduction in water body and forest) around Oyan dam as well as degraded and poor maintenance of dam and associated infrastructure have affected the dam in varying degrees vis-à-vis reduction in dam level and storage capacity, inability to meet the designed water supply capacity, and inability to generates hydro-electric power amongst others. These have had considerable adverse effects on designed functions of the dam and the basin ecosystem, and therefore, resulted to continuous reduction in the serving condition and capacity of the dam especially on irrigation scheme, livelihoods and other socio-economic potentials.

5 Conclusions

This study identified two major resource potentials of Oyan dam as at the 2016 vis-à-vis commercial water supply and irrigation scheme. It was observed that the commercial water supply varied with season. Most of the water are supply during the dry season with 2012 accounts for higher percentage of water supply. Irrigation scheme at Oyan upper section (8000 ha) and Makoloki farm settlement (9000 ha) were observed to be under-utilization of water resource potential of the dam while the 9-megawatt hydropower potential is yet to be tapped.

The continuous release of water from Oyan Dam in the recent time has increased the occurrences of downstream flooding at the lower part of the Ogun basin. For instance, there have been persistence flooding ravaging the city of Lagos over the year with significant effects on socio-economic activities. This also resulted to a situation where the irrigation project of the dam has either been under-utilized or none in existence. Series of conflicts have also triggered from other operational activities around the dam. These conflicts emanated from the inadequacy of water released for irrigation, fisheries and raw water supply for domestic water uses in part of Ogun and Lagos States. It is on this premise that the study examines the prime purpose of dam construction, evaluate its carrying capacity in terms of its depth, surface area extent, volume of water generates and agricultural potential, spatio-temporal analysis of the dam was examined between 1980 and 2015 and lastly, socio-economic impacts of the dam development were carried out among its neighboring communities.

The socio-economic impacts of the dam revealed that few people are actually benefiting or deriving their livelihoods from the dam. Thus, communities expect more from this dam in terms of employment of the local communities, compensation and relocation of the displaced communities.

Data availability. Data are available at Ogun-Oshun River Basin Development Authority (OORBDA) head office, Abeokuta, Ogun State, Nigeria.

Competing interests. The authors declare that they have no conflict of interest.

Special issue statement. This article is part of the special issue "Innovative water resources management – understanding and balancing interactions between humankind and nature". It is a result of the 8th International Water Resources Management Conference of ICWRS, Beijing, China, 13–15 June 2018.

Acknowledgements. The Authors are grateful to the Ogun-Osun River Basin Development Authority (OORBDA) for providing historical Hydrological dataset.

Edited by: Wenchao Sun
Reviewed by: two anonymous referees

References

Ayeni, A. O., Cho, M. A., Mathieu, R., and Adegoke, J. O.: The local experts' perception of environmental change and its impacts on surface water in Southwestern Nigeria, Environ. Develop., 17, 33–47, 2016.

Chander, G., Markham, B., and Helder, D.: Summary of current radiometric calibration coefficients for Landsat MSS, TM, ETM+, and EO-1 ALI sensors, Remote Sens. Environ., 113, 893–903, 2009.

Collier, M., Webb, R. H., and Schmidt, R. H.: Dams and rivers: primer on the downstream effects of dams, U.S. Geological Survey Circular 1126, Denver, Colorado, USA, 1996.

de Vries, C., Danaher, T., Denham, R., Scarth, P., and Phinn, S.: An operational invariant target sites, Remote Sens. Environ., 107, 414–429, 2007.

Imevbore, A. M. A., Ofoezie I. E., and Obot E. A.: Report on the study on snail and cyclopoid borne disease: problems of small scale water resources development Nigeria, submitted to the World Health Organization by the Institute of Ecology, O.A.U., Ile-Ife, Nigeria, 1986.

McCartney, M. and King, J.: Use of decision support systems to improve dam planning and dam operation in Africa, Colombo, Sri Lanka: CGIAR Challenge Program for Water and Food (CPWF), 74 pp. (CPWF Research for Development Series 02), 2011.

Ofoezie, I. E.: Human health and sustainable water resources development in Nigeria: schistosomiasis in artificial lakes, Natural Resources Forum, 26, 150–160, https://doi.org/10.1111/1477-8947.00015, 2002.

Ogbeide, H. E. and Uyigue, E.: Access to safe drinking water and schistosomiasis, in: Nigeria: survey on Ipogun Community, Ondo State of Nigeria, submitted to the Society for Water and Public Health Protection (SWAPHEP), 2004.

OORBDA: Ogun-Osun River Basin Development Authority (OORBDA), 2017.

Richter, B. D., Postel, S., Revenga, C., Scudder, T., Lehner, B., Churchill, A., and Chow, M.: Lost in development's shadow: The downstream human consequences of dams, Water Alternatives, 3, 14–42, 2010.

Richter, B. D. and Thomas, G. A.: Restoring environmental flows by modifying dam operations, Ecol. Soc., 12, 12, 2007.

Sivongxay, A., Greiner, R., and Garnett, S. T.: Livelihood impacts of hydropower projects on downstream communities in central Laos and mitigation measures, Water Resources and Rural Development, 9, 46–55, 2017.

Taviano, C. and Marcelo, G.: The day after tomorrow: a handbook on the future of economic policy in the developing world, The International Bank for Reconstruction and Development/The World Bank, 2010.

Webbe, G.: Schistosomiasis: Some Advances, British Medical Journal (Clinical Research Edition), 283, 1104–1106, 1981.

WHO: Human Health and Dams, WCD Working Paper, 1999.

Williams, G. P. and Wolman, M. G.: Downstream effects of dams on alluvial rivers. U.S. Geological Survey Professional Paper 1286, Denver, Colorado, USA, 1984.

World Commission on Dams: Dams and development: a new framework for decision-making, Earthscan, London, UK, 2000.

https://doi.org/10.5194/piahs-379-181-2018

Horizontal insulating barriers as a way to protect groundwater

Renata Cicha-Szot[1], Krzysztof Labus[2], Sławomir Falkowicz[1], and Norbert Madetko[3]

[1]Oil and Gas Institute – National Research Institute, Department of Petroleum Engineering, Kraków 30-001, Poland
[2]Silesian University of Technology, Department of Applied Geology, Gliwice, 44-100, Poland
[3]Soley LCC, Balice, 32-083, Poland

Correspondence: Krzysztof Labus (krzysztof.labus@polsl.pl)

Received: 25 December 2017 – Revised: 29 April 2018 – Accepted: 3 May 2018 – Published: 5 June 2018

Abstract. Trenchless Technology of Forming Horizontal Insulating Barriers (TFHB) can be considered a method of groundwater protection against inflow of pollutants. In TFHB technology, the working fluid (sodium silicate solution) and the gelling agent (CO_2) are injected separately, using one tool, to different zones of the aquifer profile. Carbon dioxide injected into the saturation zone rises due to buoyancy forces and reaches the silicate which was injected at the water table level. This initiates the process of silicate gelation, resulting in the formation of an insulating barrier. For technological purposes, the gelation time must be controlled, and the resulting gel must have certain mechanical properties. In order to apply THFB in real conditions it was necessary to identify important technological and technical parameters, as well as to define interactions between the injected fluid and the aquifer rocks. Geochemical modelling (equilibrium, reaction path and reactive transport) was used to identify potential geochemical effects of the application of TFHB in sandy aquifers. Certain petrophysical parameters and mineralogical assemblages of aquifers were addressed, taking into account both low and strongly mineralized groundwater. The simulations revealed that TFHB does not have a negative impact on the chemistry of rock-water systems described in this work.

1 Introduction

Protection of groundwater resources is a key challenge in the 21st century. New technologies are crucial in order to prevent pollution not only close to urban settings with highly congested infrastructure, but also in the case of demanding topographical relief. Trenchless Technology of Forming Horizontal Insulating Barriers (TFHB) can be considered an emergency form of immediate groundwater protection against inflow of pollutants, e.g. from hazardous landfill; however, so far there is no approved insulating technology which does not require the displacement of waste.

Trenchless methods seem to be effective; however, for the optimum solution a combination of numerous factors should be considered, including materials and geotechnical aspects. One advantage of the developed TFHB technology in shallow groundwater aquifers is the injection of a non-toxic, environmentally friendly modified water solution of sodium silicate. The set of additives which enable hybrid gelation mechanisms are determined by the composition and properties of the applied sodium silicate gels and the complex chemical reactions which cause reduced hydraulic conductivity.

2 Technology outline

The idea of injecting sodium silicate with CO_2 into geological formations has existed since 1946 (Anderson, 1946). In TFHB technology, the working fluid (sodium silicate solution) and the gelling agent (carbon dioxide) are injected separately, using one tool, to different zones of the aquifer profile. Carbon dioxide, injected into the saturation zone, rises due to buoyancy and reaches the silicate injected at the water table level.

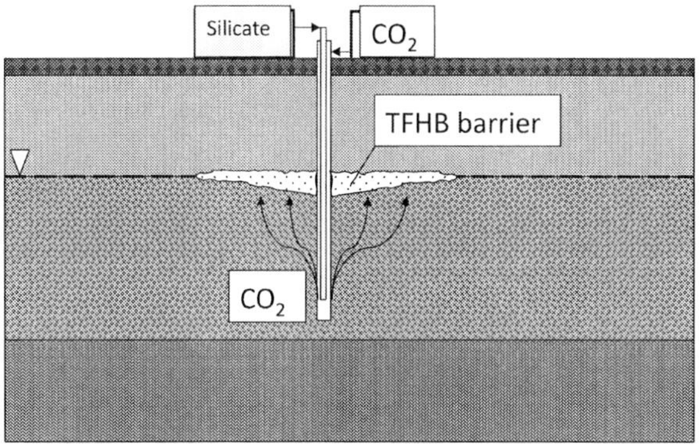

Figure 1. Creating a TFHB barrier in the roof of an aquifer.

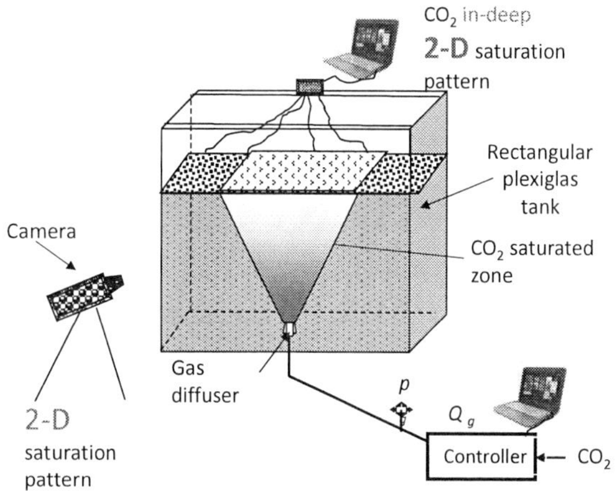

Figure 2. Diagram of the laboratory setup.

This initiates the process of silicate gelation, resulting in the formation of an insulating barrier of expected shape and size, as shown in Fig. 1. It is assumed that a barrier formed in this way will eliminate or significantly reduce vertical filtration of liquids, eliminating the paths by which pollutants migrate into aquifers. For technological purposes, the gelation time must be controlled, and the resulting gel must have certain mechanical properties.

An important element of TFHB technology is the treatment fluid, which is based on sodium silicate water solution and characterized by low viscosity. The use of special additives allows the resistance of the gel to chemical, thermal and pressure factors to be increased. Nano-additives cause the injected liquid to turn into a gel after about 48 h, even in limited contact with CO_2. Initial injection of CO_2 is, however, necessary to prevent the injected silicate migrating downwards in the aquifer. The resistance of the resulting gel against hydraulic breakthrough is high, and it can reach 56.6 MPa in sandstones (Nasr-El-Din and Taylor, 2005).

3 Laboratory tests

CO_2 injection is an important step in TFHB technology. The gas plume of carbon dioxide injected to an aquifer with a drilling tool should ideally have the shape of a paraboloid or inverted cone, with a height equal to the depth of the gas injection in the aquifer. A similar phenomenon is used in the removal of hydrocarbon-based pollutants from soil and shallow aquifers by the method of aeration (air sparging). Air sparging is technology in which air is injected below the water table through a drilling tool fitted with a special screen. The injected air migrates upwards, initiating an aerobic decontamination processes in polluted aquifers. Air sparging was implemented in the mid-1980s in Germany to enhance the process of removing chlorinated solvents from shallow aquifers. It was also used in the United States to clean up saturated zones contaminated with gasoline and other volatile contaminants (Johnson et al., 1993).

In most cases the horizontal permeability (k_h) of porous aquifer formations is higher (up to 30 times) than the vertical permeability (k_v). This has a beneficial effect on TFHB barrier formation because the greater the $k_\mathrm{h}/k_\mathrm{v}$ ratio, the easier and the more efficient the lateral injection of silicate solution into an aquifer. Moreover, the greater the permeability anisotropy, the more the shape of the injected gas cloud forms a paraboloid.

Laboratory flow visualization experiments using glass beads and sand as a porous medium were carried out to study CO_2 sparging. The aim of the experiments was to determine the effect of the gas injection method and heterogeneity of the porous media on the gas flow pattern. In order to visualize the experimental results, we built the laboratory setup shown in Fig. 2. It consists of a relatively thin rectangular tank ($5 \times 95 \times 110$ cm) made of Plexiglas and filled with porous medium and tap water.

The CO_2 gas is supplied through a cylindrical diffuser placed at bottom centre of the tank. The gas rate is operated by a flow controller in a continuous and cyclic manner.

Digital photos were taken to evaluate the geometry of the gas plume. Additionally, continuous scanning of conductivity confirmed the changes inside the porous medium. Different sands and spherical glass beads were used separately to test the effect of grain size and shape on the gas flow structure. The example results are shown in Fig. 3. Three distinct gas flow patterns were observed, depending both on grain shape and gas injection method. In the case of spherical glass beads, the channel type flow pattern shown in Fig. 3a was observed; however, for the non-spherical sands the geometry of gas flow depends primarily on the continuous or cyclic regime of the gas injection. When gas was injected in cyclic way (Fig. 3c) the gas plume was wider compared to the continuous regime (Fig. 3b).

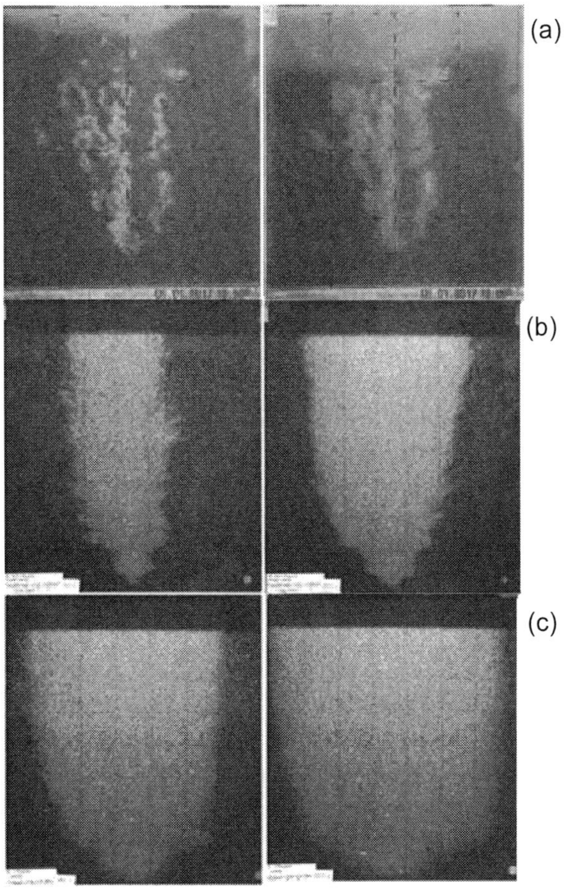

Figure 3. Gas flow patterns. Explanations in the text.

4 Tracking geochemical effects of the application of TFHB in sandy aquifers

Non-toxicity and environmental friendliness are the basic conditions for the application of TFHB technology in shallow aquifers. Moreover, it is also important that the injected working fluid (silicate solution) and the CO_2 gas (gelling catalyst) do not cause adverse or unpredictable interactions with the aquifer, including rapid and radical changes in pH; however, the dissolution of primary minerals combined with the crystallization of secondary phases can have a beneficial effect on the reduction of porosity. Therefore, the potential geochemical effects of the application of TFHB in the example aquifers were identified with the aid of simple geochemical simulations (Labus et al., 2015; Suchodolska and Labus, 2016).

Three types of dynamic modelling (equilibrium, reaction path and reactive transport) were performed on static geological aquifer models with similar petrophysical parameters but different mineralogical composition, taking into account both fresh and highly mineralized waters. The simulations performed do not relate to the effects of the pore space being blocked by silicate gels. They allow the analysis and forecasting of geochemical phenomena occurring in the aquifer systems due to the injected silicate solution.

Interactions between the injected solutions and the aquifer rock matrix and pore water were simulated by means of the Geochemists' Workbench Package (GWB, Bethke, 2008) for the following systems, each with 20 % porosity:

1. aquifer composed of quartz sand (80.0 %) with 0.01 % Fe $(OH)_3$
2. quartz sand (79.9 %) and calcite (0.1 %),
3. quartz sand (79.9 %) and calcite (0.1 %) with 0.01 % Fe $(OH)_3$,
4. quartz sand (77 %) and clay minerals (3 % illite)
5. quartz sand (80.0 %) with seawater,
6. quartz sand (80.0 %) with seawater, using pre-treatment pad fluid injection,
7. quartz sand (79.9 %) and dolomite (0.01 %).

4.1 Two-dimensional model of transport and reaction

The aim of the two-dimensional transport and reaction models was to follow the development of changes in the simulated aquifer, taking into account changes in the hydrodynamic field. GWB's X2t package was applied in this simulation. The model domain simulates a 200 m × 200 m section of an aquifer that is 0.5 m thick and is composed of 625 nodal blocks, each of which is 8 m × 8 m. The following parameters were considered: porosity, 20 %; permeability, 1-D; transverse dispersivity, 0.2; longitudinal dispersivity, 1 m; diffusion coefficient, 1×10^{-6} cm^2 s^{-1}. Groundwater flow across the domain is controlled by a hydraulic gradient of 0.05 between the left boundary (inflow) of the model and the right boundary (outflow). Injection wells (9 wells situated in a square grid) are located in the left part of the model; the silicate solution is injected within 3 h with the rate of 1.5 m^3 h^{-1} in each well. It was assumed that the injection of the working liquid into the aquifer environment would cause changes in the pore water chemistry as well as in the mineralogical composition of the aquifer matrix.

For a solute the rate of accumulation of a component in the groundwater is the sum of the rate due to transport, and

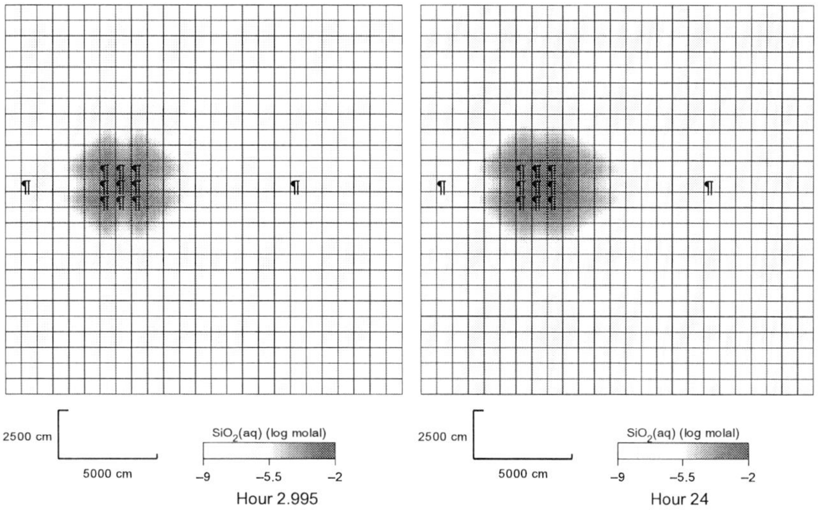

Figure 4. Plan view of $SiO_{2(aq)}$ concentrations around the group of 9 injection wells, in quartz aquifer, 3 and 24 h after the silicate fluid injection.

thechemical reactions, Eq. (1):

$$\frac{\partial(\phi C_i)}{\partial t} = \frac{\partial}{\partial x}\left(\phi D_{xx}\frac{\partial C_i}{\partial x}\right) + \frac{\partial}{\partial x}\left(\phi D_{xy}\frac{\partial C_i}{\partial y}\right) + \frac{\partial}{\partial y}\left(\phi D_{yx}\frac{\partial C_i}{\partial x}\right) + \frac{\partial}{\partial y}\left(\phi D_{yy}\frac{\partial C_i}{\partial y}\right) - \frac{\partial}{\partial x}(\phi \upsilon_x C_i) - \frac{\partial}{\partial y}(\phi \upsilon_y C_i) + \phi R_i, \quad (1)$$

where: φ is porosity, R_i is reaction rate (mol cm^{-3} s^{-1}), C_i is the component's concentration, D_{xx} is the entries in the dispersion tensor, and (υ_x, υ_y) is groundwater velocity vector (Bethke, 2008).

It was found that in all cases the injection of silicate solution triggers the pH rise within a range of about 10 m from the axis of the injection well. In the quartz sand aquifer, the quartz mineral slowly dissolves and the concentration of dissolved SiO_2 increases, as shown in Fig. 4.

In the modelled systems that included 0.1 % vol. of carbonate rocks and 0.01 % vol. of $Fe(OH)_3$, the dissolution of small volumes of calcite was also found; this precipitated back after the injection was terminated.

In the systems with illite, the effects of its hydrolysis were observed: precipitation of secondary kaolinite, release of potassium cations, and intensification of the decrease in pH.

During simulation of injection of silicate solution into the aquifer with seawater, it was noted that the pH change is limited practically to the well skin zone. Considering the actual, wider migration range of the injected liquid, this phenomenon may have consequences for the gelation time of the silicate. In order to extend the gelation time, the alkalinity of the injected solution should be increased, and/or the pad fluid should be used prior to the silicate solution injection. The injection zone located in the aquifer in equilibrium with seawater is prone to secondary dolomite precipitation, the volume of which is however small and reaches 0.0013 % vol.

4.2 Kinetic reaction path modelling

The performed reaction path simulation (simulation of reaction sequences) made it possible to determine the impact of chemical processes on the geochemical composition of porous media filled with a silicate solution. CO_2 was injected to accelerate sol-gel transition of the silicate solution. This led to geochemical composition changes which increased CO_2 partial pressure. The GWB React package was applied in this modelling. The following kinetic dissolution/precipitation rate equation was used in the calculations Eq. (2):

$$r_k = A_s k_T \left(1 - \frac{Q}{K}\right), \quad (2)$$

where r_k is reaction rate (mol s^{-1}), dissolution is $r_k>0$, precipitation is $r_k<0$), A_s is the mineral's surface area (cm^2), k_T is the rate constant (mol cm^{-2} s^{-1}) at temperature T, Q is the activity product (–), K is the equilibrium constant for the reaction (–). According to the above equation a given mineral precipitates when it is supersaturated or dissolves when it is undersaturated at a rate proportional to its rate constant and the surface area.

In all modelled aquifers, CO_2 injection causes a decrease in pH to about 5.8.

Small amounts of iron are released in the aquifer composed of quartz sand with iron (III) hydroxide due to the decomposition of $Fe(OH)_3$, as shown in Fig. 5. The solution of the treatment fluid in pore waters is oversaturated for SiO_2 phases and $Ca(FeO_2)_2$, and undersaturated for carbonates and sulphates. The crystallization of mineral phases of SiO_2 has a negligible effect on porosity.

In the presence of $Fe(OH)_3$ in the aquifer containing small amounts of calcite, a slight increase in porosity is observed, mainly related to the decomposition of $CaCO_3$. This process is responsible for the increase of Ca^{2+} concentrations and for the decrease of sulphate concentration, due to the synthesis and subsequent sorption of $CaSO_4$. Precipitation of silica and a release of small amounts of iron is also noticeable.

Decomposition of illite, amorphous silica and $Fe(OH)_3$, and the subsequent crystallization of secondary nontronite-Na has virtually no effect on the porosity in the sandy aquifer with illite.

In the aquifers saturated with seawater, the pH of the solution drops more to about 5.2 after injection of CO_2. Precipitation of quartz and decomposition of dolomite practically does not affect the porosity of the system. After the injection of CO_2, a high undersaturation with carbonate phases is observed, whereas the tendency of amorphous silica to precipitate increases.

4.3 One-dimensional transport and reaction model

The one-dimensional transport and reaction model was introduced to simulate the interactions and geochemical changes along the flow path of CO_2 injected below the injected silicate solution into the zone of saturation to accelerate the gelation process, as shown in Fig. 6; GWB's X1t package was applied. The flow field is simulated in the spherical domain with an internal radius of $r1 = 5\,\mathrm{cm}$ and an external radius of $r2 = 205\,\mathrm{cm}$; this corresponds to the gas migration path, $L = 200\,\mathrm{cm}$. The domain consists of 10 cells of equal heights.

During the injection of CO_2, the pH of pore waters in the modelled aquifer decreases. The strongest and fastest changes in pH are visible in the initial cells of the model, as shown in Fig. 7.

The increase in CO_2 fugacity and ion concentrations in pore water is most pronounced in the initial cells of the model in the proximity of the injection point. As a result of the decomposition of the dolomite in the initial cell and its crystallization in subsequent cells of the model, an increase or decrease in porosity is observed, respectively. The porosity extremes change over time, according to the direction of CO_2 transport in pore waters, see Fig. 7. The mineralogical changes of the rock matrix and the chemical composition of the pore fluid observed in all models are very small.

Under real conditions, not a single-phase solution in equilibrium with CO_2, but a two-phase $CO_2 + H_2O$ fluid will be transported through the aquifer. In this case, the migration rate will increase as the fluid moves upwards into the lower pressure zones, where the gas bubbles expand, the density of the $CO_2 + H_2O$ two-phase fluid decreases, and its transport is accelerated.

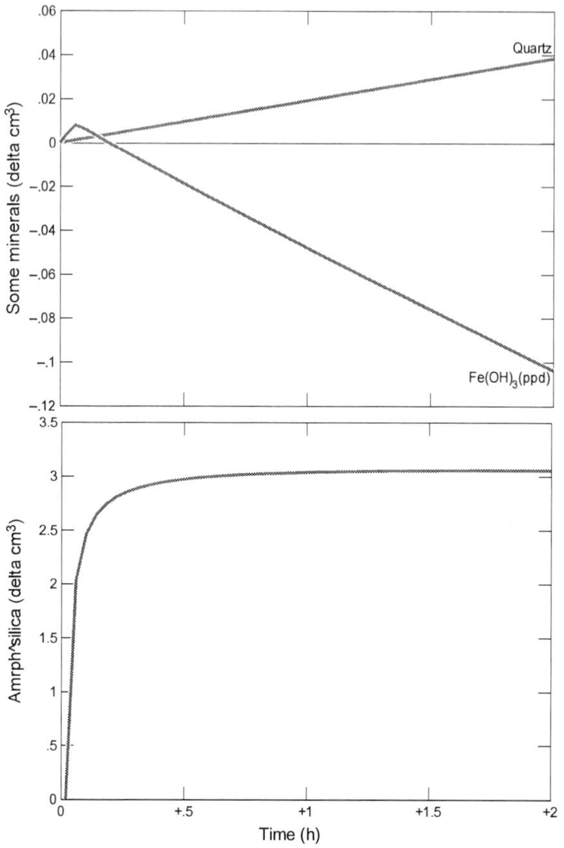

Figure 5. Recrystallization of quartz, decomposition of $Fe(OH)_3$ and precipitation of amorphous silica due to CO_2 injection – quartz sand aquifer. Compare the amounts to the model volume of $10\,000\,\mathrm{cm}^3$.

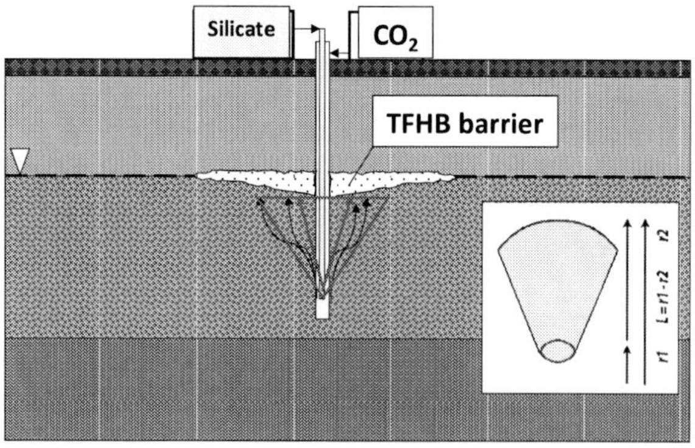

Figure 6. Domain of the 1-D transport and reaction model.

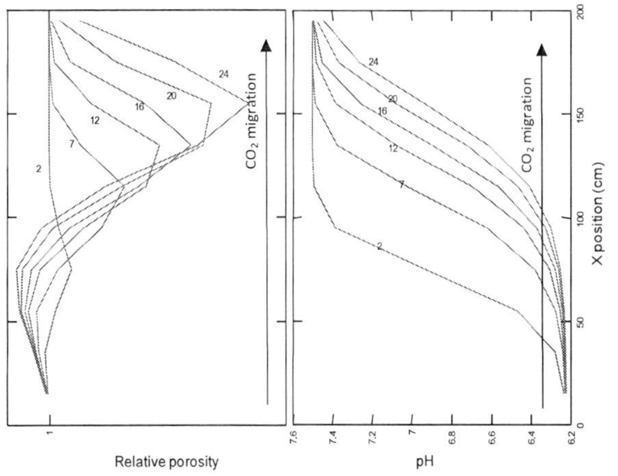

Figure 7. Relative porosity, and pH changes along the flow path of CO_2 injected into the zone of saturation.

5 Conclusions

The experiments indicated that the gas flow regime in a saturated porous medium varies greatly as result of heterogeneity and anisotropy of the porous medium.

Geochemical modelling (equilibrium, reaction path and reactive transport) was used to identify potential geochemical effects of the application of TFHB in sandy aquifers. Certain petrophysical parameters and mineralogical assemblages of aquifers were addressed, taking into account both low and highly mineralized groundwater. The simulations revealed that the TFHB does not have a negative impact on the chemistry of the rock-water systems considered.

Data availability. The data that support the findings of this study are available on request from the corresponsing author Krzysztof Labus (krzysztof.labus@polsl.pl). The data are not publicly available due to co financing of the research project by private company.

Competing interests. The authors declare that they have no conflict of interest.

Special issue statement. This article is part of the special issue "Innovative water resources management – understanding and balancing interactions between humankind and nature". It is a result of the 8th International Water Resources Management Conference of ICWRS, Beijing, China, 13–15 June 2018.

Acknowledgements. This work is the result of research conducted within the research project "A novel environmentally friendly technique of creating horizontal barriers in water-bearing zones", funded by the National Centre for Research and Development within the Smart Growth Operational Programme Priority axis 1: SUPPORT FOR R&D ACTIVITY OF ENTERPRISES Priority 1.1.1: R&D projects of enterprises, Contract No.: POIR.01.01.01-00-1038/15-00.

Edited by: Bo Pang
Reviewed by: Meifang Ren and one anonymous referee

References

Anderson, K. H.: Method of Oil Recovery, U.S. Patent 2,402,588, 1946.

Bethke C. M.: Geochemical and biogeochemical reaction modelling, Cambridge Univ. Press, Cambridge, 1–543, 2008.

Johnson, R. L., Johnson, P. C., McWhorter, D. B., Hinchee, R. E., and Goodman, I.: An overview of in situ air sparging, Ground Water Monit. Remediat., 13, 127–135, https://doi.org/10.1111/j.1745-6592.1993.tb00456.x, 1993.

Labus, K., Tarkowski, R., and Wdowin, M.: Modeling gas–rock–water interactions in carbon dioxide storage capacity assessment: a case study of Jurassic sandstones in Poland, Int. J. Environ. Sci. Technol., 12, 2493–2502, https://doi.org/10.1007/s13762-014-0652-6, 2015.

Nasr-El-Din, H. A. and Taylor, H. C.: Evaluation of sodium silicate/urea gels used for water shut-off treatments, J. Pet. Sci. Eng., 48, 141–160, 2005.

Suchodolska, K. and Labus, K.: Simplified models of transport and reactions in conditions of CO_2 storage in saline aquifers, Energy Proced., 97, 509–514, https://doi.org/10.1016/j.egypro.2016.10.062, 2016.

https://doi.org/10.5194/piahs-379-187-2018

Open Access PIAHS

Multi-scale fluctuation analysis of precipitation in Beijing by Extreme-point Symmetric Mode Decomposition

Jiqing Li, Zhipeng Duan, and Jing Huang
Renewable Energy School, North China Electric Power University, Beijing 102206, China
Correspondence: Jiqing Li (jqli6688@163.com)

Received: 28 December 2017 – Revised: 1 February 2018 – Accepted: 1 February 2018 – Published: 5 June 2018

Abstract. With the aggravation of the global climate change, the shortage of water resources in China is becoming more and more serious. Using reasonable methods to study changes in precipitation is very important for planning and management of water resources. Based on the time series of precipitation in Beijing from 1951 to 2015, the multi-scale features of precipitation are analyzed by the Extreme-point Symmetric Mode Decomposition (ESMD) method to forecast the precipitation shift. The results show that the precipitation series have periodic changes of 2.6, 4.3, 14 and 21.7 years, and the variance contribution rate of each modal component shows that the inter-annual variation dominates the precipitation in Beijing. It is predicted that precipitation in Beijing will continue to decrease in the near future.

1 Introduction

Beijing is one of the cities with the most serious water shortage in China. In the context of global warming, the annual precipitation is significantly reduced, which aggravates the water shortage crisis in Beijing. Therefore, studying the changing law of precipitation and the trend of its future development in Beijing has become a central issue of widespread concern. At present, many scholars have done a lot of researches on the variation and evolution of precipitation in Beijing. Zhu et al. (2012) used Empirical Mode Decomposition method to analyze the multi-scale oscillations of the precipitation time series from 1951 to 2009 in Beijing, the results show that the annual precipitation in Beijing will decrease continuously in the short term (Zhu et al., 2012). Through statistical calculations and frequency analysis of precipitation data from 16 rainfall stations in Beijing, Sun et al. (2007) explored the spatial and temporal distribution characteristics of precipitation in Beijing and its changing trend. It is found that continuous drought in Beijing after the 1990s and annual precipitation after 2008 may enter a relatively abundant period (Sun et al., 2007). Wang et al. (2009) analyzed the changes of temperature and precipitation over the past 48 years in Beijing and found that the annual precipitation showed a decreasing trend in both suburban and urban areas. The above research results have analyzed the cycle and variation characteristics of temperature and precipitation in Beijing from different perspectives, which is not only helpful for understanding the process and laws of climate change, but also great significance for the social and economic development of Beijing.

The climate system is a nonlinear, non-stationary, and hierarchical system (Xue et al., 2013). In the traditional trend analysis, large-scale cycles and trend changes are likely to mix together, can not tell the trend of changes or periodic oscillations. Extreme-point Symmetric Mode Decomposition (ESMD) is a locally adaptive time series analysis technique developed in recent years. Compared with the wavelet analysis method, it is free from the constraints of Fourier transform and is very suitable for analyzing non-stationary and nonlinear time series (Xia and Liu, 2017). This paper uses ESMD method to conduct multi-scale analysis of the annual precipitation series from 1951 to 2015 in Beijing, and compares the results with wavelet analysis to provide reference for future precipitation forecast and water resources planning and management in Beijing.

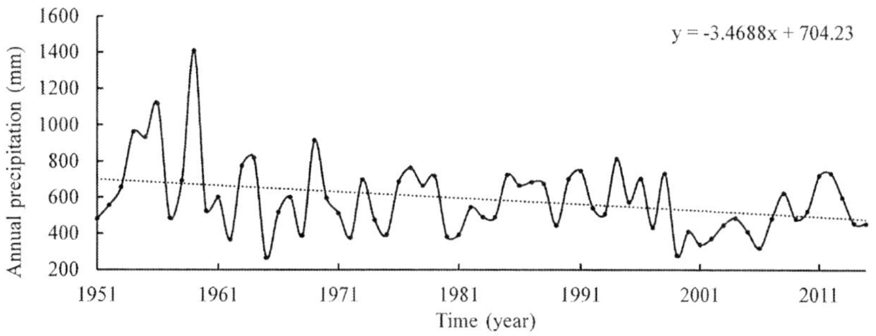

Figure 1. Annual precipitation time series from 1951 to 2015 at Beijing station.

2 Research methods

2.1 The basic principle of Empirical Mode Decomposition

The Empirical Mode Decomposition (EMD) method proposed by Huang et al. (1998), which is very suitable for dealing with nonlinear and non-stationary signals. The EMD method separates the fluctuations of different periods from the original signal and finally obtains the trend component. Fluctuations at different scales are defined as Intrinsic Mode Function (IMF). The modal decomposition of the EMD method is as follows:

1. Find all maxima and minima of the sequence $x(t)$, through the cubic spline interpolation, the upper envelope ($u_1(t)$) and the lower envelope ($u_2(t)$) of $x(t)$ are calculated.

2. Calculate the median curve ($m_1(t)$).

$$m_1(t) = \frac{1}{2}(u_1(t) + u_2(t)) \tag{1}$$

3. Subtracting the median curve from the original signal yields the remaining signal.

$$h_1(t) = x(t) - m_1(t) \tag{2}$$

4. Using $h_1(t)$ as a new signal sequence, repeat the above operation until $h_1(t)$ becomes a zero-mean process.

5. After the zero-mean process $h_1(t)$ is obtained, it is taken as the first IMF component $c_1(t)$, It represents the highest frequency component of the original signal.

6. Subtracting $c_1(t)$ from the original signal $x(t)$ results in a new signal process $r_1(t)$.

$$r_1(t) = x(t) - c_1(t) \tag{3}$$

7. The sequence $r_1(t)$ is taken as the original data, repeat steps (1)–(5) and obtain the sequence $c_2(t)$, $c_3(t)$, $c_4(t)$ and so on. The original signal is finally reconstructed.

$$x(t) = \sum_{j=1}^{n-1} c_j(t) + r_n(t) \tag{4}$$

Where: The sequence $r_n(t)$ is the trend term, representing the average trend of the signal.

2.2 The basic principle of Extreme-point Symmetric Mode Decomposition

The Extreme-point Symmetric Mode Decomposition (ESMD) method is a new development of the EMD method. This method solves the problem of "modal aliasing" in EMD and it has a big advantage in climate data analysis (Wang and Li, 2014). This method is good at finding trends and can separate inter-annual trends and the general trend of climate change in the observed sequence, helping to probe the issue of global warming. And this method can find the anomalous period and frequency band from the decomposition mode, which is good for the research of climate anomaly. The modal decomposition of the ESMD method is as follows:

1. Find all maxima and minima of the sequence $X(t)$, denoted by $E_i (1 \leq i \leq n)$.

2. Connect all the adjacent poles with line segments and mark the mid-points as $F_i (1 \leq i \leq n-1)$, and add the boundary mid-points F_0 and F_n at the left and right ends.

3. The interpolation curve $L_1, \ldots, L_p (p \geq 1)$ is constructed by using $n+1$ midpoints, and calculated the mean curve $L^* = (L_1 + \ldots + L_p)/p$.

4. The above steps are repeated for the $X(t) - L^*$ sequence until $|L^*| \leq \varepsilon$ (ε is the allowable error) or the number of screened times reaches the preset maximum value K, resulting in the first empirical mode $M_1(t)$.

5. The above four steps are repeated for the remaining sequences $X(t) - M_1(t)$, and the empirical modes $M_2(t)$–

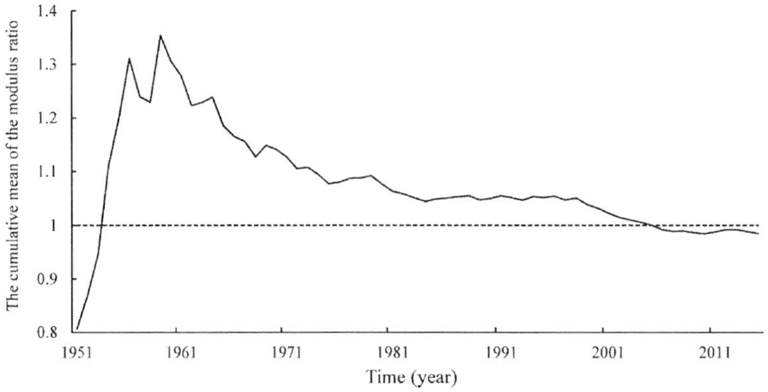

Figure 2. Beijing annual precipitation model ratio coefficient curve.

$M_3(t)$ ... are obtained respectively until the remaining sequence $R(t)$ remains only for a certain number of poles.

6. Change the maximum number of screening times K within the defined interval $[K_{\min}, K_{\max}]$ and repeat the above five steps. Then calculate the variance ratio σ/σ_0, and draw it with K changes in the map to find the minimum σ/σ_0 corresponding K_0. Repeat the above five steps again with K_0 as the limiting condition. The last remaining term $R(t)$ is the adaptive global averaging of the sequence $X(t)$.

After decomposition, the original time series $X(t)$ is reconstructed as:

$$X(t) = \sum M_i(t) + R(t) \qquad (5)$$

3 Case study

3.1 Study area profile

Beijing is located in the northwestern end of the North China Plain, the central Haihe River Basin, the land area of 16 410 km^2, is a typical northern temperate semi-humid monsoon climate. The complex and varied topography has created a diversity of weather conditions in Beijing. Summer heavy rainfall and strong convective weather are the main causes of frequent droughts and floods in Beijing. Shown in Fig. 1 for the Beijing 1951–2015 nearly 65 years of precipitation. Among them, the largest precipitation was 1407 mm in 1957, the smallest precipitation was 267 mm in 1965, the average annual rainfall was 589.75 mm, and the rate of linear precipitation decrease is 3.47 mm a^{-1}.

In this study, the annual rainfall data are provided by the National Climate Center of China Meteorological Administration. The data are released after the reorganization and review of the institution, so the reliability of the data can be basically guaranteed. As shown in Fig. 2, plot annual precipitation model ratio coefficient curve in Beijing, and analyze the representativeness of the data. It can be seen from the figure that with the increase of time, the amplitude of the curve is getting smaller and smaller, and when the annual rainfall series is over 50 years, the cumulative mean of the modulus ratio tends to 1, indicating that this series of data has certain stability.

3.2 Multi-scale analysis of precipitation changes

The ESMD method was used to decompose the annual mean precipitation series from 1951 to 2015 in Beijing. The best screening frequency was 26 times when the ratio of variance was the smallest, and 4 modes (Mode 1–4) and a trend margin R, as shown in Fig. 3.

In order to verify the reliability of the decomposition results, this paper synthesize a reconstructed sequence from the decomposed Mode 1–4 and the trend series R, and find that the reconstructed sequence is completely consistent with the original precipitation sequence, indicating that the decomposition result of ESMD method is credible of. In Fig. 3, the four mode components of ESMD decomposition each reflect the oscillations of different characteristic scales inherent in the original sequence, and the influence of signal fluctuation frequency of each scale on the overall characteristics of the original data is represented by the variance contribution rate. The Periodic Diagram Method is used to estimate the average period of each decomposition mode. Table 1 shows the mean period and the variance contribution rate of each modal component. It can be seen that annual precipitation in Beijing has annual variations of 2.6 and 4.3 years, and inter-decadal changes of 14 and 21.7 years. In addition, the main periods of the precipitation time series does not change over time during the study period.

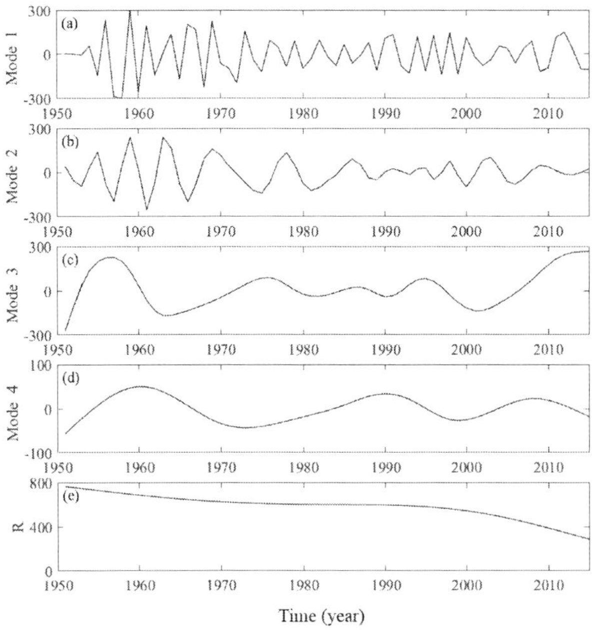

Figure 3. Modal components and trend of precipitation time series by ESMD

Combined with Table 1 and Fig. 3 quantitative analysis: The precipitation series of 65 years in Beijing contain many time-scale features, and the precipitation changes are mainly determined by the three higher-frequency oscillations of Mode 1–3. Among them, the Mode 1 (Fig. 3a) with a period of 2.6 years has the largest variance contribution rate, reaching 32.47 %. The signal oscillation of Mode 1 is very obvious and indicates the precipitation has a decreasing-increasing cycle change, which basically reflects the alternation of precipitation in Beijing before the 1970s. Mode 2 (Fig. 3b) indicates that there is a 4.3-year periodic change in the precipitation series in Beijing. Its variance contribution rate is about 16.74 %. Throughout the study period, Mode 2 shows that the amplitude of rainfall in Beijing fluctuated greatly before the 1970s and gradually decreased after the 1970s. Mode 3 (Fig. 3c) with a period of 14 years has only a smaller variance contribution rate of Mode 1, about 27.5 %. As can be seen from Fig. 3c, the amplitude changes steadily and fluctuates relatively little in the late 1960s and mid-1990s, but it significantly increased from the early 1950s to the late 1960s and after the 1990s. This shows that the precipitation variation in these two periods is great compared with other periods. Mode 4 (Fig. 3d) with a period of 21.7 years has the smallest variance contribution rate, only 1.37 %. Its amplitude is relatively stable over the entire time span. Comparing Mode 3 with Mode 4 can find that Mode 3 has a wide range of amplitude variations and the amplitude range of Mode 4 is significantly smaller. This indicates that the anomalous precipitation mainly occurs on the time scale

Table 1. Contribution rates of ESMD decomposition for mean precipitation.

Mode components	Mode 1	Mode 2	Mode 3	Mode 4	R
Period (a)	2.6	4.3	14	21.7	–
Contribution rates (%)	32.47	16.74	27.50	1.37	21.92

Table 2. Comparison of the cycle result by ESMD and Wavelet Analysis.

ESMD Analysis (a)	Wavelet Analysis Cycle (a)	Difference (a)
2.6	2	0.6
4.3	5–8	0.7–3.7
14	10–15	1–4
21.7	25	3.3

of 14 years.The variance contribution rate of trend item R (Fig. 3e) is about 16.74 %. It reflects the overall trend of precipitation in Beijing over the past 65 years. As shown in Fig. 3e, trend item show a non-linear trend of gradual decline over the entire time scale.

3.3 Wavelet analysis and verification

Wavelet analysis is a hot frontier field in recent years because of its special advantages for signal processing, it is widely used in the field of time-frequency structure analysis of weather and climate sequences (Yang and Shu, 2017). This paper uses the Morlet wavelet transform to analyzes the annual precipitation anomaly sequence. Figure 4 shows the real part time frequency distribution of wavelet coefficients of the precipitation anomaly sequence in Beijing in recent 65 years. When the wavelet coefficient is positive, it indicates that the precipitation is relatively large and the greater the value, the more precipitation. When the wavelet coefficient is negative, the precipitation is less than normal.

It can be seen from Fig. 4 that precipitation in Beijing has multi-scale changes of 2, 5–8, 10–15 and 25 years. Throughout the study period, the wavelet coefficients alternated between positive and negative, indicating that the change of precipitation with the alternation of abundance and dryness over time. Table 2 shows the comparison between the wavelet and the ESMD analysis cycle results, the results of the two methods have some differences, and the longer the cycle shows the greater the difference.

As a time-frequency analysis method, wavelet transform is the same as other time-frequency analysis methods, such as short-time Fourier transform and classical spectral estimation, which is based on Fourier transform theory as a basis. There are some limitations, such as the selection of wavelet

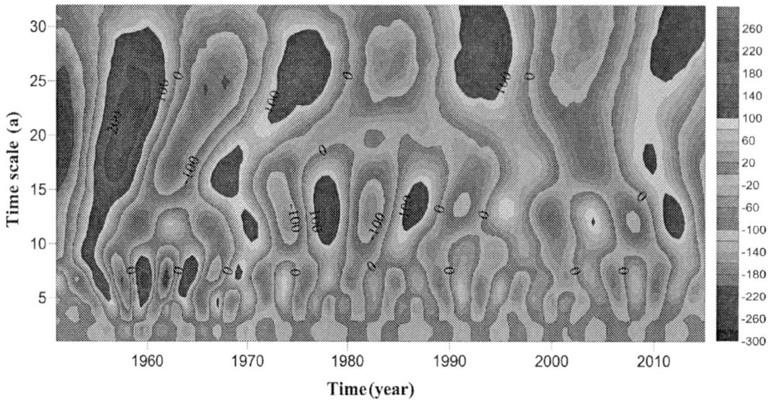

Figure 4. The real part distribution of wavelet coefficients.

basis functions, constant multi-resolution and other issues. Compared with the wavelet analysis, the ESMD method based entirely on the characteristics of data changes decomposition and got rid of the shackles of the Fourier transform theory. This method not only has the advantages of multi-resolution wavelet transform, but also breaks through restrictions of the wavelet basis functions, it has strong flexibility and adaptability. The decomposition process of this method is easier than wavelet analysis, and the components can clearly characterize the variation characteristics of signals on different time scales. It is more helpful to explore the inherent laws of things.

4 Conclusion

ESMD method is a suitable method for analyzing non-linear and non-stationary sequence signals. Applying the ESMD method to climate element time series not only separates the annual and inter-decadal trends, but also separates the general trend of climate change. This study shows that the 65-year precipitation series in Beijing are mainly composed of Mode 1, Mode 2 and Mode 3, indicating that the 2.6-, 4.3-, and 9-year scale oscillations play a major role in the change of the whole sequence. Combining several mode components and the trend term R predicts that the annual precipitation in Beijing will continue to decrease in the next few years.

Under the background of global warming, the influence of natural factors and human activities are superimposed on each other, and the variation of hydrological time series is more complicated. Based on the study of precipitation series in Beijing, this paper shows that ESMD method has better analysis effect on time series of hydrology. However, the current research is only in the initial stage. How to determine the influence factors of each mode component, calculate its weight and finally apply it to the prediction of time series will be the key work in future.

Data availability. Data are available at: http://cmdp.ncc-cma.net/cn/index.htm. This site requires user registration to download precipitation data.

Competing interests. The authors declare that they have no conflict of interest.

Special issue statement. This article is part of the special issue "Innovative water resources management – understanding and balancing interactions between humankind and nature". It is a result of the 8th International Water Resources Management Conference of ICWRS, Beijing, China, 13–15 June 2018.

Acknowledgements. This study was financially supported by National Key Projects of China "Water resources efficient development and utilization" (2017YFC0405900, 2016YFC0402208, 2016YFC0401903) and National Natural Science Foundation of China (No. 51641901). The author would like to give special thanks to anonymous reviewers.

Edited by: Dingzhi Peng
Reviewed by: two anonymous referees

References

Huang, N. E., Shen, Z., and Long, S. R.: The empirical mode decomposition and the hilbert spectrum for nonlinear and nonstationary time series analysis, Philos. T. R. Soc. A, 454, 903–995, https://doi.org/10.1098/rspa.1998.0193, 1998.

Sun, Z., Feng, S., and Yang, Z.: Precipitation characteristics in Beijing from 1950 to 2005, J. Irrig. Drain. E., 2, 12–16, 2007.

Wang, J. and Li, Z.: ESMD method for climate data analysis, Climate Change Research Letters, 3, 1–5, https://doi.org/10.12677/ccrl.2014.31001, 2014.

Wang, W., Zhang, W., and Cai, X.: Changes of temperature and precipitation in Beijing in recent 50 years, Journal of Arid Meteorology, 27, 350–353, 2009.

Xia, C. and Liu, C.: Conjoint analysis of near-fault multi-pulse ground motion based on wavelet transformation and extreme-point symmetric mode decomposition, Journal of Disaster Prevention and Mitigation Engineering, 37, 697–704, 2017.

Xue, C., Hou, W., and Zhao, J.: Application of set empirical mode decomposition in multi-scale analysis of regional precipitation change and research of climate change response, Acta Phys. Sin., 62, 109–203, 2013.

Yang, Y. and Shu, H.: Small and medium scale of climatic variation characteristics of Baiyin city – Gansu Province, Journal of Arid Land Resources and Environment, 31, 126–131, https://doi.org/10.13448/j.cnki.jalre.2017.155, 2017.

Zhu, L., Chen, Y., and Li, L.: Analysis of precipitation variation trend in Beijing City from 1951 to 2009, Water Resources Protection, 28, 42–46, https://doi.org/10.3969/j.issn.1004-6933.2012.03.008, 2012.

https://doi.org/10.5194/piahs-379-193-2018

Estimating parameter values of a socio-hydrological flood model

Marlies Holkje Barendrecht[1], Alberto Viglione[1], Heidi Kreibich[2], Sergiy Vorogushyn[2], Bruno Merz[2], and Günter Blöschl[1]

[1]Vienna University of Technology, Institute of Hydraulic Engineering and Water Resources Management, Vienna, Austria

[2]GeoForschungsZentrum Potsdam (GFZ), Section Engineering Hydrology, Telegrafenberg, Potsdam, Germany

Correspondence: Marlies Holkje Barendrecht (barendrecht@hydro.tuwien.ac.at)

Received: 31 December 2017 – Accepted: 10 February 2018 – Published: 5 June 2018

Abstract. Socio-hydrological modelling studies that have been published so far show that dynamic coupled human-flood models are a promising tool to represent the phenomena and the feedbacks in human-flood systems. So far these models are mostly generic and have not been developed and calibrated to represent specific case studies. We believe that applying and calibrating these type of models to real world case studies can help us to further develop our understanding about the phenomena that occur in these systems. In this paper we propose a method to estimate the parameter values of a socio-hydrological model and we test it by applying it to an artificial case study. We postulate a model that describes the feedbacks between floods, awareness and preparedness. After simulating hypothetical time series with a given combination of parameters, we sample few data points for our variables and try to estimate the parameters given these data points using Bayesian Inference. The results show that, if we are able to collect data for our case study, we would, in theory, be able to estimate the parameter values for our socio-hydrological flood model.

1 Introduction

Socio-hydrology aims to study the long term feedbacks between humans and hydrology and tries to explain the phenomena that occur as a result of these feedbacks (Sivapalan and Blöschl, 2015; Sivapalan et al., 2012). So far, socio-hydrological flood models are mostly based on concepts derived from personal experience/observation of how the system works (Di Baldassarre et al., 2013, 2015, 2017; Grames et al., 2016, Grelot and Barreteau, 2012; Viglione et al., 2014). Most of them are general models that are not developed to reproduce and evaluate the dynamics of a specific case study and have not been compared to data. An exception to this is the work of Ciullo et al. (2017): they made a qualitative comparison between the model results of Di Baldassarre et al. (2015) and the human-flood system of the city of Rome and Bangladesh using data on population density, flood losses and levee heights. Chen et al. (2016) developed a socio-hydrological model to reflect the human-flood dynamics in the Kissimmee River Basin, Florida. Using flood series and data about wetland area they made a qualitative assessment of the model's performance. While the socio-hydrological models and studies so far show that these models can yield valuable insights into human-flood dynamics, there is a lack of application to real world systems and none of the models so far have been calibrated to represent a specific case study. It is therefore of interest to assess whether calibrating socio-hydrological models is indeed feasible and understanding what are the data needed and their amount. Here we develop a socio-hydrological model and explore whether we are able to estimate the parameter values of this model using Bayesian Inference (Gelman et al., 2014) if we would have data available.

2 Methods

2.1 Model structure

History has shown that repeated flood events may result in lower damages if flood events happen again, because peo-

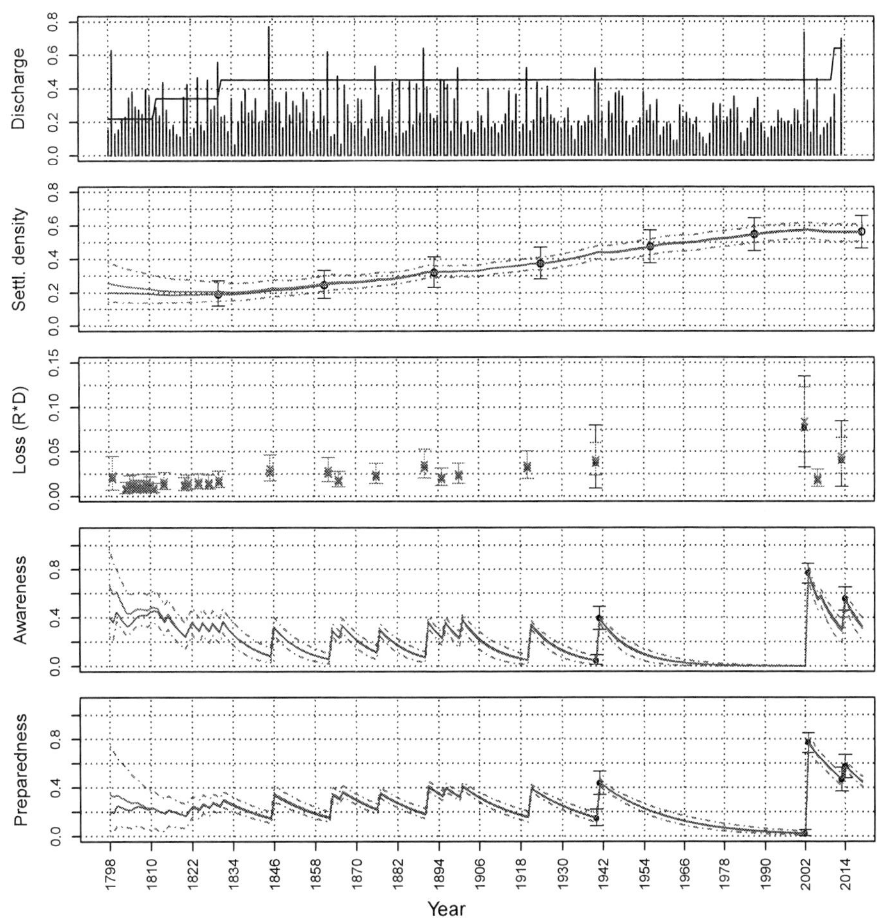

Figure 1. Generated dynamics of system variables, sampled data points and estimated time series. The generated dynamics of the system variables is represented in blue. Sampled data points are shown in black, with their 90 % uncertainty bounds. Estimated time series are plotted in red, with the mean as a solid line and the 90 % credible bounds as dotted lines.

ple become aware of the flood risk and adapt. We want our model to capture these interactions between floods, awareness and preparedness. We model these interactions with the following variables: Floods (W), Losses (L), Relative Losses (R), Settlement Density (D), Awareness (A) and Preparedness (P). The behavior of these variables over time is described with the system of differential equations in Eq. (1).

$$L = RD \tag{1a}$$

$$R = \begin{cases} R_{\max} - \beta_R \exp\left(-\alpha_R (P_{\max} - P)\dfrac{W}{W_{\max}}\right), \\ \text{if} \quad W > H \\ 0, \quad \text{otherwise} \end{cases} \tag{1b}$$

$$\frac{\mathrm{d}D}{\mathrm{d}t} = U(1 - \alpha_D A) D \left(1 - \frac{D}{D_{\max}}\right) \tag{1c}$$

$$\frac{\mathrm{d}A}{\mathrm{d}t} = \alpha_A L \left(1 - \frac{A}{A_{\max}}\right) - \mu_A A \tag{1d}$$

$$\frac{\mathrm{d}P}{\mathrm{d}t} = \alpha_P \frac{\mathrm{d}A}{\mathrm{d}t}\left(1 - \frac{P}{P_{\max}}\right) S(R) - \mu_P P \tag{1e}$$

$$S(R) = \tanh(1000R) \tag{1f}$$

High water levels (W) (here called floods) and the protection level (H) are external variables. For illustrative purposes, we use time series of the gauge at Dresden, Germany. If the water level (W) is higher than the protection level (H), this results in a relative loss (R) as given in Eq. (1b). R increases exponentially to a maximum of one for a maximum flood. Together with the settlement density (D), this results in an actual loss (L), according to Eq. (1a). The actual loss is higher if the relative loss is higher or if the settlement density (and thus the exposure) is higher.

Table 1. Model parameter description and units (n_h = number of households, n_m = number of precautionary measures)

Parameter	Definition	Unit
α_D	Risk taking attitude	$[1/(n_h/n_h)]$
α_A	Scaredness	[1/ (€/€)]
μ_A	Forgetfulness	$[1/t]$
μ_P	Decay rate of precautionary measures	$[1/t]$
α_P	Activeness	$[(n_m/n_m)t/(n_h/n_h)]$
α_R	Effectiveness of preparedness	$[1/(n_m/n_m)]$

According to the literature, awareness (Eq. 1d) increases when households experience damage (Bradford et al., 2012; Bubeck et al., 2013; Osberghaus, 2015; Owusu et al., 2015; Poussin et al., 2014; Wachinger et al., 2013), and decreases exponentially over time with rate μ_A as households forget (ICPR, 2002; Kreibich et al., 2011). We assume the increase in awareness depends on the size of the awareness as well, i.e. if awareness is high already it will increase less than when it is close to zero.

An increase in awareness may result in an increase in the uptake of precautionary measures and thus preparedness. How big this increase in preparedness is depends on other factors as well, like coping appraisal (Bubeck et al., 2012), maladaptive coping responses (Bubeck et al., 2013) or worry (Miceli et al., 2008; Raaijmakers et al., 2008). The fact that not all households that are aware of the risk take precautionary measures is represented by the parameter α_P, which determines how much of an increase in awareness results in an uptake of precautionary measures and thus an increase in preparedness. As described by Eq. (1e), preparedness increases after a flood event with an amount relative to the change in awareness and depending on how high the preparedness was before the flood event (if households have already implemented many measures, there is less room for an increase in preparedness than when they have not implemented any measures). Preparedness only increases when damage occurs, i.e. when the relative loss is higher than zero. We approximate this step function with a tangens hyperbolicus (Eq. 1f). Like the awareness, preparedness decreases exponentially over time with a rate μ_P because households forget and measures deteriorate.

According to the literature, implementing precautionary measures can greatly reduce the losses because of a flood (Kreibich et al., 2015; Poussin et al., 2015). So if preparedness is higher the relative losses will be lower. How much lower the relative losses will be depends on the parameter α_R (Eq. 1b). As explained the losses depend on the relative losses and on the settlement density. The settlement density grows with a general growth rate U, but if awareness is higher this will reduce the growth rate and thus settlement density will grow slower or may even decline as described in Eq. (1c).

Table 2. Parameter values. Real values used to generate our data points and the mean and standard deviation of the posterior distribution estimated with Bayesian inference

Parameter	Real value	Estimated
α_D	3	mean = 3.30, sd = 0.64
α_A	10	mean = 9.80, sd = 0.07
μ_A	0.1	mean = 0.11, sd = 0.02
μ_P	0.05	mean = 0.05, sd = 0.01
α_P	1	mean = 1.00, sd = 0.09
α_R	0.2	mean = 0.23, sd = 0.07
D_0	0.2	mean = 0.26, sd = 0.07
A_0	0.4	mean = 0.67, sd = 0.21
P_0	0.2	mean = 0.34, sd = 0.22

2.2 Bayesian inference

We want to estimate the values of the model parameters from data. The variables are scaled between zero and one. We assume β_R is one and try to estimate the values of the six parameters that do not have a value of one. These parameters are described in Table 1. In addition we need to estimate the starting values of the variables settlement density, awareness and preparedness, because we describe the evolution of these variables with a differential equation.

Since socio-hydrology studies both the human and the hydrological system, we have to deal with various types of data with different uncertainties. Bayesian inference allows us to include all of these different types of data. Bayes' Theorem (Eq. 2) tells us that the posterior distribution of our parameters depends on the likelihood of the data and our prior estimation of the parameter distributions:

$$p(\theta|y) = \frac{p(y|\theta)\,p(\theta)}{\int p(y|\theta)\,p(\theta)\,\mathrm{d}\theta} \tag{2}$$

$p(\theta|y)$ is the estimated posterior distribution of our parameter values. $p(\theta)$ represents our prior belief about the distribution of the parameters. If we have any information about our parameters beforehand we can specify an informative distribution, if we do not have any information we use non-informative distributions, which is what we will do here. $p(y|\theta)$ gives us the likelihood of our data given the parameter values. Here we can incorporate our data and specify the un-

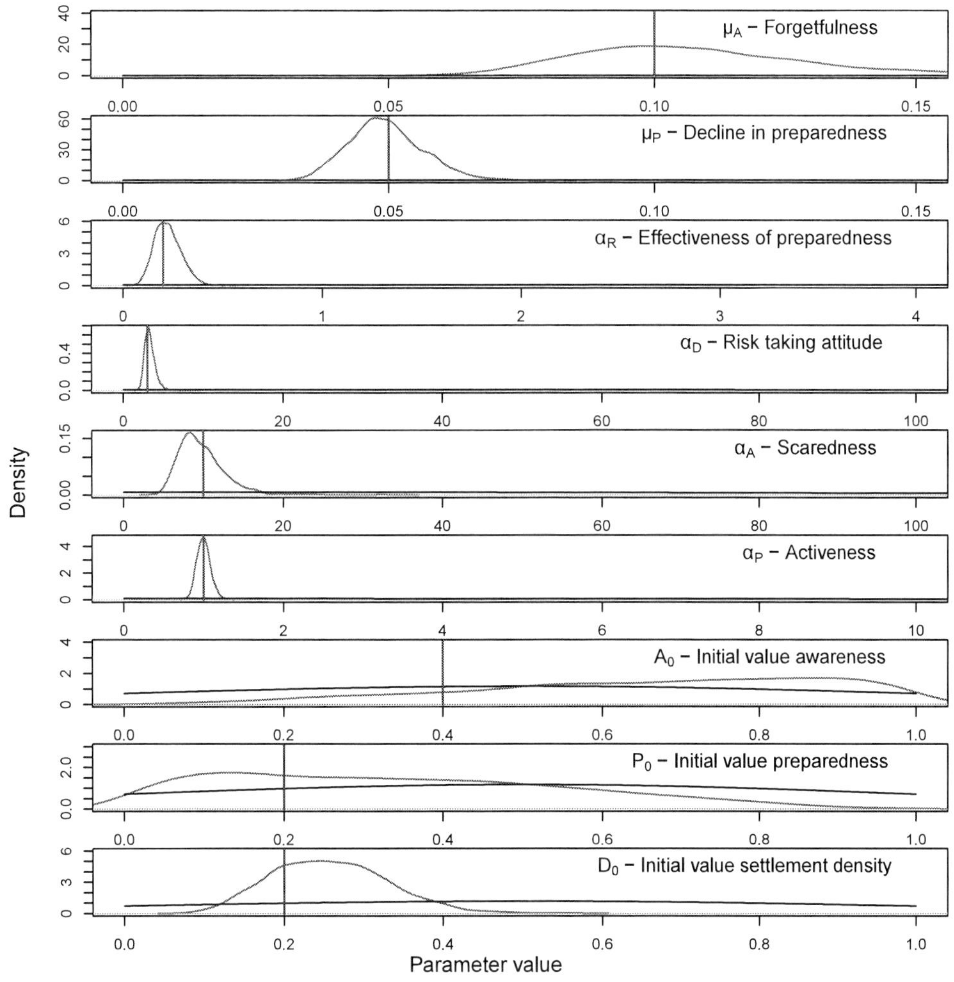

Figure 2. Prior (black) and posterior (red) distributions of the parameters. The "real" parameters to be estimated are represented by the blue vertical lines (see Table 2 for their values).

certainties. In most cases it is not possible to compute the integral in the denominator, therefore we approximate this using Markov chain Monte-Carlo (MCMC) methods (Gelman et al., 2014). In this case we use the software Stan (Carpenter et al., 2017) to perform the estimation. Stan uses Hamiltonian Monte Carlo sampling, which uses the gradient of the log probability to speed up convergence and parameter exploration (Stan Development Team, 2017).

3 Results

We assume that the parameters have the values given in Table 2 and simulate time series of all state variables (blue lines in Fig. 1). The question we ask is: if we could observe, with uncertainty, the variables only in a limited number of time points, would we be able to infer the temporal evolution of the variables or, which is analogous, would we be able to estimate the parameters of the model? To test whether this is possible, we sample some data points for each of the variables: Settlement Density (D), Loss ($L = R \cdot D$), Awareness (A) and Preparedness (P). These data points and their uncertainties are plotted in black in Fig. 1. Using these data points and uninformative priors for our parameters, we perform the Bayesian Inference using MCMC. Figure 1 shows the mean (solid red line) and 90 % credible bounds (dashed red lines) of the time series as estimated with the inference.

The top graph gives normalized discharges and protection level time series at the annual time scales. If the discharge is higher than the protection level this results in a loss as shown in the third graph. The size of this loss depends on the relative loss and the settlement density. If losses occur, this causes an increase in awareness (second graph from the bottom) and subsequently an increase in preparedness (bottom graph). Both the awareness and preparedness decrease gradually over time when no flood occurs. The awareness does slightly influence the growth of the settlement density since every time a flood occurs the settlement density grows more slowly in the following years.

The credible bounds of the simulation are quite narrow, which means that the posterior distributions of our parameter values are narrow as well and that we are confident that we almost got the parameter values right. Figure 2 shows the real parameter values (used to generate our data points) in blue, the prior distributions that we used for the Bayesian inference in black and the posterior distributions estimated with the inference in red. The posterior mean estimates are very close to the real parameter values (see also Table 2). The estimates of the initial values are bit further off from the real values, which is not surprising given that we did not pick any data points at the start of our time series and therefore we do not have any information available to estimate the variables at the start. We used flat uninformative prior distributions and we are able to determine the posterior distributions quite well. The posterior distributions are not much influenced by the prior distributions and mainly depend on the data.

4 Conclusions

We have conducted an artificial experiment where a Socio-Hydrological model is assumed whose "true parameters" are known to us, since we impose them, and observations are sampled from variables generated with the model. The aim of this paper is to assess whether the model can be calibrated to these observed data. This is not trivial because the model is highly nonlinear and it is not clear what amount of data is needed for calibration and, more importantly, whether the amount we can imagine to find in a real world case study can be enough. We demonstrate that, if the assumptions underlying the model are valid, we are able to estimate quite accurately the parameter values from relatively few data, which could be available in real case studies, using Bayesian Inference. Other estimation techniques, such as those used in rainfall-runoff modelling, could be used instead but the fact that Bayesian Inference is a flexible tool that can incorporate different types of information makes it ideal for the application to socio-hydrological models, where data are highly uncertain and prior information may be available to constrain model parameters.

The next step will be to gather data for a specific real world case study, develop and apply a model to it. This will pose a bigger challenge than it is in the case presented here, since the model will not represent correctly and fully the real world. Also, another challenge will be to map the information gathered on variables such as awareness and preparedness to the zero-to-one space adopted here. However it is promising to know that, based on what we show in this paper, if we can rely on the assumptions of the model so that the uncertainty is mainly due to not knowing its parameters, we can actually calibrate a socio-hydrological flood model to the few data available.

Data availability. The analyses in the paper are based on generated data, with the exception of the time series of water levels of the river Elbe at Dresden, which is the result of the research project "Integration von historischen und hydraulisch/hydrologischen Analysen zur Verbesserung der regionalen Gefährdungsabschätzung und zur Erhöhung des Hochwasserbewusstseins 2005–2007" (BMBF-Projekt, 2007) (Uwe Grünewald and Sabine Schümberg, Hydrology and Water Resource Management, Brandenburg University of Technology BTU Cottbus, Germany).

Competing interests. The authors declare that they have no conflict of interest.

Special issue statement. This article is part of the special issue "Innovative water resources management – understanding and balancing interactions between humankind and nature". It is a result of the 8th International Water Resources Management Conference of ICWRS, Beijing, China, 13–15 June 2018.

Acknowledgements. We would like to acknowledge funding from the Marie Sklodowska-Curie Innovative Training Network "A Large-Scale Systems Approach to Flood Risk Assessment and Management – SYSTEM-RISK" (Grant Agreement number: 676027) and the project "SozioHydroKopplung" funded by the German Bundesanstalt für Gewässerkunde. We would like to thank Hongrui Wang, Dingzhi Peng and one anonymous reviewer for their useful comments to the original version of the paper.

Edited by: Dingzhi Peng
Reviewed by: Hongrui Wang and one anonymous referee

References

BMBF-Projekt: "Integration von historischen und hydrologisch/hydraulischen Analysen zur Verbesserung der regionalen Gefährdungsabschätzung und zur Erhöhung des Hochwasserbewusstseins", BTU Cottbus, Cottbus, available at: http://www-docs.tu-cottbus.de/hydrologie/public/files/Schlussbericht_BMBF0330686.pdf, 2007.

Bradford, R. A., O'Sullivan, J. J., van der Craats, I. M., Krywkow, J., Rotko, P., Aaltonen, J., Bonaiuto, M., De Dominicis, S., Waylen, K., and Schelfaut, K.: Risk perception – issues for flood management in Europe, Nat. Hazards Earth Syst. Sci., 12, 2299–2309, https://doi.org/10.5194/nhess-12-2299-2012, 2012.

Bubeck, P., Botzen, W. J. W., and Aerts, J. C. J. H.: A review of risk perceptions and other factors that influence flood mitigation behavior, Risk Anal., 32, 1481–1495, 2012.

Bubeck, P., Botzen, W. J. W., Kreibich, H., and Aerts, J.: Detailed insights into the influence of flood-coping appraisals on mitigation behaviour, Global Environ. Chang., 23, 1327–1338, 2013.

Carpenter, B., Gelman, A., Hoffman, M., Lee, D., Goodrich, B., Betancourt, M., and Riddell, A.: Stan: A Probabilistic Programming Language, J. Stat. Softw., 76, 1–32, https://doi.org/10.18637/jss.v076.i01, 2017.

Chen, X., Wang, D., Tian, F., and Sivapalan, M.: From channelization to restoration: Sociohydrologic modeling with changing community preferences in the Kissimmee River Basin, Florida, Water Resour. Res., 52, 1227–1244, https://doi.org/10.1002/2015WR018194, 2016.

Ciullo, A., Viglione, A., Castellarin, A., Crisci, M., and Di Baldassarre, G.: Socio-hydrological modelling of flood-risk dynamics: comparing the resilience of green and technological systems, Hydrolog. Sci. J., 62, 880–891, https://doi.org/10.1080/02626667.2016.1273527, 2017.

Di Baldassarre, G., Martinez, F., Kalantari, Z., and Viglione, A.: Drought and flood in the Anthropocene: feedback mechanisms in reservoir operation, Earth Syst. Dynam., 8, 225–233, https://doi.org/10.5194/esd-8-225-2017, 2017.

Di Baldassarre, G., Viglione, A., Carr, G., Kuil, L., Salinas, J. L., and Blöschl, G.: Socio-hydrology: conceptualising human-flood interactions, Hydrol. Earth Syst. Sci., 17, 3295–3303, https://doi.org/10.5194/hess-17-3295-2013, 2013.

Di Baldassarre, G., Viglione, A., Carr, G., Kuil, L., Yan, K., Brandimarte, L., and Blöschl, G.: Debates – Perspectives on socio-hydrology: Capturing feedbacks between physical and social processes, Water Resour. Res., 51, 4770–4781, https://doi.org/10.1002/2014WR016416, 2015.

Gelman, A., Carlin, J. B., Stern, H. S., Dunson, D. B., Vehtari, A., and Rubin, D. B.: Bayesian data analysis (Vol. 2), CRC press Boca Raton, FL, 2014.

Grames, J., Prskawetz, A., Grass, D., Viglione, A., and Blöschl, G.: Modeling the interaction between flooding events and economic growth, Ecol. Econ., 129, 193–209, https://doi.org/10.1016/j.ecolecon.2016.06.014, 2016.

Grelot, F. and Barreteau, O.: Simulation of Resilience of an Insurance System to Flood Risk, International Environmental Modelling and Software Society (iEMSs) 2012 International Congress on Environmental Modelling and Software Managing Resources of a Limited Planet: Pathways and Visions under Uncertainty, Sixth Biennial Meeting, Leipzig, Germany, 2012.

ICPR (International Commission for the Protection of the Rhine): Non structural flood plain management – measures and their effectiveness, ICPR, Koblenz, 2002.

Kreibich, H., Seifert, I., Thieken, A.H., Lindquist, E., Wagner, K., and Merz, B.: Recent changes in flood preparedness of private households and businesses in Germany, Reg. Environ. Change, 11, 59–71, https://doi.org/10.1007/s10113-010-0119-3, 2011.

Kreibich, H., Bubeck, P., Van Vliet, M., and De Moel, H.: A review of damage-reducing measures to manage fluvial flood risks in a changing climate, Mitigation and Adaptation Strategies for Global Change, 20, 967–989, 2015.

Miceli, R., Sotgiu, I., and Settanni, M.: Disaster preparedness and perception of flood risk: A study in an alpine valley in Italy, J. Environ. Psychol., 28, 164–173, 2008.

Osberghaus, D.: The determinants of private flood mitigation measures in Germany – Evidence from a nationwide survey, Ecol. Econ., 110, 36–50, 2015.

Owusu, S., Wright, G., and Arthur, S.: Public attitudes towards flooding and property-level flood protection measures, Nat. Hazards, 77, 1963–1978, 2015.

Poussin, J. K., Botzen, W. J. W., and Aerts, J. C. J. H.: Factors of influence on flood damage mitigation behaviour by households, Environ. Sci. Policy, 40, 69–77, 2014.

Poussin, J. K., Botzen, W. J. W., and Aerts, J. C. J. H. Effectiveness of flood damage mitigation measures: Empirical evidence from French flood disasters, Global Environmental Change, 31, 74–84, 2015.

Raaijmakers, R., Krywkow, J., and van der Veen, A.: Flood risk perceptions and spatial multi-criteria analysis: an exploratory research for hazard mitigation, Natural Hazards, 46, 307–322, 2008.

Sivapalan, M. and Blöschl, G.: Time scale interactions and the coevolution of humans and water, Water Resour. Res., 51, 6988–7022, https://doi.org/10.1002/2015WR017896, 2015.

Sivapalan, M., Savenije, H. H. G., and Blöschl, G.: Socio-hydrology: A new science of people and water, Hydrol. Process., 26, 1270–1276, https://doi.org/10.1002/hyp.8426, 2012.

Stan Development Team: Stan Modeling Language Users Guide and Reference Manual, Version 2.17.0, http://mc-stan.org, 2017.

Viglione, A., Di Baldassarre, G., Brandimarte, L., Kuil, L., Carr, G., Salinas, J. L., and Bloeschl, G.: Insights from socio-hydrology modelling on dealing with flood risk – Roles of collective memory, risk-taking attitude and trust, J. Hydrol., 518, 71–82, https://doi.org/10.1016/j.jhydrol.2014.01.018, 2014.

Wachinger, G., Renn, O., Begg, C., and Kuhlicke, C.: The risk perception paradox – implications for governance and communication of natural hazards, Risk Analysis, 33, 1049–1065, 2013.

https://doi.org/10.5194/piahs-379-199-2018

Open Access
PIAHS

Mid and long-term optimize scheduling of cascade hydro-power stations based on modified GA-POA method

Jiqing Li and Xiong Yang

School of Renewable Energy, North China Electric Power University, Beijing 102206, China

Correspondence: Jiqing Li (jqli6688@163.com)

Received: 30 December 2017 – Revised: 2 April 2018 – Accepted: 3 April 2018 – Published: 5 June 2018

Abstract. In this paper, to explore the efficiency and rationality of the cascade combined generation, a cascade combined optimal model with the maximum generating capacity is established, and solving the model by the modified GA-POA method. It provides a useful reference for the joint development of cascade hydro-power stations in large river basins. The typical annual runoff data are selected to calculate the difference between the calculated results under different representative years. The results show that the cascade operation of cascaded hydro-power stations can significantly increase the overall power generation of cascade and ease the flood risk caused by concentration of flood season.

1 Introduction

As more and more large and medium hydro-power stations are put into use in our country, the joint dispatching of cascaded hydro-power plants has been getting more and more widespread attention. In order to encourage all hydro-power stations in cascade hydro-power stations to actively participate in the joint dispatch in order to maximize the benefits, the hydro-power industry advocates the implementation of "Cascade, rolling, integrated" mode of operation (Xu, 2015; He, 2014; Zhou, 2016). In the joint operation of cascade hydro-power stations, the leading hydro-power station in the upper reaches has good regulating performance, which can greatly increase the power generation capacity of each downstream hydro-power station and has objective power generation compensation benefits (Li, 2016; Wang, 2014). So, reflecting the compensating function of power plants with good performance in cascade joint scheduling and solving the joint operation process by highly efficient and feasible algorithms mean very much.

2 Modified GA-POA method

According to the idea of successive optimization, the T-stage optimization problem can be divided into several two-stage optimization sub-problems, while the sub-problem decision-making, using real-coded genetic algorithm to solve. Traditional genetic algorithms generate initial populations in a random way. However, randomized methods often fail to meet the requirements of "global distribution" of initial populations, which leads to their low representation and inefficiency. Based on this, the idea of uniform design is introduced into the generation of initial population of genetic algorithm to improve the initial population quality by increasing the uniform distribution of the initial population.

2.1 Coding method and initial population generation

Taking the water level of the library as a gene and coding with an integer, the population size is set as Size, $i = 1 \sim \text{Size}$. The initial population is generated as follows:

$$P_{i,t} = Z_{t,\min} + 0.1n \tag{1}$$

$$n = 10 \times \left(Z_{t,\max} - Z_{t,\min}\right) = 1, 2, \ldots, \text{size}, \tag{2}$$

where $P_{i,t}$ is the integer coding of the i individual in the parent population at time t, $t = 1, 2, \ldots, T+1$; T is the number of scheduling period; $Z_{t,\min}$, $Z_{t,\max}$ is upper and lower water level.

2.2 Fitness function setting

The fitness function is used to evaluate the individual's advantages and disadvantages. In this paper, the size of power generation is mainly considered. When the system initial solution does not meet the minimum system output requirements, the penalty function is used to adjust. The objective function value:

$$\phi(Z) = \sum_{t=1}^{T} N_t \cdot \Delta t + \alpha \cdot \min\left(\frac{N_t - N_{\min}}{N_{\min}}, 0\right), \tag{3}$$

where T is Each time period; N_t is the output of the cascade at time t; $N_{\min}$ is system minimum output.

2.3 Cross operation

Using the single point crossover method, it is assumed that i_1 and i_2 cross at the time point:

$$P'_{k,t} = \begin{cases} p_{i1,t} & t < \text{time} \\ p_{i2,t} & t \geq \text{time} \end{cases} \quad P'_{k+1,t} = \begin{cases} p_{i2,t} & t < \text{time} \\ p_{i1,t} & t \geq \text{time}, \end{cases} \tag{4}$$

where time $\in [1, 2, \ldots, T+1]$, after crossing, a population P' of the same size as Size is formed.

2.4 Variation operation

Using random mutation, the gene P_y is controlled by the probability P_y, a new gene is generated at the mutation point:

$$P''_{i,t} = \begin{cases} Z_{t,\min} + (Z_{t,\max} - Z_{t,\min}) \cdot P'_y P'_y < P_y \\ P_{i,t} P'_y \geq P_y, \end{cases} \tag{5}$$

where P_y, P'_y are all random numbers distributed between [0, 1], mutation to population P'' with the same population size of size.

2.5 Choose operation

The league method (Howson, 1975) was used to score a total of 3Size individuals in the parental group P, crossover group$P$, and mutation group P. The scoring rules of individual i were as follows: size random individuals were randomly selected without duplication, and the adaptability of i exceeded that of the competition individuals counts Score as i score:

$$\text{Score}_i = \text{Count}\{\phi_i > \phi_j \,|i, j \in \Omega_{\text{size}}\}. \tag{6}$$

3Size individuals ranked according to the score, the highest score of Size individuals as a parent group into the next generation of evolution.

2.6 Evolution termination conditions

The evolution termination condition is to reach the maximum evolutionary algebra set in advance.

The multi-stage optimization problem in the progressive optimality algorithm is split into multiple two-stage sub-problems, and then the modified genetic algorithm is applied to the decision of the sub-problems. The modified GA-POA method solving flowchart is shown as Fig. 1.

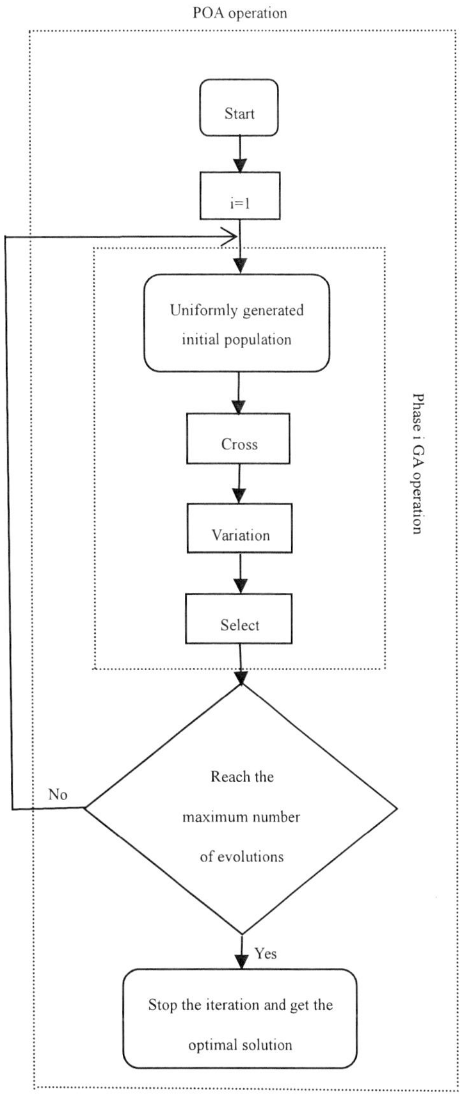

Figure 1. The modified GA-POA method solving flowchart.

3 Mid and Long-Term Optimize Scheduling of Cascade Hydro-power Stations

3.1 The objective function

Model to maximize the power generation as the optimization goal, the mathematical description of:

$$\max F = \sum_{i=1}^{n} \sum_{j=1}^{T} N_{ij} \Delta t, \tag{7}$$

where F is the goal of optimal scheduling is the total power generation of all stations in the cascade throughout the entire scheduling period; n is the number of reservoirs; T is scheduling period; N_{ij} is the i reservoir's output of j period; Δt is Time interval.

3.2 Restrictions

3.2.1 Water balance constraints

$$V_{t+1} - V_t = (Q_t - q_t) \cdot \Delta t, \quad (8)$$

where V_t, V_{t+1} is the beginning and the end of the reservoir's storage capacity of t period;

Q_t, q_t is the average inflow and the average outflow.

3.2.2 Water level constraints

$$Z_{\min} < Z_t < Z_{\max}, \quad (9)$$

where $Z_{\max}$, $Z_{\min}$ is the reservoir's upper and lower water level of t period.

3.2.3 Output constraint

$$N_{\min} < N_t < N_{\max}, \quad (10)$$

where $N_{\max}$, $N_{\min}$ is the reservoir's upper and lower output.

3.2.4 Discharge flow constraints

$$q_{\min} < q_t < q_{\max}, \quad (11)$$

where $q_{\max}$, $q_{\min}$ is the reservoir's upper and lower outflow.

3.2.5 Hydraulic equation

$$I_{i,l} = q_{i-1,l} + Q_{qj,l} - Q_{ks,l}, \quad (12)$$

where $I_{i,l}$ is the reservoir i's inflow of the l period; $q_{i-1,l}$ is the reservoir $i-1$'s outflow of the l period; $Q_{qj,l}$, $Q_{ks,l}$ is the interval flow and the loss flow of the l period.

3.2.6 Non-negative conditional constraints

All of the above variables satisfy the non-negative condition.

3.3 Output and power generation

$$N = KQ \cdot (H - \Delta H) \quad (13)$$

$$\Delta H = \theta \cdot Q^2 \quad (14)$$

$$E = N \cdot t, \quad (15)$$

where K is output coefficient; Q is used flow; H is net head; ΔH is head loss; θ is coefficient of head loss; t is time; N, E is the output and power generation of hydro-power station.

Table 1. Cascade power generation 10^8 kW h.

Generation	Flood year		Mid-year		Dry year	
	Alone	Joint	Alone	Joint	Alone	Joint
Plant A	211.97	209.21	165.43	161.32	111.95	103.75
Plant B	149.27	157.85	116.94	120.03	80.76	82.29
Plant C	124.10	139.90	93.32	102.19	63.35	73.62
Plant D	107.04	111.86	79.79	82.15	53.14	54.99
Total	592.37	618.82	455.47	465.70	309.20	314.65
Increase rate	4.46%		2.25%		1.76%	

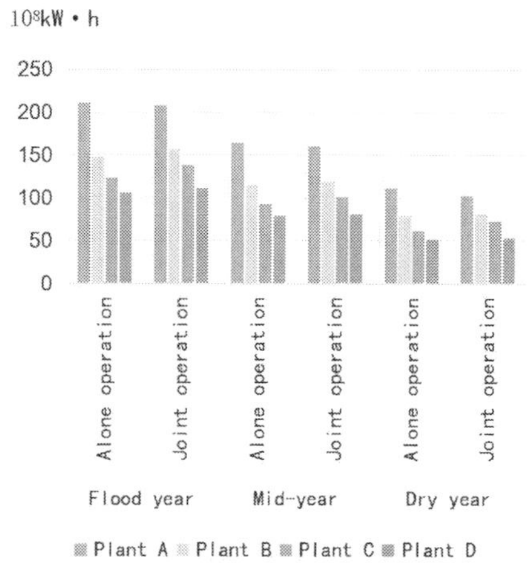

Figure 2. Cascade power generation.

4 Case study

In this study, an example is given to four hydro-power stations in a river basin cascade in China, which are represented by A, B, C and D. Among them, A is the leading power plant with huge storage capacity and significant power generation benefit which has not been completed and put into production. The adjustment capacity of B, C and D is small, and the runoff data used in this study is the monthly typical runoff which are regarded as run-of-run power plants. In this study, population size = 200, maximum iteration generation = 300, crossover time = 3 and mutation probability = 0.03. Solving the model by the established modified GA-POA method with C++ programming language, the cascade power generation under different representative years are shown as Table 1 and Fig. 2.

When cascade hydro-power station is operated in combination, the leading hydro-power station can effectively increase the power generation of the downstream hydro-power plants during the dry period, and the total generating capacity of the hydro-power station increases significantly. The power generation of joint operation growth by 4.46 % in flood year, 2.25 % in mid-year and 1.76 % in dry year than alone operation. The compensatory benefits meet the requirements of Flood Year > Mid-Year > Dry Year.

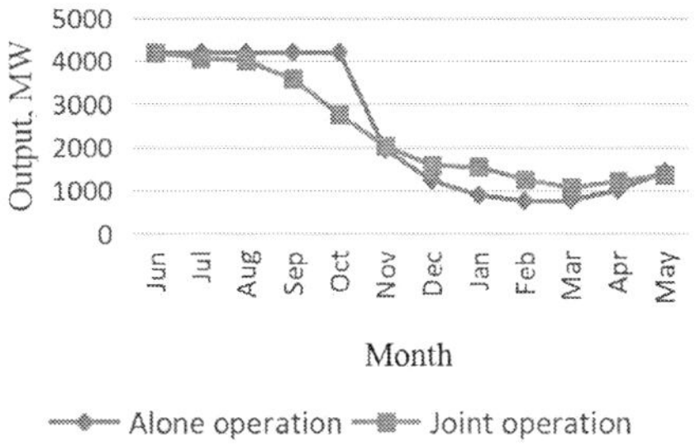

Figure 3. The output of plant A.

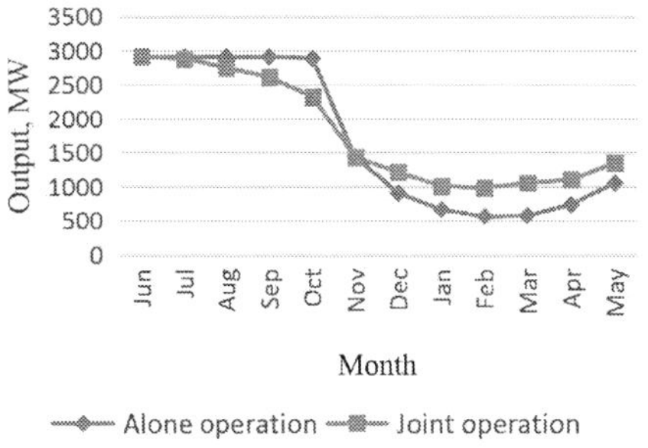

Figure 4. The output of plant B.

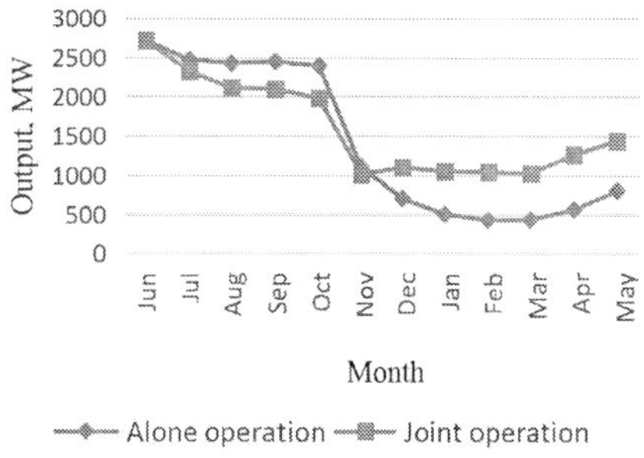

Figure 5. The output of plant C.

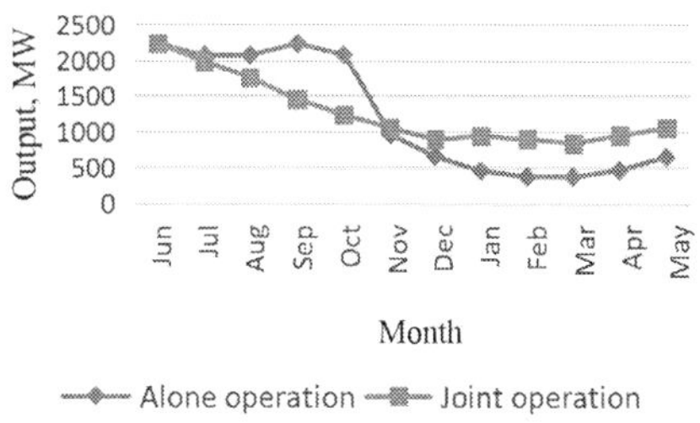

Figure 6. The output of plant D.

The joint operation can significantly improve the overall power generation of the cascade. When the water inflow is small, the joint operation has a greater significant. The upstream power station compensates the power plants of the downstream power stations and improves the output of the dry season and the seasonal power generation. It can also ease the flood risk caused by concentration of flood season. This will help mobilize the enthusiasm of the leading hydropower station and downstream hydro-power stations to participate in the joint cascade scheduling and achieve the goal of joint operation, which bring the maximum benefit in the basin.

The monthly output process of each station in flood year when it is operated independently and jointly is shown as Figs. 3 to 6. When each station of cascades operates independently, the output of the power plant fluctuates greatly. The specific performance is that when in the flood season, the output is large; when in the dry season, the output is small.

When the water supply is reduced rapidly in some months, there are efforts to make a sharp turn; when the cascade power plants are jointly operated, the power plant output process is relatively gradual. Due to the regulation and storage of the leading power station, the output of the power plant during the flood season is smaller than that of the power station alone. In the dry season, the power plant output is greater than when the power station is operated alone. Output, the overall change is smaller than that of each power station when it is running alone, and the leading power station has a significant effect on the adjustment of the steps.

5 Conclusion

In this paper, the long-term optimal scheduling model is established, and the modified GA-POA method is proposed to solve the model by C++ programming. This model is applied to cascade reservoirs in a river basin in China, which achieved good results. The study provides a beneficial reference for the joint development of large-scale cascaded hydropower stations.

Data availability. The data used in this paper are provided by the reservoir management departments, which can not be provided externally because the data are confidential.

Competing interests. The authors declare that they have no conflict of interest.

Special issue statement. This article is part of the special issue "Innovative water resources management – understanding and balancing interactions between humankind and nature". It is a result of the 8th International Water Resources Management Conference of ICWRS, Beijing, China, 13–15 June 2018.

Acknowledgements. This study was financially supported by National Key Projects of China "Water resources efficient development and utilization" (2017YFC0405900, 2016YFC0402208, 2016YFC0401903) and National Natural Science Foundation of

China (No.51641901). The author would like to give special thanks to anonymous reviewers.

Edited by: Yangbo Chen
Reviewed by: two anonymous referees

References

He, X., Wu, Y., and Jiang, G.: Profit allocation model of supply network based on revised Shapely value, Soft Science, 28, 70–73, 2014.

Howson, H. R. and Sancho, N. G. F.: A new algorithm for the solution of multi-state dynamic programming problems, Math. Program., 8, 104–106, 1975.

Li, Y., Zhou, J., Zhang, S., Huang, C., and Yan, D.: Combined dispatching chart of cascade hydropower stations on Yalong river, Jinsha river and Yangtze river, Journal of Hydroelectric Engineering, 35, 32–40, 2016.

Wang, Y., Guo, S., Zhou, Y., Hong, X., Hu, T., and Liu, D.: Uncertain analysis of long-term generation optimal operation for reservoirs, Water Resources and Power, 32, 61–65, 2014.

Xu, W.: Research on some issues in the joint optimal operation of cascaded reservoirs between Long yang xia and Liu jia xia, Tsinghua University, 2015.

Zhou, B.: Research on compensative operation of two-reservoirs and multi-cascaded hydropower plants, Da Lian University of Technology, 2016.

https://doi.org/10.5194/piahs-379-205-2018

A new method for indirectly estimating infiltration of paddy fields in situ

Yunqiang Xu, Baolin Su, Hongqi Wang, and Jingyi He
College of Water Sciences, Beijing Normal University, Beijing 100875, China
Correspondence: Baolin Su (subl@bnu.edu.cn) and Hongqi Wang (amba@bnu.edu.cn)

Received: 3 January 2018 – Revised: 9 April 2018 – Accepted: 13 April 2018 – Published: 5 June 2018

Abstract. Infiltration is one of the major procedures in water balance research and pollution load estimation in paddy fields. In this study, a new method for indirectly estimating infiltration of paddy fields in situ was proposed and implemented in Taihu Lake basin. Since when there were no rainfall, irrigation and artificial drainage, the water depth variation process of a paddy field is only influenced by evapotranspiration and infiltration $(E+F)$. Firstly, $(E+F)$ was estimated by deciding the steady decreasing rate of water depth; then the evapotranspiration (ET) of the paddy field was calculated by using the crop coefficient method with the recommended FAO-56 Penman-Monteith equation; finally, the infiltration of the paddy field was obtained by subtracting ET from $(E+F)$. Results show that the mean infiltration of the studied paddy field during rice jointing-booting period was 7.41 mm day^{-1}, and the mean vertical infiltration and lateral seepage of the paddy field were 5.46 and 1.95 mm day^{-1} respectively.

1 Introduction

Infiltration is one of the major significant factors controlling water use efficiency and influencing the transport of water and chemicals in paddy fields (Wopereis et al., 1994; Aimrun and Amin, 2009; Zhou et al., 2011). It is also an important component for water balance research and pollution load calculating (Huang et al., 2014). Infiltration in flooded paddy fields not only occurs vertically downwards, but also flows into the earthen ridges surrounding the fields (Chen and Liu, 2002; Huang et al., 2003). It means infiltration consists of vertical infiltration and lateral seepage. Vertical infiltration (vertical percolation) is the vertical movement of water beyond the root zone to the water table. In flooded rice fields, there is a continuous downward flow of water from the puddled layer to below the plow pole that basically prevents capillary rise into the root zone (Bouman et al., 2007). Lateral seepage (lateral flow/seepage) is the lateral movement of subsurface water of a paddy field. The study revealed the existence of lateral flow through the ridges to the adjacent lower fields (Janssen and Lennartz, 2007). With well-maintained ridges, lateral seepage is generally small (Bouman et al., 2007). But the study showed lateral seepage flow may occur either underneath well-maintained ridges or through and underneath ill-maintained ridges (Bouman et al., 1994).

The infiltration of paddy fields can be obtained through field experiments monitoring with instruments, such as cylinder, infiltrometer and lysimeter. But it is often prohibitive because of the excessive time and expenditure involved in the execution (Khepar et al., 2000). There are also some difficulties for measuring and calculating the infiltration of paddy fields in situ with multiple runoff outlets. The prediction of infiltration (F) is complicated. Researchers commonly assume a constant infiltration rate which is an overgeneralization of a dynamic process in modeling the water balance of rice fields, through the puddled topsoil. This assumption is at times misleading because the infiltration rate of rice fields is affected by range of soil physical and hydraulic properties, such as structure, texture, conductivity of different layers, and also by the hydrologic environment, e.g. subsoil moisture content, groundwater table depth and ponded water depth on the soil surface (Wickham and Singh, 1978; Khepar et al., 2000; Lin et al., 2014).

Based on synchronous observation of precipitation and water depth, we can use the method (Huang et al., 2014; Zhou et al., 2016a) to estimate the evapotranspiration and infiltration $(E+F)$ of paddy fields in situ. When there were no rainfall, irrigation and artificial drainage, the water depth variation process of a paddy field is only influenced by $(E+F)$. We can choose the steady decline processes of water depth to estimate $(E+F)$ by deciding the steady decreasing rate of water depth. This method can be applied to general paddy fields in situ with multiple runoff outlets and dynamic variation of lowest ridge height during growing season. Evapotranspiration (ET) of paddy fields can be calculated by using the crop coefficient method with synchronous meteorological data. We could get rice evapotranspiration (ET_c) of paddy fields with multiplying reference crop evapotranspiration (ET_0) which can be determined by the recommended FAO-56 Penman-Monteith equation, by crop coefficients (K_c) according to different rice growing stages (Allen et al., 1998). With the obtained $(E+F)$ and ET_c, the infiltration of paddy fields can be calculated by subtracting ET_c from $(E+F)$.

2 Materials and methods

2.1 Study area

The experimental paddy field (31.47361° N, 119.45020° E) located at Liyang, Changzhou, Jiangsu Province is about 900 m^2. Deep water irrigation mode under an intermittent irrigation system is adopted. Except for drying period, water depth keeps at least 30 mm. The rice is transplanted in late June, stops irrigation in mid-September, and harvests in mid-November.

This research area is located at the south bank of the lower reaches of the Yangtze River and in Taihu Lake basin, where the topography of fields is almost flat. It belongs to the subtropical humid monsoon climate. The annual mean temperature is 15.4 °C, annual precipitation is 1071.5 mm, annual sunshine duration is 2047.5 h, and the annual frost-free period is 227.6 days.

2.2 Instrument installation and data collection

Data used in this paper includes precipitation and water depth in the experimental paddy field and fundamental meteorological data of Liyang. Therein, precipitation and water depth were collected during the rice growing period in 2014 by using pluviograph and automatic mariograph. The observation frequency of water depth was 30 min, while that of precipitation was 10 min. A pluviograph was installed on a fixed tripodal platform in target plot, and a rain barrel was kept vertically 70 cm high away from ground. An automatic mariograph was installed near the ridge in the paddy plot, and the mariograph probe was installed about 2 cm high above the field bottom to avoid being blocked by mud. The probe installation height was kept unchanged during the growing period.

The fundamental meteorological data which is on daily scale, was obtained from Liyang Meteorological Station. The data includes cumulative precipitation (mm), average relative humidity (%), mean wind speed ($m\,s^{-1}$), sunshine duration (h), mean air temperature (°C), maximum air temperature (°C), minimum air temperature (°C) and net radiation ($MJ\,m^{-2}\,day^{-1}$). The mean speed is measured at 2 m above the ground level.

2.3 Data analysis methods

2.3.1 The method for estimating $(E+F)$ of paddy fields

The $(E+F)$ of paddy fields is estimated by using the water balance model (Khepar et al., 2000; Huang et al., 2014). Water balance equation is shown as follow:

$$P + R_{in} - (E+F) - R_{out} = \Delta H \tag{1}$$

where P is the daily precipitation (mm), R_{in} is the irrigation water inflow (mm), $(E+F)$ is the evapotranspiration and infiltration loss (mm), R_{out} is the water outflow (mm), and ΔH is the depth variation (mm).

Factors influencing water depth include natural factors and artificial factors. Natural factors mainly include $(E+F)$ and precipitation. Artificial factors mainly include irrigation and artificial drainage. When there were no rainfall, irrigation, and artificial drainage, the steady decline of water depth variation process line is only affected by $(E+F)$ process. Therefore, the Eq. (1) can be simplified as follow:

$$-(E+F) = \Delta H \tag{2}$$

The key procedure for estimating $(E+F)$ based on synchronous observation of precipitation and water depth is to detect steady decline processes of water depth which were selected to calibrate $(E+F)$ by deciding the steady decreasing rate of water depth. The rice growing period includes four stages, rice regreening, tillering, jointing-booting and maturation. The $(E+F)$ analysis of this paper is applied to the rice jointing-booting stage with water kept in the paddy field all the time.

2.3.2 The method for calculating ET of paddy fields

The crop coefficient method with the recommended FAO-56 Penman-Monteith equation (Allen et al., 1998) is applied to calculate the ET of the paddy field with synchronous meteorological data. In this paper, single crop coefficient method was applied. During the research period of rice jointing-booting stage, water is kept in the paddy field, indicating there is no water stress. The ET_c of the paddy field is calculated as follow:

$$ET_c = K_c \times ET_0 \tag{3}$$

Where ET_c is the daily ET ($mm\,day^{-1}$) of paddy field, ET_0 is the reference crop evapotranspiration ($mm\,day^{-1}$), and K_c is the rice coefficient.

1. Reference crop evapotranspiration (ET_0)

   The FAO Penman-Monteith method defines the reference crop as a hypothetical crop with an assumed height of 0.12 m, with a surface resistance of 70 s m^{-1}, and an albedo of 0.23, closely resembling the evaporation from an extensive surface of green grass. The reference crop evapotranspiration (ET_0) is defined and calculated using the FAO-56 Penman-Monteith equation:

$$ET_0 = \frac{0.408\Delta(R_n - G) + \gamma \frac{900}{T+273} u_2 (e_s - e_a)}{\Delta + \gamma(1 + 0.34u_2)} \quad (4)$$

   where ET_0 is reference crop evapotranspiration (mm day^{-1}), R_n is the net radiation at the crop surface (MJ m^{-2} day^{-1}), G is the soil heat flux density (MJ m^{-2} day^{-1}), T is the mean daily air temperature at 2 m height (°C), u_2 is the wind speed at 2 m height (m s^{-1}), e_s is the saturation vapour pressure (kPa), e_a is the actual vapour pressure (kPa), $e_s - e_a$ is the saturation vapour pressure deficit (kPa), Δ is the slope vapour pressure curve (kPa °C^{-1}), and γ is the psychrometric constant (kPa °C^{-1}).

2. Rice coefficient (K_c)

   The crop coefficient incorporates crop characteristics and averaged effects of evaporation from the field of the single crop coefficient method. The FAO-56 Penman-Monteith equation gives the recommended values of crop coefficient during different crop growing stages under standard condition. For rice coefficient (K_c), the recommended value was adjusted to the non-standard condition value, using the follow equation:

$$K_c = K_{c,mid} + [0.04(u_2 - 2) - 0.004(RH_{min} - 45)]\left(\frac{h}{3}\right)^{0.3} \quad (5)$$

   where $K_{c,mid}$ is the recommended value of the mid-season stage (including rice jointing-booting stage), RH_{min} is the mean value for daily minimum relative humidity (%) during the mid-season growing stage, and h is the mean maximum plant height (m).

2.3.3 A new method for indirectly estimating infiltration of paddy fields

Since when $(E + F)$ and ET_c of the paddy field have been obtained, the infiltration can be calculated by subtracting ET_c from $(E + F)$:

$$F = (E + F) - ET_c \quad (6)$$

where F is the infiltration loss (mm).

3 Results

3.1 Estimated $(E + F)$ of the paddy field

The water depth and precipitation of the paddy field during rice jointing-booting stage are shown in Fig. 1. The periods described by black dotted line indicate the steady declines of water depth with no rainfall, irrigation, and artificial drainage. Those steady decline processes can be used for estimating $(E + F)$. Meanwhile, the periods described by solid line indicate the water depth variation mutation periods. In those situation, the $(E + F)$ should be arbitrarily approximated with the mean value of adjacent growing period under the same weather condition. In this paper, we just discussed the estimated $(E + F)$ of water depth during steady decline periods.

Using the proposed method, $(E + F)$ during the rice jointing-booting stage was calculated. As shown in Fig. 2, the maximum $(E + F)$, minimum $(E + F)$ and mean $(E + F)$ were 15.63, 3.69 and 10.62 mm day^{-1} respectively. Compared with the $(E + F)$ of different paddy fields using the same method, we got the proximate parameter range of $(E + F)$ (Table 1).

3.2 Calculated ET_c of the paddy field

According to the related FAO irrigation and drainage papers, rice coefficient ($K_{c,mid}$) is 1.2 and the mean maximum rice height is 1.0 m (Allen et al., 1998). ET_c of the paddy field is shown in Fig. 2. The maximum ET_c, minimum ET_c and mean ET_c were 6.23, 1.21 and 3.21 mm day^{-1} respectively.

3.3 Estimated infiltration of the paddy field

In this study, there are 45 days of the rice jointing-booting stage in total and 27 values of infiltration of the paddy field were obtained. As shown in Fig. 2, the maximum infiltration, minimum infiltration and mean infiltration were 12.71, 2.20 and 7.41 mm day^{-1}, respectively.

4 Discussions

In practices, vertical infiltration and lateral seepage are not easily separated because transition flows can neither be classified as vertical percolation nor lateral seepage (Wickham and Singh, 1978). Typical combined values for vertical infiltration and lateral seepage vary from 1–5 mm day^{-1} in heavy clay soils to 25–30 mm day^{-1} in sandy and sandy loam soils (Bouman and Tuong, 2001). The preferred rate of infiltration is 5–15 mm day^{-1} in irrigated temperate rice lands where infiltration may be needed to leach organic toxins that can persist from year to year (Wopereis et al., 1994). The vertical infiltration and lateral seepage rates of paddy fields vary from 1 to 6 mm day^{-1}, with the mean value of 3 mm day^{-1} in Tuanlin, Hubei province, and 3.8 mm day^{-1} in Jinhua, Zhejiang

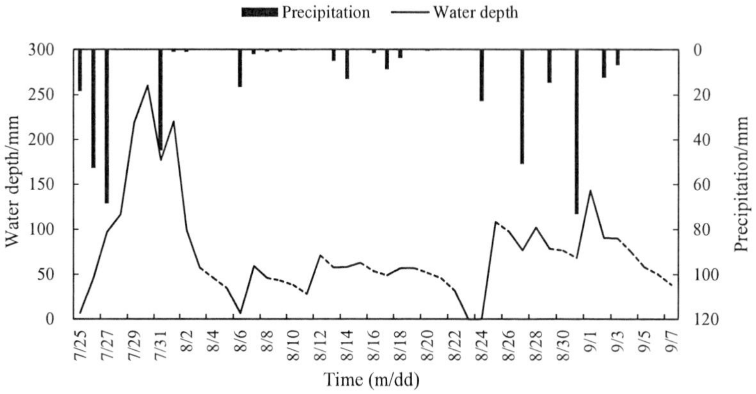

Figure 1. Water depth and precipitation of the paddy field during rice jointing-booting stage.

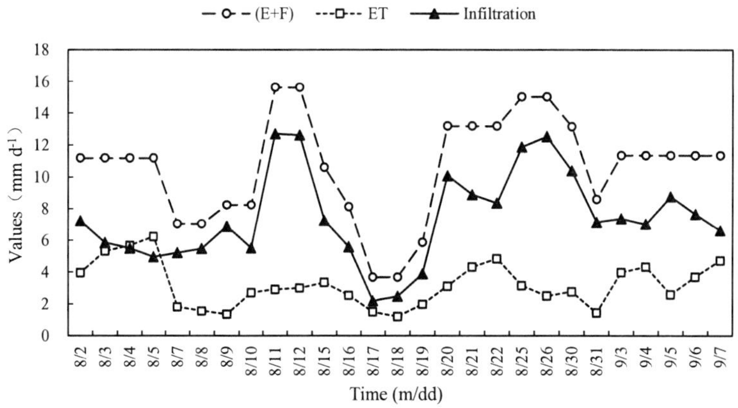

Figure 2. Schematics of $(E + F)$, ET_c and infiltration of the paddy field.

province (Cabangon et al., 2004). The research observed the vertical infiltration through the plough pan and the lateral seepage on sloping land formed into terraces and showed that the higher the position on sloping land was, the greater the potential of water losses was because of higher vertical infiltration and lateral seepage (Tsubo et al., 2005). The study carried out in Taiwan indicates the vertical infiltration and lateral seepage rates of the flat paddy field through the bottom of the ridge were 4.0 and 3.3 mm day^{-1} respectively with the plow sole underneath, and 8.5 and 4.5 mm day^{-1} respectively without the plow sole underneath (Huang et al., 2003). Therefore, in the experimental paddy field with a plow sole in this study, the infiltration range of 2.20–12.71 mm day^{-1} is moderate and credible, which demonstrates the new method for indirectly estimating infiltration of paddy fields in situ is relatively reliable.

The physical and hydraulic characteristics of the plow sole and the underlying subsoil determine the magnitude and constancy of vertical infiltration rate (Bouman et al., 1994). For many studies, there were usually assumed or measured values in vertical infiltration rates. According to the Jiangsu province actual observed statistic data, the vertical infiltration rate of single crop rice region with clay soil, loam soil and sandy loam soil are 1.2–4, 2.1–5.3 and 5.5–9.8 mm day^{-1} respectively (Wu, 2015). In the Taihu Lake basin, with approximate soil texture of loam soil, the vertical infiltration of different paddy fields is shown in Table 2, which illustrates the approximate vertical infiltration range of Jiangsu province.

The estimated infiltration values of the paddy filed were analyzed using SPSS 20.0, and the box and whisker plot were shown in Fig. 3. The first quartile (25th percentile) infiltration values of the paddy filed was 5.46 mm day^{-1}. Then the mean vertical infiltration of the paddy field was assumed to be 5.46 mm day^{-1}, according to the previous analysis of vertical infiltration in Jiangsu province and Taihu Lake basin. Subtracting the mean vertical infiltration from mean infiltration of 7.41 mm day^{-1}, the mean lateral seepage of the paddy field turned out to be 1.95 mm day^{-1}. The mean lateral infiltration accounted for 26.4 % of the infiltration, indicating the lateral seepage was also an important component of infiltration in paddy fields in situ with multiple runoff outlets. The

Table 1. $(E+F)$ of different paddy fields using the same method.

Year	Growing stage	Study area	$(E+F)$ (mm day^{-1})	Source
2014	rice jointing-booting	Liyang, Jiangsu	4–15	Zhou et al. (2016b)
2013	rice jointing-booting	Haidian, Beijing	4–12	Zhou et al. (2016a)
2012	rice jointing-booting	Haidian, Beijing	8–12	Huang (2014), Huang et al. (2014)

Table 2. Vertical infiltration of different paddy fields in Taihu Lake basin.

Year	Study area	Crop	Vertical infiltration (mm day^{-1})	Source
2007–2009	Taihu Lake basin	irrigated rice	2	Zhao et al. (2012)
2008	Danyang, Jiangsu	irrigated rice	6.0	Cheng et al. (2009)
–	Danyang, Jiangsu	irrigated rice	6.4	Li et al. (2010)
2010	Taihu Lake basin	irrigated rice	1	Lin et al. (2012)
2010	Taihu Lake basin	irrigated rice	2–3	Lin et al. (2011)
2010	Suzhou, Jiangsu	irrigated rice	4.23	Yin (2012)
2011	Suzhou, Jiangsu	irrigated rice	4.02	Yin (2012)

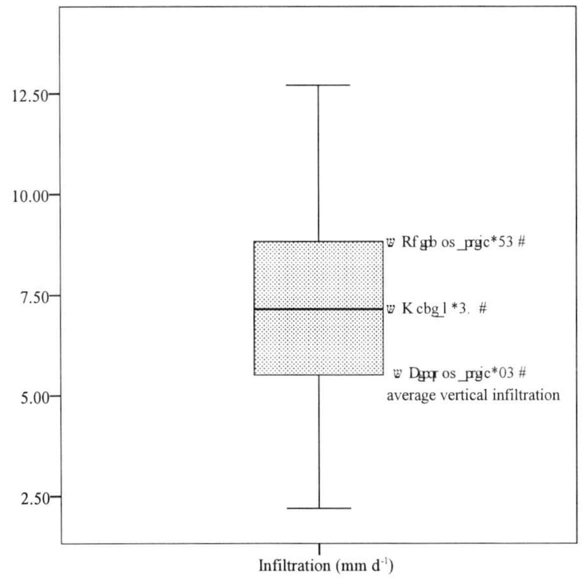

Figure 3. The box and whisker plot of infiltration of the paddy field.

lateral seepage should also be paid more attention in future paddy field research.

5 Conclusions

A new method for indirectly estimating infiltration of the paddy field was put forward. The $(E+F)$ of the paddy field were estimated by deciding the steady decreasing rate of water depth, when there were steady decline processes of water depth. The ET of the paddy field was calculated by using single crop coefficient method with the recommended FAO-56 Penman-Monteith equation. Therefore, the infiltration was calculated by subtracting calculated ET from estimated $(E+F)$. The mean infiltration of the paddy field during rice jointing-booting period was 7.41 mm day^{-1}. By analyzing the infiltration values obtained in this study and the reference vertical infiltration in Jiangsu province and Taihu Lake basin, the mean vertical infiltration and lateral seepage of the paddy field in situ turned out to be 5.46 and 1.95 mm day^{-1} respectively.

Data availability. The underlying research data belongs to the project which is not publicly accessible.

Competing interests. The authors declare that they have no conflict of interest.

Special issue statement. This article is part of the special issue “Innovative water resources management –understanding and balancing interactions between humankind and nature”. It is a result of the 8th International Water Resources Management Conference of ICWRS, Beijing, China, 13–15 June 2018.

Acknowledgements. This work is supported by Major Project on Science & Technology of Water Body Pollution Control & Treatment (2017ZX07301-003).

Edited by: Dingzhi Peng
Reviewed by: two anonymous referees

References

Aimrun, W. and Amin, M. S. M.: Pedo-transfer function for saturated hydraulic conductivity of lowland paddy soils, Paddy Water Environ., 7, 217–225, 2009.

Allen, R. G., Pereira, L. S., Raes, D., and Smith, M.: Crop evapotranspiration. Guidelines for computing crop water requirements, Fao Irrigation & Drainage Paper 56, FAO, Rome, Italy, 1998.

Bouman, B. A. M. and Tuong, T. P.: Field water management to save water and increase its productivity in irrigated lowland rice, Agr. Water Manage., 49, 11–30, 2001.

Bouman, B. A. M., Wopereis, M. C. S., Kropff, M. J., Berge, H. F. M. T., and Tuong, T. P.: Water use efficiency of flooded rice fields II. Percolation and seepage losses, Agr. Water Manage., 26, 291–304, 1994.

Bouman, M., Lampayan, R. M., and Tuong, T. P.: Water management in irrigated rice: coping with water scarcity, International Rice Research Institute, Los Baños, Philippines, 72 pp., 2007.

Cabangon, R. J., Tuong, T. P., Castillo, E. G., Bao, L. X., Lu, G., Wang, G., Cui, Y., Bouman, B. A. M., Li, Y., and Chen, C.: Effect of irrigation method and N-fertilizer management on rice yield, water productivity and nutrient-use efficiencies in typical lowland rice conditions in China, Paddy Water Environ., 2, 195–206, 2004.

Chen, S. K. and Liu, C. W.: Analysis of water movement in paddy rice fields (I) experimental studies, J. Hydrol., 260, 206–215, 2002.

Cheng, W., Li, Y., Zhu, L., Gu, J., and Yang, L. Z.: Research on the loss of nitrogen in direct-seeding rice field in Taihu Lake basin, Journal of Anhui Agricultural Sciences, 37, 2620–2621, 2717, 2009 (in Chinese).

Huang, H. C., Liu, C. W., Chen, S. K., and Chen, J. S.: Analysis of percolation and seepage through paddy bunds, J. Hydrol., 284, 13–25, 2003.

Huang, N. B.: Study on quantification of non-point source pollution and potential of pollutant emission reduction from paddy fields, MS thesis, Beijing Normal University, Beijing, China, 2014 (in Chinese).

Huang, N. B., Su, B. L., Li, R. R., Yang, W. Z., and Shen, M. M.: A field-scale observation method for non-point source pollution of paddy fields, Agr. Water Manage., 146, 305–313, 2014.

Janssen, M. and Lennartz, B.: Horizontal and vertical water and solute fluxes in paddy rice fields, Soil Till. Res., 94, 133–141, 2007.

Khepar, S. D., Yadav, A. K., Sondhi, S. K., and Siag, M.: Water balance model for paddy fields under intermittent irrigation practices, Irrigation Sci., 19, 199–208, 2000.

Li, Y., Yang, L. Z., and Yin, G. D.: Experimental study on nitrogen leaching in a direct-seeding rice paddy of Taihu Lake Basin, Plant Nutrition & Fertilizer Science, 16, 99–104, 2010 (in Chinese).

Lin, G. Y., Luo, S., Shi, Y. D., and Gao, Y.: Modification and verification of rice field runoff model under non-saturated soil condition, Water Saving Irrigation, 10, 33–36, 41, 2011 (in Chinese).

Lin, G. Y., Shen, A. C., Wang, Y. K., and Gao, Y.: Research on the irrigation and drainage water quantity under the condition of water-saving measures in the area of Taihu Basin based on numerical simulation, Water Saving Irrigation, 1, 11–14, 21, 2012 (in Chinese).

Lin, L., Zhang, Z. B., Janssen, M., and Lennartz, B.: Infiltration properties of paddy fields under intermittent irrigation, Paddy Water Environ., 12, 17–24, 2014.

Tsubo, M., Fukai, S., Basnayake, J., Tuong, T. P., Bouman, B., Harnpichitvitaya, D., Borrell, A., and Shu, F. K.: Estimating percolation and lateral water flow on sloping land in rainfed lowland rice ecosystem, Plant Prod. Sci., 8, 354–357, 2005.

Wickham, T. H. and Singh, V. P.: Water movement through wet soils, in: Soils and rice, International Rice Research Institute, Los Baños, Philippines, 337–357, 1978.

Wopereis, M. C. S., Bouman, B. A. M., Kropff, M. J., Berge, H. F. M. T., and Maligaya, A. R.: Water use efficiency of flooded rice fields I. Validation of the soil-water balance model SAWAH, Agr. Water Manage., 26, 277–289, 1994.

Wu, D.: Drainage processes and the lateral drainage effect of the main ditches on the paddy fields in the Plain River Network Regain, MS thesis, Xi'an University of Technology, Xi'an, China, 2015 (in Chinese).

Yin, H. F.: Effect of different water and nitrogen conditions on growth and nitrogen leaching of rice, MS thesis, Nanjing Agricultural University, Nanjing, China, 2012 (in Chinese).

Zhao, X., Zhou, Y., Min, J., Wang, S., Shi, W., and Xing, G.: Nitrogen runoff dominates water nitrogen pollution from rice-wheat rotation in the Taihu Lake region of China, Agr. Ecosyst. Environ., 156, 1–11, 2012.

Zhou, J. W., Su, B. L., Huang, N. B., Guan, Y. T., and Zhao, K.: In situ experiment on non-point source pollution in paddy fields, Acta Scientiae Circumstantiae, 36, 1145–1152, 2016a (in Chinese).

Zhou, J. W., Su, B. L., Huang, N. B., Guan, Y. T., and Zhao, K.: Runoff pollution experiments of paddy fields under different irrigation patterns, Environm. Sci., 37, 963–969, 2016b (in Chinese).

Zhou, S., Sugawara, S., Riya, S., Sagehashi, M., Toyota, K., Terada, A., and Hosomi, M.: Effect of infiltration rate on nitrogen dynamics in paddy soil after high-load nitrogen application containing ^{15}N tracer, Ecol. Eng., 37, 685–692, 2011.

https://doi.org/10.5194/piahs-379-211-2018

Open Access
PIAHS

Impacts of the thawing-freezing process on runoff generation in the Sources Area of the Yellow River on the northeastern Qinghai-Tibet Plateau

Xiaoling Wu[1], Xiaohua Xiang[1], Chao Qiu[2], and Li Li[3]

[1]College of Hydrology and Water Resource, Hohai University, Nanjing 210098, China
[2]Zhejiang Provincial Hydrology Burean, Hangzhou 310009, China
[3]Jiangsu Province Water Engineering Sci-tech Consulting. Corp.Ltd, Nanjing 210009, China

Correspondence: Xiaohua Xiang (freebir7237@hhu.edu.cn)

Received: 31 December 2017 – Revised: 9 April 2018 – Accepted: 14 April 2018 – Published: 5 June 2018

Abstract. In cold regions, precipitation, air temperature and snow cover significantly influence soil water, heat transfer, the freezing-thawing processes of the active soil layer, and runoff generation. Hydrological regimes of the world's major rivers in cold regions have changed remarkably since the 1960s, but the mechanisms underlying the changes have not yet been fully understood. Using the basic physical processes for water and heat balances and transfers in snow covered soil, a water-heat coupling model for snow cover and its underlying soil layers was established. We found that freezing-thawing processes can affect the thickness of the active layer, storage capacity for liquid water, and subsequent surface runoffs. Based on calculations of thawing-freezing processes, we investigated hydrological processes at Qumalai. The results show that the water-heat coupling model can be used in this region to provide an understanding of the local movement of hydrological regimes.

1 Introduction

Recently, the earth system has been influenced by permafrost and glacier evolvement. The area of China's seasonally frozen soil region is $5.137 \times 10^6\,km^2$, with an additional $2.068 \times 10^6\,km^2$ as permafrost. This accounts for 75 % of China's total land area (Xu and Deng, 1991). Investigation of permafrost hydrology is important for local ecosystem security and human life. The extensive collection of field measurement data during the last decades has increased considerably, including the available information on hydro-meteorological processes in cold regions, but most of these data are too fragmentary.

Frozen soil plays an important role in the hydrology of cold regions (Harris et al., 2009). Many studies on frozen soil have been conducted. The SHAW model incorporates land surface processes to describe the snow-melting effect and the freezing-thawing process in a one-dimensional study (Flerchinger and Saxton, 1989). The most important part of this kind of model is the connection between coupled mechanisms. Snow cover greatly affects the heat and water transfer to the underlying soil. Thus, soil water-heat movements and snow cover conditions have important theoretical value (Jorgenson et al., 2001; Harris et al., 2001; Heginbottom, 2002; Romanovsky and Osterkamp, 2001).

The coupled, dynamic model developed in this paper, using snow cover and frozen soil together, sets the connection between snow cover and soil and demonstrates an association between snow cover and soil.

2 Methods

The snow-soil coupled model established in this paper comprises a snow cover: water and heat movement model and a frozen soil: water-heat coupled movement model.

2.1 Snow cover model

On the basis of conservation of energy (Xu and Li, 2003), the net heat flux of snow cover is equal to the intersystem heat increase over the same interval. The snow exchanges energy with the atmosphere above and ground below. The net energy flux of the snowpack can be expressed as

$$Q_{\text{net}} = Q_{\text{r}} + Q_{\text{s}} + Q_{\text{l}} + Q_{\text{p}} + Q_{\text{g}} \tag{1}$$

where Q_{net} is net input energy for snowpack (J m^{-2}); Q_{r} is the net radiation flux (W m^{-2}); Q_{s} is flux of sensible heat (W m^{-2}); Q_{l} is flux of latent heat (W m^{-2}); Q_{p} is heat flux supplied by precipitation (W m^{-2}); Q_{g} is ground heat flux (W m^{-2}) always being neglected.

Based on the water balance, daily snow melt was estimated using the net energy of the snow cover.

The snow melt can be divided into three phases: warming, ripening and output (Dingman, 1994).

$$W_{\text{liq}} = W_{\text{liq}} + (P_{\text{r}} E + \Delta W) \tag{2}$$

$$W_{\text{s}} = W_{\text{s}} + (P_{\text{s}}\, S\, \Delta W) \tag{3}$$

where W_{liq} is liquid water content (m); W_{s} is snow water equivalent (m); ΔW is the energy for water transfer; P_{s} is snowfall (m); S is sublimation (m); E is evaporation (m); C_i is heat capacity of snow ($\text{J kg}^{-1\,\circ -1}$); λ_{f} is latent heat of fusion (kJ kg^{-1}).

When W_{liq} is larger than the holding capacity of the snow cover, the outflow from the snow cover is described as

$$W_{\text{out}} = W_{\text{liq}} - W_{\text{liq\_max}} \tag{4}$$

where $W_{\text{liq\_max}}$ is the holding capacity of the snow cover (m) and W_{out} is the outflow from the snow cover (m).

2.2 The frozen soil water-heat coupled movement model

We hypothesized the following assumptions could be applied to our model, including the influence of soil water vapor migration and heat flow transfer can be ignored; the soil is considered vertical and non–plastic soil; vertical heat and moisture flux; and Darcy's law applies to moisture movement in both saturated and unsaturated conditions (Shoop and Bigl, 1996). Based on these assumptions, water and heat movement in seasonally frozen soil can be treated as a one-dimensional problem.

Water movement in freezing and thawing soil can be expressed as Richard's equation and a one-dimensional heat transport equation for a freezing or thawing column of soil.

$$C_{\text{s}} \frac{\partial T}{\partial t} = \frac{\partial}{\partial Z}\left(\lambda \frac{\partial T}{\partial Z}\right) + L_{\text{f}} \rho_i \frac{\partial \theta_i}{\partial t} \tag{5}$$

$$\frac{\partial \theta_{\text{l}}}{\partial t} = \frac{\partial}{\partial Z}\left(D(\theta_{\text{l}}) \frac{\partial \theta_{\text{l}}}{\partial Z} - K(\theta_{\text{l}})\right) - \frac{\rho_i}{\rho_{\text{l}}} \frac{\partial \theta_i}{\partial t} \tag{6}$$

where $D(\theta_{\text{l}})$ is the diffusivity of soil ($\text{m}^2\,\text{s}^{-1}$); $K(\theta_{\text{l}})$ is unsaturated hydraulic conductivity of soil (m s^{-1}); ρ_{s} is density of ice (kg m^{-3}); ρ_{l} is density of liquid water (kg m^{-3}); θ_{l} is volumetric liquid water content (%); θ_{s} volumetric ice content (%); C_{s} is volumetric heat capacity of soil ($\text{J m}^{-3\,\circ -1}$); λ is thermal conductivity of soil ($\text{W m}^{-1\,\circ -1}$); L_f is latent heat of fusion (J kg^{-1}); T is soil temperature (°); Z is depth (m); t is time (s).

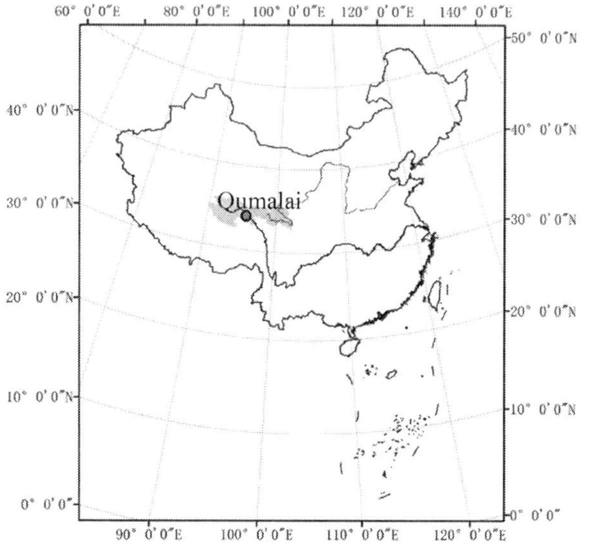

Figure 1. Distribution of observation site in the Three–River Source Region.

2.3 Coupled model

Based on the water balance, the water content $\theta_{\text{l\_top}}$ in the top layer of soil relied on the outflow from snow pack W_{out}, and evaporation from soil surface E_{soil}, the liquid content in top soil layer was given by

$$\theta_{\text{l\_top}} = \beta (W_{\text{out}} - E_{\text{soil}}) \tag{7}$$

where β is the constant coefficient.

3 Study site

The observation site selected for this study, Qumalai (34.13° N, 95.78° E), is located in the middle of the Three-River Source Region. The average elevation of Qumalai is 4500 m (Fig. 1). Controlled by the high altitude and pressure, the climate in Qumalai includes strong radiation, long light duration and seasonal variations for warm (four or five) months and cold (seven or eight) months. The mean annual air temperature is −3.3°. The mean annual precipitation is 380–470 mm. (Yin et al., 2010).

The atmospheric data used in this study includes hourly precipitation, air temperature, relative humidity, wind velocity, hours or daylight and atmospheric pressure. Daily snow thickness, soil freezing depths, soil temperature at depth of

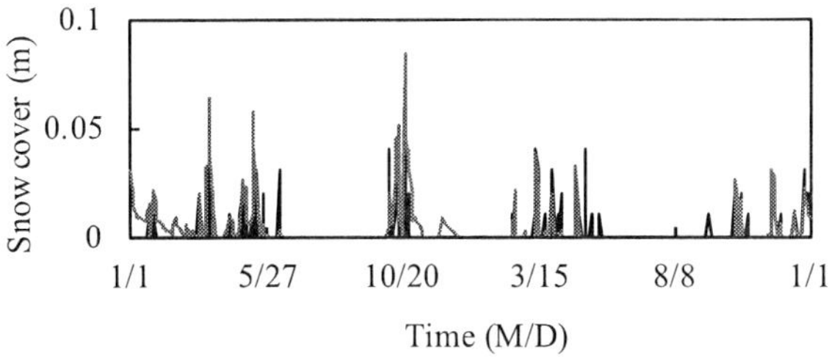

Figure 2. Comparison between simulated (red solid line) and measured (black solid line) results for the thickness of snow cover from 2005 to 2006.

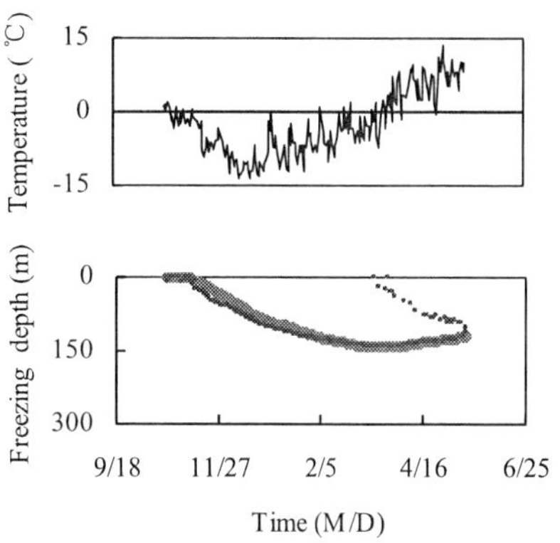

Figure 3. With the observed soil surface temperature at the corresponding times, the comparison between measured (red dashed line) and simulated (blue dashed line) freezing depth of soil from October 2005 to June 2006.

15 cm and soil water content at the depth of 50 cm were monitored from 2005 to 2006.

Permafrost and seasonal frozen soil is widely distributed throughout the study area. Because of the presence of frozen soil, water retention in the soil can be enhanced for both rain and snow melt compared to soils in non-frozen or temperate regions. The permafrost dynamics greatly influence the hydrologic processes (McGuire, 2002; Walker et al., 2003).

4 Results

4.1 Snow depth

Observation of air temperature, snow surface temperature, humidity, rainfall, wind velocity, atmospheric pressure and hours of daylight provided a good dataset for driving the snow cover model. Comparisons between simulated and measured results for snow depth in 2005–2006 are shown in Fig. 2.

According to Fig. 2, the snow depth trends were the same for both the simulated results and the measured values. At Qumalai, snow accumulation was measured from late October through the middle of May the following year, when the melting process finished. Between mid-March and early November, the snow depth reached a maximum.

4.2 Freezing depth

The daily mean freezing depth was used in the following analysis. Figure 3 shows the depth of frozen soil from October 2005 to June 2006. The simulation of the frozen soil trend for the whole period was consistent with measurements.

The daily mean soil temperature grew progressively colder in October, and the surface soil began to freeze at the same time. Over time, the soil froze deeper, and after 141 days (11 March), it was frozen to 95 cm.

The majority of the site began to thaw in April, when temperatures grew warmer. The shallow layer of soil began thawing on 15 March. By early April, the layer below 140 cm also began to thaw, though the layers between 70 and 140 cm depth remained frozen. The remaining frozen soil thawed from both the top and bottom depths.

4.3 Soil temperature

Figure 4 shows the simulated and measured daily average temperatures for every day of 2005–2006. According to Fig. 4, both the simulated and measured soil temperatures show the same trend. The soil temperature warmed from late March into late October, which corresponds to when the soil thawed (Fig. 3).

To detect the numerical model efficiency, efficiency coefficients (EC) and root mean squared errors (RMSE) were used. RMSE for simulated daily soil temperatures of 2005–2006 ranged from 1.82 to 1.9°, and EC was 0.94. There was no distinct difference for the surface soil. This means that there was no system error for the model used at these sites, and the simulated temperature for surface soil was coincident with the measured value.

4.4 Soil water content

Soil liquid water content was measured at 50 cm from April to October 2005 at Qumalai. The comparisons between simulated and measured results are shown in Fig. 5.

The simulated liquid water content decreased in early May, when there was heavy evaporation and no precipitation. As the evaporation decreased, the liquid water content increased, while precipitation created an increase in soil moisture.

From the simulation results, we found that during the thawing period, the soil liquid water content was significantly affected by the surface conditions, including precipitation and evaporation.

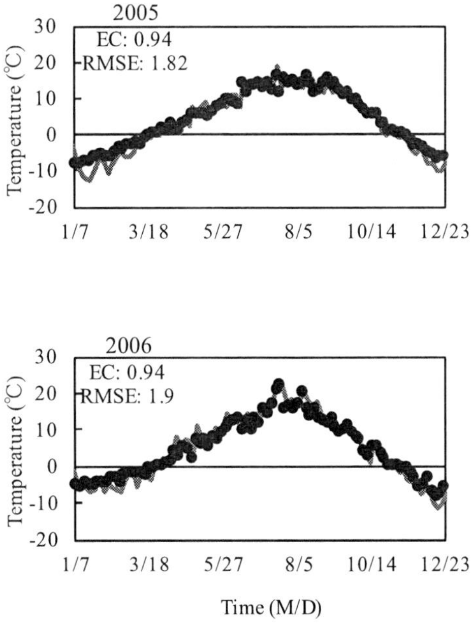

Figure 4. Comparison between simulated (red solid line) and measured (black dashed line) results for soil temperature at 15 cm depth in 2005–2006.

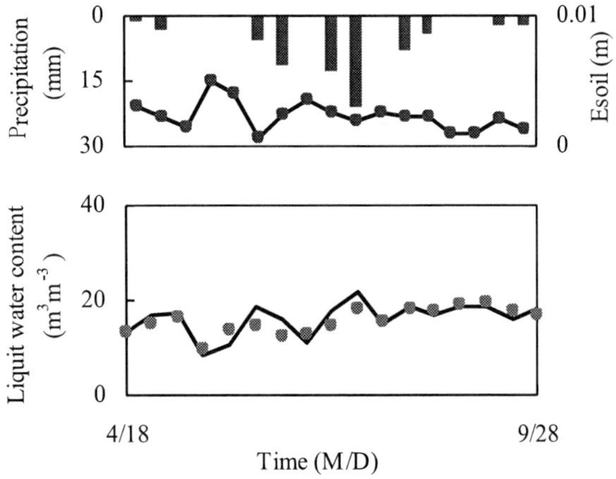

Figure 5. With the observed precipitation (blue bar) and simulated evaporation (black solid line with blue circle) at the corresponding time, comparison between simulated (black solid line) and measured (red dotted line) results for soil liquid water content at 50 cm depth in 2005.

4.5 Runoff

4.5.1 In soil thawing period

In most parts of the Three-River Source Region, the liquid water storage capacity gradually increased as the soil water phase changed from ice to liquid. The thawing process

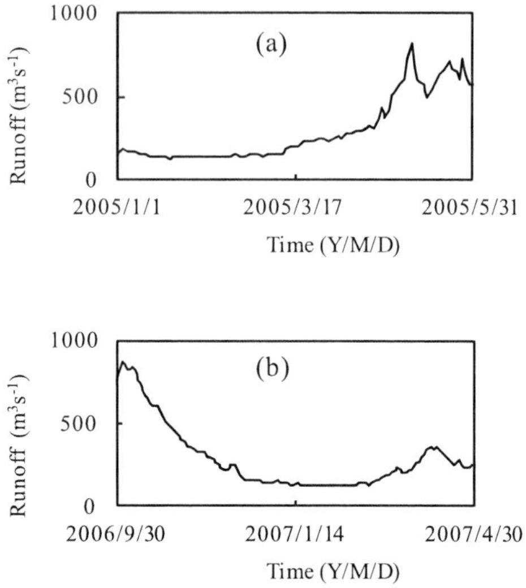

Figure 6. Runoff in the Yellow-River Source Region from 2005 to 2007. The study was carried out during the thawing (March to April) and freezing (October to March) periods.

caused the unsaturated soil status and snowmelt filtration. Runoff from the Yellow-River Source Region did not vary very much until late April, when the soil was completely thawed (Fig. 6a).

4.5.2 In soil freezing period

Runoff fluctuated between 124 and 240 $m^3 s^{-1}$, resulting from increased air temperature, evaporation and snowmelt. Runoff increased suddenly in mid-December 2006 (Fig. 6b) because of the high air temperature and deep frost depth. Based on the above analysis, the main factors affecting runoff during the frozen soil period are air temperature, evaporation and frozen soil, which stops further penetration of liquid water causing it to runoff.

During the later period of frozen soil, especially from late February to mid-March 2007, the runoff increased. This was due to the frozen soil and increasing precipitation and air temperature.

5 Conclusions

Snow cover has significant effects on water and heat in freezing and thawing soil. According to the main physical processes of water and heat transfer in snow cover and soil, a cold region snow-soil coupled model was established. Using measured data for seasonal snowmelt, water and heat status, frozen depth, liquid water content and soil temperature were simulated numerically. The simulation results were compared with measured values at corresponding times and locations to verify the reliability and precision of the model.

With the thawing and freezing of soil, runoff varied greatly. This means seasonal variations in the thawing-freezing process of the soil's active layer played a significant role in the seasonal runoff shift in the Three-River Source Region.

Data availability. The underlying research data can be accessed, although some of the data are not publicly accessible. Because some of the data used in manuscript was collected from the related office of government, some of which can not be visited from outside.

Competing interests. The authors declare that they have no conflict of interest.

Special issue statement. This article is part of the special issue "Innovative water resources management – understanding and balancing interactions between humankind and nature". It is a result of the 8th International Water Resources Management Conference of ICWRS, Beijing, China, 13–15 June 2018.

Acknowledgements. This research is supported by the National Key Research and Development Program of China (No. 2016YFA0601000) and the Fundamental Research Funds for the Central Universities (No. 2015B20314).

Edited by: Bo Pang
Reviewed by: two anonymous referees

References

Dingman, S. L.: Physical Hydrology, Prentice-Hall, Old Tappan, NJ, 1994.

Flerchinger, G. N. and Saxton, K. E.: Simulataneous heat and water model of a freezing snow-residue-soil system I. Theory and development, Trans. of ASCE, 32, 565–571, 1989.

Harris, C., Haeberli, W., Mühll, V. D., and King, L.: Permafrost monitoring in the high mountains of Europe: the PACE project in its global context, Permafrost Periglac., 12, 3–11, 2001.

Harris, C., Arenson, L. U., Christiansen, H. H., Etzemuller, B., Frauenfelder, R., Gruber, S., Haeberli, W., Hauck, C., Holzle, M., and Humlum, O.: Permafrost and climate in Europe: monitoring and modelling thermal, geomorphological and geotechnical responses, Earth-Sci. Rev., 92, 117–171, 2009.

Heginbottom, J. A.: Permafrost mapping: a review, Prog. Phys. Geog., 26, 623–642, 2002.

Jorgenson, M. T., Racine, C. H., Walters, J. C., and Osterkamp, T. E.: Permafrost degradation and ecological changes associated with a warming climate in central Alaska, Climatic Change, 48, 551–579, 2001.

McGuire, A. D.: Environmental variation, vegetation distribution, carbon dynamics and water/energy exchange at high latitudes, J. Veg. Sci., 13, 301–314, 2002.

Romanovsky, V. E. and Osterkamp, T. E.: Permafrost: changes and impacts, in: Permafrost Response on Economic Development, Environmental Security and Natural Resources, edited by: Paepe, R. and Melnikov, V., Kluwer Academic Publishers, 297–315, 2001.

Shoop, S. A. and Bigl, S. R.: Moisture migration during freeze and thaw of unsaturated soils: modeling and large scale experiments, Cold Reg. Sci. Technol., 25, 33–45, 1996.

Walker, D. A., Jia, G. J., and Epstein, H. E.: Vegetation-soil-thaw-depth relationships along a low-arctic bioclimate gradient, Alaska: synthesis of information from the ATLAS studies, Permafrost Periglac., 14, 103–123, 2003.

Xu, X. Z. and Deng, Y. S.: Experimental research of moisture migration in frozen soil, Science Press, 1991 (in Chinese).

Xu, Z. X. and Li, J. Y.: Estimating basin evapotranspiration using distributed hydrologic model, J. Hydrol. Eng., 8, 74–80, 2003.

Yin, Z. F., Ouyang, H., and Chen, H.: Simulating Soil Freezing and Thawing of Temperate Desert Ecosystem on the Qinghai-Tibet Plateau, Proc. Environ. Sci., 2, 476–485, 2010.

https://doi.org/10.5194/piahs-379-217-2018

Open Access
PIAHS

Predicting future land cover change and its impact on streamflow and sediment load in a trans-boundary river basin

Jie Wang[1], Hao Wang[2], Shaowei Ning[3], and Ishidaira Hiroshi[4]

[1]School of Hydrology and Water Resources, Nanjing University of Information Science & Technology, Nanjing, 210044, China
[2]School of Resource and Geosciences, China University of Mining and Technology, Xuzhou, 221116, China
[3]HeFei University of Technology, Hefei, 230009, China
[4]International Research Center for River Basin Environment, University of Yamanashi, Kofu, 4008511, Japan

Correspondence: Jie Wang (wangjie0775@163.com)

Received: 30 December 2017 – Revised: 30 March 2018 – Accepted: 12 April 2018 – Published: 5 June 2018

Abstract. Sediment load can provide very important perspective on erosion of river basin. The changes of human-induced vegetation cover, such as deforestation or afforestation, affect sediment yield process of a catchment. We have already evaluated that climate change and land cover change changed the historical streamflow and sediment yield, and land cover change is the main factor in Red river basin. But future streamflow and sediment yield changes under potential future land cover change scenario still have not been evaluated. For this purpose, future scenario of land cover change is developed based on historical land cover changes and land change model (LCM). In addition, future leaf area index (LAI) is simulated by ecological model (Biome-BGC) based on future land cover scenario. Then future scenarios of land cover change and LAI are used to drive hydrological model and new sediment rating curve. The results of this research provide information that decision-makers need in order to promote water resources planning efforts. Besides that, this study also contributes a basic framework for assessing climate change impacts on streamflow and sediment yield that can be applied in the other basins around the world.

1 Introduction

Studies on hydrological processes in a changing environment have been the focus of hydrological science in the 21st century. The geographic extent of land cover change worldwide has undergone tremendous change in the past years. It has been pointed out land cover change is a major driving force in degradation of natural environment, such as more frequent incidents of local flooding, increasing soil erosion (García-Ruiz et al., 2008).

The research of the impact of land cover change on hydrological condition of the catchment area has therefore received increasing considerable attention from both field observations and model simulations (Brown et al., 2005). However, most studies have been performed on the impact of land cover change on streamflow (Zheng et al., 2009) or sediment yield itself (Leh et al., 2013), few researches were related to impact of land use change on both streamflow and sediment load (Tang et al., 2011; Dao and Suetsugi, 2014). Consequently, it is of great importance to predict future land cover change and its effects on combined streamflow and sediment flow in an effort to optimize the use of water resources and prevent reservoir siltation in Red river basin.

The overall objective of this study is to investigate changes in streamflow and sediment load response to future land cover change in the target river basin. For this purpose, future scenario of land cover change is developed based on historical land cover changes and land change model (LCM).

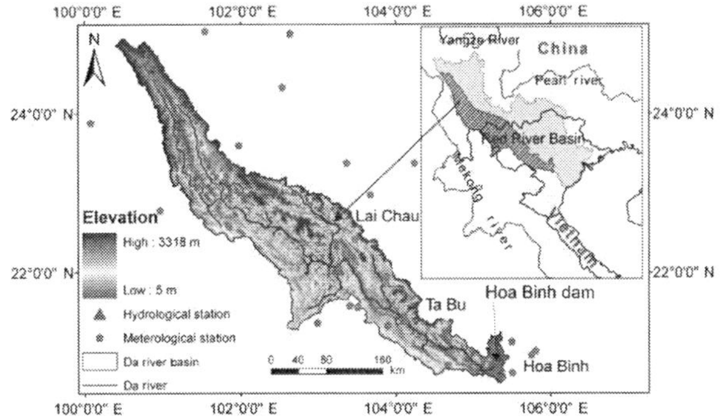

Figure 1. Location of DRB and the hydro-meteorological stations.

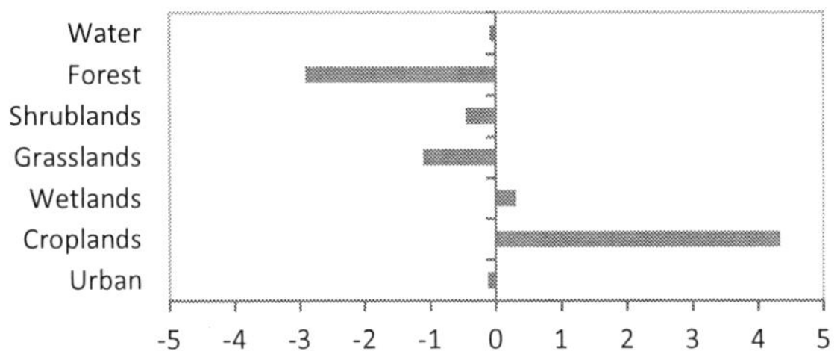

Figure 2. Net land cover changes between 2001 and 2008.

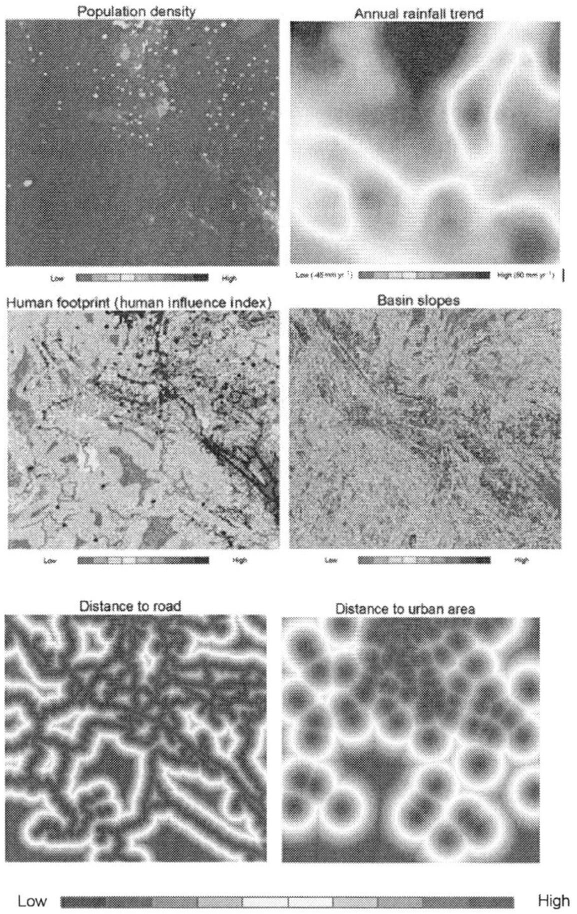

Figure 3. Selected drive factors for LCM.

And future leaf area index (LAI) is simulated by ecological model based on the future land cover scenario. Then future land cover change scenario is used to drive a distributed hydrological model to forecast the impact of the predicted land use change on streamflow. And scenarios of future land cover change and LAI and new sediment rating curve model to assess the impact of the future land use change on sediment load into Hoabinh reservoir. A methodology is developed in which the land change model and ecological model are coupled with a calibrated a distributed hydrological model and sediment simulation model. This coupling is new and essential for evaluating effects of land cover change on streamflow and sediment load, required for providing information that decision-makers need in order to promote water resources planning efforts.

2 Study Area and Data Description

2.1 Study Area Description

The Da River Basin (DRB) located in humid region is the biggest branch of the Red River which gets its name from the reddish-brown color caused by its high sediment load rich in iron dioxide. The DRB drains 55 000 km^2, originates in Yunnan Province, China (Fig. 1). The river cross sections are narrow, with a steep slope of 0.37. The annual mean runoff is about 1168 $m^3\,s^{-1}$ from 1988 to 2004 at Laichau station, associated with the total annual sediment load about $40.1 \times 10^6\,t\,yr^{-1}$. The Da River basin is characterized by a tropical monsoonal. Summer season is warm and humid, whereas the winter season is cool and dry (Le et al., 2007; Dang et al., 2010). The annual mean rainfall is about 1320 mm for the Da River basin, 85 % of which falls during wet rainy season. The HoaBinh reservoir, located on the downstream of Laichau station, is one of the largest ($V = 9.5\,km^3$) and highest (120 m) dams in South-East Asia. It was completely finished in 1993, with the main purpose of flood control, irrigation and hydropower generation (Le et al., 2007).

2.2 Dataset Description

Daily streamflow data and Monthly suspended sediment concentration (SSC) data at Laichau (LC) and Tabu (TB) stations in DRB (Fig. 1) were used in hydrological simulation. Daily meteorological data of 14th stations well distributed are obtained from the China Meteorological Data Sharing Service Center, which has been checked by the primary quality control. All datasets above covering 1991–2000 were used as inputs for hydrological simulation and sediment load calcula-

Table 1. Dataset summary for different models in the study.

Models	Spatial dataset	Point dataset
Hydrological model	Elevation, land cover, soil map	Rainfall, streamflow
New sediment rating curve		Streamflow, LAI, SSC
Land change model	Elevation, land cover, population density, road, river network, slope, human footprint, rainfall trend	No need
Ecological model	Elevation, land cover, soil map	Rainfall, maximum and minimum temperature

Figure 4. Potential map for transition from forest to croplands.

Table 2. Difference between the reference map of 2011 and the simulated map 2011.

Land cover type	Reference 2011 (pix)	Simulated 2011 (pix)	Error percent (%)	Kappa index
Water	869	842	−3.1	
Forest	184 701	177 657	−3.8	
Shrublands	60 433	69 411	14.8	
Grasslands	10 743	9126	−15.0	0.94
Wetlands	825	896	8.6	
Croplands	58 871	58 403	−0.8	
Urban	1618	1725	6.6	

tion. A series of geographical datasets (e.g., DEM, LAI, land cover type) were also employed in ecological simulation, hydrological simulation and land change prediction, summarized in this study (Table 1).

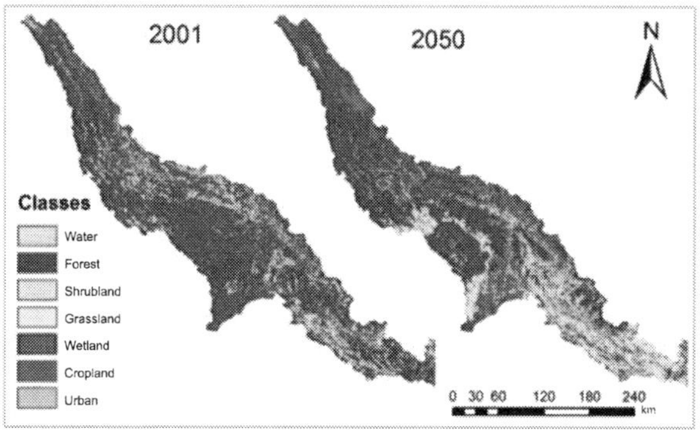

Figure 5. Land cover map of 2001 (basin line) and 2050 (predicted).

3 Methodology

In this study, a group of models coupled in a "one-way" manner framework were applied. Firstly, the Land Change Modeler (LCM) was used to predict future the land cover change in the year of 2050. LCM is a powerful tool to predict future land cover change scenarios and was widely applied (Wilson and Weng, 2011). More technical details of this model could be found in the reference (Clark Labs, 2009). Then, Biome-BGC was also applied to calculate the potential LAI under future land cover changes. In our previous work, we have already described it in detailed and validated this model well in DRB (Wang and Ishidaira, 2013). So in this research this validated Biome-BGC model would be applied to predict future LAI under future land cover scenario directly.

Finally, future land cover and LAI obtained above were used to drive hydrological and sediment simulation model. The hydrological model "Blockwise use of TOPMODEL with Muskingum–Cunge routing" (BTOPMC), a grid-based distributed hydrological model developed by University of Yamanashi, Japan (Takeuchi et al., 1999; Ao et al., 2003), was applied in river flow simulation. And sediment simulation model considering vegetation cover (LAI) which has already been developed and could give better agreement simu-

Table 3. Future land cover changes of 2050 created through neural networks compared with the baseline map of 2001 image in DRB (area percent: %).

Land cover	Baseline 2001 (%)	Predicted 2050 (%)	Changes (%)
Water	0.06	0.05	−0.01
Forest	70.91	49.57	−21.34
Shrublands	15.5	25.47	9.97
Grasslands	0.72	0.16	−0.56
Wetlands	0.19	0.34	0.15
Croplands	12.57	24.27	11.7
Urban	0.06	0.14	0.08

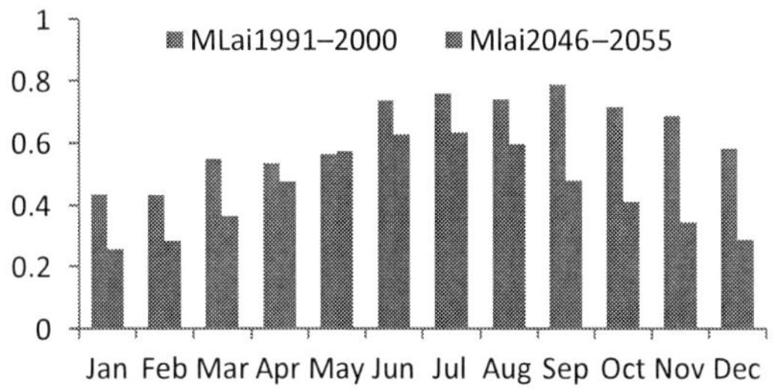

Figure 6. Comparison between current average LAI ($M_{\mathrm{lai}1991-2000}$) and future average LAI ($M_{\mathrm{lai}2046-2055}$).

lation result in several Asian basins (Wang et al., 2013; Wang and Ishidaira, 2013), was used to simulate SSC in DRB.

4 Results

4.1 Future land cover and LAI prediction

Land cover changes between year of 2001 and 2008 were calculated firstly by land change modeller, which showed that cropland increased and forest decreased obviously. So the increase in cropland mainly came from forest changes (Fig. 2). In addition, six driving factors (Fig. 3) which may affect land cover changes and land cover changes obtained above were introduced to build up potential land cover transition maps by multi-layer perceptron neural network method. For example, transition possibilities map from forest to croplands in Fig. 4 shows that there is a high possibility in the bottom and center of our basin.

Land cover map of 2011 simulated by LCM based on potential land cover transition maps showed an agreement spatially with actual land cover map of 2011. Additionally, high kappa index (Jensen, 1996) and low mean error between simulated and observed land cover maps in 2011 was calculated (Table 2), which indicated that simulated map had a good match with the reference one. So LCM was proved to have the ability to simulate land cover changes well, then we apply it to predict future land cover map of 2050 (Fig. 5 and Table 3). Comparison of the land cover maps for 2001 and 2050 reveals that that forest area would have an obvious decrease of about 26 % and croplands have an increase of 11.7 % in 2050 compared with 2001.

Finally, Biome-BGC model which has already been validated well in DRB (Wang and Ishidaira, 2013) was driven by future land cover of 2050 to predict future LAI in the period of 2046 to 2055. Results (Fig. 6) showed that future LAI would decrease especially for dry season due to deforestation.

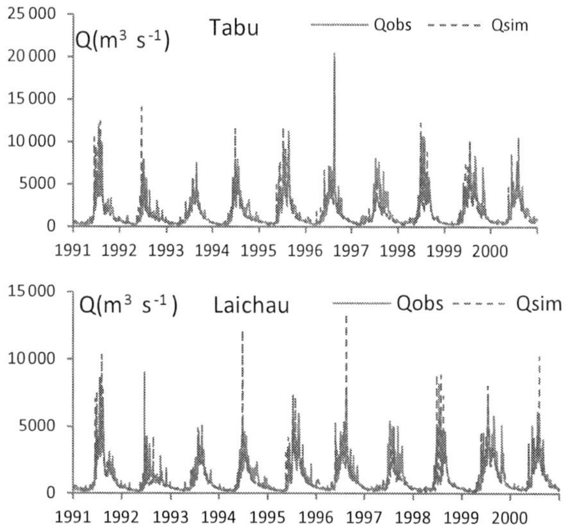

Figure 7. Comparison of observed and simulated daily streamflow in the DRB (calibration period: 1991–1995, validation period: 1996–2000).

Table 4. Evaluation of model simulation during the baseline period for the catchments controlled by Laichau and Tabu stations in the DRB.

	Laichau		Tabu	
	Calibration	Validation	Calibration	Validation
NSE	0.75	0.70	0.70	0.60
R^2	0.83	0.77	0.76	0.72
MAE (mm)	0.74	0.80	0.77	0.91
PBIAS (%)	6.2	7.2	7.0	7.8

4.2 Hydrological and sediment simulation

As shown in Tables 4 and 5, all statistics indicated that BTOPMC hydrological model and sediment simulation model could give a better simulation results in river flow and SSC, and simulated hydrograph also has a good match with the observed (Fig. 7). Results above comprehensively explained that BTOPMC and sediment model can simulate streamflow and SSC accurately.

Table 5. Performance of sediment simulation model for Laichau.

Catchment	R^2	MAE ($g\,m^{-3}$)	PBIAS (%)
Laichau	0.894	3.23	0.524

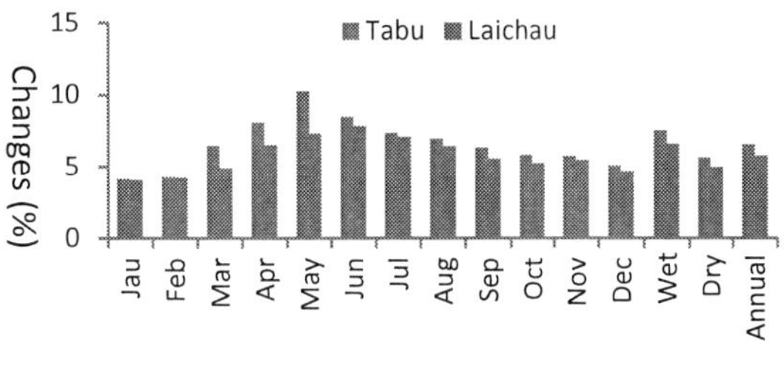

Figure 8. Changes in streamflow under land cover change in 2050 at Laichau and Tabu station.

4.3 Future land cover change impacts

Predicted land cover map, LAI data and fixing the climatic conditions were used to drive the validated hydrological and sediment simulation models to investigate future land cover change impacts on streamflow and sediment load. Changes in streamflow in upstream and downstream are illustrated in Fig. 8, which indicate that land cover change affect streamflow stronger in the downstream of DRB. As shown in Fig. 9, the effects of future land cover change would increase sediment load by 9.6 %, while the change rates in streamflow were within 5.7 %. Sediment load was found more sensitive to land cover change than streamflow. As for seasonally variations of stream flow and sediment load caused by land cover changes, changes in streamflow and sediment load were more pronounced during the wet season.

Other factors such as model parameters may lead to not realistic results in our study. To reduce uncertainty in model simulation, more field observations should be done.

5 Conclusions

In this study, one framework to evaluate future land cover change effects on streamflow and sediment load was developed and successfully applied in Da River Basin. Generally, the streamflow and sediment yield will increase in 2050s under land cover change. Streamflow and sediment load increase in 2050s emphasize the importance of building adaptation to land cover changes to avoid flood and soil erosion. The results obtained in this study could be useful for the appropriate utilization of water resources, flood control, soil conservation and ecological protection in this region by enhancing the understanding of the impact of land cover change scenarios on streamflow and sediment yield.

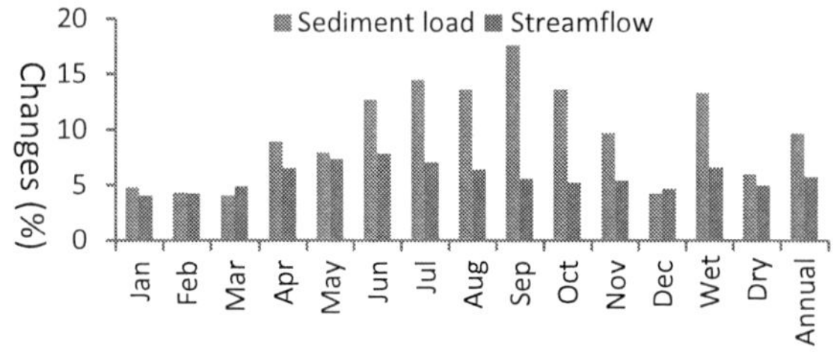

Figure 9. Changes in streamflow and sediment load under land cover changes in 2050 at Laichau station.

Data availability. 1. Daily meteorological and hydrological data from 1991 to 2000 were obtained from the hydrological stations and precipitation stations from China Meteorological Data Sharing Service Center and Vietnam Academy of Science and Technology which is considered confidential.

2. Digital elevation data (GTOPO30) is obtained from the flowing URL: https://lta.cr.usgs.gov/GTOPO30 (last access: 16 May 2017)
3. Soil type data is obtained from the flowing URL: http://www.fao.org/soils-portal/soil-survey/soil-maps-and-databases/harmonized-world-soil-database-v12 (last access: 15 December 2017)
4. Land cover data from 2001 to 2011 is obtained from the flowing URL: https://modis-land.gsfc.nasa.gov/ (last access: 10 June 2017)
5. Global roads dataset and global Human Footprint (HF) are obtained from the flowing URL: http://sedac.ciesin.columbia.edu/ (last access: 15 June 2017)
6. Monthly TRMM satellite rainfall data (TRMM_3A12) are obtained from the flowing URL: https://pmm.nasa.gov/data-access/downloads/trmm (last access: 15 May 2017).

Competing interests. The authors declare that they have no conflict of interest.

Special issue statement. This article is part of the special issue "Innovative water resources management – understanding and balancing interactions between humankind and nature". It is a result of the the 8th International Water Resources Management Conference of ICWRS, Beijing, China, 13–15 June 2018.

Acknowledgements. This research was supported by the National Natural Science Foundation of China (Grant No. 41501029) and the Startup Foundation for Introducing Talent of NUIST. The authors are also thankful to China Meteorological Data Sharing Service Center for providing the hydro-meteorological data.

Edited by: Depeng Zuo
Reviewed by: Kai Ma and one anonymous referee

References

Ao, T., Takeuchi, T., Ishidaira, H., Yoshitani, J., and Fukami, K.: Development and application of a new algorithm for automated pits removal for grid DEMs, Hydrol. Sci. J., 48, 985–997, 2003.

Brown, A. E., Zhang, L., Mcmahon, T. A., Western, A. W., and Vertessy, R. A.: A review of paired catchment studies for determining changes in water yield resulting from alterations in vegetation, J. Hydrol., 310, 28–61, 2005.

CIESIN: Center for International Earth Science Information Network, Columbia University, Environmental Treaties and Resource Indicators (ENTRI) Query Service, Palisades, NY: NASA Socioeconomic Data and Applications Center (SEDAC), available at: http://sedac.ciesin.columbia.edu (last access: 15 June 2017), 2013.

Clark Labs: The Land Change Modeler for Ecological Sustainability, IDRISI Focus Paper, Clark University, Worcester, MA, 2009.

Dang, T. H., Coynel, A., Orange, D., Blanc, G., Etcheber, H., and Lan, A. L.: Long-term monitoring (1960–2008) of the river-sediment transport in the Red River Watershed (Vietnam): temporal variability and dam-reservoir impact, Sci. Total Environ., 408, 4654–4664, 2010.

Dao, N. K. and Suetsugi, T.: The responses of hydrological processes and sediment yield to land-use and climate change in the Be River Catchment, Vietnam, Hydrol. Process., 28, 640–652, 2014.

FAO: FAO/UNESCO Soil Map of the World, available at: http://www.fao.org/soils-portal/soil-survey/soil-maps-and-databases/harmonized-world-soil-database-v12 (last access: 15 Dec 2017), 2011.

García-Ruiz, J. M., Regüés, D., Alvera, B., Lana-Renault, N., Serrano-Muela, P., Nadal-Romero, E., Navas, A., Latron, J., Martí-Bono, C., and Arnáez, J.: Flood generation and sediment transport in experimental catchments affected by land use changes in the Central Pyrenees, J. Hydrol., 356, 245–260, 2008.

Huffman, G., Bolvin, D., Braithwaite, D., Hsu, K., Joyce, R., and NASA's Precipitation Processing Center: 3B42 RT Derived Daily Product, version 7, available at: https://pmm.nasa.gov/data-access/downloads/trmm (last access: 15 May 2017), 2015.

Jensen, J. R.: Introductory Digital Image Processing: A Remote Sensing Perspective (Second edition), Prentice Hall, Upper Saddle River, New Jersey, USA, 1996.

Le, T. P. Q., Garnier, J., Gilles, B., Sylvain, T., and Minh, C. V.: The changing flow regime and sediment load of the Red River, VietNam, J. Hydrol. 334, 199–214, 2007.

Leh, M., Bajwa, S., and Chaubey, I.: Impact of land use change on erosion risk: an integrated remote sensing, geographic information system and modeling methodology, Land Degrad. Dev., 24, 409–421, 2013.

NASA Land Processes Distributed Active Archive Center: USGS Products, Digital elevation data (GTOPO30), available at: https://lta.cr.usgs.gov/GTOPO30TS1, last access: 16 May, 2017.

NASA LP DAAC: The V005 MODIS Land Cover Dynamics (MCD12Q2) product, NASA EOSDIS Land Processes DAAC, USGS Earth Resources Observation and Science (EROS) Center, Sioux Falls, South Dakota, available at: https://lpdaac.usgs.gov (accessed 10 Jun 2017), 2015

Takeuchi, K., Ao, T. Q., and Ishidaira, H.: Introduction of block-wise use of TOPMODEL and Muskingum–Cunge method for the hydro-environmental simulation of a large ungauged catchment, Hydrol. Sci. J., 44, 633–646, 1999.

Tang, L., Yang, D., Hu, H., and Gao, B.: Detecting the effect of land-use change on streamflow, sediment and nutrient losses by distributed hydrological simulation, J. Hydrol., 409, 172–182, 2011.

Wang, J. and Ishidaira, H.: Effects of human-induced vegetation cover change on sediment flow using satellite observations and terrestrial ecosystem model, J. Japan Soc. Civil Eng. B1, 69, I_205–I_210, 2013.

Wang, J., Ishidaira, H., Sun, W. C., and Ning, S. W.: Development and Interpretation of New Sediment Rating Curve Considering the Effect of Vegetation Cover for Asian Basins, The Scientific World J., 2013, 154375, https://doi.org/10.1155/2013/154375, 2013.

Wilson, C. O. and Weng, Q.: Simulating the impacts of future land use and climate changes on surface water quality in the Des Plaines River watershed, Chicago Metropolitan Statistical Area, Illinois, Sci. Total Environ., 409, 4387–4405, 2011.

Zheng, H. X., Lu, Z., Zhu, R., Liu, C., Sato, Y., and Fukushima, Y.: Responses of streamflow to climate and land surface change in the Headwaters of the Yellow River Basin, Water Resour. Res., 45, W00A19, https://doi.org/10.1029/2007WR006665, 2009.

https://doi.org/10.5194/piahs-379-223-2018

Open Access
PIAHS

Modelling the ability of source control measures to reduce inundation risk in a community-scale urban drainage system

Chao Mei[1], Jiahong Liu[1,2], Hao Wang[1,2], Weiwei Shao[1], Lin Xia[1], Chenyao Xiang[1], and Jinjun Zhou[1]

[1]China Institute of Water Resources and Hydropower Research, State Key Laboratory of Simulation and Regulation of Water Cycle in River Basin, China, Beijing 100038, China

[2]Engineering and Technology Research Center for Water resources and Hydroecology of the Ministry of Water Resources, Beijing 100044, China

Correspondence: Jiahong Liu (liujh@iwhr.com)

Received: 19 December 2017 – Revised: 2 February 2018 – Accepted: 5 February 2018 – Published: 5 June 2018

Abstract. Urban inundation is a serious challenge that increasingly confronts the residents of many cities, as well as policymakers, in the context of rapid urbanization and climate change worldwide. In recent years, source control measures (SCMs) such as green roofs, permeable pavements, rain gardens, and vegetative swales have been implemented to address flood inundation in urban settings, and proven to be cost-effective and sustainable. In order to investigate the ability of SCMs on reducing inundation in a community-scale urban drainage system, a dynamic rainfall-runoff model of a community-scale urban drainage system was developed based on SWMM. SCMs implementing scenarios were modelled under six design rainstorm events with return period ranging from 2 to 100 years, and inundation risks of the drainage system were evaluated before and after the proposed implementation of SCMs, with a risk-evaluation method based on SWMM and analytic hierarchy process (AHP). Results show that, SCMs implementation resulting in significantly reduction of hydrological indexes that related to inundation risks, range of reduction rates of average flow, peak flow, and total flooded volume of the drainage system were 28.1–72.1, 19.0–69.2, and 33.9–56.0 %, respectively, under six rainfall events with return periods ranging from 2 to 100 years. Corresponding, the inundation risks of the drainage system were significantly reduced after SCMs implementation, the risk values falling below 0.2 when the rainfall return period was less than 10 years. Simulation results confirm the effectiveness of SCMs on mitigating inundation, and quantified the potential of SCMs on reducing inundation risks in the urban drainage system, which provided scientific references for implementing SCMs for inundation control of the study area.

1 Introduction

The risk of urban inundation has been growing rapidly in recent years due to global urbanization and climate change (Zhao et al., 2016). Over the past several decades, increasing surface imperviousness caused by rapid urbanization has led to significant changes in the characteristics of the urban hydrologic cycle (Palla and Gnecco, 2015). In general, the reduction of native vegetation and the thinning of natural soils related to urbanization reduces the infiltration and storage of water (Jacobson, 2011), leading to increasing runoff rates and volumes and ultimately increasing the hydraulic load for urban drainage systems (Kim et al., 2015). However, increased surface runoff, faster runoff concentration, and higher peak flow rates can exceed an urban system's drainage capacity and cause inundation, resulting in traffic interruption, economic loss, pollution, and health issues (Dash et al., 2016). Improving the drainage capacity in rapidly urbanizing areas is a straightforward method for reducing inundation risk, and a significant improvement can be achieved in most cases (Pochwa et al., 2017). However, simple and isolated improvements to drainage capacity are now proven to be unsustainable, costly, and even impractical because of great uncertainties in the broader context of land-

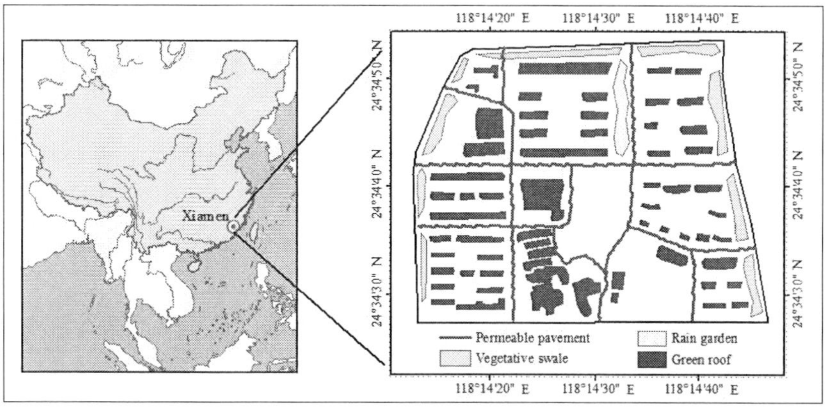

Figure 1. Location of the study area and the sketch map of SCMs settings in YN.

scape and climate change, particularly in densely urbanized areas. SCMs are facilities capable of effectively storing and infiltrating storm water runoff, usually involving green infrastructure such as green roofs, permeable pavements, vegetative swales, rain gardens, and bio-retention systems (Todeschini et al., 2012). The implementation of SCMs is usually low-cost and environmentally friendly and the benefits for runoff reduction have been confirmed by many studies at different scales. Hu et al. (2017) reported that permeable pavements are effective at mitigating flood inundation hazards in urban watersheds. Liu et al. (2014) assessed the effectiveness of green infrastructure on urban flooding reduction at a community scale and showed that total runoff reduction ranged from 85–100 % and peak flow was reduced 92.8–100 %. Pyke et al. (2011) illustrated the potential benefits of a common LID practice for increasing the resilience of communities to changing precipitation patterns. This objective of this study was to determine whether SCMs can reduce the inundation risks in urban drainage systems and assess the reduction under different rainfall conditions using scenario analysis.

2 Materials and methods

2.1 Site description and data preparation

The study area, Yangtang Neighbourhood (YN), is a residential community covering 62.87 ha (Fig. 1, 24°34′18″–24°35′ N, 118°14′02″–118°14′59″ E) in Xiamen, an important harbour city on the southeast coast of China. The average annual precipitation in YN is 1388 mm (1985–2014), and over the course of a year the distribution of precipitation is extremely uneven, with rainfall occurring mainly in the flood season, from May–September, there is an increasing trend in precipitation of YN according to local investigation. The main land uses have changed from village, bare soil, farmland, and ponds to roads, housing, and other buildings, leading to an increase in the proportion of impervious surfaces from 18 to 68 %. Rapid urbanization has therefore dramatically changed the urban water cycle, leading to urban flooding and the inundation of drainage systems.

2.2 Modelling the urban drainage system in YN

The Storm Water Management Model (SWMM), developed by the United States Environmental Protection Agency, is an integrated simulation platform for urban water systems; version 5.1 was used in this study. SWMM can be used for dynamic hydrology-hydraulic and water quality simulation and has been widely used in urban water system simulations throughout the world (Chui et al., 2016; Zhu et al., 2016). The SWMM model solved the Saint-Venant equations and modelling drainage system with dynamic wave, is suitable for urban drainage system in a community-scale in this study. The SWMM platform was used to develop an integrated simulation model of the urban drainage system in YN, using both logical parameters based on earlier experience and local knowledge while consulting the SWMM user manual and previous research. The modelled area was divided into 12 sub-catchments, 25 Junctions, and 37 conduits (Fig. 2) and used a simulation time step of 1 min with a total simulation time of 6 h.

2.3 Inundation risk evaluation methods

2.3.1 Indices for the evaluation of inundation risk

Urban drainage systems are mainly defined by catchment area, underlying surfaces, and pipe networks. A risk-evaluation index for these systems should include indices such as land use, flood amount, and drainage networks that should be built. Nine indices are selected: inundation duration, largest water flow, total flood amount, runoff coefficient, impermeable surface area, catchment area, and the diameter, gradient, and congested proportion of pipes down-

Table 1. Normalized weights of the risk evaluation indices.

Categories	Indices	Normalized weight
Hydraulic	Inundation duration	0.1955
	Largest water flow	0.1135
	Total flood amount	0.3076
Environmental	Runoff coefficient	0.0553
	Impervious area proportion	0.0481
	Catchment area	0.0993
Drainage	Diameter of downstream pipe	0.0944
	Gradient of downstream pipe	0.0479
	Congested proportion of downstream pipe	0.0444

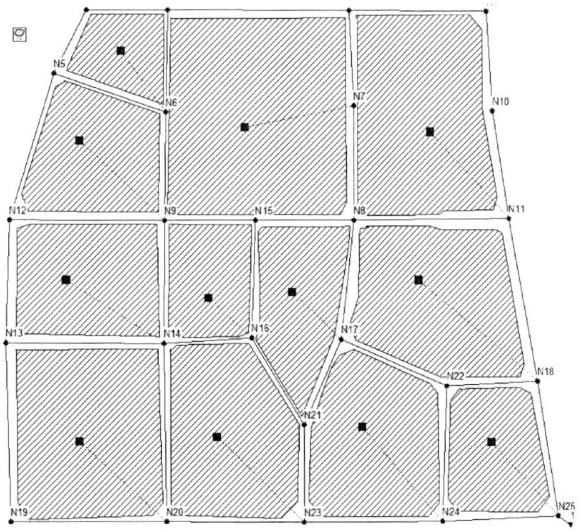

Figure 2. Drainage system modeled on YN.

stream of each manhole (Wang, 2010). We then classified these into three independent groups: hydraulic, environmental, and drainage.

2.3.2 Calculation of inundation risks

In this study, the inundation risk of each manhole in the YN drainage system was defined as a weighted sum as follows:

$$R = \sum_{i=1}^{m} X_i W_i$$

Where R is the inundation risk value of each manhole, m is the number of evaluation indices, i is the serial number of the indices, X_i is the normalized value of each index, and W_i is the normalized weight of the index X_i. In this study, the greater the value of R, the higher the inundation risk of the manhole. Thus, the calculated R value of each manhole under different rainfall scenarios could be used to evaluate inundation risk (Islam et al., 2013).

To properly evaluate the inundation risk, appropriate weights for the indices needed to be determined. We used analytic hierarchy process (AHP) in this study due to its unique advantages when the elements of certain decisions are difficult to quantify or compare (Jiao et al., 2017). The detailed principles and procedures of AHP can be found in the literature (Jenifer and Jha, 2017). We consulted 10 high-level experts in the fields of urban planning, city management, environment risk, water supply, and drainage when considering the proper weights for indices used in the AHP method, settling on the weights for inundation risk evaluation indices presented in Table 1.

2.4 Simulation scenarios

2.4.1 SCM scenarios

Although a variety of SCMs can achieve runoff reduction, not all can be adapted to a specific area; therefore, the selection of site-appropriate SCMs is important. After considering these factors with relation to YN, green roofs, rain gardens, permeable pavements, and vegetative swales were selected as those that could be implemented most effectively. In this plan, all SCMs are work together to reduce surface runoff and storm water load in the urban drainage system. The design and planning work for the YN SCMs was conducted by the Xiamen Municipal Engineering Design Institute Co., Ltd. Most roofs will be converted into green roofs, with a total area of $125\,000\,\mathrm{m}^2$; rain gardens with a total area of $30\,000\,\mathrm{m}^2$ will collect runoff; most roads will be converted to pervious surfaces such as permeable asphalt, concrete, and brick with a total area of $129\,000\,\mathrm{m}^2$; and green space surrounding YN will be converted to vegetative swales with a total area of $42\,000\,\mathrm{m}^2$ (Fig. 1).

2.4.2 Storm scenarios

The SWMM model developed for this study can be used to simulate annual rainfall or a singular event. As the main objective of this study was to assess the effects of SCM implementation on the YN urban drainage systems, we chose to model short-term rainfall using 2 h design storm rainfalls

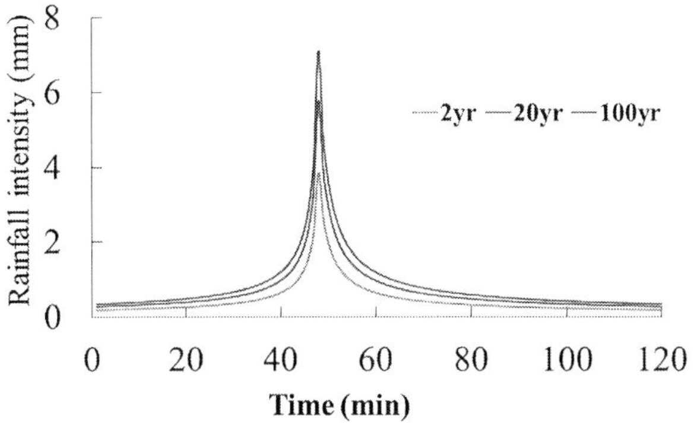

Figure 3. Synthetic rainfall events with different return periods modelled for YN.

with a 1 min time step and different return periods as calculated by the storm intensity formula for Xiamen, which is a local standard established by the department of municipal administration:

$$q = \frac{1432.348(1 + 0.582 \lg P)}{(t + 4.56)^{0.633}} \quad (1)$$

Where i is the storm intensity in $\mathrm{L\,min^{-1}\,ha^{-1}}$), P is the storm return period in years, and t is the convergence time in min.

In order to evaluate the reduction effects under different rainfall intensities, we generated six 2 h synthetic storms with return periods of 2, 5, 10, 20, 50, and 100 years (partially showed in Fig. 3) and used these six events to model the effects on YN's urban drainage system. Due to a lack of research on rainfall patterns in Xiamen, the widely used Chicago-storms synthetic rainfall pattern was selected to simulate these rainfall events with different return periods, with the peak occurrence of all events set to be 0.4. Finally, the drainage system of YN are modelled under two scenarios: before implementation of the SCMs (baseline, BL) and after implementation (SCM).

3 Results and discussion

3.1 Overall hydrologic performance

The implementation of SCMs changes the hydrological cycle of an area by increasing the storage of rainwater, thus delaying peak flow, reducing flood volume, and mitigating flood effects. In this study, we used hydrographs and hydrologic performance indices to evaluate the impacts of SCM implementation; Fig. 4 shows the outfall runoff hydrographs for YN under the BL and SCM scenarios.

As shown in Fig. 4, SCMs implementation resulting in significantly reduction of hydrological indexes that related to inundation risks, range of reduction rates of average flow, peak flow, and total flooded volume of the drainage system

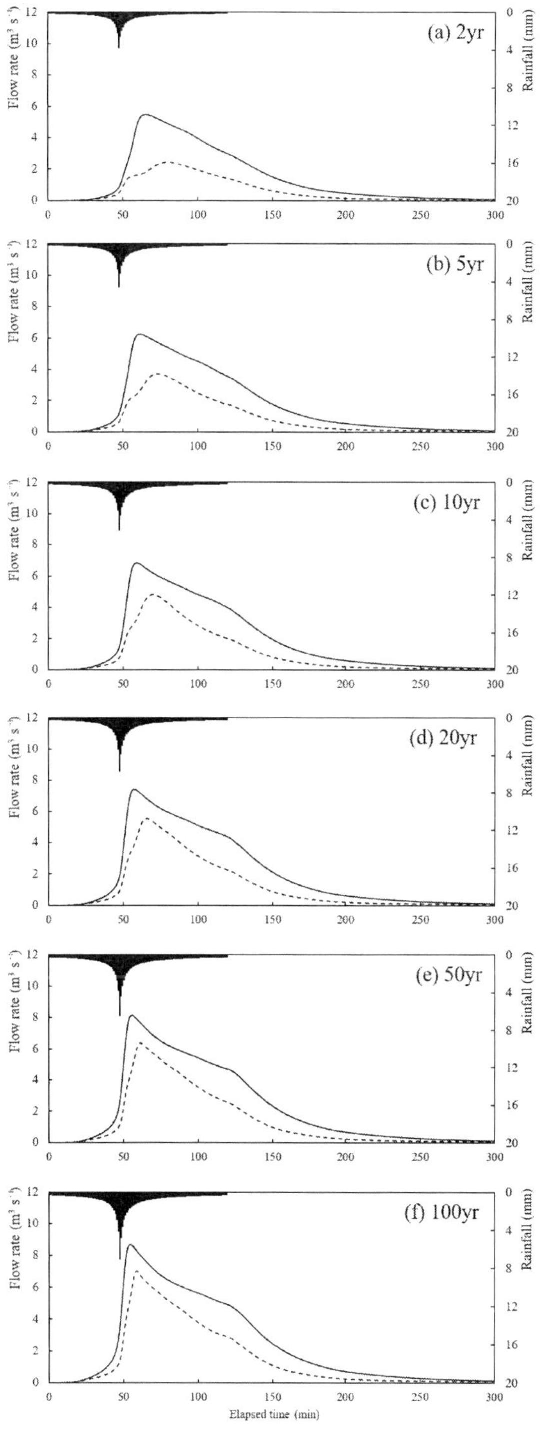

Figure 4. Runoff hydrographs for YN under baseline (solid) and SCM (dotted) scenarios for 6 modeled rainfall events (black-shaded curve).

were 28.1–72.1, 19.0–69.2, and 33.9–56.0 %, respectively, under six rainfall events with return periods ranging from 2 to 100 years, and the mean reduction rates of the average flow, peak flow, and total flood volume are 44.2, 37.4, and 42 % for the six rainfall return periods, there are significant differences in flow process between the BL and SCM scenarios.

For all events, the peak flow is reduced under the SCM scenario, though this effect is most pronounced at lower rainfall intensities, indicating that SCM implementation relieves the runoff concentration of storm water and reduces the flood load of YN's urban drainage system.

3.2 Detailed hydrologic performance

We next considered the inundation risk value for each manhole in YN (Fig. 5) for each rainfall period in both scenarios. As the risk value increased, the inundation risk of the manholes also increased, allowing the evaluation of relative risk. Inundation risk factors for each specific manhole are shown in Fig. 5. As the rainfall return period increases, inundation risk rises under both the BL and SCM scenarios, though the SCM implementation clearly reduces this risk dramatically even under intense rainfall events.

Manholes N7, N14, N17, N20, and N21 consistently show the highest inundation risk because of poor drainage systems upstream. According to on-site investigations, N7 was connected to a large, mostly impervious area, and the others were located downstream of a community area with low gradients in downstream pipes. In addition, these five manholes are at the junctions of three drainage pipes that receive water from more than 30 % of the catchment area. Thus these manholes have higher risks under the BL scenario and SCM implementation is effective at reducing inundation under heavy rains with a return period smaller than 10 years. For rainfall return periods above 10 years, the SCMs still show distinct reduction effects, but the total storm water cannot be completely handled by these installations, leading to flows exceeding the capacity of YN's urban drainage system. Because that the infiltration capacity of green infrastructure would diminish under continuous rainfall, then there should be enough green infrastructures in order to completely eliminate inundation in the drainage system. Therefore, when there is not enough space to implement green infrastructures, some gray infrastructures such as underground reservoirs and deep tunnels should be implemented in conjunction with green infrastructures.

The inundation risks of the drainage system were significantly reduced after SCMs implementation, the risk values falling below 0.2 when the rainfall return period was less than 10 years. The inundation risk increases with the increasing rainfall return periods, but increasing speed are different to each other between BL and SCM scenarios. Under the BL scenario, the inundation risk of the urban drainage system shows a continuous and sharply increase with the increasing rainfall return periods, and under SCM scenario, the inunda-

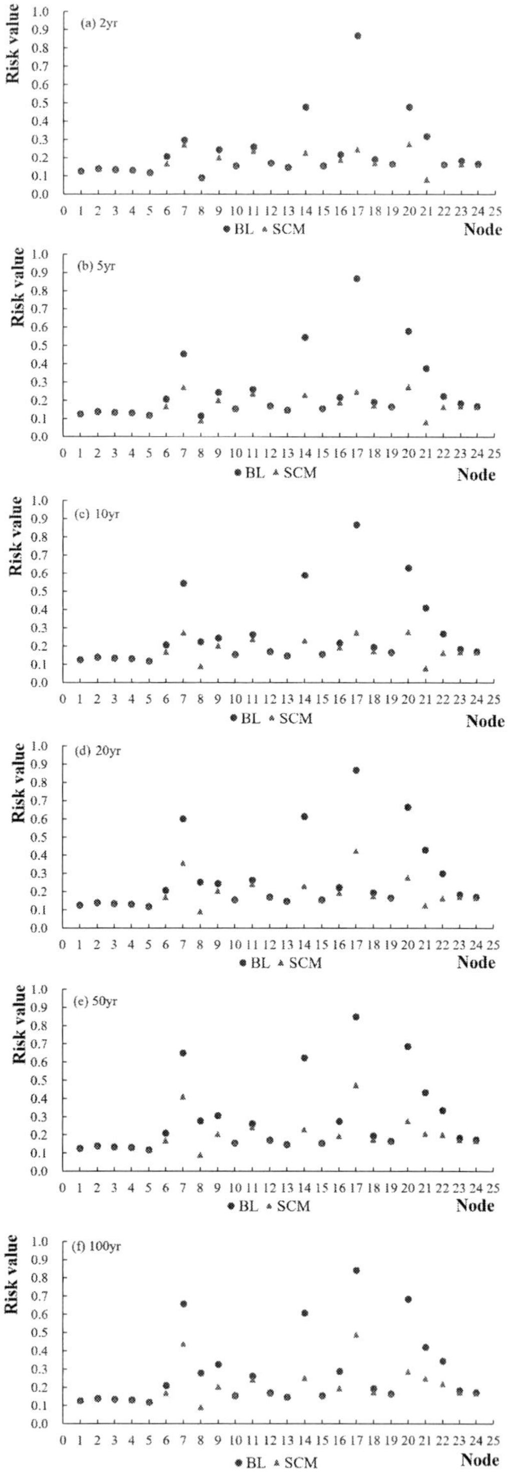

Figure 5. Risk values for each manhole in YN under BL and SCM scenarios with different rainfall return periods.

tion risk shows no increase when the rainfall return period is lower than 5, and a slow and continuous increase when the rainfall return period is higher than 10. The different increasing characteristics of the inundation risk for BL and SCM scenarios indicates that, the SCMs implementing acquires better effects under relatively light rains, and with increasing of the rainfall intensity, the mitigation effects of urban flood and inundation of urban drainage systems decrease gradually, which is a proof that the SCMs implementing enhances the resilience of urban drainage systems of YN to extreme rainfalls.

4 Conclusions

1. SCMs have significant impacts on three hydrologic performance indices related to inundation risk. The highest reduction rates of the average flow, peak flow, and total flooded volume were 72.1, 69.2, and 56.0 %, respectively, and the reduction rates for these decreased with increasing rainfall return period. In addition, the SCMs caused a hydrograph delay of the outfall flow from the urban drainage system; the lag time also decreased with increasing rainfall return period.

2. SCMs implementation showed an effective reduction of inundation risk in the urban drainage system. A comparison of risk values for each manhole between the BL and SCM scenarios showed that the number of inundation-affected manholes diminished from 7 to 0 under a 5 years return period rainfall, with the risk values for all manholes falling below 0.2 when the rainfall return period was below 10 years. At rainfall return periods higher than 10 years, the risk values for all manholes still fell to below 0.5.

3. The inundation risk increases with increasing rainfall return period, but shows different characteristics between the BL and SCM scenarios. Under the former, the inundation risk shows a continuous increase with increasing rainfall return period, but under the latter, the inundation risk shows no increase when the rainfall return period is below 5 years, with a slow and continuous increase beyond that.

It could be concluded that the proper design and implementation of SCMs settings could reduce inundation risks in urban drainage systems and achieve the sustainability objectives of storm water management at a community scale. These conclusions can guide planners and decision-makers in future considerations related to inundation risk management in community-scale urban drainage systems.

Data availability. The data are not publicly accessible.

Competing interests. The authors declare that they have no conflict of interest.

Special issue statement. This article is part of the special issue "Innovative water resources management – understanding and balancing interactions between humankind and nature". It is a result of the 8th International Water Resources Management Conference of ICWRS, Beijing, China, 13–15 June 2018.

Acknowledgements. The authors would like to extend their thanks to the Chinese National Natural Science Foundation (No. 51522907 & 51739011), and the National Key Research and Development Program of China (2016YFC0401401). This work was supported by the Research Fund of the State Key Laboratory of Simulation and Regulation of Water Cycle in River Basin, China Institute of Water Resources and Hydropower Research (grant number 2017ZY02).

Edited by: Wenchao Sun
Reviewed by: two anonymous referees

References

Chui, T. F. M., Liu, X., and Zhan, W.: Assessing cost-effectiveness of specific LID practice designs in response to large storm events, J. Hydrol., 533, 353–364, https://doi.org/10.1016/j.jhydrol.2015.12.011, 2016.

Dash, A. K., Bhattcharjee, R. M., and Paul, P. S.: Lessons Learnt from Indian Inundation Disasters: An Analysis of Case Studies, Int. J. Disaster Risk Red., 20, 93–102, https://doi.org/10.1016/j.ijdrr.2016.10.013, 2016.

Hu, M., Sayama, T., Zhang, X., Tanaka, K. K., Takara, H., and Yang, H.: Evaluation of low impact development approach for mitigating flood inundation at a watershed scale in China, J. Environ. Manag., 193, 430–438, https://doi.org/10.1016/j.jenvman.2017.02.020, 2017.

Islam, M. S., Swapan, M. S. H., and Haque, S. M.: Disaster risk index: How far should it take account of local attributes?, Int. J. Disaster Risk Red., 3, 76–87, https://doi.org/10.1016/j.ijdrr.2012.10.001, 2013.

Jacobson, C. R.: Identification and quantification of the hydrological impacts of imperviousness in urban catchments: a review, J. Environ. Manag., 92, 1438–1448, https://doi.org/10.1016/j.jenvman.2011.01.018, 2011.

Jenifer, M. A. and Jha, M. K.: Comparison of Analytic Hierarchy Process, Catastrophe and Entropy techniques for evaluating groundwater prospect of hard-rock aquifer systems, J. Hydrol., 54, 605–624, https://doi.org/10.1016/j.jhydrol.2017.03.023, 2017.

Jiao, S., Zhang, X., and Xu, Y.: A review of Chinese land suitability assessment from the rainfall-waterlogging perspective: Evidence from the Su Yu Yuan area, J. Clean. Prod., 144, 100–106, https://doi.org/10.1016/j.jclepro.2016.12.162, 2017.

Kim, Y., Kim, T., Park, H., and Han, M.: Design method for determining rainwater tank retention volumes to control runoff from building rooftops, KSCE J. Civ. Eng., 19, 1585–1590, https://doi.org/10.1007/s12205-013-0269-1, 2015.

Liu, W., Chen, W., and Peng, C.: Assessing the effectiveness of green infrastructures on urban flooding reduction: A community scale study, Ecol. Model., 291, 6–14, https://doi.org/10.1016/j.ecolmodel.2014.07.012, 2014.

Palla, A. and Gnecco, I.: Hydrologic modeling of Low Impact Development systems at the urban catchment scale, J. Hydrol., 528, 361–368, https://doi.org/10.1016/j.jhydrol.2015.06.050, 2015.

Pochwa, K., Słys, D., and Kordana, S.: The temporal variability of a rainfall synthetic hyetograph for the dimensioning of stormwater retention tanks in small urban catchments, J. Hydrol., 549, 501–511, https://doi.org/10.1016/j.jhydrol.2017.04.026, 2017.

Pyke, C., Warren, M. P., Johnson, T., LaGro Jr., J., Scharfenburg, J., Groth, P., Freed, R., Schroeer, W., and Main, E.: Assessment of low impact development for managing stormwater with changing precipitation due to climate change, Landsc. Urban Plan., 103, 166–173, https://doi.org/10.1016/j.landurbplan.2011.07.006, 2011.

Todeschini, S., Papiri, S., and Ciaponi, C.: Performance of storm water detention tanks for urban drainage systems in northern Italy, J. Environ. Manag., 101, 33–45, https://doi.org/10.1016/j.jenvman.2012.02.003, 2012.

Wang, L.: The Research on Urban Drainage Network Inundation Evaluation and Control Based on Modeling Technology, Beijing University of Technology, Beijing, 2010.

Zhao, G., Shi, R., Pang, B., Xu, Z., Du, L., and Chang, X.: Impact of rapid urbanization on rainfall-runoff processes in urban catchment: Case study for Liangshui River basin, J. Hydroelectr. Eng., 35, 55–64, 2016 (in Chinese).

Zhu, Z., Chen, Z., Chen, X., and He, P.: Approach for evaluating inundation risks in urban drainage systems, Sci. Total Environ., 553, https://doi.org/10.1016/j.scitotenv.2016.02.025, 2016.

https://doi.org/10.5194/piahs-379-231-2018

Open Access
PIAHS

Response of streamflow to climate change in a sub-basin of the source region of the Yellow River based on a tank model

Pan Wu, Xu-Sheng Wang, and Sihai Liang

School of Water Resources and Environment, China University of Geosciences, Beijing, China

Correspondence: Xu-Sheng Wang (wxsh@cugb.edu.cn) and Sihai Liang (liangsh@cugb.edu.cn)

Received: 19 December 2017 – Revised: 6 February 2018 – Accepted: 11 February 2018 – Published: 5 June 2018

Abstract. Though extensive researches were conducted in the source region of the Yellow River (SRYR) to analyse climate change influence on streamflow, however, few researches concentrate on streamflow of the sub-basin above the Huangheyan station in the SRYR (HSRYR) where a water retaining dam was built in the outlet in 1999. To improve the reservoir regulation strategies, this study analysed streamflow change of the HSRYR in a mesoscale. A tank model (TM) was proposed and calibrated with monthly observation streamflow from 1991 to 1998. In the validation period, though there is a simulation deviation during the water storage and power generation period, simulated streamflow agrees favourably with observation data from 2008 to 2013. The model was further validated by two inside lakes area obtained from Landsat 5, 7, 8 datasets from 2000 to 2014, and significant correlations were found between the simulated lake outlet runoff and respective lake area. Then 21 Global Climate Models (GCM) ensembled data of three emission scenarios (SRA2, SRA1B and SRB1) were downscaled and used as input to the TM to simulate the runoff change of three benchmark periods 2011–2030 (2020s), 2046–2065 (2050s), 2080–2099 (2090s), respectively. Though temperature increase dramatically, these projected results similarly indicated that streamflow shows an increase trend in the long term. Runoff increase is mainly caused by increasing precipitation and decreasing evaporation. Water resources distribution is projected to change from summer-autumn dominant to autumn winter dominant. Annual lowest runoff will occur in May caused by earlier snow melting and increasing evaporation in March. According to the obtained results, winter runoff should be artificially stored by reservoir regulation in the future to prevent zero-flow occurrent in May. This research is helpful for water resources management and provides a better understand of streamflow change caused by climate change in the future.

1 Introduction

Recently, climate change (IPCC, 2007) and uncertainties of global water resources change caused draw extensive attention all over the world (Taylor et al., 2013; Bae et al., 2008; Lan et al., 2009). As the water tower of Asia (Immerzeel et al., 2010), the Tibetan Plateau is sensitive to climate change (Su et al., 2016; Li et al., 2014; Yang et al. 2012; Li et al., 2013). The Yellow River, originates from the Tibetan Plateau, is the second longest river of China and the fifth longest river of the world. Water resources and runoff response to climate change of the Yellow River basin have been extensively researched (Yang et al., 2004; Fu et al., 2004; Chang et al., 2007), especially in the head water of the Yellow River (Wang and Cheng, 2000; Tang et al., 2008; Zheng et al., 2009, 2010; Liang et al., 2010; Lan et al., 2010, 2013). Above Huangheyan station, two lakes which play a role as nature reservoirs for the basin located in the topside source region of the Yellow River (HSRYR) with area about 21 000 km^2 (Liang et al., 2010; Brierley et al., 2016). Because of water shortage and zero-flow phenomenon in 1990s, a 20 m-high Huangheyan Dam was built at the Ngoring Lake downstream, during 1999 and early 21st century (Liang et al., 2010; Brierley et al., 2016), to conserve enough water for the ecological environment maintenance. Though projections of runoff change were proposed by inputting the GCMs data to a calibrated model (Immerzeel et al., 2010; Xu et al., 2009), these researches conducted at a largescale more than $100\,000\,\text{km}^2$ area is not suitable for water resources management of a mesoscale basin like the HSRYR. Reasonable regulation of the Huangheyan dam in the climate changing future is important. Tank model were first proposed by Sugawara and Maruyama (1956), a typical runoff model with a brief model structure requires less data and gets a good performance in simulation and forecast (Franchini and Pacciani, 1991).

This study aims to establish a semi-distributed model by connecting several typical tank models. GCMs data are input into the built and calibrated model to assess the runoff in 21st century. This study will offer a reference for dam regulation strategy and water resources management in the sub-basin above Huangheyan station.

2 Study area background and data processing

2.1 Study area background

The HSRYR with drainage area about $21\,000\,\text{km}^2$ is a topside sub-basin of the source region of the Yellow River above Huangheyan hydrological station (HSRYR) which is embraced by high mountains from the north, south, to west sides (Fig. 1). Three meteorological stations are around or inside this area, from southwest to northeast is Qumalai, Qingshuihe and Maduo, respectively. This is a cold and arid area with an annual mean air temperature of $-4\,°\text{C}$ and annual mean precipitation of 310 mm. The potential evapotranspiration is about 1300–1400 mm year^{-1} (Liang et al., 2010).

The two major lakes, Gyaring and Ngoring, show an expansion trend in the past decade (Duan et al., 2015). Huangheyan dam is about 20 m-height which located at 1km away from Ngoring lake outlet (Brierley et al., 2016). Previously, the dam had been used for hydroelectric generation and water regulation, but recently it has only been used to store water due to low power generation efficiency. However, if the cut off phenomenon will show again in this area is still unknown. Notably, there is a runoff disturbed period caused by dam construction activity and power generation during 1999 to early 21st century.

2.2 Data processing

Monthly runoff data from 1990–2013 are obtained from the Yellow River Conservancy Commission (YRCC). Daily precipitation, air temperature and wind speed at 10 m height, relative humidity and sunshine duration data between 1961 and 2013 at the three meteorological stations are collected from the China Meteorological Administration (CMA). These data are applied to derive monthly mean air temperature (T), monthly positive accumulated temperature ($\sum T^{+}$) monthly precipitation (P) and snowfall (P_{s}). The potential evapotranspiration is calculated with the FAO Penman-Monteith equation (Allen et al., 1998). 21 Global Climate Models (GCMs) are selected as shown in Table 1. Due to uncertainty of each model projection, mean value of 21 GCMs output data under three emission scenarios (SRA2, SRA1B and SRB1), including shortwave flux at surface (W m^{-2}) (ΔR_{s}), precipitation rate change (mm day^{-1}) and air temperature change(ΔT) are used as ensembled data in this study. Mean value of respective 20C3M output precipitation rate P_0' (mm day^{-1}) during 1961 to 1990 is used as reference value. Landsat 5, 7, 8 datasets of July or August from 2000 to 2014 are used to extract lake area for further validation of model distribution ability.

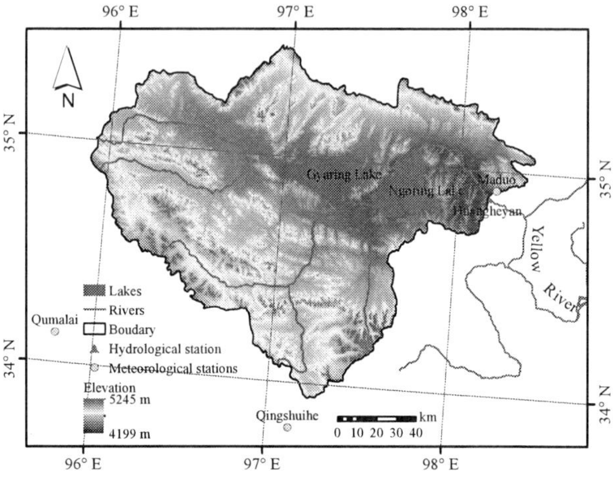

Figure 1. Stations distribution and study area topography.

3 Methods

3.1 Model construction and mechanism

As shown in Fig. 2, snow, surface runoff and baseflow are considered in each land surface hydrologic response unit (HRU). Parameters described in Fig. 2 and used in the following equations are shown in Table 2.

Snowfall and rainfall are distinguished based on critical temperature T_{c}, expressed as:

$$P_{\text{s}}(i) = \begin{cases} 0, & T(i) > T_{\text{c}} \\ P(i), & T(i) > T_{\text{c}} \end{cases} \tag{1}$$

Snow accumulation is indicated by following equations:

$$h_{\text{s}}(i) = \tag{2}$$

$$\begin{cases} T(i) > T_{\text{c}}: & \begin{cases} h_{\text{s}}(i-1) \geq M_{\text{s}}(i) + E_{\text{s}}(i): \\ \quad h_{\text{s}}(i-1) - M_{\text{s}}(i) - E_{\text{s}}(i), \\ h_{\text{s}}(i-1) < M_{\text{s}}(i) + E_{\text{s}}(i): 0, \end{cases} \\ T(i) \leq T_{\text{c}}: & h_{\text{s}}(i-1) + P_{\text{s}}(i) - E_{\text{s}}(i), \end{cases}$$

h_{s} is accumulated snow relative depth (mm) in snow water equivalent (SWE) form.

Table 1. Details of selected model and respective output scenarios.

Institute ID	Model name	Output scenarios of each model			
BCCR	BCM2	20C3M	SRA2	SRA1B	SRB1
CCCMA	CGCM3_1-T47	20C3M	SRA2	SRA1B	SRB1
CNRM	CM3	20C3M	SRA2	SRA1B	SRB1
CSIRO	MK3	20C3M	SRA2	SRA1B	SRB1
CONS	ECHO-G	20C3M	SRA2	SRA1B	–
LASG	FGOALS-G1_0	20C3M	–	SRA1B	SRB1
GFDL	CM2	20C3M	SRA2	SRA1B	SRB1
GFDL	CM2_1	20C3M	SRA2	SRA1B	SRB1
NASA	GISS-AOM	–	–	SRA1B	SRB1
NASA	GISS-EH	–	–	SRA1B	–
NASA	GISS-ER	20C3M	SRA2	SRA1B	SRB1
UKMO	HADCM3	20C3M	SRA2	SRA1B	SRB1
UKMO	HADGEM1	20C3M	SRA2	SRA1B	–
INM	CM3	20C3M	SRA2	SRA1B	SRB1
IPSL	CM4	–	SRA2	SRA1B	SRB1
NIES	MIROC3_2-HI	20C3M	–	SRA1B	SRB1
NIES	MIROC3_2-MED	20C3M	SRA2	SRA1B	SRB1
MPIM	ECHAM5	20C3M	SRA2	SRA1B	SRB1
MRI	CGCM2_3_2	20C3M	SRA2	SRA1B	SRB1
NCAR	CCSM3	20C3M	SRA2	SRA1B	SRB1
NCAR	PCM	20C3M	SRA2	SRA1B	–

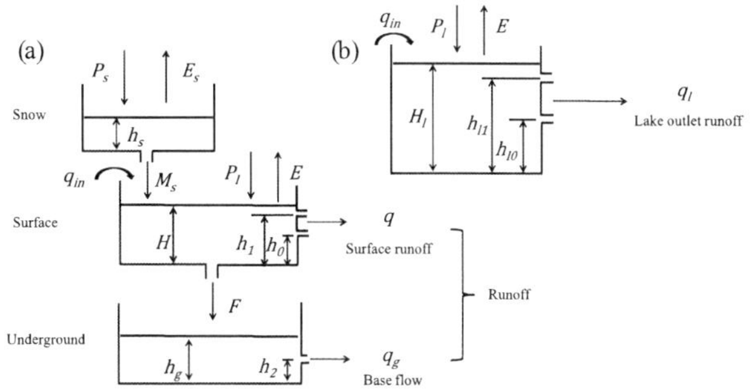

Figure 2. Typical tank model constructions, panels **(a)** and **(b)** indicate land surface HRU and lake HRU, respectively.

Snowmelt is calculated by Eq. (3)

$$M_s(i) = \begin{cases} T(i) > T_c: & \begin{cases} h_s(i-1) \geq M_{sp}(i) + E_{sp}(i): M_{sp}(i), \\ h_s(i-1) < M_{sp}(i) + E_{sp}(i): h_s(i-1), \end{cases} \\ T(i) \leq T_c: & 0 \end{cases} \tag{3}$$

$M_{sp}(i)$ is potential snowmelt water as indicated in Eq. (4). And the potential snow sublimation is given by Eq. (5):

$$M_{sp}(i) = \alpha \cdot \sum (T^+ - T_c) \tag{4}$$

$$E_{sp}(i) = \beta \cdot \sum T^+ \tag{5}$$

Surface runoff is calculated by Eq. (6) based on different water depth.

$$q(i) = \begin{cases} 0, & H(i-1) < 0 \\ b_1 \cdot H(i-1), & 0 < H(i-1) \leq h_0 \\ b_1 \cdot H(i-1) \\ \quad + b_2 \cdot (H(i-1) - h_0), & H(i-1) > h_0 \end{cases} \tag{6}$$

Surface water infiltration is indicated as follow:

$$F(i) = b_0 \times H(i) \tag{7}$$

Baseflow is given by Eqs. (8) and (9):

$$q_g(i) = \begin{cases} d \cdot \left(h_g(i) - h_2\right), & h_g(i) > h_2 \\ 0, & h_g(i) > h_2 \end{cases} \tag{8}$$

As Fig. 2 indicated lake HRU plays a role as a regulate reservoir which outlet runoff is calculated by following equations:

$$q_l(i) = \tag{9}$$

$$\begin{cases} T(i) > T_c: & \begin{cases} h_l(i-1) < h_{l1}: 0, \\ h_{l1} \leq h_l(i-1) < h_{l2}: \\ \quad c_0(h_l(i-1) - h_{l1}), \\ h_{l2} \leq h_l(i-1): c_0 + (h_l(i-1) - h_{l1}) \\ \quad + c_1(h_l(i-1) - h_{l2}),) \end{cases} \\ T(i) \leq T_c: & \begin{cases} h_l(i-1) < h_{l1}: 0, \\ h_l(i-1) \geq h_{l1}: c_0(h_l(i-1) - h_{l1}), \end{cases} \end{cases}$$

Snow accumulated depth, water depth of each step can be calculated by solving following equation by implicit finite difference method.

Time-varying values of snow depth, land surface water depth, baseflow and lake water depth are given by Eqs. (10), (11), (12) and (13) respectively:

$$\frac{dh_s}{dt} = P_s - E_s - M_s \tag{10}$$

$$\frac{dH}{dt} = P_l + M_s - E_{ls} - F + q_{in} - q \tag{11}$$

$$\frac{dh_g}{dt} = F - q_g \tag{12}$$

$$\frac{dH_l}{dt} = P - E_l + q_{in} - q_l \tag{13}$$

Land surface actual evaporation and lake water evaporation are calculated from:

$$E = \begin{cases} E_{ls} = k_1 \cdot E_0 \\ E_l = k_2 \cdot E_0 \end{cases} \tag{14}$$

Then total outlet runoff is transformed into $m^3\ s^{-1}$ unit.

3.2 Model performance evaluation

Volume difference (D_v), and Nash–Sutcliffe efficiency (R^2) are used to evaluate the model performance defined as following:

$$R^2 = 1 - \frac{\sum_{i=1}^{n} (Q_i - Q_i')^2}{\sum_{i=1}^{n} (Q_i - \overline{Q})} \tag{15}$$

$$D_v = \frac{V_R - V_R'}{V_R} \cdot 100\,\% \tag{16}$$

Table 2. Parameters and descriptions for the built model.

Model constructions	Parameters	Unit	Descriptions
Snow model (Eqs. 1, 2, 3, 4, 5, 10)	P	mm	Precipitation
	P_s		Snowfall in SWE form
	P_l		Rainfall
	h_s		Snow depth in SWE form
	M_{sp}		Potential snowmelt in SWE form
	M_s		Snowmelt in SWE form
	E_{sp}		Potential snow sublimation in SWE form
	E_s		Snow sublimation in SWE form
	α	mm (°C)$^{-1}$ day^{-1}	Degree-day factor
	β		Snow sublimation coefficient
Surface Runoff (Eqs. 6, 11, 14)	q	mm	Surface runoff
	H		Water depth of surface tank
	E_{ls}		Land surface actual evaporation
	h_0		Critical value for surface runoff generation
	h_1		Critical value for peak runoff
	k_2	–	Land surface evaporation conversion coefficient
	b_1		Runoff coefficient for normal runoff generation
	b_2		Runoff coefficient for peak value
Infiltration and Baseflow (Eqs. 7, 8, 12)	F	mm	Infiltration water
	h_g		Water depth of underground tank
	q_g		Base flow
	h_2		Critical value for base flow generation
	b_0	–	Infiltration coefficient
	d		Base flow coefficient
Lake model (Eqs. 9, 13, 14)	q_l	mm	Lake outlet runoff
	E_l		Lake water evaporation
	H_l		Water depth of lake tank
	h_{l1}		Critical value for lake outlet runoff generation
	h_{l2}		Lake outlet runoff coefficient for peak value
	k_2	–	Water evaporation conversion coefficient
	c_0		Lake outlet runoff coefficient for normal runoff generation
	c_1		Lake outlet runoff coefficient for peak value
Common Parameters	T^+	°C	Positive temperature
	T_c		Critical temperature for melt
	i	month	Time step indicator

R^2 is Nash–Sutcliffe efficiency Q_i is observed monthly runoff, Q_i' simulated monthly runoff, $\overline{Q}$ is observed annual mean runoff, n is the total number of simulated months. D_v is volume difference between observed runoff and simulated runoff. V_R indicates total observed runoff volume. V_R' indicates total simulated runoff volume.

There are five HRUs included in this study, land surface HRUs: HRU 1, HRU 2 and HRU 3. Two Lakes HRUs: HRU G and HRU N. Each HRU is represented by one typical tank model, the model connection is based on flow direction as shown in Fig. 3.

3.3 GCMs data downscaling

There are several normally used downscaling methods for GCMs data: statistical downscaling method, dynamic downscaling method and delta method (Xu et al., 2009). But different method has different shortage and advantage (Fowler et al. 2007). Delta method is a simple method commonly applied in hydrological studies of climate change (Merritt et al., 2006). In this study delta method is selected instead of complex spatial downscaling mothed.

Precipitation is projected by:

$$P = P_0 \cdot \ 1 + \Delta P / P_0' \tag{17}$$

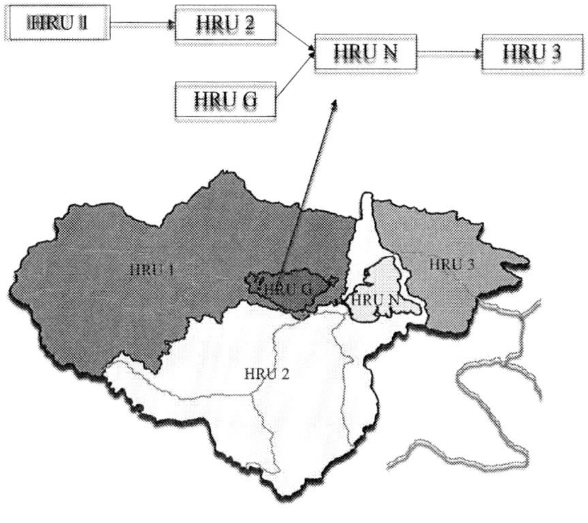

Figure 3. Connection of typical tank models in built model.

P_0 (mm) is observed monthly precipitation of pre-change period (1961–1990), P'_0 reference precipitation rate (mm day^{-1}) of period 1961–1990 obtained from mean value of GCMs under scenario 20C3M, ΔP (mm day^{-1}) is the mean value of 21 GCMs projected change of precipitation rate under respective scenarios.

For temperature projection:

$$T = T_0 + \Delta T \cdot \frac{°\mathrm{C}}{\mathrm{K}} \tag{18}$$

T_0 (°C) is monthly pre-change value of temperature (mean value of 1961–1990), ΔT(K) is the projection change of temperature.

$$R_s = \frac{\Delta R_s \times 86400}{1000000} + R_{s0} \tag{19}$$

R_{s0} (MJ (m^2 × day)$^{-1}$) is monthly mean value of pre-change period (1961–1990) obtained from FAO Penman-Monteith equation (Allen et al., 1998). ΔR_s (W m^{-2}) is mean value of 21 GCMs projected shortwave flux change at surface under respective scenarios.

3.4 Projection of $\sum T^+$ and E_0

Monthly accumulated positive temperature $\sum T^+$ is significantly correlated with monthly mean temperature as shown in Fig. 4. By using the piecewise equation shown in Fig. 4, $\sum T^+$ can be obtained from monthly mean temperature (T). Potential evaporation(E_0) is calculated by linear correlation with projected monthly mean temperature, and solar radiation (Table 4).

4 Results and discussions

4.1 Results of simulation

Monthly comparisons between the simulated runoff and observed runoff in the calibration period (1990–1998) and validation period (2000–2013) are shown in Fig. 5. As the red line in Fig. 5a and red points in Fig. 5c indicated, the runoff is obviously disturbed during dam construction and hydropower generation period. After disturbed period, simulated runoff variation agrees favourably with the observation runoff except slight overestimation (Fig. 5a and b). The model performance in calibration period: R^2: 0.84, D_v: −0.5 %, in period 2008–2013: R^2: 0.67, D_v: −24.88 %. Generally model performance is very good if $R^2 > 0.75$, satisfactory if $0.36 < R^2 < 0.75$, and unsatisfactory if $R^2 < 0.36$ (Nash and Sutcliffe, 1970; Krause et al., 2005; Moriasi et al., 2007). The D_v is about −25 % which indicates overestimation. But it is reasonable, because there will be more water conserved in the reservoir after the a 20 m-height dam built.

4.2 Lakes expansion and relationship with respective outlet runoff

Figure 6 indicates variation of the lake area (Fig. 6b), in the past decade expansion of Ngoring lake is more obvious than Gyaring lake (Fig. 6a). The two lakes show a similar change pattern, lake area shrank at first (2000–2004) and then expand (2004–2012). It is caused by water resources change in this area, the last zero-flow phenomenon happened in 2004 (Chang et al., 2007), and runoff rise again from 2004 to 2012. Though this model is only calibrated by runoff of Huangheyan station, simulated outlet runoff of the two lakes is significantly correlated with the lake area change (Fig. 7) that indicate a distribution characteristic of this model.

4.3 Results of projection

Change of projected ΔP (%), ΔE_0 (%), ΔR_s (W m^{-2}) and ΔT (°) in the three respective periods under scenarios SRA2, SRA1B and SRB1 are displayed in Fig. 8. Three scenarios similarly indicate that temperature and precipitation will increase, but radiation will decrease in the future. Temperature and precipitation is projected to increase for all seasons but with largest increase appearing in winter and spring. Radiation is projected to decease all the time but with largest decrease appearing in spring. Impacted by change of temperature and radiation, potential evapotranspiration is projected to decrease in summer and winter but increase in spring and autumn. Though projected variations under different scenarios show a similar seasonality, the projected increment and decrement are different resulted from different emission scenarios. Radiation and temperature projected by SRA2 show the largest change among the three scenarios, but precipitation and potential evapotranspiration projected by SRB1 show a largest change compare to the other two scenarios.

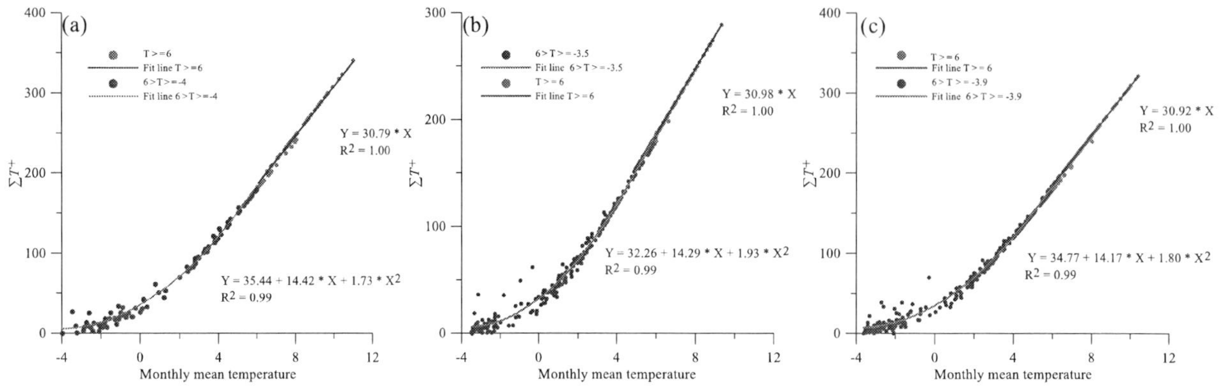

Figure 4. Relationship between monthly mean temperature and monthly positive accumulated temperature (1961–2013). Panels **(a, b, c)** indicate Qumalai, Qingshuihe and Maduo station, respectively.

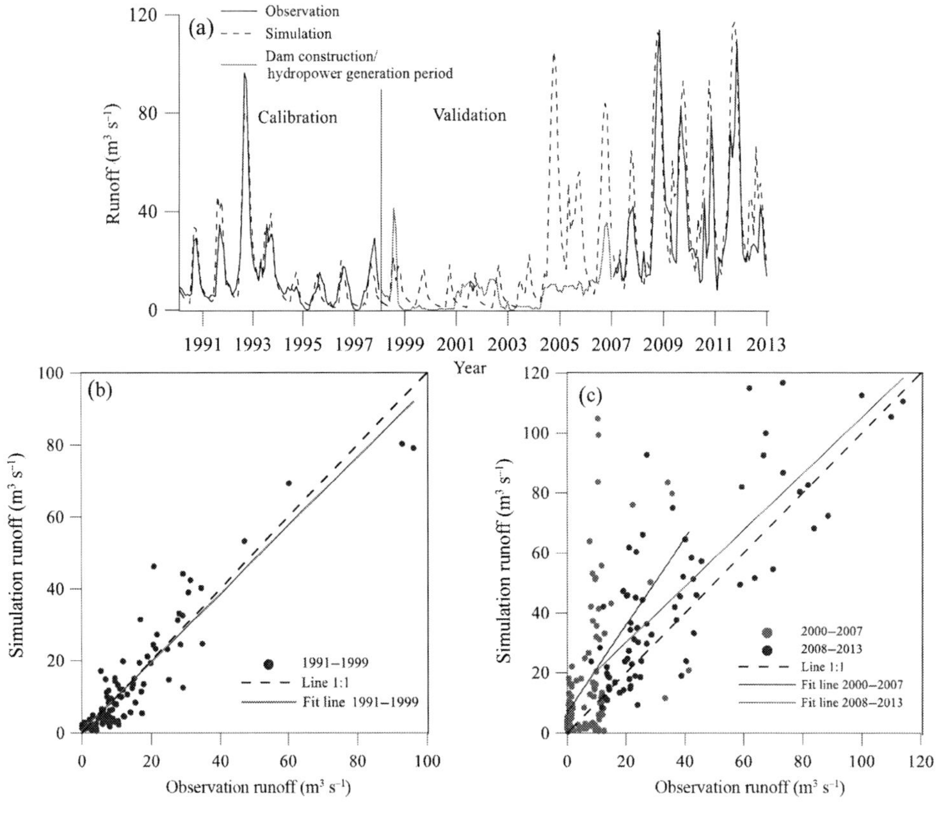

Figure 5. Comparison between simulation and observation runoff.

Variables projected by SRA1B show a moderate change. These are resulted from different scenarios setting different future greenhouse gas emissions accompanied by storylines of social, economic and technological development (Parry and Cox, 2007).

Projected runoff in 2011–2030 (2020s), 2046–2065 (2050s), 2080–2099 (2090s) under the three scenarios SRA2, SRA1B and SRB1 are shown in Fig. 9a, b and c, respectively. There are two dash line, blue and purple, indicate pre-change period (1961–1990) and low-flow (1990s) period, respectively. Results in Fig. 9 similarly indicated that runoff shows an increase trend in the long term. Increasing runoff is mainly caused by dramatically increase precipitation. Projected runoff of SRA1B and SRB1 are lower than SRA2 but

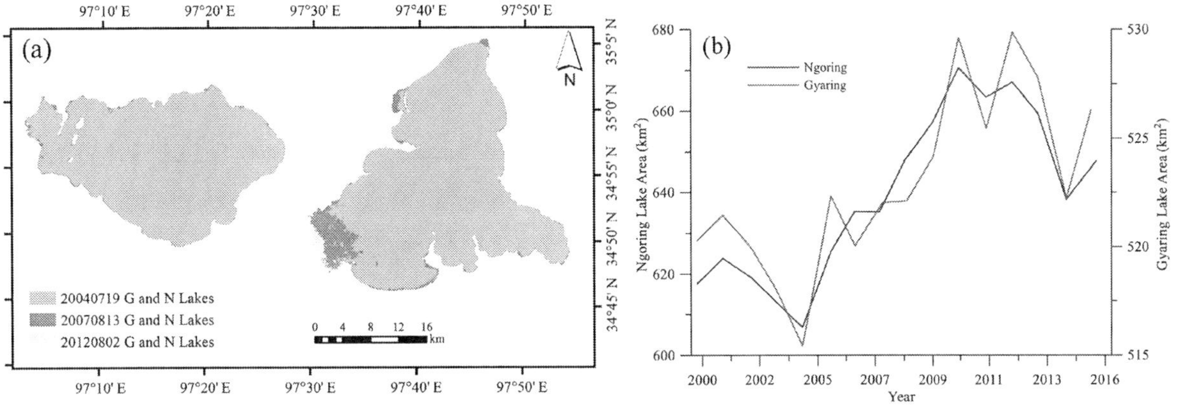

Figure 6. Lakes expansion and area change shown in panels **(a)** and **(b)**, respectively.

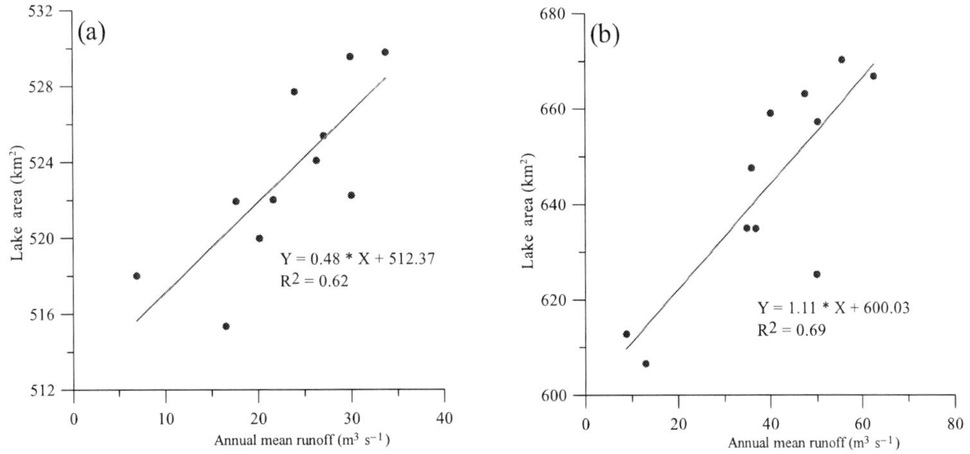

Figure 7. Relationship between lake area and respective outlet runoff.

showed a similar trend as SRA2 in the three periods (Fig. 9c). Runoff is projected to increase eventually in the future but won't return to a pre-change level. As indicated by results of SRA2 (Fig. 9b), runoff of August to December will reach to a pre-change level in 2090s, but runoff of January to July in 2090s is lower than pre-change period. Annual runoff is projected to return in 2090s, but runoff in spring and summer is projected to be lower than pre-change period. Water resources distribution is projected to change from summer-autumn dominant to autumn winter dominant. Temperature increase in winter-spring and potential evapotranspiration increase in spring cause earlier snowmelt runoff and increase evapotranspiration in spring. Annual lowest runoff is projected to occur in May as different from the winter months, as seen in the 1990s (Fig. 9).

4.4 Will zero-flow show again?

The last zero-flow phenomenon occurred in 2004 January to March (Chang et al., 2007). In the future, precipitation is projected to increase in winter and earlier snow melting will occur which result in higher winter runoff than before (Fig. 8), the zero flow won't occur in winter again. And there is no zero-flow phenomenon in projected results of different periods under the three scenarios. Due to increase temperature in winter, frozen water resources will release earlier than before which will cause less melt water recharge in May runoff and a relative lower runoff appearing in May. If the dam impacts are taken into consideration, zero-flow phenomenon is possible show again in May in the warming future. To prevent appearing of zero-flow phenomenon, artificially regulate the dam to store water in winter is imperative.

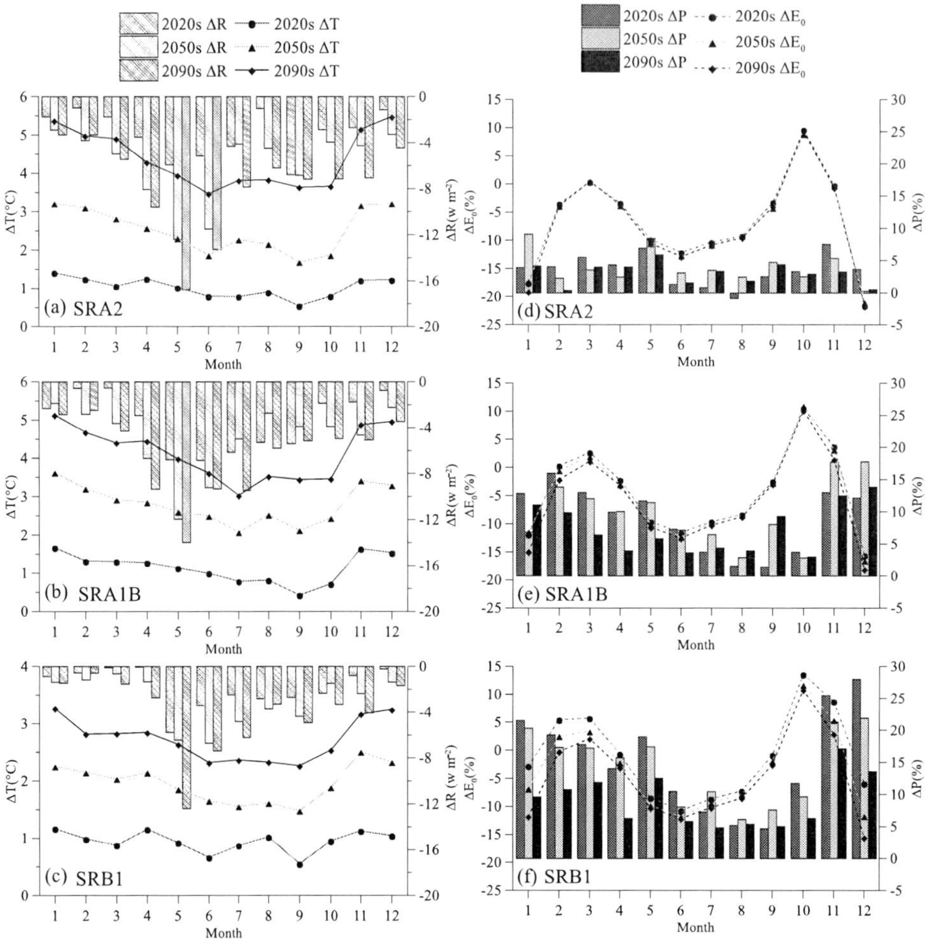

Figure 8. Projected climatic elements of the three periods.

Table 3. Calibrated parameters.

Parameters		Area	$H_{(l)}(0)$	h_0	h_1	h_2	$h_{g(s)}(0)$	h_{10}	h_{11}	T_c	α	β	$k_{1(2)}$	b_0	b_1	b_2	d	c_0	c_1
HRU	Unit	km^2	mm							°C	mm (°C)$^{-1}$ day^{-1}		–						
Land	HRU 1	9003.5	10	10	0	10	10	–	–	0	0.12	0.18	0.55	0.05	0.01	0.1	0.1	–	–
	HRU 2	8059	10	10	0	10	10	–	–	0	0.12	0.18		0.06	0.01	0.1	0.4	–	–
	HRU 3	2821	10	10	0	10	10	–	–	0	0.12	0.18		0.31	0.1	0.22	0.2	–	–
Lake	HRU G	526	7000	–	–	–	0	6000	8400	0	–	–	0.75	–	–	–	–	0.01 0.001*	0.9
	HRU N	610.7	10 000	–	–	–	0	10 000	17 000	0	–	–		–	–	–	–	0.01 0.001*	0.9

* Coefficient for frozen lake water seepage in winter ($T < 0$).

4.5 Limitation remarks

Due to low people density (0.34 person km^{-2}) (Liang et al., 2010) and difficulty of estimating dam construction and hydropower generation impacts, observed runoff used in this study was not naturalized first. However, this model is well calibrated by undisturbed runoff from 1990–1998 and then validated by 2008–2013 runoff and lake area obtained from Landsat datasets.

Only two variables are involved in E_0 projection in this study, but as shown in Table 3, a strong relationship existed between R_s and T. For $\sum T^+$ projection, when daily air temperature is positive for in the whole month, monthly positive

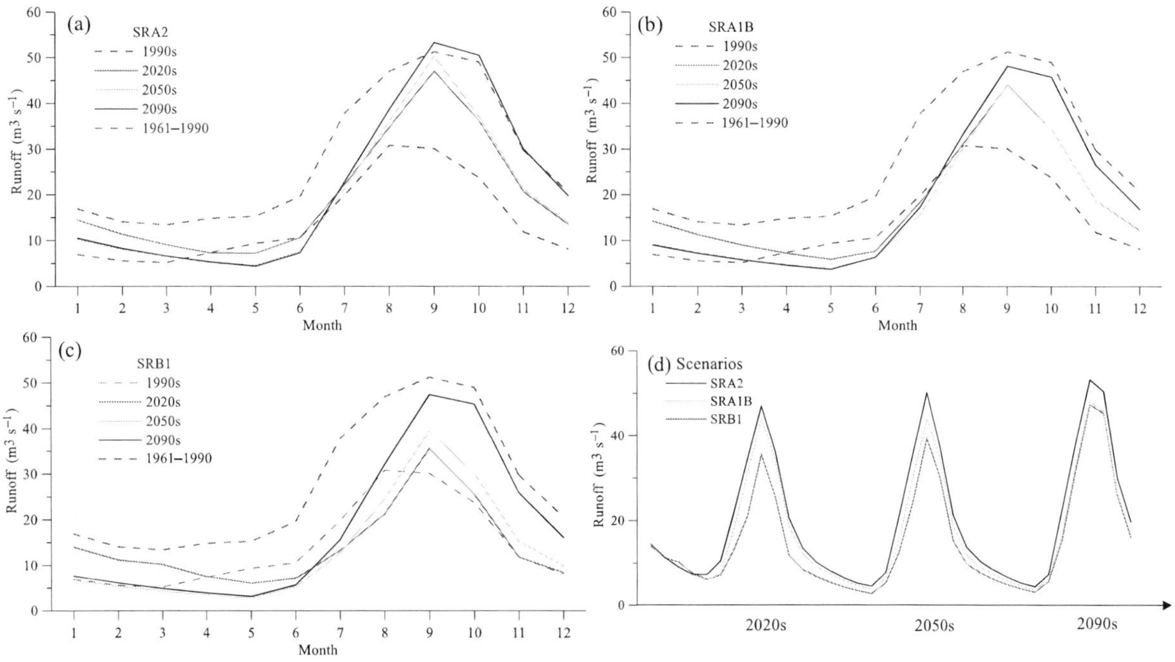

Figure 9. Projected runoff of different scenarios.

Table 4. Regression Model for potential evaporation projection.

Stations	Intercept	Monthly T	R_s	Significant F	R^2
Qumalai	0.100	0.065	0.126	0	0.99
Qingshuihe	0.238	0.062	0.112	0	0.99
Maduo	0.214	0.067	0.118	0	0.99

accumulated temperature should be calculated from:

$$\sum T^{+} = \text{days} \cdot T \tag{20}$$

But the deviation is negligible, the results will be same as which obtained the linear equation of piecewise equations in Fig. 4.

Frozen ground is widely distributed in this area (Jin et al., 2010) which degradation in the warming future will impact runoff generation (Hayashi et al., 2003; Shanley and Chalmers, 1999; Wang et al., 2009). But few conception models consider impacts of frozen ground and its degradation on runoff generation. It needs a further and deeply study to conceptualize the physical process into conception model.

5 Conclusions

In this study, the built model connected several typical tank models showed a good performance in runoff simulation and projection. And it can be used as a semi-distribution model proved by significantly correlation with the lake area change. Different projections obtained by different scenarios that improve uncertainties of water resources change in the future, but there are similarities:

a. Compared to low-flow period (1990s), the projected runoff shows a recovery trend in the long term.

b. Runoff distribution will change from summer-autumn dominant to autumn-winter dominant. In the future, annual lowest runoff will occur in May rather than in the winter.

c. Though there is no zero-flow showed among these projections, but zero-flow is possible to occur in May by considering dam impacts. Reservoir regulation strategies should be made for storing winter runoff to prevent zero-flow phenomenon in May.

This study is helpful for water resources management and provides a better understand of streamflow change caused by climate change in the future

Data availability. GCMs data are obtained from IPCC Data Distribution Centre: http://www.ipcc-data.org/sim/gcm_clim/SRES_AR4/index.html.

The availability of other data are mentioned in Sect. 2.2.

Competing interests. The authors declare that they have no conflict of interest.

Special issue statement. This article is part of the special issue "Innovative water resources management – understanding and balancing interactions between humankind and nature". It is a result of the 8th International Water Resources Management Conference of ICWRS, Beijing, China, 13–15 June 2018.

Acknowledgements. This research was supported by the China Geological Survey from 2000, the National Natural Science Foundation of China (41330634, 41072191, 91125011) and the Fundamental Research Funds for Central Universities. The authors are grateful to the two reviewers for their insightful and constructive comments which have greatly improved the quality of the paper.

Edited by: Wenchao Sun
Reviewed by: Shaowei Ning and one anonymous referee

References

Allen, R., Pereira, L. S., Raes, D., and Smith, M.: Crop evapotranspiration – Guidelines for computing crop water requirements, FAO Irrigation and Drainage Paper no. 56, Rome, Italy, 1998.

Bae, D. H., Jung, I. W., and Chang, H.: Long-term trend of precipitation and runoff in korean river basins, Hydrol. Process., 22, 2644–2656, 2008.

Brierley, G. J., Li, X., Cullum, C., and Gao, J.: Landscape and ecosystem diversity, dynamics and management in the Yellow River source zone, Springer Geography, available at: https://link.springer.com/book/10.1007/978-3-319-30475-5 (last access: 28 February 2018), 2016.

Chang, G., Li, L., Zhu, X., Wang, Z., Xiao, J., and Li, F.: Influencing factors of water resources in the source region of the Yellow River, J. Geogr. Sci., 17, 131–140, 2007.

Duan, S., Fan, S., Cao, G., Liu, X., and Sun, Y.: The changing features and cause analysis of the lakes in the source regions of the yellow river from 1976 to 2014, J. Glaciol. Geocryol., 37, 745–756, 2015.

Fowler, H. J., Blenkinsop, S., and Tebaldi, C.: Linking climate change modelling to impacts studies: recent advances in downscaling techniques for hydrological modelling, Int. J. Climatol., 27, 1547–1578, 2007.

Franchini, M. and Pacciani, M.: Comparative analysis of several conceptual rainfall-runoff models, J. Hydrol., 122, 161–219, 1991.

Fu, G., Chen, S., Liu, C., and Shepard, D.: Hydro-climatic trends of the yellow river basin for the last 50 years, Climatic Change, 65, 149–178, 2004.

Hayashi, M., Kamp, G. V. D., and Schmidt, R.: Focused infiltration of snowmelt water in partially frozen soil under small depressions, J. Hydrol., 270, 214–229, 2003.

Immerzeel, W. W., van Beek, L. P., and Bierkens, M. F.: Climate change will affect the asian water towers, Science, 328, 1382–1385, 2010.

IPCC: Climate change 2007: The Physical Science Basis, in: Contribution of Working Group I to the Fourth Assessment Report of the Intergovernmental Panel on Climate Change, Cambridge University Press, Cambridge, UK, 996 pp., 2007.

Jin, H. J., Wang, S. L., Lan-Zhi, L., Rui-Xia, H. E., Chang, X. L., and Luo, D. L.: Features and degradation of frozen ground in the sources area of the yellow river, China, J. Glaciol. Geocryol., 49, 5522–5529, 2010.

Krause, P., Boyle, D. P., and Bäse, F.: Comparison of different efficiency criteria for hydrological model assessment, Adv. Geosci., 5, 89–97, https://doi.org/10.5194/adgeo-5-89-2005, 2005.

Lan, C., Lettenmaier, D. P., Alberti, M., and Richey, J. E.: Effects of a century of land cover and climate change on the hydrology of the puget sound basin, Hydrol. Process., 23, 907–933, 2009.

Lan, C., Zhang, Y., Gao, Y., Hao, Z., and Cairang, L.: The impacts of climate change and land cover/use transition on the hydrology in the upper yellow river basin, China, J. Hydrol., 502, 37–52, 2013.

Lan, Y., Zhao, G., Zhang, Y., Wen, J., Liu, J. Q., and Hu, X.: Response of runoff in the source region of the yellow river to climate warming, Quatern. Int., 226, 60–65, 2010.

Li, B., Yu, Z., Liang, Z., and Acharya, K.: Hydrologic response of a high altitude glacierized basin in the central tibetan plateau, Global Planet. Change, 118, 69–84, 2014.

Li, F., Zhang, Y., Xu, Z., Teng, J., Liu, C., Liu, W., and Mpelasoka, F. The impact of climate change on runoff in the southeastern Tibetan Plateau, J. Hydrol., 505, 188–201, 2013.

Liang, S., Ge, S., Wan, L., and Zhang, J.: Can climate change cause the yellow river to dry up?, Water Resour. Res., 46, 228–236, 2010.

Merritt, W. S., Alila, Y., Barton, M., Taylor, B., Cohen, S., and Neilsen, D.: Hydrologic response to scenarios of climate change in subwatersheds of the Okanagan basin, British Columbia, J. Hydrol. 326, 79–108, 2006.

Moriasi, D. N., Arnold, J. G., Van Liew, M. W., Bingner, R. L., Harmel, R. D., and Veith, T. L.: Model evaluation guidelines for systematic quantification of accuracy in watershed simulations, Am. Soc. Agric. Biol. Eng., 50, 885–890, 2007.

Nash, J. E. and Sutcliffe, J. V.: River flow forecasting through conceptual models: Part I. A discussion of principles, J. Hydrol., 10, 282–290, 1970.

Parry, M. L. and Cox, B. S.: Technical Summary, in: Climate Change 2007: Impacts, Adaptation and Vulnerability. Contribution of Working Group II to the Fourth Assessment Report of the Intergovernmental Panel on Climate Change, Cambridge University Press, Cambridge, UK, 23–789, 2007.

Shanley, J. B. and Chalmers, A.: The effect of frozen soil on snowmelt runoff at Sleepers River, Vermont, Hydrol. Process., 13, 1843–1857, 1999.

Su, F., Zhang, L., Ou, T., Chen, D., Yao, T., Tong, K., and Qi, Y.: Hydrological response to future climate changes for the major upstream river basins in the Tibetan Plateau, Global Planet. Change, 136, 82–95, 2016.

Sugawara, M. and Maruyama, F.: A method of revision of the river discharge by means of a rainfall model. Collection of research papers about forecasting hydrologic variables, The Geosphere Research Institute of Saitama University, Saitama, Japan, 14–18, 1956.

Tang, Q., Oki, T., Kanae, S., and Hu, H.: Hydrological cycles change in the yellow river basin during the last half of the twentieth century, J. Climate, 21, 1790–1806, 2008.

Taylor, R. G., Scanlon, B., Döll, P., Rodell, M., Van Beek, R., Wada, Y., Longuevergne, L., Leblanc, M., Famiglietti, J. S., Edmunds,

M., and Konikow, L.: Ground water and climate change, Nature Clim. Change, 3, 322–332, 2013.

Wang, G. and Cheng, G.: Eco-environmental changes and causative analysis in the source regions of the yangtze and yellow rivers, China, Environmentalist, 20, 221–232, 2000.

Wang, G. X., Hu, H. C., and Li, T. B.: The influence of freeze-thaw cycles of active soil layer on surface runoff in a permafrost watershed, J. Hydrol., 375, 438–449, 2009.

Xu, Z. X., Zhao, F. F., and Li, J. Y.: Response of streamflow to climate change in the headwater catchment of the yellow river basin, Quatern. Int., 208, 62–75, 2009.

Yang, D., Li, C., Hu, H., Lei, Z., Yang, S., Kusuda, T., Koike, T., and Musiake, K.: Analysis of water resources variability in the Yellow River of China during the last half century using historical data, Water Resour. Res., 40, 308–322, 2004.

Yang, T., Hao, X., Shao, Q., Xu, C. Y., Zhao, C., Chen, X., and Wang, W.: Multi-model ensemble projections in temperature and precipitation extremes of the tibetan plateau in the 21st century, Global Planet. Change, 80–81, 1–13, 2012.

Zheng, H., Lu, Z., Zhu, R., Liu, C., Sato, Y., and Fukushima, Y.: Responses of streamflow to climate and land surface change in the headwaters of the yellow river basin, Water Resour. Res., 45, 641–648, 2009.

Zheng, H., Zhang, L., Liu, C., Shao, Q., and Fukushima, Y.: Changes in stream flow regime in headwater catchments of the yellow river basin since the 1950s, Hydrol. Process., 21, 886–893, 2010.

https://doi.org/10.5194/piahs-379-243-2018

Land Use/Land Cover Changes and Its Response to Hydrological Characteristics in the Upper Reaches of Minjiang River

Kai Ma[1,2], **Xiaorong Huang**[1,2], **Biying Guo**[1,2], **Yanqiu Wang**[1,2], **and Linyun Gao**[1,2]

[1]State Key Laboratory of Hydraulics and Mountain River Engineering, Chengdu, Sichuan, 610065, China
[2]College of Water Resource & Hydropower, Sichuan University, Chengdu, Sichuan 610065, China

Correspondence: Xiaorong Huang (hxiaorong@scu.edu.cn)

Received: 30 December 2017 – Revised: 23 March 2018 – Accepted: 23 March 2018 – Published: 5 June 2018

Abstract. Land use changes alter the hydrological characteristics of the land surface, and have significant impacts on hydrological cycle and water balance, the analysis of complex effects on natural systems has become one of the main concerns. In this study, we generated the land use conversion matrixes using ArcGIS and selected several landscape indexes (contagion index, CONTAG, Shannon's diversity index, SHDI, etc.) to evaluate the impact of land use/cover changes on hydrological process in the upper reaches of Minjiang River. We also used a statistical regression model which was established based on hydrology and precipitation data during the period of 1959–2008 to simulate the impacts of different land use conditions on rainfall and runoff in different periods. Our results showed that the simulated annual mean flow from 1985 to 1995 and 1995 to 2008 are 9.19 and 1.04 $m^3 s^{-1}$ lower than the measured values, respectively, which implied that the ecological protection measures should be strengthened in the study area. Our study could provide a scientific basis for water resource management and proper land use planning of upper reaches of Minjiang River.

1 Introduction

With the continuous development of human society, the land use pattern has significantly changed from the basin scale to the global scale, and influenced the biological energy and water cycle processes on earth's surface (Kalnay and Cai, 2003; Lambin et al., 2002). This has been one of most popular scientific issues in this century, which focused on the potential impacts of land use changes on regional resources, environment and sustainable development (Potter, 1991). At the basin scale, the impacts of climate and land use changes on hydrological process would directly induce the changes in water resource supply and demand, and further significantly impact the local ecology, environment and economic development (Wang et al., 2005). Therefore, it is important to have a better understanding about the land use changes and their influence on regional water cycle. However, the impacts of human activities in different time periods and the different geographical features reflected the large inconsistency in response of land use pattern to hydrological process. Thus, a large number of case studies in typical areas are needed to explore the general application theory (Huang and Zhang, 2004).

2 The Study Area

As one of main tributaries at upper reach of the Yangtze River, Minjiang River originates from the border of Sichuan and Gansu Provinces on northern foot of Minshan Mountain (Fig. 1). Within the 31–33° N and 102–104° E, its upper reaches are located in the transition zone between hilly Sichuan Basin and Northwest Sichuan Plateau and under the administrative jurisdiction of Wenchuan, Maoxian, Heishui and Songpan Counties in Sichuan Province. The basin area is 22 950 km^2, with the total river length being 340 km and average slope being 8.85 ‰. The mean annual discharge and the annual mean flow are 462 $m^3 s^{-1}$ and 14.57 billion m^3, respectively. As upper reaches of Minjiang River are of complicated geological structure, the overall terrain descends from the northwest to the southeast. There are great differ-

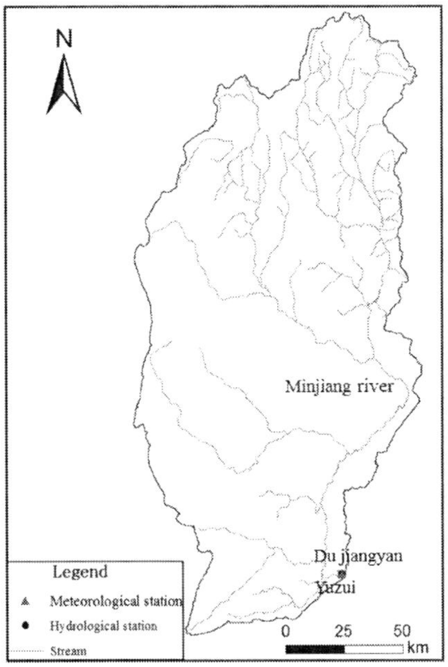

Figure 1. Location of study area and hydrometeorological stations.

ences between proportions of high and low mountains in this area, with average height difference being over 1000 m. For a long time, this area is relatively economically underdeveloped due to great differences in production and life styles of multiple peoples, which also results in diversity of agricultural land use. These provide goods disturbance gradient conditions for study on this land use impacts.

3 Data

3.1 Land use data

Satellite-based image data in 1985, 1995, 2005 and 2014 are used in this study, with the support of EDARS IMAGINE and ArcGIS software. In this study, land use is divided into 6 types, i.e. arable land, forest land, grass land, water area, land for urban and rural residents and unused land, and subdivided into 20 types, i.e. paddy field, dry land, forest land, shrub, open forest land, other forest land, high coverage grass land, moderate coverage grass land, low coverage grass land, canal, lake, permanent glaciers and snowfield, urban land, rural residential land and swamp.

3.2 Water and weather data

The monthly precipitation used in this study was observed by the China Meteorological Administration at Dujiangyan station during 1959–2010, while the monthly runoff data of Yuzui hydrological station in Dujiangyan from 1959 to 2008 were derived from the Survey Bureau of The Hydrology and Water Resources.

3.3 Analysis on land use pattern changes

The land use changes in the last two decades was analyzed based on the image data in 1985, 1995, 2005 and 2014. Firstly, The conversion matrix of land use change that established by superimposing the land use image during two different periods was employed to analyze the land use changes (Liu and Zhu, 2010). Then, 7 indexes were selected based on the theory of landscape ecology (Ravan and Roy, 1997; Liu et al., 2005; Xiao and Bu, 1997; Bu et al., 2005) to analyze spatial pattern of land use for upper reaches of Minjiang River, including Patch Density (PD), Edge density (ED), Landscape Shape Index (LSI), Mean Fractal Dimension (FRAC_MN), Contagion Index (CONTAG), Landscape Diversity Index (DIVISION) and Shannon's Diversity Index (SHDI). The detailed physical meaning and calculation of each index could be seen in Bu et al. (2005) and Liu et al. (2005).

3.4 Response of land use changes to hydrological process

There are many studies (Zhou et al., 2015; Bewket and Sterk, 2005; Hao and Su, 2015) about the impacts of land use pattern changes on hydrological process, which are mostly statistical analysis for hydrological data and distributed hydrological model simulation analysis based on soil physical and hydrological data. However, it is difficult to collect the data (such as soil physical data) that are necessary for hydrological model establishment. Meanwhile, it is hard to identify the relevant parameters affecting rainfall and runoff process due to different types of vegetation cover (Napoli et al., 2017; Bütof et al., 2015). So while the model analysis with higher simulation precision and physical mechanism, but it is difficult to be suitable in practice, especially in Information deficiency areas. In this study, the statistical regression model is established based on rainfall and runoff data of Dujiangyan Station during 1959–2008 to simulate the impacts different land use conditions on runoff under same precipitation condition.

4 Results

4.1 Characteristics of dynamic changes of land use

The result from land use data in 1985 and 2014 showed that during the period of 1985–2014, the land was mainly used as grass land and forest land (accounting for over 96 % of total catchment area) at upper reaches of Minjiang River, and then followed by arable land, and others were water area, land for urban and rural residents and unused land (Table 1). In recent 20 years, the grass land area was decreased sharply with the average speed of 351.29 km^2 during 1985–2014, while the others land areas were increased. The areas of water area and arable land accounted for 2.56 and 3.38 % of total catchment area in 1985 and 2014, respectively. The areas of arable land

Table 1. Land use conversion matrix from 1985 to 2014.

		2014						Total (%)	Decrement (%)
		Arable land	forest land	Grass land	Water area	Land for urban & rural residents	Used land		
1985	Arable land	56.80	20.83	18.08	2.79	1.41	0.08	2.39	1.03
	Forest land	1.50	82.48	15.23	0.49	0.15	0.15	46.88	8.21
	Grass land	1.45	16.34	81.74	0.29	0.11	0.07	50.47	9.21
	Water area	0.45	3.57	15.07	80.82	0.01	0.08	0.17	0.03
	Land for urban & rural residents	16.82	13.43	10.00	16.22	43.53	0.00	0.04	0.02
	Used land	0.00	22.34	31.50	0.84	0.00	45.32	0.04	0.02
Total (%)		2.80	47.43	48.87	0.58	0.18	0.13	100	
Decrement (%)		1.44	8.77	7.62	0.45	0.16	0.11	1.44	
Added area (km^2)		89.92	121.21	−351.29	91.17	30.21	18.79	89.92	

and water area were increased by 89.91 and 91.16 km^2, respectively, and their total increment exceeded the increment of forest land in this period. The increased arable land was mainly forest land and grass land, with large ratio for conversion between forest land and grass land. With social development of society, land for urban and rural residents was increased gradually.

The Table 2 showed the conversion matrixes for typical land use types at upper reaches of Minjiang River at different time periods. During the period of 1985–1995, the converted land area accounted for 19.04 % of the total land use area, mainly involving conversion between high and moderate coverage grass land. Some shrub areas were converted to moderate coverage grass land, and the moderate coverage grass land was mainly converted to high coverage grass land, which accounted for 5.46 % of the total area. During the period of 1995–2005, the land use conversion area was 35.5 %, still dominated by the conversion between high and moderate coverage grass land. The high coverage grass land was converted to low coverage grass land, with a ratio of 5.09 % and the shrub area was converted to moderate coverage grass land at the same time. However, the land use conversion area was only 1.16 % during the period of 2005–2014, mainly involving conversion from shrub to other land use types, i.e. forest land (0.31 %), low coverage grass land (0.04 %) and dry land (0.04 %).

4.2 Changes in land use pattern at landscape level

The land use pattern at landscape level in the upper reaches of Minjiang River was shown in Table 3. The landscape density (PD, ED) at the whole basin was characterized by decreasing at first and then increasing. The landscape density and the number of patches reached the maximum in 2014. The landscape shape index (LSI) and mean fractal dimension (FRAC_MN) could reflect the type and shape of landscape and the complexity of boundary. The increasing LSI and FRAC_MN during 1985–1995 indicated that during the land development period, there was a lack of effective planning, and the large-scale tree cutting and grazing would induce continuous deterioration of natural vegetation and fragmentation of land type. However, the increasing trends of LSI and FRAC_MN during 1995–2014 might be highly related to local grain for green policy implemented in later 1990s. The CONTAG reflects the cluster degree of different patch type in landscape. A higher value indicates better connectivity of patches dominated by few large patches or of same type in the landscape. With space information included, the CONTAG is the most important index to describe landscape pattern. The high CONTAG in the study indicated the good connection the dominant patch types of forest land and grass land, while the gradually decreasing trend implied the growing interference effect of human activities, and large patches dominated by forest land and grass land were disturbed in social development, causing the connectivity decreased. The DIVISION refers to separation degree of different patch individuals distributed in some landscape. The entire study area had a distinct nature gradient and limited changes in DIVISION. The SHDI can reflect the landscape heterogeneity, and was especially sensitive to unbalanced distribution of various patches types in the landscape, which emphasized the contribution of rare patch types. The SHDI in the study area was relatively large and showed land-use types are increasingly enriched during 1985–1995, but it was followed closely by decreasing.

In conclusion, changes in land use of upper reaches of Minjiang River were mainly characterized by the exchange of land use types between natural forest land and grass land since 1985. The policy of grain for green promoted the increase in arable land area, but these change in arable land were less obvious compared to the changes in forest and grass land. The social development in the early 20th century protected the ecological environment, but also expanded the influence of human activities.

Table 2. Conversion matrixes for typical land use types from 1985 to 2014 (%).

		Dry land	Forest land	Shrub	Open forest land	High coverage grass land	Moderate coverage grass land	Low coverage grass land	Urban land	Rural residential land
1985–1995	Dry land	71.37	3.11	12.34	1.53	2.09	8.56	0.59	0.03	0.14
	Forest land	0.25	88.25	5.24	0.28	1.23	4.58	0.17	0.00	0.00
	Shrub	1.01	3.10	85.93	0.25	1.29	7.98	0.41	0.00	0.01
	Open forest land	1.46	1.75	3.10	85.27	1.88	5.89	0.57	0.00	0.01
	High coverage grass land	1.52	3.41	5.60	0.33	51.09	37.96	0.08	0.00	0.01
	Moderate coverage grass land	0.34	2.23	3.88	0.35	13.10	79.58	0.42	0.00	0.01
	Low coverage grass land	0.71	1.83	4.22	0.41	6.12	0.63	86.05	0.00	0.03
	Urban land	7.57	0.72	6.25	1.49	3.47	0.00	3.87	0.00	0.00
	Rural residential land	11.93	2.50	9.05	0.79	3.24	1.35	8.53	0.98	0.13
1995–2005	Dry land	58.65	4.45	10.76	3.15	7.34	9.65	1.57	0.11	0.70
	Forest land	0.59	74.27	9.94	2.51	5.34	6.34	0.47	0.00	0.01
	Shrub	2.23	10.86	64.29	4.32	8.09	8.09	1.19	0.01	0.04
	Open forest land	2.53	9.89	9.22	61.93	7.96	7.05	0.30	0.03	0.19
	High coverage grass land	1.69	7.32	9.14	3.36	23.64	52.22	1.79	0.00	0.05
	Moderate coverage grass land	1.42	4.81	7.97	2.19	11.40	70.84	0.89	0.01	0.02
	Low coverage grass land	1.10	3.86	14.23	4.54	4.25	19.92	50.11	0.14	1.19
	Urban land	25.55	0.01	2.20	2.11	0.60	4.36	0.00	52.07	0.00
	Rural residential land	22.12	3.66	9.25	0.67	4.03	13.00	3.29	2.18	23.79
2005–2014	Dry land	95.48	0.09	1.06	0.32	1.09	0.19	0.02	0.03	0.44
	Forest land	0.03	99.45	0.08	0.12	0.06	0.05	0.06	0.00	0.01
	Shrub	0.16	1.39	97.83	0.07	0.06	0.15	0.17	0.00	0.03
	Open forest land	0.07	0.15	0.09	98.89	0.03	0.03	0.54	0.00	0.01
	High coverage grass land	0.16	0.15	0.23	0.21	99.07	0.09	0.03	0.00	0.01
	Moderate coverage grass land	0.05	0.06	0.07	0.07	0.07	99.56	0.01	0.00	0.01
	Low coverage grass land	0.49	0.06	0.37	0.54	0.79	0.17	97.49	0.00	0.02
	Urban land	1.98	0.00	0.00	0.00	0.00	0.00	0.00	97.10	0.92
	Rural residential land	10.44	0.25	1.99	0.00	0.00	0.04	0.09	0.00	80.50

Table 3. Landscape indexes of land use from 1985 to 2014.

Year	PD	ED	LSI	FRAC_MN	CONTAG	DIVISION	SHDI
1985	0.1474	14.6405	68.2270	1.0941	68.6634	0.9496	1.6549
1995	0.1432	14.6823	68.4190	1.0948	67.3052	0.9465	1.6821
2005	0.4018	24.5776	93.2695	1.0898	63.2793	0.9470	1.6380
2014	0.4749	25.3151	96.0111	1.0838	62.8380	0.9487	1.6491

4.3 Hydrological response to land use changes

The hydrological and meteorological data in Dujiangyan Station were used in this study to analyze annual trend of precipitation and runoff during 1959–2008. The results were shown in Fig. 2. The precipitation was overall decreased by 1.9 mm per year; meanwhile, the annual mean flow was also showed a decreasing trend, with average decreasing speed of 1.5 $m^3 s^{-1}$ per year, and the annual runoff volume decreased at a rate of $47.5 \times 10^6 m^3$ per year. The changes in precipitation and runoff were nearly synchronous, because precipitation is the main factor to constitute runoff. Therefore, based on the precipitation and runoff data from 1959 to 2008, we established statistical equations (Table 4) for precipitation and runoff during the period 1959–2008, 1985–1995 and 1996–2008 through regression analysis, respectively. These equations represent the relationship between rainfall and runoff under different land use impacts.

The R^2 of regression equation at each period were larger than 0.6, indicating the close relationship between runoff and precipitation. Therefore, the statistical model could relatively accurately describe this relation. The monomial coefficient in Eq. (3) is 0.537, smaller than 0.623 in Eq. (2), indicating the runoff generation capacity in the 1996–2008 was decreased compared with that in the previous 10 years.

Since 1970s, the precipitation and runoff process in the basin were gradually weakened, and these may partially due to the land use changes. The predicted runoff from 1985–1995 was calculated with the statistical model in Eq. (3) to obtain the comparative predicted values under the same pre-

Table 4. Statistical regression analysis and statistical models for runoff and precipitation in different periods.

Period	Correlation regression equation*	R^2
1959–2008	(1) $Q = -1.329E-05P^2 + 0.588P + 12.673$	0.747
1985–1995	(2) $Q = 0.623P - 24.545$	0.798
1996–2008	(3) $Q = 0.537P + 34.646$	0.670

* Q and P in the equations are annual mean flow and precipitation.

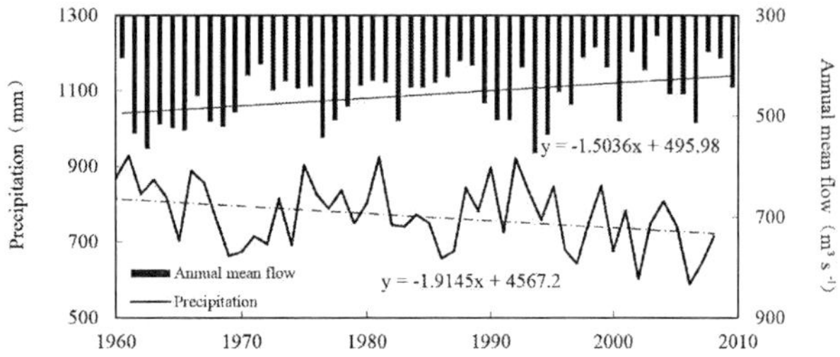

Figure 2. Precipitation and annual mean flow in Dujiangyan from 1959 to 2008.

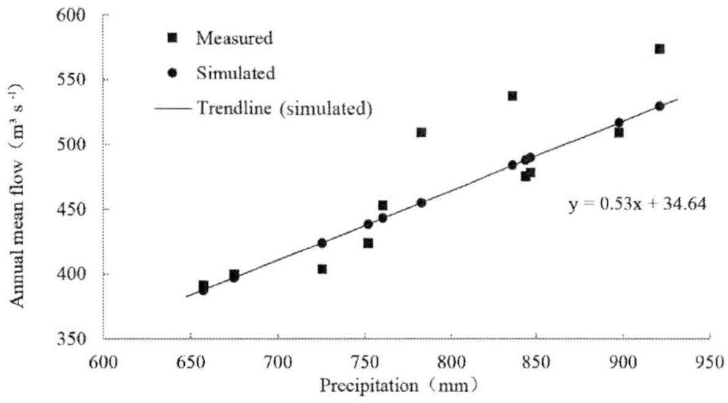

Figure 3. Comparison of simulated and measured annual mean flow from 1985 to 1995.

cipitation and different land use conditions. The comparison of simulated and measured flow during 1985–1995 was shown in Fig. 3. The simulated annual mean flow was highly consisted with the measured flow, but there were still some differences. The runoff generation capacity was relatively low under the same land use conditions, and the simulation decreased by $9.19\,m^3\,s^{-1}$ compared with observation. The land use pattern in the study area was dominated by grass land and forest land. Since the boom of forest resource exploitation appeared in the upper reaches of Minjiang River in 1980s and the "wood finance" appeared in the counties and districts, the forest land types were damaged. Previous research showed that vegetation condition influenced the surface runoff, which is an important part (Xiao et al., 2006) of the eco-hydrological functions, such as water conservation, preserving water and soil, reducing peak-flow and flood, and reducing sediment yield. Forest land has flood intercepting, low water supply, water purifying and other functions (Abari et al., 2017). The destruction of forest land reduced the water and soil reservation capacity, which was the main reason of the higher measured flow.

The comparison between simulated and measured annual mean flow from 1996–2008 was shown in Fig. 4. Similar with the comparison trend in the period of 1985–1995, the simulated flow from statistical model was still lower than the measured value under the same land use conditions of forest land destroyed. The simulated flow was $1.04\,m^3\,s^{-1}$

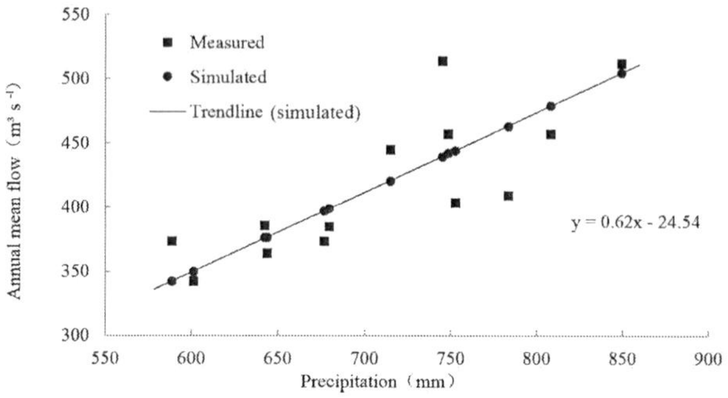

Figure 4. Comparison of simulated *r* and measured annual mean flow from 1996 to 2008.

decreased than the measured value on average. This result was related to the increase in arable area and decrease in grass land area in the study area. The growth rate of population in the upper reaches of Minjiang River kept at about 10 % since 1995, while the agriculture population was only increased by 5000 during 1995–2000. Special geomorphic features, less cultivable land area and unchanged traditional cultivation method in the study area compelled the people to reclaim the hillside land (Zhao et al., 2006; Sichuan Census Office and Sichuan Provisional Bureau of Statistics, 2012), expand animal husbandry and turn forest land into grass land. In such cases, grass land was severely degraded due to influence from animal husbandry. Baneful influence was brought to ecological environment, water and soil loss was intensified, and measured annual mean flow was increased. The restoration of ecological environment needed to be further reinforced.

5 Conclusions

The development in economy and society changed people's way of life and land use, which affected the water cycle process between rainfall and surface runoff in the basin. According to the analysis of land use pattern in last 30 years through hydrological and precipitation data and establishing of statistical regression model, the conclusion is as follows:

1. During the period of 1985–2014, dynamic changes in land use were mainly dominated by the change of forest and grass land and characterized by the exchanges of various types between natural forest and grass land. Arable land was changed remarkably due to corresponding policies. During the period of 1985–1995, the land use pattern and the types of forest and grass land were fragmented. The ecological environment was degraded and arable land area was decreased. During the period of 1995–2014, a large number of shrub wood was changed into grass land and the arable area was increased due to the impact of human activities, resulting in higher heterogeneity of land use pattern.

2. The decrease in runoff at the whole basin was attributed to the decrease in rainfall, which was closely related to forest and grass land patterns, while the complex physical mechanism should be further studied. Statistic models at different times in this study show that: from 1985–1995, simulated flow was $9.19\,m^3\,s^{-1}$ lower than the measured value. As for the land use, the forest and grass land were mainly disturbed and the ecosystem maintenance as well as the water supply function for the forest were seriously damaged. From 1995–2008, the simulated value was $1.04\,m^3\,s^{-1}$ lower than the measured value. Due to the contradiction between the development in human society and the limited resources in agriculture was gradually intensified, the arable land area was increased and the shrub wood as well as grass land were degraded. It is still necessary to increase efforts on protecting the ecological environment and maintaining the sustainable development of ecological functions.

Data availability. Precipitation data is provided by the China Meteorological Administration and can be downloaded from its official website. Hydrological data and land use data are not publicly accessible. Runoff data are provided by the Survey Bureau of The Hydrology and Water Resources, and land use data are provided by China Institute of Water Resources and Hydropower Research.

Competing interests. The authors declare that they have no conflict of interest.

Special issue statement. This article is part of the special issue "Innovative water resources management – understanding and balancing interactions between humankind and nature". It is a result of the 8th International Water Resources Management Conference of ICWRS, Beijing, China, 13–15 June 2018.

Acknowledgements. This work is supported by the National Science Foundation of China (No. 51579161, 51779160) and National Program on Key Basic Research Project (2015CB452701).

Edited by: Zongxue Xu
Reviewed by: two anonymous referees

References

Abari, M. E., Majnounian, B., Malekian, A., and Jourgholami, M.: Effects of forest harvesting on runoff and sediment characteristics in the Hyrcanian forests, northern Iran, Eur. J. Forest Res., 136, 375–386, 2017.

Bewket, W. and Sterk, G.: Dynamics in land cover and its effect on stream flow in the Chemoga watershed, Blue Nile basin, Ethiopia, Hydrol. Process., 19, 445–458, 2005.

Bu, R., Hu, M., Chang, Y., Li, X., and He, H.: Correlation Analysis between Landscape Indexes, Acta Ecologica Sinica, 25, 2764–2775, 2005.

Bütof, A., Von Riedmatten, L. R., Dormann, C. F., Scherer-Lorenzen, M., Welk, E., and Bruelheide, H.: The responses of grassland plants to experimentally simulated climate change depend on land use and region, Glob. Change Biol., 18, 127–137, 2015.

Hao, Z. and Su, Z.: Effects of land use change on runoff in the typical areas in Haihe River basin, Advances in Water Science, 26, 491–499, 2015.

Huang, M. and Zhang, L.: Hydrological responses to conservation practices in a catchment of the Loess Plateau, China, Hydrol. Process., 18, 1885–1898, 2004.

Kalnay, E. and Cai, M.: Impact of urbanization and land use change on climate, Nature, 423, 528–531, 2003.

Lambin, E. F., Baulies, X., Bockstael, N. E., Fischer, G., Krug, T., Leemans, R., Moran, E. F., Rindfuss, R. R., Skole, D. L., and Turner, I. I.: Land use and Land-Cover Change: Implementation Strategy, Shock, 30, 10–29, 2002.

Liu, J., Liu, M., Tian, H., Zhuang, D., Zhang, Z., Zhang, W., Tang, X., and Deng, X.: Spatial and temporal patterns of China's cropland during 1990–2000: An analysis based on Landsat TM data, Remote Sens. Environ., 98, 442–456, 2005.

Liu, R. and Zhu, D.: Methods for Detecting Land Use Changes Based on the Land Use Transition Matrix, Resources Science, 32, 1544–1550, 2010.

Napoli, M., Massetti, L., and Orlandini, S.: Hydrological response to land use and climate changes in a rural hilly basin in Italy, Catena, 157, 1–11, 2017.

Potter, K. W.: Hydrological impacts of changing land management practices in a moderate-sized agricultural catchment, Water Resour. Res., 27, 845–855, 1991.

Ravan, S. A. and Roy, P. S.: Satellite Remote Sensing for Ecological Analysis of Forested Landscape, Plant Ecol., 131, 129–141, 1997.

Sichuan Census Office, and Sichuan Provisional Bureau of Statistics: Tabulation on the 2010 Population census of sichuan province, China Statistics Press, Sichuan, 2012.

Wang, G., Zhang, Y., Liu, G., and Chen, L.: Impact of Land Use Change During 1967–2000 in Ma Ying River Basin on River Runoff, Science China, 35, 671–681, 2005.

Xiao, D. and Bu, C.: Spatial Ecology and Landscape Heterogeneity, Acta Ecologica Sinica, 17, 453–461, 1997.

Xiao, J., Zhang, Z., and Wu, J.: Preliminary Study about Impact of Slope Scale and Forested Land Vegetation on Surface Runoff and Soil Moisture, Research of soil and Water Conservation, 13, 227–231, 2006.

Zhao, Y., He, X., Hu, Y., and Chang, Y.: Land Use/Land Cover Changes and driving forces in the upper reaches of Minjiang River, Chinese Journal of Applied Ecology, 17, 862–866, 2006.

Zhou, J., Lei, L., Shi, P., Wang, L., Wei, W., Liu, H., Amp, G., and University, N. N.: Response of runoff to the climate and land use pattern changes in Shiyang River Basin, Acta Ecologica Sinica, 35, 3788–3796, 2015.

https://doi.org/10.5194/piahs-379-249-2018

Open Access
PIAHS

The use of an integrated variable fuzzy sets in water resources management

Qingtai Qiu[1], Jia Liu[1], Chuanzhe Li[1], Xinzhe Yu[2], and Yang Wang[1]

[1]China Institute of Water Resources and Hydropower Research, State Key Laboratory of Simulation and Regulation of Water Cycle in River Basin, China, Beijing 100038, China

[2]Collage of Environment, Hohai University, No.1 Xikang Road, Nanjing 210098, China

Correspondence: Jia Liu (hettyliu@126.com)

Received: 31 December 2017 – Accepted: 12 February 2018 – Published: 5 June 2018

Abstract. Based on the evaluation of the present situation of water resources and the development of water conservancy projects and social economy, optimal allocation of regional water resources presents an increasing need in the water resources management. Meanwhile it is also the most effective way to promote the harmonic relationship between human and water. In view of the own limitations of the traditional evaluations of which always choose a single index model using in optimal allocation of regional water resources, on the basis of the theory of variable fuzzy sets (VFS) and system dynamics (SD), an integrated variable fuzzy sets model (IVFS) is proposed to address dynamically complex problems in regional water resources management in this paper. The model is applied to evaluate the level of the optimal allocation of regional water resources of Zoucheng in China. Results show that the level of allocation schemes of water resources ranging from 2.5 to 3.5, generally showing a trend of lower level. To achieve optimal regional management of water resources, this model conveys a certain degree of accessing water resources management, which prominently improve the authentic assessment of water resources management by using the eigenvector of level H.

1 Introduction

Regional rational allocation of water resources means to perform rational allocation of water resources available in the region by various means following certain principles and corresponding criteria (Wang et al., 2008). With the improvement achieved in water resources development and utilization, China has performed water allocation based on macroscopic economy, ecological water consumption and reservoir regulation etc. successively, and realized water allocation-based channel water quality distribution; water resource allocation played a vital role in guiding the rational utilization of water resources and the rapid social and economic development in China during the implementation of the Tenth Five-year Plan, the Eleventh Five-year Plan, and the Twelfth Five-year Plan (Wang and You, 2008; Wang et al., 2008). Due to the aged urban water facilities and the rural water consumption that was difficult to make statistics of in the past, industrial water utilization was far from being consistent with the most stringent system of water resources management; with the update and perfection of urban pipe network facilities, the significant improvement of rural water facilities, and the enhancement of water resources development and utilization in the northern part of China in recent years, it's essential to further improve the water resources management measures to achieve the development of industrial structure featuring "industry determination by water"; in view of the current water resources development and utilization, dynamic industrial water allocation should be a major measure for management improvement.

With Zoucheng County as an example, this paper analyzes and predicts the water consumption structure and water consumption level of Zoucheng County in the future level year through system dynamics model (SD), builds and evaluates the water use index system for the planned level year using variable fuzzy set evaluation model (VFS), performs qualitative and quantitative analysis of the water resources consumption and the extent to which water supports social and economic development in Zoucheng County, and proposes the measures for improving water resource management in Zoucheng County based on evaluation result.

2 Theory and methodology

2.1 System dynamics (SD) model

System dynamics (SD) model suits for the management of modern social system since it's based on OP and control theory. SD model divides the research object into a number of subsystems, establishes the cause-and-effect network between subsystems based on the information feedback mechanism in each subsystem, and probes into the relationship between the whole and subsystems, and between subsystems (Rehan et al., 2013). As well , SD model was applied as a decision support tool to evaluate and analyze ingapore's integrated water resources. The results shows it is a useful support tool for deciding sustainable water resources (Xi and Poh, 2013). In nature, SD model is a first order differential equation group that describes the dependence of the variation rate of each state variable in the system on state variables or certain inputs etc., further takes into account some factors contributing to the change of state variables, and resolves the description of variation rate into the description of some flow rates based on the actual conditions of system and the research requirements. Mathematical equation of SD model (Simonovic and Fahmy, 1999):

1. State variable: Input and output variables or the variables accumulated by either thereof.

   $$L_k = L_j + \mathrm{DT}\left(\mathrm{IR}_{jk} - \mathrm{OR}_{jk}\right) \quad (1)$$

   Where, L_k, R_{jk}-Output and input rates (variation rate); DT-Time interval.

2. Rate equation: Input and output parallel variable.

   $$\frac{L_k - L_j}{\mathrm{DT}} = \frac{\mathrm{DL}}{\mathrm{DT}} = \mathrm{IR}_{jk} - \mathrm{OR}_{jk} \quad (2)$$

3. Auxiliary equation:

   In SD model, the algebra of rate equation requires additional algebraic calculation, i.e. auxiliary equation. It can be used to solve the problems of nonlinear, high order and complex time-varying systems. It can accurately understand and show the relationship between internal and external factors.

4. Table function:

   Since simple algebraic combination-based auxiliary variable can't accurately describe nonlinear relationship, graphical representation of non-liner function is needed, i.e. through table function.

System dynamics method features fast analysis and simple model building, as well as the use of non-linear equation, but this in turn represents the weakness of system dynamics: Even a tiny perturbation of nonlinear equation parameters that describe the intrinsic relation of system may bring about the absurdity of long-term analysis result. A novel integrated decision support tool, which synergized System Dynamics (SD) and Analytic Hierarchy Process (AHP) was proposed in Singapore's sources of water supply, shows that the limited capacity and long construction time of underground water have a profound impact on Singapore's water resources (Xi and Poh, 2015). As well, variable set fuzzy evaluation is suitable for the study on complex system since it gives overall evaluation of objectives constrained by various factors using fuzzy mathematics as its theoretical basis, thereby giving a quantitative description of water resources bearing capacity; furthermore, variable set fuzzy evaluation can be used with scenario analysis to provide decision makers with technical support.

2.2 Variable fuzzy sets (VFS)

Defining the concept of VFS, we set that U is a fuzzy concept, A and Ac reflect the attractability and repellency respectively. To any elements u ($u \in U$), μ_A and μ_{Ac} are the relative membership function of element u to A and Ac that express the degrees of attractability and repellency (Chen, 2012, 2013; Wang et al., 2014).

$$\mu_A + \mu_{\mathrm{Ac}} = 1, 0 \leq \mu_A \leq 1, 0 \leq \mu_{\mathrm{Ac}} \leq 1 \quad (3)$$

Where, μA is defined as relative difference degree u to A.

Suppose the regional water resources utilized level has m indices eigenvalue as follows:

$$I_{ab} = \begin{bmatrix} [a_{11}, b_{11}] & [a_{12}, b_{12}] & \dots & [a_{1c}, b_{1c}] \\ [a_{21}, b_{21}] & [a_{22}, b_{22}] & \dots & [a_{2c}, b_{2c}] \\ \vdots & \vdots & \dots & \vdots \\ [a_{m1}, b_{m1}] & [a_{m2}, b_{m2}] & \dots & [a_{mc}, b_{mc}] \end{bmatrix} = ([a_{ih}, b_{ih}]) \quad (4)$$

Where, I_{ab} represents the index function; a means the index; b stands for the index level.

First, the subjective weight of index is determined through analytic hierarchy process; then, entropy method is used to obtain the expert weight entropy model based on the pairwise discrimination matrices established by experts for evaluating the quality of information presented by experts; finally, combination is performed to figure out the index fusion weight.

$$W = \sum_{j=1}^{m} w_j \times S (j = 1, 2, \cdots, m) \quad (5)$$

Where, W means the final fusion weight; w_j represents the index weight proposed by each expert; S stands for expert's weight.

Integrated relative membership degree vector of the scheme for Level h:

$$V_h(u_j) = \frac{1}{1+\left\{\frac{\sum_{i=1}^{m}[w_i(1-\mu_{ih}(u_j))]^p}{\sum_{i=1}^{m}[w_i\mu_{ih}(u_j)]^p}\right\}} \tag{6}$$

Where, $V_h\ (u_j)$ represents the relative membership degree of each index of Scheme j for Level h; w_i means the weight of index i, while $w_1 + w_2 + \ldots w_m = 1$; $\alpha = 1$, and p represents distance parameter: In this paper, Hamming distance $p = 1$ is assumed to achieve a linear model of integrated relative membership degree of the scheme for Level h, namely

$$V_h(u_j) = \sum_{i=1}^{m} w_i\mu_{ih}(u_j) \tag{7}$$

Since the distribution sequence of relative membership degree of u for Level C in the domain of discourse and the Weight h of Variable c are given, their sum is called "the eigenvalue of level variable":

$$H_{(u)} = \sum_{h=1}^{c} \mu_A(u_o)h \tag{8}$$

2.3 Integrated variable fuzzy sets model (IVFS)

System dynamics (SD) model works out the water allocation that reflects the present regional situation and future industrial structure through the establishment of water use index system; the variable fuzzy sets choose the evaluation index system that reflect water consumption level, compute the water resource utilization level eigenvalue, and identify the focus of regional water allocation and management in future based on eigenvalue variation analysis. IVFS is relatively low in parameter accuracy and pays more attention to dynamic behavior and system structure.

3 Case study

3.1 water utilization in the index system

Located in Jining, Shandong Province, Zoucheng County is one of top 100 counties in China. Industry-specific water consumption in Zoucheng County in 2015: The per capita water consumption was 256.93 $m^3\,a^{-1}$, approx. 1/2 of the per capita water consumption in China; the water consumption per RMB 10 000 of industrial GDP amounted to 15.71 m^3, approx. 1/8 of the national average; the urban domestic water consumption was 106 $L\,p^{-1}$ d, approx. 1/2 of the national average; the rural residents' living water consumption was 83 $L\,p^{-1}$ d, which was approximately equal to the national average; the water consumption per Mu of farmland irrigated was 258 m^3, approx. 1/2 of the national average. Taken together, Zoucheng County leads in China in terms of water utilization.

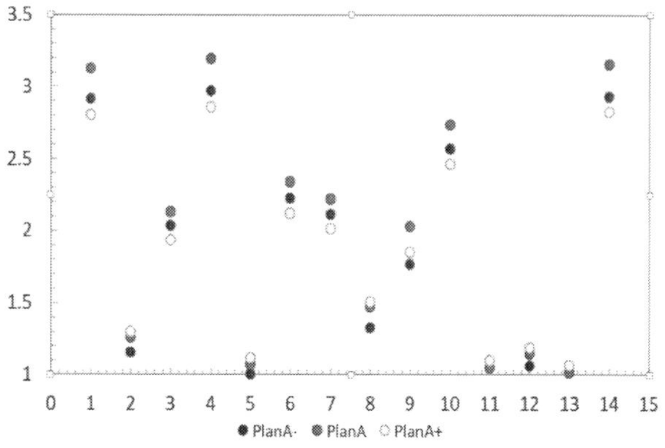

Figure 1. The Water Utilization level of Each town

In order to suit the comparison for the performance of this approach, evaluated index system is chosen to keep in line with the regional actual situation. In this paper, a series of indices are chosen to input SD model in Table 1 which can be divided into two types according to the status of the index: one is a constant and the other change with the function.

After the system dynamics model to calculate the region water resources allocation, according to the actual area analysis, analysis level indicators of water resources, and according to the analysis of regional water level change management. From the output of SD model, this paper Selects the evaluation index: water resources per capita (WRC) (m^3), development and utilization rate of water resources (DURW) (%), regional economic growth rate (REG) (%), water shortage ratio (WRS) (%), per-10000-yuan-GDP water (PG) (m^3), proportion of ecological water (%) (PEW), overlimit rate of water pollution (OWP) (%).

3.2 Water level based on the IVFS

According to the SD model analysis, matrix $\mathbf{I}_{ab}$ can be determined as follows.

$$\mathbf{I}_{ab} = \begin{bmatrix} [550,450] & [450,350] & [350,250] & [250,150] & [150.50] \\ [0.55,0.6] & [0.6,0.65] & [0.65.0.7] & [0.7,0.75] & [0.75,0.8] \\ [11,10] & [10,9] & [9,8] & [7,6] & [5,4] \\ [0,0.03] & [0.03,0.06] & [0.06,0.09] & [0.09,0.12] & [0.12,0.15] \\ [10,12] & [12,14] & [14,16] & [16,18] & [18,20] \\ [0.5,0.48] & [0.48,0.46] & [0.46,0.44] & [0.44.0.42] & [0.42,0.4] \\ [0,0.1] & [0.1,0.2] & [0.2,0.3] & [0.3,0.4] & [0.4,0.5] \end{bmatrix}$$

In order to ensure the consistency of region, this article selects AHP and entropy method to build the index weight (Yu et al., 2009)

In consideration of clear expression, this paper order the town of Zoucheng: Chengqu, Xiangcheng, Chengqian, Dasu, Beishu, Zhongxindian, Tangcun, Taiping, Shiqiang, Zeshan, Kanzhuang, Zhangzhuang,Taihuang and Guoli from 1 to 14.

As Fig. 1 shows the water level of each town in Zoucheng performs a big changes, mainly because the present situation

Table 1. The regional variable of evaluating index system for Zoucheng.

Index	Variable	Index	Variable	Index	Variable
Invest 1990–2030	*C*	WatSupply	*C*	AllocateRatio	*T*
Local invest	*T*	GrowthCurve	*T*	Water Invest1990–2030	*T*
RatioIncre	*T*	InflationRate	*T*	Limitedrecye	*T*
InvToWatIncre	*T*	IrrigatedAreaCash	*T*	RatAgrWatSavInv	*T*

Table 2. The evaluation index system based on VFS.

Index	Change (%)	Max (%)
development and utilization rate of water resources	±1	±4
regional economic growth rate	2.5	10
per-10000-yuan-GDP water	1.5	14
proportion of ecological water	±4	±16

Table 3. The weights of indices by AHP and Entropy Model.

Index	WRC	DURW	REG	WRS	PG	PEW	OWP
Weight	0.091	0.096	0.221	0.134	0.44	0.184	0.14

of towns economy, and the regional water resources utilized level. As well, these towns which get a higher water level shows an order of Plan A, Plan A−, Plan A+. This implies that, although the water level level is higher, but these towns already present a full capacity. These towns that get a low water level, are still has certain development potential.

4 Conclusions

According to the model output result:

a. The water resources consumption of Zoucheng County in current and planned level years presents an increasing trend; the consumption will increase fast between current year and 2020 (short-range planned level year), and increase slowly between 2020 and 2030 (long-term planned level year), which conforms to the overall development trend of Zoucheng County. The variation of growth rate mainly results from the fact that the high water consumption in Zoucheng County narrows its growth potential.

b. In Zoucheng County, the water consumption is high in urban area and its surroundings, and at general level in remote towns, which reflects that fact that the water consumption decreases with the increase of the percentage of agricultural water; in consideration of the overall industrial layout and rural population in Zoucheng County, a threshold for agricultural water shall be defined for water resource management in the future.

It can be concluded that the evaluation of Water resources using settings can play a part in water management. These conclusions can guide decision makers and government in future considerations related to water-lack management in regional development.

Data availability. The main basic data of this study comes from official statistical materials such as "Zoucheng Statistical Bulletin" and "Zoucheng Water Resources Bulletin".

Competing interests. The authors declare that they have no conflict of interest.

Special issue statement. This article is part of the special issue "Innovative water resources management – understanding and balancing interactions between humankind and nature". It is a result of the 8th International Water Resources Management Conference of ICWRS, Beijing, China, 13–15 June 2018.

Acknowledgements. This study was supported by the National Key Research and Development Program (Grant No. 2017YFC1502405, 2016YFA0601503), the Major Science and Technology Program for Water Pollution Control and Treatment (Grant No. 2018ZX07110001), the National Natural Science Foundation of China (Grant No. 51409270), the Hebei Province Water Scientific Research Project (Grant No. 2015-16), and the IWHR Research & Development Support Program (Grant No. WR0145B732017).

Edited by: Dingzhi Peng
Reviewed by: Dehua Zhu and one anonymous referee

References

Chen, S.-Y.: Variable sets and the theorem and method of optimal decision making for water resource system, J. Hydraul. Eng., 43, 1066–1074, 2012 (in Chinese).

Chen, S.-Y.: Variable sets assessment theory and method of water resource system, J. Hydraul. Eng., 44, 134–142, 2013 (in Chinese).

Rehan, R., Knight, M. A., Unger, A. J. A., and Haas, C. Tl.: Development of a system dynamics model for financially sustainable management of municipal watermain networks, J. Water Res.,

47, 7184–205, https://doi.org/10.1016/j.ecoleng.2015.11.008, 2013.

Simonovic, S. P. and Fahmy, H.: A new modeling approach for water resources policy analysi, J. Water Resour. Res., 35, 295–304, https://doi.org/10.1029/1998WR900023, 1999.

Slobodan, P. S. and Hussamm, F.: A new modeling approach for water resources policy analysi, J. Water Resour. Res., 35, 295–304, doi:10.1029/1998WR900023, 1999.

Wang, H. and You, J.-J.: Advancements and development course of research on water resources deployment, J. Hydraul. Eng., 39, 1168–1175, 2008 (in Chinese).

Wang, L., Fang, L., and Hipel, K. W.: Basin-wide cooperative water resources allocation, J. Eur. J. Operat. Res., 190, 798–817, https://doi.org/10.1016/j.ejor.2007.06.045, 2008.

Wang, Y., Dong, S., Wang, D., and Feng, X.: Variable Fuzzy Set Theory to Assess Water Quality of the Meiliang Bay in Taihu Lake Basin, J. Water Resour. Manage., 28, 867–880, https://doi.org/10.1007/s11269-014-0521-6 2014.

Xi, X. and Poh, K. L.: Using System Dynamics for Sustainable Water Resources Management in Singapore, J. Proc. Comput. Sci., 16, 157–166, https://doi.org/10.1016/j.procs.2013.01.017, 2013.

Xi, X. and Poh, K. L.: A Novel Integrated Decision Support Tool for Sustainable Water Resources Management in Singapore: Synergies Between System Dynamics and Analytic Hierarchy Process, J. Water Resour. Manage., 29, 1329–1350, https://doi.org/10.1007/s11269-014-0876-8, 2015.

Yu, J.-X., Jiang, X.-G., and Lian, J.-J.: Comprehensive evaluation model for optimal deployment of water resources based on fuzzy theory and information entropy, J. Hydraul. Eng., 40, 729–735, 2009 (in Chinese).

https://doi.org/10.5194/piahs-379-255-2018

Long-term changes in river system hydrology in Texas

Yiwen Zhang[1] **and Ralph Wurbs**[2]

[1]Water Management and Hydrologic Science Program, Texas A&M University, College Station, Texas, 77840, USA

[2]Zachry Department of Civil Engineering, Texas A&M University, College Station, Texas, 77840, USA

Correspondence: Yiwen Zhang (zhang1990@tamu.edu) and Ralph Wurbs (r-wurbs@tamu.edu)

Received: 15 December 2017 – Accepted: 16 January 2018 – Published: 5 June 2018

Abstract. Climate change and human actives are recognized as a topical issue that change long-term water budget, flow-frequency, and storage-frequency characteristics of different river systems. Texas is characterized by extreme hydrologic variability both spatially and temporally. Meanwhile, population and economic growth and accompanying water resources development projects have greatly impacted river flows throughout Texas. The relative effects of climate change, water resources development, water use, and other factors on long-term changes in river flow, reservoir storage, evaporation, water use, and other components of the water budgets of different river basins of Texas have been simulated in this research using the monthly version of the Water Rights Analysis Package (WRAP) modelling system with input databases sets from the Texas Commission on Environmental Quality (TCEQ) and Texas Water Development Board (TWDB). The results show that long-term changes are minimal from analysis monthly precipitation depths. Evaporation rates vary greatly seasonally and for much of the state appear to have a gradually upward trend. River/reservoir system water budgets and river flow characteristics have changed significantly during the past 75 years in response to water resources development and use.

1 Introduction

Texas is a large state located in the south-central United States that is representative of both the drier western and wetter eastern regions of the country. Climate, hydrology, geography, economic development, and water management vary dramatically across the 15 major river basins and 8 coastal basins of Texas shown in Fig. 1. The flows in Texas rivers are highly variable with daily, seasonal, and multiple-year fluctuations reflecting the extremes of floods and droughts as well as less severe variations. Mean annual precipitation ranges from 20 cm at El Paso on the Rio Grande in west Texas to 142 cm in the Sabine River Basin on the eastern border. The population of the state increased from 5.8 million people in 1930 to 25.4 million in 2010 and is projected to increase to 29.6 million in 2020, 46.3 million by 2060 and 51 million by 2070 which is nearly twice the population (25.4 million people) of the year 2010 (Texas Water Development Board, 2012).

Hydrology is extremely variable, subject to major floods and severe droughts along with seasonal and continuous fluctuations. The hydrologically most severe drought on record for much of the state began gradually in 1950 and ended in April 1957 with a major widespread flood event. Major droughts during the 1910s and 1930s were also extended by multiple-year dry periods over large areas. Droughts since the 1950s have been much costlier than the earlier droughts due to population and economic growth. More recent dry periods, for more than half of Texas, in 2011 experienced the smallest annual precipitation since the beginning of official precipitation records in 1895. On the other extreme, the year 2015 was one of the wettest on record throughout the year and included multiple major floods.

Many studies have investigated the impacts of land use changes, water resource management, and climate change on stream flow. Zhang and Schilling (2006) analysed the trend of increasing base flow in the Mississippi River attributed to land use changes. Reductions in the flow of the Wei River in China are attributable to a combination of multiple factors

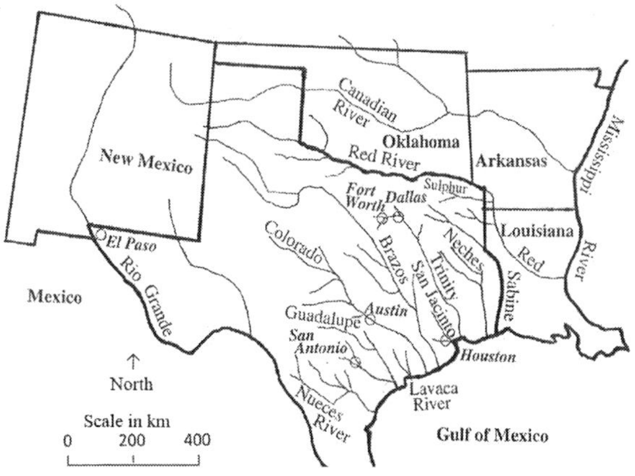

Figure 1. Major rivers and largest cities in Texas.

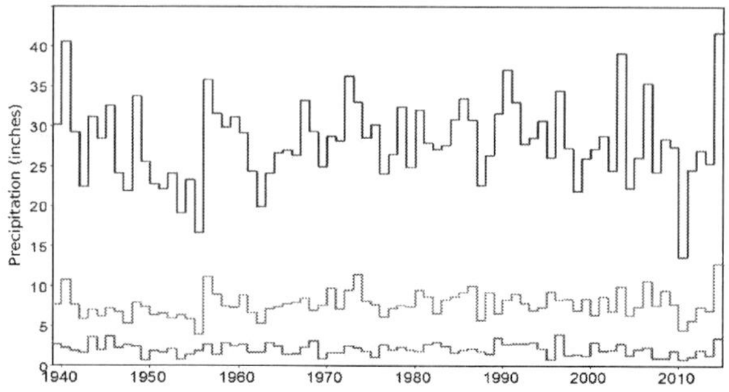

Figure 2. Statewide annual mean (top plot), two-month maximum (middle plot), and two-month minimum (bottom plot) precipitation during each year of 1940–2014.

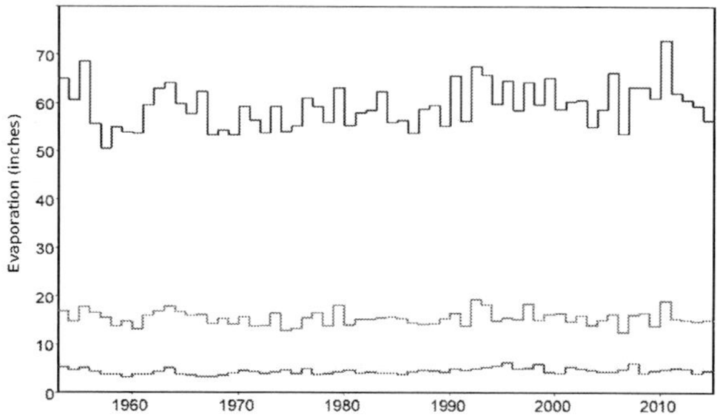

Figure 3. Statewide annual mean (top plot), two-month maximum (middle plot), and two-month minimum (bottom plot) reservoir evaporation rates during each year of 1954–2014.

(Gao et al., 2013). Huang et al. (2015) and Gao et al. (2012) analysed flow regime changes resulting from constructing large dams on the Yangtze and Qingyi Rivers in China. This paper explores river flows and related quantities with a focus on the response mechanisms to climate change, water resource development, and other factors on long-term changes in hydrologic variables for the river systems of Texas (Wurbs and Zhang, 2014).

2 Observed precipitation and reservoir evaporation rates in Texas

Databases maintained by the Texas Water Development Board (TWDB) contain the data of 1940–2014 monthly precipitation and 1954–2014 monthly reservoir evaporation rates that were used in studies at Texas A&M University (TAMU) to explore variability and past long-term trends (Wurbs and Zhang, 2014). The investigation included plots and statistical analyses of the precipitation and evaporation sequences for each of the 92 quadrangles. Plots of annual depth versus two-month maximum and two-month minimum depths precipitation for 1940–2014 and reservoir evaporation rates for 1954–2014 are presented in Figs. 2 and 3. The area-weighted 1940–2014 mean annual precipitation for the state is 709 mm, with individual quads varying from 295 mm in the extreme west increasing to 1430 mm in the southeast. The 1954–2014 state-wide mean annual reservoir evaporation rate is 1510 mm ranging from 1100 mm in east Texas to 1800 mm in the west. Tables 1 and 2 provide a summary of slopes from 1940–2014 annual precipitation, minimum two-month precipitation during each year, maximum two-month precipitation during each year. likewise, 1954–2014 trend analyses were performed for reservoir evaporation annual totals and annual two-month minima and maxima. As indicated in Table 1, a linear least squares regression line through the 75 years of annual state-wide precipitation depths has a slope of $-0.0766\,\mathrm{mm\,year^{-1}}$ ($-0.0108\,\%$ of 709 mm). The trend slopes for the total annual precipitation for the 92 individual quads are negative for 38 of the quads and positive for the other 54 and average -0.363 percent of their respective means (Zhang, 2014). The same analyses have been applied to monthly reservoir evaporation rates. The trend slope for state-wide evaporation implies an increase averaging $1.97\,\mathrm{mm\,year^{-1}}$ (Zhang, 2014). However, long-term changes possibly implied by these metrics are very small relative to the great continuous variability.

3 Observed river flows

Wurbs and Zhang (2014) analysed observed flows at many selected U.S. Geological Survey (USGS) gauging stations on major rivers with records of at least 70 years. All sites exhibit drastic fluctuations including severe droughts and floods as well as continuous variability. The data were downloaded from the USGS NWIS website using HEC-DSSVue. Observed mean daily flows in cubic feet per second (cfs) at a gauging station on the Brazos River near Waco are plotted in Fig. 4 while annual volumes and annual minimum monthly flow are plotted in Fig. 5. The dataset con-

Table 1. Statistics for mean statewide precipitation and evaporation.

	Precipitation			Reservoir Evaporation		
Quantity	Mean (mm)	Slope (mm year^{-1})	Slope (% mean)	Mean (mm)	Slope (mm year^{-1})	Slope % mean
12-month total	709	−0.0766	−0.0108	1510	1.97	0.130
2-month minimum	53.6	−0.142	−0.265	114	0.777	0.417
2-month maximum	195	0.235	0.120	393	0.0440	0.0112

Table 2. Statistics for mean statewide precipitation and evaporation.

	Precipitation			Reservoir Evaporation		
	Average	Number	Number	Average	Number	Number
Quantity	92 Quads	Negative	Positive	92 Quads	Negative	Positive
12-month total	−0.363	38	54	0.138	24	58
2-month minimum	−0.413	69	23	0.412	7	85
2-month maximum	0.0378	29	63	0.0262	39	53

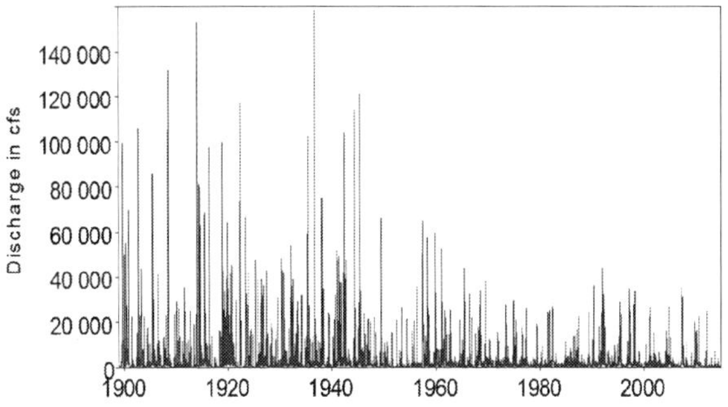

Figure 4. Daily Gauged Flows of Brazos River at Waco during 1 October 1898–4 July 2016.

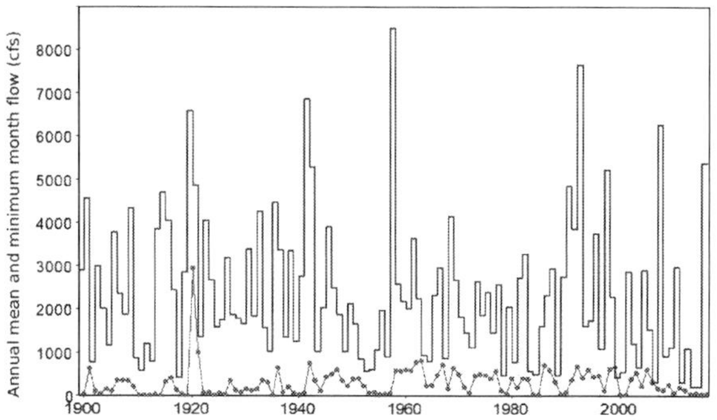

Figure 5. 1899–2015 Annual Mean Flow (blue solid line) and Annual Minimum Monthly Flow (red dots) of the Brazos River at Richmond.

sists of the 117 years from January 1899 through December 2015. The 42 782 flows from the dataset range from 0 to 158 000 cfs and have a mean and standard deviation (SD) of 2337 and 5835 cfs, respectively. The hydrographs at the Waco gauge on the Brazos River have presented the possibility that long-term changes in flow characteristics may differ significantly between daily and annual means. However, long-term changes in observed monthly and annual flows at most gauge sites appear to be relatively minimal.

4 IHA Analysis of Observed Daily Flows

Comparative analyses of statistical metrics of observed gauged flows during historical periods prior to significant water resource development versus during more recent periods reflecting development are performed for daily flows at selected gauge sites. Indicators of Hydrologic Alteration (IAH) software, developed by The Nature Conservancy (TNC), calculate the characteristics of natural and altered hydrologic regimes by separating one period-of-record into two-time periods usually known as Pre-Impact and Post-Impact periods. All of the observed daily flows exhibit the great variability that characterizes stream flow throughout Texas. Long-term trends in river flow characteristics especially in high flows and low flows are very dependent on both location and the period-of-analysis adopted. Over the past several decades, long-term trends of decreases in high flows and increases of low flows have been evident at some gauges, due to construction and operation of reservoir projects. At a number of other sites both high flows and low flows have increased due to discharged return flows and increased rainfall runoff due to urbanization. Meanwhile, some sites exhibit no evident long-term changes at all.

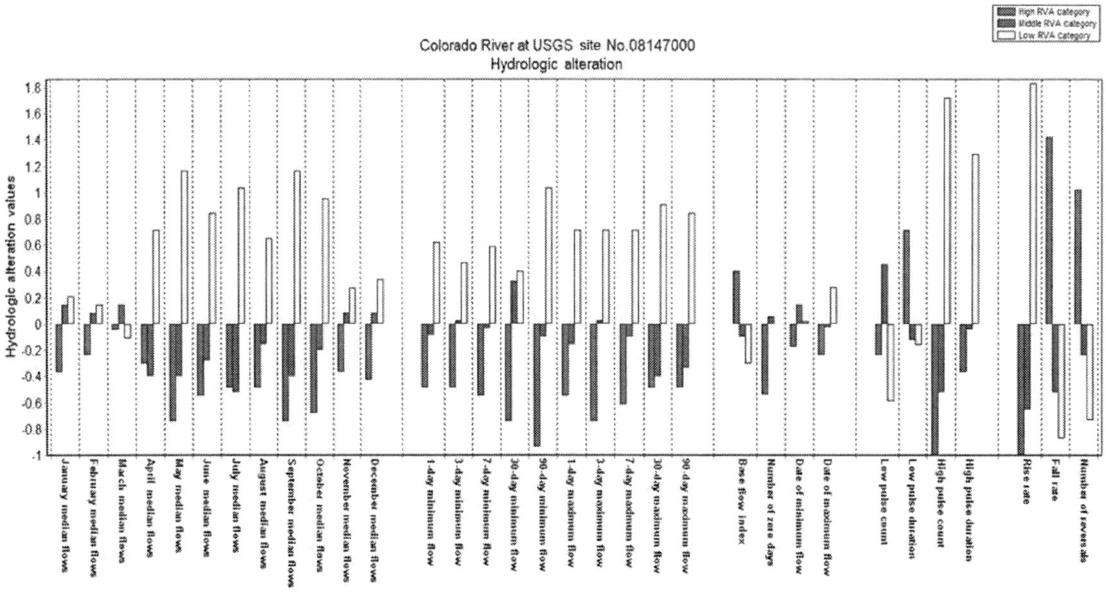

Figure 6. Hydrologic Alteration factors for Colorado River near San Saba.

Table 3. River basin characteristics.

WAM Basin	Watershed Area Total	Watershed Area Texas	Annual Precip	Naturalized Flow Outflow	Naturalized Flow Outflow	Reservoirs Number	Reservoirs Storage Capacity	Reservoirs Storage Capacity	Contents
	(10^6 km^2)	(10^6 km^2)	(mm)	(10^6 m^3)	(mm)		(10^6 m^3)	(% Flow)	(% cap)
Brazos	121 760	114 750	748	8942	77.9	716	4956	55.4	83.0
Canadian	123 560	33 320	495	268	8.1	47	1086	405	69.4
Colorado	107 430	106 910	622	3849	3.60	489	5738	149	74.3
Cypress	9200	7590	120	2068	272	91	1083	52.4	85.9
GSA	26 240	26 244	820	2740	104	241	934	34.1	79.8
Lavaca	5980	5980	1009	1062	177	21	207	19.5	92.6
Neches	25 740	25 740	1236	7680	298	203	4512	58.8	98.2
Nueces	43 250	43 250	630	800	18.5	125	1184	148	53.0
Red	242 040	62 930	649	12 455	198	237	4542	36.5	89.1
Rio Grande	471 940	127 910	408	1357	10.6	80	7359	542	37.0
Sabine	25 270	19 610	1213	8185	417	213	7728	94.4	97.6
San Jacinto	10 194	10 190	1185	2801	324	114	725	25.9	91.2
Sulphur	9760	9270	1184	3197	345	57	887	27.7	86.9
Trinity	46 400	46 400	1000	8182	177	686	9078	111	79.1
Col-Lavaca	2430	2430	1017	371	13.7	8	9	2.43	82.6
Lavaca-Guad	2590	2590	1006	489	201	0	0	0.0	0.0
Neches-Trin	1990	1990	1259	502	194	31	72	14.3	34.2
Nueces-RG	27 050	27 050	642	1423	714	65	140	9.84	34.5
SA-Nueces	6870	6870	890	697	101	9	2	0.29	76.8
Trinity-San J	640	640	1222	223	349	13	6	2.69	65.5
Entire State		681 660	709	67 290	98.7	3446	50 250	74.7	79.8

Table 4. Frequency metrics for naturalized flows at basin outlets in $m^3 s^{-1}$.

WAM Basin	Mean	Standard Deviation	Exceedance Frequency 90 %	75 %	50 %	25 %	10 %
Major River Basins							
Brazos	235	305	25.2	53.1	123	284	605
Canadian	8.50	18.5	0.31	0.75	2.44	8.27	20.0
Colorado	108	131	21.3	34.2	61.0	135	243
Cypress	65.5	83.9	0.71	6.72	30.4	95.3	182
GSA	85.5	107	10.9	26.6	49.1	106	201
Lavaca	33.6	58.0	1.31	3.84	10.4	35.2	97.8
Neches	242	274	20.3	46.7	133	354	621
Nueces	25.3	59.3	0.68	1.92	5.82	22.4	66.6
Red	395	427	59.4	112	247	544	880
Rio Grande	43.0	39.6	18.8	23.9	31.9	48.6	74.5
Sabine	257	264	27.0	60.1	161	386	625
San Jacinto	88.7	119	6.84	15.2	40.8	117	246
Sulphur	101	139	0.96	6.10	43.0	144	285
Trinity	236	294	12.1	43.2	119	323	625
Coastal Basins							
Colorado-Lav	15.3	25.2	0.4	1.5	5.9	17.1	44.0
Lavaca-Guad	15.9	33.0	0.19	0.75	3.49	14.5	46.8
Neches-Trinity	45.1	55.2	2.14	9.51	26.9	61.8	103
Nueces-Rio G	11.7	38.6	0.00	0.00	0.00	3.7	31.0
San A-Nueces	22.1	72.9	0.26	0.56	1.79	11.4	48.7
Trinity-San Jac	7.07	11.0	0.39	0.90	2.65	8.64	19.4

Hydrologic Alteration (HA) factors are used in the RVA analyses in the IHA program. If the HA factor is positive, the frequency of values in the category (lowest third, middle third, highest third) has increased in the post-impact period; in contrast, if the HA factor is negative, it means that the frequency of values in the category has decreased in the post-impact period (Conservancy, 2009). Hydrologic Alteration factors at a gauging station on the Colorado River near San Saba are plotted in Fig. 6. The gauge located on the upper Colorado River near San Saba (5.2 mi downstream from San Saba River, 9.2 mi east of San Saba) has a daily recorded flow for the period from 1915 to present. The J. B. Thomas, E. V. Spence, and O. H. Ivie Reservoir are the three water suppliers in this upper Colorado River basin. The unimpacted period for this gauge is from 1915 to 1969, and the impacted period is from 1969 to 2017. The graph at this site illustrates the extreme variability characteristic of river flows throughout Texas on different periods. Figure 6 shows that the changes of runoff in the Colorado River near San Saba varies widely from month to month. The greatest Hydrologic Alteration (HA) factors were the increase in magnitude of baseflows, low-flow events, and monthly flows from April to October. Monthly flows from January to March are essentially unchanged for the impacted period. Although the magnitude and duration are similar, both 1-day, 7-day, 30-day, and 90-day minimum and maximum flow events tend to show a gradual decline of post-impact period.

5 Hydrology and water management in Texas

The TCEQ WAMs combine historical natural river basin hydrology with specified scenarios of water resources development, allocation, management, and use. Hydrologic period-of-analysis sequences of naturalized, regulated, and unappropriated flows are generated (Wurbs, 2005, 2013, 2014). The approximately 200 major reservoirs in Texas with water supply storage capacities of at least 6.17 million m^3 (5000 acre-feet) are all impounded by constructed dams. The WAMs datasets are available for simulation of the 3446 permitted reservoirs and other constructed water control facilities. The WAM System was applied to explore changes in river flows and the water budgets of the river basins. Simulation results include naturalized river flows that represent natural hydrology without human water development and use. Sequences of regulated flows in WAM are computed by altering homogenous naturalized flows to reflect a specified scenario of water resource development and use. Results reported by Wurbs and Zhang (2014) are summarized as the following: river basin characteristics are in Table 3 while frequency metrics for monthly naturalized flows are tabulated in 4 and 5.

Table 5. Frequency metrics for regulated flows at basin outlets as a percent of naturalized flows.

Basin		Standard	Exceedance Frequency				
WAM	Mean	Deviation	90 %	75 %	50 %	25 %	10 %
Major River Basins							
Brazos	79.3 %	96.8 %	1.1 %	32.1 %	54.2 %	80.7 %	89.3 %
Canadian	59.0 %	75.0 %	47.4 %	46.1 %	55.6 %	48.9 %	50.1 %
Colorado	54.9 %	84.2 %	0.08 %	14.2 %	26.0 %	51.8 %	63.9 %
Cypress	87.9 %	97.6 %	7.0 %	1.0 %	64.1 %	91.1 %	96.3 %
GSA	92.9 %	98.3 %	72.1 %	78.6 %	85.9 %	93.3 %	96.4 %
Lavaca	93.7 %	97.9 %	99.5 %	90.9 %	81.5 %	92.0 %	97.0 %
Neches	89.4 %	101.2 %	4.0 %	31.4 %	76.9 %	92.8 %	96.5 %
Nueces	68.0 %	71.3 %	658.8 %	255.4 %	106.7 %	51.4 %	48.7 %
Red	90.3 %	97.3 %	70.2 %	72.0 %	85.1 %	94.5 %	93.4 %
Rio Grande	6.84 %	42.58 %	0.33 %	0.66 %	1.02 %	1.41 %	3.02 %
Sabine	93.3 %	101.5 %	53.5 %	61.4 %	84.8 %	95.8 %	99.5 %
San Jacinto	106.4 %	96.1 %	332.2 %	183.7 %	115.0 %	95.8 %	98.7 %
Sulphur	96.9 %	95.2 %	737.0 %	161.4 %	95.8 %	92.8 %	96.2 %
Trinity	82.0 %	86.7 %	247 %	91.6 %	67.7 %	76.2 %	81.4 %
Coastal Basins							
Colorado-Lav	97.1 %	98.9 %	168.2 %	107.3 %	82.7 %	94.0 %	98.5 %
Lavaca-Guad	102.6 %	100.0 %	302.5 %	159.9 %	112.9 %	102.3 %	100.8 %
Neches-Trinity	91.4 %	96.3 %	112.6 %	79.2 %	86.4 %	89.2 %	93.2 %
Nueces-Rio G	105.9 %	99.1 %	0.83 $m^3\,s^{-1}$	0.95 $m^3\,s^{-1}$	1.03 $m^3\,s^{-1}$	116.4 %	99.2 %
San A-Nueces	100.0 %	100.0 %	101.6 %	100.2 %	100.2 %	99.3 %	100.0 %
Trinity-San Jac	105.1 %	99.5 %	234.1 %	148.7 %	110.6 %	101.0 %	102.2 %

The state-wide river basin characteristics information presented in Table 3 was developed by compiling information from the individual river basins derived from WAM simulation results. The long-term mean naturalized flow of 67.3 billion m^3 represents the total of all stream flows under undeveloped conditions discharging into the Gulf of Mexico or across state borders under natural conditions. The estimated mean naturalized flow of 67.3 billion m^3 is an annual volume equivalent to covering the state to a depth of 98.7 mm or 12.4 % of the mean annual precipitation of 709 mm $year^{-1}$ falling on the state.

The frequency metrics in units of $m^3\,s^{-1}$ are tabulated in Tables 4 and 5 and provide a comparison of river flows for current actual conditions of development versus natural undeveloped conditions. The impacts of development on river flows vary greatly between the river basins and also between different sites in the same river basin. The Brazos River Basin is used as an example for interpreting Tables 4 and 5. The mean and median (50 % exceedance) naturalized flows at the Brazos River outlet are 235 and 123 $m^3\,s^{-1}$ (Table 4). The corresponding mean and median regulated flows are 210 and 8.7 $m^3\,s^{-1}$, which are 79.3 and 54.2 % (Table 5) of the naturalized flows.

6 Summary and conclusions

River basin hydrology in Texas, like elsewhere, is characterized by tremendous variability that includes continual short-term fluctuations, seasonality, and the extremes of severe multiple-year droughts and intense floods. Water resource development and management are governed largely by needs for dealing with the extremes of floods and droughts. Large volumes of reservoir storage are essential. Numerous reservoirs have been constructed in Texas, most since the 1940s, and are in operation today. Global warming has impacts on hydrology and population growth causing increased water use to be an important consideration in river system management over the past several decades.

The databases maintained by water agencies in Texas provide excellent opportunities to explore and facilitate an improved understanding of flow characteristics and long-term changes in flow characteristics. Long-term changes in precipitation during 1940–2014 are not evident, while reservoir evaporation rates in areas of Texas appear to have gradually increased over the past several decades. The IHA software program was to process daily hydrologic records to enable characterization of natural water conditions and facilitate evaluations of human-induced changes to flow regimes. Changes in flow characteristics exhibit dramatic fluctuations on all sites. Long-term trends of decreases in flows are evident at some gauges; increases are evident at others. Some exhibit both increases and decreases, and some sites exhibit no evident long-term changes. Analysis results are sensitive to the length of the period-of-analysis. Long-term changes in daily flows may differ greatly from changes in monthly and annual flows at the same site. Changes in low flow characteristics are very different than changes in flood flows. Compared with Climate Change, the impact contributed by human activity is relatively a greater influence on streamflow.

The WAM system provided water supply reliability and flow frequency metrics that are fundamental to water resources planning and administration of water rights permit system. The naturalized flow results from WAM exhibit essentially no long-term trends. Long-term changes in river system water budgets and flow characteristics are quantified by comparing WAM regulated versus naturalized flows. The comparative evaluations of simulation flow can be used to support effective water management between diverse types of human water users and various aspects of environmental requirements.

Data availability. The monthly WRAP input datasets for all the river basins of the Texas are available at the TCEQ WAM website: https://www.tceq.texas.gov/permitting/water_rights/wr_technical-resources/wam.html (Texas Commission on Environmental Quality, 2015) observed stream flows from the USGS National Water Information System (http://waterdata.usgs.gov/tx/nwis/nwis, USGS National Water Information System, 2015).

Competing interests. The authors declare that they have no conflict of interest.

Special issue statement. This article is part of the special issue "Innovative water resources management – understanding and balancing interactions between humankind and nature". It is a result of the 8th International Water Resources Management Conference of ICWRS, Beijing, China, 13–15 June 2018.

Edited by: Dingzhi Peng
Reviewed by: two anonymous referees

References

Conservancy, N.: IHA Software Version 7.1 User's Manual, The Nature Conservancy, http://www.nature.org, 2009.

Gao, B., Yang, D., Zhao, T., and Yang, H.: Changes in the eco-flow metrics of the Upper Yangtze River from 1961 to 2008, J. Hydrology., 448–449, 30–38, 2012.

Gao, P., Geissen, V., Ritsema, C. J., Mu, X.-M., and Wang, F.: Impact of climate change and anthropogenic activities on stream flow and sediment discharge in the Wei River basin, China, J. Hydrol. Earth Syst. Sci., 17, 961–972, 2013.

Huang, F., Chen, Q., Li, F., Zhang, X., Chen, Y., Xia, Z., and Qiu, L.: Reservoir-induced changes in flow fluctuations at monthly and hourly scales: Case study of the Qingyi River, China, J. Hydrol. Eng., 20, https://doi.org/10.1061/(ASCE)HE.1943-5584.0001233, 2015.

Texas Water Development Board: Water for Texas 2012, Austin, TX, 2012.

Texas Commission on Environmental Quality: Water Availability Models, https://www.tceq.texas.gov/permitting/water_rights/wr_technical-resources/wam.html, last access: August 2015.

USGS National Water Information System: USGS Water Data for Texas, http://waterdata.usgs.gov/tx/nwis/nwis, last access: 31 December 2015.

Wurbs, R. A.: Texas water availability modeling system, J. Water Res. Pl., 131, 270–279, 2005.

Wurbs, R. A.: Water Rights Analysis Package (WRAP) Modeling System Reference Manual, Technical Report 255, Texas Water Resources Institute, 10th Edition, College Station, TX, 2013.

Wurbs, R. A.: Sustainable statewide water management in Texas, J. Water Res. Pl., 141, https://doi.org/10.1061/(ASCE)WR.1943-5452.0000499, 2014.

Wurbs, R. A. and Zhang, Y.: River System Hydrology in Texas, Technical Report 461, Texas Water Resources Institute, College Station, TX, 2014.

Zhang, Y.: Long-Term Changes in River System Water Budget in Texas, Master's thesis, Texas A & M University, available at: http://hdl.handle.net/1969.1/153334, 2014.

Zhang, Y. K. and Schilling, K. E.: Increasing streamflow and baseflow in Mississippi River since the 1940s: Effect of land use change, J. Hydrol., 324, 412–422, 2006.

https://doi.org/10.5194/piahs-379-263-2018

Analysis of Spring Flow Change in the Jinan City under Influences of Recent Human Activities

Xiaomeng Liu[1,2,3], Litang Hu[1,2,3], and Kangning Sun[1,2,3]

[1]College of Water Sciences, Beijing Normal University, Beijing 100875, P.R. China
[2]Beijing Key Laboratory of Urban Hydrological Cycle and Sponge City Technology, Beijing 100875, P.R. China
[3]Engineering Research Center of Groundwater Pollution Control and Remediation of Ministry of Education, Beijing Normal University, Beijing 100875, P.R. China

Correspondence: Litang Hu (litanghu@bnu.edu.cn)

Received: 29 December 2017 – Revised: 9 May 2018 – Accepted: 23 May 2018 – Published: 5 June 2018

Abstract. Jinan city, the capital of Shandong Province in China, is famous for its beautiful springs. With the rapid development of the economy in recent years, water demand in Jinan city has been increasing rapidly. The over-exploitation of groundwater has caused a decline in groundwater level and, notably, dried up springs under extreme climate conditions. To keep the springs gushing perennially and sustainably use groundwater resources, the local government has implemented many measures to restore the water table, such as the Sponge City Construction Project in Jinan. Focusing on changes in spring flow and its impact factors in Jinan, this paper analyzes the changes in observed spring flow in the most recent 50 years and then discusses the causes of decreases in the spring flow with the consideration of climate and human activities. Spring flow in the study area was changed from the natural state to a period of multiwater source management. The artificial neural network (ANN) model was developed to demonstrate the relationship among spring flow, precipitation, and groundwater abstraction to predict the variations of spring flow under the conditions of climate change and human activities. The good agreement between the simulated and observed results indicates that both precipitation and exploitation are important influence factors. However the effective infiltration of precipitation into groundwater is the most influential factor. The results can provide guidance for groundwater resource protection in the Jinan spring catchment.

1 Introduction

Covering an area of approximately 7–12 % of the earth's land area, karst regions are an important part of the planet, with karst aquifers supplying more than 25 % of the drinking water for the world's population. Karst water is an indispensable water source for the world (Vries and Simmers, 2002; Bakalowicz, 2005; Hartmann et al., 2014). In northern China, there are abundant karst catchments in Shanxi, Shandong, Henan, and Hebei provinces, where karst aquifers also supply the major water resources for domestic, municipal and industrial uses and agricultural irrigation. The karst systems in these areas are dominated by fissures, fractures and conduits, which provide extensive storage for precipitation and enough water resources for karst springs (Ma et al., 2004; Qian et al., 2006; Hu et al., 2008; Kang et al., 2011).

Jinan is the provincial capital of Shandong province in China and is famous for numerous karst springs, such as Baotu springs and Heihu springs. According to the Jinan Statistical Yearbook, Jinan's gross domestic product (GDP) was approximately CNY 9.83 billion (USD 1.552 billion) in 1988, then reached CNY 95.22 billion (USD 15.04 billion) in 2000, and reached CNY 610.02 billion (USD 96.33 billion) in 2015, thus increasing more than 62 times in 27 years. As a result of the rapid development of the economy, water demand quickly increased, resulting in an increase in groundwater abstraction. The over-exploitation of groundwater in Jinan has caused many problems, such as the decline of the

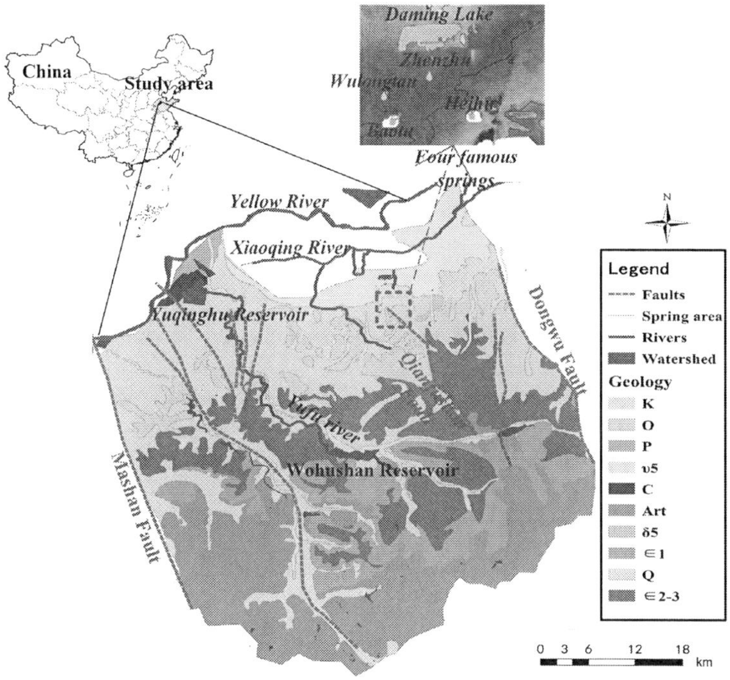

Figure 1. Map of the study area.

groundwater table, the drying up of springs, and the degradation of groundwater quality.

Since 1972, springs in Jinan have suffered from drying up. The longest duration for drying up of the Baotu Spring was approximately 926 days at the beginning of the 21st century. Since 1981, much research has been carried out to study the recharges and discharges of the groundwater system in Jinan. Since 2001, a series of measures have been implemented by the Jinan government to restore the flow rate of springs to sustain the demand for economic growth. For example, many pumping wells in the spring catchment were closed, especially in the urban district; groundwater usage was partially replaced by the Yellow River water and the Yangtze River water, and the infiltration of surface water and rains were added (Xi et al., 1987; Cai et al., 2015). The springs started to regush from 2003 and have continued gushing until now. Since 2015, the Sponge City Construction Project has been another important measure put forward by the local government. However, the threat of springs drying up still exists, and the groundwater level is still in a lower position. How to keep the springs gushing perennially and maintain a virtuous circle of water environment in the spring catchment is still the focus of the society.

Considering the dynamic changes in the spring flow and its influence factors in the Jinan spring catchment, this paper will analyze the changes in the spring flow rate in the most recent 50 years and then discuss the causes of decreases in the spring flow in view of climate change and human activities. An artificial neural network method is employed to represent the relationship among spring flow, precipitation and groundwater use. The results can provide suggestions for the protection of the springs in Jinan city.

2 Study area

2.1 Background

Covering an area of approximately 1500 km^2, the Jinan spring catchment (116°40′–117°14′ E, 36°28′–36°46′ N) is situated in the middle of Jinan, Shandong province, China. The northern boundary of the area is the contact zone of metamorphic and magmatic activity, mainly along the Yellow River and the Xiaoqing River. The southern boundary divides the watershed, and the eastern and western boundaries of the spring catchment are constrained by two water-barrier faults, the Dongwu fault and the Mashan fault, respectively (as shown in Fig. 1). The catchment is a monoclinic structure, ranging in elevation from 700 m above sea level (m a.s.l.) in the south to 26 m a.s.l. in the north. The study area has the continental climate of a warm temperate zone. The annual mean precipitation is approximately 670.47 mm, while the annual mean evaporation is approximately 1475.6 mm. More than 70 % of the precipitation falls from June to September, which leads to serious problems of urban waterlogging in Jinan city during the season. The average annual temperature is 14.2 °C, and the frost-free period is approximately 192–

238 days. The Yellow River, Xiaoqing River, Yufu River and Beidasha River are the major rivers in the study area.

2.2 Geology and hydrogeology

The study area is situated in the transitional zone of the Luzhong Mountains and the piedmont plain. Topographically, the karst aquifer system is complex and diverse and is divided into three areas from south to north: the southern part consists of steep mountains and deep eroded valleys with an absolute elevation of 500–700 m; the central part is a hilly area, with broad valleys and developed gullies with an elevation of 250–500 m; the northern part includes the Yellow River Piedmont and an alluvial plain with an absolute elevation of approximately 20–50 m. The strata exposed in the study area are mostly complete, with Upper Cambrian and Lower-Middle Ordovician limestone as the major aquifers. The basement of the catchment is the Archean Taishan Group, consisting of metamorphic rocks exposed in the south area. The Cambrian strata are well exposed from the south to the north and are characterized by the interbedding of limestone and shale. The Ordovician strata are mostly distributed in the middle of the study area and are composed by thick-bedded limestone. All of the Paleozoic strata are gradually inclined to the north with a dip direction of 20° NE and then buried under the Quaternary. The Quaternary is well exposed in the area from the lower Pleistocene to the Holocene. Among them, some faults are widely developed throughout the study area, such as the Mashan fault, Dongwu fault and Qianfoshan fault. The faults with north-west and north-east directions are dominating features, providing possible formations for spring water. Subject to the geology and topography conditions, the southern area has become the indirect recharge area of the spring catchment, the middle part is the direct recharge area and the runoff area, and the northern part is the drainage area where most springs are located.

Based on the topography, lithology, geology and other characteristics, the aquifers in the study area can be divided into four types: pore water in a Quaternary aquifer; fractured karst water in an Ordovician aquifer; karst fissure water in Cambrian aquifers; and fractures water in an Archaean aquifer (Qian et al., 2006). Among them, pore water and karst water are the major groundwater resources in the study area, accounting for more than 80 % of the total groundwater discharge, with the karst aquifer being the major supplier of water resources for springs. The Archaean aquifer only exists in some special formations and contributes little to the springs. This paper mainly analyses the former three types of groundwater aquifers.

3 Analysis of the spring flow and its influence factors

Groundwater level is the most fundamental indicator of the status of groundwater. In the Jinan spring catchment, the most important influence factors are precipitation and groundwater abstraction. Other aspects, such as artificial recharge, also affect the dynamics of the Jinan spring flow, but with limited effect. The dynamic changes in precipitation, groundwater abstraction and spring flow will be analyzed below.

3.1 Recharge items of groundwater

3.1.1 Changes in the precipitation

The annual precipitation in the spring catchment is shown in Fig. 2. It can be seen that the precipitation from 1956–2013 showed strong variations, i.e., the highest precipitation is 1196.6 mm in 1964, while the lowest is just 340 mm in 1989. There are several significant periods: 1961–1964 and 2003–2005 are the wet years; 1969–1980, 1982–1985, and 2007–2010 are normal years; 1965–1968, 1986–1989 and 1999–2002 are the dry years.

3.1.2 Other recharge sources to the aquifer

In addition to precipitation, the infiltration from the Yufu River and the Beidasha River and agricultural irrigation are also important recharge sources of the karst aquifers. However, after reservoirs were built upstream and in branches of the Yufu River and the Beidasha River, the river channels were dry most of the time. Though the artificial recharge in the strong leakage areas causes a short-term increase in groundwater level, the influence is limited and the effect is fugacious. For the agricultural irrigation, the amount is much less than the artificial recharge and thus has much less effect on the rise of groundwater level and the regushing of spring flow. As a consequence, the major recharge item of the karst aquifer is precipitation and is considered an important factor in this paper.

3.2 Groundwater discharges

It is evident in Fig. 2 that the spring flow and groundwater abstraction correlated negatively. The spring flow showed a decreasing trend before 2003, while the groundwater abstraction increased gradually. The groundwater abstraction in the urban district increased in the earlier years and then decreased rapidly in 1980 because most of the pumping wells were shut down by the local government for the protection of the springs. These wells have maintained a rate of approximately $10 \times 10^4\,m^3\,d^{-1}$ until now. As supplements, the groundwater abstraction in suburban areas increased rapidly to meet the demand of the economic development and the requirements of protecting springs. The total groundwater abstraction in the spring catchment also increased sharply and has maintained a rate of approximately $70 \times 10^4\,m^3\,d^{-1}$ from 2003 until parts of the groundwater supplement were replaced by the Yellow River and the Yangtze River water.

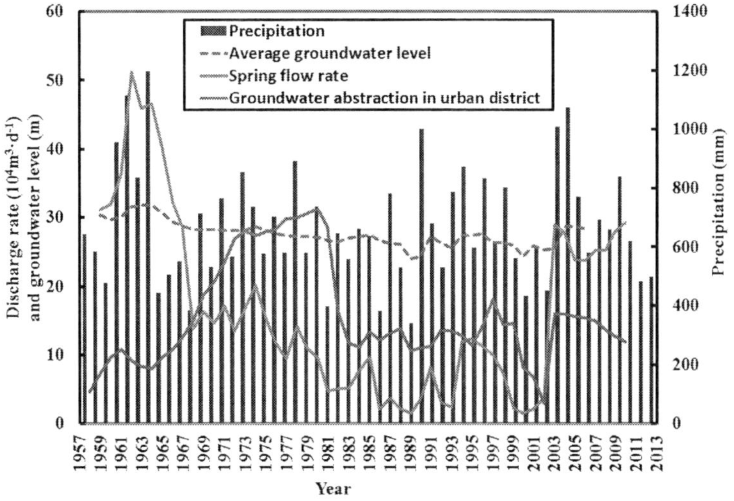

Figure 2. Changes in precipitation, the spring flow rate, groundwater abstraction and groundwater level.

3.3 Relation between groundwater level and spring flow

Only when the groundwater level is above the ground surface will springs have the chance to pour out. In other words, spring discharge is strongly affected by groundwater level. It can be seen from Fig. 2 that groundwater level has a similar trend as spring flow. However, spring flow is more sensitive, as it changes faster than groundwater level. A correlation analysis was performed to describe the relation between them. The result shows that groundwater level and spring flow have a very close relation with the correlation coefficient greater than 0.885.

3.4 Responses of spring flow changes to human activities

The changes in the spring flow rate, average groundwater level, groundwater abstraction in the urban district and precipitation are displayed in Fig. 2. It is obvious that spring flow has a positive relation with precipitation and an inverse relation with groundwater abstraction. To further analyze the characteristics of spring flow and its response to natural and human activities, the Mann-Kenall (M-K) test and Pettitt change point test were applied (Li et al., 2014). The result from the M-K test indicated an increasing trend before 1967 and a decreasing trend after 1968. The change point in the spring flow series was 1980 and was calculated using the Pettitt change point analysis. Based on the statistical analysis, comprehensively considering the human activities and natural conditions, the characteristics of spring discharge were divided into four periods:

1. The natural period (before 1967): There was little groundwater abstraction before 1967. The average spring discharge rate varied from 30×10^4 to $40 \times 10^4 m^3 d^{-1}$. The highest spring flow rate was $50.18 \times 10^4 m^3 d^{-1}$ in 1962. In addition, the spring flow varied similarly to the trend of precipitation. In this period, precipitation was the major influence factor of spring flow, as the coefficient parameter between them was 0.75.
2. The period of exploiting groundwater seriously (1968–1980): since 1965, a large number of pumping wells have been drilled in the spring discharge area. The rate of groundwater abstraction in the urban district increased from 10×10^4 to $30 \times 10^4 m^3 d^{-1}$ in the first five years. The rate of total groundwater abstraction in the Jinan spring catchment reached $60 \times 10^4 m^3 d^{-1}$ in 1972, and then increased to $70 \times 10^4 m^3 d^{-1}$ by 1980. As a result of the over-exploitation of groundwater, the springs stopped gushing during the dry seasons since 1972, and the number of dried-up days increased.
3. The period of controlling groundwater use (1981–2000): in order to protect springs, the Jinan government put forward a series of measures and the most important one was transferring groundwater abstraction in urban areas to suburban areas. The exploitation rate in urban areas (spring discharge areas) decreased suddenly from 30×10^4 to $12 \times 10^4 m^3 d^{-1}$, and the groundwater level was restored gradually. However, the spring flow still decreased gradually, and the number of dried-up days increased, as the spring catchment is a groundwater system.
4. The period of multiwater source use (after 2001): since 2001, water from the Yellow River, the Yangtze River and two reservoirs (i.e., Wohushan Reservoir and Yuqinghu Reservoir) were gradually used jointly in the supply water system, thus reducing much of the abstraction of groundwater. Additionally, the superfluous sur-

face water was artificially recharged into the groundwater system since 2001. These factors have led to the gushing of springs from September 2003 until now, and the spring flow was recovered gradually.

4 Relationship among spring flow and influence factors

The spring flow is strongly affected by natural and human activities, and common models cannot reflect the changes in spring flow very well. To further understand the dynamic changes in spring flow and the relationship among spring flow, precipitation and human activities, an artificial neural network (ANN) model was developed. An ANN model is a nonlinear method inspired by the structure and functions of the human nervous system. ANN is self-adaptive and has high accuracy. ANN has been proven to be an efficient method in powerful computations of complex nonlinear relationships. Back-propagation (BP) neural networks are the most popular ANN models and are widely used (Kurtulus and Razack, 2006; Hu et al., 2008; Li et al., 2012).

The precipitation and exploitation rate of groundwater in the current and previous year were used as the input factors; and the spring flow rate served as the output item. The period from 1960–2007 is taken as the calibration period to train the model, and the period from 2008–2010 is regarded as the verification period to check the model. The result is shown in Fig. 3. The correlation coefficient between the measured and simulated values is 0.99, which indicates that the ANN model can reflect the changes of spring flow and can be used to predict the spring flow in the Jinan spring catchment.

5 Concluding remarks

Karst aquifers play an important role in supplying the drinking water and supporting the economic development of Jinan city, and the spring flow is a significant indicator of the changes in the groundwater system and famous landscape. Consequently, analyzing the changes of spring flow and its response to the influence factors is highly necessary for the managers of water resources. The spring flow in the Jinan spring catchment is strongly influenced by precipitation and human activities, such as groundwater abstraction, artificial recharge and water transfer projects. A partial correlation analysis was performed to quantitatively demonstrate its importance. The results showed that the partial correlation coefficient between spring flow rate and precipitation is 0.666, and the partial correlation coefficient between spring flow rate and the groundwater abstraction in the previous year is −0.259. This means that 92.5 % of the dynamic changes in the spring flow are affected by precipitation and groundwater abstraction. Precipitation plays a major role in affecting the spring flow.

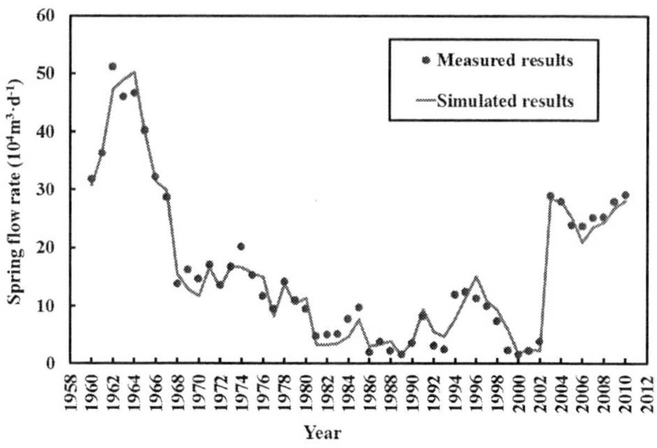

Figure 3. Comparison of measured and simulated spring flow rate.

The characteristics of spring flow were divided into four periods: (1) the natural period; (2) the period of exploiting groundwater seriously; (3) the period of controlling groundwater abstraction; and (4) the period of multiwater source use. After a series of measurements were proposed and implemented by hydrogeologists and the local government in Jinan, the spring flow almost remained in a new dynamic balance. An ANN model was developed to simulate the changes of spring flow, and the good agreement of the results further suggests precipitation and exploitation are the most important influence factors.

Though the groundwater level is recovered to a certain degree and the spring flow has been regushing for more than fourteen years since 2003, the spring flow in Jinan is still threatened by human activities, mainly because the water resources are scarce and an imbalance between water storage and demand exits. Considering the expensive investment for transferring water from an outside catchment in Jinan, the current best suggestion to protect the famous springs in the city is to improve water utilization efficiency and the recycling utilization efficiency, especially the spring water utilization. This paper mainly discusses the changes in spring flow, precipitation, and groundwater abstraction and the relationship among them. The detailed analysis of rational groundwater use and protection of springs will be further studied in the near future.

Data availability. Part of the raw data required to reproduce these findings are available to download from http://jnswj.net/jsp/sw/jnsw.jsp and http://jnsww.com.cn:9090/jnsw/ (last access: 30 May 2018). Other data are maintained secrecy according to Chinese law and not publicly accessible. The processed data required to reproduce these findings cannot be shared at this time as the data also forms part of an ongoing study.

Author contributions. XL and KS collected the data. XL and LH analyzed the results. KS developed the ANN model. XL, LH and KS wrote the manuscript.

Competing interests. The authors declare that they have no conflict of interest.

Special issue statement. This article is part of the special issue "Innovative water resources management – understanding and balancing interactions between humankind and nature". It is a result of the 8th International Water Resources Management Conference of ICWRS, Beijing, China, 13–15 June 2018.

Acknowledgements. This study was supported by the National Nature Science Foundation of China (grant number: 41572220). The authors would also like to acknowledge the financial support of the project named Water Cycle and Hydrological Process Modeling of Jinan "Sponge City".

Edited by: Yangbo Chen
Reviewed by: two anonymous referees

References

Bakalowicz, M.: Karst groundwater: a challenge for new resources, Hydrogeol. J., 13, 148–160, https://doi.org/10.1007/s10040-004-0402-9, 2005.

Cai, Y., Yu, D., Lu, S., Cheng, H., and Wu, D.: The Water Supply Hydrogeology Report of High Quality in Jinan Area (in Chinese), Report, Shandong Provincial Geo-mineral Engineering Exploration Institute, Jinan, China, 223 pp., 2015.

Hartmann, A., Goldscheider, N., Wagener, T., Lange, J., and Weiler, M.: Karst water resources in a changing world: Review of hydrological modeling approaches, Rev. Geophys., 52, 218–242, https://doi.org/10.1002/2013RG000443, 2014.

Hu, C., Hao, Y., Yeh, T.-C. J., Pang, B., and Wu, Z.: Simulation of spring flows from a karst aquifer with an artificial neural network, Hydrol. Process., 22, 596–604, https://doi.org/10.1002/hyp.6625, 2008.

Kang, F., Jin, M., and Qin, P.: Sustainable yield of a karst aquifer system: a case study of Jinan springs in northern China, Hydrogeol. J., 19, 851–863, https://doi.org/10.1007/s10040-011-0725-2, 2011.

Kurtulus, B. and Razack, M.: Evaluation of the ability of an artificial neural network model to simulate the input-output responses of a large karstic aquifer: the La Rochefoucauld aquifer (Charente, France), Hydrogeol. J., 15, 241–254, https://doi.org/10.1007/s10040-006-0077-5, 2006.

Li, B., Yu, Z., Liang, Z., Song, K., Li, H., Wang, Y., Zhang, W., and Acharya, K.: Effects of Climate Variations and Human Activities on Runoff in the Zoige Alpine Wetland in the Eastern Edge of the Tibetan Plateau, J. Hydrol. Eng., 19, 1026–1035, 2014.

Li, X., Shu, L., Liu, L., Yin, D., and Wen, J.: Sensitivity analysis of groundwater level in Jinci Spring Basin (China) based on artificial neural network modeling, Hydrogeol. J., 20, 727–738, https://doi.org/10.1007/s10040-012-0843-5, 2012.

Ma, T., Wang, Y., and Guo, Q.: Response of carbonate aquifer to climate change in northern China: a case study at the Shentou karst springs, J. Hydrol., 297, 274–284, https://doi.org/10.1016/j.jhydrol.2004.04.020, 2004.

Qian, J., Zhan, H., Wu, Y., Li, F., and Wang, J.: Fractured-karst spring-flow protections: a case study in Jinan, China, Hydrogeol. J., 14, 1192, https://doi.org/10.1007/s10040-006-0061-0, 2006.

Vries, J. J. D. and Simmers, I.: Groundwater recharge: an overview of processes and challenges, Hydrogeol. J., 10, 5–17, 2002.

Xi, D., Cao, Y., Li, Z., Fang, P., Shao, Z., and Jia, G.: The Water Supply Hydrogeology Exploitation Report in Jinan City, Shandong Province (in Chinese), Report, Shandong Provincial Geomineral Engineering Exploration Institute, Jinan, China, 165 pp., 1987.

https://doi.org/10.5194/piahs-379-269-2018

Preliminary research on quantitative methods of water resources carrying capacity based on water resources balance sheet

Yanqiu Wang[1], **Xiaorong Huang**[1,2], **Linyun Gao**[1], **Biying Guo**[1], **and Kai Ma**[1]

[1]College of Water Resource & Hydropower, Sichuan University, Chengdu 610065, China
[2]State Key Laboratory of Hydraulics and Mountain River Engineering, Sichuan University, Chengdu 610065, China

Correspondence: Xiaorong Huang (hxiaorong@scu.edu.cn)

Received: 30 December 2017 – Revised: 4 March 2018 – Accepted: 5 March 2018 – Published: 5 June 2018

Abstract. Water resources are not only basic natural resources, but also strategic economic resources and ecological control factors. Water resources carrying capacity constrains the sustainable development of regional economy and society. Studies of water resources carrying capacity can provide helpful information about how the socioeconomic system is both supported and restrained by the water resources system. Based on the research of different scholars, major problems in the study of water resources carrying capacity were summarized as follows: the definition of water resources carrying capacity is not yet unified; the methods of carrying capacity quantification based on the definition of inconsistency are poor in operability; the current quantitative research methods of water resources carrying capacity did not fully reflect the principles of sustainable development; it is difficult to quantify the relationship among the "water resources, economic society and ecological environment". Therefore, it is necessary to develop a better quantitative evaluation method to determine the regional water resources carrying capacity. This paper proposes a new approach to quantifying water resources carrying capacity (that is, through the compilation of the water resources balance sheet) to get a grasp of the regional water resources depletion and water environmental degradation (as well as regional water resources stock assets and liabilities), figure out the squeeze of socioeconomic activities on the environment, and discuss the quantitative calculation methods and technical route of water resources carrying capacity which are able to embody the substance of sustainable development.

1 Introduction

As a sort of irreplaceable natural resource, water resources are an essential element for human existence and an important guarantee for sustainable socioeconomic development. In China, a country experiencing an acute shortage of water, nearly two-thirds of cities are short of water to varying degrees (Chen, 2016) and the contradiction between water resources constraint and socioeconomic development is becoming increasingly prominent. As a result, control over the socioeconomic development and utilization intensity within the range of carrying capacity has become a core part of national efforts to guarantee water resources security in the future (Wang, 2017).

Most of the existing studies incorporate the carrying capacity of water resources into the theory of sustainable development, believing that economic development should be based on a virtuous cycle of the ecological environment, and the research should focus on the hydrology cycle process and mechanism under the natural – artificial binary mode with human interventions. Further, systematic consideration should be given to the relationship among "water resources, socioeconomy and ecological environment" (Zhu et al., 2002; Feng and Huang, 2008). The conclusions of several studies considered the fact that the carrying capacity of water resources signifies that water resources are needed to sustain a healthy social system (Cerutti et al., 2010; Bakker, 2012). However, current studies in this area lack depth. There are great difficulties in how to systematically quantify the relationship between the environment and the economy, as well as natural resources depletion and environmental degradation caused by economic activities (Gan et al., 2014). In recent years researchers have been paying more attention to the calculation methods of water resources carrying capacity rather than the theoretical research of water resources carrying capacity (Fu et al., 2012; Kang and Xu, 2012). Hence, this paper proposes to bring water resources depletion and water environmental damage into the traditional national economic accounting system and establish a regional water resources balance sheet relying on studies on regional water resources assets and liabilities, so as to expand the quantitative channels for water resources carrying capacity.

2 Concepts and research progress of water resources carrying capacity

Carrying capacity, with its limit thought to be first derived from Aristotle and later applied as a mechanics index, refers to the maximum (extreme) carrying capacity of an object without causing any damage (Long et al., 2004). The earliest study on the carrying capacity of water resources in China was conducted in 1989 when Xinjiang Water Resources Soft Science Research Group carried out studies on Xinjiang's water resources carrying capacity and development strategies for the first time. Since 1990, many scholars have put forward numerous viewpoints, concepts, methods, evaluation index systems and calculation models related to the water resources carrying capacity, but no unified understanding has been formed so far (Zhao, 2009).

2.1 Status quo of the research on water resources carrying capacity

In general, definitions of water resources carrying capacity can be summarized into two schools of thought. The maximum support capacity theory represented by Shi Yafeng argued that water resources carrying capacity refers to the maximum capacity of water resources in a region to carry (accommodate) the agriculture, industry, city size and population without disruptions to the society and ecosystem, at a certain stage of the social, historic and technological development. It is a comprehensive objective changing with the development of the society, economy, and science and technology (Liu et al., 2011).

The maximum development capacity theory represented by Xu Youpeng argued that water resources carrying capacity refers to the maximum capacity of water resources to supply water for industrial and agricultural production, people's life, and the ecological environment under certain technical and economic level and social production conditions, i.e., the maximum development capacity of water resources. Water resources can be renewed naturally and constantly utilized by people, and bring benefits to mankind without causing any deterioration under this maximum development capacity (Xu, 1993).

Calculation methods for water resources carrying capacity are broadly divided into two categories. The first one is the evaluation method, which was designed to evaluate the degree of support for water resources to a certain scale of development through building an indicator system, such as the principal component analysis method (Fu and Ji, 1997), aggregative indicator method (Wang et al., 2000) and fuzzy evaluation method (Li and Jin, 2009); the second one is the planning method, which was designed to take a large number of complex factors including economy, society, resources and environment as a whole to find out internal relations between them, and apply mathematical equations to simulate or build models, so as to figure out the optimal development plan, such as the system dynamics method (Yang et al., 2015) and multiobjective analysis method (Xue et al., 2000). In addition to these traditional quantitative methods, many scholars have made different attempts. For example, Feng et al. (2008) applied the water shortage risk assessment model based on information diffusion technologies to studies on the water resources carrying capacity and analyzed the water resources carrying capacity of Jinhua City, Zhejiang Province (Feng and Huang, 2008). Ren et al. (2016) put forward the regional water resources metabolism theory, which introduced the concept of biological metabolism to the water resources carrying capacity (Ren et al., 2016).

2.2 Existing problems

2.2.1 The definition of water resources carrying capacity is not yet unified

Water resources carrying capacity has been defined in two different ways. The former, taking the size of population and social economy that can be carried as an indicator of the water resources carrying capacity from the perspective of a water resources carrying object, namely, the socioeconomic system, is complete in terms of the physical meaning but comparatively abstract; while the latter, taking a specific volume such as water supply capacity as an indicator of the water resources carrying capacity from the perspective of water resources carrying subject, namely, the water resources system (Yuan et al., 2006), is incomplete in terms of the physical meaning but more intuitive. It is considered here that the former statement can better embody the substance of water resources carrying capacity and also have more practical significance.

2.2.2 The methods of carrying capacity quantification based on the definition of inconsistency are poor in operability

Two categories of calculation methods summed up in the above section can correspond respectively to two kinds of different definitions of the water resources carrying capacity.

Due to the nonunified definitions, the results from the quantitative methods raised on the basis of different definitions are difficult to quantify and compare in reality, which leads to poor operability of these methods.

2.2.3 Failure to fully reflect the principle of sustainable development

The emphasis of research on water resources carrying capacity needs to persevere in the principle of sustainable development, but in general, the principle cannot be clearly reflected in the research process. The principle can only be implicitly reflected in researchers' research process according to their understanding of the connotation of sustainable development (Zhang et al., 2003). Consequently, the calculated results of the water resources carrying capacity can hardly reflect the depletion of resources and the loss of the environment.

2.2.4 Bottlenecks, which are difficult to quantify, exist in the relationship among water resources, socioeconomy and ecological environment

Studies on water resources carrying capacity at the present stage are generally conducted on water resources or social and economic activities as the main research objects only, which ignore the ecological environment and the internal relations between the trios. To do this, an optimal choice is to explore and compile a water resources balance sheet.

3 Carrying capacity and water resources balance sheet

3.1 Concept of water resources balance sheet

In 1946, John Richard Hicks, a British economist, put forward the green GDP idea for the first time, and he placed emphasis on changing the perception of pure economic growth (Zhang, 2006). In the mid-1990s, the loss caused by ecological environment destruction every year in China accounted for more than 8 % of GDP (Chen, 2007), indicating that the economic development comes at the expense of the environment and the extensive development model has brought serious negative impact to the ecological environment. When economic prosperity comes with serious environmental damage and health hazards, the false image of national growth emerges, which will cause distortion or twist to the evaluation of resources carrying capacity. Hence, clarification of the creditor relationship between the environment and the economy can play a key role in calculating water resources carrying capacity.

What is water resources liability? Generalized water resources liabilities can be defined as the negative impact on water resources volume, environmental quality and ecosystem function due to excessive improper economic activities. It could be further described as the overuse of water resources leading to the damage of water cycling process and renewability of water, the overdischarge of wastewater and pollutants to natural water bodies leading to the reduction of environmental capacity, and the excessive consumption of ecological water use leading to the decrease of ecosystem service (Jia et al., 2017). As the human society attaches importance to the ecological environment and sustainable development, the US, Canada, Japan, Australia and other countries have conducted research on the water resources balance sheet. Since China carried out the green GDP accounting study in the early 21st century, natural resources assets accounting has begun to be put into practice. National balance sheet is subject to the System of National Accounts (SNA), and so the System of Integrated Environmental and Economic Accounting (SEEA) jointly launched by the United Nations and the World Bank as an auxiliary accounting system to SNA, is considered to be comprised of the current international norms which are closest to the compilation of China's natural resources balance sheet (NRBS) (Geng et al., 2015). On this basis, the United Nations Statistics Division cooperated with the London Group on Environmental Accounting in compiling the System of Environmental and Economic Accounting for Water (SEEAW) based on SEEA, and China also formed the general framework of Comprehensive System of Environmental and Economic Accounting for Water (CSEEAW) (Chai et al., 2016) in 2009.

However, the SEEAW did not set specific liability items, and that is why it differs from the water balance sheet (Huang, 2017). Therefore, the decision on whether or not to recognize water resources liabilities determines the compilation of the statement of water resources assets and liabilities or water resources assets balance. So, is there any water resources liability? It can be seen that the resources depletion, environmental damage and ecological disruption caused by current socioeconomic development represent a sort of "debt" to natural resources, which could clearly reflect the "accounting" based on the concept of sustainable development. The environment should be treated as a virtual subject to specify water resources equities and liabilities of the economy, so as to restore and compensate the environment (Jia et al., 2017).

The system of accounting for the water resources balance sheet shall include various statement forms and design the order in accordance with the statement forms as follows: physical object first, value second; stock first, flow second; classification first, consolidation second, which includes the physical accounts of water resources, the quality accounts of water resources, and the hybrid supply and use table. Finally, the compilation of water resources balance sheet could be completed on the basis of the accounting for water resources magnitude of value. In the *System of Trial Preparation of Natural Resources Balance Sheet (Preparation Guide)* published in November 2015, China has come up with the standard statement forms of the physical accounts of water resources and the quality accounts of water resources. How-

Table 1. Basic statement form of the water resources equity balance sheet.

Item		Opening stock	Equity distribution	Equity use	Current net change			Closing stock
					Equity assets	Equity liabilities	Net equity assets	
Water resources	1 Surface water							
	2 Groundwater							
	3 ...							
	4 Total							
Water environment	5 COD discharge							
	6 Ammonia nitrogen discharge							
	7 ...							
	8 Total							
Water ecology	9 Ecological water demand							
	10 Sea inflow							
	11 ...							
	12 Total							

Note: there is a balance relationship in the table, i.e., equity assets − equity liabilities = net equity assets; the Three Red Lines index can be used as a reference for equity distribution. Due to data limitations, the compilation of water environment and water ecology is very difficult at the present stage.
Explanation of some conceptions: (1) opening stock – can be set to zero in the beginning of the accounting year; meanwhile the ending stock represents the changes in the current period. (2) Equity distribution – distribution quota that based on the Three Red Lines index or assessment indicators, which generally refers to the multiyear average. (3) Equity use – water consumption during the accounting period. (4) Equity assets – the rights and benefits of water users under the legal protection during water consumption. (5) Equity liabilities – obligation of water users to compensate or repair the environment due to exceeding the rights stipulated by law during the water consumption.

ever, current studies on the value of water resources are not well developed. Hence the standards for statement forms of water resources balance sheet have not yet been unified. The research achievements of Hong Gan, Lin Wang, Chang-Hai Qin, and Lin Jia are referred here to list the basic statement form of the water resources equity balance sheet (Chai et al., 2016; Zhou et al., 2017); see Table 1.

It can be seen that the inclusion of resources depletion, environmental damage and ecological benefits in the system of evaluation for socioeconomic development can establish a system of national economy accounting in favor of green and low-carbon cyclic development. Problems about the water resources carrying capacity at this stage are due to the lack of quantification of resource depletion and environmental damage. Hence, if these two can be connected, then deficiencies in the quantification of water resources carrying capacity may be improved.

3.2 Discussion on quantification methods

The concept of water resources carrying capacity is related to a huge and complex system with many factors such as population, resource, environment, ecology, society, economy and technique, while the previous calculation of the carrying capacity usually failed to consider the interaction among various factors in the complex system. In order to solve the problem that the current research has ignored the inherent relationship among water resources, economic society and ecological environment, the feedback relationship among the water resources system, socioeconomic system and ecological environment system is analyzed and summarized as shown in Fig. 1.

It is observed that the physical accounts of water resources clearly reflects the increase (source) and decrease (outflow) in stock of regional water resources in each year, in accordance with the feedback relationship between the water resources system (supply) and socioeconomic system (depletion). And the accounting of assets and liabilities for water resources, water environment and water ecology in the water resources balance sheet can reflect the depletion and degradation of the ecological environment caused by regional socioeconomic water utilization. Therefore, an attempt is made to combine these two with the quantification of the carrying capacity, to avoid the twisting or distortion of the calculation of water resources carrying capacity, and propose the technical route for quantification of the carrying capacity as shown in Fig. 2.

Studies on water resources valuation at the present stage are inadequate. Consequently, the discussion here is based on the physical water assets account, in which the consideration may be given to deduct water liabilities (i.e., the equity liability item in Fig. 1) from the available water resources for compensation for water liabilities. It can be seen that the combination with the water resources balance sheet can more completely reflect the endowment conditions of a regional water resources circulation system, the resource threshold

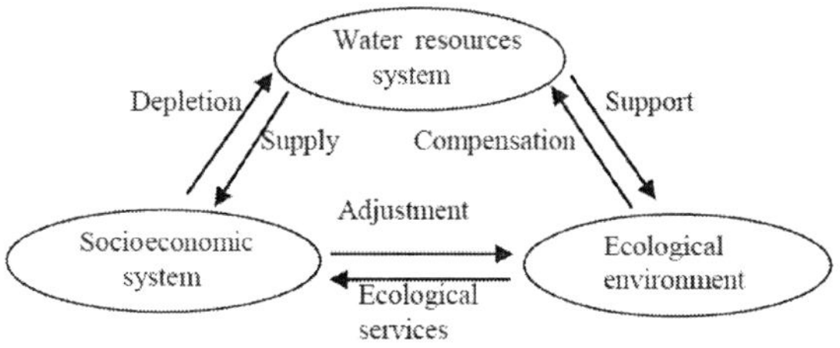

Figure 1. System feedback diagram.

that can be provided facing the economic development pressure and the water resources constraint after generating the liabilities, as well as avoid the distortion of evaluation results. Meanwhile, water resources liability items reflect the resources depletion and environmental degradation resulting from social and economic activities and lay a foundation for maintaining a virtuous ecological cycle. And the quantification of water resources carrying capacity on this basis is able to fully reflect the sustainable development theory.

4 Case study and discussion

4.1 Calculation of water resources carrying capacity in Chengdu

Due to its abundant amount of data and the good foundation of previous research, Chengdu was selected as the study area. Chengdu, located in central Sichuan at 102°54′–104°53′ E and 30°05′–31°26′ N, covers an area of 12 390 km^2 in total. Chengdu has a moderate subtropical monsoon climate and abundant precipitation, with the annual average precipitation ranging from 900 to 1300 mm and multiple years of total water resources reaching 8.607 billion m^3. In general, Chengdu boasts abundant water resources, which mainly come from the plentiful transit water resources in the region, while Chengdu does not have rich local water resources due to a dense population and relatively little per capita water resources.

According to the above quantization technology route, the water resources capacity of Chengdu city in 2015 is calculated as follows. The data are from the *Chengdu Water Resources Bulletin* (Chengdu Water Authority, 2015) and some relevant reports. Taking the multiyear average of 1980–2005 groundwater in Chengdu at 9.720 billion m^3 as the beginning stock and the year of 2015 as the beginning year, the physical accounts of water resources of Chengdu in 2015 (see Table 2) is compiled with reference to the trial compilation achievement of Xi Yuanyuan (Xi, 2017) and the compilation requirements of the *System of Trial Preparation of Natural Resources Balance Sheet (Preparation Guide)*. Meanwhile, the "Three Red Lines" control objectives of all districts in Chengdu are combined to compile the water resources items in the statement of water resources equity balance sheet of Chengdu in 2015, see Table 3. Due to data limitations, the water environment and water ecology items are not filled in here.

Total available regional water resources are determined according to the method set out in the *Technical Details of National IWRM Planning*. The formula is as follows:

$$W = W_0 - W_S + W_T, \tag{1}$$

where W refers to the total available regional water resources, W_0 refers to the total water resources, W_s refers to the ecological water demand and W_T refers to the water resources diverted in from outside the region, all measured in 100 million m^3. According to the meaning of each item in the physical accounts of water resources, it can be seen that the total water resources corresponds to the value in the third row of the first column in the Table 2, expressed as [3, 1], meanwhile there is a balance relationship, i.e., [3, 1] = [3, 2] + [3, 3]. The water resources diverted in from the outside region corresponds to the value in the sixth row of the first column in the Table 2, expressed as [6, 1].

The ecological water demand can be calculated in accordance with the method set out in the *Estimation Method of Available Water Resources (Trial)* by the Ministry of Water Resources of China. With the percentage of multiyear average regional runoff (10–20 % for northern areas and 20–30 % for southern areas generally) as the minimum ecological water demand of rivers, here the ecological water demand is calculated at 20 %, deriving the result of ecological water demand of 2.996 billion m^3.

$$W = [3, 1] + [6, 1] - 29.96 \tag{2}$$

Thus, the total available water resources of Chengdu in 2015 stood at 3.853 billion m^3 by calculation.

According to the *Statistical Yearbook of Chengdu* (Chengdu Bureau of Statistics, 2016), the per capita water consumption A_1 is calculated by extracting water consumption of domestic, industrial, agricultural and ecological water outside the channel from the physical accounts of water resources.

$$A_1 = \frac{[12, 1]}{P(t)} = \frac{[13, 1] + [14, 1] + [15, 1] + [16, 1]}{P(t)}, \tag{3}$$

where $P(t)$ corresponds to the total population in the corresponding year; the value of numerators derive from Table 2. Thus the per capita water consumption of Chengdu in 2015 is 358 m^3 by calculation. The total available water resources in the corresponding year divided by the per capita water consumption derives the water resources carrying capacity of that year. The overcarrying is calculated using the following formula (Wang et al., 2007) to measure the overcarrying of regional water resources.

$$R = \frac{A_1 \times P(t)}{W} \tag{4}$$

When $R < 1$, it is not overcarrying, and the smaller the value is, the greater the potential carrying capacity is; when $R > 1$,

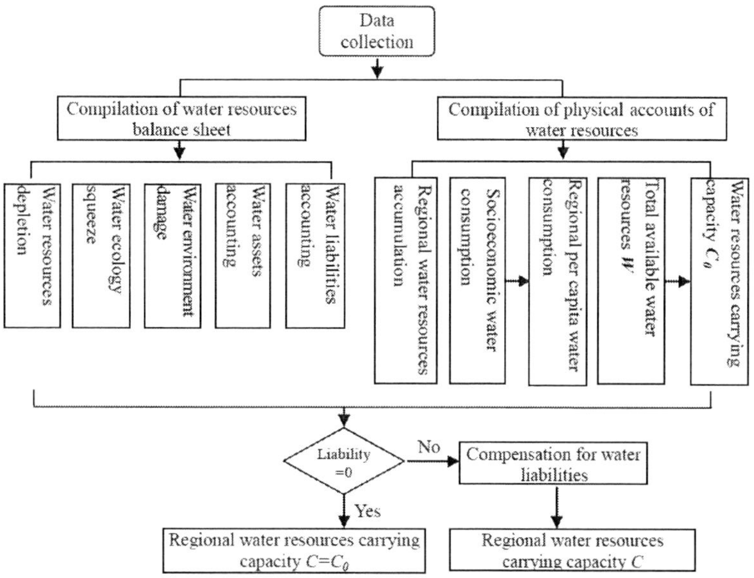

Figure 2. Technical route of carrying capacity quantification.

Table 2. The physical accounts of water resources of Chengdu in 2015, measured in units of 100 million m^3.

Indicator	Code	Total	Surface water	Groundwater
A	B	1	2	3
Opening stock	1	99.13	1.93	97.20
Increase in stocks	2	324.35	296.57	27.78
Water resources from rainfall	3	66.32	41.48	24.84
Inflows and diverted in	4	235.36	232.42	2.94
Inflows from outside region	5	205.92	205.92	–
Diverted in from outside region	6	2.17	2.17	–
Inflows from other water bodies	7	27.27	24.33	2.94
Socioeconomic return water	8	22.67	22.67	–
Irrigation return water	9	11.62	11.62	–
Nonirrigation return water	10	11.05	11.05	–
Decrease in stocks	11	326.05	296.59	29.46
Water consumption	12	52.44	49.7	2.74
Domestic	13	9.26	–	–
Industrial	14	15.96	–	–
Agricultural	15	25.69	–	–
Ecological water outside the channel	16	1.53	1.53	–
Outflows and diverted out	17	268.99	244.66	24.33
Outflows to outside region	18	213.73	213.73	–
Diverted out to outside region	20	27.99	27.99	–
Outflows to other water bodies	21	27.27	2.94	24.33
Nonwater consumption	22	4.62	2.23	2.39
Closing stock	23	97.43	1.91	95.52

Note: the nonwater consumption in the table refers to the evaporation losses of the lakes and reservoirs, the loss of drainage evaporation and the evaporation of underground water. However, due to the current state of monitoring technology, this indicator cannot be obtained by calculation or evaluation. Therefore it is treated as a balancing item to keep the entire table in balance. Definitions of other indicators refer to *Chengdu Water Resources Bulletin*.

Table 3. Water resources equity balance sheet of Chengdu in 2015, measured in units of 100 million m^3.

Item		Opening stock	Equity distribution	Equity use	Current net change			Closing stock
					Equity assets	Equity liabilities	Net equity assets	
Water Resources	1 Surface water	—	55.14	49.7	5.44	0	5.44	5.44
	2 Groundwater	—	3.76	2.74	1.02	0	1.02	1.02
	3 Total	—	59.9	52.44	6.46	0	6.46	6.46

the region is in the overcarrying state, and the greater the value is, the more serious the overcarrying is.

The water resources carrying capacity of Chengdu in 2015 is calculated as shown in Table 4. It can be seen that Chengdu did not generate water resources liability in 2015, but water resources were in the overcarrying state. Therefore, it is necessary to strengthen water conservation, tighten up pollution control and speed up the industrial restructuring, so as to guarantee the ecological environmental water demand and improve the water resources carrying capacity appropriately.

Table 4. Water resources carrying capacity of Chengdu in 2015.

Item	2015
Total water resources (100 million m^3)	66.32
Total population (10 000 persons)	1465.8
Total available water resources W (100 million m^3)	38.53
Per capita water consumption (m^3)	358
Water resources carrying capacity C (10 000 persons)	1076.26
Water resources liabilities (100 million m^3)	0
Water resources carrying capacity C (10 000 persons)	1076.26
Overcarrying	1.36

4.2 Discussion

The results of the discussion are as follows.

1. Table 2 shows that the groundwater extraction in 2015 was 274 million m^3, less than the target of 376 million m^3 controlled by the Three Red Lines index in 2015 in Chengdu. And the total amount of water used was 5.244 billion m^3, which did not exceed the target of 5.990 billion m^3 in the Three Red Lines index in 2015 in Chengdu. Therefore, it can be seen from Table 3 that no water equity liabilities have been generated in Chengdu in 2015.

2. The proportion of the total amount of agricultural water in Table 2 is the largest, accounting for 48.99 % of the total, followed by industrial water consumption, accounting for 30.43 %. Ecological water outside the channel takes up 2.92 %. It can be seen that most of the water used in Chengdu was agricultural water, followed by industrial water consumption, which is only about half of agricultural water.

3. Table 4 shows that the water resources carrying capacity of Chengdu in 2015 was 10 762 600 persons and the overcarrying state was 1.36. Through observing the structure of water consumption in Chengdu, it is suggested that the water resources carrying capacity of Chengdu should be improved appropriately by increasing the utilization rate of industrial and agricultural water in Chengdu.

4. Compared with other research (Guo, 2013), which calculated the water resources carrying capacity of Chengdu in 2015 was 11 949 500 persons by the system dynamics, its results are close to the results calculated through the above technical route. Since the above technical route takes into account whether the economic society will generate liabilities in terms of water resources depletion, it can better reflect the connotation of water resources carrying capacity and embody the sustainability theory compared with the traditional methods. In addition to calculating the regional water resources carrying capacity, the idea of this article can also be used to calculate the water environmental capacity of the area in combination with the quality accounts of water resources.

5 Conclusions

The conclusions to this study are as follows.

1. Compared to the past achievements, Chengdu's water resources carrying capacity calculated by the above technical route is reasonable and practical. However, the water environment and water ecological liabilities were not taken into account due to the limitation of data at this stage, and thus the result can not reflect water environmental damage and water ecology squeeze. Subsequent further studies are needed for improvement.

2. The approach to quantifying water resources carrying capacity through the compilation of the water resources balance sheet is feasible. However, the compilation of the water resources balance sheet needs further

research, focusing on studying how to value the water resources, and replenishing two indicators in terms of water environment and water ecology, so as to optimize and expand the quantitative methods of water resources carrying capacity.

Data availability. The data used in this article are from the Chengdu Water Authority (2015; http://www.cdwater.gov.cn/wsbs/bmcx/sl/system/2017/03/1/000002497.html) and the Chengdu Bureau of Statistics (2016; http://www.cdstats.chengdu.gov.cn/htm/detail_63038.html)

Competing interests. The authors declare that they have no conflict of interest.

Special issue statement. This article is part of the special issue "Innovative water resources management – understanding and balancing interactions between humankind and nature". It is a result of the 8th International Water Resources Management Conference of ICWRS, Beijing, China, 13–15 June 2018.

Acknowledgements. This study was supported by the National Science Foundation of China grants 51779160 and 51579161.

Edited by: Zongxue Xu
Reviewed by: two anonymous referees

References

Bakker, K.: Water management, Water security: research challenges and opportunities, Science, 337, 914, https://doi.org/10.1126/science.1226337, 2012.

Cerutti, A. K., Bagliani, M., Beccaro, G. L., and Bounous, G.: Application of Ecological Footprint Analysis on nectarine production: methodological issues and results from a case study in Italy, J. Clean. Prod., 18, 771–776, 2010.

Chai, X. R., Huang, X. R., Xi, Y. Y., and Yang, P. P.: Analysis on preparation of water resources balance sheet, Journal of Water Resources and Water Engineering, 27, 44–49, 2016.

Chen, L.: National implementation of double-control action on total water consumption and intensity strive to ensure sustained and healthy economic and social development, China Water Resources, 23, 21–24, 2016.

Chen, Z. Z.: Promote ecological science with enthusiasm and truth-seeking spirit, Resources Environment and Development, 2, 1–2, 2007.

Chengdu Bureau of Statistics: The Statistical Yearbook of Chengdu, available at: http://www.cdstats.chengdu.gov.cn/htm/detail_63038.html (last access: 26 March 2018), 2016.

Chengdu Water Authority: Chengdu Water Resources Bulletin, available at: http://www.cdwater.gov.cn/wsbs/bmcx/sl/system/2017/03/1/000002497.html (last access: 26 March 2018), 2015.

Feng, L. H. and Huang, C. F.: A risk assessment model of water shortage based on information diffusion technology and its application in analyzing carrying capacity of water resources, Water Resources Management, 22, 621, https://doi.org/10.1007/s11269-007-9182-z, 2008.

Fu, Q., Jiang, Q., and Wang, Z.: Comprehensive Evaluation of Regional Agricultural Water and Land Resources Carrying Capacity Based on DPSIR Concept Framework and PP Mode, Springer Berlin Heidelberg, Germany, 370, 391–398, 2012.

Fu, X. and Ji, C. M.: A multi-dimension DP Model and application, Water Resources and Power, 4, 1–6, 1997.

Gan, H., Wang, L., Qing, C. H., and Jia, L.: Understanding of balance sheet of water resources, China Water Resources, 14, 1–7, 2014.

Geng, J. X., Hu, T. Y., and Liu, Z. J.: The research on preparation and application of Chinese National Balance Sheet and Natural Resources Balance Sheet – An Analysis Based on SNA 2008 and SEEA 2012, Accounting Research, 1, 15–24, 2015.

Guo, W.: Dynamic Simulation Analysis of Water Resources Population Carrying Capacity Based on System Dynamics – A Case Study of Chengdu, Journal of Henan Science and Technology, 16, 176–177, 2013.

Huang, X. R.: Perspective on theories and methods study the compilation of water resources balance sheet, Journal of Water Resources and Water Engineering, 28, 1–5, 2017.

Jia, L., Gan, H., and Wang, L.: Discussion on water resources liabilities, Journal of Natural Resources, 32, 1–11, 2017.

Kang, P. and Xu, L.: Water Environmental Carrying Capacity Assessment of an Industrial Park, Procedia Environ. Sci., 13, 879–890, 2012.

Li, G. and Jin, C. L.: Fuzzy comprehensive evaluation for carrying capacity of regional water resources, Water Resour. Manag., 23, 2505–2513, 2009.

Long, T. R., Jiang, W. C., and He, Q.: Water resources carrying capacity: new perspectives based on eco-economic analysis and sustainable development, J. Hydraul. Eng., 35, 38–45, 2004.

Ren, C., Guo, P., and Li, M.: An innovative method for water resources carrying capacity research–metabolic theory of regional water resources, J. Environ. Manage., 167, 139–146, 2016.

Liu, J. J., Dong, S. C., and Li, Z. H.: Comprehensive Evaluation of Chinese Water Resources Carrying Capacity, J. Nat. Resour., 26, 258–269, 2011.

Wang, D. X., Wang, H., and Ma, J.: Water resources supporting capacity for regional development in China, J. Hydraul. Eng., 11, 21–26, 2000.

Wang, J. H.: Evaluation and strategic allocation of national water resources carrying capacity, Chin. J. Environment Management, 9, 111–112, 2017.

Wang, Z. J., Liao, S. H., Wu, X. F., Zhao, J. S., and Gan, H.: Water resources carrying capacity in Datong city, South-to-North Water Diversion and Water Science & Technology, 5, 47–50, 2007.

Xi, Y. Y.: The research of physical asset account for water resources in Chengdu, Sichuan University, Sichuan University Press, 2017.

Xu, Y. P.: A study of comprehensive evaluation of the water resources carrying capacity in the arid area-a case study in the Hetian river basin of Xinjiang, Journal of Natural Resources, 8, 229–237, 1993.

Xue, X. J., Hui, Y. H., and Huang, Q.: Research on city water resources capacity and its practical example, Journal of Northwest A&F University, 28, 135–139, 2000.

Yang, J., Lei, K., and Khu, S.: Assessment of water environmental carrying capacity for sustainable development using a coupled system dynamics approach applied to the Tieling of the Liao river basin, Environ. Earth Sci., 73, 5173–5183, 2015.

Yuan, Y., Gan, H., Wang, Z. J., and Bao, W. F.: Discussion on progress and development trend of the research on water resources carrying capacity, Journal of China Institute of Water Resources and Hydropower Research, 4, 62–67, 2006.

Zhang, L.: Research on forest resources accounting of green GDP, Harbin Institute of Technology, 2006.

Zhang, L., Dong, Z. C., and Zhang, W.: Progress and prospect of research of water resources capacity, Water Resources and Hydropower Engineering, 34, 1–4, 2003.

Zhao, N.: The empirical research of Beijing water resources carrying capacity and pressure state, Urban Stud., 8, 148–152, 2009.

Zhou, P., Jia, L., and Gan, H.: The framework of physical water resources accounting for water entities, Accounting Research, 2017, 24–31, 2017.

Zhu, Y. Z., Xia, J., and Tan, G.: A primary study on the theories and process of water resources carrying capacity, Progress in Geography, 21, 180–188, 2002.

https://doi.org/10.5194/piahs-379-279-2018

Discussion on water resources value accounting and its application

Biying Guo[1], Xiaorong Huang[1,2], Kai Ma[1], Linyun Gao[1], and Yanqiu Wang[1]

[1]College of Water Resource & Hydropower, Sichuan University, Chengdu 610065, China
[2]State Key Laboratory of Hydraulics and Mountain River Engineering, Sichuan University, Chengdu 610065, China

Correspondence: Xiaorong Huang (hxiaorong@scu.edu.cn)

Received: 30 December 2017 – Revised: 17 April 2018 – Accepted: 19 April 2018 – Published: 5 June 2018

Abstract. The exploration of the compilation of natural resources balance sheet has been proposed since 2013. Several elements of water resources balance sheet have been discussed positively in China, including basic concept, framework and accounting methods, which focused on calculating the amount of water resources with statistical methods but lacked the analysis of the interrelationship between physical volume and magnitude of value. Based on the study of physical accounting of water resources balance sheet, the connotation of water resources value is analyzed in combination with research on the value of water resources in the world. What's more, the theoretical framework, form of measurement and research methods of water resources value accounting are further explored. Taking Chengdu, China as an example, the index system of water resources balance sheet in Chengdu which includes both physical and valuable volume is established to account the depletion of water resources, environmental damage and ecological water occupation caused by economic and social water use. Moreover, the water resources balance sheet in this region which reflects the negative impact of the economy on the environment is established. It provides a reference for advancing water resources management, improving government and social investment, realizing scientific and rational allocation of water resources.

1 Introduction

Influenced by congenital conditions of water resources and scale of economic and social development, there are prominent water resource problems in China. Water shortage, severe water pollution and degradation of water ecology have become significant constraints on regional economic development. In 2013, The Third Plenary Session of the eighteen first proposed to explore the compilation of natural resources balance sheet in order to get a clear picture of natural resource assets and an accurate grasp of occupation, use, consumption, recovery and value addition, among other activities, of natural resource assets by the economy (Jia et al., 2017), which was of great significance for China to comprehensively deepen the reform. Water resources balance sheet, an integral part of natural resources, is an accounting statement that calculates depletion of water resources, water environmental damage and ecological water occupation due to social and economic water consumption with indicators including water quantity, water quality and water ecology, and reflects negative impacts of economy and society on environment. Gan et al. (2014) have put forward the basic positioning about compilation of water resources balance sheet. That is, it should serve the most rigorous water resources management and the compilation of natural resources balance sheet, focus on accounting physical volume of regional economy and the environment, and target the water value accounting and integrated environmental and economic accounting. At present, insufficient research on water resources value accounting not only influences the compilation of water resources balance sheet but also restricts the reflection of the economic value of natural resources in national economic accounting. Therefore, it is necessary to discuss water resources value accounting in an in-depth way.

2 Research status

Since the introduction of the concept "natural resources balance sheet", there has been a lot of research on this theory in China. Wang (2017) considered that natural resources balance sheet serves as a bridge and a bond between System of National Accounts (SNA) and System of Environmental-Economic Accounting (SEEA), therefore, it should be deemed as an independent system based on SEEA and supplementing SNA. Geng et al. (2017) introduced the design, balance relationship and compilation purpose of balance sheet, put forward two possible statement compilation schemes, and tried to compile land resources balance sheet and water resources balance sheet of Yongning County, Ningxia by collecting data with questionnaire. Feng et al. (2014) proposed the compilation principle of physical volume, namely first in physical and then in monetary terms, first in stock and then in flow terms and first in category and then in integration terms, and a conceptual framework of natural resources balance sheet which comprises categorized physical sheet and integrated monetary sheet of natural resources was developed. Now there remain some challenges faced by natural resources accounting, including insufficient theoretical research, lack of relatively unified, standard and mature statistic accounting system across the world. In addition, at present stage, studies mostly focus on the compilation of natural resources assets balance based on physical volume. If accounting is merely based on physical volume, natural resources assets account can only be used as a satellite account and statistically serves as a ledger. It can't reflect the utilization, consumption and protection benefits of natural resources in combination with factors including economic growth and can't realize the shift from balance sheet of natural resources to balance sheet of asset (Jiao et al., 2015; Huang et al., 2017). Thus, the shift from physical volume of natural resources to their magnitude of value is the key technical issue for advancing resources accounting.

2.1 Research status of water resources value

Water resources, as a kind of basic resource, provide means of production and means of livelihood for our human society, keep ecological balance, and are valuable assets without a doubt. In 1988, the research group of resources accounting and its inclusion in SNA proposed theoretical accounting framework regarding such three aspects as physical volume accounting, value magnitude accounting and water quality accounting for water resources, though the group hasn't put forward mature theory and approach of integrating it into SNA (Xu et al., 2006; Jiang et al., 1999). Shen et al. (1998) defined value of water resources from varying aspects including land rent theory, labor theory of value, theory of marginal utility and non-use value theory by specific characteristics of water resources, and proposed that the value connotation of water resources was mainly reflected by such three aspects as scarcity, resource property right and labor value. Xu et al. (2006) further developed the theory "payment for water quality resources use" which means that the payment for water use depends on water quality, and advanced the principle of "quality water at premium price". Huang et al. (2017) suggested a technology route for research on water resources valuation, i.e., the application of shadow prices approach to evaluate water resources depletion, cost method to evaluate water environmental damage, and emergy analysis method to evaluate ecological water occupation, respectively. Water resource value accounting, the key to promoting compilation of water resources balance sheet, lays a theoretical foundation for formation of water market mechanism, facilitates efficient and reasonable allocation of water resources and is of great significance for building a water-conservation society. Nevertheless, water resources not only offer means of livelihood for economic society as a commodity, but also are characterized by its mobility, renewability and duality of advantage and disadvantage as a type of natural resources. So it is difficult to quantify the value of water resources and there is little research on water resources value accounting. In light of the above problems, this study discusses the value of water resources, studies existing water resources value accounting models, and tries to use various methods to calculate the value and price of water resources in such three aspects as water resources, water environment and water eco-system, so as to provide a reference for establishing a reasonable accounting approach to water resources value.

2.2 The connotation of water resources value

Value is the foundation of price while price is a monetary expression of value. Natural resources property right system in China has defined national ownership of natural resources including water resources. Water price is fundamentally determined by the government rather than the market. Thus, the value of water resources is not fully reflected in market price and can't fully reflect the scarcity of resources. Wang et al. (1999) divided water resources into three categories, i.e., resource-oriented ones, engineering ones and renewable ones, to discuss the value content and price structure of water resources and proposed that water resources value accounting was based on shadow price. Gan et al. (2012) held the view that, water resources value, a special relationship with the appraisal subject established in the socio-economic and environmental system cyclic process, was represented by use value, property value, labor value, and compensated value, and was reflected by equilibrium price in the market-oriented economy. At national level, water resources value shows the attributes of property value and compensated value; national activities reflect actions of managers and exploitation and maintenance of water resources embody labor value attribute of water resources; as for users, their activities represent use value attribute of water resources (Chinese Environmental Economic Accounting Report, 2009).

Water resources value accounting should start with definition of accounting subject and should cover both unpolluted and directly available water resources and polluted and worthless water resources, appraise value of water resources regarding water resources depletion, water environmental damage and ecological water occupation in an integrated manner, and reflect water flow processes in socio-economic and environmental system in the form of economic information. The value of water resources is different from commodity value. Whether water resources are in their natural state or have agglutinated indiscriminate human labor, the production of water resources value is closely linked to their scarcity and economically manifests the ownership (Jiang and Wang, 1996). With the help of water resources value accounting, we can derive water resources price reflecting resource scarcity, use economic leverage to protect and improve water quality, increase water price reasonably to lessen wasted water resources, in order to avoid water resources crisis and realize sustainable development.

3 Research on water resources value accounting approach

3.1 Theories of water resources value

In a traditional sense, water resources are "inexhaustible" and views like "priceless resources" give rise to the free utilization without pay of water resource (Jiang, 1998a) and restrict sustainable economic and social development. With the development of western economics, land rent theory, Marxism labor theory of value, marginal utility theory of value, environmental value theory, ecological value theory, existential value view or non-use value view and other views have gradually emerged. According to land rent theory of Marx, land rent is a form in which land ownership is realized economically and represents revenues obtained by land owners merely depending on their ownership of land. In China, water resources are owned by the State and land rent paid for water resources is an important part of water resources price. According to Marx's labor theory of value, labor consumption of water resources invested by our human beings determines the price of water resources. This prompts discussion on whether natural resources without necessary social labor are valuable or not. The labor theory of value overemphasizes human's social labor time but ignores consumption compensation of water resources itself. Therefore, this theory is unable to address value issues related to natural resources including water resources. Marginal utility theory of value holds that (Wang et al., 2003), utility is the source of price and a necessary condition for the formation of value; the combination of utility and scarcity generates commodity value; marginal utility is a measure to gauge the magnitude of value. However, marginal utility theory of value has its flaw. It confuses commodity value with use value or goods utility, and is subjective in the evaluation of utility, thus eliminates the economic component of water resources value (Jiang, 1998a). Environmental value theory and ecological value theory examine the interaction between environment and water resources, dependence of economic development on nature, respectively. However, both of them merely focus on framework building and haven't developed any complete theoretical system. According to the above analysis of water resources value theories we can find that, there are no unified views on the sources of water resources value among the varied value theories discussed above. Therefore, the value accounting approaches differ significantly from each other, leading to different conclusions and difficult to get applied and promoted.

3.2 Value accounting model of water resources

Now, models and approaches of water resources value accounting mainly include shadow prices method, marginal opportunity cost model, supply-demand pricing model, fuzzy mathematics model, emergy analysis method, etc. Shadow price can reflect the scarcity of resources and is derived through solving linear programming, adjustments based on market prices at home and abroad or opportunity cost approach. But, shadow price is difficult to calculate and can't represent the value of water resources. Hence, there is a great restriction on model application. Marginal opportunity cost model provides definite calculation formula:

$$\mathrm{MOC} = \mathrm{MPC} + \mathrm{MUC} + \mathrm{MEC} \quad (1)$$

where MOC is marginal opportunity cost, MPC is marginal production cost, MUC is marginal user cost, MEC is marginal external cost.

The downside is that, it is difficult to determine MUC and MEC in the formula, and MOC only takes into consideration volume of water resources and ignores quality. With simple calculation formula and easily available data, supply-demand pricing model is suitable for the market-oriented economy environment. However, the disadvantage of this model is that, value of water resources is only determined by volume, ignoring function and ecological influence of water resources. Though fuzzy mathematics model considers multiple uncertain complex factors involved in water resources problems to establish function and matrix calculation formula and has relatively comprehensive analysis coverage, in the course of synthesize evaluation of water resources value, certain experience is required to determine the weight (Jiang et al., 1998b), which increases operational difficulty. Started from the emergy cycle and conversion of water resources eco-economic system, water resources emergy theory, which integrates the connotation of water resources value with emergy theory based on concepts including emergy conversion rate, converts different emergy flows, materials flows and currency flows in eco-system or socio-economic system into emergy in unified standards for

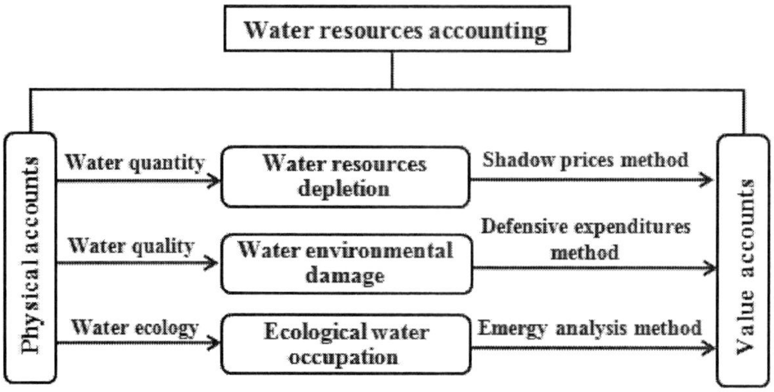

Figure 1. Technical routes of water resources accounting.

measurement and analysis. As a result, it can provide reasonable explanations for value sources, formation mechanism and other problems regarding water resources value research, and it is the new development direction for water resources value theory (Lv et al., 2009; Wang and He, 2015).

According to this paper, because of multiple purposes and renewability of water resources, we should use different methods to account water resources value from varied perspectives including water resources depletion, environmental damage and ecological occupation resulting from social and economic water consumption as shown in Fig. 1. Depletion of water resources represents a process of change in volume of water resources. The severer the depletion, the more prominent the scarcity is. Scarcity can be measured by shadow price and greater scarcity indicates higher shadow price. Water environmental damage refers to unfavorable changes in water resources environment and functional degradation of water eco-system. The deterioration of water quality is an important cause for water environmental damage and it also leads to reduced ability of water resources functional utility to satisfy subject needs, i.e., weakened utility value of water resources. Restoration cost method is applicable to non-market price appraisal of resource environment without direct market performance pattern and it uses cost required for restoration of damaged environment into its original state as the value of the original resource environment (Wang and Qu, 2001). The method can be used to measure damage of water environment. Overuse of water resources takes up existing water resources necessary for normal development and keeping a relatively stable status of eco-system. In turn, eco-environmental degradation imposes huge pressures on normal social and economic operation and exerts multiple impacts on socio-economic-eco-system. Emergy analysis approach unifies varied energies of quality and quantity in eco-system into emergy, reflects the actual price and contribution of emergy at different levels, and can assess value indicators of eco-system in a more objective manner by taking factors difficult to account into consideration.

4 Research on water resources accounting in Chengdu

4.1 Study area

Chengdu, located in central Sichuan Province and the western part of Sichuan Basin, belongs to Minjiang and Tuojiang basins in Yangtze River system and boasts a land of 12 121 km^2 and a permanent population of 14 657 500. Blessed with abundant rainfalls and sufficient water resources, it has built up the water supply infrastructure system project mainly backed by Dujiangyan and combining storage, diversion and pumping together. In 2015, Chengdu's average precipitation, total precipitation, total water resources reached 882.0 mm, 10 690 and 6632 million m^3 (including 6581 million m^3 of surface water and 2484 million m^3 of groundwater), respectively. Chengdu has high-quality water as it lies at the upstream of Yangtze River Basin, most of its river resources are composed of meteoric waters, underground undercurrents and melt snow and rivers are far away from man-made pollution before they flow into Chengdu Plain.

4.2 Physical accounts of waters resources assets in Chengdu

Physical account of water resources is the first step towards compilation of water resources balance sheet and also provides data basis for value accounting of water resources assets. Physical assets accounting reflects information about total water resources of a region and their changes based on statistics of water resources and measured in physical unit (Jian et al., 2016). Statistics of physical accounts of waters resources assets in Chengdu cover three sub-indicators, i.e., water resources, water environment and water eco-system.

Table 1. The physical accounts of water resources assets in Chengdu.

Type	Item	Unit	Surface water	Groundwater	Total water resources
Water Resources	Opening stocks	100 million m^3	1.93	93.59	95.52
	Increases in stocks		296.57	27.78	324.35
	Precipitation		41.48	24.84	66.32
	Inflows		232.42	2.94	235.36
	Returns		22.67	0	22.67
	Decreases in stocks		296.59	29.46	326.05
	Abstraction		49.7	2.74	52.44
	Outflows		244.66	24.33	268.99
	Ecological water consumption of river and lake		2.23	2.39	4.62
	Closing stocks		1.91	91.91	93.82
	Available amount		56.14	3.76	59.90
	Depletion		19.38	0	19.38
Water environment		10 000 tons	Wastewater	COD	Ammonia nitrogen
	Discharge		135 848.04	17.7	2.0
	Treatment		110 539.75	–	–
	Reduction		25 308.29	5.69	0.60
Water ecology		100 million m^3	Volume based ecological water occupation	Quality based ecological water occupation	Total
	Opening value		40.46	25.99	66.45
	Closing value		39.85	21.39	61.24
	Changes		−0.61	−4.60	−5.21

Note: Depletion of surface water = Abstraction of surface water − Available surface water resources (2)
Depletion of groundwater = Abstraction of shallow groundwater − Available shallow groundwater + Abstraction of deepgroundwater (3)
Volume based ecological water occupation = local water withdrawal − returned water − environmental wateruse (4)
Quality based ecological water occupation = amount of unqualified rivers + amount of unqualified lakes + amount of unqualified reservoirs (Bai et al., 2001) (5)

Table 2. Emergy assessment of water resources value in research area.

Item	Chemical emergy of rainwater	Solar emergy of rainwater	Unit water resources solar emergy	Unit water resources value	
	$\times 10^{16}$ J^{-1}	$\times 10^{20}$ sej^{-1}	$\times 10^{11}$ (sej m^{-3})	/(USD m^{-3})	/(CNY m^{-3})
	5.28	8.13	1.23	0.12	0.80

Water resources depletion, waste water discharge and ecological water occupation in Chengdu are estimated based on *Chengdu Water Resources Bulletin*, *Chengdu Environment Quality Bulletin*, *Chengdu Statistical Yearbook (2014–2015)* and other materials as well as "The Three Red Lines" control indicators. Considering the availability of data, Chengdu physical accounts of water resources assets in 2015 are established as shown in Table 1.

It can be seen that, in terms of water resources, in 2015, beginning stock and ending stock of water resources in Chengdu were 9552 and 9382 million m^3, respectively, a decrease of 170 million m^3, and total water resources depletion reached 1938 million m^3; in terms of water environment, untreated wastewater was 253 million tons, and as total COD and ammonia nitrogen emissions in 2015 should be limited to 120 100 and 14 000 tons according to implementation opinions on cutting down total amount of major pollutants under the Twelfth Five-year Plan of Chengdu; emissions of COD and ammonia nitrogen should be reduced by 56 900 and 6000 tons; in terms of water eco-system, ecological water occupation caused by human activities in 2015 reached 6124 million m^3, a decrease of 521 million m^3 compared to that at the end of 2014.

Table 3. The value accounts of water resources assets in Chengdu.

Item				
1. Water resources	Opening stock value	Net value change (100 million CNY)	Closing stock value	Depletion cost
1.1. Surface water	6.45	−0.07	6.38	64.73
1.2. Groundwater	312.59	−5.61	306.98	0
1.3. Total water resources	319.04	−5.68	313.36	64.73
2. Water environment	Amount of pollutants to be reduced (10 000 tons)		Environmental degradation cost (100 million CNY)	
2.1. Wastewater	25 308.29		3.97	
2.2. COD	5.69		5.46	
2.3. Ammonia nitrogen	0.60		10.02	
3. Water eco-system		Opening water occupation value	Closing water occupation value (100 million CNY)	Value changes in current year
3.1. Ecological water occupation		52.89	48.75	−4.14

4.3 Water price in Chengdu

Price is a bridge between physical volume and magnitude of water resources value. Three different methods are adopted to calculate the value for water resources, water environment and water eco-system in physical account.

1. Firstly, shadow prices method is used to calculate water resources value. By reference to shadow price calculation made by He and Chen (2005) for nine major basins in China in important years by applying non-linear dynamic input-occupancy-output model, in 2015, shadow price in Chengdu was about 3.34 CNY m^{-3}, slightly higher than local domestic water price (2.96 CNY m^{-3}).

2. Defensive expenditures method is adopted to calculate water environment degradation cost. The calculation formula is:

$$\begin{aligned}\text{defensive expenditures} &= \text{reduced cost of unit pollutants}\\ &\times \text{amount of pollutants to be reduced}\\ &= \text{treatment cost of wastewater}\\ &\times \text{wastewater to be treated}\end{aligned} \quad (2)$$

   According to investigation, Chengdu's unit wastewater treatment cost is about 1.57 CNY m^{-3}, including 0.17 CNY m^{-3} of fixed wastewater treatment cost, 1.07 CNY m^{-3} of unit wastewater treatment operation expense and 0.33 CNY of unit wastewater treatment statutory tax. Based on results of national environmental and economic accounting for water, reduced expenses of COD and ammonia nitrogen were 9.6 and 167.0 CNY kg^{-1}, respectively.

3. Emergy analysis method is used to account ecological water occupation which mainly considers the water resources necessary for various types of ecosystems occupied by economic and social overuse. This part of water originally belongs to natural water resources, and therefore the value of natural water resources is taken as the value of ecological water occupation. According to the concept analysis of the energy conversion process of water resources, the rainwater in the study area is used as the energy source of the water resources, assuming that solar emergy of rainwater is equal to that of water resources. Relevant calculation formulas are (Chen et al., 2006):

$$\begin{aligned}\text{chemical emergy of rainwater} &= \text{water volume}\\ &\times \text{Gibbs free emergy} \times \text{density}\end{aligned} \quad (3)$$

$$\begin{aligned}\text{solar emergy of rainwater}\\ &= \text{chemical emergy of rainwater}\\ &\times \text{solar emergy conversion rate of rainwater}\end{aligned} \quad (4)$$

$$\begin{aligned}\text{unit water resources solar emergy}\\ &= \text{solar emergy of water resources}\\ &\div \text{amount of water resources}\end{aligned} \quad (5)$$

$$\begin{aligned}\text{unit water resources value}\\ &= \text{unit water resources solar emergy}\\ &\div \text{emergy/currency rate}\end{aligned} \quad (6)$$

Relevant parameters are: Gibbs free emergy is 4.94 J g^{-1} (Odum, 1996); solar emergy conversion rate of rainwater is 1.54×10^4 sej J^{-1} (Zeng et al., 2011); emergy currency^{-1} rate in research area in 2015 is 1.00×10^{12} sej USD^{-1} (Zeng et al., 2011); RMB exchange rate is 6.49 CNY USD^{-1}. Results of water resources emergy in Chengdu are shown in Table 2.

4.4 Value accounts of waters resources assets in Chengdu

Value accounts of water resources assets derive corresponding aggregate indicators through overall accounting of water resources assets based on physical account. Figure of each item in value accounts of water resources is equal to the price of water resource and their quantity in physical account. We can learn from Table 3 that, at the end of 2015, the value of water resources stock in Chengdu totaled 31 336 million CNY including 638 million CNY of surface water stock, 30 698 million CNY of groundwater stock and water resources depletion cost reached 6473 million CNY; environmental degradation costs based on wastewater discharges was 397 million CNY, and amounts of COD and ammonia nitrogen to be reduced were 546 million CNY and 1002 million CNY, respectively; Based on emergy analysis method, unit water resources value in Chengdu in 2015 is 0.80 $\mathrm{CNY\,m^{-3}}$, then we multiply it with ecological water occupation in the current year and then derive ecological occupation cost at 4875 million CNY.

4.5 Discussion

In this paper, the water resources accounting of Chengdu was carried out based on the idea of first physical, then value. The research results are discussed as follows:

1. In the physical account, the calculation method of ecological water occupation is not perfect which has a duplicate part with the water resource depletion. In the future research, in consideration of the minimum ecological base flow occupied by the process of water abstraction, the process impact of the decline in ecological service function can be analyzed dynamically.

2. The practice of water resources accounting indicated that, accurate shadow price in the calculation of water resources depletion cost requires support from complete data and calculation should be made at municipal level or above; as pollutant control expense calculated by protection expense method is less than environmental damage caused by pollutant emission, such expense can only be used as the lower limit for environmental degradation cost; now, water eco-system in water resources balance sheet haven't been well studied in China and both physical volume of ecological water occupation and calculation method for ecological value of water resources still await further research and practice.

5 Conclusions

Beginning with the connotation of water resources value, this paper analyzed basic principles and existing flaws of water resources value theories and accounting methods, proposed to primarily account water resources depletion cost, water environment degradation cost and ecological water occupation cost from three aspects, i.e. water resources, water environment and water eco-system. Physical and value accounts of water resources in Chengdu, China were established based on shadow prices method, defensive expenditures method and emergy analysis method. Although these methods have certain deficiencies which need to be improved in future research, this paper makes a preliminary attempt to shift from physical accounts of water resources to value accounts and advances the compilation of water resources balance sheet.

Data availability. The data used in this article are from the Chengdu Water Authority (2015; http://www.cdwater.gov.cn/wsbs/bmcx/sl/system/2017/03/1/000002497.html); the Chengdu Bureau of Statistics (2016; http://www.cdstats.chengdu.gov.cn/htm/detail_63038.html); and the Chengdu Environmental Protection Bureau (2015; http://www.cdepb.gov.cn/cdepbws/Web/Template/GovDefaultInfo.aspx?cid=65205&aid=B1C16A95EB0242198B5D67870E8BE084).

Competing interests. The authors declare that they have no conflict of interest.

Special issue statement. This article is part of the special issue "Innovative water resources management – understanding and balancing interactions between humankind and nature". It is a result of the 8th International Water Resources Management Conference of ICWRS, Beijing, China, 13–15 June 2018.

Acknowledgements. This study is funded by the National Science Foundation of China (no. 51779160, 51579161).

Edited by: Zongxue Xu
Reviewed by: two anonymous referees

References

Bai, Y. L., Wang, R. S., and Yao, L.: Ecological water depletion by human use in Beijing City, Acta Ecologica Sinica, 31, 4415–4426, 2011.

Chen, D., Chen, J., and Luo, C. H.: Evaluation method of natural water resources based on emergy theory and its application, J. Hydraul. Engin., 37, 1188–1192, https://doi.org/10.13243/j.cnki.slxb.2006.10.006, 2006.

China Institute of Water Resources and Hydropower Research: Chinese Environmental Economic Accounting Report, 110–114, 2009.

Feng, Z. M., Yang, Y. Z., and Li, P.: From Natural Resources Accounting to Balance-sheet of Natural Resources Asset Compilation, Ecosystem Service and Assessment, 29, 449–456, 2014.

Gan, H., Qin, C. H., Wang, L., and Zhang, X. J.: Study on water pricing method and practice – Discussion on the connotation of water resources value, J. Hydraul. Eng., 43, 289–295, 2012.
Gan, H., Wang, L., Qin, C. H., and Jia, L.: Understanding of balance sheet of water resources, China Water Resources, 14, 1–7, 2014.
Geng, J. X., Li, Z. J., Hu, T. Y., and Tang, J. L.: Practice and Exploration of Balance Sheet of Natural Resources – A Case Study of Yongning, Ningxia, Friends of Accounting, 5, 9–25, 2017.
He, J. and Chen, X. K.: Researches on sub price calculation of dynamic water resources of nine river basins in China, Journal of Economics of Water Resources, 23, 14–19, 2005.
Huang, X. R., Guo, B. Y., Xi, Y. Y., Wang, Y. Q., and Gao, L. Y.: Perspective on theories and methods study of the compilation of water resources balance sheet, Journal of Water Resource and Water Engineering, 28, 1–5, 2017.
Jia, L., Gan, H., Wang, L., and Qin, C. H.: Discussion on Water Resources Liabilities, Journal of Natural Resources, 32, 1–11, 2017.
Jian, F. H., Song, X. Y., Yu, W. B., and Cai, G. Y.: Establishment and Analysis of Water Assets Accounts for Water Balance Sheet Compilation: a case study of Zhangye City in Heihe River Basin, J. Desert Res., 36, 851–856, 2016.
Jiang, W. L.: Theory of water resource value, Beijing, Science Press, 1998a.
Jiang, W. L.: A Study on Model of Water Resources Value, Resour. Sci., 20, 36–37, 1998b.
Jiang, W. L.: The evolution course problems and expectation on check and calculation of natural resources in China, Territory and Natural Resource Study, 3, 43–46, 1999.
Jiang, W. L. and Wang, H. D.: Current situation and prospection research of water resource value in China, Geography and Territorial Research, 12, 1–5, 1996.
Jiao, R. J., Geng, J. X., and Wu, X. Y.: Discussion on Method of Preparing Water Balance Sheet in China, Wa. Sci. Technol., 1, 214–220, 2015.
Lv, C. M., Wu, Z. N., and Hu, C. H.: Progress and prospection theory research of water resource value, Resources and Environment in the Yangtze Basin, 18, 545–549, 2009.
Odum, H. T.: Environmental Accounting: Emergy and Environmental Decision Making, New York, John Wiley & Sons, 1996.
Shen, D. J., Liang, R. J., Wang, H., and Jiang, Y. Z.: The Water Resources Value, J. Hydraul. Eng., 5, 54–59, 1998.
Wang, Y. D.: The Positioning Strategy of the Balance Sheet of Natural Resources in China, Statistics and Decision, 12, 52–55, 2017.
Wang, L. and He, Q.: A Review of the Ecosystem Value Based on Emergy Theory, Ecological Economy, 31, 133–136, 2015.
Wang, S. M. and Qu, F. T.: Water resource accounting and adjustment of GDP – A case study of the developed areas in the eastern part of China, Journal of Nanjing Agricultural University, 24, 115–118, 2001.
Wang, D. X., Wang, H., and Yin, M. W.: Implication of water resources and its value, Adv. Water Sci., 10, 195–200, 1999.
Wang, H., Ruan, B. Q., and Shen, D. J.: Water Price Theory and Practice for Sustainable Development, Beijing, Science Press, 2003.
Xu, Z. C., Ye, Y. X., and Peng, X. C.: Research progress in water resource value accounting, Ecol. Environ., 15, 1117–1121, 2006.
Zeng, X., Yao, J., and Sun, H.: Sustainability assessment on eco-economic system of Chengdu City based on emergy analysis, Chinese J. Ecol., 30, 2875–2880, 2011.

https://doi.org/10.5194/piahs-379-287-2018

Open Access
PIAHS

Analysis of the spatial-temporal change of the vegetation index in the upper reach of Han River Basin in 2000–2016

Jinkai Luan[1], Dengfeng Liu[1,2], Lianpeng Zhang[1], Qiang Huang[1], Jiuliang Feng[3], Mu Lin[4], and Guobao Li[5]

[1]State Key Laboratory of Eco-hydraulics in Northwest Arid Region of China, School of Water Resources and Hydropower, Xi'an University of Technology, Xi'an 710048, China
[2]Department of Land Resources and Environmental Sciences, Montana State University, Bozeman, MT 59717, USA
[3]Shanxi Provincal Water and Soil Conservation and Ecological Environment Construction Center, Taiyuan 030002, China
[4]School of statistics and Mathematics, Central University of Finance and Economics, Beijing 100081, China
[5]Work team of hydraulic of Yulin City, Yulin 719000, China

Correspondence: Dengfeng Liu (liudf@xaut.edu.cn)

Received: 29 December 2017 – Revised: 25 March 2018 – Accepted: 26 March 2018 – Published: 5 June 2018

Abstract. Han River is the water source region of the middle route of South-to-North Water Diversion in China and the ecological projects were implemented since many years ago. In order to monitor the change of vegetation in Han River and evaluate the effect of ecological projects, it is needed to reveal the spatial-temporal change of the vegetation in the upper reach of Han River quantitatively. The study is based on MODIS/Terra NDVI remote sensing data, and analyzes the spatial-temporal changes of the NDVI in August from 2000 to 2016 at pixel scale in the upper reach of Han River Basin. The results show that, the area with increasing NDVI between 0 and 0.005 per year accounts for 62.07 % of the area of upper reach of Han River Basin, and the area with changing rate between −0.005 and 0 per year accounts for 26.65 % of the research area. The area with significant decreasing trend only accounts for 2.76 %, while area significant increasing trend accounts for 13.47 %, and the area with increasing NDVI is much larger than the area with reducing NDVI. The vegetation index of each county is evaluated and found that, the areal proportion with significant decreasing trend in Hantai is the biggest, reaching 35.57 %. The areal proportion with significant increasing trend in Zhenba County, Ziyang County, Xunyang County, Zhashui County, Shangzhou District, Shanyang County and Yun County is larger than the others, and the areal proportions are more than 20 %. The largest areal proportion with significant increasing trend is in Shangzhou District and it reaches 31.11 %. On the whole, the area ratio in all districts and counties with increasing NDVI is much larger than the area ratio with decreasing NDVI.

1 Introduction

Vegetation is a link between soil, atmosphere and water (Li et al., 2011), and plays a coordinating role in the balance of the earth ecosystem, climate change and water cycle (Sun et al., 1998; Mao et al., 2012). At the same time, vegetation is one of the most important and sensitive components of global land cover, which can reflect the impacts of climate change and human activities in a short period (Zhang et al., 2008). As the best indicator of vegetation growth and coverage, Normalized difference vegetation index (NDVI) is the surface characteristic parameter of remote sensing data, which reflects the distribution characteristics and changes of vegetation, and is the basic data of vegetation research (Zhao, 2012; Marchetti et al., 2016; Wu et al., 2017; Birtwistle et al., 2016).

Based on MODIS NDVI remote sensing data, Liu et al. (2017) analyzed the vegetation change in the southern Mu Us desert from 2000 to 2013, and concluded that the average annual NDVI showed an upward trend with an annual growth rate of 2.69×10^{-3} per year, and also analyzed the level of NDVI year by year transfer. Luo et al. (2017) used the trend analysis, R/S analysis, partial correlation analysis and residual analysis methods to analyze the spatial and temporal variation characteristics of NDVI in the past 2000–2013 years. Based on the MODIS NDVI vegetation data from 2001 to 2010 in Qaidam Basin, Xu and Yang (2014) used linear trend analysis method, and analyzed the variation characteristics of vegetation in the Qaidam Basin in a decade. Liu et al. (2015) found that the vegetation NDVI showed a significant upward trend in North China in 1981–2006 years, with an increase rate of 0.009/10a, and pointed out that there was obvious spatial difference in this change. Zhang et al. (2013) found that the annual growth rate of NDVI in Inner Mongolia grassland growth season was 0.735 %, among which the typical grassland growth was the most obvious, and 31.807 % of grassland NDVI showed a downward trend. After studying the vegetation index of Yulin in Shaanxi Province, the authors found that the vegetation index of most areas of Yulin has shown a significant increase in recent 17 years (Luan et al., 2017). Using the trend analysis method, Lv et al. (2014) analyzed the temporal and spatial variation of vegetation cover in the Yarlung Zangbo River Basin based on the pixel size and found that the improved area of NDVI in the watershed was larger than the degraded area, showing a slowly rising trend overall, but the persistence of this trend showing a significant spatial difference. Bai et al. (2014) used correlation analysis and spatial statistical analysis to analyze the temporal and spatial variation characteristics of 2000-2010 vegetation NDVI in Northern Shaanxi Province. It is found that the vegetation NDVI in Northern Shaanxi has experienced 3 stages, from relatively rapid growth to small fluctuations, and then to steady growth. Li and Chen (2014) analyzed the temporal and spatial distribution of NDVI in the arid region of northwest China from 1982 to 2006 and found that the NDVI of vegetation coverage and growing season increased significantly before 2000 but decreased after 2000. The obvious decrease is in Yili Valley, Zhongtian Mountain and plain areas, showing an increasing trend in the mountainous areas or at the source of rivers and on some banks of some rivers. Generally, the analysis of NDVI spatial and temporal variation is based on the linear regression analysis, which can be used to analyze the variation and spatial distribution of NDVI.

Han River Basin is the water source of the Middle Route Project of South-to-North Water Diversion and the water diversion project from Han to Wei River. In recent years, the ecological protection projects such as comprehensive watershed management and vegetation conservation have been carried out extensively. The implementation of these projects, especially the "Changzhi" and "Danzhi" project has achieved a certain effect (Gao, 2012; Wang, 2012; Wang et al., 2011). With the implementation of the ecological protection project, the vegetation in the Han River Basin has been improved to some extent. In order to reveal the temporal and spatial characteristics of vegetation change and evaluate the implementation effect of ecological engineering, it is an effective method to quantify the vegetation index by remote sensing.

In the related research, Ma (2013) carried on the trend line analysis to the annual average NDVI of the whole Han River Basin in Shaanxi Province in 2000–2010 years, compared the NDVI in 2000 and 2010, but did not carry on the spatial and temporal variation analysis and the change degree evaluation based on the pixel scale. At present, there are few studies on the temporal and spatial variation characteristics of vegetation index in the upper reach of Han River Basin, especially the spatial variability of vegetation index at pixel scale. Therefore, taking the upper reach of the Han River Basin as the research area, NDVI of August (growing season) from 2000 to 2016 was analyzed by linear regression trend line on the pixel scale to evaluate the NDVI variation of each pixel. Finally, the vegetation index of each district in the upper reach of the Han River Basin was evaluated to reveal the changes of vegetation on each scale in each county in recent years. This may provide reference for policy making and effect evaluation of ecological protection of water sources in the middle route of South-to-North Water Diversion Project.

2 Research area overview

Han River is the largest tributary of the Yangtze River, with the total length of 1577 km, the basin area of 159 000 km^2, and the average annual water resources 56.6 billion m^3. Taking Danjiangkou Reservoir and Zhongxiang as the boundaries of upper, middle and lower reach, it flows into the Yangtze River in Wuhan, and the upper reach of the Han River Basin are located above Danjiangkou Reservoir (Fig. 1). The basin includes Hanzhong City in Shaanxi Province, Ankang City in Shaanxi Province, Shangluo City in Shaanxi Province, Hubei Province Shiyan City, in addition to a small part of Gansu Province and Henan Province. The western region of the study area belongs to the north subtropical monsoon climate, and the eastern part is the transition climate between the north subtropical zone and the warm temperate zone. The climate is mild, the heat is relatively sufficient, and the four seasons are distinct (Zhou et al., 2011).

3 Data and methods

3.1 Data

Using MODIS NDVI data to analyze vegetation changes is one of the most important ways to study vegetation evolution (Liu and Tian, 2015). So this article uses NDVI remote sensing data from the MODIS/Terra website, the full name of the dataset is MODIS/Terra Vegetation Indices 16-Day L3 Global 250 m SIN Grid V005, referred to as MOD13Q1.

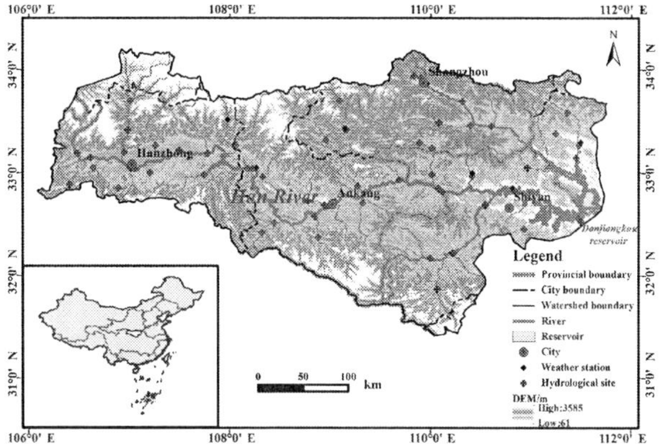

Figure 1. Location and topographic map of upper reach of Han River Basin.

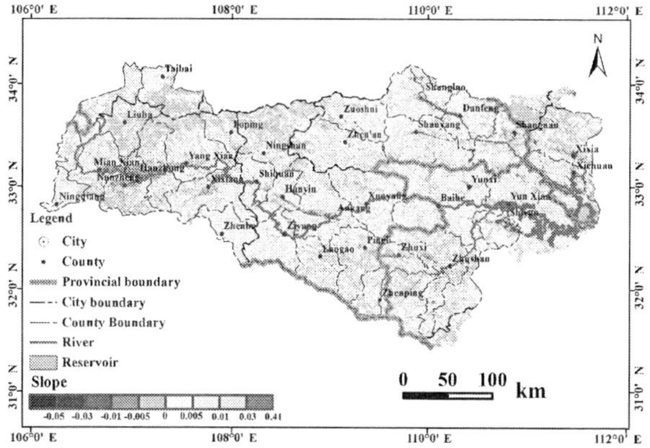

Figure 2. Distribution of NDVI linear regression slope in upper reach of Han River Basin from 2000 to 2016.

MOD13Q1 data is a 16-day composite with a spatial resolution of 250 m. The data file contains data for 12 fields. Valid NDVI values are between −1 and 1.

In this paper, two images of h26v5 and h27v5 are used. Firstly, the downloaded images are combined, cropped and re-projected by using the software MODIS Reprojection Tool (MRT). In this study, only NDVI images were analyzed in August, MOD13Q1 data had 2 images in August, and the images were maximized and composited. So we get an image every year, and 17 images in 2000–2016 years.

3.2 Linear regression trend line analysis

Linear regression trend line analysis is a method of regression analysis for a set of variable with time. The method can analyze the change trend of each pixel in the study area, and simulate the Greenness Rate of Change (GRC) of vegetation. GRC is defined as the slope of the linear regression equation for the inter-annual variability of the Seasonally Integrated Normalized Difference Vegetation Index (SINDVI) over a period of time (Song and Ma, 2008; Stow et al., 2003). In this paper, the NDVI of each pixel for 17 years can be used to simulate the change trend of NDVI of this pixel during the 17 years and to estimate the variation range. The regression slope is calculated by the least square method. The formula is as follows:

$$\text{slope} = \frac{17 \cdot \sum_{i=1}^{17} i \cdot \text{NDVI}_i - \left(\sum_{i=1}^{17} i\right)\left(\sum_{i=1}^{17} \text{NDVI}_i\right)}{17 \cdot \sum_{i=1}^{17} i^2 - \left(\sum_{i=1}^{17} i\right)^2}, \quad (1)$$

where i is the ith year from 1 to 17 years; NDVI_i is the NDVI value in ith year. If slope > 0, NDVI has shown an increasing trend; slope = 0, NDVI has not changed; slope < 0, NDVI showed a decreasing trend.

Table 1. NDVI variance linear regression trend line analysis of changes in upper reach of Han River Basin from 2000 to 2016.

NDVI tread	Area (km^2)	Proportion (%)
slope ≤ −0.05	0.43	0.00
−0.05 < slope ≤ −0.03	21.84	0.03
−0.03 < slope ≤ −0.01	616.71	0.83
−0.01 < slope ≤ −0.005	1555.74	2.10
−0.005 < slope ≤ 0	19695.95	26.65
0 < slope ≤ 0.005	45871.65	62.07
0.005 < slope ≤ 0.01	5674.56	7.68
0.01 < slope ≤ 0.03	468.17	0.63
slope > 0.03	2.74	0.00

4 Results

4.1 Temporal and spatial variation characteristics of NDVI

By means of linear regression trend line analysis, the pixel NDVI spatial distribution slope map (Fig. 2) and different levels of slope area in 2000–2016 years in August (Table 1) are obtained. The slope of the whole basin was between −0.054 and 0.041, and the vegetation index increased more than the decrease range. The area of green region in Fig. 2 is larger than the area of red region, indicating that the area of increased vegetation index in recent 17 years is larger than the area of reduced vegetation index. There was no obvious regularity in the distribution of the areas where vegetation index increased and decreased.

Table 1 is a linear regression trend line analysis of vegetation index in Han River Basin, the area of reduced NDVI is 21 890.67 km^2, accounting for 29.61 % of the total area of the whole basin. The area of slope between 0 and −0.005 is 19 695.95 km^2, accounting for 26.65 % of the total area of Han River Basin (above Danjiangkou). The area of increased

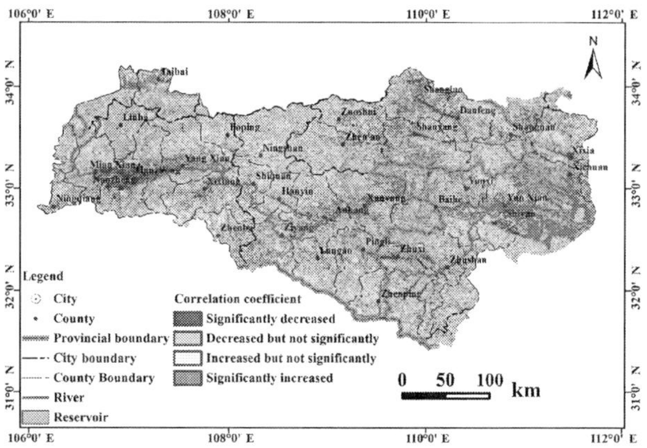

Figure 3. The significance test of NDVI dynamic change from 2000 to 2016.

NDVI is 52 017.12 km^2, accounting for 70.39 % of the total area of the river basin, in which the area of slope between 0 and 0.005/a accounts for 62.07 %. It shows that over 70 % of the Han River basin (above the Danjiangkou) has shown an increasing trend in NDVI in recent years, and the increasing value is basically between 0 and 0.005/a. The vegetation coverage of the Han River basin (above the Danjiangkou) is gradually increasing.

4.2 Evaluation of variation degree about NDVI

In order to quantitatively analyze the degree of NDVI change in the study area, correlation coefficient analysis is conducted between annual NDVI images and time series to obtain the regression coefficient of each NDVI value and year, that is r_{xy}, and the coefficient was tested at the significant level of 0.05, and the distribution is shown in Fig. 3.

If the r_{xy} value is greater than 0.05 significance level ($p < 0.05$) $r_{0.05} = 0.482$, the trend of NDVI increase or decrease is considered significant. As shown in Fig. 3, the distribution of area significantly decreased, decreased but not significantly, increased but not significantly and significantly increased was disorganized. So it is necessary to quantitatively analyze the area and proportion of each region (Table 2). In Table 2, the areas with significant reduction accounted for only 2.76 % of the whole basin, the areas with lower but less significant accounted for 26.86 %, the areas with increased but not significant accounted for 56.91 %, and the areas with significant increase accounted for 13.47 %. By comparing with Table 1, it is found that the region of slope ≤ -0.005/a is substantially reduced. In general, the index of vegetation in the upper reaches of the Han River shows an increasing trend, which means that the vegetation coverage gradually increases. According to the Soil and Water Conservation Bulletin of the Yangtze River Basin, it was found that the total area of water and soil erosion in the Danjiangkou reservoir area and the upper reaches of Danjiangkou was 20 762 km^2.

Table 2. NDVI change evaluation in upper reach of Han River Basin from 2000 to 2016.

NDVI change evaluation	Area (km^2)	Proportion (%)
Significantly decreased	2038.61	2.76
Decreased but not significantly	19852.06	26.86
Increased but not significantly	42063.06	56.91
Significantly increased	9954.04	13.47

The Danjiangkou reservoir area and upstream is a key region of the national soil and water conservation area, and the soil erosion is serious before. The soil and water conservation project has achieved a certain effect and the vegetation coverage increased.

4.3 Evaluation of the change degree of NDVI in each county

Because the spatial variation of each level of change is irregular, in order to find out the changes of vegetation in each county in recent years, we evaluate the NDVI of each county in the upper reach of the Han River basin, which will provide a reference for better environmental policy. Since there are many counties in the upper reach of the Han River basin and only a small part of some counties are located in the upper reach of the Han River basin, Therefore, this paper selects Taibai County of Baoji City, Hantai District, Lueyang County, Ningqiang County, Mianxian County, Nanzheng County, Liuba County, Chenggu County, Yang County, Xixiang County, Zhenba County and Foping County; Ningshan County, Shiquan County, Hanyin County, Ziyang County, Hanbin District, Langao County, Xunyang County, Pingli County, Zhenping County and Baihe County; Shangzhou District, Zhashui County, Zhen'an County, Sanyang County, Danfeng County and Shangnan County, Xixia County in Henan Province, Zhuxi County, Zhushan County, Yunxi County, Fang County and Yun County in Hubei Province County and Shiyan City (Maojian District, Zhangwan District, Shiyan Economic Development Zone), a total of 35 counties, including almost all the upper reach of the entire basin, according to Fig. 1. To evaluate the changes of NDVI in these counties, and determine the proportions of each change degree of each county and city in the upper reach of Han River, as the results are shown in Table 3.

As shown in Table 3, as long as the proportion of Hantai District area is significantly reduced by more than 10 %, and even reached 35.57 %, the proportion of significant reduction in other districts and counties is below 10 %. The proportion of increased but not significant accounted for the largest proportion, basically all districts and counties have reached more than 50 %, and some counties close to 70 %. The proportion of decreasing but not significant proportion was the second, but fluctuated among different dis-

Table 3. NDVI change evaluation of every counties and district in upper reach of Han River Basin from 2000 to 2016.

Name	Area proportion of			
	decreased significant	decreased but not significantly	increased but not significantly	increased significantly
	(%)			
Taibai County	1.71	41.42	41.91	14.97
Hantai District	35.57	38.26	24.46	2.19
Lueyang County	0.79	28.61	65.04	5.56
Ningqiang County	1.29	16.38	67.22	15.11
Mianxian County	6.93	30.18	56.89	6.00
Nanzheng County	7.34	50.32	40.31	2.03
Liuba County	1.00	51.52	42.96	4.51
Chenggu County	4.12	42.78	50.30	2.80
Yang County	2.18	33.55	55.79	8.48
Xixiang County	1.87	24.45	62.52	11.17
Zhenba County	0.12	8.26	67.15	24.47
Foping County	2.22	38.66	53.83	5.30
Ningshan County	0.72	43.51	50.34	5.42
Shiquan County	0.85	15.15	68.72	15.28
Hanyin County	1.57	19.96	68.33	10.13
Ziyang County	0.28	10.27	61.35	28.10
Hanbin District	1.53	18.89	62.91	16.66
Langao County	1.41	31.03	58.18	9.39
Xunyang County	0.19	15.85	63.83	20.12
Pingli County	2.40	30.28	58.10	9.22
Zhenping County	1.38	30.24	61.15	7.24
Baihe County	0.46	15.82	66.71	17.01
Shangzhou District	5.51	14.13	49.25	31.11
Zhashui County	0.68	18.88	56.18	24.26
Zhen'an County	0.27	24.62	61.11	14.00
Shanyang County	2.00	18.55	57.77	21.68
Danfeng County	2.71	22.11	59.29	15.89
Shangnan County	3.64	28.06	58.07	10.23
Xixia County	6.46	36.93	50.13	6.48
Zhuxi County	4.55	29.37	55.67	10.40
Zhushan County	3.74	34.74	54.65	6.87
Yunxi County	1.11	15.02	65.03	18.84
Yun County	2.56	13.11	59.69	24.64
Fang County	1.00	29.37	58.76	10.88
Shiyan City	7.77	22.23	60.63	9.37

tricts and counties, between 8.26 ∼ 51.52 %. Significant increase in a larger proportion for Zhenba County, Ziyang County, Xunyang County, Zhashui County, Shangzhou District, Shanyang county and Yun county, were more than 20 %. The largest is in Shangzhou District, and it reached 31.11 %. Overall, the NDVI increased proportion in each county is far greater than the proportion of reduced. The reason for the decrease of vegetation index in Hantai is that in recent years, Hantai district has been actively developing and building industrial parks, and the total area of cultivated land has decreased by 16.56 % from 2000 to 2016.

5 Conclusion

This paper uses MODIS/Terra NDVI time series data to analyse the temporal and spatial variation characteristics of vegetation index in the upper reach of Han River during growing season (August) in 2000–2016, and the conclusions are the following.

1. According to the linear regression trend line analysis, it was found that 70.39 % of the upper reach of the Han River basin had an increased regional vegetation index, and the area with increasing rate of 0–0.005/a accounts for 62.07 % of the total area; the area with slope between 0 and −0.005 accounts for 26.65 % of the total area in the upper reach of the Han River Basin; the variation of NDVI in the upper reach of the basin is basically between −0.005 and 0.01/a.

2. After the assessment of the change of vegetation index in the upper reach of the Han River Basin, it was found that the significant reduction area accounted for only 2.76 % of the whole basin, and the significant increased area accounted for 13.47 %. The increased area is far greater than the reduced area.

3. In order to reveal the distribution of significant changes in the region, the vegetation index in each county was evaluated and found that the largest proportion of significantly decreased is Hantai and it reached 35.57 %, while the percentage of other districts and counties significantly decreased below 10 %. The Zhaoyang County, Xiyang County, Xunyang County, Shangzhou District, Zhashui County, Shanyang County and Yun County were all significantly increased by more than 20 %, and the largest is Shangzhou District, reaching 31.11 %. On the whole, the area ratio in all districts and counties with increasing NDVI is much larger than the area ratio with decreasing NDVI.

Data availability. For specific basic data,please contact the author Luan Jinkai (ljk_940202@163.com).

Competing interests. The authors declare that they have no conflict of interest.

Special issue statement. This article is part of the special issue "Innovative water resources management – understanding and balancing interactions between humankind and nature". It is a result of the 8th International Water Resources Management Conference of ICWRS, Beijing, China, 13–15 June 2018.

Acknowledgements. This work was partially supported by the National Natural Science Foundation of China (NSFC) (Grant No. 51779203, 51609270, 41330858). Funding was also provided by the Basic Research Plan of Natural Science of Shaanxi Province (2016JQ5105) and Project of the plan of science and technology of Yulin (2016-21). The authors would like to thank the anonymous reviewers for their comments to improve the paper.

Edited by: Zongxue Xu
Reviewed by: two anonymous referees

References

Bai, J.-j., Bai, J.-t., and Wang, L.: Spatio-temporal Change of Vegetation NDVI and Its Relations with Regional Climate in Northern Shaanxi Province in 2000–2010, Scientia Geographica Sinica, 34, 882–888, 2014.

Birtwistle, A. N., Laituri, M., Bledsoe, B., and Friedman, J. M.: Using NDVI to measure precipitation in semi-arid landscapes, J. Arid Environ., 131, 15–24, 2016.

Gao, Q.: Problems and Countermeasures of ecological management in Hanjiang River Basin, Shaanxi Agricultural Sciences, 3, 192–195, 2012.

Li, M. J., Hou, X. Y., Ying, L. L., Lu, X., and Zhu, M. M.: Spatial-Temporal Dynamics of NDVI and Its Response to Temperature and Precipitation in the Yellow River Delta during the Period 1998–2008, Resource Science, 33, 322–327, 2011.

Li, Q. and Chen, Y.: Response of spatial and temporal distribution of NDVI to hydrothermal condition variation in arid regions of Northwest China during 1981–2006, Journal of Glaciology and Geocryology, 36, 327–334, 2014.

Liu, B., Sun, Y., Wang, Z. H., and Zhao, T.: Analysis of the vegetation cover change and the relative role of its influencing factors in North China, Journal of Natural Resource, 30, 12–23, 2015.

Liu, D. and Tian, F.: Ecohydrological Evolution Model On Riparian Vegetation In Lower Tarim River, Science Press, Beijing, 37–46, 2015.

Liu, D., Wang, X., Huang, Q., Meng X., and Lin M.: Study on the evolution law of the vegetation index in the southern Maowusu Desert, Journal of Water Resources and Water Engineering, 28, 5–9, 2017.

Luo, M., Jiapaer, G., Guo, H., Zhang, P.-f., Meng, F.-h., and Liu, T.: Spatial-temporal Variation of Growing-season NDVI and Its Responses to Hydrothermal Condition in the Tarim River Basin from 2000 to 2013, Journal of Natural Resource, 32, 50–63, 2017.

Luan, J., Liu, D., Huang, Q., Feng, J., Lin M., and Li G.: Analysis of the spatial-temporal change and impact factors of the vegetation index in Yulin of Shaanxi Province in the last 17 years, J. Ecol., 38, 1–11, 2018.

Lv, Y., Dong, G., Yang, Sh., Zhou, Q., and Cai, M.: Spatio-Temporal Variation in NDVI in the Yarlung Zangbo River Basin and Its Relationship with Precipitation and Elevation, Resource Science, 36, 0603–0611, 2014.

Ma, C.: Multi-scale vegetation cover changes in Shaanxi Hanjiang river basin, Journal of Shaanxi University of Technology (Natural Science Edition), 29, 66–70, 2013.

Mao, D., Wang, Z., Luo, L., and Guang, Y.: Correlation Analysis between NDVI and Climate in Northeast China based on AVHRR and GIMMS Data Sources, Remote Sensing Technology and Application, 27, 77–85, 2012.

Marchetti, Z. Y., Minotti, P. G., Ramonell, C. G., Schivo, F., and Kandus, P.: NDVI patterns as indicator of morphodynamic activity in the middle Paraná River floodplain, Geomorphology, 253, 146–158, 2016.

Song, Y. and Ma, M.: Variation of AVHRR NDVI and its Relationship with Climate in Chinese Arid and Cold Regions, Journal of Remote Sensing, 12, 499–505, 2008.

Sun, H., Wang, Ch., Niu, Zh., Bu, He., and Li, B.: Analysis of the Vegetation Cover Change and the Relationship between NDVI and Environmental Factors by Using NOAA Time Series Data, Journal of Remote Sensing, 2, 204–210, 1998.

Stow, D., Daeschner, S., Hope, A., Douglas, D., Petersen, A., Myneni, R., Zhou, L., and Oechel, W.: Variability of the seasonally integrated Normalized Difference Vegetation Index across the north slope of Alaska in the 1990s, Int. J. Remote Sens., 24, 1111–1117, 2003.

Xu, H.-j. and Yang, T.-b.: Spatial-temporal variations of Vegetation Activities and Its Responses to Climatic Factors in the Qaidam Basin, Journal of Natural Resource, 29, 398–409, 2014.

Wang, X.: Technology and practice of ecological clean small watershed construction in Dan Han River Basin of Shaanxi Province, Soil and water conservation in China, 2, 11–13, 2012.

Wang, X., Li, Zh., Li, P., and Zhang, L.: Present situation and Control Countermeasures of non-point source pollution in Dan Han River Basin of Shaanxi Province, Soil and Water Conservation Bulletin, 31, 186–189, 2011.

Wu, C., Peng, D., Soudani, K., Siebicke, L., Gough, C. M., Arain, M. A., Bohrer, G., Lafleur, P. M., Peichl, M., Gonsamo, A., Xu, S., Fang, B., and Ge, Q.: Land surface phenology derived from normalized differencevegetation index (NDVI) at global FLUXNET sites, Agr. Forest Meteorol., 233, 171–182, 2017.

Zhang, Q., Wu, Sh., Zhao, D., and Dai, E.: Responses of Growing Season Vegetation Changes to Climatic Factors in Inner Mongolia Grassland, Journal of Natural Resource, 28, 754–764, 2013.

Zhang, Y., Zhao, Zh., Li, Sh., and Meng, X.: Indicating variation of surface vegetation cover using SPOT NDVI in the northern part of North China, Geographical Research, 27, 745–755, 2008.

Zhao, Y.: Principles and Methods of Remote Sensing Application Analysis, Science Press, Beijing, 366–375, 2012.

Zhou, Q., Bian, J., and Zheng, J.: Variation of Air Temperature and Thermal Resources in the Northern and Southern Regions of the Qinling Mountains from 1951 to 2009, Acta Geographica Sinica, 66, 1211–1218, 2011.

https://doi.org/10.5194/piahs-379-293-2018

Impact of possible climate changes on river runoff under different natural conditions

Yeugeniy M. Gusev, Olga N. Nasonova, Evgeny E. Kovalev, and Georgy V. Ayzel

Institute of Water Problems, Russian Academy of Sciences, Moscow, 119333, Russian Federation

Correspondence: Yeugeniy M. Gusev (sowaso@yandex.ru)

Received: 28 December 2017 – Revised: 13 April 2018 – Accepted: 19 April 2018 – Published: 5 June 2018

Abstract. The present study was carried out within the framework of the International Inter-Sectoral Impact Model Intercomparison Project (ISI-MIP) for 11 large river basins located in different continents of the globe under a wide variety of natural conditions. The aim of the study was to investigate possible changes in various characteristics of annual river runoff (mean values, standard deviations, frequency of extreme annual runoff) up to 2100 on the basis of application of the land surface model SWAP and meteorological projections simulated by five General Circulation Models (GCMs) according to four RCP scenarios. Analysis of the obtained results has shown that changes in climatic runoff are different (both in magnitude and sign) for the river basins located in different regions of the planet due to differences in natural (primarily climatic) conditions. The climatic elasticities of river runoff to changes in air temperature and precipitation were estimated that makes it possible, as the first approximation, to project changes in climatic values of annual runoff, using the projected changes in mean annual air temperature and annual precipitation for the river basins. It was found that for most rivers under study, the frequency of occurrence of extreme runoff values increases. This is true both for extremely high runoff (when the projected climatic runoff increases) and for extremely low values (when the projected climatic runoff decreases).

1 Introduction

In the reports of the IPCC (Intergovernmental Panel on Climate Change, 2001, 2014), there are estimates of the increment in the mean annual air temperature and annual precipitation both for the entire globe and for various regions of the planet by the end of the 21st century. They show that for the past 100 years the global values of air temperature and precipitation have increased, respectively, by 0.6 °C and 2 % with large differences between the continents (IPCC, 2001). These climatic changes cause changes in characteristics of hydrological regime in different regions of the world. Investigation of possible changes in hydrological characteristics of various objects is an urgent problem in the environmental sciences.

In 2013, the International Inter-Sectoral Impact Model Intercomparison Project (ISI-MIP) (Warszawski et al., 2013) was launched. Its Water Sector was aimed at validation and improvement of hydrological models on regional and global scales and application of the models for hydrological projections of possible changes in water resources due to expected climatic changes. The present work was carried out within the framework of the ISI-MIP2 and its purpose was to study possible changes in characteristics of runoff from 11 large river basins (located on different continents under different natural conditions) by the end of 21st century caused by projected changes of climate. Similar large-scale multi-basin modelling studies were performed by nine hydrological models participating in the ISI-MIP (Krysanova and Hattermann, 2017). Here, in contrast to the mentioned paper, the Land Surface Model (LSM) SWAP (Soil Water-Atmoshere-Plants) (Gusev and Nasonova, 2010), rather than a hydrological model, is used as a main calculational tool.

Changes in such characteristics of river runoff as climatic runoff (runoff averaged over the period of 30–35 years), standard deviation of annual runoff within the specified climatic period, as well as a frequency of extreme values of annual runoff were calculated. The study of the two latter characteristics was motivated by the fact that increase in the frequency and magnitude of extreme climate events are stated in a number of publications (e.g., IPCC, 2001, 2014). As such, river flows are also expected to become more variable in the future, with more flash floods and lower minimum flows. This is confirmed by some fragmentary studies of changes in river runoff variability (e.g., Lin et al., 2008). The present paper is devoted to investigation of this statement by means of studying changes in standard deviation of annual runoff (caused by its natural variability) and frequency of annual runoff extreme values in the 21st century. In addition, the authors tried to generalize the calculated changes in climatic runoff in terms of the so-called climate elasticity of runoff in relation to changes in climatic values of air temperature and precipitation.

2 The study river basins

Objects of this study represent 11 large river basins located in different continents under different natural conditions: the Rhine and Tagus in Europe; the Ganges, Lena, Upper Yellow and Upper Yangtze in Asia, the Niger in Africa; the Mackenzie and Upper Mississippi in North America, the Upper Amazon in South America and the Darling in Australia (Fig. 1). The main characteristics of the selected river basins are given in Table 1.

The basins were schematized as a set of regular grid cells with a spatial resolution $0.5° \times 0.5°$ in latitude and longitude connected by a river network (Gusev et al., 2017b).

3 Methodology of investigation: SWAP model and data

3.1 SWAP model

The main tool for the present study is the SWAP model developed in the Institute of Water Problems of the Russian Academy of Sciences (Gusev and Nasonova, 1998, 2010). It is a physically-based land surface model with distributed parameters describing the processes of heat and water exchange in a soil – vegetation/snowpack – atmosphere system (SVAS). For a large-scale modelling, a selected river basin is divided into a number of calculational grid cells connected by a river network. The model simulations are performed for each calculational grid cell. After that the river routing model is used to simulate runoff at the basin outlet.

The SWAP model outputs are several dozens characteristics of thermal and water regimes of a river basin. However, here, only river runoff is considered. Model inputs include the land surface parameters and meteorological forcing data.

Different versions of the model SWAP were validated against observations including characteristics both related to energy balance or thermal regime of SVAS (sensible and latent heat fluxes, ground heat flux, net radiation, upward longwave and shortwave radiation, surface temperature, soil freezing and thawing depths) and related to hydrological cycle or water regime of SVAS (surface and total runoff from a catchment, river discharge, soil water storage in different layers, evapotranspiration, snow evaporation, intercepted precipitation, water table depth, snow density, snow depth and snow water equivalent, water yield of snow cover). The validations were performed for "point" experimental sites and for catchments and river basins of different areas (from 10^{-1} to 10^5 km^2) on a long-term basis and under different natural conditions (from arid to humid and from non-frozen soils to permafrost). The results of validation (e.g., Boone et al., 2004; Grippa et al., 2017; Gusev et al., 2017a; Gusev and Nasonova, 2010; Nasonova et al., 2009) have shown that SWAP is able to reproduce annual and interannual dynamics of the mentioned characteristics fairly well.

3.2 Data

Here, for model calibration and validation, daily values of meteorological forcing data for the period of 1970–2001 to drive the SWAP model were derived from the Water and Global Change (WATCH) data set (Weedon et al., 2011) which was created on the base of the ERA-40 reanalysis product hybridized with monthly values of ground-based measurements (including air temperature, the number of days with precipitation, cloudiness and precipitation) taken from the Global Precipitation Climatology Center (GPCC) and the Climatic Research Unit (CRU of University of East England) data sets.

For hydrological projections, daily meteorological forcing data simulated by five General Circulation Models (GCMs: HadGEM2-ES, IPSL-CM5A-LR, MIROC-ESM-CHEM, GFDL-ESM2M, and NorESM1-M) for the period from 2006 to 2099 and for the four RCP (Representative Concentration Pathways) scenarios (RCP2.6, RCP4.5, RCP6.0 and RCP8.5) were used (Krysanova and Hattermann, 2017). These scenarios were designed for the 5th assessment report of IPCC (2014). The numbers in the names of scenarios correspond to increment of incoming radiation (in W m^{-2}) in 2100 caused by increased greenhouse gas emission into the atmosphere as compared to a pre-industrial period. The larger numbers correspond to more aggressive anthropogenic scenarios due to increase in greenhouse gas emission and weak measures for its limitation. To increase the quality of hydrological projections a post-processing bias correction of meteorological projections to the WATCH data was performed within the framework of the ISI-MIP project.

The land surface parameters for each calculational grid cell were derived from the global ECOCLIMAP parameter data set (Champeaux et al., 2005) as described in Gusev et al. (2017b).

River runoff data for gauges listed in Table 1 were provided within the framework of the ISI-MIP project (see Huang et al., 2017 for details). Additional data for subbasins of the Lena and MacKenzie rivers were taken from the Global Runoff Data Centre (GRDC).

Table 1. River basins with their area; gauge stations with coordinates; averaged over 1971–2000 air temperature T and precipitation P; change in the probability of extremely high runoff due to climate change (in accordance with RCP scenarios). The values of T and P are taken from (Krysanova and Hattermann, 2017); F_3 is the probability of extremely high runoff for the period P3, F_0 is the probability of extremely high runoff for the period P0, equaled to 2.5 %.

River	Streamflow gauge station	Latitude	Longitude	Area (km^2)	T, °C	P, mm yr^{-1}	F_3, %	F_3/F_0
Lena	Stolb	72.37	126.80	2 460 000	−10.2	384	67.6	27.0
U. Amazon	Sao Paulo de Olivenca	−3.45	−68.75	990 781	21.7	2122	2.4	0.9
MacKenzie	Arctic Red River	67.46	−133.74	1 660 000	−4.3	435	44.0	17.6
U. Yangtze	Cuntan	29.61	106.60	804 859	6.8	768	5.2	2.1
Ganges	Farakka	25.00	87.92	835 000	21.1	1173	9.6	3.8
U. Yellow	Tangnaihai	35.50	100.15	121 000	−2	506	3.1	1.3
Niger	Lokoja	7.80	6.77	2 074 171	27.7	625	2.2	0.9
Rhine	Lobith	51.84	6.11	160 800	8.7	1038	4.8	1.9
U. Mississippi	Alton	38.89	−90.18	444 185	7.3	967	4.2	1.7
Tagus	Almourol	39.47	−8.37	67 490	14	671	1.8	0.7
Darling	Louth	−30.53	145.11	489 300	19.2	590	0.9	0.4

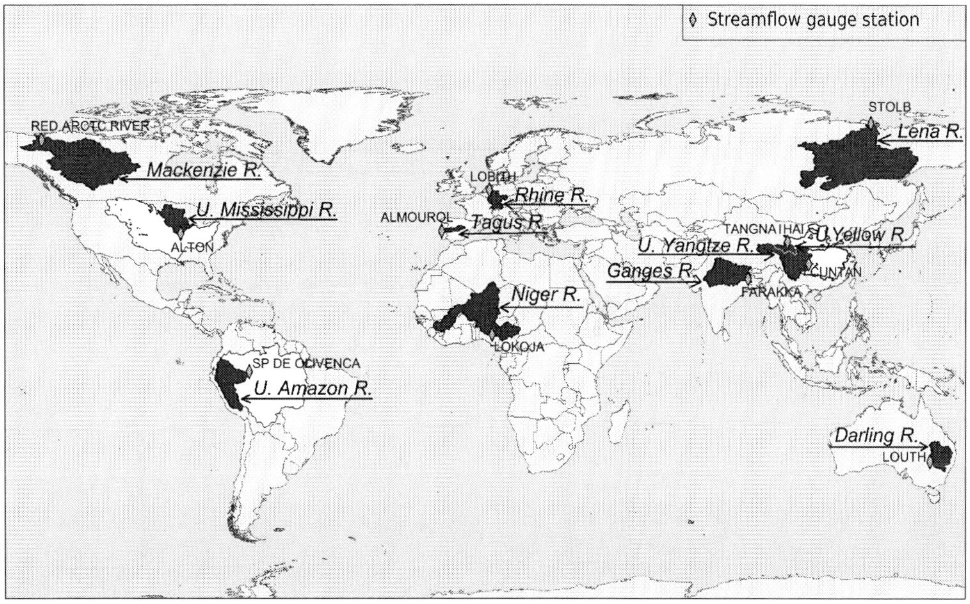

Figure 1. Location of the selected river basins (Gusev et al., 2017b).

3.3 SWAP model calibration and validation

To improve the quality of hydrological calculations an automatic calibration of the most important model parameters based on the algorithm SCE-UA for searching the global extremum of an objective function (Duan et al., 1992) under the additional condition that the systematic error of runoff calculation cannot exceed 5 % (Nasonova et al., 2009) was carried out. The Nash and Sutcliffe coefficient of efficiency of simulating monthly river runoff was used as an objective function. For each river basin, the calibration was carried out for 8-year period (different for different rivers). Validation of the obtained optimal values of model parameters was carried out for the period of 1970–2001 and showed a satisfactory agreement between measured and simulated monthly runoff (Gusev et al., 2017b).

4 Results and discussion

The obtained optimal values of model parameters were used for simulating daily runoff for the selected rivers by the SWAP model with meteorological forcing data from the above five GCMs. The runoff simulations were performed for two periods: historical P0 (1962–2005) and prognostic

(2006–2099). The latter was divided into three parts: P1 (2006–2035), P2 (2036–2065) and P3 (2066–2099). For the prognostic period, runoff was simulated for the above four climate scenarios of the RCP family.

4.1 Projections of changes in climatic river runoff

Some examples of the projected river runoff by the end of the 21st century are given in Fig. 2, where the dynamics of annual values of river runoff (obtained as a result of averaging of five runoff simulations with meteorological projections from five GCMs) for six river basins and four climate scenarios are shown. Here, for the historical period, river runoff was averaged over the period, while, for the prognostic periods, runoff values were calculated as 30-year moving averages to obtain an evolution of climatic runoff. Besides, Fig. 2 also shows the uncertainty of runoff projections caused by a scatter in GCMs' results and estimated as a root mean square error of an ensemble average.

The obtained results confirmed an expected conclusion that more aggressive scenarios lead to larger changes in river runoff from the selected basins, both in the positive and negative directions (for different basins). As to the relative error of climatic runoff (equaled to the absolute error divided by average runoff), it is the greatest for the Darling and Tagus rivers. It should be noted that according to Gusev et al. (2017b) these basins are also characterized by the greatest natural uncertainty of runoff caused by weather noise. For the northern rivers, Lena and Mackenzie, the relative error is the smallest that corresponds to the lowest values the natural uncertainty of river runoff from these basins (Gusev et al., 2017b). Apparently, for basins with the greatest amplitude of weather noise, the scatter of the dynamics of meteorological outputs from different GCMs is also the greatest, which, in turn, leads to greater differences between the values of river runoff calculated on their basis.

As to the changes in river runoff ΔR, they are evidently different (both in magnitude and sign) for basins in different regions of the planet due to differences in natural conditions, primarily climatic ones. We tried to generalize the obtained results, using the concept of climate elasticity of runoff.

Climate elasticity, introduced by Fu (1981), is used to determine the response of any variable (e.g., runoff) to climate change. For example, a precipitation elasticity of annual runoff of 2.2 suggests that annual runoff would increase with 22 % due to an increase in annual precipitation with 10 % (22 %/10 % = 2.2). Negative elasticity means that changes of these two variables are of an opposite sign. The temperature elasticity (ε_T) and precipitation elasticity (ε_P) of runoff from each river basin was calculated as follows (Grijsen and Patel, 2014):

$$\Delta R/R0 = \varepsilon_T \Delta T/T0 + \varepsilon_P \Delta P/P0, \quad (1)$$

where $R0$, $T0$ and $P0$ are climatic values of annual runoff, air temperature and precipitation, respectively; ΔR, ΔT and ΔP are changes of these variables (here, P and R are in mm year^{-1}, while T is in K).

The calculated values of climate elasticities of runoff ε_T and ε_P for the selected basins were mapped, using the kriging procedure in the phase coordinates of $T0$ and $P0$ (Fig. 3). As can be seen in Fig. 3, the temperature elasticity of runoff ε_T is negative, because an increase in temperature raises evapotranspiration, leading to a decrease in runoff. The values of precipitation elasticity of runoff ε_P are positive, because an increase in precipitation results in an increment of runoff.

The values of climate elasticities in Fig. 3 give an opportunity to project, as the first approximation, changes in climatic annual river runoff, provided that projections of changes in mean annual air temperature and annual precipitation for a river basin are available.

4.2 Projections of changes in natural variability of river runoff

Now let us consider the second statistical moment – the standard deviation of annual river runoff σ_R within a 30-year range associated with a natural variability of runoff (caused by a weather noise). An error of estimation of the second moment is higher than that of the mean value of runoff and equals to approximately 13 % of σ_R for a sample of annual runoff values of length ≈ 30 (such an error is of the order of averaged over the basins change in σ_R in the 21st century). As such, to increase the accuracy of estimation of changes in runoff variability the values of σ_R obtained for each of the four climatic scenarios and five GCMs were averaged over the scenarios and GCMs for the periods P0, P1, P2 and P3. After that, their temporal dynamics was approximated by a linear or parabolic function. The error of the obtained value of σ_R did not exceed 2 %.

The results of estimation of the dynamics of σ_R in the 21st century obtained for different rivers are presented in Fig. 4. They show that for all rivers, with the exception of the Amazon, Darling and Tagus, the standard deviation of annual runoff σ_R increases by the end of the 21st century. The values of σ_R for the Amazon and Tagus rivers decrease, but insignificantly.

As to climatic values of annual runoff R, averaged over all climatic scenarios and GCMs, they have a noticeable downward trend only for the Tagus and Darling rivers. (Note, that the relative error of such estimates for different rivers is within the range of 0.5–1.5 %). For the Lena, Mackenzie and Ganges rivers, the climatic values of runoff demonstrate its apparent increase. The climatic runoff of the remaining rivers has an unremarkable growth by the end of the century (the Mississippi, Niger, and Rhine rivers) or does not change (the Amazon, Yellow and Yangtze rivers).

Figure 5 shows the relationship (for the climatic periods P1, P2 and P3 and for 11 rivers) between the relative standard deviation of annual runoff from its climatic value $\varepsilon_\sigma = \frac{\sigma_{\mathrm{R,pr}}}{\sigma_{\mathrm{R,P0}}}$

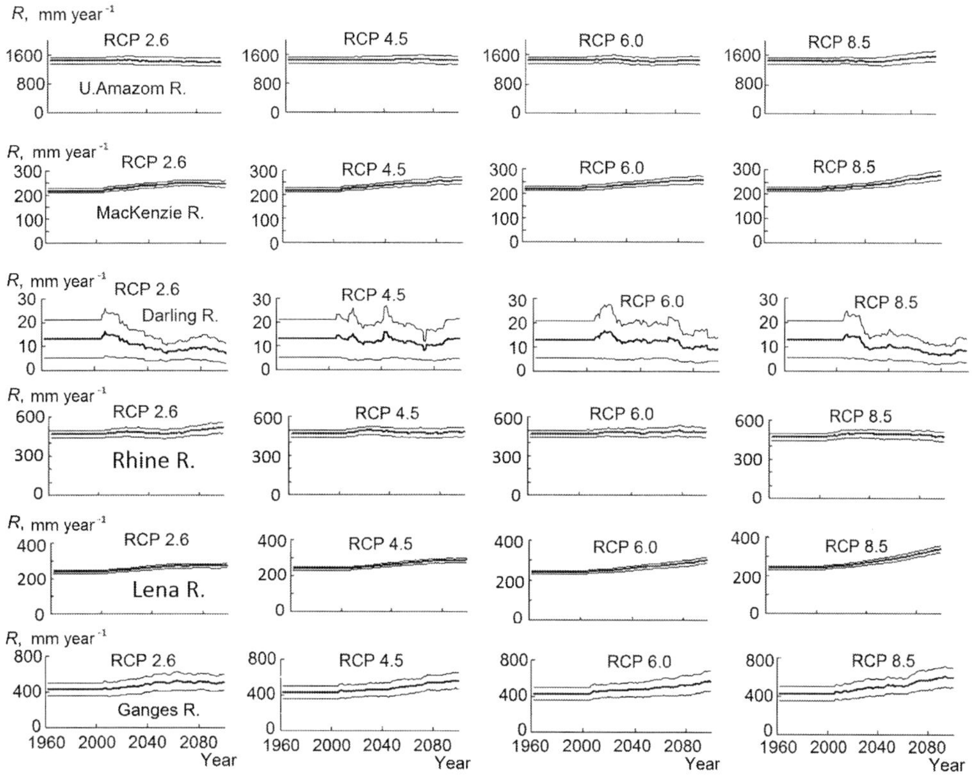

Figure 2. Dynamics of multi-modal climatic river runoff (bold line) by the end of the 21st century in accordance with four RCP scenarios, as well as root mean square error of multi-model mean values (thin line).

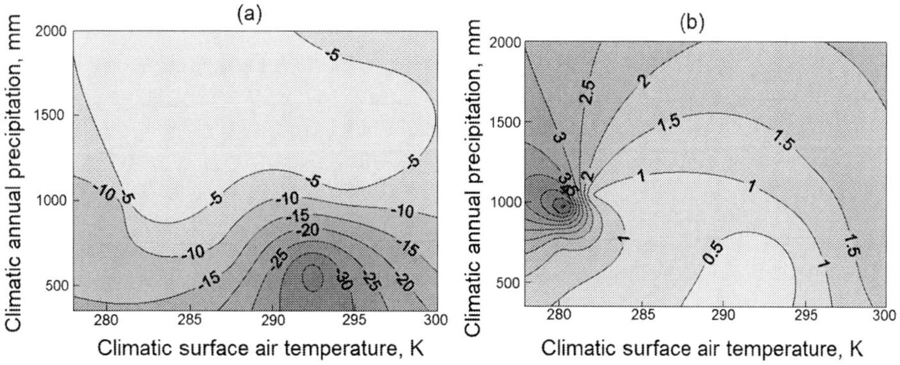

Figure 3. Temperature elasticity of runoff ε_T **(a)** and precipitation elasticity of runoff ε_P **(b)**. The elasticities are shown in isolines.

(where pr means any of prognostic periods: P1, P2 or P3; $\sigma_{\mathrm{R,pr}}$ and $\sigma_{\mathrm{R,P0}}$ are standard deviations of annual runoff R for any of the prognostic period pr and for the historical period P0, respectively) and the relative value of climatic runoff $\varepsilon_{\mathrm{R}} = \frac{R_{\mathrm{pr}}}{R_{\mathrm{P0}}}$. It is seen that, in the first approximation (with a coefficient of correlation ≈ 0.8), the change in the relative standard deviation of annual runoff ε_σ is proportional to the change in the relative climatic runoff ε_{R}. This means that for a particular river the variability of its annual runoff changes in proportion to changes in climatic runoff. Thus, as the first approximation, the value of $\sigma_{\mathrm{R,pr}}$ can be obtained on the basis of the following equation:

$$\sigma_{\mathrm{R,pr}} \approx \sigma_{\mathrm{R,P0}} \frac{R_{\mathrm{pr}}}{R_{\mathrm{P0}}}. \tag{2}$$

However, the obtained relation can be treated only very approximately, which is confirmed by the deviations of the circles in Fig. 5 from the line 1 : 1. This means that the ratio of

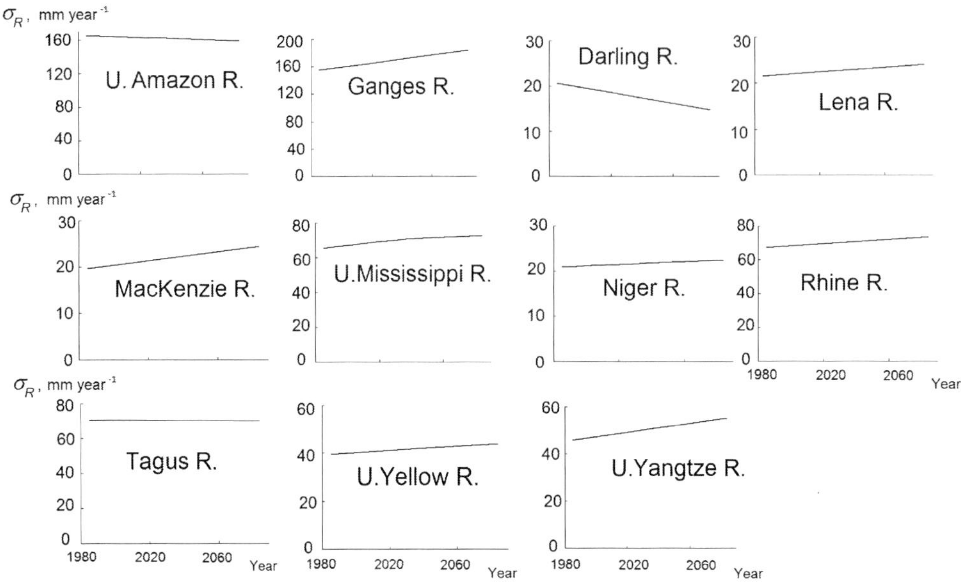

Figure 4. The dynamics of standard deviation of annual runoff σ_R during the 21st century.

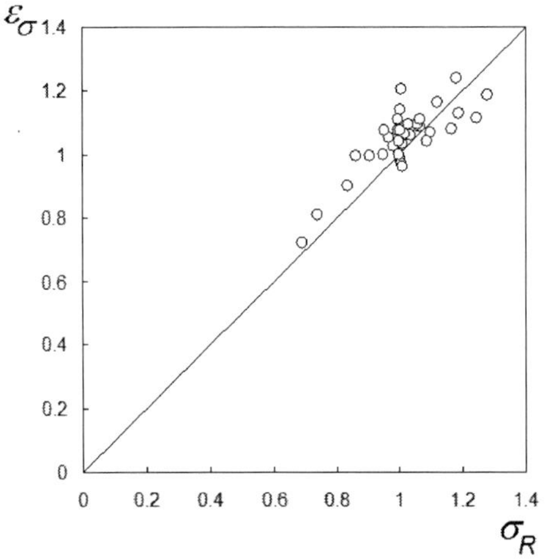

Figure 5. Relative standard deviation of annual river runoff ε_σ versus relative value of climatic runoff ε_R.

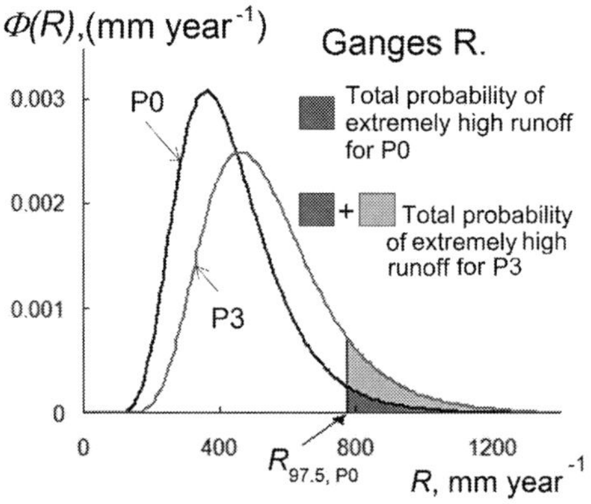

Figure 6. Probability density Φ of annual runoff R for the Ganges River for the historical P0 and prognostic P3 periods. $R_{97.5,\mathrm{P0}}$ is the quantile of 97.5 % probability for the period P0 (the cut point for the extremely high runoff).

deterministic and stochastic components in runoff of a particular river (Gusev et al., 2017b) can vary with climate change and, evidently, the value of $\sigma_{\mathrm{R,pr}}$ can be more accurately obtained by means of model sumulations.

4.3 Changing the frequency of extreme runoff

First of all, it should be noted that, here, runoff values that exceed quantile of annual runoff with 97.5 % probability for the historical period P0 are treated as extremely high annual river runoff, while runoff values that are lower than quantile of annual runoff with 2.5 % probability for the period P0 are considered as extremely low ones. Here, the main attention will be paid to changing the frequency of occurrence of extremely high runoff.

To solve this problem, we need to assess the impact of climate change on the distribution function of annual runoff. Different variants for approximating the distribution of river runoff on the basis of application of different distribution functions (the lognormal distribution, the Pearson distribution, the gamma distribution, the distribution by Kritskii and Menckel) are considered in Sikan (2007). An analysis of these variants in (Sikan, 2007) and the authors' own experience (Gusev et al., 2017a, b) shows that empirical distribution of annual runoff (at least within the range of probabilities from 0.025 to 0.975) can be satisfactorily approximated by the function of the two-parameter lognormal distribution. Application of more precise three-parameter functions of distribution in this case is not possible, because the accuracy of estimation of the third statistical moment, which is necessary to obtain the third parameter of the distribution function and reflects the degree of its asymmetry, is rather low (in contrast to the above-mentioned accuracy of the first two statistical moments) in the case of application of the samples of annual river runoff available for us. Therefore, in the present paper, a hypothesis about the possibility of approximation of the empirical distribution functions of annual runoff for the rivers under study by the two-parameter lognormal distribution is used.

The parameters of the lognormal distribution function of annual runoff for a particular river and time period (P0, ..., P3) were derived from the empirical distribution of annual runoff calculated using forcing data from each GCM and for each RCP scenario. Then, these parameters were averaged over all GCMs and scenarios, and the obtained averages were used for constructing the final distribution function of runoff for each river and time period.

The obtained distribution functions of annual river runoff were used for estimating the change in the frequency of its extreme values due to possible climate change.

Figure 6 shows the calculated functions of distribution of annual runoff of the Ganges River for the historical P0 and prognostic P3 periods. The probability of appearance of extremely high runoff changes from 2.5 % for the historical period (the shaded area under the black curve) to 9.6 % for the prognostic period (the shaded area under the grey curve).

Similar calculations were carried out for all rivers under study. The results are shown in Table 1 (two last columns). Analysis of the results shows that changes in the frequency of occurrence of extremely high annual runoff f_{max} depend on the natural conditions of a particular river basin. For most rivers the frequency increases due to climate change (especially for rivers in the boreal zone). As a rule, for such rivers, according to the RCP scenarios, the climatic runoff also increases. For the U. Amazon, U. Yellow and Niger rivers, changes in f_{max} are within the error of its estimation (with accounting for errors of estimation of parameters of the approximating functions). For these rivers, changes in climatic runoff are also small. For the Darling and Tagus rivers, the frequency of extremely high runoff is noticeably reduced. And climatic runoff of these rivers also decreases. As to the frequency of extremely low annual runoff f_{min}, for the period P3, it is 4.1 and 1.2 times larger (for the Darling and Tagus rivers, respectively) than for the historical period P0, i.e. the frequency of the appearance of extremely low annual runoff is increasing.

5 Conclusions

For 11 large river basins, characterized by different natural conditions, possible changes in characteristics of river runoff caused by the expected climate changes according to climatic scenarios of the RCP-family up to 2100 are investigated. Changes in climatic runoff are different (both in magnitude and sign) for the river basins located in different regions of the planet due to differences in natural (primarily climatic) conditions. Climatic runoff has a noticeable downward trend only for the Tagus and Darling rivers. For the Lena, Mackenzie and Ganges rivers, the climatic values of runoff demonstrate its apparent increase. The climatic runoff of the remaining rivers has an unremarkable growth by the end of the century (the Mississippi, Niger, and Rhine rivers) or does not change (the Amazon, Yellow and Yangtze rivers).

As to variability of annual runoff, changes in the standard deviation of annual runoff from its climatic value, in the first (fairly coarse) approximation, are proportional to changes in climatic runoff. It is shown that for all rivers, with the exception of the Amazon, Darling and Tagus, the standard deviation of annual runoff σ_{R} increases by the end of the 21st century. The values of σ_{R} for the Amazon and Tagus rivers decrease, but insignificantly.

For most rivers under study, the frequency of occurrence of extreme runoff values increases. This is true both for extremely high runoff (when the projected climatic runoff increases) and for extremely low values (when the projected climatic runoff decreases).

Data availability. Data were provided by ISI-MIP regional-scale water sector organizers, and by GRDC (the Global Runoff Data Centre, 56068 Koblenz, Germany).

Competing interests. The authors declare that they have no conflict of interest.

Special issue statement. This article is part of the special issue "Innovative water resources management – understanding and balancing interactions between humankind and nature". It is a result of the 8th International Water Resources Management Conference of ICWRS, Beijing, China, 13–15 June 2018.

Acknowledgements. The study was supported by the Russian Science Foundation (grant no. 14-17-00700 – Sect. 4.2 and 16-17-10039 all the other sections). We are grateful to ISI-MIP regional-scale water sector organizers for providing with data and materials, and to GRDC for providing discharge data.

Edited by: Zongxue Xu
Reviewed by: two anonymous referees

References

Boone, A., Habets, F., Noilhan, J., Clark, D., Dirmeyer, P., Fox, S., Gusev, Y., Haddeland, I., Koster, R., Lohmann, D., Mahanama, S., Mitchell, K., Nasonova, O., Niu, G.-Y., Pitman, A., Polcher, J., Shmakin, A. B., Tanaka, K., van den Hurk, B., Verant, S., Verseghy, D., Viterbo, P., and Yang, Z.-L.: The Rhone-aggregation land surface scheme intercomparison project: An overview, J. Climate, 17, 187–208, 2004.

Champeaux, J. L., Masson, V., and Chauvin, F.: ECOCLIMAP: a global database of land surface parameters at 1 km resolution, Meteorol. Appl., 12, 29–32, 2005.

Duan, Q., Sorooshian, S., and Gupta, V. K.: Effective and efficient global optimization for conceptual rainfall runoff models, Water Resour. Res., 28, 1015–1031, 1992.

Fu, B.: On the calculation of the evaporation from land surface, Sci. Atmos. Sin., 5, 23–31, 1981.

Grijsen, J. and Patel, H.: Understanding the impact of climate change on hydropower: the case of Cameroon – climate risk assessment for hydropower generation in Cameroon, World Bank Group, AFTEG, Washington DC, Report 87913, 151, 2014.

Grippa, M., Kergoat, L., Boone, A., Peugeot, C., Demarty, J., Cappelaere, B., Gal, L., Hiernaux, P., Mougin, E., Ducharne, A., Dutra, E., Anderson, M., Hain, C., Ait-Mesbah, S., Balsamo, G., Boussetta, S., Chaffard, V., Chan, E., Cohard, JM, Favot, F., Gascon, T., Gascoin, S., Getirana, A., Galle, S., Gosset, M., Guichard, F., Gusev, Y., Harris, P., He, X., Hector, B., Kaptue, A., Kazuaki, Y., Kotsuki, S., Kim, H., Kumar, S., Lebel, T., M-H, Lo, Mahanama, S., Magand, C., Maignan, F., Mamadou, O., Nasonova, O., Oki, T., Ottlé, C., Pellarin, T., Pierre, C., Richard, A., Roujean, J.-L., Polcher, J., Quantin, G., Séguis, L., Shmakin, A., Sokratov, V., Tanaka, K., Turkov, D., Velluet, C., Verseghy, D., and Vischel, T.: Modelling surface runoff and water fluxes over contrasted soils in pastoral Sahel: evaluation of the ALMIP2 land surface models over the Gourma region in Mali, J. Hydrometeorol., 18, 1847–1866, 2017.

Gusev, E. M. and Nasonova, O. N.: Modelling heat and water exchange between the land surface and the atmosphere, Nauka, Moscow, 2010.

Gusev, Y. M. and Nasonova, O. N.: The Land Surface Parameterization scheme SWAP: description and partial validation, Global Planet. Change, 19, 63–86, 1998.

Gusev, Y. M., Semenov, V. A., Nasonova, O. N., and Kovalev, E. E.: Weather noise impact on the uncertainty of simulated water balance components of river basins, Hydrol. Sci. J., 62, 1181–1199, 2017a.

Gusev, Y. M., Nasonova, O. N., Kovalev, E. E., and Aizel, G. V.: Modelling river runoff and estimating its weather-related uncertainty for 11 large-scale rivers located in different regions of the globe, Hydrol. Res., https://doi.org/10.2166/nh.2017.015, in press, 2017b.

Huang, S., Kumar, R., Florke, M., Yang, T., Hundecha, Y., Kraft, P., Gao, C., Gelfan, A., Liersch, S., Lobanova, A., Strauch, M., Ogtrop, F., Reinhardt, J., Haberlandt, U., and Krysanova, V.: Evaluation of an ensemble of regional hydrological models in 12 large-scale river basins worldwide, Climatic Change, 141, 381–397, 2017.

Intergovernmental Panel on Climate Change (IPCC): Climate Change 2001: The scientific basis, Contribution of Working Group I to the Third Assessment Report of the Intergovernmental Panel on Climate Change, Cambridge University Press, Cambridge, UK and New York, 881 pp., 2001.

Intergovernmental Panel on Climate Change (IPCC): Climate Change 2014: Synthesis Report, Contribution of Working Groups I, II and III to the Fifh Assessment Report of the Intergovernmental Panel on Climate Change, edited by: Core Writing Team, Pachauri, R. K., and Meyer, L. A., IPCC, Geneva, Switzerland, 151 pp., 2014.

Krysanova, V. and Hattermann, F. F.: Intercomparison of climate change impacts in 12 large river basins: overview of methods and summary of results, Climatic Change, 141, 363–379, 2017.

Lin, X., Zhang, Y., Yao, Z., Gong, T., Wang, H., Chu, D., Liu, L., and Zhang, F.: The trend on runoff variations in the Lhasa River Basin, J. Geogr. Sci., 18, 95–106, doi.org/10.1007/s11442-008-0095-4, 2008.

Nasonova, O. N., Gusev, Y. M., and Kovalev, Y. E.: Investigating the ability of a land surface model to simulate streamflow with the accuracy of hydrological models: A case study using MOPEX materials, J. Hydrometeorol., 10, 1128–1150, 2009.

Warszawski, L., Frieler, K., Huber, V., Piontek, F., Serdeczny, O., and Schewe, J.: The inter-sectoral impact model intercomparison project (ISI–MIP): project framework, P. Natl. Acad. Sci. USA, 111, 3228–3232, 2013.

Weedon, G. P., Gomes, S., Viterbo, P., Shuttleworth, W. J., Blyth, E., Österle, H., Adam, J. C., Bellouin, O., and Best, M.: Creation of the WATCH Forcing Data and its use to assess global and regional reference crop evaporation over land during the twentieth century, J. Hydrometeol., 12, 823–848, 2011.

Sikan, A. V.: Methods of statistical processing of hydrometeorological information, Textbook, Speciality "Hydrology", RGGMU, St. Petersburg, 2007.

https://doi.org/10.5194/piahs-379-301-2018

Open Access
PIAHS

The impacts of climate change on irrigation and crop production in Northeast China and implications for energy use and GHG Emission

Tingting Yan[1], **Jinxia Wang**[2], **Jikun Huang**[2], **Wei Xie**[2], **and Tingju Zhu**[3]

[1]Development Research Center of the Ministry of Water Resources, Beijing 100038, China
[2]School of Advanced Agricultural Sciences, Peking University, Beijing 100871, China
[3]International Food Policy Research Institute, 2033 K St, NW, Washington, DC 20006, USA

Correspondence: Jinxia Wang (jxwang.ccap@pku.edu.cn)

Received: 31 December 2017 – Revised: 26 March 2018 – Accepted: 26 March 2018 – Published: 5 June 2018

Abstract. The water-food-energy-GHG nexus under climate change has been gaining increasing attention from both the research and policy communities, especially over the past several years. However, most existing nexus studies are qualitative and explorative in nature. So far, very few studies provide integrated analysis of this nexus across all the four sectors. The purpose of this paper is to examine this nexus by assessing the effects of climate change on agricultural production through the change in water availability, evaluating the adjustment responses and resulting energy consumption and GHG emission, with the Northeast China as a case study. Based on our simulation results, by 2030, climate change is projected to increase water supply and demand gap for irrigation in Northeast China. Due to the increase in water scarcity, irrigated areas will decrease, and the cropping pattern will be adjusted by increasing maize sown areas and decreasing rice sown areas. As a result, the total output of crops and profits will clearly be reduced. Finally, energy consumption and GHG emission from irrigation will be reduced. This study suggests that climate change impact assessment fully consider the nexus among water, food, energy and GHG; however, more studies need to be conducted in the future.

1 Introduction

Increasing evidence of climate change intensifies our concern about its impacts on agricultural production and water availability. With high confidence, IPCC (2014) concluded that global temperature will continue to increase in the future. Numerous agronomic and economic studies found that agricultural production is vulnerable to climate change; and its vulnerability is highly related to adaptation capacity (Mendelsohn and Dinar, 2009; IPCC, 2014). Besides direct impact on crop productivity, climate change also indirectly influences agricultural production through water availability changes. Recent studies reveal that climate change may enlarge the gap between irrigation demand and supply, therefore jeopardizing irrigated agriculture (Rosenzweig et al., 2004; Wang et al., 2013). Obviously, to understand the impacts of climate change on agricultural production requires integrated analysis that combines climate change impacts on both the agricultural and water sectors. In addition, farmers' adaptive responses in adjusting agricultural production also need to be included into the impact assessment (Wang et al., 2013). However, so far very few such integrated analyses have been conducted for China.

Importantly, the change of water availability and agricultural production will also influence energy consumption and GHG emission. Theoretically, water, food, energy and GHG form a complex web of interlinked components, which is more complicated under the background of climate change. Food production highly depends on irrigation; irrigation water supply consumes energy in pumping, delivering water, or other activities. Finally, energy production results in GHG emission, which indirectly influence agricultural production and irrigation through climate change feedbacks. In 2011, the Water-Energy-Food Security Nexus Conference held in Bonn, Germany, which started using the nexus terminology for the first time and concluded that the issues of water, food and energy security are highly related, and thus must be addressed with a comprehensive insight and a multidisciplinary approach into the nexus (Waughray, 2011). Since then, the water-food-energy-GHG nexus under climate change has become a hot topic.

However, most of those published literatures focused on the two-way relationships, such as Food-Energy Nexus (Wise et al., 2014; Cobuloglu and Büyüktahtakın, 2015), Energy-Water Nexus (Wang et al., 2012; Kao et al., 2015) or Food-Water Nexus (Waughray, 2011; Wang et al., 2013), little research analysed the relationship among water, energy and food as a whole. According to Keairns et al. (2016), as of December 2016, there were only about 40 literatures, whose subject and abstract contain water, energy, food and nexus at the same time. Interestingly, although not only the concept of the water-energy-food nexus and its important role on decision-making under climate change but also the systematic approach to the nexus have been increasingly addressed by many scholars (Bazilian et al., 2011; Martin-Gorriz et al., 2014; Rasul and Sharma, 2015), little study has quantified the water-food-energy nexus for the future. Even in few work with estimation of future water-food-energy nexus (Nelson et al., 2009; Finley and Seiber, 2014), climate change scenarios published by IPCC (Van Vuuren et al., 2011a) are not used.

Therefore, to have a better understanding of the water, food, energy and GHG nexus with forthcoming climate change, we must answer the following questions. How will climate change affect water availability for agricultural production? If water availability changes, then, what types of adjustments in agricultural production will likely to be made by farmers? That is, what changes will be made in irrigated and rain-fed areas? How will cropping patterns be adjusted? Based on these adjustments, how will farmers' profits from agricultural production change? Due to changes in water availability and the adjustments in agricultural production, how will energy consumption and GHG emission be influenced?

As a major grain production region in China, the Northeast depends to a large extent on irrigation for crop production. The region encompasses four provinces (Liaoning, Jilin, Heilongjiang and the northeastern part of Inner Mongolia) and covers two large river basins (Songhuajiang and Liaohe River Basins, SRB and LRB). In China, 41 % of soybean, 34 % of maize and 40 % of japonica rice are produced in the Northeast (NBSC, 2013). About 30 % of the grain produced in this region depends on irrigation. According to national and regional plans, to further increase agricultural production in Northeast China requires further expansion of irrigated areas. However, the water endowment in this region is limited and it tends to decrease. Per capita water availability is 1500 m^3 in the region, lower than the national average (2100 m^3 per capita). In the LRB, per capita water availability is only approximately 630 m^3 (NBSC, 2013). Over the past 50 years (1961–2011), runoff in the SRB and LRB has declined by 3 and 8 %, respectively (ECSNCCA, 2011). In the future, water is expected to become more scare in these two river basins (Wang et al., 2013).

The overall goal of this study is to illuminate the water-food-energy-GHG nexus under climate change in Northeast China by assessing the effects of climate change on agricultural production through the change in water availability, evaluating the adjustment responses by farmers and resulted changes in energy consumption and GHG emission. We first simulate the impacts of climate change on crop yield through its effects on the water availability without considering farmers' adaptive adjustments. Second, after considering farmers' autonomous adaptations to climate change driven by profit maximization (and risk minimization), we analyse changes in cropping pattern, irrigation intensity, crop yield and output, and profits. Finally, the energy consumption and GHG emission from irrigation are estimated at both the regional and the national levels.

2 Methodology and data

It takes process to estimate the effects of climate change on irrigation water supply and agricultural production, energy consumption and GHG emission. First, we introduce the China Water Simulation Model (CWSM) that is used to simulate the impacts of climate change on water and agricultural production. Second, we describe the steps and methods to estimate the energy consumption and GHG emission due to the changed water and agricultural production. Finally, we clarify the specifications of three simulation scenarios, including a reference scenario and two climate change scenarios.

2.1 Specification of the China Water Simulation Model

The model used in this study is the CWSM model developed by the first and second authors, together with their team. The CWSM is a model that integrates climate, hydrology, crop and water allocation optimization to simulate the regional water balance, water allocation and their impacts on agricultural production under socioeconomic, policy, and climate change scenarios (Wang et al., 2013). The model includes two main components: the water balance and the water allocation. When the balance between the water supply and water demand is affected by climate change, water is reallocated among sectors (i.e. agricultural, industrial and domestic water users) and within the agricultural sector. For a detailed description of the CWSM, see Wang et al. (2013). In this paper, an improvement is made for allocation within the agricultural sector.

The major approach used in the water allocation within the agricultural sector in the CWSM is the Positive Mathematical Programming (PMP) model (Wang et al., 2013). The PMP model is a method for calibrating models of agricultural production with resource constraints by using nonlinear production or cost functions, avoiding the corner solution in a linear optimization model. The method is implemented in three stages. First, calibration constraints are added to the primal problem. Second, the shadow values from the first step are used to estimate the parameters of a new cost function. Third, a non-linear objective function is specified, without the calibration constraints such that it can reproduce the base-year activity levels, closely or exactly, thus verifying the calibrated unconstrained optimization model.

Based on the work of Wang et al. (2013), our model has made the following major improvements. First, a production function with a constant elasticity of substitution (CES) is used, considering the substitution between irrigated land and irrigation water, the return-to-scale effect, and the relationship between crop yield and irrigation water. Second, the shadow costs of the inputs are introduced into the cost function, which are calibrated by using the method specified by Graveline and Mérel (2014). More details about parameters specification can be found in Yan (2015). Compared with our previous study (Wang et al., 2013), the CES PMP model used in this paper is more flexible, ensuring realistic responses to price changes, replicating the observed profits and taking into account yield response to irrigation water.

The model is given as follows for the SRB and LRB.

$$\max_{\left\{\begin{array}{l}A_{r,c,ir}\\A_{r,c,rf}\\W_{r,c,ir}\end{array}\right\}} \prod = \sum_{r}\sum_{c}\left[P_{r,c}\left(Q_{r,c,ir}+Q_{r,c,rf}\right) - \left(C_{r,c,a}+\mu_{r,c,a,ir}\right)A_{r,c,ir} - \left(C_{r,c,a}+\mu_{r,c,a,rf}\right)A_{r,c,rf} - \left(C_{r,c,w}+\mu_{r,c,w}\right)W_{r,c,ir}\right] \quad (1)$$

$$\text{st.}\left\{\begin{array}{ll}\sum_{c}\left(A_{r,c,ir}+A_{r,c,rf}\right)\leq\sum_{c}\left(\bar{A}_{r,c,ir}+\bar{A}_{r,c,rf}\right) & (2)\\ \sum_{c}W_{r,c,ir}\leq I_{cc} & (3)\\ Q_{r,c,ir}=\alpha_{r,c,ir}\left(\beta_{r,c,ir,a}A_{r,c,ir}^{\rho_c}+\beta_{r,c,ir,w}W_{r,c,ir}^{\rho_c}\right)^{\frac{\delta_{r,c,ir}}{\rho_c}} & (4)\\ Q_{r,c,rf}=\alpha_{r,c,rf}A_{r,c,rf}^{\delta_{r,c,rf}} & (5)\end{array}\right.$$

The subscript "r" refers to the SRB and LRB; "c" refers to crop group (i.e. rice, wheat, maize, soybean, sugar crops, edible oil crops, cotton, vegetable and other crops); "ir" refers to irrigated production, and "rf" refers to rainfed production; the subscript "a" refers to the parameter related to land, and the subscript "w" refers to the parameter related to irrigation water; "cc" refers to climate change scenarios (we used the average values from 26 cmip5 GCMs under different representative concentration pathways (RCP), Ma et al., 2015); the reference information is indicated with bars.

The following are the definitions of the variables in the model systems:

- $A_{r,c,ir}$ and $A_{r,c,rf}$: choice variables in this model, representing sown areas of irrigated and rainfed crops by river basin, respectively;
- $W_{r,c,ir}$: a choice variable in this model, indicating the irrigation water use by crop and by river basin;
- $P_{r,c}$: the crop commodity price by river basin;
- $Q_{r,c,ir}$ and $Q_{r,c,rf}$: irrigated and rainfed crop production by crop and river basin, respectively;
- $C_{r,c,a}$: the observed cost of inputs other than irrigation water, including seeds, fertilizer, pesticide, etc.;
- $C_{r,c,w}$: the observed cost of the irrigation water used by crop by river basin;
- $\mu_{r,c,a,ir}$ and $\mu_{r,c,a,rf}$: the shadow cost of irrigated and rainfed land, respectively;
- $\mu_{r,c,w}$: the shadow cost of irrigation water;
- I_{cc}: the available water supply for irrigation by river basin under climate change;
- $\alpha_{r,c,ir}$ and $\alpha_{r,c,rf}$: the scaling parameter of the production function, $\alpha > 0$;
- $\beta_{r,c,ir,a}$ and $\beta_{r,c,ir,w}$: the share parameters of production function, $\left(\beta_{r,c,ir,a}, \beta_{r,c,ir,w}\right) \in [0, 1]$ and $\beta_{r,c,ir,a} + \beta_{r,c,ir,w} = 1$;
- $\delta_{r,c,ir}$ and $\delta_{r,c,rf}$: the return-to-scale parameters of irrigated and rain-fed production, $\delta_{r,c} \in (0, 1)$;
- ρ_c: the substitution parameter, $\rho_c \in (-\infty, 0) \cup (0, 1)$.

In the above program, Eq. (1) is the objective function, which is to maximize the producer's profit Π_r subject to the land constraint (2), water constraint (3). In the objective function (1), a crop's profit is defined as the difference between gross revenue and the cost of crop production. A crop's gross revenue is the product of its output and its selling price. The cost of crop production includes the costs of using irrigated land, rainfed land and irrigation water, each of which consists of two parts, the observed cost and the shadow cost. The observed cost of irrigated land and rainfed land include the cost of all types of input cost except irrigation water, such as seeds, pesticides, fertilizers and labor, the observed cost of irrigation water is the water fees. The land constraint (2) requires that the total area allocated to all crops not exceed the total land available in the base year. The water constraint (3) implies that the agricultural water use in each river basin cannot be more than the irrigation water supply under each climate change scenario. In the objective function, crop's total output changes with its sown area and irrigation water. Equation (4) is a production function for irrigated crop with a constant elasticity of substitution (CES) between irrigated land and irrigation water. When crop planted on rainfed land, no irrigation water is used, and the production function can be simplified as Eq. (5).

2.2 Estimation of Energy consumption and GHG emission

The Energy consumption of and GHG emission from irrigation activities mainly depend on pumping, conveying and pressurizing water for application to the field (Rothausen and Conway, 2011). Due to data limit, we calculate emission from groundwater pumping for irrigation and then derive the emission from pumping surface water by using a conversion factor. According to the method introduced by Wang et al. (2012), the emission from pumping groundwater for irrigation can be calculated in the following steps.

First, the energy use rate is calculated by using

$$\text{Energy use rate}\left(\text{kWh}\,\text{m}^{-3}\right)=\frac{9.8\,\text{m}\,\text{s}^{-2}\times\text{Lift (m)}}{3.6\times10^{3}\times\text{Efficiency (\%)}} \tag{6}$$

which means that the energy required to lift 1 m^3 of water 1 m up at 100 % efficiency is 0.0027 kWh (Rothausen and Conway, 2011). Lift, the most crucial factor in the energy use rate, relates to the vertical distance over which water is raised before its application to the field. The average pump lift for the SRB and LRB is from provincial data, and the estimated relationship based on groundwater level is in Wang et al. (2012) (see Table 7). Another key factor in Eq. (6) is pumping efficiency. In China, there are two types of pump engines, namely, electric and diesel; their shares are 76 and 24 %, respectively, and their pumping efficiency are set at 40 % (with an additional 15 % transmission and distribution loss) and 15 %, respectively, according to Wang et al. (2012).

Second, a conversion factor is used to derive the GHG emission rate from the energy use rate, which is 0.32021 kg CO_2 e kWh^{-1} for diesel and 0.94773 kg CO_2 e kWh^{-1} for electricity produced in China (Wang et al., 2012). Equation (7) illustrates this calculation:

$$\begin{aligned}&\text{GHG emissionrate}\left(\text{kg}\,\text{CO}_2\,\text{e}\,\text{m}^{-3}\right)= \qquad (7)\\&\quad\text{Energy use rate}_{\text{Electric}}\left(\text{kWh}\,\text{m}^{-3}\right)\times\text{Share}_{\text{Electric}}\ (\%)\\&\quad\times 0.94773\ \text{kg}\,\text{CO}_2\ \text{e}\,\text{kWh}^{-1}\\&\quad+\text{Energy use rate}_{\text{Diesel}}\left(\text{kWh}\,\text{m}^{-3}\right)\\&\quad\times\text{Share}_{\text{Diesel}}\ (\%)\times 0.32021\ \text{kg}\,\text{CO}_2\ \text{e}\,\text{kWh}^{-1},\end{aligned}$$

Third, with the agricultural water use simulated by the CWSM model and the share of groundwater use (calculated based on proportion of groundwater-fed irrigated area from Wang et al. (2012) and the change rate of the groundwater supply from the CWSM model) under different scenarios, the groundwater pumped for irrigation can be derived. Then, the resulting energy consumption and GHG emission from groundwater irrigation can be calculated as the product of the energy use rate and GHG emission rate and groundwater use in agriculture.

The above steps are used to calculate the GHG emission from pumping groundwater for irrigation. For the total GHG emission from both surface water and groundwater pumping irrigation, a conversion coefficient of 0.58 is used. In China, the installed pumping capacity for surface water and groundwater are 22 million and 37.7 million kW respectively (Wang et al., 2012); thus, the ratio between emission from pumping surface water and groundwater is 0.58. That is, the total emission from irrigation are 1.58 times the emission from groundwater pumping.

2.3 Scenario specification

To simulate the impact of climate change in the future, we set a reference scenario without climate change and two alternative climate change scenarios. The base year in our study is 2010, represented by the average level of temperature and precipitation from 1990 to 2019, and the projection year is 2030, represented by the average level from 2020 to 2049. Historical data are mainly calculated from the China Statistical Yearbook published by National Bureau of Statistics of China and China Water Resources Bulletin published by the Ministry of Water Resources and calibrated by using survey data from the CCAP (Chinese Center for Agricultural Policy, Peking University).

2.3.1 Reference scenario

There are three main water use sectors in the system, namely, the agricultural, industrial and domestic sectors. Without climate change, their water demands of water are determined by principally socioeconomic development and water use efficiency, with the total supply of water in 2030 remaining unchanged. For more details about the water balance in our model, see Wang et al. (2013).

In the agricultural sector, we assume that the sown area, cropping pattern and yield stay the same as in 2010. The development of agriculture is represented by the increase in irrigated areas. According to information in the National Integrated Water Resources Plan 2010–2030 from the Ministry of Water Resources (MWR) in China, irrigated areas in SRB and LRB will expand by 13 and 17 %, respectively. For the industrial sector, the annual growth rate of GDP in Northeast China will be 4.4 % but with gradually declining growth rates from 5.1 % in 2010–2020 to 3.8 % in 2020–2030 (Sun et al., 2008). Regarding the domestic sector, the key factors are population and the urbanization rate. Population data are from the recent projection by the United Nations, which shows that the average annual population growth rate will be 0.34 % in 2011–2030, ranging from approximately 0.5 % in 2011 to nearly zero in 2030. As projected by IIASA, the urbanization rate in Northeast China will increase by approximately 22 % from 2010 to 2030 (Toth et al., 2008).

To prevent the impact of climate change from being overwhelmed, we assume that the water use efficiency in various sectors (agricultural, industrial and domestic sectors) in 2030 will be equal to that in the base year of 2010. Water use efficiency in the agricultural sector is measured by the share of crop water consumption over the total water withdrawal in the agricultural sector at the basin level. Industrial water use efficiency or the industrial water use quota is measured by water consumption per 10 000 RMB Yuan of GDP in the industrial sector. Water use efficiency in the domestic sector has been divided into rural and urban areas, and it is measured by the per capita daily water use quota in rural and urban areas, respectively.

2.3.2 Climate change scenarios

Representative Concentration Pathways (RCPs) are used in this paper to provide future changes in climate variables, temperature and precipitation. Compared with the SRES scenarios (Nakicenovic et al., 2000), RCPs are a new set of climate scenarios, and they cover a wide range of year-2100 global radiative forcing levels (from 2.6 to 8.5 $W\,m^{-2}$) that were introduced in a special issue of Climatic Change in 2011 (Van Vuuren et al., 2011a).

RCP2.6 represents the lowest emission and radiative forcing. Under this scenario, the increase in the global mean temperature is limited to 2 °C, and emission from energy use will be negative in the second half of the 21st century. It requires the full participation of all nations, the rapid implantation of new technologies, and sufficient political and societal support (Van Vuuren et al., 2011b). Under RCP8.5, it is assumed that the population growth rate is high, the income growth rate is relatively low, technological change and energy intensity improvements are at modest rates, and climate mitigation policies are absent. As a result, RCP8.5 will have the highest greenhouse gas emission among all RCPs (Riahi et al., 2011).

To obtain high-resolution changes in temperature and precipitation in the SRB and LRB, 26 CMIP5 models are used (Ma et al., 2014). Given the temperature and precipitation under the two RCPs, the future changes in runoff are then estimated by using the CWSM. As shown in Table 1, both the average temperature and precipitation in Northeast China under the RCPs in 2030 will be higher than those in 2010. The future average temperature will increase by up to 2.19 °C, and the change in precipitation will vary from 3.41 to 7.07 % in the SRB and LRB. As a result, the runoff of Northeast China under the RCP scenarios will be 3.30 to 5.98 % lower than it is in 2010. Clearly, the LRB will be affected more negatively by climate change than the SRB. Comparing all four RCP scenarios, RCP8.5 has the most negative effects on runoff (simulated by using climate elasticity, Fu et al., 2007) in both the SRB and the LRB, whereas RCP6.0 has the mildest effects. To capture a wide range of future climate change, RCP6.0 and RCP8.5 are used for further simulations.

Climate change will directly affect the crop yield and water balance. The direct impacts of climate change on the crop

Table 1. Change in the average annual temperature, precipitation and runoff under alternative climate change scenarios from 2010 to 2030.

	Climate Scenarios			
	RCP2.6	RCP4.5	RCP6.0	RCP8.5
Change in temperature (°C)				
SRB	1.8	1.71	1.82	2.19
LRB	1.5	1.43	1.36	1.9
Change in precipitation (%)				
SRB	5.58	5.34	7.07	5.59
LRB	4.05	4.26	6.16	3.41
Change in runoff (%)				
SRB	−3.3	−3.12	−1.74	−5.44
LRB	−4.21	−3.83	−0.55	−5.98

Sources: Changes in temperature and precipitation are simulated by Beijing Normal University based on 26 CMIP5 models. Changes in runoff are simulated by the authors based on the CWSM model.

yield are not considered in our study because convincing estimations of the effects on the yield of many crops under the RCPs are not currently available (Ju et al., 2013). Thus, in this paper, we focus on the effects of climate change on crop production through its impacts on the water balance. As discussed above, the SRB and LRB under the two climate scenarios differ in temperature, precipitation and runoff, leading to different impacts on the water supply and water demand in 2030. According to the simulation of the CWSM (Table 2), compared with the reference scenario, the water supply in the SRB will drop by 1.44 to 4.49 % under the RCPs, whereas the water supply in the LRB will drop by as much as 7.88 % (under RCP8.5). Meanwhile, the water demand increases by approximately 1 % mainly because of the increasing irrigation demand of crops. Consequently, climate change will further widen the gap between the water supply and the water demand. In the LRB under the RCP8.5, the water gap rate (share of water shortage in demand) will be almost 7 % higher than the reference scenario in 2030.

3 Results and Discussions

3.1 Agricultural production under climate change without optimization adjustments

Suppose that farmers facing climate change will not optimize their production, which means that cropping patterns will remain exactly the same and irrigation water will be allocated to each crop according to its share of irrigation demand under the reference scenario. Thus, the percentage change of irrigated area, rainfed area and sown area will be zero and that of irrigation intensity equal the reduction ratio of irrigation supply. Since crops' total output and their yield, as well

Table 2. Change in the water balance under alternative climate change scenarios compared with the reference scenario in 2030 (%).

	Climate Scenarios			
	RCP2.6	RCP4.5	RCP6.0	RCP8.5
Water supply				
SRB	−2.73	−2.58	−1.44	−4.49
LRB	−5.55	−5.05	−0.73	−7.88
Water demand				
SRB	0.81	0.44	0.39	1.34
LRB	0.39	0.11	0.02	1.68
Water gap rate				
SRB	−2.54	−2.19	−1.36	−4.11
LRB	−4.36	−3.82	−0.76	−6.79

Sources: Simulation results based on the CWSM model.

as total profits and profit per ha, will vary to the same extent, we only show the results of the average crop yield in Table 3 and the total profits in Table 4.

In Northeast China, the average yield of all crops decreases under both two climate change scenarios. In both the SRB and the LRB, rice and wheat suffer the most negative impacts of climate change and vegetables will also experience a considerable yield loss. These crops have a relatively high water sensitivity under irrigation, which means that they will be more affected when the irrigation water supply decreases because of climate change. The total profits at the basin level also decline, but the profits of crops widely vary. Cotton and sugar crops will experience a huge increase in profits, whereas the other seven crops will face losses in the profit. RCP8.5 in the LRB has the most profound impact on agricultural production, especially for rice. All of these changes are highly in line with the fluctuation in the irrigation water supply.

Sources: Simulation results based on the CWSM model

Table 4 Percentage change in total profits under two climate scenarios without optimization in 2030 (compared with the reference scenario, %)

Sources: Simulation results based on the CWSM model

3.2 Adjustment of agricultural production to climate change

Considering the adjustments made by farmers, the simulated effects of climate change on agricultural production under two climate change scenarios are presented in Tables 5 and 6, including the cropping pattern, irrigation intensity, average crop yield (of both irrigated and rainfed crops), crop outputs and agricultural profits.

Table 3. Percent change in yield under two climate scenarios autonomous adaptation in 2030 (compared with the reference scenario, %).

	SRB		LRB	
	RCP6.0	RCP8.5	RCP6.0	RCP8.5
Rice	−1.14	−3.67	−0.41	−5.55
Wheat	−0.82	−2.64	−0.31	−4.22
Maize	−0.23	−0.75	−0.13	−1.79
Soybean	−0.10	−0.34	−0.05	−0.64
Sugar crops	−0.35	−1.13	−0.12	−1.60
Oil crops	−0.14	−0.47	−0.06	−0.79
Cotton	−0.25	−0.83	−0.08	−1.11
Vegetable	−0.69	−2.24	−0.17	−2.39
Other crops	−0.18	−0.58	−0.08	−1.09

Sources: Simulation results based on the CWSM model.

Table 4. Percentage change in total profits under two climate scenarios without optimization in 2030 (compared with the reference scenario, %).

	SRB		LRB	
	RCP6.0	RCP8.5	RCP6.0	RCP8.5
Total	−1.12	−3.58	−0.37	−5.06
Rice	−3.39	−10.72	−1.06	−13.89
Wheat	−1.56	−4.98	−0.57	−7.62
Maize	−0.61	−1.98	−0.27	−3.73
Soybean	−0.46	−1.49	−0.20	−2.75
Sugar crops	−0.98	−3.15	−0.33	−4.52
Oil crops	−0.50	−1.60	−0.23	−3.12
Cotton	−0.79	−2.59	−0.26	−3.71
Vegetable	−1.52	−4.88	−0.41	−5.63

Sources: Simulation results based on the CWSM model.

3.2.1 Cropping pattern and irrigation intensity

The reallocation of land and water within crops in the PMP model is realized by the adjustments of the cropping pattern and irrigation intensity. Theoretically, when water scarcity in Northeast China becomes more serious, the irrigated area of each river basin and crop, in addition to the irrigation water intensity, will decrease, and the rainfed area will increase. Because RCP8.5 represents the climate change scenario with the highest emission, we will focus on its impacts on agricultural production through water availability in the SRB and LRB below.

The shares of irrigated area to the total sown area in the SRB and LRB are 32.24 and 42.91 %, respectively, under the reference scenario and they fall to 31.86 and 42.59 %, respectively, under RCP8.5. Due to the relatively high sensitivity of the yield response to water and the relatively low gross revenues per hectare, the most striking effects of the decrease in the irrigated area will occur in rice, soybean and sugar crops,

Table 5. Percent changes of key factors in the SRB River Basin under climate in 2030 (%).

	Sown area	Irrigated area	Rain-fed area	Irrigation intensity
RCP6.0				
Total	0.00	−0.36	0.17	−0.57
Rice	−0.87	−0.87	0.00	−0.46
Wheat	−0.03	−0.12	0.17	−0.88
Maize	0.19	0.45	0.12	−0.53
Soybean	0.09	−0.90	0.25	−0.41
Sugar crops	−0.25	−0.92	0.10	−0.71
Oil crops	0.22	−0.05	0.30	−0.33
Cotton	−0.10	−0.25	0.07	−0.19
Vegetable	0.06	0.07	0.02	−0.11
Other crops	0.08	0.10	0.08	−0.32
RCP8.5				
Total	0.00	−1.17	0.56	−7.59
Rice	−2.84	−2.84	0.00	−6.19
Wheat	−0.04	−0.29	0.56	−11.87
Maize	0.62	1.48	0.38	−7.23
Soybean	0.28	−2.91	0.79	−5.47
Sugar crops	−0.78	−2.93	0.33	−9.50
Oil crops	0.75	0.00	0.97	−4.42
Cotton	−0.30	−0.83	0.24	−2.59
Vegetable	0.18	0.20	0.07	−1.53
Other crops	0.26	0.30	0.25	−4.38
	Output	Yield	Total Profits	Profits per ha
RCP6.0				
Total	–	–	−0.95	−0.95
Rice	−1.31	−0.44	−3.93	−3.08
Wheat	−1.16	−1.13	−2.00	−1.97
Maize	−0.09	−0.28	−0.41	−0.60
Soybean	−0.05	−0.14	−0.40	−0.49
Sugar crops	−0.53	−0.29	−1.17	−0.93
Oil crops	−0.15	−0.37	−0.35	−0.57
Cotton	−0.09	0.01	−0.61	−0.51
Vegetable	−0.08	−0.14	−0.61	−0.67
Other crops	−0.07	−0.15	−0.63	−0.71
RCP8.5				
Total	–	–	−3.02	−3.02
Rice	−4.21	−1.41	−12.39	−9.83
Wheat	−3.72	−3.68	−6.30	−6.26
Maize	−0.28	−0.90	−1.33	−1.94
Soybean	−0.17	−0.44	−1.30	−1.57
Sugar crops	−1.70	−0.93	−3.75	−2.99
Oil crops	−0.49	−1.23	−1.12	−1.85
Cotton	−0.31	−0.01	−1.97	−1.68
Vegetable	−0.26	−0.44	−1.96	−2.14
Other crops	−0.23	−0.48	−2.05	−2.30

Sources: Simulation results based on the CWSM model.

by approximately 2.9 % in the SRB and 6.5 % in the LRB. By contrast, maize and vegetables will experience an expansion in the irrigated area, in addition to wheat (by 1.48 and 1.92 % in the SRB and the LRB, respectively).

The rainfed areas in both river basins will expand, by 0.56 %, with the growth rate crop varying from 0.07 % (vegetable) to 0.97 % (oil crops) in the SRB and 0.11 % (vegetable) to 0.98 % (soybean) in the LRB. Note that there are no rainfed rice in Northeast China under any scenarios due to insufficient rainfall.

Correspondingly, the sown area of each crop will change with the decrease in the irrigated area and the increase in the rainfed area. The crops that will expand and contract the most are maize and rice, respectively, in both the SRB (0.62 and −2.84 %, respectively) and the LRB (1.13 and −6.72 %, respectively). However, maize, soybean and rice will still be the dominant crops in Northeast China, and their shares of sown area to the entire basin's will hardly change, based on the calibration methods we used and the assumptions of stability in the sown area of the entire basin under all scenarios.

Under the reference scenario, the irrigation intensity in the SRB and LRB will be 3522 and 3650 $m^3\ ha^{-1}$, respectively. With the reduction in water availability under climate change scenario RCP8.5, the irrigation intensity in these two river basins will decrease by 4.07 and 7.59 %, respectively. The extent to which the irrigation intensity of crops will decrease depends on the water-yield response and gross revenue per hectare. For example, in the LRB, the irrigation intensities of oil crops will decline the most (by 8.41 %) as a consequence of being the least profitable crops.

3.2.2 Crop outputs and yield

Despite the considerable reduction in water availability under RCP8.5, the crop outputs of the SRB and LRB will not change significantly. Thanks to the adjustments in the cropping pattern and irrigation intensity, the resulted output are higher than those without adaptation measures. For example, the reduction of the output of grain crops (rice, wheat and maize) in the LRB declines from 2.35 to 2.16 %. Rice and wheat are the most water-sensitive crops; their output in the LRB drops the most (by approximately 8.5 and 7.2 %, respectively). In Northeast China, wheat production is marginal, but the promotion of rice should be considered carefully given the sharp drop in the water supply as a result of climate change.

Regarding the average crop yield, the only crop whose yield increases is cotton in the LRB, by 0.2 %. A large drop in the wheat production of the LRB will occur (5.65 %). Generally, crop yields does not change as much as those of irrigated area or water intensity in our model. The reason is partly related to the assumption that the production of crop is diminishing marginal return. However, the most important reason is, unlike many studies on the impact of climate change on yield (Wang et al., 2009; Tao and Zhang, 2010), we consider

Table 6. Percent changes of key factors in the LRB under climate scenarios in 2030 (%).

	Sown area	Irrigated area	Rainfed area	Irrigation intensity
RCP6.0				
Total	0.00	−0.06	0.04	−1.26
Rice	−0.49	−0.49	0.00	−0.99
Wheat	−0.14	−0.20	0.06	−2.21
Maize	0.08	0.14	0.04	−1.29
Soybean	−0.04	−0.46	0.07	−0.90
Sugar crops	−0.15	−0.52	0.04	−1.67
Oil crops	0.00	−0.19	0.05	−2.64
Cotton	−0.07	−0.16	0.04	−0.29
Vegetable	0.01	0.01	0.01	−0.26
Other crops	0.01	−0.05	0.02	−0.98
RCP8.5				
Total	0.00	−0.75	0.56	−4.07
Rice	−6.72	−6.72	0.00	−3.18
Wheat	−1.64	−2.49	0.87	−7.14
Maize	1.13	1.92	0.60	−4.18
Soybean	−0.53	−6.26	0.98	−2.89
Sugar crops	−1.89	−6.84	0.52	−5.34
Oil crops	−0.10	−2.69	0.63	−8.41
Cotton	−0.97	−2.26	0.52	−0.94
Vegetable	0.10	0.09	0.11	−0.83
Other crops	0.06	−0.83	0.30	−3.17
	Output	Yield	Total Profits	Profits per ha
RCP6.0				
Total	–	–	−0.35	−0.35
Rice	−0.63	−0.14	−1.68	−1.19
Wheat	−0.54	−0.40	−0.95	−0.81
Maize	−0.07	−0.15	−0.23	−0.31
Soybean	−0.06	−0.02	−0.26	−0.22
Sugar crops	−0.23	−0.08	−0.52	−0.37
Oil crops	−0.04	−0.03	−0.23	−0.22
Cotton	−0.05	0.01	−0.29	−0.22
Vegetable	−0.03	−0.04	−0.25	−0.26
Other crops	−0.04	−0.05	−0.32	−0.33
RCP8.5				
Total	–	–	−4.67	−4.67
Rice	−8.49	−1.90	−21.42	−15.76
Wheat	−7.20	−5.65	−12.16	−10.70
Maize	−1.04	−2.14	−3.19	−4.27
Soybean	−0.78	−0.25	−3.58	−3.07
Sugar crops	−2.99	−1.12	−6.90	−5.11
Oil crops	−0.51	−0.41	−3.11	−3.01
Cotton	−0.78	0.20	−3.98	−3.04
Vegetable	−0.39	−0.49	−3.44	−3.54
Other crops	−0.58	−0.64	−4.37	−4.43

Sources: Simulation results based on the CWSM model.

only the indirect effect of climate change through the changing water stress and the effect of the optimization adjustment measures.

3.2.3 Agricultural profits

Agricultural profits decrease by 3.02 and 4.67 % for the SRB and the LRB, respectively, under RCP8.5; these reductions are smaller than losses without optimization adjustment measures (3.58 and 5.06 %, respectively). This finding shows that farmers' autonomous adaptation measures such as adjusting the cropping pattern and irrigation intensity can effectively protect farmers' profits. Moreover, the profit loss at the basin level is much milder than the reduction in the irrigation water supply, in relative term. It can be understood in three ways. First, irrigated activities are not dominate in agricultural productions, only accounting for approximately 35 % of the total sown area in Northeast China under the reference scenario, and will remain stable under climate change (as discussed above). Second, farmers can lower their irrigation water intensity, spending less on water fees and energy cost to adapt to declining water availability. Third, crop diversification at the basin level plays an important role in stabilizing agricultural profits. Although the total profits of producing a single-crop drops by as much as 21.42 % (rice in the LRB), such climate-related risks could be mitigated by planting a combination of diverse crops (for example, maize, oil crops and vegetables). Thus, confronting changing climatic conditions, farmers are recommended to diversify their crop portfolios (Huang et al., 2014).

3.2.4 Sensitivity analysis

The robustness of our model to the most important parameters, the substitution elasticity σ and the supply elasticity η, are tested. The results show although the output and net profits are not sensitive to both parameters, the cropping pattern and irrigation intensity are directly related to them. For a better simulation, estimating the key parameters of the model by using survey data and econometric methods is highly recommended (Heckelei et al., 2012), as what we did in this study.

3.3 Energy consumption and GHG emission

As shown in Table 7, the groundwater level in the SRB and LRB is approximately 15 m below ground. According to the relationship between the pump lifts and the groundwater levels defined by Wang et al. (2012), the estimated pump lift is approximately 35 m in Northeast China. Using Eqs. (6) and (7), the energy use rate and GHG emission rate are estimated as 0.32 $kWh\,m^{-3}$ and 0.25 $kg\,CO_2\,e\,m^{-3}$, respectively, in the SRB and 0.34 $kWh\,m^{-3}$ and 0.27 $kg\,CO_2\,e\,m^{-3}$, respectively, in the LRB.

Under the reference scenario in 2030, the energy consumption in the SRB will be 4.91 million $kWh\,m^{-3}$ and

Table 7. Energy use rate and GHG emission rate from irrigation in the SRB and LRB.

	SRB	LRB
Groundwater level (m)	13.82	16.43
Estimated pump lift (m)	34.27	36.64
Energy use rate ($kWh\,m^{-3}$)	0.32	0.34
GHG emission rate ($kg\,CO_2e\,m^{-3}$)	0.25	0.27

Sources: Calculated by the authors based on Wang et al. (2012).

Table 8. Total GHG emission from irrigation in the SRB and LRB in 2030.

	Reference	RCP6.0	RCP8.5
SRB			
Agricultural water use[a] (billion $m^3\,yr^{-1}$)	23.30	22.95	22.19
Share of GW use[b] (%)	51.16	50.71	49.75
Agricultural GW use (billion $m^3\,yr^{-1}$)	11.92	11.64	11.04
Energy consumption from GW pumping (million $kWh\,m^{-3}$)	3.78	3.69	3.5
GHG emission from GW ($MtCO_2e$)	3.01	2.94	2.79
Total GHG emission[c] ($Mt\,CO_2\,e$)	4.76	4.65	4.41
LRB			
Agricultural water use[a] (billion $m^3\,yr^{-1}$)	11.96	11.89	11.05
Share of GW use[b] (%)	48.19	48.07	46.88
Agricultural GW use (billion $m^3\,yr^{-1}$)	5.76	5.71	5.18
Energy consumption from GW pumping (million $kWh\,m^{-3}$)	1.95	1.94	1.76
GHG emission from GW ($Mt\,CO_2\,e$)	1.56	1.54	1.40
Total GHG emission[c] ($Mt\,CO_2\,e$)	2.46	2.44	2.21

that in the LRB is projected to be 50 % less, 2.45 million $kWh\,m^{-3}$ (see Table 8). The total GHG emission from irrigation under the reference scenario will be 4.76 and 2.46 $Mt\,CO_2$ e in the SRB and LRB, respectively. Under the RCP scenarios, GHG emission are projected to decrease in both river basins, with change rate ranging from −2.36 to −7.41 % in the SRB and from −0.81 to −10.07 % in the LRB. This decrease is related to the reduction in the agricultural water supply as a result of climate change. If an increasing number of water-saving technologies, such as low-pressure pipelines, sprinklers and micro-irrigation, are used as planned by the government, the future trend of GHG emission from irrigation will be unclear because low-pressure pipelines are more energy consuming-efficient (Zou et al., 2013) whereas sprinklers and micro-irrigation are more energy consuming-intensive.

To have a general idea at the national level, the results of the other eight major river basins in China are calculated in the same manner. Under all four RCP scenarios, the total GHG emission from irrigation in China varies from 42.58 $Mt\,CO_2$ e (RCP8.5) to 45.12 $Mt\,CO_2$ e (RCP4.5), of which Northeast China accounts for approximately 16 %. If China is responsible for 20 % of global GHG emission (IPCC, 2014), then the ratio of GHG emission from irrigation to total national emission will range from 0.4 % (RCP8.5) to 0.75 % (RCP2.6) in 2030. This ratio is calculated to be 0.58 % in 2006 (Wang et al., 2012), and it will be higher under both the RCP2.6 and the RCP6.0 (0.62 %) scenarios. The significance of irrigation as a source of GHG emission should not be ignored but instead should draw more attention from both researchers and political leaders in China.

4 Conclusion and policy implications

The nexus among water, food, energy and GHG under climate change has become a hot topic in recent years. However, few integrated analysis of their relationships have been done so far. The purpose of this paper is to examine the nexus by assessing the effects of climate change on agricultural production through the change in water, evaluating the adjustment responses and the resulting energy consumption and GHG emission. The study area focuses on Northeast China, which is a major grain production region and highly dependent on irrigation with a limited water endowment.

Based on our simulation results, by 2030, climate change will result in an increase in the water supply and demand gap for irrigation from 1.36 to 4.11 % in the SRB and from 0.76 to 6.79 % in the LRB in Northeast China under two climate scenarios, RCP6.0 and RCP8.5. Due to the increase in water scarcity, under the same climate scenarios, the irrigated areas in these two river basins decrease from 0.36 to 1.17 % and from 0.06 to 0.75 %, respectively. In addition to the change in irrigated areas, the cropping pattern will be adjusted by farmers. The major adjustment in these two river basins is the increase in the maize sown areas and the decrease in the rice sown areas. As a result, the total output of crops will decline. For example, in the SRB, major grain production (wheat, maize and rice) will be reduced by from 1.26 to 1.47 %, while in the LRB, major grain production will be reduced by from 2.16 to 2.35 %. Further analysis also indicates a reduction in agricultural production profits by from 0.95 to 3.02 % in the SRB and from 0.35 to 4.67 % in the LRB. Finally, we find that, under RCP6.0 and RCP8.5, energy consumption and GHG emission will decrease by from

2.36 to 7.41 % in the SRB and from 0.81 to 10.07 % in the LRB.

Our results have policy implications. First, in assessing the impacts of climate change, the nexus among water, food, energy and GHG emission should be fully considered; otherwise, the overall impacts of climate change will be underestimated. In addition, further research at the regional level in China is needed, especially for those regions that are expected to experience more serious water shortage under climate change, such as the Haihe, Huaihe and Yellow RBs.

Second, we found that the adaptation measures adopted by farmers will reduce the risk of climate change for agricultural production, which is a good news for China's food security. To mitigate the negative effects of climate change on agricultural production, implementing irrigation measures (such as extending water-saving technologies), optimizing the cropping pattern (reducing the sown area of water-intensive crops) and improving water productivities (such as planting drought-resilient varieties) are highly recommended.

Finally, energy consumption and GHG emission both decrease due to the reduction in irrigation water caused by climate change. That is, although climate change negatively influences water and agricultural production, it will not lead to a further increase in energy consumption and GHG emission. However, the estimation of energy consumption and GHG emission in this paper is mainly based on the change in the irrigation water supply. There are other factors that may affect future emission, such as the application of water-saving technologies, the limitations of pumping groundwater and the use of renewable energy sources for irrigation. How these factors influence energy consumption and GHG emission is an interesting topic for future research.

Data availability. The original data set can be accessed: http://www.mwr.gov.cn/english/publs/ and http://data.stats.gov.cn/english/easyquery.htm (last access: 13 April 2018).

Competing interests. The authors declare that they have no conflict of interest.

Special issue statement. This article is part of the special issue "Innovative water resources management – understanding and balancing interactions between humankind and nature". It is a result of the 8th International Water Resources Management Conference of ICWRS, Beijing, China, 13–15 June 2018.

Acknowledgements. We acknowledge financial support from the National Natural Sciences Foundation in China (71161140351).

Edited by: Zongxue Xu
Reviewed by: two anonymous referees

References

Bazilian, M., Rogner, H., Howells, M., Hermann, S., Arent, D., Gielen, D., Steduto, P., Mueller, A., Komor, P., Tol, R. S. J., and Yumkella, K. K.: Considering the energy, water and food nexus: Towards an integrated modelling approach, Energ. Policy, 39, 7896–7906, 2011.

Cobuloglu, H. I. and Büyüktahtakın, İ. E.: Food vs. biofuel: An optimization approach to the spatio-temporal analysis of land-use competition and environmental impacts, Appl. Energ., 140, 418–434, 2015.

ECSNCCA (Edition Committee of the Second National Climate Change Assessment): Second National Climate Change Assessment Report, Science Press, 2011.

Finley, J. W. and Seiber, J. N.: The nexus of food, energy, and water, J. Agr. Food Chem., 62, 6255–6262, 2014.

Fu, G., Charles, S. P., and Chiew, F. H. S.: A two-parameter climate elasticity of streamflow index to assess climate change effects on 80 annual streamflow, Water Resour. Res., 43, W11419, https://doi.org/10.1029/2007WR005890, 2007.

Graveline, N. and Mérel, P.: Intensive and extensive margin adjustments to water scarcity in France's Cereal Belt, European Review of Agricultural Economics, 41, 707–743, https://doi.org/10.1093/erae/jbt039, 2014.

Heckelei, T., Britz, W., and Zhang, Y.: Positive Mathematical Programming Approaches–Recent Developments in Literature and Applied Modelling, Bio-based and Applied Economics, 1, 109–124, 2012.

Huang, J., Jiang, J., Wang, J., and Hou, L.: Crop Diversification in Coping with Extreme Weather Events in China, J. Integr Agr, 13, 677–686, 2014.

IPCC: Climate Change 2013: The Physical Science Basis, Contribution of Working Group I to the Fifth Assessment Report of the Intergovernmental Panel on Climate Change, Cambridge University Press, 1395–1445, 2014.

Ju, H., Lin, E., Wheeler, T., Challinor, A., and Jiang, S.: Climate Change Modelling and Its Roles to Chinese Crops Yield, J. Integr Agr, 12, 892–902, 2013.

Kao, S., Sale, M. J., Ashfaq, M., Uria Martinez, R., Kaiser, D. P., Wei, Y., and Diffenbaugh, N. S.: Projecting changes in annual hydropower generation using regional runoff data: An assessment of the United States federal hydropower plants, Energy, 80, 239–250, 2015.

Keairns, D. L., Darton, R. C., and Irabien, A.: The Energy-Water-Food Nexus, Annual Review of Chemical & Biomolecular Engineering, 7, 239, 2016.

Ma, J., Dong, W., Wei, Z., and Yan, X.: Evaluating daily surface maximum temperature interpolation error by adding climate stations near border areas over China, Int. J. Climatol., 35, 2798–2808, 2015.

Martin-Gorriz, B., Soto-García, M., and Martínez-Alvarez, V.: Energy and greenhouse-gas emissions in irrigated agriculture of SE (southeast) Spain. Effects of alternative water supply scenarios, Energy, 77, 478–488, 2014.

Mendelsohn, R. and Dinar, A.: Climate Change and Agriculture: An Economic Analysis of Global Impacts, Adaptation, and Distributional Effects, Edward Elgar Publishing, 2009.

Nakicenovic, N., Alcamo, J., Davis, G., De Vries, B., Fenhann, J., Gaffin, S., Gregory, K., Grubler, A., Jung, T. Y., Kram, T.,

La Rovere, E. L., Michaelis, L., Mori, S., Morita, T., Pepper, W., Pitcher, H. M., Price, L., Riahi, K., Roehrl, A., Rogner, H., Sankovski, A., Schlesinger, M., Shukla, P., Smith, S.J., Swart, R., Van Rooijen, S., Victor, N., and Dadi, Z.: Special Report on Emissions Scenarios: a special report of Working Group III of the Intergovernmental Panel on Climate Change, Cambridge University Press, 2000.

NBSC (National Bureau of Statistics of China): China statistical yearbook 2013, China Statistics Press, 2013.

Nelson, G. C., Robertson, R., Msangi, S., Zhu, T. J., Liao, X. L., and Jawajar, P.: Greenhouse gas mitigation: Issues for Indian agriculture, International Food Policy Research Institute, Environment and Production Technology Division, Washington, DC, IFPRI Discussion Papers 00900, 2009.

Rasul, G. and Sharma, B.: The nexus approach to water-energy-food security: an option for adaptation to climate change, Climate Policy, 16, 682–702, 2015.

Riahi, K., Rao, S., Krey, V., Cho, C., Chirkov, V., Fischer, G., Kindermann, G., Nakicenovic, N., and Rafaj, P.: RCP 8.5 – A scenario of comparatively high greenhouse gas emissions, Climatic Change, 109, 33–57, 2011.

Rosenzweig, C., Strzepek, K. M., Major, D. C., Iglesias, A., Yates, D. N., McCluskey, A., and Hillel, D.: Water resources for agriculture in a changing climate: international case studies, Global Environ. Change, 14, 345–360, 2004.

Rothausen, S. G. S. A. and Conway, D.: Greenhouse-gas emissions from energy use in the water sector, Nat. Clim. Change, 1, 210–219, 2011.

Sun, L., Veen, W. V., Zhang, X., Huang, J., and Qiu, H.: China's Economic Growth outside the Farm Sector: Plausible Trends towards 2030, CATSEI Working Paper, available at: http://www.catsei.org/upload/2009/320093512113891.pdf (last access: 13 April 2018), 2008.

Tao, F. and Zhang, Z.: Adaptation of maize production to climate change in North China Plain: Quantify the relative contributions of adaptation options, Eur. J. Agron., 33, 103–116, 2010.

Toth, F. L., Cao, G., and Hizsnyik, E.: Regional population projections for China, International Institute for Applied Systems Analysis, Laxenburg, Austria, Interim Report IR-03-042, 1–55, 2008.

Van Vuuren, D. P., Edmonds, J. A., Kainuma, M., Riahi, K., and Weyant, J.: A special issue on the RCPs, Climatic Change, 109, 1–4, 2011a.

Van Vuuren, D. P., Stehfest, E., Den Elzen, M. G. J., Kram, T., Van Vliet, J., Deetman, S., Isaac, M., Klein Goldewijk, K., Hof, A., Mendoza Beltran, A., Oostenrijk, R., and van Ruijven, B.: RCP2.6: exploring the possibility to keep global mean temperature increase below 2 °C, Climatic Change, 109, 95–116, 2011b.

Wang, J., Mendelsohn, R., Dinar, A., Huang, J., Rozelle, S., and Zhang, L.: The impact of climate change on China's agriculture, Agr. Econ.-Blackwell, 40, 323–337, 2009.

Wang, J., Rothausen, S. G. S. A., Conway, D., Zhang, L., Xiong, W., Holman, I. P., and Li, Y.: China's wate-energy nexus: greenhouse-gas emissions from groundwater use for agriculture, Environ. Res. Lett., 7, 014035, https://doi.org/10.1088/1748-9326/7/1/014035, 2012.

Wang, J., Huang, J., and Yan, T.: Impacts of climate change on water and agricultural production in ten large river basins in China, J. Integr. Agr., 12, 1267–1278, 2013.

Waughray, D.: Water security: the water-food-energy-climate nexus, World Economic Forum water initiative, island press, Washington, 2011.

Wise, M., Dooley, J., Luckow, P., Calvin, K., and Kyle, P.: Agriculture, land use, energy and carbon emission impacts of global biofuel mandates to mid-century, Appl. Energ., 114, 763–773, 2014.

Yan, T.: Optimal Allocation of Agricultural Water Use and Its Impacts on Crop Production under Climate Change: Study in Five Large River Basins in Northern China, The University of Chinese Academy of Sciences, 2015.

Zou, X., Li, Y., Cremades, R., Gao, Q., Wan, Y., and Qin, X.: Cost-effectiveness analysis of water-saving irrigation technologies based on climate change response: A case study of China, Agr. Water Manage., 129, 9–20, 2013.

https://doi.org/10.5194/piahs-379-313-2018

Temporal and spatial variation of hydrological condition in the Ziwu River Basin of the Han River in China

Ziyan Li[1], Dengfeng Liu[1,2], Qiang Huang[1], Tao Bai[1], Shuai Zhou[1], and Mu Lin[3]

[1]State Key Laboratory of Eco-hydraulics in Northwest Arid Region of China, School of Water Resources and Hydropower Engineering, Xi'an University of Technology, Xi'an, 710048, China
[2]Department of Land Resources and Environmental Sciences, Montana State University, Bozeman, MT 59717, USA
[3]School of statistics and Mathematics, Central University of Finance and Economics, Beijing 100081, China

Correspondence: Dengfeng Liu (liudf@xaut.edu.cn)

Received: 11 December 2017 – Revised: 5 March 2018 – Accepted: 12 March 2018 – Published: 5 June 2018

Abstract. The middle route of South-To-North Water Diversion in China transfers water from the Han River and Han-To-Wei Water Diversion project of Shaanxi Province will transfer water from the Ziwu River, which is a tributary of the Han River. In order to gain a better understanding of future changes in the hydrological conditions within the Ziwu River basin, a Mann–Kendall (M–K) trend analysis is coupled with a persistence analysis using the rescaled range analysis (R/S) method. The future change in the hydrological characteristics of the Ziwu River basin is obtained by analysing the change of meteorological factors. The results show that, the future precipitation and potential evaporation are seasonal, and the spatial variation is significant. The proportion of basin area where the spring, summer, autumn and winter precipitation is predicted to continue increase is 0.00, 100.00, 19.00 and 16.00 %, meanwhile, the proportion of basin area that will continue to decrease in the future respectively will be 100.00, 0.00, 81.00 and 74.00 %.The future potential evapotranspiration of the four seasons in the basin shows a decreasing trend. The future water supply situation in the spring and autumn of the Ziwu River basin will degrade, and the future water supply situation in the summer and winter will improve. In addition, the areas with the same water supply situation are relatively concentrated. The results will provide scientific basis for the planning and management of river basin water resources and socio-hydrological processes analysis.

1 Introduction

According to the Intergovernmental Panel Climate Change report (IPCC, 2014), climate change acts as the driving factor of water cycle in hydrological processes and precipitation and evaporation are the main climatic factors affecting fresh water resources. The interaction of climate change with other factors has had irreversible negative impacts on the ecological and hydrological cycle of the basin. The Ziwu River basin, which is a tributary of the upper reach of the Han River, is located in the warm humid and semi-humid climate region of the north subtropics. The precipitation in the river basin changes significantly between seasons. Summer has the most precipitation, accounting for 46 % of the annual precipitation averaged across the basin, and winter the least at about 3 % (Zhang and Zhang, 2004). The seasonal variation of precipitation will inevitably lead to differences in the temporal and spatial distribution of hydrological conditions, and therefore it is very important to study the temporal and spatial variation of hydrological condition in the basin of the climate change.

In the context of the growing shortage of water resources in the world, the impact of water resources in the changing environment has received widespread attention (Xia et al., 2015). In view of the study of changing environment, domestic and foreign scholars mainly study the characteristics of river basin climate element change (Zhu et al., 2005; Chebana et al., 2013; Ma et al., 2012). Qiu et al. (2003) used the 20 cm caliber evaporator data from 123 meteorological stations in the Yellow River basin and its surrounding areas from 1961 to 2000 to analyse the climatic trend of pan evaporation. The study found that the average annual temperature of the Yellow River basin increased by 0.6 °C during the period from 1960 to 2000, but the evaporation of the evaporating pan showed a decreasing trend, especially during spring and summer. Shi et al. (2006) established an exponential formula for estimating evaporation capacity according to the relationship between evaporation capacity and temperature in the Yellow River basin, to analyse the influence of temperature change on the evaporation capacity. The results showed that the basin evaporation capacity increased by 5.0–7.0 % when the temperature increased by 1 °C.

Generally, the methods of M–K mutation point and trend test, rescaled range analysis (R/S) and wavelet analysis are used to analyse the change characteristics of climate elements in the basin (Rzeszotko, 2012; Wallis and Matalas, 1970). For the trend analysis of meteorological elements, most scholars adopt the mature Mann–Kendall and rescaled range analysis method (Whitfield and Hendrata, 2006; Polemio et al., 2008). Ling et al. (2011) tested the temperature, precipitation and evaporation in the Manas River basin for nearly 52 years and used the R/S analysis method to predict the future trend. The results showed that the temperature and precipitation increased significantly from 1956 to 2007 and the evaporation is significantly reduced. Using the R/S analysis method to test for long-range temporal dependence in the data, Dinh and Molnr (1999) showed that the R/S analysis method has good robustness. It can be seen that in the time series, a trend has a significant influence on the estimated Hurst value. Using global inventory modeling and mapping studies (GIMMS) data, Bisognin and Lopes (2007) selected three periods to study seasonal vegetation trends, resulting in a nonlinear variation in vegetation changes caused by drought. Zhang et al. (2012) used R/S analysis to predict the dynamic trend of future groundwater level in the Loess Platform area of Xingping City. The results show that the annual precipitation in Xingping decreases slightly, and the quantity of irrigation water decreases significantly, which is consistent with the dynamic trend of groundwater level change in the same period. Previous studies have mostly analyzed the temporal and spatial evolution characteristics of the meteorological elements in the basin under the historical conditions, while there are relatively fewer studies of the impact of meteorological factors on the future hydrological conditions of the basin.

In summary, this paper takes the Ziwu River basin of the upper reach of the Han River as the study area. In order to gain a better understanding of the temporal and spatial distribution of future water resources changes in the Ziwu River basin, the Mann–Kendall (M–K) trend analysis method is combined with a persistence analysis using the rescaled range analysis (R/S) method in this study. The temporal and spatial variation characteristics of precipitation and evaporation trends and persistence in future is obtained by setting indicators set for determining the future change of regional meteorological elements and setting situations of regional future water supply. This will provide scientific basis for the planning and management of river basin water resources and socio-hydrological processes analysis.

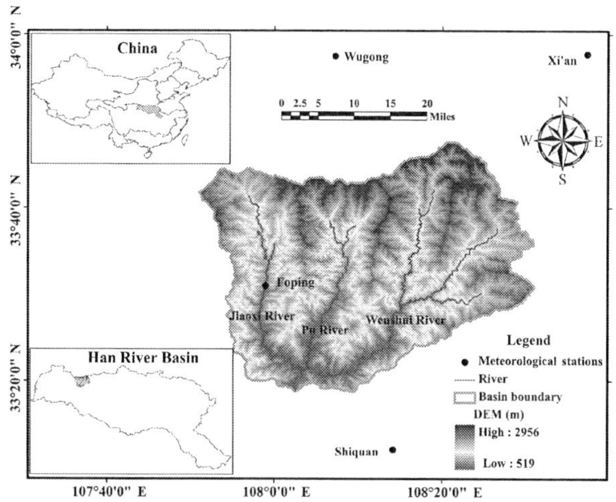

Figure 1. The location of the meteorological stations and the river network of Ziwu River.

2 Study area and data

The Ziwu River, which is a tributary of the Han River, originates in the south of the Qinling Mountain of Gangtie Town, Ningshan County, flows through Ningshan County, Shiquan County, Yangxian County and Xixiang County, and enters into Han River at Xixiang County. In Fig. 1, the Wenshui River above Sanhekou reservoir is 106 km, the Jiaoxi River is 70 km and the Pu River is 58 km. The basin area is 3028 km^2, and the altitude ranges from 360–2960 m. It is in the warm humid and semi-humid climate region of the north subtropics, where summer is warm and humid, winter is cold and dry and the spatial difference of seasonal precipitation in the basin is significant. The average annual temperature is between 7.7–15.7 °C and annual average precipitation is 900 mm.

In this paper, the precipitation and potential evapotranspiration data of the four meteorological stations in the Ziwu River basin from 1960 to 2013 are used and the potential evaporation data are calculated by the Penman–Monteith formula revised by the Food and Agricultural Organization (FAO). The Penman–Monteith formula is used to calculate potential evaporation data because it considered the influence of meteorological factors such as temperature, relative humidity, sunshine hours and wind speed and it is also a commonly used formula for calculating the potential evaporation in hydrological, meteorological and agriculture fields (Yang et al., 2012).

3 Methodology

3.1 Mann–Kendall trend analysis

The nonparametric M–K test (Mann, 1945; Kendall, 1975) has been commonly used to assess the significance of monotonic trends in meteorological and hydrological data time series, but it can't predict the future trend of the sequence (Hamed, 2008). Suppose a set of hydro-meteorological time series $X = (x_1, x_2, x_3 \cdot, x_n)$ (n is the number of variables in the sequence), and then establish the standard normal distribution statistic Z.

The M–K test statistic (S) is calculated in the following Eqs. (1) and (2):

$$\mathrm{sgn}\,(x_j - x_i) = \begin{cases} +1; & \text{If } x_j > x_i \\ 0; & \text{If } x_j = x_i \\ -1; & \text{If } x_j < x_i \end{cases}, \tag{1}$$

$$S = \sum\nolimits_{i=1}^{n-1} \sum\nolimits_{j=i+1}^{n} \mathrm{sgn}\,(x_j - x_i), \tag{2}$$

where x_i and x_j are the data values at times i and j, n indicates the length of the data set. While a positive value of S indicates an increasing trend, a negative of S indicates a decreasing trend. The following expression, as an assumption, is used for the series where the data length $n > 10$ and data are approximately normally distributed (variance ($\sigma^2 = 1$) and mean ($\mu = 0$) value).

$$\mathrm{Var}(S) = \frac{n(n-1)(2n+5) - \sum_{k=1}^{P}(t_k - 1)(2t_k + 5)}{18}, \tag{3}$$

in this equation, P is the number of tied groups, and the summary sign ($\sum$) indicates the summation over all tied groups. t_k is the number of data values in the "Pth" group. If there are not the tied groups, this summary process can be ignored.

After the calculation of the variance of time series data with Eq. (3), the standard Z value is calculated according to the following Eq. (4).

$$Z = \begin{cases} \dfrac{S-1}{\sqrt{\mathrm{Va}(S)}}; & \text{If } S > 0 \\ 0; & \text{If } S = 0 \\ \dfrac{S+1}{\sqrt{\mathrm{Var}(S)}}; & \text{If } S < 0 \end{cases}, \tag{4}$$

the changing trend of judging the standard: (1) If the statistics Z is greater than 0, the past trend is increase, on the contrary, if the statistics Z is less than 0, the past trend is reduced; (2) if the statistics Z is greater than or equal to 1.96, the trend is increase significantly, on the contrary, if the statistics Z is less than or equal to -1.96, the trend is decrease significantly.

3.2 Hurst Exponent

The R/S analysis (Rescaled Range Analysis) is a statistical method proposed by Hurst, a British hydrologist in 1951 to deal with hydrologic time series from the Nile (Hurst, 1951). This method is used to quantitatively characterize the persistence or long-range dependency of a time series. The statistic Hurst exponent can well reveal the relationship between the future and past trends of the time series so that future trend changes can be predicted based on the past trends of the time series. Up until now, it has been widely applied to the qualitative prediction of meteorological and hydrology time series such as rainfall, evaporation and runoff. And the magnitude of the H value can be used to determine the strength of persistence and anti-persistence of future trend components (Feng et al., 2008). The Hurst index rating table is showed in Table 1. Although the H value can show the trend of change, it can't determine the actual historical trends. The Hurst exponent calculation process is as follows:

Suppose a set of hydro-meteorological time series $\{X(t)\}$ (t is the number of variables in the sequence). For any positive integer ($\tau \geq 1$), the mean value series ($\overline{X}$) is calculated in the following Eq. (5).

$$\overline{X}(\tau) = \frac{1}{\tau}\sum\nolimits_{\tau=1}^{n} X(\tau) \quad \tau = 1, 2, \cdots, n, \tag{5}$$

then the cumulative deviation can be calculated in the following Eq. (6).

$$X(t,\tau) = \sum\nolimits_{t=1}^{\tau}\left[X(t) - \overline{X}(\tau)\right] \quad 1 \leq t \leq \tau. \tag{6}$$

The range $R(\tau)$ and standard deviation $S(\tau)$ are calculated in the following Eqs. (7) and (8) respectively.

$$R(\tau) = \underset{1\leq t\leq \tau}{\mathrm{Max}}\, X(t,\tau) - \underset{1\leq t\leq \tau}{\mathrm{Min}}\, X(t,\tau) \tag{7}$$

$$S(\tau) = \sqrt{\frac{1}{\tau}\sum\nolimits_{t=1}^{\tau}[X(t) - X(\tau)]^2}. \tag{8}$$

Therefore, the Hurst exponent can be expressed by following Eq. (9).

$$\frac{R(\tau)}{S(\tau)} = (\alpha\tau)^H, \tag{9}$$

where α is constant and H is the Hurst index.

Eventually, the Hurst exponent is calculated in the following Eq. (10). It can be obtained by logarithm of Eq. (9).

$$H = \left\{\ln\left[\frac{R(\tau)}{S(\tau)}\right]\right\} \Big/ (\ln\tau + \ln\alpha). \tag{10}$$

The criterion is as follows: if H value is greater than 0.50 and less than 1, the variables are positively correlated and the trend of future changes is consistent with the past; the H value is closer to 1, the persistence is stronger. If H value is greater than 0 and less than 0.50, the variables are negatively correlated and the trend of future changes is opposite to the past; the H value is closer to 0, the anti-persistence is stronger. If H value is equal to 0.50, the time series is random and future trends have nothing to do with the past.

Table 1. Hurst index rating table.

grade	the range of H value	persistence strength	grade	the range of H value	persistence strength
1	$0.50 < H < 0.55$	very weak	1	$0.45 < H < 0.50$	very weak
2	$0.55 < H < 0.65$	weak	2	$0.35 < H < 0.45$	weak
3	$0.65 < H < 0.75$	a little strong	3	$0.25 < H < 0.35$	a little strong
4	$0.75 < H < 0.80$	strong	4	$0.20 < H < 0.25$	strong
5	$0.80 < H < 1.00$	very strong	5	$0.00 < H < 0.20$	very strong

3.3 Discriminant analysis of hydrological condition

Precipitation and evaporation are important links in the process of water cycle and also important indicators of the local hydrological conditions. Because the test method of Mann–Kendall and R/S are unable to independently obtain the past and future trends of hydrological sequences, this paper uses the method of Mann–Kendall test and R/S test coupled to determine the indicators set for determining the future change of regional meteorological elements, which are shown in Table 2.

Regional hydrological conditions are greatly influenced by evaporation and precipitation under natural conditions. If one of the following three situations can happen, the regional water supply situation will improve in the future. In the situation one, the future regional precipitation increases and the evaporation decreases. In the situation two, the trend of precipitation increase is greater than that of evaporation increase. In the situation three, the trend of decrease in precipitation is smaller than that of evaporation reduction. If one of the following three situations can happen, the regional water supply situation will degrade in the future. In the situation one, the future regional precipitation decreases and the evaporation increases. In the situation two, the trend of precipitation increase is smaller than that of evaporation increase. In the situation three, the trend of decrease in precipitation is greater than that of evaporation reduction. Scientifically and reasonably evaluating the evolvement characteristics of the water supply situation in the future will have important practical significance for the sustainable development of the regional economy and society. This paper sets the situations of regional future water supply, which are seen in Table 3.

Whether the future water supply situation will be advantageous or disadvantageous depends on the relative differences in the change degree of precipitation and evaporation. This paper considers a total of six situations of advantageous (situation one, situation two and situation three) and disadvantageous (situation one, situation two and situation three) situations.

In summary, the advantageous/disadvantageous area for future water supply in the river basin is calculated as follows:

The Thiessen polygon method is used to calculate the area weights f_i of basin features represented by the meteorological stations in the river basin and calculate the area represented by each station.

The Mann–Kendall analysis method and R/S analysis method is used to calculate the Z and H values of rainfall statistics at each station.

Based on the water supply evaluation indicators, the future water supply characteristics of the areas represented by each site are analysed.

Collect advantageous and disadvantageous sites for future water supply and calculate the area of advantageous or disadvantageous area for future water supply in the basin by combining the area weights of advantageous or disadvantageous sites calculated in step (1) with Eq. (11):

$$F_{\text{advantageous/disadvantageous}} = \sum f_i / F, \tag{11}$$

where F is the total area of the river basin and f_i is the area weights of advantageous or disadvantageous area for future water supply.

4 Results and analysis

4.1 Analysis of tendency and sustainability of precipitation change

4.1.1 Precipitation historical change trend

The Mann–Kendall nonparametric test method is used to calculate past trends of precipitation for all 4 weather stations considered in this study. The results shows that the proportion of sites with decreasing trend respectively in spring, summer, autumn and winter is 100.00, 0.00, 100.00 and 84.00 %, respectively, and the proportion of sites with increasing trend in spring, summer, autumn and winter is 0.00, 100.00, 0.00 and 16.00 %, respectively. It can be seen from historical trends that the supply of water in spring and autumn is in a form of harshness and the water supply in summer is in a good form.

4.1.2 Precipitation change sustainability

The precipitation H statistic value of each weather station is calculated by the R/S analysis method, and spatial interpolation is carried out by kriging interpolation method based on the ArcGIS environment. The result is shown in Fig. 2. According to Fig. 2, the H values of the four seasons are mainly concentrated in the range of 0.50–0.75 and the precipitation H value is significantly different in the spatial dis-

Table 2. Indicators set for determining the future change of regional meteorological elements.

Z	Hurst value	variation characteristics of meteorological elements
$Z \leq -1.96$	$H > 0.50$	The trend of reduction in past is significant, and it will continue to decrease in the future.
$Z \leq -1.96$	$H < 0.50$	The trend of reduction in past is significant, but it is likely to reverse in the future.
$-1.96 < Z \leq 0$	$H > 0.50$	The trend of reduction in past is not significant, and it will continue to decrease in the future.
$-1.96 < Z \leq 0$	$H < 0.50$	The trend of reduction in past is not significant, but it is likely to reverse in the future.
$0 < Z \leq 1.96$	$H > 0.50$	The trend of increase in past is not significant, and it will continue to increase in the future.
$0 < Z \leq 1.96$	$H < 0.50$	The trend of increase in past is not significant, but it is likely to reverse in the future.
$Z > 1.96$	$H > 0.50$	The trend of increase in past is significant, and it will continue to increase in the future.
$Z > 1.96$	$H < 0.50$	The trend of increase in past is significant, but it is likely to reverse in the future.

Table 3. Situations of regional future water supply.

future water supply situation	situation	state of water resources	statistical reflection Z and H
advantage	situation one	precipitation increases and the evaporation decreases	$Z_{\text{precipitation}} > 0$, $H_{\text{precipitation}} > 0.5$ and $Z_{\text{evaporation}} < 0$, $H_{\text{evaporation}} > 0.5$
	situation two	the trend of precipitation increase is greater than that of evaporation increase	$\lvert Z\rvert_{\text{precipitation}} > \lvert Z\rvert_{\text{evaporation}}$, and $H_{\text{precipitation}} > 0.5$, $H_{\text{evaporation}} > 0.5$
	situation three	the trend of decrease in precipitation is smaller than that of evaporation reduction	$\lvert Z\rvert_{\text{precipitation}} < \lvert Z\rvert_{\text{evaporation}}$, and $H_{\text{precipitation}} > 0.5$, $H_{\text{evaporation}} > 0.5$
disadvantage	situation one	precipitation decreases and the evaporation increases	$Z_{\text{precipitation}} < 0$, $H_{\text{precipitation}} > 0.5$ and $Z_{\text{evaporation}} > 0$, $H_{\text{evaporation}} > 0.5$
	situation two	the trend of precipitation increase is smaller than that of evaporation increase	$\lvert Z\rvert_{\text{precipitation}} < \lvert Z\rvert_{\text{evaporation}}$, and $H_{\text{precipitation}} > 0.5$, $H_{\text{evaporation}} > 0.5$
	situation three	the trend of decrease in precipitation is greater than that of evaporation reduction	$\lvert Z\rvert_{\text{precipitation}} > \lvert Z\rvert_{\text{evaporation}}$, and $H_{\text{precipitation}} > 0.5$, $H_{\text{evaporation}} > 0.5$

tribution. The spatial distribution trend of the H value in the four seasons is as follows: In spring, the H value of precipitation gradually increases from southwest to northeast, that is to say, the intensity of future sustainability gradually increases from southwest to northeast; In summer, the H value of precipitation is the smallest in the northwest local region and it is very large in the southeast part; In autumn, the H value of precipitation increases from south to north, that is to say, the intensity of future sustainability gradually increases from south to north; In winter, the H value of precipitation gradually increases from west to east, that is to say, the intensity of future sustainability gradually increases from west to east. Each season has a strong sustainability and regularity, which is conducive to the study of future water supply and development.

4.1.3 Precipitation trends in the future

The results of the trend of future precipitation in the Ziwu River basin with time are shown in Table 4. The proportion of basin area where the spring, summer, autumn and winter precipitation is predicted to continue increase is 0.00, 100.00, 19.00 and 16.00 %, corresponding to the average annual precipitation of 0.00, 360.60, 279.69 and 21.15 mm, respectively; The proportion of basin area that will continue to decrease in the future respectively will be 100.00, 0.00, 81.00 and 74.00 %, corresponding to the average annual precipitation of 161.02, 0.00, 308.43 and 27.44 mm, respectively. The proportion of area where precipitation will be reversed in the future will be smaller in the seasonal scale.

4.2 Analysis of tendency and sustainability of potential evaporation change

4.2.1 Potential evaporation historical change trend

The Mann–Kendall nonparametric test method is used to calculate past trends of potential evaporation in all weather stations in the basin. In theory, the reduction of potential evaporation is conducive to the change of water supply at all seasons from poor to favorable trend.

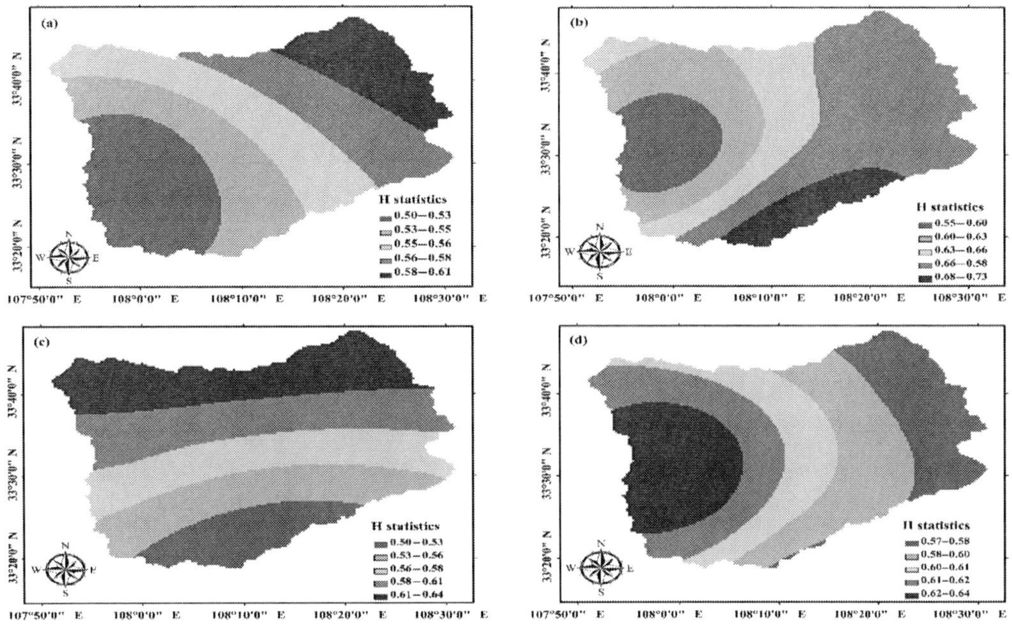

Figure 2. Spatial distribution of H value of precipitation in the four seasons **(a)** spring **(b)** summer **(c)** autumn **(d)** winter.

Table 4. Changes of the future precipitation trend in the Ziwu River basin.

season	future trend of increase ($Z > 0$, $H > 0.5$)		future trend of decrease ($Z < 0$, $H > 0.5$)		future trend reverse as increase ($Z < 0$, $H < 0.5$)		future trend reverse as decrease ($Z > 0$, $H < 0.5$)	
	proportion of occupied area	historical average annual precipitation	proportion of occupied area	historical average annual precipitation	proportion of occupied area	historical average annual precipitation	proportion of occupied area	historical average annual precipitation
	%	mm	%	mm	%	mm	%	mm
spring	0.00	0.00	100.00	161.02	0.00	0.00	0.00	0.00
summer	100.00	360.60	0.00	0.00	0.00	0.00	0.00	0.00
autumn	0.00	0.00	81.00	308.43	19.00	279.69	0.00	0.00
winter	16.00	21.15	74.00	27.44	0.00	0.00	0.00	0.00

4.2.2 Potential evaporation change sustainability

The potential evaporation H statistic value of each weather station was calculated by the R/S analysis method and spatial interpolation was carried out by kriging interpolation method based on the ArcGIS environment. The result is shown in Fig. 3. According to Fig. 3, the spatial distribution of potential evaporation H is significantly different. The potential evaporation H is significantly different. The potential evaporation in summer is the largest, even in some areas reached 0.88; it in winter take second place; and it in spring is the smallest. Each season has a strong sustainability and regularity, which is conducive to the study of future water supply and development.

4.2.3 Potential evaporation trends in the future

The results of a comprehensive analysis of potential evaporation past trends (Mann–Kendall test) and future sustainability (R/S test) are shown in Table 5. The results show that the potential evapotranspiration of the four seasons in the basin shows a decreasing trend. The historical potential evapotranspiration in spring, summer, autumn and winter respectively are 246.40, 354.68, 150.45 and 89.40 mm, which indicates that the basin will be humid in the future and is beneficial to the sustainable development of the local water resources.

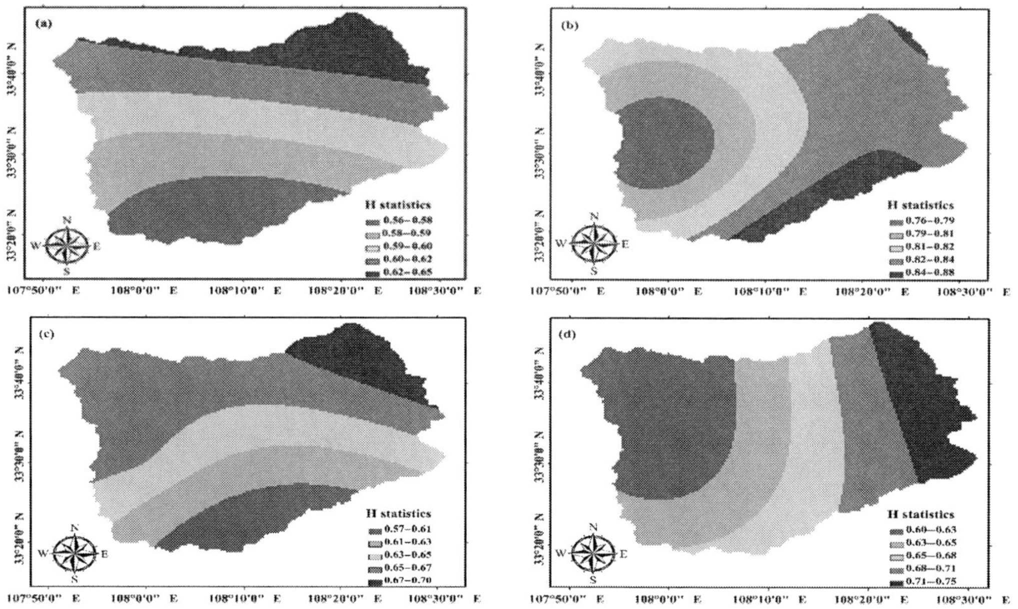

Figure 3. Spatial distribution of H value of potential evaporation in the four seasons **(a)** spring **(b)** summer **(c)** autumn **(d)** winter.

Table 5. Changes of the future potential evaporation trend in the Ziwu River basin.

season	future trend of increase ($Z > 0$, $H > 0.5$)		future trend of decrease ($Z < 0$, $H > 0.5$)		future trend reverse as increase ($Z < 0$, $H < 0.5$)		future trend reverse as decrease ($Z > 0$, $H < 0.5$)	
	proportion of occupied area	historical average annual precipitation	proportion of occupied area	historical average annual precipitation	proportion of occupied area	historical average annual precipitation	proportion of occupied area	historical average annual precipitation
	%	mm	%	mm	%	mm	%	mm
spring	0.00	0.00	100.00	246.40	0.00	0.00	0.00	0.00
summer	0.00	0.00	0.00	354.68	0.00	0.00	0.00	0.00
autumn	0.00	0.00	81.00	150.45	19.00	148.83	0.00	0.00
winter	0.00	0.00	100.00	89.40	0.00	0.00	0.00	0.00

4.3 Variation characteristics of future hydrological conditions in the basin

According to the trend of precipitation and evaporation in the future, the precipitation and evaporation in the basin have seasonal variation, which lays a good foundation for the study of the future hydrological conditions of the basin. Table 6 and Fig. 4 show that the future water supply situation of the basin is severe (disadvantageous) and good (advantageous) in regional spatial and temporal distribution.

The results show that: the area of severe water supply in spring, summer, autumn and winter respectively are 100.00, 0.00, 81.00 and 0.00 %. The future of the water supply in the spring and autumn is predicted to be more limited than in the past. The trend of decreasing precipitation in spring and autumn is greater than that in potential evaporation, and most of the water supply in the autumn basin is severe excluding area which is 19 % in the southeast. These indicate that the future water supply is severe and are not conducive to the good development of regional hydrological conditions.

The area of water supply in spring, summer, autumn and winter good areas respectively account for 0.00, 100.00, 19.00 and 100.00 %. The summer water supply is situation one and the contradiction between supply and demand of water resources will be eased in the season; The water supply in autumn is situation two, which mainly distribute in some areas of the southeast of the basin; The water supply in winter are situation one and three. The trend of decreasing precipitation is less than that of potential evaporation in most areas of the river basin, and the contradiction between the future water supply in summer and winter will be alleviated.

To sum up, the water supply in the future will be severe in spring and autumn, and the water supply in the future will be

Table 6. Changes of future water supply characteristics of the basin.

season	advantageous situation proportion %			disadvantageous situation proportion %		
	Situation one	Situation two	Situation three	Situation one	Situation two	Situation three
spring	0.00	0.00	0.00	0.00	0.00	100.00
summer	100.00	0.00	0.00	0.00	0.00	0.00
autumn	0.00	19.00	0.00	0.00	0.00	81.00
winter	16.00	0.00	84.00	0.00	0.00	0.00

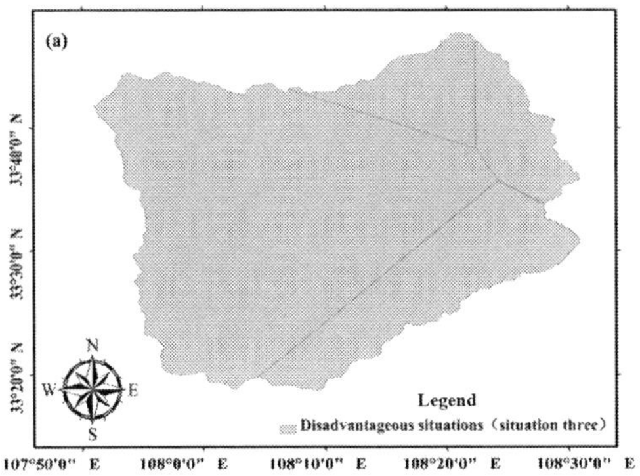

(a) Space distribution of spring water supply

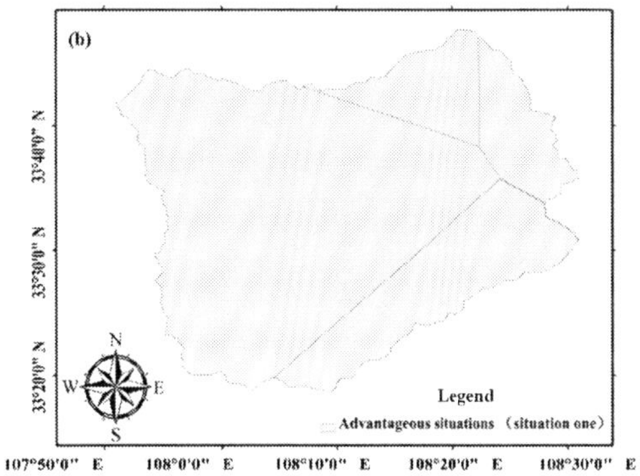

(b) Space distribution of summer water supply

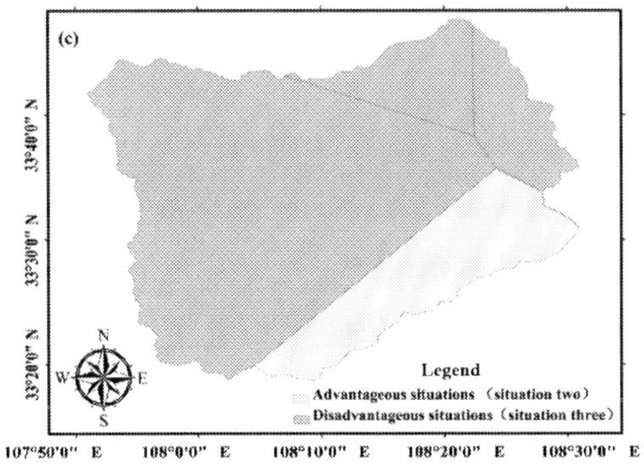

(c) Space distribution of autumn water supply

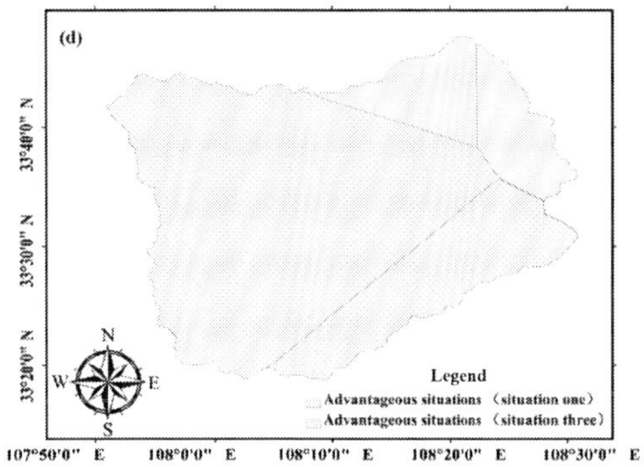

(d) Space distribution of winter water supply

Figure 4. Spatial distribution of future water supply in the basin.

good in summer and winter, especially in the summer with a good space for development which precipitation will increase and evaporation will decrease.

5 Conclusions

In this paper, the Mann–Kendall non-parametric test and R/S test methods were combined to analyse the historical trend and future trend of precipitation and potential evaporation in the river basin. After combining the characteristics of future hydrological conditions, the following conclusions are obtained:

The future precipitation and potential evaporation are seasonal, and the spatial variation is significant. In addition, the areas with the same water supply situation are relatively concentrated. The proportion of basin area where the spring, summer, autumn and winter precipitation is predicted to continue increase is 0.00, 100.00, 19.00 and 16.00 %, meanwhile, the proportion of basin area that will continue to decrease in the future respectively will be 100.00, 0.00, 81.00 and 74.00 %.The future potential evapotranspiration of the four seasons in the basin shows a decreasing trend.

The area of water supply in spring, summer, autumn and winter good areas respectively account for 0.00, 100.00, 19.00 and 100.00 % and the area of severe water supply in spring, summer, autumn and winter respectively are 100.00, 0.00, 81.00 and 0.00 %. The future water supply in the spring and autumn of the Ziwu River basin will be severe, and the future water supply in the summer and winter will be better, especially in the summer.

Data availability. The meteorological data in the paper are provided by China Meteorological Data Service Center, from their web site at http://data.cma.cn/en.

Competing interests. The authors declare that they have no conflict of interest.

Special issue statement. This article is part of the special issue "Innovative water resources management – understanding and balancing interactions between humankind and nature". It is a result of the 8th International Water Resources Management Conference of ICWRS, Beijing, China, 13–15 June 2018.

Acknowledgements. This work was partially supported by National Key R&D Program of China (2017YFC0405900) and National Natural Science Foundation of China (NSFC) (Grant No. 51779203, 51609270, 41330858). Funding was also provided by the Basic Research Plan of Natural Science of Shaanxi Province (2016JQ5105). The authors would like to thank Andreas Schumann and the anonymous reviewers for their comments to improve the paper.

Edited by: Andreas Schumann
Reviewed by: two anonymous referees

References

Bisognin, C. and Lopes, S. R. C.: Estimating and forecasting the long-memory parameter in the presence of periodicity, J. Forecast., 26, 405–427, https://doi.org/10.1002/for.1030, 2007.

Chebana, F., Ouarda, T. B. M. J., and Duong, T. C.: Testing for multivariate trends in hydrologic frequency analysis, J. Hydrol., 486, 519–530, https://doi.org/10.1016/j.jhydrol.2013.01.007, 2013.

China Meteorological Data Service Center: available at: http://data.cma.cn/en (last access: 21 May 2017), 2017.

Dang, T. D. and Molnr, S.: On the effects of non-stationarity in long-range dependence tests, Periodica Polytechnica Ser.el.eng, 43, 227–250, 1999.

Feng, X. L., Feng, Z. L., Luo, L. C., Qin, L. L., and Lin, P.: R/S analysis and Hurst exponential test of climate change trend of Qinghai-Tibet Plateau, J. Arid Land Geogr., 31, 175–181, https://doi.org/10.13826/j.cnki.cn65-1103/x.2008.02.002, 2008.

Hamed, K. H.: Trend detection in hydrologic data: The Mann–Kendall trend test under the scaling hypothesis, J. Hydrol., 349, 350–363, https://doi.org/10.1016/j.jhydrol.2007.11.009, 2008.

Hurst, H. E.: Long term storage capacities of reservoirs, Trans. Am. Soc. Civil Eng., 116, 776–808, 1951.

IPCC: Synthesis Report, Contribution of Working Groups I, II and III to the Fifth Assessment Report of the Intergovernmental Panel on Climate Change, edited by: Core Writing Team, Pachauri, R. K., and Meyer, L. A., Geneva, Switzerland, 151 pp., 2014.

Kendall, M. G.: Rank Correlation Methods, M, 4th ed, Charles Griffin, London, 1975.

Ling, H. B., Xu, H. L., Zhang, Q. Q., and Shi, W.: Climate Change in the Manas River Basin, Xinjiang during 1956–2007, J. Glaciol. Geocryol., 33, 64–71, 2011.

Ma, X. N., Zhang, M. J., Huang, X. Y., Ma, Q., and Pan, S. K.: Characteristics of climate change and future trend in the upper Yellow River basin from 1960 to 2008, Arid Land Resour. Environ., 26, 17–23, https://doi.org/10.13448/j.cnki.jalre.2012.06.029, 2012.

Mann, H. B.: Nonparametric tests against trend, J. Econometrica, 13, 245–259, 1945.

Polemio, M., Casarano, D., Dragoni, W., and Sukhija, B. S.: Climate change, drought and groundwater availability in southern Italy, Geol. Soc. London Special Pub., 288, 39–51, https://doi.org/10.1144/SP288.4, 2008.

Qiu, X. F., Liu, C. M., and Zeng, Y.: Changes of pan evaporation in the recent 40 years over the Yellow River Basin, J. Nat. Res., 18, 437–442, 2003.

Rzeszotko, Z.: Estimation of the Hurst exponent for market indices by use of rescaled range method and CAS Mathematica. Case study: S&P500, Electro. J. Mathe. Technol., 6, 256–267, 2012.

Shi, Z. H., Wang, G. Q., Yv, H., Jiang, Y. F., Jin, X. A., and Li, H. B.: Impact of Temperature change on Potential Evaporation in Yellow River Basin, Meteorol. J. of Henan, 1, 31–32, https://doi.org/10.3969/j.issn.1673-7148.2006.01.015, 2006.

Wallis, J. R. and Matalas, N. C.: Small Sample Properties of H and K Estimators of the Hurst Coefficient h, Water Resour. Res., 6, 1583–1594, 1970.

Whitfield, P. H. and Hendrata, M.: Assessing Detectability of Change in Low Flows in Future Climates from Stage Discharge Measurements, Canadian Water Res. J., 31, 1–12, https://doi.org/10.4296/cwrj3101001, 2006.

Xia, J., Shi, W., Luo, X., Hong, S., Ning, L., and Chiristopher, J. G.: Revisions on water resources vulnerability and adaption measures under climate change, J. Adv. Water Sci., 26, 279–286, https://doi.org/10.14042/j.cnki.32.1309.2015.02.019, 2015.

Yang, H. B., Lv, H. F., Yang, D. W., and Hu, Q. F.: Seasonality of precipitation and potential evaporation and its impact on catchment water-energy balance, J. Hydroelec. Eng., 31, 54–59, 2012.

Zhang, K. and Zhang, S. T.: Analyses of "02.6" Ziwu River Extraordinary Flood in South of Shaanxi Province, J. Water Resour. Archit. Eng., 2, 54–56, 2004.

Zhang, Y. Q., Hu, W., and Liu, J. M.: Groundwater level regime variation trend on basis of rescaled range analysis, J. Central South Univ., 43, 4912–4916, 2012.

Zhu, H. H., Qin, D. Y., Zhou, Z. H., Yan, D. H., and Yang, G. Y.: Study on Precipitation Evolution Law of the Yellow River Basin, J. Yellow River, 27, 21–24, 2005.

https://doi.org/10.5194/piahs-379-323-2018

Open Access
PIAHS

Distributed source pollutant transport module based on BTOPMC: a case study of the Laixi River basin in the Sichuan province of southwest China

Hongbo Zhang[1], Tianqi Ao[2], Maksym Gusyev[3], Hiroshi Ishidaira[4], Jun Magome[4], and Kuniyoshi Takeuchi[4]

[1]Faculty of Electric Power Engineering, Kunming University of Science and Technology, Kunming 650500, China
[2]College of Water Resources and Hydropower, Sichuan University, Chengdu 610065, China
[3]International Centre for Water Hazard and Risk Management (ICHARM), Public Works Research Institute (PWRI), Tsukuba 305-8516, Japan
[4]International Research Center for River Basin Environment, University of Yamanashi, Kofu 400-8511, Japan

Correspondence: Hongbo Zhang (zhanghongbo@kmust.edu.cn)

Received: 30 December 2017 – Accepted: 12 February 2018 – Published: 5 June 2018

Abstract. Nitrogen and phosphorus concentrations in Chinese river catchments are contributed by agricultural non-point and industrial point sources causing deterioration of river water quality and degradation of ecosystem functioning for a long distance downstream. To evaluate these impacts, a distributed pollutant transport module was developed on the basis of BTOPMC (Block-Wise Use of TOPMODEL with Muskingum-Cunge Method), a grid-based distributed hydrological model, using the water flow routing process of BTOPMC as the carrier of pollutant transport due a direct runoff. The pollutant flux at each grid is simulated based on mass balance of pollutants within the grid and surface water transport of these pollutants occurs between grids in the direction of the water flow on daily time steps. The model was tested in the study area of the Lu county area situated in the Laixi River basin in the Sichuan province of southwest China. The simulated concentrations of nitrogen and phosphorus are compared with the available monthly data at several water quality stations. These results demonstrate a greater pollutant concentration in the beginning of high flow period indicating the main mechanism of pollution transport. From these preliminary results, we suggest that the distributed pollutant transport model can reflect the characteristics of the pollutant transport and reach the expected target.

1 Introduction

The assessment and treatment of watershed water pollution, especially non-point source (NPS) pollution, is currently one of the key environmental protection subjects being addressed by scientists, engineers, and local governments. This is especially true in developing countries like China. According to *The Reports on the State of the Environment of China*, between 1989–2011, the treatment rate of industrial wastewater and domestic sewage had reached over 90 and 75 %, respectively, until 2010. However, the proportion of monitored sections for water quality beyond Class III in the 7 stream networks (including 469 state-level sections) and main lakes (including 26 state-level sections) of China was still as high as 40.1 and 80.8 %. These data indicates that PS pollution has been controlled effectively in China during the last more than ten years but NPS pollution are still very serious and has become the dominant pollution source leading to over-standard water quality. This is caused by the characteristics of NPS, such as the dispersion, uneven spatial and temporal distribution, randomness, long incubation period, and the difficulty for monitoring and controlling relative to PS pollution (Wu et al., 2006; Nigussie et al., 2003; Hao et al., 2006). The complicated mechanism of NPS decides that it is very difficult to simulate and evaluate NPS quantitatively and accurately.

The assessment models for NPS pollution can be categorized into two types – the empirical model and the mechanism model (Nash et al., 1970; Beven et al., 1979; Jiang et al., 2011). The empirical models are built on the basis of empirical relationships between watershed physiographic characteristics and pollutant exports, such as the Export Coefficient Method (Johnes and Heathwaite, 1996; Johnes and O'Sullivan, 1989) and Source Strength Coefficient Method (SSCM) (Chinese Academy for Environmental Planning, 2003). In general the empirical models have a lower simulating accuracy, but they are usually applicable enough for many practical applications like the watershed water pollution treatment, water environmental impact assessments, water environmental planning and etc. In comparison, the mechanism models generally have stronger simulating ability than the empirical models. There are many good NPS models developed on some mechanism of pollutant transport and transformation, such as the Hydrological Simulation Program–Fortran (HSPF), Soil and Water Analysis Tools (SWAT), Better Assessment Science Integrating Point and Non Point Sources (BASIN) (Nasr et al., 2007; Anderton et al., 2002; Orth et al., 2015) and etc., which can simulate the physical, chemical, and biochemical processes in some extent and to estimate the effects of agricultural management measures and practices (USEPA, 2000). Mechanism model has some advantages compared with empirical model like the higher accuracy, clear principles, parameter portability between different catchments and great development potential; however, its complex mechanism requires the more accurate and larger amount of input data which makes it very difficult to apply in the small basins or the basins without enough monitoring data (Andréassian et al., 2009; Cho et al., 2008; Jiang et al., 2011), especially in the developing countries which has poor monitoring data of hydrological and meteorological observation data (Kobierska et al., 2013; Singh et al., 2014; Ying et al., 2010).

Certainly, the mechanism model has become the main area of NPS study and will has great development and application in the future accompanying with our deeper and more understanding to the NPS mechanism, the increasing of the observation data and the development of observation techniques and equipments. However, due to the scarcity of observation data and sites, the empirical NPS model is still widely applied in the ungauged basins (PUB) and regions (Bookera and Woods, 2014; Viviroli et al., 2009; Ying et al., 2010). And with the economic progress, more and more developing countries and regions have the serious water pollution issue, which have urgent and pressing need to evaluate and control the NPS although they have poor observation data of hydrological, meteorological and water quality observation data and equipments.

Based on the advantages of mechanism model and the actual situation of ungauged basins and regions: the scarcity of observation data and sites and the wide application of empirical models, this study developed a distributed source pollutant transport module (DSPTM) based on a grid-based distributed hydrological model, BTOPMC and the empirical NPS assessment model, the Export Coefficient Method, to simulate the NPS pollutant transport process with the direct runoff and evaluate the contribution of NPS and the impact of the direct runoff on the N and P concentration at the monitored sections and applied it in the study area of the Lu county area situated in the Laixi River basin in the Sichuan province of southwest China.

2 Methodology

2.1 The NPS transport mechanism in DSPTM

BTOPMC (referred to as the Yamanashi Hydrological Model (YHyM) in Japan) is a grid-based distributed hydrological model, developed in Yamanashi University in the 1990s (Ao et al., 2003; Takeuchi et al., 2008) and applied widely around the world, with good impacts and research value (Bao et al., 2010; Takeuchi et al., 2013; Yoshimura et al., 2010). It has been applied particularly in areas with warm, humid climates and with high elevations and sloping landscapes, such as Japan and Southeast Asian countries (Zhou et al., 2006; Yoshimura et al., 2007, 2010; Reichert et al., 2001; Ao et al., 2006; Takeuchi et al., 1999). BTOPMC has five calibrating and validating model parameters–saturated soil hydraulic conductivity (T_0), decay factor of T_0 (m), maximum root zone capacity (S_{rmax}), initial value of the average soil saturation deficit (S_{bar0}), and Manning's rough coefficient (n_0) (Ao et al., 2003; Takeuchi et al., 1999).

The DSPTM is a distributed source pollutant module developed on the BTOPMC and implanted into the BTOPMC as an extension of its function (Fig. 1). The basic principles of the model include the mass balance of pollutants within the grid and the pollutant transmission between grids along the runoff direction. According to the research findings, the transport and transformation of pollutants are carried by runoff. In this study, the concentration process of BTOPMC was treated as the carrier of the pollutant transport. The BTOPMC model is a grid-based distributed hydrological model. The basin is segmented into a rectangular grid net and the runoff yield and concentration processes are calculated in each rectangular grid for each time interval. The concentration mechanism of BTOPMC is based on the Muskingum-Cunge Method which is used for the concentration calculation of each grid. The concentration calculation was applied grid by grid, successively along the runoff direction, until the control section grids were reached. Simultaneously, the pollutant was transferred along the runoff route of the concentration for each time interval, and the mass balance for pollutants within each grid was calculated (Eq. 1). In this way, the pollutant flux through each grid including the control section grid was calculated for each time interval.

In BTOPMC, the basin grids are divided into two types – the slop grid and the river grid by setting the area threshold of the upstream catchment of each grid. The NPS pollutant transport module for the slop grid and river grid is built on the basis of two basic principles – the mass balance for each pollutant within each grid and the pollutant transmission between grids along the runoff route of the concentration. The module includes 5 variables of pollutant input and output: (1) pollutant mass within the grid at the beginning of the period $WP_{ini}(t)$; (2) pollutant mass within the grid at the end of the period $WP_{end}(t)$; (3) pollutant input mass from the upstream grids for each time interval $\sum WP_{in}(t)$. (Because it is similar to the concentration process, a grid might receive pollutant input from several upstream grids.) (4) Total pollutant mass from the direct input (NPS pollutants for slop grids and PS pollutant for river grids which were set as the PS input grids) $\sum WP_{direct\_in}(t)$; (5) pollutant output mass to the downstream grid $TN_{out}(t)$. Pollutant mass conservation within a grid is shown in the following formula:

$$\begin{aligned}\Delta WP &= WP_{end} - WP_{ini} \\ &= C_{WP\_end} V_{end} - C_{WP\_ini} V_{ini} \\ &= \sum WP_{in} + \sum WP_{direct\_in} - WP_{out}\end{aligned} \tag{1}$$

where ΔWP: Pollutant mass increase within the grid for each time interval, t; C_{WP_ini}: Pollutant concentration at the beginning of the time interval, $mg\,L^{-1}$; C_{WP_end}: Pollutant concentration at the end of the time interval, $mg\,L^{-1}$; V_{ini}: Water storage within grids at the beginning of the time interval, m^3; V_{end}: Water storage within grids at the end of the time interval, m^3.

2.2 Assumptions of the pollutant load in grids

Due to the scarcity of observation data, the annual total nitrogen (TN) and total phosphorous (TP) load of different land-use area of study area has been estimated by the Export Coefficient Method in the earlier study (Zhang et al., 2013). Therefore, based on the characteristic of local residents treating the domestic wastes, the starting-stage situation of model study and the NPS accumulation law, this study has three assumptions for the pollutant load in grids.

In the rural and town area of southern China, the domestic sewages like the livestock and poultry waste are usually treated with the biogas tank or septic tanks then returned to the farmland as a fertilizer anytime and it is difficult to understand the accurate time law of wastes entering the fields. So this study assumed that the load entering into the study area is uniform daily in the study period (Assumption (1): load temporal distribution), the load into the same land -use grids is uniform in the spatial distribution (Assumption (2): load spatial distribution) and the load will be accumulated when no runoff in this grid and brought to the downstream grid when there is runoff in the grid (Assumption (3): load accumulation).

3 Data of study area

3.1 Basic information of study area

The study area, the Lu County area of Laixi River basin, with a drainage area of 728 km^2, is located in Sichuan province in southwest China at 105°14′57″–105°41′51″ E, 28°59′56″–29°20′3″ N (Fig. 2). It includes three main rivers. The Lu County reach of the Laixi River and its two branches – Jiuqu River and Maxi River, have a total length of 130 km. The Laixi River has a basin area of 3257 km^2, is one of the two main tributaries of the Tuo River of the upper Yangtze stream networks, and originates from Dazu district of Chongqing city and flows into the Tuo River in Hushi town of Luzhou city, Sichuan province.

The study basin has a humid subtropical climate with an average annual precipitation of 991.7 mm and reference evaporation of 980 mm. The mean annual temperature is 17.7°, and there is a large mean temperature difference of 19.2° between winter and summer. There is heavy forest coverage, and, until 2010, it covered 39.5 % of the area. The dominant terrain type is hills, which cover nearly 87.5 % of the study area, and incline towards the southwest with elevations ranging from 218 to 757.5 m and an average slope of 8 %. The dominant land use types are the paddy field and dry land, with a total ratio of 69 % of study area, which means that most of the area is used as agricultural land and agricultural NPS pollution will be a serious a result of the heavy utilization of the fertilizer and pesticides.

The Study area includes 10 towns and 157 villages, with an urban population of 73 775 and an agricultural population of 472 504 (about 85 % of the total population). There is only one sewage treatment plant until 2011, located in Fuji Town, which has a daily treatment capacity of 6000 t (equal to the daily domestic sewage amount for about 17 911 persons). Therefore, except a part of the domestic sewage from Fuji Town, which is treated by the Fuji sewage treatment plant and discharged into the stream networks as the Discharge Standard of the Pollutants for Municipal Wastewater Treatment Plant (GB18918-2002), the other domestic sewage from Fuji and the other towns (including both the agricultural population and nonagricultural population) is only treated with septic tanks or biogas tanks. These results in livestock and poultry waste being discharged into the simple septic tanks, which are usually dug out in the ground, and then returned to the farmland as a fertilizer when the septic tank is full. In large-scale livestock and poultry farms waste will be treated with a biogas tank anaerobic system, and then the biogas slurry and residue from the waste will similarly be returned to farmland. In the study area, there are many industrial plants, including wine plants, paper mills, chemical plants, and glass factories, which are the major source of PS pollution.

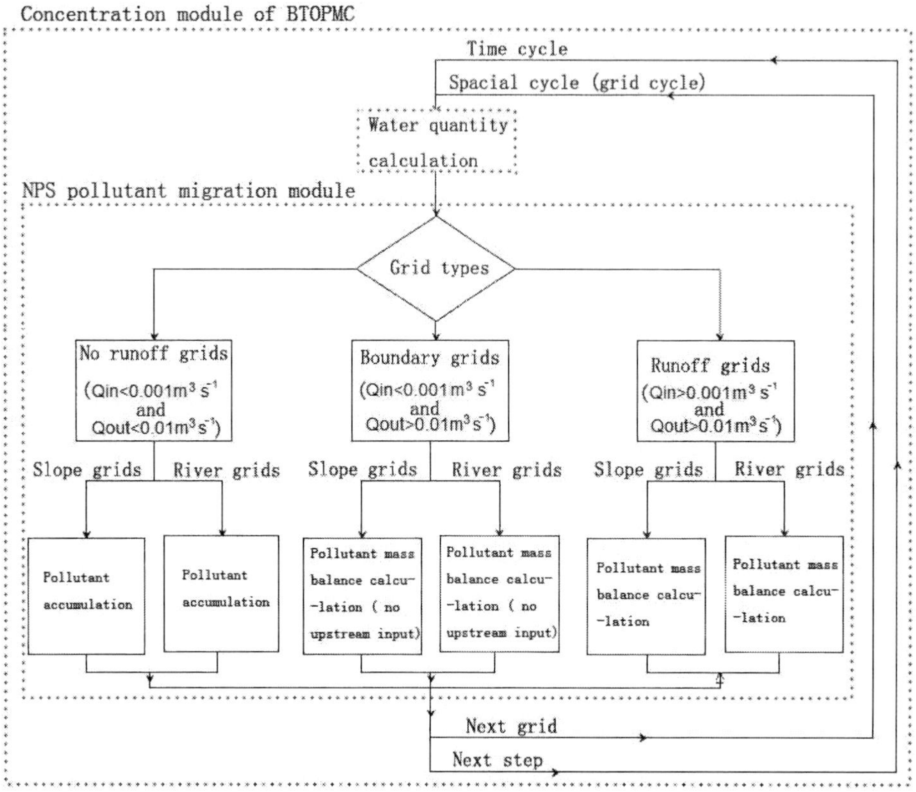

Figure 1. Framework of NPS pollutant transport module.

3.2 The related data for DSPTM application

1. DEM, land use/land cover and soil data
   The Digital Elevation Model (DEM) of the study area was obtained from the Soil and Sources Bureau of Sichuan Lu County, with the accuracy of 30 m. Soil types and their distribution data (1000 m) were download from the Food and Agriculture Organization of the United Nations (FAO). Land use/land cover data (30 m) were collected from the Second Land and Resources Investigation of China.

2. Hydrological and meteorological data
   Hydrological and meteorological data were collected from local hydrological and meteorological governmental sectors. The eight Rainfall stations that were used were Xihu, Duobao, Yutan, Fangdong, Longshi, Shinian, Shuanghe, and Fuji (Fig. 2), with daily rainfall from 1979–1987 and 2011. The two evaporation stations that were used were Dengying and Guanyin(2*), with monthly evaporation data from 1979–1987 and 2011. The two runoff stations used were the Fuji station and Yutan station, both with daily rainfall data from 1979–1987 and 2011 (Fig. 2).

3. Water quality, PS and NPS pollution data
   Population and animal data were derived from local government sectors and a field survey of the local committees of the village, farmers, and scientific and technical personnel. Water quality data were collected from the environmental protection administration of Lu County. The six water quaity monitoring sections that were used were the Tianzhusi, E'xiangjing and Guandu Bridge station in the Laixi River (3 stations); Shuangshengyan and Nan Bridge in Jiuqu River (2 stations) and Dawutan in Maxi River (1 station) (Fig. 2). The study area has the monthly water quality data of 2011, with 22 monitoring factors including ammonia nitrogen (NH_4^+-N), TN, and TP, among others. Two pollutant indicators were TN and TP and the six pollution sources were natural woodland source, rural NPS (agricultural NPS, rural and town domestic sewage, livestock and poultry source), construction land source and industry wastewater source (PS).

   Based on the annual pollutant load of each land-use type estimated in the earlier study and the assumption (1) and (2), the daily load of each grid from three land use of natural land, farmland and impervious surface is calculated (Table 1 and Fig. 3). The industry waste-water

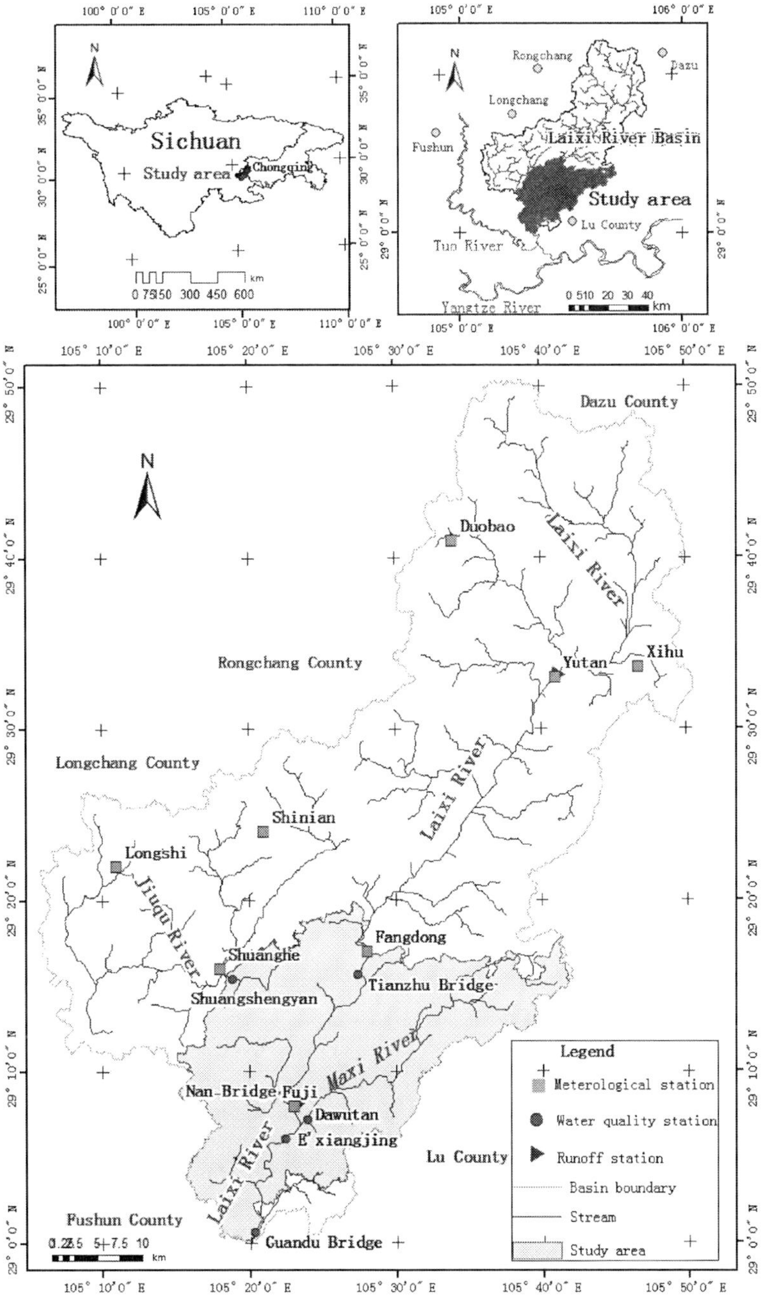

Figure 2. Study area and the spatial distribution of hydrological, meteorological, and water quality stations.

source (PS) is allocated into the PS input grids. As there was no TP data for the industry waste-water source, the PS input only included TN input (Table 2 and Fig. 3).

4 The application of DSPTM in study area

The application for DSPTM includes two steps – the calibration and validation stage for BTOPMC and the application stage for DSPTM. As the carrier of the pollutant transport, the BTOPMC was calibrated and validated first, and then the DSPTM was applied in the study area.

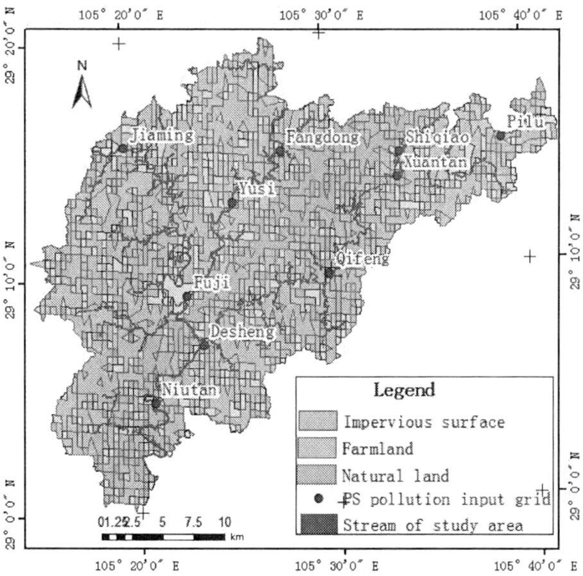

Figure 3. Reclassified land use and PS pollution input grid distribution.

Table 1. Pollutant daily input amount for each land-use grid Units: kg/grid/day.

Land use type	Natural land	Farmland	Impervious surface
Grid quantity (grids)	1708	5262	1122
TN	0.07	2.1817	0.2761
TP	0.0177	0.7887	0.0064

Natural land is the land type of shallow-root and deep-root vegetation and farmland is the land type of shallow-root vegetation and irrigation land in the land use classification of BTOPMC.

4.1 The calibration and validation for hydrological model, BTOPMC

The calibration and validation information for the BTOPMC model parameters can be seen in Tables 3 and 4. The calibration and validation effect meets the requirements, with the Nash-Sutcliffe efficient as 57.46 % for calibration and 65.36 % for validation.

4.2 The application of DSPTM in study area

The DSPTM was applied in the Lu county area of Laixi river basin with the data of 2011. According to the spatial distribution of the water quality monitoring sections, the three sections of Tianzhusi, Shuangshengyan, and Dawutan were set as the inflow water quality monitoring sections of the study area (Fig. 2). Therefore, in the NPS pollutant transport simulation, the measured pollutant concentration of the three inflow sections was used to replace the pollutant concentration of the corresponding river grids, to simulate the pollutant concentration of Nan Bridge, E'xiangjing, and Guandu Bridge for each simulation interval. The time step for simulating pollutant concentration in this study is daily, but the monitoring time step of the measured data is monthly. The change of the water quality is gradual when there are no sudden water quality events. Therefore, the monthly measured values of water quality monitoring sections were converted into daily pollutant concentrations by the temporal linear interpolation method (Table 5).

In addtion, the model was intialized like this (Starting time: 1 January 2011): (1) Initial concentration of river channel grids. Pollutant initial concentration of river channel grids can be calculated through the spatial linear interpolation of the measured concentration values of water quality monitoring sections on the starting time. (2) Initial concentration of slope grids. The study assumed that initial pollutant concentration of slope grids was zero and the pollutant concentration of slope grids was zero when there was no runoff (The grid outflow is smaller than 0.01 m^3/s).

Table 2. Daily river pollution load and PS input grid No. for industry waste-water source (PS) of each town.

Town	TN ($kg\,day^{-1}$)	Grid No. for PS input
Fuji	36.85	64 076
Fangdong	0.08	54 779
Qifeng	0.47	63 078
Desheng	12.22	66 929
Niutan	7.84	69 764
Yusi	1.97	58 391
Xuantan	0.03	57 140
Jiaming	0.19	54 477
Shiqiao	0.14	55 847
Pilu	0.38	55 098

Table 3. Calibration and validation information for BTOPMC model parameters.

Items	Basic information
Model name	Modified version of BTOPMC
Runoff simulation area	Interval watershed area between Fuji and Yutan station (1456.56 km^2)
Calibration period	1979–1984
Validation period	1985–1987
Time step	daily
Inflow section of simulation area	Yutan station (Control area: 747.4 km^2)
Section for calibration and validation	Fuji station (Control area: 2203.96 km^2)

Table 4. Calibrated model parameters for BTOPMC.

Parameter	Name and physical meaning	Basin physical characteristics Classification	Calibrated parameters
T_0 ($m^2 s^{-1}$)	Saturated soil hydraulic conductivity	Clay area	1.938
		Sand area	1.421
		Silt area	1.424
m (m)	Decay factor of T_0	Clay area	0.099
		Sand area	0.299
		Silt area	0.2
S_{rmax} (m)	Maximum root zone capacity	Deep rooted area	0.029
		Shallow rooted area	0.021
		Shallow rooted and irrigated area	0.02
		Impervious area	0.004
S_{bar0} (m)	Initial value of the average soil saturation deficit	$0 \leq \lambda \leq 10$	0.288
		$11 \leq \lambda \leq 20$	0.798
		$21 \leq \lambda \leq 30$	0.166
		$31 \leq \lambda \leq 40$	0.155
n_0	Manning's rough coefficient	Stream network	0.004
		Building area	0.12
		Clay area	0.157
		Sand area	0.117
		Silt area	0.277

Notes: λ is the topographic index (Beven and Kirkby, 1979; Quinn et al., 1995).

Table 5. Basic information for DSPTM application.

Items	Basic information
Model name	DSPTM based on BTOPMC
Study area	Lu county area of the Laixi River basin (728 km^2)
Simulation period	2011
Time step	Daily
Upstream inflow sections for water quality monitoring	Tianzhusi, Shuangshengyan, and Dawutan
Validation sections	Nan Bridge, E'xiangjing, and Guandu Bridge

4.3 Results

In this study, the 2011 daily observed water quality data of Nan Bridge, E'xiangjing, and Guandu Bridge sections were used to test the performance of DSPTM with two pollution indicators of TN and TP. The TN and TP graphs of the three control sections are shown in Fig. 4a–f. Meanwhile, the model prints the spatial distribution of TN on 1 July (wet season) and 1 February (dry season), see Fig. 6. Overall, the simulated results are as expected.

1. The simulated and measured values for annual pollutant flux of the control sections (Nan Bridge, E'xiangjing, and Guandu Bridge) are close to each other. The results of comparisons between the simulated and measured value showed that for 2011 TN and TP pollutant flux of the control sections, the relative errors are within 10 % of one another and close to 10 %, respectively. The simulation relative error for the TN pollutant flux of the Guandu Bridge is slightly larger, at nearly 11.34 %. The relative error of the other control sections can be seen in Table 6. This indicates that the simulation effect for the annual pollutant load amount of study area is generally not bad.

2. The statistics indicators of Nash-suttcliffe coefficient and Mean relative errors shows that the simulation effect of the concentration variation process is bad (Table 6). As shown in Fig. 4a–f, for the 2011 TN and TP daily concentrations of the three control sections, the simulation and measured process graphs display a uniform variance trend and the simulated values fall around the measured values, within a certain range. This indicates that the simulation results can reflect the temporal change trend of pollutant concentration for control sections but the simulation values have poor performance, especially at peak concentration. This is because assumptions (1) and (2) cannot reflect the actual concentration distribution. In further studies the temporal and spatial distribution simulation for NPS pollutants will be the research emphasis.

3. The simulation results show that the pollutants concentration in the beginning of the flood period is slightly greater, which is in accordance with the characteristics of the NPS pollution generating process. This indicates that assumption (3) is reasonable. By analyzing the concentration peak and the process grphs for TN and TP of the control sections, the following features can be found: (a) Pollutant concentration peaks typically occur at the end of April or the beginning of May, which marks the beginning of the flood period; (b) With the flood period continuing, the pollutant concentration decreases gradually and tends to be stable; (c) At the beginning of November, the dry period, the pollutant concentration increases slightly. The above features reflect the characteristics of the occurrence of NPS pollution. Before the flood period, pollutants accumulate on the ground surface due to the weak driving force from rainfall and runoff. With the beginning of the flood period, the accumulated pollutants are carried off and migrated by the runoff and afflux into the river. That is why river pollutant concentration increases at the beginning of the flood period. NPS pollution refers to the phenomenon that the dispersive pollutants on ground surface are carried and migrated into river. With the continuous scouring for pollutants on the ground surface with runoff, the accumulated pollutant amount decreases gradually, which cause the reduction in the generating intensity of NPS. Meanwhile, the river pollutant concentration also decreases gradually and tends to be stable. During the dry period, although there is no NPS pollution and PS pollutants are the main pollution sources, the river pollutant concentration increases slightly because the river flow is relatively small. The above analysis demonstrates that assumption (3) is reasonable to some extent.

4. The spatial distribution of TN on the wet and dry season shows that a large amount of NPS pollutant is accumulated on the slope grids of the study area on the dry season and the pollutant is brought away with the direct runoff on the wet season (Fig. 5). That meets the transport feature of NPS.

5 Summary and conclusion

This study aimed to develop a distributed NPS simulation model based on BTOPMC. This study developed a DSPTM based on the concentration process of BTOPMC. Runoff is the carrier of pollutant transport and transformation in catchment. The basic principles of DSPTM include the mass balance of pollutants within a grid and the pollutant transmission between grids along the runoff direction in catchment. The grids in catchment were divided into slope grid and river channel grid, therefore, the transport process was divided into two stages of slope pollutant transport and river channel

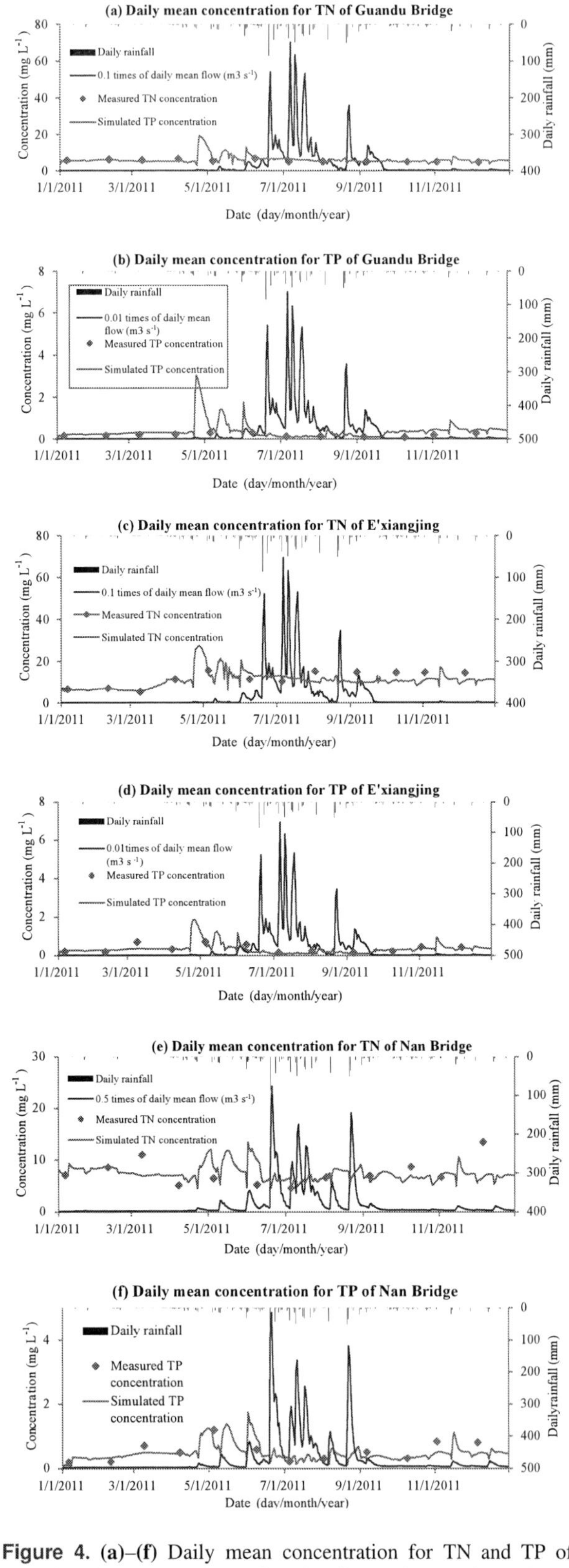

Figure 4. **(a)–(f)** Daily mean concentration for TN and TP of Guandu Bridge, TN and TP of E'xiangjing and TN and TP of Nan Bridge.

Table 6. Pollutant flux for TN and TP of controlling sections and their relative errors.

Section	TN					TP				
	Measured (ton)	Simulated (ton)	Relative error (%)	Nash-Suttcliffe Coefficient	Mean relative error	Measured (ton)	Simulated (ton)	Relative error (%)	Nash-Suttcliffe Coefficient	Mean relative error
Nan Bridge	3502.03	3356.62	−4.15	−0.34	26 %	49.29	54.39	10.36	0.30	38 %
E'xiangjing	13 106.16	11 771.14	−10.19	0.29	18 %	381.03	352.24	−7.55	0.42	35 %
Guandu Bridge	6012.58	6694.67	11.34	−0.82	14 %	205.03	199.94	−2.48	−3.3	75 %

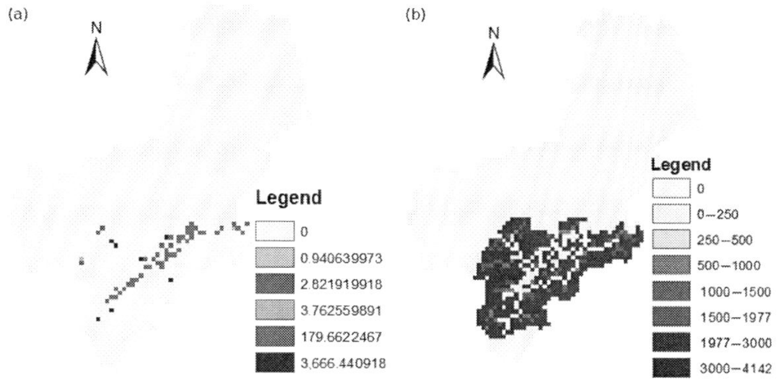

Figure 5. The spatial distribution of TN on 1 July (**a**, wet season) and 1 February (**b**, dry season), kg.

pollutant transport. For slope transport process, NPS, such as natural woodland source, rural NPS and construction land source, are the dominant pollution sources. For river channel grids, the dominant pollution source is PS, i.e. industrial waste-water source.

In this model application, the pollution source data derived from earlier results from the study area, which were calculated using the Export Coefficient Method. Any errors in the earlier study results would influence the accuracy of the DSPTM simulated results. Because of the shortage of hydrological and environmental data, the Lu County area of the Laixi river basin was selected as the study area; this area is not an integrated watershed. Three upstream sections of the study area, Tianzhusi, Shuangshengyan, and Dawutan, were set as the pollutant input sections, which made the modeling process complicated. The measured errors of the water quality data of the three sections could have increased errors in the simulated results. The runoff parameters of BTOPMC were calibrated and validated in the integrated Laixi basin, so they would not influence the results of water quality simulation.

To fit the DSPTM test, three assumptions – the even distribution of NPS pollutants in temporal and spatial scale and the pollutant accumulation within the grids where there is no runoff, were proposed. We found that the first two assumptions could not reflect the actual pollutant distribution in the basin, which lead to a bad fitting effect of the simulated and measured pollutant concentration. However, according to the simulated results, including the time accordance of pollutant concentration peak with the flood peak assumption 3 was confirmed to be reasonable.

Overall, the simulated results show that DSPTM can, to some extent, reflect the characteristics of NPS pollutant transport and reach the expected target. In further studies, the pollutant distribution mode in space and time and the pollutant generating model will be developed to improve the simulated results.

Data availability. The Digital Elevation Model (DEM) can be acquired from USGS (Lehner et al., 2008), the Soil data from FAO (Food and Agriculture Organization of the United Nations, 2003) and the Land use/land cover data from GlobeLand30 (Chen et al., 2017). Hydrological and meteorological data can be collected from the related hdrological and meteorological sections of Sichuan and Chinese ministry of water resources (Sichuan Provincial Water Resources Department, 2017; Sichuan Meteorological Service, 2017; China Hydrology, 2017; Hydrological and Meteorological Information of China Hydrology) and the water quality data has been attached. Other related data can be found in the the manuscript.

Competing interests. The authors declare that they have no conflict of interest.

Special issue statement. This article is part of the special issue "Innovative water resources management – understanding and balancing interactions between humankind and nature". It is a result of the 8th International Water Resources Management Conference of ICWRS, Beijing, China, 13–15 June 2018.

Acknowledgements. The work was funded by Personnel Training Project of Yunnan Province "The development and application for non-point source pollutant transport model for slope and river channel and its transformation model for river channel based on BTOPMC" (Grant number: KKSY201404107), Program of international S&T cooperation of China "Joint Research on the Key Techniques of Prevention and Remedy for Watershed Non-point Source Pollution" (Grant number: 2012DFG21780) and National Natural Science Foundation of China "Mechanism Study on Yunnan Laterite Reservoir Bank Instability Caused by Erosion" (Grant number: 51269006). Thanks are expressed to Wenqing Chen, Department of Environmental Engineering, Sichuan University, China and the members of her research team – Lusheng Ye and Haixia Luo. Appreciation is given to my lab fellows: Xiaodong Li, Xing Liu, Yali Ma, Yuhong Dong, Li Zhou, Min Ke, Biqiong Wu, and Xiaoxia Wang of Digital Information Engineering Center for Resources and Environment. Gratitude is expressed to the governmental sectors for for providing data for the hydrological environmental, meteorological, soil, and sources of Lu County, Sichuan, China.

Edited by: Dingzhi Peng
Reviewed by: two anonymous referees

References

Anderton, S. P., Latron, J., White, S. M., Llorens, P., Gallart, F., Salvany, C., and O'Connell, P. E.: Internal evaluation of a physically-based distributed model using data from a Mediterranean mountain catchment, Hydrol. Earth Syst. Sci., 6, 67–84, https://doi.org/10.5194/hess-6-67-2002, 2002.

Andréassian, V., Perrin, C., Berthet, L., Le Moine, N., Lerat, J., Loumagne, C., Oudin, L., Mathevet, T., Ramos, M.-H., and Valéry, A.: HESS Opinions "Crash tests for a standardized evaluation of hydrological models", Hydrol. Earth Syst. Sci., 13, 1757–1764, https://doi.org/10.5194/hess-13-1757-2009, 2009.

Ao, T. Q., Yoshitani, J., Takeuchi, K., Fukami, K., Mutsura, T., and Ishidaira, H.: Effects of block scale on runoff simulation in distributed hydrological model: BTOPMC, in: Weather Radar Information and Distributed Hydrological Modelling, edited by: Tachikawa, Y., Vieux, B. E., Georgakakos, K. P., and Nakakita, E., IAHS Publication, 282, 227–234, 2003.

Ao, T. Q., Ishidaira, H., Takeuchi, K., Kiem, A. S., Yoshitari, J., Fukami, K., and Magome, J.: Relating BTOPMC model parameters to physical features of MOPEX basins, J. Hydrol., 320, 84–102, 2006.

Bao, H. J., Wang, L. L., Li, Z. J., Zhao, L. N., and Zhang, G. P.: Hydrological daily rainfall-runoff simulation with BTOPMC model and comparison with Xin'anjiang model, Water Science and Engineering, 3, 121–131, 2010.

Beven, K. J. and Kirkby, M. J.: A physically based, variable contributing area model of hydrology, Hydrol. Sci. B., 24, 43–69, 1979.

Bookera, D. J. and Woods, R. A.: Comparing and combining physically-based and empirically-based approaches for estimating the hydrology of ungauged catchments, J. Hydrol., 508, 227–239, 2014.

Chen J., Liao A., and Chen J., et al.: 30-meter GlobalLand cover data product- GLobeLand30 [J]. Geomatics World, available at: http://www.globallandcover.com/user/login.aspx?para=124, 1–8, 2017. (in Chinese)

China Hydrology: The Daily Hdrological Monitoring Data, available at: http://swgl.mwr.gov.cn/, 1979–1987, 2017.

Chinese Academy for Environmental Planning: National Water Environmental Capacity Validation Manual, Beijing, 2003.

Cho, J., Park, S. W., and Im, S. J.: Evaluation of Agricultural Non-point Source (AGNPS) model for small watersheds in Korea applying irregular cell delineation, Agr. Water Manage., 9, 400–408, 2008.

Food and Agriculture Organization of the United Nations: THE DIGITAL SOIL MAP OF THE WORLD Version 3.6, available at: http://www.fao.org/soils-portal/soil-survey/soil-maps-and-databases/faounesco-soil-map-of-the-world/en/, 2003.

Hao, F. H., Cheng, H. G., and Yang, S. T.: Theoretical methods and application of non-point source pollution Environmental Science Press of China, Beijing, 1, 4–5, 2006.

Hydrological and Meteorological Information of China Hydrology: The Daily Hdrological and Meteorological Monitoring Data, 1979–1987, available at: http://xxfb.hydroinfo.gov.cn/, 2017.

Jiang, H. Y., Kyung, S. M., and Youngchul, K.: Development of EMC-based empirical model for estimating spatial distribution of pollutant loads, Desalin. Water Treat., 27, 1–14, 2011.

Johnes, P. J. and Heathwaite, A. L.: Modelling the impact of land use change on water quality in agricultural catchments, Hydrol. Process., 11, 269–286, 1996.

Johnes, P. J. and O'Sullivan, P. E.: Nitrogen and phosphorus losses from the catchment of Slapton Ley, Devon-an export coefficient approach, Field Stud., 7, 285–309, 1989.

Kobierska, T., Jonas, M., Zappa, M., Bavay, J., Magnusson, S. M., and Bernasconi, S. M.: Future runoff from a partly glacierized watershed in Central Switzerland: a 2 model approach, Adv. Water Res., 55, 204–214, 2013.

Lehner, B., Verdin, K., and Jarvis, A.: New global hydrography derived from spaceborne elevation data, Eos, Transactions, AGU, available at: https://hydrosheds.cr.usgs.gov/datadownload.php?reqdata=3demg89, 93–94, 2008.

Nash, J. E. and Sutcliffe, J. V.: River flow forecasting through conceptual models. Part 1: A discussion of principals, J. Hydrol., 10, 282–290, 1970.

Nasr, A., Bruen, M., Jordan, P., Moles, R., Kiely, G., and Byrne, P.: A comparison of SWAT, HSPF and SHETRAN/GOPC for modeling phosphorus export from three catchments in Ireland, Water Res., 41, 1065–1073, 2007.

Nigussie, H. and Fekadu, Y.: Testing and evaluation of the agricultural non-point source pollution model (AGNPS) on Augucho catchment, western Hararghe, Ethiopia, Agric. Ecosyst. Environ., 19, 201–212, 2003.

Orth, R., Staudinger, M., Seneviratne, S. I., Seibert, J., and Zappa, M.: Does model performance improve with complexity? A case study with three hydrological models, J. Hydrol., 523, 147–159, 2015.

Quinn, P. F, Beven, K. J., and Lamb, R.: The ln(a/tan/beita) index: How to calculate it and how to use it within the topmodel framework, Hydro. Process., 9, 161–182, 1995.

Reichert, P., Borchardt, D., Henze, M., Rauch, W., Shanahan, P., Somlyody, L., and Vanrolleghem, P. A.: River Water Quality Model No. 1. Scientific and Technical Report No. 12. International Water Association, London, UK, 2001.

Sichuan Meteorological Service: The Daily Meteorological Monitoring Data, available at: http://www.scqx.gov.cn/info/?1269, 1979–1987, 2017.

Sichuan Provincial Water Resources Department: The Daily Hdrological Monitoring Data, available at: http://www.scwater.gov.cn/, 1979–1987, 2017.

Singh, R., Archfield, S. A., and Wagener, T.: Identifying dominant controls on hydrologic parameter transfer from gauged to ungauged catchments – A comparative hydrology approach, J. Hydrol., 517, 985–996, 2014.

Takeuchi, K., Ao, T. Q., and Ishidaira, H.: Introduction of blockwise use of TOPMODEL and Muskingum – Cunge method for the hydroenvironmental simulation of a large ungauged basin, Hydrol. Sci. J., 44, 633–646, 1999.

Takeuchi, K., Hapuarachchi, H. A. P., Zhou, M. C., Ishidaira, H., and Magome, J.: A BTOP model to extend TOPMODEL for distributed hydrological simulation of large basins, Hydrol. Process., 22, 3236–3251, 2008.

Takeuchi, K., Hapuarachchi, H. A. P., Kiem, A. S., Ishidaira, H., Ao, T. Q., Magome, J., Zhou, M. C., Georgievski, M., Wang, G., and Yoshimura, C.: Distributed runoff predictions in the Mekong River basin, in: Runoff prediction in ungauged basins – Synthesis across processes, places and scales, Chapter 11, Cambridge University Press, Cambridge, UK, 2013.

USEPA: National management measures for the control of nonpoint pollution from agriculture, Washington DC, USA, 2000.

Wu, C. L. and Chau, K. W.: Mathematical model of water quality rehabilitation with rainwater utilization – cases study at Haigang, Int. J. Environ. Pollut., 28, 534–545, 2006.

Ying, L.-L., Hou, X.-Y., Lu, X., and Zhu, M.-M.: Discussion on the export coefficient method in non-point source pollution studies in China, J. Water Res. Water Eng., 21, 90–95, 2010.

Yoshimura, C. and Takeuchi, K.: Estimation of nutrient runoff processes in the Mekong River Basin using a distributed hydrological model, Journal of Japan Society of Hydrology & Water Resources, 20, 493–504, 2007 (in Japanese with English abstract).

Yoshimura, C., Zhou, M. C., Kiem, A. S., Fukami, K., Hapuarachchi, H. A. P., Ishidaira, H., and Takeuchi, K.: 2020s scenario analysis of nutrient load in the Mekong River Basin using a distributed hydrological model, Sci. Total Environ., 407, 5356–5366, 2010.

Viviroli, D., Zappa, M., Gurtz, J., and Weingartner, R.: An introduction to the hydrological modelling system PREVAH and its pre- and post-processing-tools, Environ. Modell. Softw., 24, 1209–1222, 2009.

Zhang, H., Li, J., Li, X.-D., Chen, X., Liu, X., and Ao, T.-Q.: Study on Rural Non-Point Source Pollution Assessment Method of Regions with Sparse Data, Journal of Sichuan University, Engineering Science Edition, 45, 58–66, 2013.

Zhou, M. C., Ishidaira, H., Hapuarachchi, H. A. P., Magome, J., Kiem, A. S., and Takeuchi, K.: Estimating potential evapotranspiration using the Shuttleworth – Wallace model and NOAA-AVHRR NDVI to feed the hydrological modeling over the Mekong River Basin, J. Hydrol., 327, 151–173, 2006.

https://doi.org/10.5194/piahs-379-335-2018

Multi-model ensemble hydrological simulation using a BP Neural Network for the upper Yalongjiang River Basin, China

Zhanjie Li[1,2], Jingshan Yu[1,2], Xinyi Xu[1], Wenchao Sun[1,2], Bo Pang[1,2], and Jiajia Yue[3]

[1]College of Water Sciences, Beijing Normal University, Beijing 100875, China
[2]Beijing Key Laboratory of Urban Hydrological Cycle and Sponge City Technology, Beijing 100875, China
[3]School of Geographical Science, Qinghai Normal University, Xining 810016, China

Correspondence: Jingshan Yu (jingshan@bnu.edu.cn)

Received: 20 November 2017 – Revised: 31 March 2018 – Accepted: 3 April 2018 – Published: 5 June 2018

Abstract. Hydrological models are important and effective tools for detecting complex hydrological processes. Different models have different strengths when capturing the various aspects of hydrological processes. Relying on a single model usually leads to simulation uncertainties. Ensemble approaches, based on multi-model hydrological simulations, can improve application performance over single models. In this study, the upper Yalongjiang River Basin was selected for a case study. Three commonly used hydrological models (SWAT, VIC, and BTOPMC) were selected and used for independent simulations with the same input and initial values. Then, the BP neural network method was employed to combine the results from the three models. The results show that the accuracy of BP ensemble simulation is better than that of the single models.

1 Introduction

Hydrological processes are complex and are affected by many factors, including climate, weather, topography, and the sub-surface. Hydrological models provide important and effective methods of hydrological research for describing complex water cycle processes. Because of the different structure of each model, their simulation ability for a hydrological process is also different. Each hydrological model has its own characteristics and basins where it can be more effectively applied. Relying on a single model often leads to predictions that capture some phenomena at the expense of others (Duan et al., 2007). Multi-model ensemble hydrological simulation has been an effective method for improving simulation accuracy. Many researchers have applied multi-model simulation methods to improve the simulation and prediction accuracy of hydrological models (Ajami et al., 2007; Devineni et al., 2008; Najafi and Moradkhani, 2016; Razavi and Coulibaly, 2016; Zhu et al., 2016).

Artificial neural networks (ANNs), as an abstraction and simulation of some of the basic characteristics of human brain or natural neural networks, have received great attention because of their advantages in solving highly nonlinear problems of great social concern. The "feed-forward, back propagation" neural network (BPNN), which is currently the most popular network architecture, can be applied in a variety of fields according to the characteristics of the model (Hu et al., 2005; Yaseen et al., 2015). Wei et al. (2013) established a predictive model, based on a wavelet-neural network hybrid modelling approach, for monthly river flow estimation and prediction (Wei et al., 2013). Wang et al. (2017) proposed a new back-propagation neural network algorithm and applied it in the semi-distributed Xinanjiang (XAJ) model. The improved hydrological model was capable of updating the flow forecasting error without losing the lead time (Wang et al., 2017).

This study explores the use of BP for hydrological streamflow predictions. We investigated how the BP scheme can be used to improve the accuracy of streamflow predictions. This paper is organized as follows. Section 2 presents the study area and data collection. Section 3 introduces the methodology, including hydrological models and the BP neural network. Section 4 describes the streamflow simulation of each ensemble member and the validation results of multi-model

Table 1. Weather stations in the basin.

Station name	Latitude (N)	Longitude (E)	Elevation (m)
Qingshuihe	97°08′	33°48′	4415.4
Shiqu	98°06′	32°59′	4200.0
Shiquluoxu	98°00′	32°28′	3242.1
Ganzi	100°00′	31°37′	3393.5
Seda	100°20′	32°17′	3893.9
Daofu	101°07′	30°59′	2957.2
Xinlong	100°19′	30°56′	3000.0
Litang	100°16′	30°00′	3948.9
Qianning	101°29′	30°29′	3449.0

predictions using the BP schemes. Section 5 provides summaries and conclusions.

2 Study area and data collection

2.1 Study area

The Yalongjiang River (YLJR) is located in the Sichuan Province in southwestern China, in the eastern Tibetan Plateau, and is the largest tributary of the Jinshajiang River in the upper Yangtze River. The total length of the main stream of YLJR is 1571 km, and the YLJR Basin area is about 128 440 km^2, accounting for 13 % of the total area of the upper reaches of the Yangtze River. The average annual runoff in the YLJR Basin is about 58 billion m^3. The upper reaches of the YLJR Basin, above the Yajiang in situ gauging station, were selected as the study area.

The terrain elevation of the upper YLJR Basin varies greatly and from north to south ranges from 2600 to 6111 m. Because of the influence of the westerly atmospheric circulation and monsoons, the north-south climate change is also very obvious, with a dry continental climate in the northern plateau and a subtropical climate in the central and southern parts of the basin. The winters are long and cold, and the summers are cool and wet, with strong radiation all year round. The average annual precipitation during the last 50 years was about 500–2470 mm, of which 73 % occurred from June to September. The inter-annual change, and regional distribution trend, of streamflow in this basin is similar to the precipitation.

2.2 Data description

The Digital Elevation Model (DEM) dataset was taken from the Computer Network Information Centre, Chinese Academy of Sciences (http://www.gscloud.cn, last access: 17 April 2018), was used to derive the mean elevation and slope of the region at a resolution of 90 m. The land cover datasets in this study were determined using MODIS (MOD12Q1-051) data (https://lpdaac.usgs.gov, last access: 17 April 2018), and included grassland, woodland, cropland, urban and built-up areas, water, and unused land. In 2010,

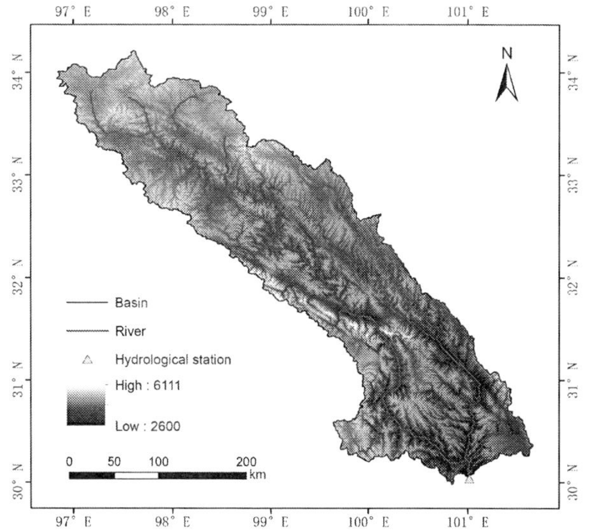

Figure 1. Topography, river networks, and streamflow gauging station of the basin.

approximately 89 % of the entire study area was covered by grassland; woodland covered approximately 9 %, and the remainder was covered by other land cover types. The soil map for the basin was derived from the Chinese national 1 : 1 000 000-scale soil map. The physical properties of these soils were obtained from the China Soil Scientific Database (http://www.soil.csdb.cn/, last access: 17 April 2018). The other properties used as parameters for the models were computed using empirical equations (Saxton, 2006). The main soil type is plateau meadow soil.

There are nine national weather stations in the basin, shown in Table 1. Daily meteorological records from these stations for 1960 to 2013 were used to assess the performance of a range of probability distribution models (http://data.cma.cn/, last access: 17 April 2018). Daily Streamflow data for the basin (Yajiang station) for 2007 to 2011 were obtained from the Institute of Geographic Sciences and Natural Resources Research, Chinese Academy of Sciences (CAS). These data were used to calibrate the model parameters. The topography, river networks, and streamflow gauging station of the basin are presented in Fig. 1.

3 Methodology

3.1 Hydrological models

For this study, we employed three kinds of common distributed hydrological models (SWAT, VIC, and BTOPMC) as the ensemble members of multi-model simulation.

SWAT (Soil and Water Assessment Tool) is a popular, physically-based, distributed hydrological model developed by the United States Department of Agriculture (Arnold et al., 1998). It is capable of simulating the water cycle

Table 2. SWAT model parameters calibrated.

Name	Description	Initial range
CN2	SCS runoff curve number	20–90
SLSUBBSN	Average slope length (m)	10–150
ALPHA_BF	Baseflow recession coefficient	0–1
GW_DELAY	Groundwater delay time (days)	30–450
EPCO	Plant uptake compensation factor	0.01–1
OV_N	Manning coefficient for overland flow	0–0.8
ESCO	Soil evaporation compensation coefficient	0.8–1
CH_K2	Hydraulic conductivity in main channel ($mm\,h^{-1}$)	5–130
SOL_AWC	Available soil water capacity ($mm\,H_2O\,mm\,Soil^{-1}$)	0–1
SOL_K	Soil saturated hydraulic conductivity ($mm\,h^{-1}$)	0–2000
SNOCOVMX	Threshold depth of snow, above which there is 100 % cover	0–500
SNO50COV	Fraction of snow volume represented by SNOCOVMX that corresponds to 50 % snow cover	0–1

Table 3. Parameter values for the VIC model.

Name	Description	Initial range
B	Exponent of variable infiltration capacity curve	0–10.0
D	Fraction of maximum base flow	0–1.0
D_{max}	Maximum velocity of base flow ($mm\,day^{-1}$)	0–30
W_s	Fraction of maximum soil moisture content of the third layer	0–1.0
d_1, d_2, d_3	three soil layer thicknesses (m)	0.05–2.0

Table 4. Parameters of BTOPMC model.

Name	Description	Range
m	Decay factor	0.005–0.1
α	Drying function parameters	−10–10
S_{bar0}	Average soil saturation deficit (m)	0.001–0.9
D_0	Coefficients of discharge ability	0.1–200
S_{rmax}	Maximum storage capacity of root zone (m)	0.001–0.8
n_0	Block average Manning's coefficient	−2.0–2.0

and transportation of sediment and pollutants at the basin scale. The hydrological processes considered in the model include precipitation, interception, infiltration, evapotranspiration, snowmelt, surface runoff, percolation, baseflow, and flow movement in river channels. The model divides a basin into multiple sub-basins that are further subdivided into hydrological response units (HRUs), which have homogeneous land use and soil characteristics. Three steps are involved in simulating the hydrological processes: (1) preparing the meteorological input and GIS data of the soil type and land cover, (2) constructing the model, and (3) the calibration/validation process. In the present study, 12 sensitive parameters were identified for model calibration using the GLUE (Beven and Binley, 1992) method. The parameters are described and listed in Table 2.

VIC (Variable Infiltration Capacity) (Liang et al., 1994) is a large-scale, semi-distributed land surface hydrological model that solves full water and energy balances, and was originally developed at the University of Washington. It is a grid-based, soil-vegetation-atmosphere transfer scheme that clearly reflects the effects of infiltration, precipitation, and the spatial variability of vegetation on water fluxes through the landscape. The model takes into account the spatial, subgrid scale, variability of infiltration, precipitation, and vegetation. The newest version of the VIC model consists of three layers that allow for explicitly depicting the dynamics of surface and groundwater interactions, and calculating the groundwater table (Liang et al., 2003). The model is driven by data on precipitation, maximum and minimum daily temperature (daily-time step), vegetation type, and soil texture. The generated runoff is routed laterally using simulated topology and stream networks to the basin outlet. The parameters and their initial prior ranges are briefly described in Table 3.

The BTOPMC (blockwise use of TOPMODEL with Muskingum-Cunge routing) model is a physically-based, distributed, hydrological model developed at the University of Yamanashi (Japan). This model extends the concepts of the original TOPMODEL to distributed hydrological simulation in large river basins (Takeuchi et al., 2008). In the BTOPMC model, the hydrological processes are considered in grid cells to represent spatial heterogeneity. The grid cells are then divided into several sub-catchments or sub-basins, where each sub-basin is treated as a block or a unit to limit the total number of parameters. The vertical column is divided into four zones: vegetation, root, unsaturated, and saturated. In the BTOPMC model, there are six basic model parameters to be identified. A detailed description of these parameters is given in Table 4.

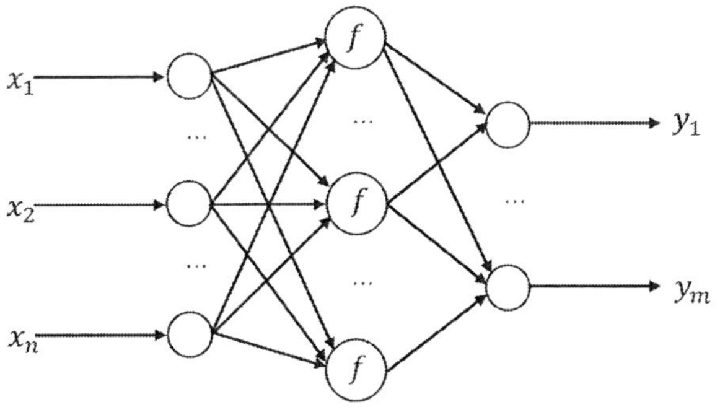

Figure 2. The topology of the BPNN model.

Table 5. Comparison between ensemble simulation and ensemble members' optimal simulating series.

		SWAT	VIC	BTOPMC	BP Ensemble
E_{NS}	Calibration	0.77	0.75	0.82	0.95
	Validation	0.66	0.65	0.75	0.90

3.2 BP neural network

The BPNN is a multilayer structure and feed forward mapping model trained by an error back propagation algorithm. The topology of the BPNN model mainly comprises the input, hidden, and output layers. The BP algorithm consists of forward propagation and error back propagation. In the forward propagation, the input information is processed from the input layer to the output layer through the hidden layer. The state of each layer is only affected by the state of the next layer. If the expected output is not obtained in the output layer, then it is transferred back to the error back propagation. The error signal is returned along the original connection path. The error is minimized by modifying the weights of each layer node. The topology of the BPNN model is shown in Fig. 2.

In the present study, the inputs of the BP network are the simulation results of three hydrological models, and the output is the streamflow at the Yajiang station.

3.3 Model performance indicators

Before the model can be applied for evaluating the method proposed in this study, the model robustness in the basin must be examined by assessing the model's performance in terms of benchmark calibrations. The Nash–Sutcliffe efficiency coefficient was selected to evaluate the simulation performance of the three individual models as well as multi-model ensemble simulation:

$$E_{NS} = 1 - \frac{\sum_{i=1}^{n} \left(Q_{\mathrm{obs},i} - Q_{\mathrm{sim},i}\right)^2}{\sum_{i=1}^{n} \left(Q_{\mathrm{obs},i} - Q_{\mathrm{obs,avg}}\right)^2}, \qquad (1)$$

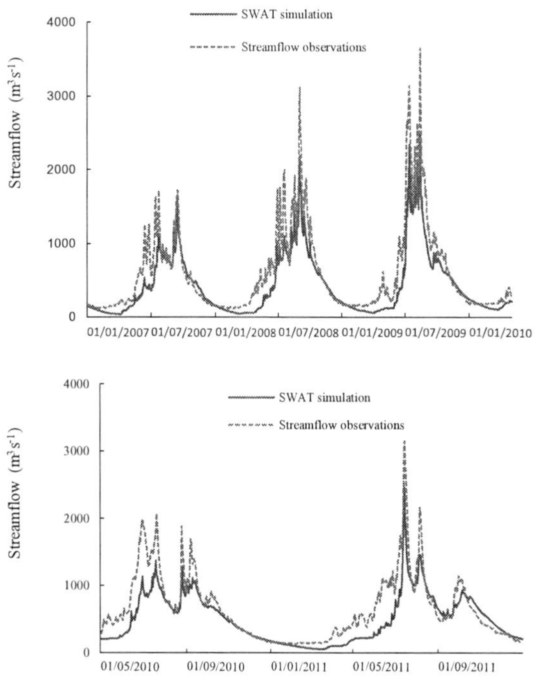

Figure 3. SWAT simulated streamflow for the basin in both the calibration and validation periods.

where $Q_{\mathrm{obs},i}$ ($\mathrm{m^3\,s^{-1}}$) and $Q_{\mathrm{sim},i}$ ($\mathrm{m^3\,s^{-1}}$) represent the observed and simulated streamflow, respectively, at time step i, and $Q_{\mathrm{obs,avg}}$ ($\mathrm{m^3\,s^{-1}}$) is the average value of the streamflow observations. The integer n is the number of samples. E_{NS} measures the agreement between modeled and observed values, with $E_{NS} = 1.0$ indicating a perfect agreement between modeled and observed streamflow for a given basin.

4 Results and discussion

4.1 Streamflow simulation of each ensemble member

Using available streamflow records, the benchmark calibration for the basin was conducted using continuous daily streamflow observations for 1 January 2007–30 April 2010, and was validated using data for 1 May 2010–31 December 2011. The results of the benchmark calibrations of the three ensemble members are listed in Table 5, and the streamflow simulations are shown in Figs. 3–5.

During the calibration period, each E_{NS} coefficient of the three ensemble members was above 0.75. In the validation period, the E_{NS} coefficient for the BTOPMC model reached 0.75, which was slightly better than the SWAT and VIC models. The results of these calibrations were taken as the benchmarks for the basin.

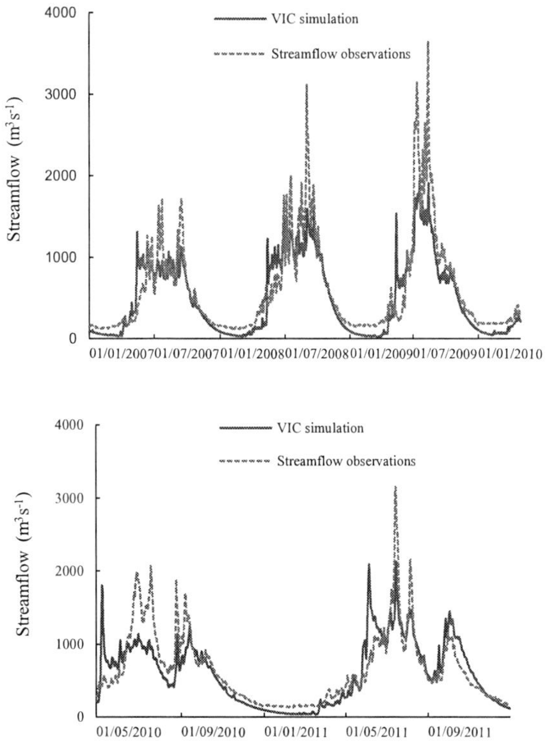

Figure 4. VIC simulated streamflow for the basin in both the calibration and validation periods.

Figures 3, 4, and 5 compare the observed and simulated daily streamflow series derived from the three ensemble members in the calibration and validation periods. It can be concluded from these figures that the simulations of daily streamflow series from three models are acceptable on the whole. However, the representations of some low flows are not as satisfactory as those of normal flows, with a noticeable underestimation of much of the low flows. It was also difficult to accurately obtain the peak flow and peak time using the three models. One of the reasons why the peak flow was difficult to simulate accurately was because the daily scale model usually converted short-term rainstorms (a few hours or even a few minutes) to long duration (1 day) events. This smoothing obviously has an effect on the simulation results for high flow volumes.

Considered as a whole, the simulated streamflow values from the three models are in good agreement with the measured data, which shows that these models can be used in the basin.

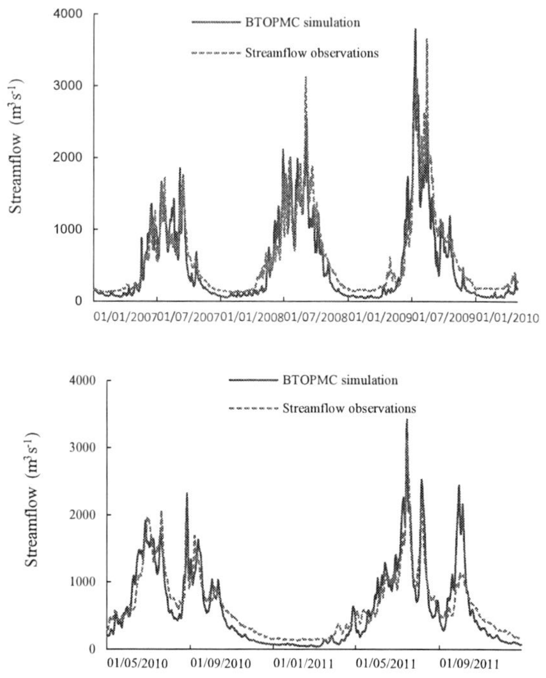

Figure 5. BTOPMC simulated streamflow for the basin in both the calibration and validation periods.

4.2 Streamflow simulation using the BPNN multi-model ensemble

The performance of the multi-model ensemble simulations in both the calibration and validation periods are shown in Fig. 6. From the simulations, it can be seen that the BPNN multi-model ensemble greatly improves the results of a single member. The BPNN multi-model ensemble adequately reproduces the observed daily streamflow series, with E_{NS} values of 0.95 and 0.90 during the calibration and validation periods, respectively. The E_{NS} of the simulation in the ensemble model is satisfactory, and it reproduces the observed hydrographs well. The fitting of the flow process in the BPNN multi-model ensemble is clearly better than that of a single member, especially for the relatively low flow period (from November to April). Moreover, the simulation accuracy of the ensemble model for peak flows is also much higher than that of a single model.

5 Conclusions

This study explored the feasibility of applying a multi-model ensemble simulation to the upper YLJR Basin. The SWAT, VIC, and BTOPMC models have good simulation results as a whole, but there are some errors for low flows and peak flows. The results of the BPNN multi-model ensemble simulation are better than that of a single model. The BPNN

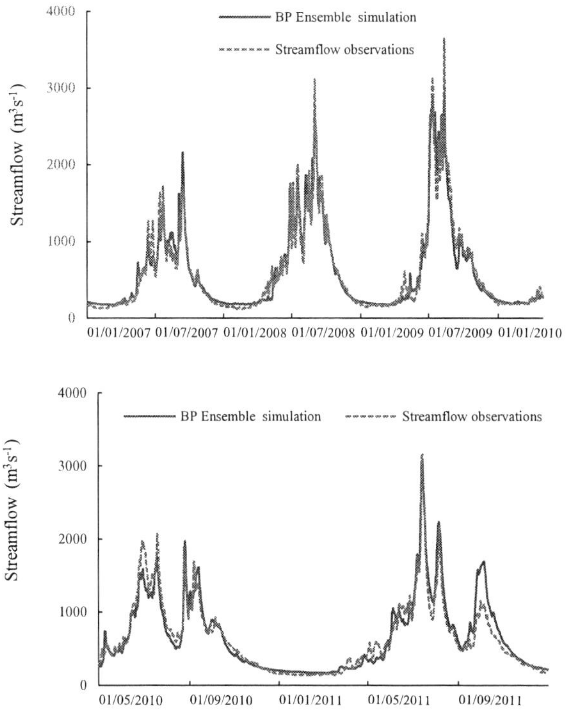

Figure 6. BP ensemble simulated streamflow for the basin in both the calibration and validation periods.

multi-model ensemble takes into account the accuracy of each model, maximizes the use of model information, and has higher precision and stability. The multi-model simulation results are more consistent with the observations, especially in the low flow and peak flow periods. Multi-model ensemble simulation should become an important direction in hydrological simulation research.

Data availability. The raw data required to reproduce these findings are available to download through URLs provided in Sect. 2.2. The processed data required to reproduce these findings cannot be shared at this time as the data also forms part of an ongoing study.

Competing interests. The authors declare that they have no conflict of interest.

Special issue statement. This article is part of the special issue "Innovative water resources management – understanding and balancing interactions between humankind and nature". It is a result of the 8th International Water Resources Management Conference of ICWRS, Beijing, China, 13–15 June 2018.

Acknowledgements. This research was supported by the National Key Technology Research and Development Program of China (Grant No. 2016YFC0401308), the National Natural Science Foundation of China (Grant Nos. 41671018, 51779007) and Fundamental Research Funds for the Central Universities.

Edited by: Yangbo Chen
Reviewed by: two anonymous referees

References

Ajami, N., Duan, Q., and Sorooshian, S.: An integrated hydrologic Bayesian multimodel combination framework: Confronting input, parameter, and model structural uncertainty in hydrologic prediction, Water Resour. Res., 43, W014031, https://doi.org/10.1029/2005WR004745, 2007.

Arnold, J., Srinivasan, R., Muttiah, S., and Williams, J.: Large area hydrologic modeling and assessment – Part 1: Model development, J. Am. Water Resour. As., 34, 73–89, 1998.

Beven, K. and Binley, A.: The future of distributed models: Model calibration and uncertainty prediction, Hydrol. Process., 6, 279–298, 1992.

Devineni, N., Sankarasubramanian, A., and Ghosh, S.: Multimodel ensembles of streamflow forecasts: Role of predictor state in developing optimal combinations, Water Resour. Res., 44, W094049, https://doi.org/10.1029/2006WR005855, 2008.

Devineni, N., Sankarasubramanian, A., and Ghosh, S.: Multi-model ensemble hydrologic prediction using Bayesian model averaging, Adv. Water Resour., 30, 1371–1386, 2007.

Hu, T., Lam, K., and Ng, S.: A modified neural network for improving river flow prediction, Hydrol. Sci. J., 50, 299–318, 2005.

Liang, X., Lettenmaier, D. P., Wood, E., and Burges, S.: A simple hydrologically based model of land surface water and energy fluxes for general circulation models, J. Geophys. Res.-Atmos., 99, 14415–14428, 1994.

Liang, X., Xie, Z., and Huang, M.: A new parameterization for surface and groundwater interactions and its impact on water budgets with the variable infiltration capacity (VIC) land surface model, J. Geophys. Res.-Atmos., 108, 8613, https://doi.org/10.1029/2002JD003090, 2003.

Najafi, M. R. and Moradkhani, H.: Ensemble Combination of Seasonal Streamflow Forecasts, J. Hydrol. Eng., 21, 040150431, https://doi.org/10.1061/(ASCE)HE.1943-5584.0001250, 2016.

Razavi, T. and Coulibaly, P.: Improving streamflow estimation in ungauged basins using a multi-modelling approach, Hydrol. Sci. J., 61, 2668–2679, 2016.

Saxton, K. R. W.: Soil water characteristic estimates by texture andorganic matter for hydrological solutions, Soil Sci. Soc. Am. J., 70, 1569–1578, 2006.

Takeuchi, K., Hapuarachchi, P., Zhou, M., Ishidaira, H., and Magome, J.: A BTOP model to extend TOPMODEL for distributed hydrological simulation of large basins, Hydrol. Process., 22, 3236–3251, 2008.

Wang, J., Shi, P., Jiang, P., Hu, J., Qu, S., Chen, X., Chen, Y., Dai, Y., and Xiao, Z.: Application of BP Neural Network Algorithm in Traditional Hydrological Model for Flood Forecasting, Water, 9, 48, https://doi.org/10.3390/w9010048, 2017.

Wei, S., Yang, H., Song, J., Abbaspour, K., and Xu, Z.: A wavelet-neural network hybrid modelling approach for estimating and predicting river monthly flows, Hydrol. Sci. J., 58, 374–389, 2013.

Yaseen, Z., El-Shafie, A., Jaafar, O., Afan, H., and Sayl, M.: Artificial intelligence based models for stream-flow forecasting: 2000–2015, J. Hydrol., 530, 829–844, 2015.

Zhu, G., Li, X., Zhang, K., Ding, Z., Han, T., Ma, J., Huang, C., He, J., and Ma, T.: Multi-model ensemble prediction of terrestrial evapotranspiration across north China using Bayesian model averaging, Hydrol. Process., 30, 2861–2879, 2016.

https://doi.org/10.5194/piahs-379-343-2018

Trading the Economic Value of Unsatisfied Municipal Water Demand

Dua'a B. Telfah[1], **Riccardo Minciardi**[2], **and Giorgio Roth**[1]

[1]Department of Civil, Chemical and Environmental Engineering, University of Genoa, Genoa, Italy
[2]Department of Informatics, Bioengineering, Robotics, and Systems Engineering, University of Genoa, Genoa, Italy

Correspondence: Gorgio Roth (giorgio.roth@unige.it)

Received: 29 December 2017 – Revised: 19 March 2018 – Accepted: 20 March 2018 – Published: 5 June 2018

Abstract. Modelling and optimization techniques for water resources allocation are proposed to identify the economic value of the unsatisfied municipal water demand against demands emerging from other sectors. While this is always an important step in integrated water resource management perspective, it became crucial for water scarce Countries. In fact, since the competition for the resource is high, they are in crucial need to trade values which will help them in satisfying their policies and needs. In this framework, hydro-economic, social equity and environmental constraints need to be satisfied. In the present study, a hydro-economic decision model based on optimization schemes has been developed for water resources allocation, that enable the evaluation of the economic cost of a deficiency in fulfilling the municipal demand. Moreover, the model enables efficient water resources management, satisfying the demand and proposing additional water resources options. The formulated model is designed to maximize the demand satisfaction and minimize water production cost subject to system priorities, preferences and constraints. The demand priorities are defined based on the effect of demand dissatisfaction, while hydrogeological and physical characteristics of the resources are embedded as constraints in the optimization problem. The application to the City of Amman is presented. Amman is the Capital City of the Hashemite Kingdom of Jordan, a Country located in the south-eastern area of the Mediterranean, on the East Bank of the Jordan River. The main challenge for Jordan, that threat the development and prosperity of all sectors, is the extreme water scarcity. In fact, Jordan is classified as semi-arid to arid region with limited financial resources and unprecedented population growth. While the easy solution directly goes to the simple but expensive approach to cover the demand, case study results show that the proposed model plays a major role in providing directions to decision makers to orient their policies and strategies in order to achieve sustainability of scarce water resources, satisfaction of the minimum required demand as well as financial sustainability. In addition, results map out national needs and priorities that are crucial in understanding and controlling the complexity of Jordan's water sector, mainly for the city of Amman.

1 Introduction

In view of water scarcity, there is a growing strain on surface and groundwater to guarantee socio-economic development and maintain healthy ecosystems. Rapid population growth, climate change, and socio-economic development generate an increasing pressure on water resources to meet domestic, agriculture and industrial water demand and maintain water quality (Loucks and van Beek, 2005). The competition over scarce water resources is increasing. Water demand centres for municipal, industrial, agricultural, hydropower, and environmental uses compete in water stressed countries over conventional (surface and groundwater) as well as unconventional (desalination, brackish and treated wastewater) water. This increase both the importance of reallocating existing water resources more efficiently, and the need for developing new unconventional resources. In view of that, water mangers address the concept of Integrated Water Resource Management (IWRM). Effective water resource management focus on managing different water sources and uses while conserving or improving water quantity and quality. Identification and development of water resources project rely on maximizing net benefit or minimizing costs while considering social, political and environmental impacts. It contains physical, political, and economic measures such as water pricing or marketing policies (Mayer and Muñoz-Hernandez, 2009).

Different approaches combining optimization and simulation tools have been applied on a wide range of water resource management problems. On the basis of objective functions and accompanying constraints, optimization models search for the best solution among the countless options for the infrastructure shape and the water allocation. These models, usually built on economic and/or hydrologic bases, present the advantage of incorporating social value systems in the allocation of the water resource. Other criteria, such as equity, political, or environmental quality can also be used (Loucks and van Beek, 2005); generally speaking, these are more useful if improvement of the system performance is the main goal.

The present contribution exploits the knowledge needed to understand and manage water resources with special emphasis for the capital city of Amman, home for almost forty-two percent of the population of Jordan. The importance of this study derived from the tremendous pressures in service delivery faced by the water uses. This is due to an extraordinary municipal demand growth, linked to the increasing of both the number of refugees entering the Country and the municipal demand, connected with economic development, improvement of living standards and health.

2 The case study

Amman is the capital and the largest city of Jordan. It is the economic, political and cultural centre. Its population is 4 million, 36 % foreign nationals, over an area of 7579 km^2.

The model case study divides the Amman area in: two domestic demand centres (Amman main and Deep South), one agricultural demand centre, and one industrial demand centre. The model identifies both existing and potential water supply sources from different resources able to satisfy the demand for these centres.

Amman is mainly supplied by renewable groundwater, fossil groundwater, surface water, and treated wastewater. All resources are allowed to supply municipal, agricultural and industrial uses, except treated wastewater that is only used for agriculture and industrial uses. In the actual system, private groundwater wells are owned by the farmer or factory. The Amman main water system is supplied by both internal, including twelve renewable groundwater wellfields, and external water sources imported from other governorates, including 13 sources in addition to the Jafer basin, namely the DISI, through the newly constructed conveyance pipeline on Build Operate and Transfer basis. Deep South water system is supplied by local resources that belong to five wellfields in the Dead Sea basin.

Renewable groundwater resources supply 126 [$Mm^3\ yr^{-1}$], fossil groundwater supply 114 [$Mm^3\ yr^{-1}$] and surface water supply an average of 135 [$Mm^3\ yr^{-1}$]. In the study period (13 years), treated wastewater supplies a total of 296 [Mm^3].

3 Methodology

An optimization model is here proposed to allocate water in regions with absolute resource scarcity. It is applied for the period 2018–2030 at Amman governorate level to: maximize social benefits; minimize the dissatisfaction of water demand; minimize the economic cost of water production, that include investment, operation and maintenance, and conveyance costs. Constraints were designed to confine the model according to technical, environmental and institutional aspects and local resource conditions. This analysis extends similar works (Wang et al., 2015; Al-zahrani et al., 2016; Davijani et al., 2016) in which multi-objective optimization models are proposed for water resources allocation considering economic and social objectives. Two objectives are finalized. The first is an economic objective that includes water production and conveyance costs, and agriculture and industrial demand dissatisfaction cost. The second is the dissatisfaction of municipal demand. The model was solved to find the trade-off value assigned to the dissatisfaction of municipal demand objective by founding Pareto front and selecting one of the preferred Pareto optimal choices (Yapo et al., 1998). In the present work, the model is formulated and solved with reference to an annual time-step. Both fixed and variable unit costs related to water production and conveyance are assumed constant in time (no inflation rate), and municipal demand comprises the physical losses.

3.1 Water Demand Prediction

The model considers three kinds of demand points (each demand is associated to a given topographical location):

a. Domestic demand points. These points are mainly in cities and refer to water used for living (drinking, cooking, washing, etc.). This kind of demand is expected to increase in time. In the statement of the problem, it is assumed that such increase is perfectly identified and connected with population growth. Physical losses are considered and estimated using the concept of Non-Revenue Water (NRW), where NRW is assumed to decrease by 1 % each year over the study period. The domestic demand has the highest priority and must be satisfied only by water coming from groundwater, surface water, and desalinated water.

b. Agriculture industrial demand points: it is assumed to be approximately constant over the years and can be satisfied by all water sources.

Each water use is characterized by: (1) the normal demand, representing the target that reflects the amount of water needed to reach economic development and to sustain living standards; (2) the minimum demand, the amount required by a sector that must be satisfied; (3) the maximum demand level (aspiration level – continuous supply). Different kinds of water demands were extracted from utility records. Water demand forecast is focused on the prediction of the normal demand level.

3.2 Allocation model

An optimal water resource allocation model is proposed to achieve the minimum economic cost and maximum social benefit for the capital city of Amman allocating limited water resources under physical and environmental constraints. The optimal water resources allocation model consists of two parts: the objective function, including economic and social aspects, and the constraints, which embed physical, environmental, and social limitations. The allocation model is built on two objectives through:

$$\mathrm{MIN} = \mathrm{F1} + \mathrm{DELTA} \cdot \mathrm{F2} \tag{1}$$

where F1 is the economic term and F2 the social term. F1 comprises production and investment costs for the development of potential resources, water conveyance cost, and cost of dissatisfaction of agriculture and industrial demands. The social term expresses the dissatisfaction of the municipal demand, associated to a trade-off factor (delta) that reflects the economic cost of a deficiency in fulfilling the municipal demand.

The model is solved with different values of delta to find the Pareto front. The preferred Pareto optimal point is then selected based on the desired importance for the two objectives.

3.2.1 Available and Potential Resources

The purpose of this section is to present costs and constraints related to the extraction of water from existing and potential resources and to develop a potential resource.

Existing Resources/Renewable Groundwater (RGW)

Let IRGW be the number of wellfields already existing. The cost term to be minimized for this kind of resources is

$$\mathrm{RGWC} = \sum_{t=1}^{T} \sum_{i=1}^{\mathrm{IRGW}} K_i^{\mathrm{RGW}} \cdot y_i^{\mathrm{RGW}}(t) + c_i^{\mathrm{RGW}} \cdot q_i^{\mathrm{RGW}}(t) \tag{2}$$

where K_i^{RGW} is the fixed cost [JD] associated with the abstraction from the ith wellfield; c_i^{RGW} is the variable cost coefficient for water from the ith wellfield [JD Mm^{-3}]; $y_i^{\mathrm{RGW}}(t)$ is a binary variable (BV) that is: 1 if the i-th wellfield is used in year t, and zero otherwise; $q_i^{\mathrm{RGW}}(t)$ is the water pumped from the ith renewable well field in year t [Mm^3 yr^{-1}].

The following constraints must be fulfilled:

$$0 \le q_i^{\mathrm{RGW}}(t) \le q_i^{\mathrm{RGW,MAX}}$$
$$i = 1, \ldots \mathrm{IRGW} \text{ and } t = 1, 2, \ldots T \tag{3}$$

Disjunctive constraint:

$$\mu \cdot y_i^{\mathrm{RGW}}(t) - q_i^{\mathrm{RGW}}(t) \ge 0$$
$$i = 1, 2, \ldots \mathrm{IRGW} \text{ and } t = 1, 2, \ldots T \tag{4}$$

where $q_i^{\mathrm{RGW,MAX}}$ is the maximum wellfield capacity of ith renewable wellfield [Mm^3 yr^{-1}]; its estimation should be based on the groundwater aquifers safe yield; μ is an arbitrary very large number (50×10^9).

Constraints (4) is introduce to ensure that the fixed cost is paid whenever the wellfield is exploited. Note that there is no necessity of ensuring the opposite as it is automatically ensured by the minimization of the cost function.

Other Existing Resources

The same applies also when considering the cost for water production from existing fossil groundwater (in this case $i = 1,$...IFW), from existing surface water source (in this case $i = 1,$...ISW), and from existing wastewater treatment. One should notice that for the case study under analysis, surface water resource uses are subject to political agreements between Jordan and Israel, that determine abstraction limits for both side, considering the ecosystem demand.

Potential Resources (not yet existing)

Besides optimizing the existing resources, the model suggests when new plants should be built and how "powerful" each new plant should be. The two cost terms are related to the resources exploitation, analogous to the term considered for existing resources, and to construction costs.

Potential Fossil Ground Water (PFW)

Let IPFW be the number of potential wellfields extracting water from the same source – in the case study the potential fossil groundwater resources of Hisban Kafrain, Sheediyya-Al Hasa, and Additional Disi. Then, the decision model minimizes both operational, PFWC1, and investment, PFWC2, costs:

$$\mathrm{PFWC1} = \sum_{t=1}^{T} \sum_{i=1}^{\mathrm{IPFW}} K_i^{\mathrm{PFW}} \cdot y_i^{\mathrm{PFW}}(t)$$
$$\cdot z_i^{\mathrm{PFW}}(t) + c_i^{\mathrm{PFW}}$$

$$\cdot q_i^{\mathrm{PFW}}(t) \cdot z_i^{\mathrm{PFW}}(t) \quad (5)$$

$$\mathrm{PFWC2} = \sum_{I=1}^{\mathrm{IPFW}} \left(\sum_{t=1}^{T} V_i^{\mathrm{PFW}}(t) \right) \cdot \left(B_i^{\mathrm{PFW}} + D_i^{\mathrm{PFW}} \cdot q_i^{\mathrm{PFW,MAX}} \right) \quad (6)$$

where K_i^{PFW} is the fixed cost of abstraction from the ith PFW wellfield [JD]; $y_i^{\mathrm{PFW}}(t)$ is a binary variable (BV): 1 if the ith PFW wellfield is used in year t, and zero otherwise; $z_i^{\mathrm{PFW}}(t)$ is a BV: 1 if the i-th PFW wellfield is fully operated at (t), zero otherwise; $c_i^{\mathrm{PFW}}(t)$ is the variable cost coefficient for quantity of the abstracted PFW from the ith wellfield [JD]; $q_i^{\mathrm{PFW}}(t)$ is water flow pumped from the ith wellfield [$\mathrm{Mm^3\,yr^{-1}}$], this constrained to the size of the new plant; $V_i^{\mathrm{PFW}}(t)$ is BV: 1 if construction the ith wellfield start in year t, and zero otherwise; $q_i^{\mathrm{PFW,MAX}}$ is maximum water flow that maybe pumped from the ith PFW wellfield [$\mathrm{Mm^3\,yr^{-1}}$]. It is a decision variable that determine the size of the new plant; D_i^{PFW} is the variable cost coefficient associated to the ith potential wellfield [JD]; B_i^{PFW} is the fixed cost associated with constructing the planned ith wellfield [JD]; L_i^{PFW} is the period for constructing PFW wellfield.

The following constraints have to be introduced:

The quantity of water extracted from the ith wellfield should be less than or equal the maximum capacity.

$$0 \le q_i^{\mathrm{PFW}}(t) \le q_i^{\mathrm{PFW,MAX}}$$
$$i = 1, \ldots \mathrm{IPFW} \text{ and } t = 1, 2 \ldots T \quad (7)$$

Disjunctive constraint ensure that the wellfield has been already established when it is exploited:

$$z_i^{\mathrm{PFW}}(t) = \begin{cases} 1 \text{ if } V_i^{\mathrm{PFW}}\left(t - L_i^{\mathrm{PFW}}\right) = 1 \\ \text{Or} \\ z_i^{\mathrm{PFW}}(t-1) = 1 \\ 0 \text{ otherwise} \end{cases}$$
$$i = 1, \ldots \mathrm{IPFW} \text{ and } t \ge L_i^{\mathrm{PFW}} \quad (8)$$

$$z_i^{\mathrm{PFW}}(t) = 0$$
$$i = 1, \ldots \mathrm{IPFW} \text{ and } t \le L_i^{\mathrm{PFW}} \quad (9)$$

Imposing fixed exploitation cost constraint if necessary.

$$\mu \cdot y_i^{\mathrm{PFW}}(t) - q_i^{\mathrm{PFW}}(t) \ge 0$$
$$i = 1, \ldots \mathrm{IPFW} \text{ and } t = 1, 2, \ldots T \quad (10)$$

Disjunctive constraint to insure the use of the wellfield when it is build and fully operated.

$$\mu \cdot z_i^{\mathrm{PFW}}(t) - y_i^{\mathrm{PFW}}(t) \ge 0$$
$$i = 1, \ldots \mathrm{IPFW} \text{ and } t = 1, 2, \ldots T \quad (11)$$

Preventing multiple start of construction of a wellfield.

$$\sum_{t=1}^{T} V_i^{\mathrm{PFW}}(t) \le 1$$
$$i = 1, \ldots \mathrm{IPFW} \text{ and } t = 1, 2, \ldots T \quad (12)$$

L_i^{PFW} is the number of time interval for the ith new FW wellfield.

A constraint limiting the size of the plant.

$$q_i^{\mathrm{PFW,MAX}} \le \mathrm{MAXSIZE}^{\mathrm{PFW}}$$
$$i = 1, \ldots \mathrm{IPFW} \text{and } t = 1, 2, \ldots T \quad (13)$$

where $\mathrm{MAXSIZE}^{\mathrm{PFW}}$ is the maximum admissible size of the plant in site i; assumed technical known.

Other Potential Resources

The above applies also when considering the cost for building plants and produce water from potential wastewater treatment plants ($i = 1,\ldots\mathrm{IPWWT}$), and for the potential desalination plants ($i = 1,\ldots\mathrm{IDP}$).

3.2.2 Transmission cost

The transmission cost is dependent on the topology of the transmission network which depend on topography. This require deep knowledge of the topographic characteristic of the region. Simplification of point to point connection costs is used based on the fact that the transmission network is (almost entirely) already existing.

Transmission Cost from Renewable Groundwater (RGW) to Agriculture Demand (AD) Points

Introducing JAD as the number of agriculture demand points (in our model one agriculture centre) and IRGW as the number of renewable groundwater wellfields, then the cost of transmitting water from a renewable groundwater source to an agricultural demand points is provided by

$$\mathrm{TC}_{\mathrm{AD}}^{\mathrm{RGW}} = \sum_{t=1}^{T} \sum_{j=1}^{\mathrm{JAD}} \sum_{i=1}^{\mathrm{IRGW}} \frac{\gamma \cdot q_{ij}^{\mathrm{AD\_RGW}}(t)(\mathrm{TDH})_{ij}}{E_{\mathrm{pi}}} \cdot C_{\mathrm{Ep}} \cdot \Delta t \cdot y_{ij}^{\mathrm{AD\_RGW}} \quad (14)$$

where γ is the specific weight of water = [9.81 $\mathrm{KN\,m^{-3}}$ at 20 °C]; $q_{ij}^{\mathrm{AD\_RGW}}$(t) is the water flow from the ith RGW wellfield in time interval (t) to jth agriculture demand point expressed [$\mathrm{Mm^3\,yr^{-1}}$]; TDH_{ij} is the total dynamic head; H_{Stat} is the difference in head between suction and discharge sides of pump in the absence of flow; equals difference in elevation of free surfaces of the fluid source and destination; H_{f} is the head loss; E_{pi} is pump efficiency; C_{Ep} is the cost of pumping water from the ith source based on the efficiency of the pump to any demand center point [JD Kwh^{-1}]; $y_{\text{ij}}^{\text{AD\_RGW}}$ is a binary parameter with two values, 1 if the H_{Stat} greater than zero, zero otherwise; L_{ij} is the distance between the ith RGW wellfield and jth agriculture demand point [m]; The flow that can be conveyed from ith source to the jth demand points is constrained by

$$\leq q_{ij}^{\text{AD\_RGW}}(t) \leq q_{ij}^{\text{AD\_RGW, MAX}} \tag{15}$$

where $q_{ij}^{\text{AD\_RGW,MAX}}$ is the maximum flow from the ith source to the jth demand point [Mm^3 yr^{-1}].

Transmission Cost from Other sources

Similar transmission cost equations and constraints are used for other sources: existing fossil groundwater ($i = 1,...\text{IFW}$), existing surface water ($i = 1,...\text{ISW}$), existing treated wastewater ($i = 1,...\text{IWWT}$), PFG ($i = 1,...\text{IPFW}$), potential wastewater treatment plants ($i = 1,...\text{IPWWT}$), and potential desalination plants ($i = 1,...\text{IDP}$).

Flows summation from all resources to jth agriculture demand points namely $\text{FLA}_J(t)$ in year t, is given by:

$$\sum_{I=0}^{\text{IRGW}} q_{ij}^{\text{AD\_RGW}}(t) + \sum_{I}^{\text{IFW}} q_{ij}^{\text{AD\_FW}}(t) + \sum_{I=1}^{\text{ISW}} q_{ij}^{\text{AD\_SW}}(t) + \sum_{I=1}^{\text{IWWT}} q_{ij}^{\text{AD\_WWT}}(t) + \sum_{I=1}^{\text{IPFW}} q_{ij}^{\text{AD\_PFW}}(t) + \sum_{I=1}^{\text{IPWWT}} q_{ij}^{\text{AD\_PWWT}}(t) + \sum_{I=1}^{\text{IDP}} q_{ij}^{\text{AD\_DP}}(t) = \text{FLA}_J(t) \tag{16}$$

Transmission Cost-Industrial and Municipal Demand

Transmission cost to industrial and municipal demand points are introduced, with index j ranges from 1 to JID or 1 to JDD (number of industrial and municipal points). $\text{FLI}_J(t)$ and $\text{FLD}_J(t)$ presents flows summation from all resources to the jth industrial and municipal demand point, excluding treated wastewater.

3.2.3 Constraints over the resource availability

Constraints on resources overall amount are introduced as

$$\sum_{j=1}^{\text{JID}} q_{ij}^{\text{ID\_RGW}}(t) + \sum_{j=1}^{\text{JAD}} q_{ij}^{\text{AD\_RGW}}(t) + \sum_{j=1}^{\text{JDD}} q_{ij}^{\text{DD\_RGW}}(t) \leq q_I^{\text{RGW}}(t)$$
$$i = 1,\ldots \text{IRGW and} t = 0, 1, \ldots T \tag{17}$$

The above is a constraint for a typical (presently existing) renewable groundwater plant. The same applies for the other sources: existing fossil groundwater ($i = 1,...\text{IFW}$); existing surface water ($i = 1,...\text{ISW}$); existing wastewater source ($i = 1,...\text{IWWT}$); potential FGW ($i = 1,...\text{IPFW}$); potential wastewater treatment plants ($i = 1,...\text{IPWWT}$); and potential desalination plants ($i = 1,...\text{IDP}$).

3.2.4 Dissatisfied Agriculture/Industrial Demand

Dissatisfied agriculture and industrial objective functions are expressed as:

$$\text{Min} \sum_{t=0}^{T} \sum_{j=1}^{\text{JID}} \epsilon_j \text{Max}(\text{ID}_j(t) - \text{FLI}_j(t), 0) + \sum_{t=0}^{T} \sum_{j=1}^{\text{JAD}} \theta_j \text{Max}(\text{AD}_j(t) - \text{FLA}_j(t), 0)) \tag{18}$$

where $\text{ID}_j(t), \text{AD}_j(t)$ is the industrial and Agricultural demand [Mm^3 yr^{-1}] for the demand centre [JID] and [JAD] at year t respectively; $\text{FLI}_j(t)$, $\text{FLA}_j(t)$ is the supplied water for industrial and Agricultural uses [Mm^3 yr^{-1}] for the demand centre [JID] and [JAD] at year t respectively; ϵ_j, θ_j are the unit costs of unsatisfied industrial and agriculture demand for the demand centre [JID] and [JAD] in [JD Mm^{-3}] respectively.

3.2.5 Municipal Demand Dissatisfaction

The unsatisfied municipal demand, is expressed as:

$$\text{Min} \sum_{t=0}^{T} \sum_{j=1}^{\text{JDD}} \text{Max}(\text{DD}_j(t) - \text{FLD}_j(t), 0) \tag{19}$$

were $\text{DD}_j(t)$, $\text{FLD}_j(t)$ are the municipal demand and supplied water [Mm^3 yr^{-1}] for the [JDD] centre at year t.

4 Results and Discussion

The present work is focused on trading the economic cost of the service with the municipal water demand satisfaction. Results here presented are based on the following assumptions: (1) municipal water demand includes the physical losses determined by Non-Revenue Water (NRW), which decline 1 % each year starting from the 2017 level of 38 %; (2) population growth rate is 1.7 %; (3) industrial and agricultural demands are assumed constant over the study period; and (4) fixed and variable unit costs do not account for inflation.

For the given conditions, a trade-off curve can be constructed to describe the relationship between the two objectives (i.e., the economic cost of the service vs. the municipal water demand satisfaction) by changing the cost of unsatisfying the municipal demand by one million cubic meters of water. Figure 1 shows the Pareto front. A negative relation between the dis-satisfaction of the municipal demand and the cost of the service is depicted.

The choice of the trade-off factor (delta) may take place, theoretically, at any discretization step. Figure 1 depicts also

Table 1. Pareto optimal selection.

Point	Economic Cost F1 [BJD]	Dissatisfaction Cost [MJD]	Per capita demand share [$L\,c^{-1}\,d^{-1}$]	DELTA [MJD]	Total cost [BJD]
1	2.9982	74	118	1.2	3.07
2	2.9985	248	118.1	2.2	3.25
3	3.38	0	120	6.09	3.38

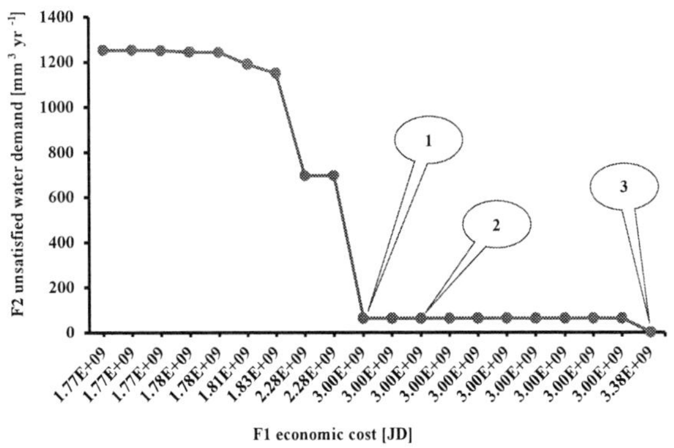

Figure 1. Pareto optimal front.

the amount of unsatisfied demand associated with different cost levels.

Selected possible solutions for the allocation problem (see Fig. 1) are represented by:

- *Point 1.* Unsatisfied municipal demand equal 62.15 [Mm^3], and trade-off factor is 1 200 000 [JD];

- *Point 2.* Un-satisfied municipal demand equal 62.03 [Mm^3], and trade-off factor is 2 200 000 [JD];

- *Point 3.* Un-satisfied municipal demand equal 0, and trade-off factor is 6 090 000 [JD].

Points 1 and 2 satisfy both social and economic objectives but are slightly biased to economic objectives. For Point 3 priority is given to the social objective and the economic objective was ignored. Based on the results described in Fig. 1 and Table 1, the selected Pareto optimal solution is Point 2 with 62.03 [Mm^3] unsatisfied municipal demand, now equivalent to 118.1 $L\,c^{-1}\,d^{-1}$, and an economic cost equal to 3.00 [BJD]. Although Point 3 totally satisfies the normal municipal water demand of 120 [$L\,c^{-1}\,d^{-1}$], the economic cost is 3.38 [BJD] exceeding Point 2 costs by 377 [MJD]. In other words, at Point 3 a cost of 377 MJD should be paid to increase the water availability by 1.9 $L\,d^{-1}$ per person. When compared to Point 1, Point 2 adds 120 000 m^3 with a total additional expenditure of 268 000 JD.

5 Conclusions

The estimation of the economic cost associated to a deficiency in fulfilling municipal water demand against demands emerging from other sectors can be obtained through modelling and optimization techniques for water resources allocation under the viewpoint of integrated water resource management (IWRM). This estimation is crucial for water scarce Countries. In fact, in these Countries the high competition for the resource call for the need to trade values which will help them in satisfying their policies and needs.

This model can be replicated and applied to other areas, especially those characterized by an increasing demand-supply deficit or dissatisfaction of municipal water consumers. It will be beneficial and will facilitate their endeavour for sustainable water management.

Data availability. Some of the data used are extracted from Miyahyna records – including the annual reports and yearly budget – which is available at: http://www.miyahuna.com.jo/YearlyReports.aspx (Miyahyna, 2018). Data on population and refugees are available at: http://dosweb.dos.gov.jo/DataBank/Population_Estimares/2017/PopulationEstimatesbyLocality_Eng.pdf (Jordan Department of Statistics, 2017). Other data were obtaided through a non-disclosure agreement from the Ministry of Water and Irrigation (MWI) and Miyahuna.

Competing interests. The authors declare that they have no conflict of interest.

Special issue statement. This article is part of the special issue "Innovative water resources management – understanding and balancing interactions between humankind and nature". It is a result of the 8th International Water Resources Management Conference of ICWRS, Beijing, China, 13–15 June 2018.

Edited by: Zongxue Xu
Reviewed by: two anonymous referees

References

Al-zahrani, M. A., Musa, A., and Chowdhury, S.: Multi-objective optimization model for water resource management: a case study for Riyadh, Saudi Arabia, Environ. Dev. Sustain., 18, 777–798, https://doi.org/10.1007/s10668-015-9677-3, 2016.

Davijani, M. H., Banihabib, M. E., Anvar, A., and Hashemi, S. R.: Multi-objective optimization model for the allocation of water resources in arid regions based on the maximization of socioeconomic efficiency, Water Resour. Manage., 30, 927–946, https://doi.org/10.1007/s11269-015-1200-y, 2016.

Jordan Department of Statistics: Estimated Population of the Kingdom by Governorate, Locality, Sex and Households, available at: http://dosweb.dos.gov.jo/DataBank/Population_Estimares/2017/PopulationEstimatesbyLocality_Eng.pdf (last access: 11 April 2018), 2017.

Loucks, D. P. and van Beek, E.: Water resources systems planning and management: an introduction to methods, models and applications, Springer, Cham, Switzerland, https://doi.org/10.1007/978-3-319-44234-1, 2005.

Mayer, A. and Muñoz-Hernandez, A.: Integrated water resources optimization models: an assessment of a multidisciplinary tool for sustainable water resources management strategies, Geography Compass, 3, 1176–1195, https://doi.org/10.1111/j.1749-8198.2009.00239.x, 2009.

Miyahyna: Yearly Reports, available at: http://www.miyahuna.com.jo/YearlyReports.aspx, last access: 11 April 2018.

Wang, Z., Yang, J., Deng, X., and Lan, X.: Optimal water resources allocation under the constraint of land use in the Heihe river basin of China, Sustainability, 7, 1558–1575, https://doi.org/10.3390/su7021558, 2015.

Yapo, P. O., Gupta, H. V., and Sorooshian, S.: Multi-objective global optimization for hydrologic models, J. Hydrol., 204, 83–97, https://doi.org/10.1016/S0022-1694(97)00107-8, 1998.

https://doi.org/10.5194/piahs-379-351-2018

Fluoride in groundwater: a case study in Precambrian terranes of Ambaji region, North Gujarat, India

Rudra Mohan Pradhan and Tapas Kumar Biswal

Department of Earth Sciences, Indian Institute of Technology Bombay, Powai-400076, India

Correspondence: Rudra Mohan Pradhan (rudra.pradhan@iitb.ac.in)

Received: 31 December 2017 – Revised: 20 February 2018 – Accepted: 26 February 2018 – Published: 5 June 2018

Abstract. Fluoride is one of the critical ions that influence the groundwater quality. World Health Organization (WHO, 1970) and Bureau of Indian Standards (BIS, 1991) set an upper limit of 1.5 mg L^{-1} in F^- concentration for drinking water purpose and above affects teeth and bones of humans. The presence of fluoride in groundwater is due to an interaction of groundwater and fluoride bearing rocks. Fluoride rich groundwater is well known in granitic aquifers in India and elsewhere. Generally, the concentration of F^- in groundwater is controlled by local geological setting; leaching and weathering of bedrock and climatic condition of an area. The main objective of the present study is to assess the hydrogeochemistry of groundwater and to understand the abundance of F^- in groundwater in hard rock terranes of Ambaji region, North Gujarat. A total of forty-three representative groundwater samples were collected and analyzed for major cations and anions using ICP-AES, Ion Chromatograph (Metrohm 883 Basic IC Plus) and titration methods. The F^- concentration in groundwater of this study area ranges from 0.17 to 2.7 mg L^{-1}. Among, twenty groundwater samples have fluoride exceeding the maximum permissible limit as per the BIS (1.5 mg L^{-1}). It is also noticed that residents of this region are affected by dental fluorosis. The general order of the dominance of major cations and anions are $Ca^{2+} > Mg^{2+} > Na^+ > K^+$ and $HCO_3^- > Cl^- > F^-$ respectively. Geochemical classification of groundwater shows most of the samples are the alkaline earth-bicarbonate type. The semi-arid climatic conditions of the region, the dominance of granitoid-granulite suite rocks and the fracture network in the disturbed and brittle zone has facilitated the development of potential aquifers and enrichment in F^- concentration in this area. The concentration of fluoride is due to high evaporation rate, longer residence time in the aquifer zone, intensive and long term pumping for irrigation.

1 Introduction

The usefulness of groundwater to human is great extent. Groundwater is considered to be the major source of drinking water in India. Groundwater chemistry is largely a function of the mineral composition of the aquifer through which it flows (Brindha and Elango, 2011). It depends upon physical and geological factors such as evaporation rate, residence time, aquifer media, rock–water interaction, recharge capacity, and anthropogenic activities. Fluoride in groundwater may be of geogenic (natural) or anthropogenic (Human induced). Fluoride is widely distributed in the environment and it ranks 13th among other elements in order of abundance in the earth's crust. Its natural abundance in the earth's crust is 0.06 to 0.09 % and 300 mg kg^{-1}. Generally, fluoride does not exhibit any colour, taste or smell when dissolved in water and hence, it is difficult to determine through physical examination. Fluoride is naturally present due to weathering of rocks rich in fluoride. Groundwater in crystalline rocks, mainly granites/granite gneisses are particularly vulnerable to fluoride because they often contain abundant fluoride bearing minerals (fluorite, apatite, hornblende, and biotite). Fluoride rich groundwater is well-known in different types of geological terranes and aquifers (Reddy et al., 2010). Several studies have mentioned an increase in dissolved F^- concentrations with increasing groundwater residence time in the aquifer systems (Nordstrom and Jenne, 1977; Apambire et al., 1997; Edmunds and Smedley, 2013). Prolonged rock-water interactions through weathering of fluoride-rich rocks enrich fluoride concentration in the groundwater systems. Consequently, fluoride is leached out and dissolves

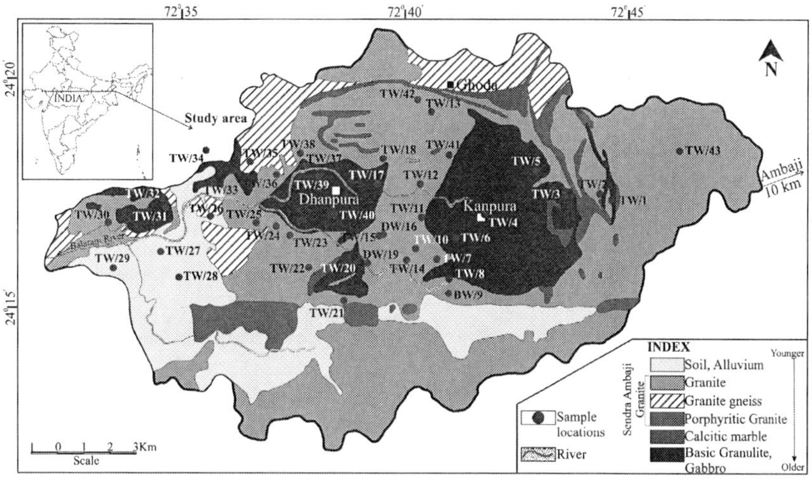

Figure 1. Geological map and sample locations of the study area (modified after Singh et al., 2010).

in groundwater and thermal gases (Deshmukh et al., 1995). Fluorite, apatite, mica (biotite) and hornblende are the minerals, which have the most effect on the geochemistry of F^- in groundwater (Dar et al., 2011) when they are interacted with groundwater and dissolved, leads to increasing F^- in groundwater (Rao, 2017). Groundwater interacting with crystalline rocks, mainly (alkaline) granites/granite gneisses (deficient in calcium) and hydrothermal solutions are probably to have reasonably high fluoride concentrations (Brunt et al., 2004). Ayenew (2008) studied that the higher concentration of F^- is due to low concentration of Ca^{2+} under an alkaline environment in Ethiopian groundwater. In different parts of China, Li et al. (2014, 2017) found that the high F^- was controlled by hydrogeological conditions, but also man-made activities and mixing of different recharge waters were also key reasons. High fluoride groundwater has been recorded as one of the most critical natural groundwater quality problems associated with hard rock areas in India (Handa, 1975; Reddy et al., 2010; Raj and Shaji, 2017; Shekhar et al., 2017). More than 200 million people from different nations suffer health problems due to high fluoride in groundwater (Ayoob and Gupta, 2006; Wu and Sun, 2016). In India, there has been an increase in the incidence of dental and skeletal fluorosis due to high fluoride concentration in drinking water. Dental fluorosis is common in 17 states of India with most pronounced in the states of Andhra Pradesh, Gujarat, Madhya Pradesh, Bihar, Punjab, Rajasthan, Tamil Nadu, Telangana, Uttar Pradesh and West Bengal (CGWB, 2015). World Health Organization (WHO, 1970) and Bureau of Indian Standards (BIS, 1991) set an upper limit of $1.5\,mg\,L^{-1}$ in F^- concentration for drinking water purpose and above affects teeth and bones of humans. In the literature, early researchers (Handa, 1988; Gupta, 1991; Sharma et al., 2005; Chinoy et al., 2005; Brindha and Elango, 2011; Raj and Shaji, 2017) discussed that fluoride and fluorosis were correlated with high concentration of fluoride ion in drinking water in different parts of India. In this paper, the data pertaining to fluoride in groundwater of Ambaji region of North Gujarat, India has been presented and discussed.

2 Materials and methods

2.1 Study area

The present area of study covers a part of Ambaji region (Ambaji-Amirgard-Dhanpura-Kanpura) situated in Banaskantha District of Gujarat state, India (Fig. 1). On the South, it is bounded by Mehsana district where high Fluoride in groundwater is reported. The area belongs to the Survey of India toposheet no. 45D/11, 45D/12 and 45D/15 (scale 1 : 50 000). The area has a diverse landscape and characterized by hilly upland with intermountain valleys, followed by piedmont zone with alluvium and residual hills and gently sloping vast alluvial plain. The area falls under semi-arid climate. Extreme temperatures (hot summer and dryness in the non-rainy seasons), erratic rainfall and high evaporation are the typical characteristic features of this type of climate. The annual rainfall of this region is 578.8 mm and is mostly received during the south-west monsoon season from June to September. Since it falls in a semi-arid type of climate, the rivers flowing through it are of ephemeral nature i.e. have water during monsoon only and dry up after monsoon period. The drainage network is constituted mainly by the Banas River and their tributaries (Balaram River). Geologically, the area belongs to the Meso-proterozoic South Delhi Terrane (1100–900 Ma), Aravalli craton. The rocks are pelitic-, calcareous-, calcareous and basic-granulites; these are intruded by three phases of Ambaji granites (G1, G2, G3). The rocks are extremely fractured along large-scale faults and shear zones (Singh et al., 2010).

2.2 Sampling and instrumentation

The area has been subjected to a high degree of deformation, which gives rise to fracture networks and this can be traced surficially and through remote sensing and GIS study. These intersecting fracture networks suggest that the water-yielding zones for phreatic and deeper aquifers are interconnected in hydraulic continuity. A total of forty-three ground water samples (Tube well/Bore well/Dug well) from different aquifers (Fig. 1) were collected in the month of May and June 2017 (Table 1) for the present study. In this area the depth of dug well ranges from 10 to 20 m b.g.l. with a diameter of 3 to 5 m and water level ranges from 3 to 15 m b.g.l. The depths of tube well are ranging from 60 to 90 m with 20 to 25 cm in diameter. Water samples were collected in HDPE bottles of 500 mL capacity. Samples collected in the field were filtered using 0.45 µm Millipore filter paper and Alkalinity of groundwater samples measured through titration methods. Water samples were brought to the laboratory for geochemical analysis. Groundwater field parameters (pH, EC, TDS, Temperature) were measured in the field using portable HANNA pH meter. Major cations and anions of water samples were determined using ICP-AES and Ion chromatograph (Metrohm 883 Basic IC Plus) with appropriate standards.

3 Results and discussion

Fluoride is one of the important ions that influence the groundwater quality. Fluorite (CaF_2) is the only principal fluorine mineral, mostly present as an accessory mineral in granites. The results of hydrogeochemical parameters of groundwater samples including fluoride concentration are presented in Table 1. Fluoride concentration in groundwater of Ambaji regions varied between 0.17 mg L^{-1} (Bhyla village) to 2.7 mg L^{-1} (Padni village). It was revealed that twenty groundwater samples have fluoride exceeding the maximum permissible limit as per the BIS (1.5 mg L^{-1}) and twenty-three groundwater samples are within the permissible limit (Table 2). Fluoride concentration in seven villages is very alarming in this region. The maximum concentration of fluoride was recorded in Padni (2.7 mg L^{-1}), Kengora (2.3 mg L^{-1}), Padliya (2.2 mg L^{-1}), Kothiavas (1.8 mg L^{-1}),Virampur (2 and 1.6 mg L^{-1}), Khemrajya (1.9 mg L^{-1}), Kanpura (1.5 (mg L^{-1}), Ajapur Mota (2 mg L^{-1}) and Dhanpura (1.8 mg L^{-1}). Wide variation of fluoride concentration is probably due to variation in chemical strata of different rocks.

Figure 2 shows the spatial distribution of fluoride concentration in different aquifers in Ambaji region. It was also found that high fluoride groundwater is concentrated mostly

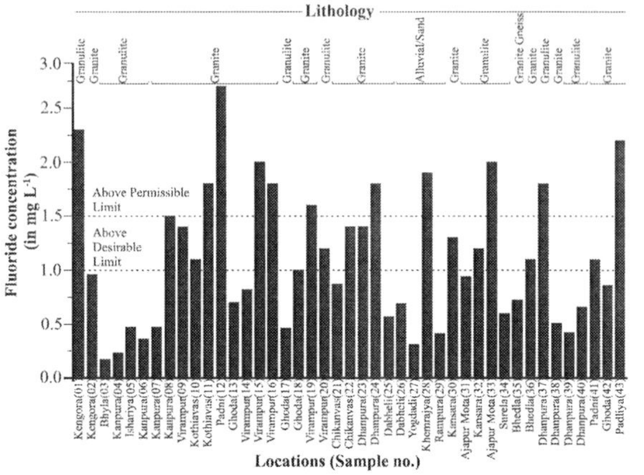

Figure 2. Spatial distribution of fluoride in groundwater of Ambaji region.

in the discharge areas than in the recharge areas with a trend of fluoride enrichment along the direction of flow. Other physico-chemical parameters such as pH, TDS, EC, major cations and anions of groundwater were analysed. Minimum, maximum and statistical values of water parameters are given in Table 3. The pH (H^+ ion in the solution) value is ranging from 6.44 to 7.45 in the study area and most of the groundwater samples found to be within the permissible limit. In the study area, Electrical conductivity (EC) ranged between 400 µS cm^{-1} (Dhanpura) to 2170 µS cm^{-1} (Padni) except TW-39 (Dhanpura) which is 4370 µS cm^{-1}. The high EC values of samples may be due to the primary minerals of earth's crust dissolved in water samples. All natural water contains different concentrations of Total Dissolved Salts (TDS) as a result of the dissolution of minerals within the rocks and soils. The essential ions that contribute TDS are carbonate, bicarbonate, sodium, potassium, calcium, magnesium, fluoride, chloride, nitrates and sulphates. TDS values varied from 200 to 1085 mg L^{-1} (Padni) except Dhanpura village (2185 mg L^{-1}) which is much higher than the permissible limit. The occurrence of other ions, mainly bicarbonate (HCO_3^-) and calcium (Ca^{2+}) ions also affects the concentration of fluoride in groundwater. Most of the samples are collected from aquifers of weathered and fractured zones occurring in the unconfined/semi unconfined conditions. Jacks (1990) studied that chemical weathering of minerals results in the formation of Ca- and Mg-carbonates which helps as good sinks for fluoride ions. But, it is the leachable state of fluoride ions that determine the water fluoride levels which is mainly due to (i) pH of the draining solutions and (ii) dissolved carbon dioxide in the soil. Dissolved fluoride in groundwater is possible only under favorable physico-chemical environments and with enough residence time (Handa, 1975).

Table 1. Physical and chemical analysis of groundwater of Ambaji region, North Gujarat, India.

Sample no.	Location	pH	Temp (°C)	EC (μS)	TDS ($mg L^{-1}$)	Fluoride ($mg L^{-1}$)	Lithology
TW/1	Kengora	6.87	27.6	780	390	2.3	Granulite
TW/2	Kengora	6.89	28.2	860	430	0.96	Granite
TW/3	Bhyla	6.92	29.7	1060	530	0.17	Granulite
TW/4	Kanpura	7.44	30	1090	545	0.23	Granulite
TW/5	Ishariya	6.96	30.1	1070	535	0.47	Granulite
TW/6	Kanpura	7.18	29.1	1820	910	0.36	Granulite
TW/7	Kanpura (Pedcholi)	6.88	29.8	1080	540	0.47	Granite
TW/8	Kanpura	6.97	29.7	1810	905	1.5	Granite
BW/9	Virampur (Golia)	7.04	29.9	1580	790	1.4	Granite
TW/10	Kothiavas	6.65	29.6	1580	790	1.1	Granite
TW/11	Kothiavas	7.11	28.7	940	470	1.8	Granite
TW/12	Padni	7.34	29.9	2170	1085	2.7	Granite
TW/13	Ghoda	6.89	29.5	770	385	0.7	Granite
TW/14	Virampur	6.77	29.2	1650	825	0.82	Granite
TW/15	Virampur	7.05	29.6	1010	505	2	Granite
DW/16	Virampur	6.85	28.8	1360	680	1.8	Granite
TW/17	Ghoda	6.86	30.1	1100	550	0.46	Granulite
TW/18	Ghoda	6.84	29.3	730	365	1	Granite
DW/19	Virampur	6.86	27.1	1980	990	1.6	Granite
TW/20	Virampur	7.15	28.6	1450	725	1.2	Granulite
TW/21	Chikanvas (Hadmana)	6.97	29	1100	550	0.87	Granite
TW/22	Chikanvas	6.84	29.4	950	475	1.4	Granite
TW/23	Dhanpura	6.66	29.3	1000	500	1.4	Granite
TW/24	Dhanpura	6.92	29.8	1160	580	1.8	Granite
TW/25	Dabheli	7.1	29.3	910	455	0.57	Granite
TW/26	Dabheli	7.07	30	1110	555	0.69	Alluvial/Sand
TW/27	Yogdadi	7.08	30.4	1050	525	0.31	Alluvial/Sand
TW/28	Khemrajya	7.08	29.6	1730	865	1.9	Alluvial/Sand
TW/29	Rampura	7.24	29.5	930	465	0.41	Alluvial/Sand
TW/30	Kansara	6.93	29.8	1360	680	1.3	Granite
TW/31	Ajapur Mota	7.25	30.7	1380	690	0.94	Granulite
TW/32	Kansara	7.08	29.9	1100	550	1.2	Granulite
TW/33	Ajapur Mota	7.45	28.6	1520	760	2	Granulite
TW/34	Surela	7.05	29.5	940	470	0.6	Granulite
TW/35	Bhedla	6.63	30.7	760	380	0.72	Granite Gneiss
TW/36	Bhedla	7.09	30	1050	525	1.1	Granite
TW/37	Dhanpura	7.24	29.6	1260	630	1.8	Granulite
TW/38	Dhanpura	6.44	30.2	400	200	0.51	Granite
TW/39	Dhanpura	7.09	29.6	4370	2185	0.42	Granulite
TW/40	Dhanpura	6.89	30.3	1010	505	0.66	Granulite
TW/41	Padni (Near School)	7.14	27.4	620	310	1.1	Granite
TW/42	Ghoda	7.27	29	790	395	0.86	Granite
TW/43	Padliya	6.77	29.2	900	450	2.2	Granite

Table 2. Comparison of fluoride content with drinking water standards (BIS, 1991 and WHO, 1970).

Parameter	BIS standard		WHO standard		Number of samples exceeding permissible limit	% of sample exceeding permissible limit
	Highest desirable ($mg L^{-1}$)	Maximum permissible ($mg L^{-1}$)	Highest desirable ($mg L^{-1}$)	Maximum permissible ($mg L^{-1}$)		
Fluoride	0.6–1.2	1.5	0.5	1.5	20	46

Table 3. Minimum, Maximum and Statistical values of water parameters.

Parameters	Units	Minimum	Maximum	Mean	Standard Deviation
EC	$\mu S\,cm^{-1}$	400	4370	1239.3	618.24
pH		6.44	7.45	6.99	0.2127
Ca^{2+}	$mg\,L^{-1}$	27	292.18	89.25	43.411
Mg^{2+}	$mg\,L^{-1}$	16.21	177.42	54.13	30.767
Na^{+}	$mg\,L^{-1}$	23.81	332.63	86.3	63.278
K^{+}	$mg\,L^{-1}$	0.16	14.62	3.22	2.478
CO_3^{-}	$mg\,L^{-1}$	0	0	0	0
HCO_3^{-}	$mg\,L^{-1}$	135	570	393.37	104.47
Cl^{-}	$mg\,L^{-1}$	29.99	769.76	121.47	128.63
F^{-}	$mg\,L^{-1}$	0.17	2.7	1.11	0.634
SO_4^{2-}	$mg\,L^{-1}$	13.05	266.97	44.16	42.223

In the present study, the general order of the dominance of major cations and anions are $Ca^{2+} > Mg^{2+} > Na^{+} > K^{+}$ and $HCO_3^{-} > Cl^{-} > F^{-}$ respectively. Geochemical classification of groundwater shows most of the samples are of alkaline earth-bicarbonate type. These waters had temporary hardness because the concentration of Ca^{2+} and Mg^{2+} ions exceeds over Na^{+} and K^{+} ions. Simultaneously, values of CO_3^{2-} and HCO_3^{-} ions together were high as compared to SO_4^{2-} and Cl^{-} ions together. The groundwater samples are classified into Ca^{2+}-Mg^{2+}-HCO_3^{-}, Ca-Cl and Na-Cl hydrogeochemical facies which are due to lithology. The possible reason for the presence of fluoride-rich groundwater in the granitic area is due to release of fluoride from the fluoride-bearing minerals in the rocks. The alkaline water depleted in calcium is mainly responsible for the high concentration of F^{-} (Shaji et al., 2007). The long residence time of water in the aquifer system, caused by a high rate of evapotranspiration and a weathered zone of low hydraulic conductivity, are the additional factors that trigger to the dissolution of fluorine-bearing minerals for further increase of F^{-} content in groundwater. Geochemical behavior of groundwater from the study suggests that high fluoride groundwater contain low levels of Calcium and high alkalinity. High pH and HCO_3^{-}/Ca^{2+} suggest favorable chemical conditions for fluoride dissolution process.

Generally, the concentration of F^{-} in groundwater is controlled by local geological setting, leaching, and weathering of bedrock and climatic condition of an area. As the area falls under semi-arid climatic conditions, the dominance of granitoid-granulite suite rocks and the fracture network in the disturbed and brittle zone has facilitated the development of potential aquifers and enrichment in F^{-} concentration. Also, it is noticed that the concentration of fluoride is due to high evaporation rate, longer residence time and comparatively low rainfall in the aquifer zone, intensive and long term pumping for irrigation.

Therefore, to maintain the quality of ground water, it is recommended for continuous monitoring of physico-chemical parameters of water and can be used for drinking and other uses only after prior treatment.

Data availability. The data presented here are part of Rudra Mohan Pradhan's doctoral thesis work and yet, Ph.D. is not completed. However, individual data are available from the author(s) on request. Those interested in using this data may contact the corresponding author.

Competing interests. The authors declare that they have no conflict of interest.

Special issue statement. This article is part of the special issue "Innovative water resources management – understanding and balancing interactions between humankind and nature". It is a result of the 8th International Water Resources Management Conference of ICWRS, Beijing, China, 13–15 June 2018.

Acknowledgements. Authors are acknowledges to Ministry of Earth Sciences (MoES), Govt. of India and Department of Earth Sciences, Indian Institute of Technology Bombay for financial assistance. Rudra Mohan Pradhan thankful to Bhuban Mohan Behera and Abdul Rajik Khan for their help during field work. Thanks are due to the two anonymous reviewers for their valuable comments that improved the quality of this manuscript.

Edited by: Wenchao Sun
Reviewed by: two anonymous referees

References

Apambire, W. B., Boyle, D. R., and Michel, F. A.: Geochemistry, genesis, and health implications of fluoriferous groundwaters in the upper regions of Ghana, Environ. Geol., 33, 13–24, 1997.

Ayenew, T.: The distribution and hydrogeological controls of fluoride in the groundwater of central Ethiopian rift and adjacent highlands, Environ. Geol., 54, 1313–1324, 2008.

Ayoob, S. and Gupta, A. K.: Fluoride in drinking water: a review on the status and stress effects, Crit. Rev. Env. Sci. Tech., 36, 433–487, 2006.

BIS: 10500: Specification for Drinking Water, Indian Standard Institution (Indian Bureau of Standard), New Delhi, 1–4, 1991.

Brindha, K. and Elango, L.: Fluoride in Groundwater: Causes, Implications and Mitigation Measures, in: Fluoride Properties, Applications and Environmental Management, edited by: Monroy, S. D., 111–136, 2011.

Brunt, R., Vasak, L., and Griffioen, J.: Fluoride in groundwater: probability of occurrence of excessive concentration on global scale, Igrac: International Groundwater Resources Assessment Centre, Report nr. SP 2004-2, 1–12, 2004.

CGWB: Groundwater quality scenario in India, Central Ground Water Board, Government of India, New Delhi, 2015.

Chinoy, N. J., Sequeira, E., Narayana, M. V., Mathews, M., Barot, V. V., Kandoi, P. R., and Jhala, D. D.: A survey of fluoride in 90 endemic villages of Mehsana and Banaskantha districts of North Gujarat, India, Fluoride, 38.3, 224, 2005.

Dar, M. A., Sankar, K., and Dar, I. A.: Fluorine contamination in groundwater: a major challenge, Environ. Monit. Assess., 173, 955–968, 2011.

Deshmukh, A. N., Wadaskar, P. M., and Malpe, D. B.: Fluorine in environment: A review, Gondwana Geol. Mag., 9, 1–20, 1995.

Edmunds, W. M. and Smedley, P. L.: Fluoride in natural waters, in: Essentials of medical geology, 311–336, Springer, Dordrecht, 2013.

Gupta, S. C.: Chemical character of groundwater in Nagaur district, Rajasthan, Ind. J. Environ. Hlth., 33, 341–349, 1991.

Handa, B. K.: Geochemistry and genesis of Fluoride-Containing ground waters in India, Groundwater, 13, 275–281, 1975.

Handa, B. K.: Fluoride occurrences in natural water in India and its significance, BHU-Jal News, 3, 21–24, 1988.

Jacks, G.: Mineral weathering studies in Scandinavia, in: The Surface Waters Acidification Programme, 215–222, Cambridge University Press, 1990.

Li, P., Qian, H., Wu, J., Chen, J., Zhang, Y., and Zhang, H.: Occurrence and hydrogeochemistry of fluoride in alluvial aquifer of Weihe River, China, Environ. Earth Sci., 71, 3133–3145, 2014.

Li, P., Tian, R., Xue, C., and Wu, J.: Progress, opportunities, and key fields for groundwater quality research under the impacts of human activities in China with a special focus on western China, Environ. Sci. Pollut. R., 24, 3224–13234, 2017.

Nordstrom, D. K. and Jenne, E. A.: Fluorite solubility equilibria in selected geothermal waters, Geochim. Cosmochim. Ac., 41, 175–188, 1977.

Raj, D. and Shaji, E.: Fluoride contamination in groundwater resources of Alleppey, southern India, Geosci. Front., 8, 117–124, 2017.

Rao, N. S.: Controlling factors of fluoride in groundwater in a part of South India, Arab. J. Geosci., 10, p. 524, 2017.

Reddy, D. V., Nagabhushanam, P., Sukhija, B. S., Reddy, A. G. S., and Smedley, P. L.: Fluoride dynamics in the granitic aquifer of the Wailapally watershed, Nalgonda District, India, Chem. Geol., 269, 278–289, 2010.

Shaji, E., Viju, J., and Thambi, D. S.: High fluoride in groundwater of Palghat District, Kerala, Current Science, 240–245, 2007.

Sharma, J. D., Parul, J., and Deepika, S.: Geological study of fluoride in groundwater of Sanganer tehsil of Jaipur district, Rajasthan, India, Fluoride, 38, p. 249, 2005.

Shekhar, S., Ghosh, M., Pandey, A. C., and Tirkey, A. S.: Impact of geology and geomorphology on fluoride contaminated groundwater in hard rock terrain of India using geoinformatics approach, Applied Water Science, 7, 2943–2956, 2017.

Singh, Y. K., De Waele, B., Karmakar, S., Sarkar, S., and Biswal, T. K.: Tectonic setting of the Balaram-Kui-Surpagla-Kengora granulites of the South Delhi Terrane of the Aravalli Mobile Belt, NW India and its implication on correlation with the East African Orogen in the Gondwana assembly, Precambrian Res., 183, 669–688, 2010.

World Health Organization (WHO): Fluoride and human health, WHO monograph series no. 59, 1970.

Wu, J. and Sun, Z.: Evaluation of shallow groundwater contamination and associated human health risk in an alluvial plain impacted by agricultural and industrial activities, mid-west China, Expos. Health, 8, 311–329, 2016.

https://doi.org/10.5194/piahs-379-357-2018

Multiple time scale analysis of sediment and runoff changes in the Lower Yellow River

Kaige Chi[1], **Zhao Gang**[2], **Bo Pang**[1], **and Ziqian Huang**[1]

[1]Beijing Normal University, Beijing, 100875, China
[2]School of Geographical Sciences, University of Bristol, Bristol BS8 1SS, UK

Correspondence: Bo Pang (pb@pb.bnu.edu.cn)

Received: 3 January 2018 – Accepted: 30 January 2018 – Published: 5 June 2018

Abstract. Sediment and runoff changes of seven hydrological stations along the Lower Yellow River (LYR) (Huayuankou Station, Jiahetan Station, Gaocun Station, Sunkou Station, Ai Shan Station, Qikou Station and Lijin Station) from 1980 to 2003 were alanyzed at multiple time scale. The maximum value of monthly, daily and hourly sediment load and runoff conservations were also analyzed with the annually mean value. Mann–Kendall non-parametric mathematics correlation test and Hurst coefficient method were adopted in the study. Research results indicate that (1) the runoff of seven hydrological stations was significantly reduced in the study period at different time scales. However, the trends of sediment load in these stations were not obvious. The sediment load of Huayuankou, Jiahetan and Aishan stations even slightly increased with the runoff decrease. (2) The trends of the sediment load with different time scale showed differences at Luokou and Lijin stations. Although the annually and monthly sediment load were broadly flat, the maximum hourly sediment load showed decrease trend. (3) According to the Hurst coefficients, the trend of sediment and runoff will be continue without taking measures, which proved the necessary of runoff-sediment regulation scheme.

1 Introduction

Climate changes and human activities have great influence on fluvial geomorphology (Zhang et al., 2006). It is of great importance to understand the sediment and runoff changes for reservoir operation, flood mitigation and fluvial management.

The Yellow River is not only the second largest river of China, but also the second largest river of the world in terms of the sediment load (Kong et al., 2015). Heavily sedimentation resulted in that the river bed of lower yellow river (LYR) is higher than the surrounding lands, which posed a huge flood risk of the residents. Since 2002, a Runoff-Sediment Regulation Scheme (WSRS) has been implemented by the Yellow River Conservancy Commission (YRCC) to reduce the siltation in LYR and transport sediment to sea (Xia et al., 2016). Many researchers tried to analyse the influence of the WSRS. Analysis of the sediment and runoff changes of LYR before WSRS is a critical step for these researches.

Many efforts are devoted to explore the sediment and runoff changes of Yellow River. Liu et al. (2011) analysed the annually runoff and sediment changes of the upper, middle and lower Yellow River from 1950 to 2006. The potential impacts of human activities on the runoff and sediment change in the basin were also discussed in the paper. Kong et al. (2015) analysed the influence of runoff–sediment regulation on the channel scouring and delta maintenance in LYR. Most of these studies were focused on the annually and monthly scale. Causing the most intensive transport processes in rivers occur during the passage of a flood wave. Research on the changes of sediment and runoff at daily and hourly scale, especially for the peak value, is even more important to management practice.

In the research, the sediment and runoff changes of seven stations along LYR from 1983 to 2002 are analysed at multiple time scale. The maximum value of monthly, daily and hourly sediment load and runoff observations were also analysed along with the annually value. The research tried to explore the sediment and runoff changes of LYR before WSRS, and built a foundation for understanding its influences.

2 Study area and data description

2.1 Study area description

The Yellow River originates from the northern Bayan Kala Mountain in the Qinghai Plateau and it through the Huanghuaihai Plain, finally into the Bohai Sea. The total length of the mainstream is 5464 km, the surface of the runoff drop is 4480 m, and the basin total area is 795 000 km^2. The downstream reaches in Taohuayu, Zhengzhou City, Henan Province is the Lower Yellow River (LYR), with length of 786 km and catchment area of it is about 23 000 km^2, accounting for 3 % of the total drainage area. The total drop of the downstream reaches is 93.6 m with average gradient of 0.12 ‰. The increase of the amount of water accounted for 3.5 % of the Yellow River water. At present, the estuary of the Yellow River is located at the junction of Bohai Bay and Laizhou Bay, it was a new channel shaped by Qingshuigou artificial diversion in 1976 (Yang et al., 2004).

LYR has a wide upper section and a narrow lower section. The reach upstream of Gaocun, with a typical braided channel pattern, is called the braided reach, with a length of 284 km. And the deposition of upside are seriously, the lower section is more stable. The reach lying approximately between Gaocun and Aishan with a channel pattern transitioning from braided to meandering is called the transitional reach, with a length of 184 km, and the reach roughly between Aishan and Lijin with a stable and well-restricted meandering channel pattern is called the meandering reach, with a length of 272 km. Floods in the LYR mainly come from three areas in the middle region. They are Helongjian, Longsanjian and Sanhuajian area. The middle reaches of the Yellow River formed Huayuankou floods. Floods from the upper reaches constitute the base flow of floods in the LYR. In the recent 40 years, the quantity of sediment which transported by flood to the estuary is averaged about 1 billion tons per year (Miao et al., 2010). Heavy sedimentation makes the surface of the Yellow River rise above 2–3 m on both side, even up to 10 m above sea level, which threaten the safety of the people on the coast (Xu et al., 2004). Heavy sedimentation also results in an obvious shrinkage of the main channel with a decrease of flood discharge capacity (Li et al., 2014). The transverse slope of water surface of parts of the Yellow river has already reached 1–2 ‰, which much larger than the vertical drop of 0.14 ‰ (Liu, 2011).

2.2 Data description

The hourly runoff and sediment data of Huayuankou (HYK), Jiahetan (JHT), Gaocun (GC), Sunkou (SK), Aishan (AS), Luokou (LK), and Lijin (LJ), which are the main hydrological stations along the LYR, are adopted in the study. The data is collected from 1980 to 2002. The annually sediment load and runoff are calculated by summing up the hourly data. The maximum value of the monthly, daily and hourly sediment load and runoff are also obtained by statistics of the hourly data.

3 Methods

In this paper, non-parametric Mann–Kendall (MK method) trend test and Hurst coefficient method are used to analysis the sediment and runoff changes of LYR.

3.1 Mann–Kendall trend test

The Mann–Kendall trend test is a rank correlation test between the ranks of observations and their time sequence (Yue et al., 2002). It has been used by many researchers to analyse the trends of precipitation, runoff, temperature, water quality and other elements in time series. The Mann–Kendall test does not require samples to follow a certain distribution (Zhang et al., 2007). It is suitable for non-normal distribution such as hydrology and meteorology. It is a frequently-used time series test method in hydrological and meteorological research (Guo et al., 2010). This method takes the sequence as a prerequisite. And it requires the sequences to be independent and random.

In the Mann–Kendall test, the hypothesis H_0 is time-series data $(X_1 ... X_n)$, which are n independent samples with the same distribution of random variables. The alternative hypothesis H_1 is a bilateral test. For all k, $j \leq n$, and $k \neq j$, then the distribution of X_k and X_j are not the same. The test statistic S is calculated as follows:

$$S = \sum_{k=1}^{n-1} \sum_{j=k+1}^{n} \operatorname{sgn}(X_j - X_k) \tag{1}$$

$$\operatorname{sgn}(X_j - X_k) = \begin{Bmatrix} +1(X_j - X_k) > 0 \\ 0(X_j - X_k) = 0 \\ -1(X_j - X_k) < 0 \end{Bmatrix} \tag{2}$$

Under the assumption that the data are independent and identically distributed, S is normal distribution. The mean is 0. The variance is $V_{or}(S) = n(n-1)(2n+5)/18$. When $n > 10$, the standard normal system variable is calculated by:

$$\operatorname{Var}(S) = \frac{n(n-1)(2n+5) - \sum_{i=1}^{n} t_i(i-1)(2i+5)}{18} \tag{3}$$

t is the range of any given node, and when $n > 10$, Z_c converges to a normal distribution and is calculated by:

$$Z_c = \begin{cases} \dfrac{S-1}{[\operatorname{Var}(S)]^{1/2}} & S > 0 \\ 0 & S = 0 \\ \dfrac{S+1}{[\operatorname{Var}(S)]^{1/2}} & S < 0 \end{cases} \tag{4}$$

In the bilateral trend test, at a given alpha significance level, if $Z_c \geq Z_{1-\alpha/2}$, rejecting the hypothesis H_0, that is, at the significance level of α, the time series data have obvious upward or downward trend. $\pm Z_{1-\alpha/2}$ is the standard normal distribution $(1-\alpha/2)$ quantiles, and α is the test's significance level. The degree of the trend can be expressed as Kendall inclination β, which is calculated as follows:

$$\beta = \text{Median}\left[\frac{x_i - x_j}{i-j}\right], \quad \forall j < i \tag{5}$$

In the formula, $1<j<i<n$. When $\beta>0$, it means the rising trend, otherwise it means the falling trend.

3.2 Hurst coefficient

Hurst found that the data did not follow the Brownian motion and the normal distribution through the Nile hydrological data, if there was a large amount of water a year, the amount of water in the following year would also tend to be larger. The Hurst coefficient was proposed in 1951. The Hurst coefficient can be used to characterize the persistence or long-term correlation of time series quantitatively.

The Hurst coefficient h means: $h<0.5$, the sequence is anti-persistent, and the trend of the future is opposite to the trend of the present; $h=0.5$, it indicates the sequence is random. And the future trend will not be affected by the present; $h>0.5$, it indicates that the sequence has a positive continuity, the trend of the future change is the same as the current trend; h is closer to 1, it shows that the sequence is more persistent. When h is the closer to 0, the stronger the anti-persistence of sequences.

When $0.5<h<0.719$, the degree of variation is low; when $0.719<h<0.839$ the degree of variation is middle; when $0.839<h<0.924$, the degree of variation is high; when $0.924<h<1.0$, the degree of variation is acute (Xie et al., 2008). The time series with long-term correlation in nature are ubiquitous. The Hurst coefficient has been widely used in hydrology, geochemistry, climate, geology and earthquake and other fields.

4 Results

4.1 Mann–Kendall test analysis

The Mann–Kendall test was used to analyse the changes of the runoff and sediment of the seven hydrological stations, and the Z_c of runoff at different time scale is shown in Table 1.

According to Table 1, Z_c are far greater than the significance level ($\alpha=0.01$), which absolute value is 2.56. We can conclude that the runoff of the seven hydrological stations are shown obviously downward trend at different time scale. We can also observe that the Z_c is larger in smaller time scale from Table 1, which indicates that the runoff peak value reduced more obviously.

The Kendall inclination β of runoff at different time scale is shown in Fig. 1.

Table 1. Z_c of runoff in different time scale.

Name	Hour	Daily	Monthly	annually
HYK	−3.67	−3.54	−3.06	−3.22
JT	−3.96	−3.49	−3.12	−3.17
GC	−3.75	−3.43	−3.12	−2.85
SK	−3.72	−3.43	−3.12	−2.85
AS	−3.96	−3.59	−3.01	−2.85
LK	−3.67	−3.38	−2.96	−2.91
LJ	−3.88	−3.59	−3.01	−2.96

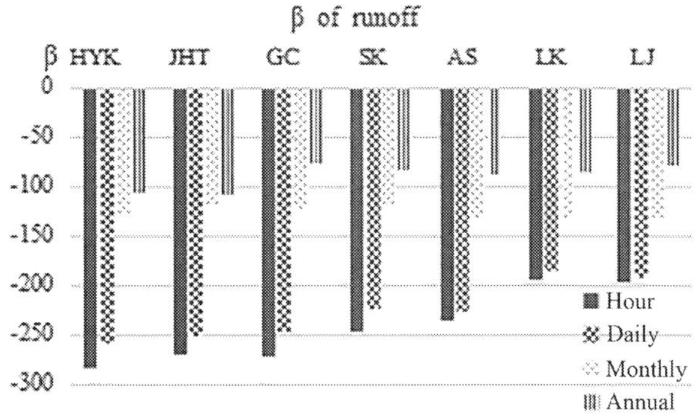

Figure 1. β of runoff in different time scale (HYK: Huayuankou; JHT: Jiahetan; GC: Gaocun; SK: Sunkou; AS: Aishan; LK: Luokou; LJ: Lijin).

It can be observed that the trend of the runoff changes at seven stations in the LYR are relatively consistent. Based on the calculated β value, the β of the runoff are all negative which shows that the trends of runoff are all decreasing. Since the middle of 1980s, runoff has decreased significantly, the largest decrease occurred in 2000. These results are consistent with the former research (Yao et al., 2015). In different time scales, at the same horological station, the smaller the scale is, the more obvious changes are, β tends to zero. Among them, the change of HYK Station is the most obvious one, β is -285 in hour scale, however, β turns out to be -106 in annually scale. At the same time scale, decline trends are gradually reduced from upstream to downstream.

The Z_c of sediment at different time scale is shown in Table 2.

According to the Z_c result (Table 2), $|Z_c|<Z_{1-\alpha/2}$, Z_c are far smaller than the significance level ($\alpha=0.01$), which absolute value is 2.56. We can accept the original hypothesis H_0 at $\alpha=0.01$ significance level, there is no clear upward or downward trend in time series data.

The Kendall inclination β of sediment at different time scale is shown in Fig. 2.

It can be observed that although the trend has not reached a significant level, β is always greater than zero (Fig. 2) which means the amount of sediment in the LYR has slightly increased since the 1980s.

Table 2. Z_c of sediment in different time scale.

Name	Hour	Daily	Monthly	annually
HYK	2.03	0.53	0.48	0.58
JT	2.06	1.11	0.63	0.21
GC	1.34	0.48	0.53	−0.21
SK	1.37	0.42	1.06	−0.05
AS	1.32	0.05	1.27	0.16
LK	−1.11	−0.11	0.69	0.00
LJ	−1.74	−0.42	0.26	−0.11

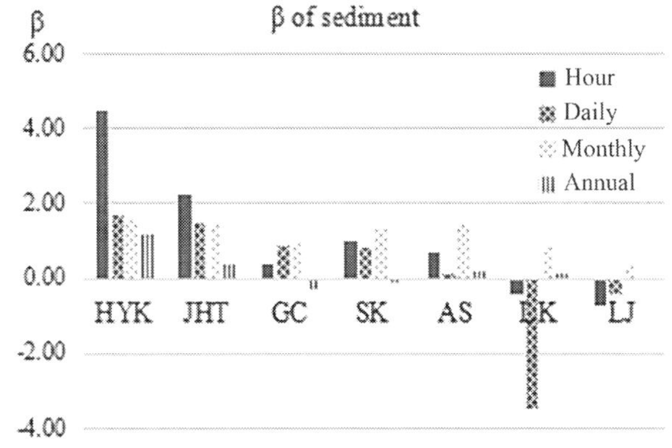

Figure 2. β of sediment in different time scale (HYK: Huayuankou; JHT: Jiahetan; GC: Gaocun; SK: Sunkou; AS: Aishan; LK: Luokou; LJ: Lijin).

The Z_c of upstream hydrological stations are close to the significance level, and the β of HYK and JHT are all greater than 0, which means their trends of sediment are slightly increased. However, the β value of downstream is more complicated, and it does not continue

the trend of upstream, and the β value decreases, even less than 0. This shows that in the flood season, a large amount of sediment are carried by the floodwater from upstream to downstream, resulting an upward trend of sediment in the upstream, However, due to the sedimentation effect, sediment has not been carried into the downstream by floods, which result in there is no trend of increase of sediment in the downstream.

According to the β of sediment in different time scales, we can find that most of the β values of sediment in the LYG over the past 20 years are greater than zero. Take HYK Station as an example, the value of β in each time scale is greater than zero. And the β value decreases in turn from the hour scale to the year scale. At the hour scale, the β of sediment in HYK station is 4.46, however, at the annually scale, the β becomes 1.16. Under different time scales, the scale becomes larger, the β gets smaller, closer to 0.

To sum up, it can be seen from the β value, whether it is in daily scale, monthly scale, or annually scale, the changes

Table 3. Hurst value of runoff.

Name	Hour	Daily	Monthly	annually
HYK	0.86^b	0.88^b	0.87^b	0.82
JHT	0.89^b	0.89^b	0.86^b	0.77
GC	0.87^b	0.89^b	0.89^b	0.78
SK	0.92^b	0.92^b	0.90^b	0.79
AS	0.95^a	0.94^a	0.90^b	0.79
LK	0.95^a	0.93^a	0.91^b	0.79
LJ	0.96^a	0.94^a	0.92^b	0.82

[a] acute degree, [b] high degree
HYK: Huayuankou; JHT: Jiahetan; GC: Gaocun; SK: Sunkou; AS: Aishan; LK:Luokou; LJ: Lijin

of sediment in the LYR in flood season all show the opposite trend of the change of runoff from 1980 to 2003. The trend of the runoff is decreased, the trend of sediment is slightly increased.The trends of different scales are similar. But the smaller the scale is, the more obvious the change will be. The changes of runoff and sediment are the response to the natural factors of climate change and human activities. Relevant studies show that in recent years, except the upstream regions where the trend of precipitation changed over time, in other regions, the annual precipitation series generally show a downward trend and the trend of downstream was more obvious (Xu and Zhang, 2006). In the past 20 years, the runoff into the LYR reduced, this trend is consistent with the trend of precipitation in the Yellow River Basin. But the sediment into the LYR has no obvious increased trend, there is no synchronized phenomenon between runoff sequence and sediment sequence change in the same station of the LYR.

The rapid change of runoff did not result to a response of the sediment. This is due to the factors that affect the change of sediment are multicomponent (Fan and Yang, 2010). The slope cropland in the Yellow River Basin is an important sediment source, rainfall erosion is the main driving force of sediment erosion, so heavy rains in the flood season can lead to the amount of sediment entering the Yellow River increased, while there is a large number of reservoirs built in the Yellow River Basin, which intercept a lot of sediment, has a great effect on reducing the amount of sediment (Xu and Zhang, 2006). Ultimately, the data show that, even the change of the flow rate is obvious increase, sediment deposition in the lower reaches did not show an obvious change.

4.2 Hurst coefficient analysis

Hurst Coefficient revealed the variation on dimension of time series, hence the H values demonstrated whether or not the data sets have changed over time. The Hurst value of runoff at different time scale is shown in Table 3.

According to Hurst results of runoff, at the same hydrological station, the time scales get smaller, the Hurst value of the runoff get higher. Take LJ station as an example,at

Table 4. Hurst value of sediment.

Name	Hour	Daily	Monthly	annually
HYK	0.97[a]	0.99[a]	0.86[b]	0.81
JHT	0.96[a]	0.94[a]	0.81	0.87[b]
GC	0.89	0.89[b]	0.87[b]	0.97[a]
SK	0.89[b]	0.9[b]	0.92[b]	0.97[a]
AS	0.88[b]	0.87[b]	0.97[a]	0.96[a]
LK	0.87[b]	0.87[b]	0.93[a]	0.89[b]
LJ	0.84	0.81	0.89[b]	0.81

[a] acute degree, [b] high degree
HYK: Huayuankou; JHT: Jiahetan; GC: Gaocun; SK: Sunkou; AS: Aishan; LK: Luokou; LJ: Lijin

the hour, daily, monthly and annually scale, Hurst values are 0.96, 0.94, 0.92, and 0.82 respectively. At the same time scale, the Hurst value of downstream hydrological stations are higher than these of the upstream hydrological stations, which could prove that the persistence of upstream is higher than that of downstream.

The Hurst value of sediment at different time scale is shown in Table 4. According to the Hurst results of sediment (Table 4), it can be seen that Hurst results of 7 stations in different time scales are all very high ,all of them are greater than 0.8. The future trend of sediment in the LYR exists state persistence and has similar changing characteristics with past features.

Generally, both runoff and sediment series during the flood season correlated well in long term, and the trend of the future changes would be affected by the current regime. The future trend of runoff and sediment are positively related to the past situation, these results are consistent with the study of Yan et al. (2007) and Ma and Li (2008). Of all runoff variations, the moderate and strong alteration appeared in annually scale, monthly scale respectively. And all the sediment series altered sharply and existed significant jump variation.

5 Discussion and conclusion

Based on the sediment and runoff data from 1980 to 2003, the characteristics of runoff and sediment sequence of 7 hydrological stations in the LYR were analysed in detail. Respectively, in the time scale of hour, daily, monthly and annually, using MK test method and Hurst coefficient method to analysis the degree of variation of runoff and sediment. We draw these conclusions as follows:

1. The runoff and sediment of the LYR during 20 years show the different trend, the runoff of seven hydrological stations was significantly reduced, However, the trends of sediment load in these stations were not obvious.
2. The time scale has big effect on the analysis results. In general, the runoff of the seven stations show more obvious trend in smaller time scale. According to the sediment load, the opposite results can be found in different time scale. According to the LK an LJ station, the annually and monthly sediment load were broadly flat, the maximum hourly sediment load showed decrease trend.
3. The changes of runoff and sediment show high persistence without additional measures, which proved the necessary of WSRS.

Data availability. The data used is the data of China's Yellow River. According to national secret regulations, only for research use and not publicly available.

Competing interests. The authors declare that they have no conflict of interest.

Special issue statement. This article is part of the special issue "Innovative water resources management – understanding and balancing interactions between humankind and nature". It is a result of the 8th International Water Resources Management Conference of ICWRS, Beijing, China, 13–15 June 2018.

Acknowledgements. This research is supported by one research programs: National Key Research and Development Program (during the 13th Five-year Plan) under Grant No. 2016YFC0401309, Ministry of Science and Technology, PRC.

Edited by: Dingzhi Peng
Reviewed by: two anonymous referees

References

Fan, H. and Yang, X. Y.: Interannual change characteristics of runoff and sediment load from the mainstream and major tributaries of the Yellow River, J. Sediment Res., 4, 11–15, 2010.

Guo, X. Q., Liu, M. C., Qian, L., and Zhou, W. X.: Evolution regularity of precipitation in the Shiyang River Basin on Mann-Kendall features, Arid Land Geography, 33, 593–599, 2010.

Kong, D. X., Miao, C. Y., Wu, J. G., Jiang, L., and Duan, Q. Y.: Bi-objective analysis of water–sediment regulation for channel scouring and delta maintenance: A study of the lower Yellow River, Global Planet. Change, 133, 27–34, 2015.

Li, W., Fu, X., Wu, W., and Wu, B.: Study on runoff and sediment process variation in the lower Yellow River, Journal of Hydroelectric Engineering, 33, 108–113, 2014.

Liu, G. W.: Geological Basis for the Management of the Lower Yellow River, Sci. China Earth Sci., 41, 1511–1523, 2011.

Ma, J. H. and Li, X. G.: Predictable time and trend analysis of the sediment time series at some control sections along the Yellow River, J. Sediment. Res., 1, 41–45, 2008.

Miao, C. Y., Ni, J. R., and Borthwick, A. G. L.: Recent changes of water discharge and sediment load in the Yellow River basin, China, Prog. Phys. Geogr., 34, 541–561, 2010.

Xia, D., Yu, Z., Xu, B., Gao, M., Mi, T., Jiang, X., and Yao, P.: Variations of Hydrodynamics and Submarine Groundwater Discharge in the Yellow River Estuary Under the Influence of the Water-Sediment Regulation Scheme, Estuar. Coast., 39, 333–343, 2016.

Xie, P., Lei, H. F., Chen, G. C., and Li, J.: A Spatial and Temporal Variation Analysis Method of Watershed Rainfall Based on Hurst Coefficient, Journal of China Hydrology, 28, 6–10, 2008.

Xu, Q. H., Yang, X. L., Zheng, Z. H., Wang, R. J., Ding, X. Y., and Liang, W. D.: Course Change and Course Realignment of the Yellow River, Geography and Geo-Information Science, 22, 77–80, 2004.

Xu, Z. X. and Zhang, N.: Long-term trend of precipitation in the Yellow River basin during the past 50 years, Geogr. Res., 25, 27–34, 2006.

Yan, A. L., Huang, Q., and Wang, Y. M.: Persistence Measure on Runoff Evolution of the Yellow River, Journal of Arid Land Resources and Environment, 21, 27–30, 2007.

Yang, D., Li, C., Hu, H., Lei, Z., Yang, S., Kusuda, T., Koike, T., and Musiake, K.: Analysis of water resources variability in the Yellow River of China during the last half century using historical data, Water Resour. Res., 40, W065021–W0650212, 2004.

Yao, W. Y., Gao, Y. J., An, C. H., and Jiao, P.: Analysis of trend of runoff and sediment load in upper and middle reaches of Yellow River at century scale, Advances in Science and Technology of Water Resources, 35, 112–120, 2015.

Yue, S., Pilon, P., and Cavadis, G.: Power of the Mann–Kendall and Spearman's rho tests for detecting monotonic trends in hydrological series, J. Hydronaut., 259, 254–271, 2002.

Zhang, J. Y., Zhang, S. L., Wang, J. X., and Li, Y.: Study on runoff trends of the six larger basins in China over the past 50 years, Adv. Water Sci., 18, 230–234, 2007.

Zhang, Q., Xu, C. Y., Becker, S., and Jiang, T.: Sediment and runoff changes in the Yangtze River basin during past 50 years, J. Hydrol., 331, 511–523, 2006.

https://doi.org/10.5194/piahs-379-363-2018

Open Access
PIAHS

Study on the influence on water ecosystem by a lake inflow filtration system

Sushu Wu, Shipei Gao, Xiaodong Hu, Songgan Weng, and Liuchao Guo
Jiangsu Hydraulic Research Institute, Nanjing 210017, China
Correspondence: Wu Sushu (wsspp8@163.com)

Received: 29 December 2017 – Revised: 2 March 2018 – Accepted: 7 March 2018 – Published: 5 June 2018

Abstract. Lakes play important roles in the economic-social sustainable development. However, due to unreasonable development and urbanization in recent years, lake water pollution and ecological degradation have occurred in China. The improvement of the lake inflow water quality is very important. A filtration system includes Gravel filtering system, Aquatic plant area and Ecological bag area was established. The test river is one of the typical lake inflow river and located in the river network in the Chang Dang lake, China. Water quality, zooplankton and phytoplankton in the inflow river were observed form July to mid-August in order to analyze the general process. The average removal rate of NH_3-N (ammonia nitrogen) TN (total nitrogen) and TP (total phosphorus) is 28.33, 25.76 and 24.43 %, respectively. The Pantle-Buck method was used to evaluate the water quality and the B/T index was used to evaluate the nutrition situation. The B/T values were reduced by 20 % and the SI pollution index was reduced by 11.8 %. Therefore, a positive effect on the water's ecological restoration was achieved by the filtration system.

1 Introduction

Lakes play important roles in the nature material circulation and economic-social sustainable development (Jun et al., 2006; Wei et al., 2006; Yang et al., 2008). As a type of important natural resource, it have multiple functions, such as flood diversion and storage, water supply, habitat, biodiversity maintenance, water quality purifying, material production, climate regulation, breeding, shipping, and leisure tourism (Houtven et al., 2014). Lakes are also important guarantees of the ecological environment, by providing the maintenance of regional ecological balances. However, due to unreasonable development of natural resources and urbanization in recent years, lake water pollution and ecological degradation have occurred (Wei and Chuan, 2003). These factors have even affected the drinking water safety, causing a threat to the sustainable use of lake water resources (Hesheng, 2001; Huixia and Jibo, 2011). Lake eutrophication has become the most serious problem of the water environments in economically developed areas. Therefore, effectively governing lake eutrophication has become the focus of the current water environment governance (Cohen et al., 2016; Li et al., 2006; Tian et al., 2006; Wang and Li, 2003).

2 Materials and methods

2.1 Test site

The contaminants are mainly flowing into the lakes through the inflow rivers and the degradation of river flow water pollution is very important to the lake (Du et al., 2017; Naftz, 2017; Jiang et al., 2006; Jiang and Xiao, 2011). The test sites of the ecological filtering system were located in Dongfeng River, one of the typical inflow rivers of Changdang Lake, Jiangsu province, China (Fig. 1). In this study, approximately 150 m of the channel was selected to apply the ecological filtration system.

2.2 Establishment of the ecological filtering system

2.2.1 Gravel filtering system

A gravel filtering system was set up at the entrance of the ecological filtering system (Fig. 2). The gravel with particle sizes of 30 to 70 mm was used to pave a three layer section in the center of the channel, and each tile area was 5 m wide × 20 m long. Submerged plants, such as ceratophyllum demersum and myriophyllum verticillatum, were alternatively planted

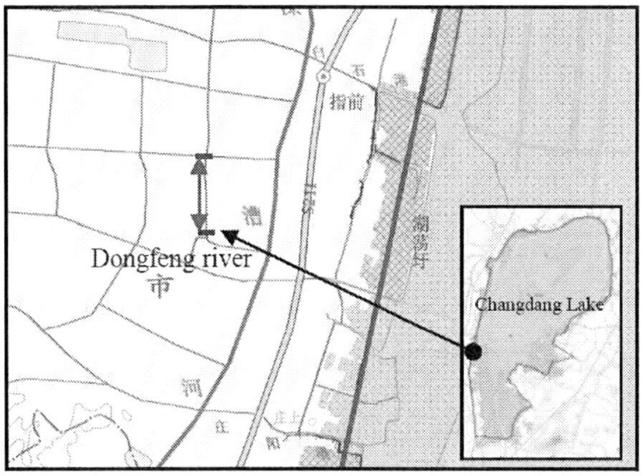

Figure 1. Location of test site, Dongfeng river.

on both sides of gravel areas, and each planting area was 5 m × 2 m, with a total of 8 pieces.

2.2.2 Aquatic plant area

In the downstream of the gravel filtering area, reeds, typha, thalia dealbata, and pontederia cordata were alternatively planted in blocks, with a planting area of 5 m × 3 m for each block, and a block interval of approximately 3 m. At the corners on the land of each block, a clump of iris wilsonii (total of 8 strains) was planted to form a landscape area.

2.2.3 Mire ecological bag area

Mire ecological bags were allocated in the downstream of the aquatic plant area, and a mire bag ecological recovery area was arranged in a block on both sides of the rivers, with each area measuring 7 m × 20 m.

2.3 Sampling and monitoring design

Five sampling points were set in this study's test. These were located in the entrance section center of the ecological filtering system (Point 1); the terminal section center of the gravel ecological filtering system (Point 2); the terminal section center of the plant ecological filtering system (Point 3); the terminal section center of the ecological bag filtering system (Point 4); and the section center 10 m outside the exit of the ecological filtering system (Point 5), respectively. The influences of an ecological filtering system on the quality and ecology of the water were primarily analyzed. The test items included the TP, TN, ammonia nitrogen, permanganate index, flow velocity, water level, and the microscopic examinations of phytoplankton and zooplankton.

The sampling time of this study's test was between 5 June and 5 September, with a weekly sampling frequency schedule, and a total of 13 samplings taken. The total nitrogen,

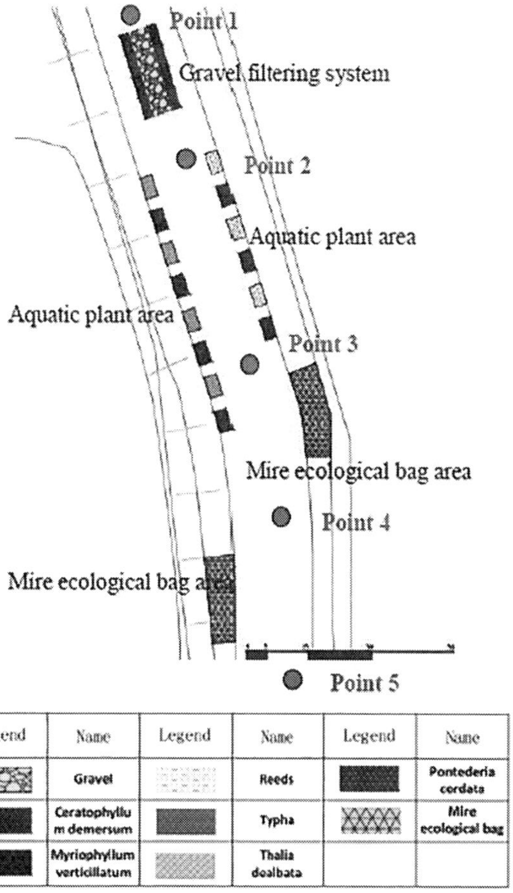

Figure 2. Filtering system composition.

total phosphorus, ammonia nitrogen and permanganate index were measured. The other hydrological data were determined by field instruments. The phytoplankton was collected according to the requirements of the Lake Ecosystem Observation Method for the purposes of classification, identification, and counting during this study's experimental testing process.

2.4 Evaluation of water quality eutrophication

Five parameters were selected, including chlorophyll *a* (chl *a*), total phosphorus (TP), total nitrogen (TN), transparency (SD), and potassium permanganate index (COD_{Mn}). The calculation equation of the comprehensive nutrition index is shown as follows:

$$\mathrm{TLI}(\Sigma) = \sum_{j=1}^{m} W_j \times \mathrm{TLI}(j) \tag{1}$$

In Eq. (1), TLI (Σ) indicates the comprehensive nutrition state index; TLI (j) represents the nutrition state index of the jth parameter; and W_j indicates the relative weight of nutrition state index of the jth parameter.

When chl a is taken as a basic parameter, then the calculation equation of the normalized relative weight of the jth parameter is as shown below:

$$W_j = \frac{r_{ij}^2}{\sum_{j=1}^{m} r_{ij}^2} \tag{2}$$

where r_{ij} is the correlation coefficient of the jth parameter and basic parameter chl a; and m is the number of evaluation parameters.

2.5 Ecological pollution index

Two pollution indexes are used to indicate the water quality by ecological indicators. The Pantle-Buck method was used to evaluate the water quality by phytoplankton and the B/T index was used to evaluate the nutrition situation by species of zooplankton.

A rotifers indicator B/T index (Sladeck., 1983) was used to evaluate the nutrition situation of the water quality as follows: $B/T = B$ (species of brachionus) / T (species of trichocerca).

A Pantle-Buck method (Pantle et al., 1995) was used to calculate the pollution index in order to evaluate the water quality. The equation is shown as follows:

$$\mathrm{SI} = \sum s \times h / \sum h \tag{3}$$

In the equation, SI is the pollution index; s is the algae pollution indication level; and h is the estimated number grade of this type of algae.

3 Results and discussions

3.1 Analysis of the water quality

3.1.1 Changes in the ammonia nitrogen

It can be seen from Fig. 3 that the ecological filtering system in the tested river had a certain intercept of ammonia nitrogen. At the beginning of the ecological filtering system, the intercept rate was quite low, at 18.70 %. However, as time went on, as of 31 July, the intercept rate was found to been increased to 33.03 %. It can be seen that as the self-improvement process of the ecological filtering system progressed, the intercept rate of the ammonia nitrogen was also gradually increased. It was found that, between 7 August and 5 September, the intercept rate of the ammonia nitrogen for the eco · logical filtering system was maintained at a higher level, with a change range of between 29.46 and 33.33 %. A peak was reached on 5 September, and the ecological filtering system was considered to be stable during this period. As viewed from the perspective of the entire process, the average intercept rate of the ammonia nitrogen was

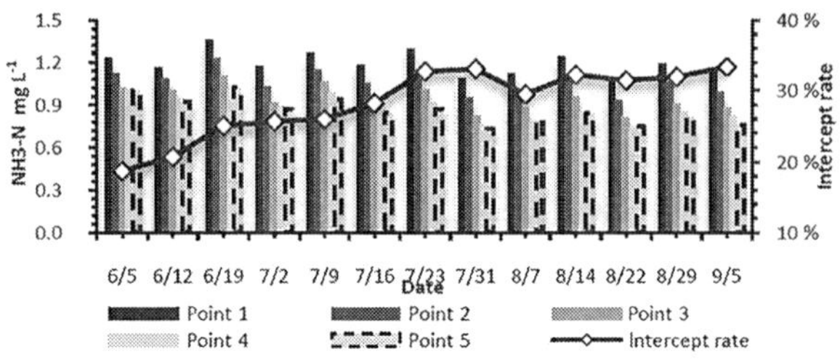

Figure 3. Change laws of the ammonia nitrogen.

28.33 %, and reached 29.90 % when the ecosystem's stability was achieved.

As viewed from the perspective of the change laws of the ammonia nitrogen between the points after the stability of ecological filtering system was achieved, the changes between Point 1 (section center 10 m in front of the ecological filtering system's entrance), and Point 2 (terminal section center of the gravel filtering system) were found to be larger than the other points. These results indicated that the gravel ecological filtering system had the most obvious incept effect on the ammonia nitrogen in the entire system. This may have been due to effects of the biomembrane system which was formed by the gravel, and also that the ceratophyllum demersum and myriophyllum verticillatum growing on the gravel absorbed some of the ammonia nitrogen (Kim et al., 2000). The overall inflow water containing ammonia nitrogen in the tested river was classified as a Grade-4 (1.0–1.5 mg L^{-1}) water quality level. In the inflow water which was filtered by the river's ecological filtering system, the overall outflow water reached a Grade-3 (0.5–1.0 mg L^{-1}) quality level.

3.1.2 Changes in the total nitrogen

It can be seen from Fig. 4 that the change laws of intercept rate of the total nitrogen were roughly the same as that of the ammonia nitrogen in the tested rivers. The minimum was found to be 17.50 % at the beginning of the ecological filtering system (12 June), and the maximum was 31.03 % (31 July). During the period from 5 June to 31 July, the intercept rate of the total nitrogen displayed a gradual rise, which indicated that the ecological filtering system of the river channel was continuously being self-improved. During the period from 31 July to 5 September, the intercept rate of the total nitrogen of the ecological filtering system maintained a higher level, and fluctuated between 27.59 and 31.03 %. These results proved that the ecological filtering system tended to be stable during this time, with an average intercept rate of total nitrogen of approximately 28.42 %, which was higher than that of the entire process (25.76 %). The overall intercept rate showed a process of a rise-volatility-stability pattern.

As viewed from the perspective of the stable ecological filtering system, the changes between Point 1 and Point 2 were determined to be larger. Therefore, it could be concluded that

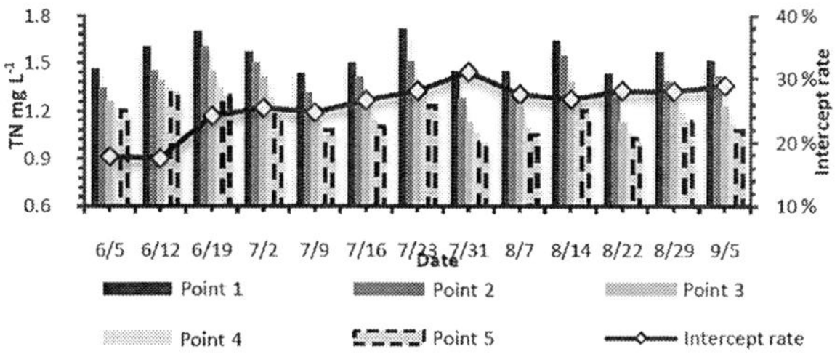

Figure 4. Change laws of the total nitrogen.

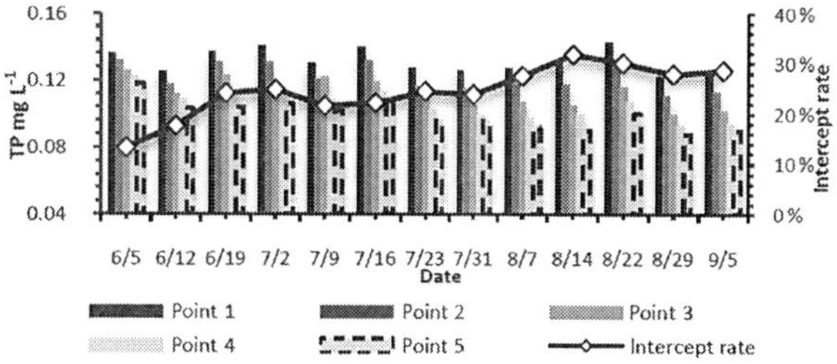

Figure 5. Change laws of the total phosphorus.

the gravel ecosystem played the largest role in the entire system, followed by that between Point 2 and Point 3, and so on, in the plant ecological filtering system.

3.1.3 Changes in the total phosphorus

It can be seen from Fig. 5 that the ecological filtering system had a good effect on the interception of the total phosphorus. The minimum of the intercept rate was determined to be 13.24 %, and the maximum was 31.82 %. The overall intercept rate showed a rising process of volatility, and the average intercept rate of the total phosphorus was 24.43 %. In June, due to the ecological filtering system at the early stage being posterior to the establishment, the interception rate of the total phosphorus was found to be lower only due to the intercept effects of the gravel and the absorption of plants and microorganisms, and thereby the intercept was at a lower level. From July to mid-August, the interception rate of the total phosphorus increased in volatility. In terms of the several samplings taken after mid-August, the interception rate of the total nitrogen presented a low-amplitude decrease, while displaying a higher level as a whole. The specific reasons for these results require further study.

As viewed from the perspective of the monitoring data after the ecosystem achieved stability, the total phosphorus degradation between Point 1 and Point 3 was higher, which further proved the previous conclusions, for example that the gravel ecological filtering and plant filtering systems had good effects on the improvements in the water quality. It was determined that biomembranes were formed on the gravel during this period, when the plants grew vigorously to form a barrier in the water, and intercepted and adsorbed insoluble phosphorus pollutants in the water (Xinyan et al., 2006). Also, a portion of the phosphorus was absorbed by plant growth.

3.1.4 Changes in permanganate index

Figure 6 illustrates that the interception of the permanganate index presented a larger volatility, of which the minimum was 6.90 % at the beginning of ecological filtering system, and the maximum reached 20.00 % on 14 August. In terms of the entire sampling period, the interception rate of the

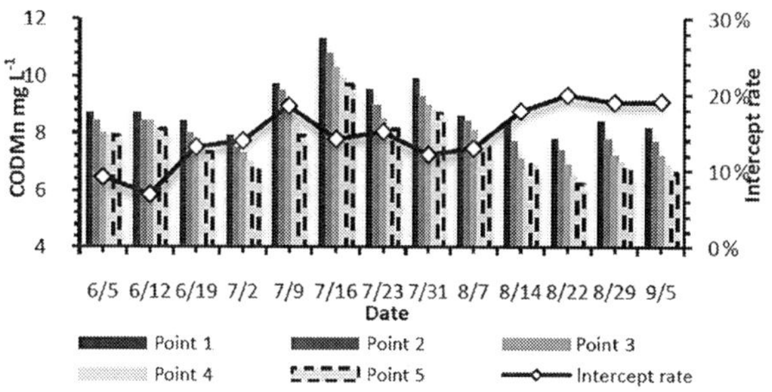

Figure 6. Change laws of the permanganate index.

ecological filtering system still showed a rising trend, and the average interception rate of the permanganate index was 14.77 %.

The ecological filtering system had a lower intercept rate of COD_{Mn}, mainly due to the fact that the COD_{Mn} belonged to the organic matter, was not easily absorbed and used directly by the plants and microbes, and its removal mainly relied on the intercept effect of the biomembranes formed by the gravel and plants. Following the stability of the ecological filter system, the changes between Point 1 and Point 3 were larger, while those between Point 3 and Point 5 were slightly smaller. These results indicated that the gravel and plant ecological systems had a better purification effect on the permanganate index than the regions with the ecological bag filtering system.

3.1.5 Changes in water quality eutrophication

During the three-month duration of the sampling period, the monthly averages of chlorophyll *a*, total phosphorus, total nitrogen, transparency, and permanganate index from June to July, 2015 were selected in order to calculate according to the calculation equations of the comprehensive nutrition state indices, and to obtain the evaluation results of the nutritional status. The results are shown in Table 1.

It can be seen that the levels of chlorophyll *a*, total nitrogen, total phosphorus, and the permanganate index were reduced month by month. Meanwhile, the transparency of the water gradually rose. However, the nutrition state index of water mass gradually declined. Therefore, it was shown that

Table 1. Evaluation results of the outflow water eutrophication in the ecological filtering system.

Month	6	7	8
TIL (chl *a*)	73.1	70.03	63.49
TIL (SD)	79.69	73.9	66.67
TIL (TP)	58.22	57.29	48.07
TIL (TN)	58.58	56.45	55.1
TIL (COD)	58.04	55.75	52.01
Nutrition state index	65.98	65.3	57.57
Nutrition level	Moderate eutrophication	Moderate eutrophication	Slight eutrophication

the existence of an ecological filtering system improved the intercept effect of the pollutants, as well as the inflow water quality.

3.2 Changes in the water ecology

3.2.1 Changes in zooplankton

In this study, from June to September 2015, an ecological sampling was carried out in the tested area. According to the quantitative water sample analysis of the zooplankton, the zooplankton species were not found to be reduced in the tested areas. These species included copepod nauplius and copepodite. There were 57 zooplankton species determined during the microscopic examination from June to September. These were made up of 19 species of protozoa, which accounted for 33.3 % of the total species; 25 species of rotifers, which accounted for 43.9 %; 7 species of cladocera, which accounted for 12.3 %; and 6 species of copepods, which made up 10.5 %.

During the period from June to September, the zooplankton species with higher occurrence frequencies in the tested areas were found to include such native animals as histiobalantium, actinophrys, and difflugia globulosa; rotifers, such as keratella cochlearis, brachionus angularis, polyarthra trigla, polyarthra dolichoptera, and trichocerca capucina; cladocera, such as diaphanosoma brachyurum and bosmina coregoni; and copepods, such as mesocyclops leuckarti. In addition, nauplius and copepodite were also determined to have existed. The zooplankton in the tested areas belonged to a cosmopolitanism.

The two sites which were allocated for ecological environment monitoring in the tested areas included Point 1 in the upstream of the tested area, and Point 5 in the downstream of the tested area. Figure 7 illustrates the microscopic difflugiaoblonga number which was determined through quantitative microscopy of the two sites from June to September. As the test river length is 150 m, we assume the temperatures of Point 1 and Point 5 are same, the changes between these two points caused by temperatures are not discussed here.

The B/T values in the two sites within the tested areas were 2.3 and 1.8 respectively. These results indicated

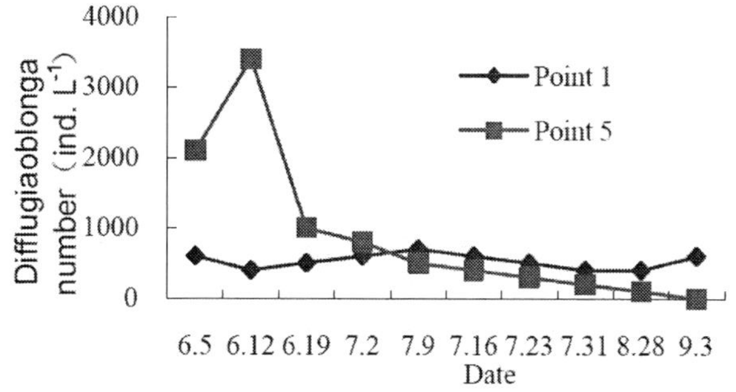

Figure 7. Correlation analysis of the microscopic difflugiaoblonga number.

that Point 1 was a eutrophic water mass, and Point 5 was a mesotrophic water mass. It was shown that the establishment of an ecological filtering system played a certain positive role in improving the downstream ecological environment, with a lesser effect on the upstream ecological environment.

3.2.2 Changes in phytoplancton

Within the two sample points of the Changdang Lake tested area, 55 genera and 96 species of phytoplankton were observed. These were dominated by chlorophyta with 28 genera and 52 species; followed by cyanophyta with 11 genera and 13 species; bacillariophyta with 8 genera and 12 species; euglenophyta with 4 genera and 10 species; chrysophyte with 3 genera and 3 species; pyrrhophyta with 3 genera and 3 species; and cryptophyta with 2 genera and 3 species. The dominant species of phytoplankton were chlorella sp., scenedeamus quadricauda and monoraphidium contortum of chlorophyta; cyclotella sp. and melosira granulata var. angustissima of bacillariophyta; pseudanabaena catenata and phormidium sp. of cyanophyta; and crypomonas erosa of cryptophyta. In addition, chlamydomonas sp., pandorina morum and oocystis parva of chlorophyta; and chroomonas acuta of cryptophyta, microcystis aeruginosa, aphanizomenon issatschenkoi and merismopedia minima of cyanophyta also displayed a higher abundance.

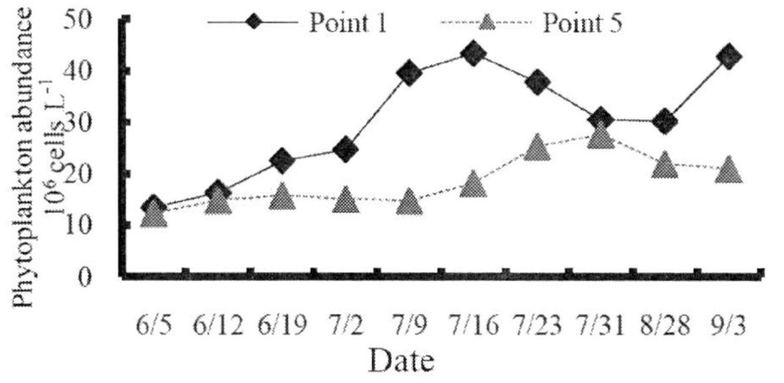

Figure 8. Changes in the phytoplankton abundance in the two sites during the test period.

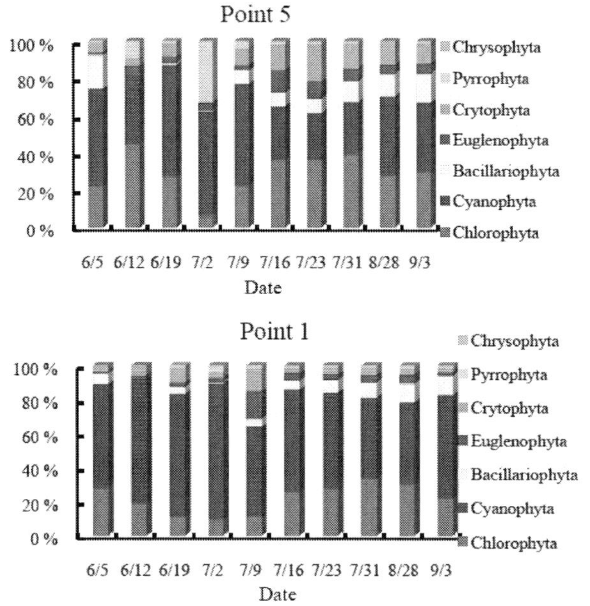

Figure 9. Changes in each phytoplankton species' proportion at Point 1 and Point 5.

It can be seen in Fig. 8 that the phytoplankton abundance was basically equal in two points during the early test period. However, two weeks later, the phytoplankton abundance at Point 5 was lower than that at Point 1 by 29.6 %. As the temperatures rose, the phytoplankton abundance in the two sites showed different degrees of increase. However, those at Point 5 presented a gentle increase trend. The phytoplankton abundance at Point 5 was lower than Point 1 by 50.9 %, which indicated that the aquatic plants at Point 5 caused a certain inhibition in the proliferation of the phytoplankton.

Table 2 shows that the statistical results of the quantities of eutrophic, meso-eutrophic, and mesotrophic indicator species of the phytoplankton at Point 1 were greater than those at Point 5 in the tested areas. Meanwhile, the quantity of oligotrophic indicator species at Point 5 was found to be greater than those at Point 1 by 3 types. As viewed from the perspective of the pollution index, both Point 1 and Point 5 belonged to a meso-eutrophic level. However, the pollution index at Point 5 was reduced by 11.8 % over that at Point 1. These results indicated that the planting of aquatic plant reduced the nutrient concentration in the water mass, inhibiting the quantities and abundances of the eutrophic and meso-eutrophic indicator species to a certain extent, and thereby reducing the water pollution degree. Therefore, a positive effect on the water's ecological restoration was achieved.

Table 2. Nutrition index number and pollution index of the phytoplankton.

	Site	Point 1	Point 5
Nutrition indicator species	Eutrophication	20	16
	Meso-eutrophication	34	31
	Mesotrophication	19	17
	Oligotrohpiccation	3	6
Pollution index		**2.2**	**1.94**

4 Conclusions

In this study, the water quality, zooplankton, and phytoplankton of the ecological filtering system were examined in order to analyze the general degradation process. It was considered that ecological filtering could potentially have a certain improvement effect on the water quality. According to the research regarding the changes in the water's quality and ecology, the ecological filtering system in the Dongfeng River was found to significantly improve the water quality, as well as the inflow water quality. Through the monitoring of the selected indicators, it could be seen that the ecological filtering system of this study displayed purification effects on such pollutants as ammonia nitrogen, total nitrogen, total phosphorus, and permanganate index. The research achievements played an exemplary role in the governance of the inflow water in the Changdang Lake Basin, as well as similar rivers.

Data availability. Data sets are available upon request by contacting the correspondence author.

Competing interests. The authors declare that they have no conflict of interest.

Special issue statement. This article is part of the special issue "Innovative water resources management – understanding and balancing interactions between humankind and nature". It is a result of the 8th International Water Resources Management Conference of ICWRS, Beijing, China, 13–15 June 2018.

Acknowledgements. This work was supported by Foundation Research Project of Jiangsu Province (Grant No., BK20141013) and the Hydraulic Science & Technology Project of Jiangsu Province (2015033).

Edited by: Depeng Zuo
Reviewed by: two anonymous referees

References

Cohen, A. S., Gergurich, E. L., Kraemer, B. M., McGlue, M. M., McIntyre, P. B., Russell, J. M., Simmons, J. D., and Swarzenski, P. W.: Climate warming reduces fish production and benthic habitat in Lake Tanganyika, one of the most biodiverse freshwater ecosystems, P. Natl. Acad. Sci. USA, 113, 9563, https://doi.org/10.1073/pnas.1603237113, 2016.

Chen, H. S.: Restoration project of the ecosystem in Tai lake, J. Resources and Environment in the Yangtze Basin, 10, 173–178, 2001.

Du, C. G., Li, Y. M., Wang, Q., Liu, G., Zheng, Z. B., Mu, M., and Li, Y.: Tempo-spatial dynamics of water quality and its response to river flow in estuary of Taihu Lake based on GOCI imagery, J. Environ. Sci. Poll. Res., 24, 28079, https://doi.org/10.1007/s11356-017-0305-7, 2017.

Hesheng, C.: Restoration project of the ecosystem in Tai lake, J. Resour. Environ. Yangtae Basin, 10, 173–178, 2001.

Houtven, G. V., Mansfield, C., Phaneuf, D. J., Haefen, R., Milstead, B., Kenney, M. A., and Reckhow, K. H.: Combining expert elicitation and stated preference methods to value ecosystem services from improved lake water quality, J. Ecol. Econ., 99, 40–52, 2014.

Huixia, J. and Jibo, X.: Eco-remediation of Polluted Rivers: Status Quo and Progress, J. Enviro. Sci. Technol., 34, 138–143, https://doi.org/10.3969/j.issn.1003-6504.2011.03.032, 2011.

Jiang, H. X. and Xiao, J. B.: Eco-remediation of Polluted Rivers: Status Quo and Progress, Environ. Sci. Technol., 34, 138–143, 2011.

Jiang, X. Y., Zhai, J. P., Huang, L., Rong, N.: Experimental study on enrichment of nitrogen and phosphor in several hydrophytes, J. Environ. Protect. Sci., 6, 13–16, 2006.

Jun, L., Liu, C. Q., Xiao, H. Y., Liu, X. Y., Li, Y., and Wang, S. L.: Summer phytoplankton diversity and water quality evaluation in northern part of Taihu Lake, China, J. Ecol. Environ., 15, 453–456, 2006.

Kim, Y. and Kim, W.: Role of water hyacinths and their roots for reducing algal concentration in the effluent from waste stabilization ponds, J. Water Res., 34, 3285–3294, 2000.

Li, J., Liu, C., Xiao, H., Liu, X., and Li, Y. Y.: Summer phytoplankton diversity and water quality evaluation in northern part of Taihu Lake, China, J. Ecol. Environ., 15, 453–456, 2006.

Naftz, D.: Inputs and Internal Cycling of Nitrogen to a Causeway Influenced, Hypersaline Lake, Great Salt Lake, Utah, USA, J. Aquat. Geochem., 23, 199–216, 2017.

Pantle, R. and Buck, H.: Die biologische uberwachung der gewaser und die darstellung ergebinsse, J. Gas und Wasserfach, 96, 604, 1995.

Sladeck, V.: Rotifers as indicators of water quality, J. Hydrobiol., 100, 169–201, 1983.

Tian, W. J., Hao, F. H., Wang, C., and Zhai, J. B.: Ammonia-nitrogen degradation in the typical streams entering Taihu Lake, J. Ecol. Environ., 15, 1138–1141, 2006.

Wang, W. and Li, C. Q.: Stream Corridor and Ecological Rehabilitation, J. Water Resources And Hydropower Engineering, 359, 56–58, 2003.

Wei, T. J., Hao, F. H., Wang, C., and Zhai, J. B.: Ammonia-nitrogen degradation in the typical streams entering Taihu Lake, J. Ecol. Environ., 15, 1138–1141, 2006.

Wei, W. and Chuan, Q. L., Stream Corridor and Ecological Rehabilitation, J. Water Resour. Hydropower Eng., 359, 56–58, 2003.

Xinyan, J., Zhai, J. P., and Huang, L.: Experimental Study on Enrichment of Nitrogen and Phosphor in Several Hydrophytes, J. Environ. Protect. Sci., 6, 13–16, https://doi.org/10.3969/j.issn.1004-6216.2006.06.005, 2006.

Yang, X., Yu, S., Ma, Y., Zhao, Y., Yan, X. J., and Xiu, C. H.: Study on the influence of wetland media on the purifying the micro-polluted raw water, J. Am. Sci., 4, 95–100, 2008.

https://doi.org/10.5194/piahs-379-371-2018

Open Access
PIAHS

Quantitative analysis on sensitive factors of runoff change in Fenhe watershed based on integration approach

Deng Wang, Shengqi Jian, Zening Wu, Zhaoxi Zhang, and Caihong Hu

College of Water Conservancy & Environmental, Zhengzhou University. Science road 100, Zhengzhou, China

Correspondence: Caihong Hu (hucaihong@zzu.edu.cn)

Received: 30 December 2017 – Revised: 7 April 2018 – Accepted: 14 April 2018 – Published: 5 June 2018

Abstract. The runoff of the Fenhe River flowed into the Yellow River (RRY) is reducing significantly due to the influence of climate change and human activities. It is generating bad situation of shortage of water resources and led to the deterioration of ecological environment of Shanxi Province. At the same time, the reduction in RRY causes the runoff reduction in Yellow River and exacerbated the water resources shortage of the middle area of the Yellow River. Therefore, it is important to alleviate water shortage and develop the soil and water conservation measurements and regional water policy by analyzing the influence of human activities and climate change on the RRY.

The existing study quantified the reduction in amount of RRY which caused by human activities and climate change using statistical methods and watershed hydrological model. The main results of the study were as follow:

1. Using hydrological variation diagnosis system, the variation characteristics of long time series of measured annual runoff were analyzed in Hejin station that is the Fenhe River control station. The results showed that the runoff of Fenhe River run into Yellow River declined year by year, in 1971, fell the most obviously.
2. The impact of LUCC on runoff was calculated using the method of area ratio in the Fenhe River basin. Human activities were major factor in the reduction of RRY than the climate change, contributed 83.09 % of the total reduction in RRY, Groundwater exploitation gave the greatest contribution to the decrease in RRY in the scope of several kinds of human activity (30.09 %), followed by coal mining (26.03 %), climate changed contributed 19.17 % of the total reduction of RRY, and the decrease of precipitation contributed 20.81 %. But the variation of air temperature and wind speed would result in the increase of the amount of RRY.

1 Introduction

The Yellow River Basin is the origin of Chinese civilization, is the second largest river after the Yangtze River, but the Yellow River Basin water shortage is extremely short, accounting for about 2 % of China's runoff is about 12 % of China's population production and living water, The basin's gross domestic product (GDP) accounts for 9 % of the country (He et al., 2013). Zhang Xuecheng and other pointed out that the trend of runoff from Fen River Basin to the yellow river was significantly reduced, is one of the major causes of runoff reduction in the Yellow River (Zhang and Wang, 2001). However, after entering the 1970s, the measured flow of the Yellow River and its main tributaries showed a significant reduction, while the main stream and the main tributaries were repeatedly broken up (Zhang et al., 2000; Mu et al., 2007; Jia et al., 2007; Xu et al., 2012). Fenhe River is the second largest tributary of the Weihe River in the middle reaches of the Yellow River. It has an important influence on the process of water and sediment evolution in the middle and lower reaches of the Yellow River. At the same time, the Fen River is the largest river in Shanxi Province. Its basin area is 39 826 km^2, the basin area of 1738.7 million km^2 of arable land, accounting for 30 % of the total cultivated area of the province. Fenhe River is the main water source in Shanxi Province. The average annual runoff of river is 2.067 billion m^3, the amount of groundwater resources is 2.50 billion m^3, and the total amount of water resources is 3.358 billion m^3, accounting for 27.2 % of the total water resources in Shanxi Province. However, under the influence of climate change and human activities, the terrestrial water circulation pattern and water resources situation in the Fenhe River Basin have changed significantly, and a series of adverse consequences have been caused, such as the annual runoff of the Fenhe River and the decrease of the groundwater level. Especially the Fen River into the Yellow River runoff dropped, as the second largest tributary of the middle reaches of the Yellow River, Fen River into the Yellow River is the middle reaches of the Yellow River water resources, an important source of supply, Fen River runoff dropped, not only increased the Fen River Basin water Resources supply and demand contradiction, but also to the middle and lower reaches of the Yellow River can be a significant reduction in water resources, increased between the provinces and the water sector water distribution work difficult. In addition, the decrease of runoff of the Yellow River caused by the decrease of the runoff of the Fenjiang River has a significant impact on the river bed evolution in the middle and lower reaches of the Yellow River, which has caused the Yellow River main channel to shrink further and thus adversely affect the river health of the Yellow River. Yuan Zhihua's study shows that (1950–2005) into the the Yellow River Fenhe River runoff reduction, human activities are the main factors leading to the decrease of runoff in the Fenhe River Basin, runoff reduction accounted for 75.9 % of the total reduction of runoff (Yuan et al., 2008); Wang Guoqing on the impact of environmental change on Runoff in the Fenhe River Basin using SIMHYD conceptual rainfall runoff model, the results of the study indicate that on average, 1970–1999 years, the impact of climate factors and human activities on runoff accounted for 35.19 % of the total reduction of runoff and 64.11 %, human activities are the main factors of Fenhe River Runoff reduction (Wang et al., 2008).

There are four methods to evaluate the hydrological effects of environmental changes: similar watershed comparative test, hydrological model simulation, sensitivity analysis based on hydrothermal coupled equilibrium equation and statistical regression analysis (Li et al., 2009). Similar watershed comparison test method is to select two experimental conditions with similar climatic conditions, topography, area and other characteristics. Under the same gradient and climatic conditions, change the climatic conditions of another experimental basin or the watershed characteristics, the differences between hydrological variables in these two experimental watersheds analyze the impact of human activities on hydrological processes in the basin. However, this method has limitations in the watershed scale for high cost and difficult to find similar watershed (Zhao et al., 2010). Statistical regression analysis is based on the analysis of hydrological and meteorological data using statistical principles. Through the analysis of the factors influencing the process of runoff formation, the main influencing factors of runoff variation are established, and the regression equation of runoff and main influencing factors is established. The advantages of the method are relatively easy to implement, but do not reflect the nonlinear relationship between rainfall and runoff (Ahn and Merwade, 2014), and the accuracy of the results is not high (Jiang et al., 2011). Sensitivity analysis is mainly used to quantitatively analyze the impact of climate change on runoff, which evaluates the impact of climate change on runoff by determining the degree of change in the runoff of the basin due to changes in the climate factors of the unit, and is widely used in quantitative analysis of environmental change Runoff effects, and achieved satisfactory results (Wang et al., 2013; Elfert and Bornmann, 2010; Awan and Ismaeel, 2014; Ahn and Merwade, 2014; Zeng et al., 2014). Based on the understanding of hydrological phenomena, the hydrological model of watershed through establishes a hydrological model by mathematical method to analyzes its causes and the relationship between hydrological elements, and. With the development of computer and GIS technology, hydrological model method has become a powerful tool to study hydrological problems (Awan and Ismaeel, 2014; Ahn and Merwade, 2014; Li et al., 2009). Although the impact of climate change and human activities on runoff, domestic and foreign scholars have done a lot of research and determine the very rich results, there are still some problems: Most research of the impact of the quantitative analysis of climate change and human activities on runoff research, only take the factors that impact runoff changes as climate factors and human factors, and on this basis, quantitative analysis the impact of both changes in runoff, the impact of human activities to the next level, a detailed division of the various climate and human activities on runoff is not included.

According to the available data, the main factors influencing the runoff in the Fenhe River Basin are: the annual rainfall of the Fenhe River Basin is reduced by 38 800 m^3 per year from 1956 to 2012. The potential evapotranspiration of the Fenhe River Basin shows a decreasing trend. The average annual growth rate of surface water in the Fenhe River Basin increased to 18.06 million m^3 per year in between 1980 and 2012, the water supply in the Fenhe River Basin accounted for The total amount of water supply was 63 %; With the development of economy, the amount of coal mining increased year by year, especially in the 20th century After the age of 1991, the province's total output of 288.57 million t, to 1993 exceeded 300 million t, the highest in 1996 to 34.9460 million t, the subsequent mining volume gradually reduced in 2000 to 246 114 500 t; so far, Fen River (61), including the large and medium-sized reservoirs 17, the total control basin area of 14 736 km^2; small (I) and small (II) type of reservoirs 165, the total storage capacity of 17.2016 billion m^3, Total control basin area 15 317 km^2; as of the end of 2006, Fenhe River Basin soil and water conservation measures accumulated area of 22 230 km^2, of which terraced fields (including beach and dry plains) 4970 km^2, water conservancy and forestry forest 11 800 km^2, economic forest 1595 km^2, grassland 830 km^2, dam 58 800 km^2, and another closed hillside area of 245 000 km^2. The exploitation of surface water affects the runoff of the surface water by the increase or decrease of surface water consumption. The groundwater exploitation affects the runoff through the groundwater consumption and the amount of submerged water. The water conservancy project affects the regional runoff through the increase or decrease of the regional evaporation and the soil and water conservation measures change the land use type area to affect runoff. Therefore, based on the principle of water balance, this paper constructs a unified framework for the reduction of gravity flow into the Yellow River. In this framework, different methods are used to quantitatively analyze the changes of rainfall, the change of potential evapotranspiration, Groundwater exploitation, coal mining, water conservancy projects and soil and water conservation measures on the Fen River into the yellow runoff rate of contribution. The purpose of this paper is as follows: (1) Analyze the trend of runoff change in Fenhe River Basin during 1956–2012 and analyze the factors influencing the change. (2) Quantitative analysis of the contribution of climate change and river basin attribute change to runoff changes.

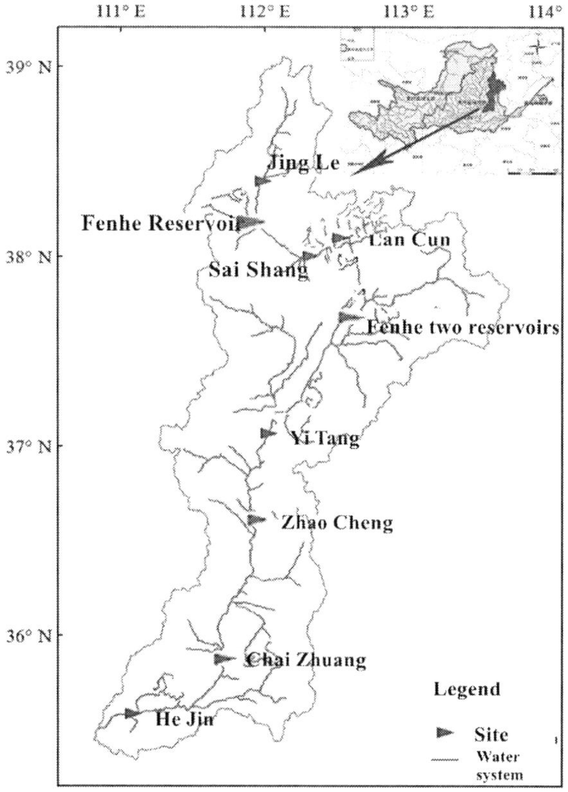

Figure 1. Fenhe River Basin.

2 Materials and methods

2.1 Overview of the study area

Fenhe River Basin is located at 110.5–113.5° E, 35.3–39.0° N, the terrain is north high, south low, longitudinal watershed topography is north-south length 413 km, 188 km long from east to west. The elevation of the basin continues to decline, and the elevation of the mountains on both sides of the east and west increases, resulting in a disparity in the height of the ground, and the transitional zone between the valley basin and the mountain is the loess plateau. Watershed landforms can be divided into three types of area: loess hilly and gully region, rocky mountainous area and river channel area. The total area of the river basin is 39 471 square kilometers (Fig. 1), accounting for 5.2 % of the Yellow River basin area, which is the second largest tributary of the Yellow River and the main source of the runoff of the Yellow River (Du and Zhang, 2016).

Fenhe River Basin belongs to mid latitude continental monsoon climate, the climate is characterized by four distinctive seasons: short day and little rain and more sand wind in spring (Liang et al., 2010), high temperature and more rainstorm in summer, the difference in start time is more, mild and sunny in autumn, scarce precipitation and cold and dry in winter. The basin is rich in light and heat resources, lack of water resources. The average temperature of Fenhe River Basin changed from 13 to 4 °C between south and north, and frost-free period was 80–205 days, which also showed a decreasing trend from south to north.

The nine hydrological stations on Fenhe River, from upstream to downstream be: Jingle, Fenhe Reservoir, Zhaishang, Lanchun, Erba, Yitang, Zhaocheng, Chaizhuang and Hejin. Among them, the Hejin Station is the Fenhe entering Yellow River control station, whose control basin area is 38 728 square kilometers, accounting for 98 % of the total area of Fenhe River basin.

The average precipitation in the Fenhe River Basin (1956–2012 series) is 503 mm, and the distribution of precipitation is decreasing from the southeast to the northwest, but it is closely related to the terrain elevation. The warm and humid airflow brought by the summer monsoon is the main source of water vapor for precipitation. Precipitation from June to September accounts for more than 70 % of the total annual precipitation.

2.2 Data selection

This study collected the annual runoff from 1956 to 2012 in the Fen River Basin into the Yellow Control Station (Hejin station), and the rainfall data from 1956 to 2012 in the Fenhe River Basin. The meteorological data of the 10 major weather stations in the basin (Hejin, Jiexiu, Lok, Linfen, Qixian, Qingxu, Taigu, Taiyuan and Yicheng) from 1956 to 2012 including mean daily wind speed, daily average temperature, daily maximum temperature, daily minimum temperature, sunshine hours, relative humidity and daily mean vapor pressure. At the same time, the data of surface water consumption, the results of soil and water conservation measures in Fenhe River Basin, the data of human water consumption, the data of water conservancy construction and the data of coal mining were collected. Runoff data and rainfall data were collected from the Hydrological Bureau of the Ministry of water resources compiled by the hydrological data series of the Yellow River. Meteorological data from China Meteorological Data network.

In addition, this study also collected the Fenhe River Basin groundwater depth data in Fenhe River Basin in 2001–2012, which include a total of 199 groundwater depth monitoring station (Fig. 2), It can be seen from the figure that the Fenhe River Basin groundwater depth monitoring stations in this study are concentrated in Taiyuan and Linfen basins. The soil data of the Fen River basin in this paper are derived from the HSWD soil database (Fig. 3) constructed by the International Food and Agriculture Organization (FAO), and the soil classification is FAO-90. And the obtain data is correct by the reference to "Shanxi Soil" edited by Liu Yaosong.

2.3 Methods

Based on the analysis of the characteristics of measured annual runoff of Hejin River in Hejin River Basin, the main climatic and human influencing factors of Fenhe River entering the Yellow River runoff were determined. Based on the annual variation of annual runoff of Hejin River. The measured annual runoff of Hejin station is divided into two periods: "basis period" and "change period", the difference between the mean measured annual runoff of "base period" and "change period" is the total reduction, The total reduction in the Fen River into the Yellow River runoff includes the reductions caused by climate change and the reductions caused by major human activities. On the basis of estimating the total reduction of the Yellow River runoff from the Fen River, a unified framework for the reduction of the diameter of the Fenhe River into the Yellow River is established. In this framework, the separation and evaluation of the effects of climate change and main human activities on runoff are as follows (Wang et al., 2008):

$$\Delta W = W_{\mathrm{hc}} - W_{\mathrm{b}} \quad (1)$$

$$\Delta W = \Delta W_{\mathrm{c}} + \Delta W_{\mathrm{hum}} \quad (2)$$

$$\Delta W_{\mathrm{c}} = \Delta W_{\mathrm{p}} + \Delta W_{E_0} \quad (3)$$

$$\Delta W_{\mathrm{hum}} = \Delta W_{\mathrm{wc}} + \Delta W_{\mathrm{gw}} + \Delta W_{\mathrm{wcp}} + \Delta W_{\mathrm{coal}} + \Delta W_{\mathrm{LUCC}} \quad (4)$$

$$\eta_{\mathrm{p}} = \frac{\Delta W_{\mathrm{p}}}{\Delta W} \times 100\,\% \quad (5)$$

$$\eta_{E_0} = \frac{\Delta W_{E_0}}{\Delta W} \times 100\,\% \quad (6)$$

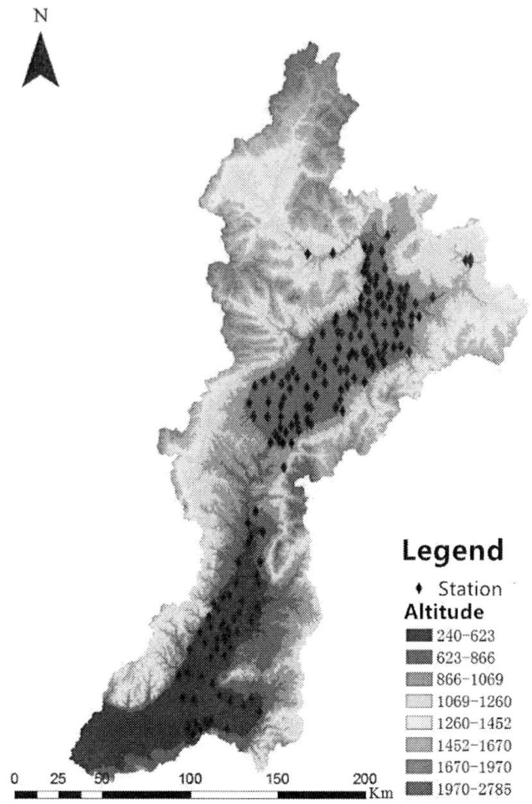

Figure 2. Distribution of Groundwater Level and Buried Survey in Fenhe River Basin.

$$\eta_{\mathrm{c}} = \eta_{\mathrm{p}} + \eta_{E_0} \quad (7)$$

$$\eta_{\mathrm{wc}} = \frac{\Delta W_{\mathrm{wc}}}{\Delta W} \times 100\,\% \quad (8)$$

$$\eta_{\mathrm{gw}} = \frac{\Delta W_{\mathrm{gw}}}{\Delta W} \times 100\,\% \quad (9)$$

$$\eta_{\mathrm{wcp}} = \frac{\Delta W_{\mathrm{wcp}}}{\Delta W} \times 100\,\% \quad (10)$$

$$\eta_{\mathrm{coal}} = \frac{\Delta W_{\mathrm{coal}}}{\Delta W} \times 100\,\% \quad (11)$$

$$\eta_{\mathrm{LUCC}} = \frac{\Delta W_{\mathrm{LUCC}}}{\Delta W} \times 100\,\% \quad (12)$$

$$\eta_{\mathrm{hum}} = \eta_{\mathrm{wc}} + \eta_{\mathrm{gw}} + \eta_{\mathrm{wcp}} + \eta_{\mathrm{coal}} + \eta_{\mathrm{LUCC}} \quad (13)$$

In the formula, ΔW is the total reduction of the Yellow River runoff into the Yellow River, the unit is $10^8\,\mathrm{m}^3$; W_{hc} is the measured annual flow rate of change, the unit is $10^8\,\mathrm{m}^3$; W_{b} is the base year measured annual runoff, the unit is $10^8\,\mathrm{m}^3$; ΔW_{c} is the amount of runoff reduction caused by climate change, the unit is $10^8\,\mathrm{m}^3$; ΔW_{hum} is the amount of runoff reduction caused by human activity, the unit is $10^8\,\mathrm{m}^3$; ΔW_{p} is the amount of runoff reduction caused by reduced rainfall, the unit is $10^8\,\mathrm{m}^3$; ΔW_{E_0} is the amount of runoff reduction caused by the decrease in potential evapotranspiration,

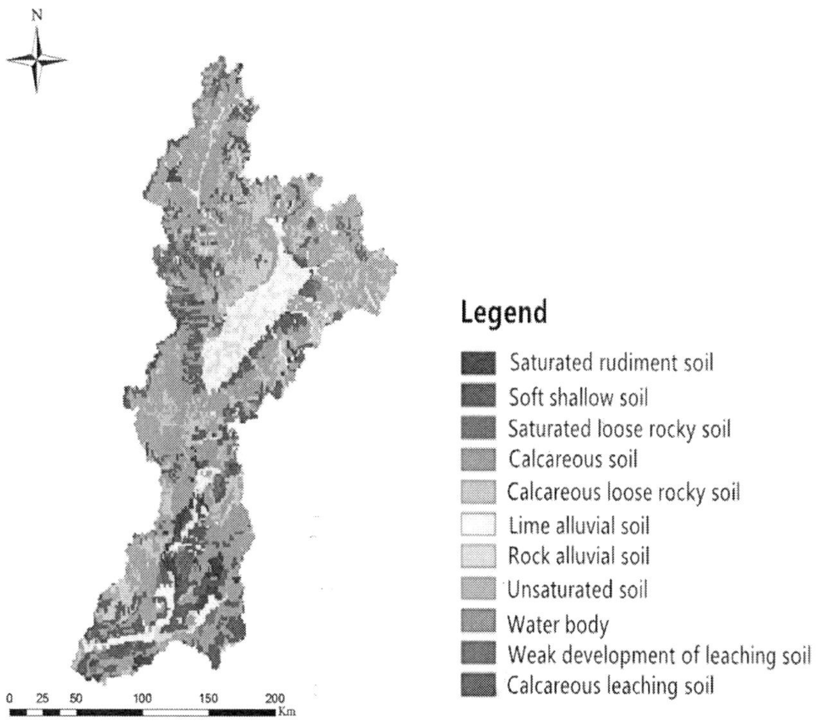

Figure 3. Distribution of Soil Types in Fenhe River Basin.

the unit is 10^8 m^3; ΔW_{wc} is the amount of runoff reduction caused by the increase in surface water consumption, the unit is 10^8 m^3; ΔW_{gw} is the amount of runoff reduction caused by groundwater exploitation, the unit is 10^8 m^3; ΔW_{wcp} is the amount of runoff reduction caused by the new evaporation of water conservancy projects, the unit is 10^8 m^3; ΔW_{coal} is the amount of runoff reduction caused by coal mining, the unit is 10^8 m^3; ΔW_{LUCC} is the amount of runoff reduction caused by land use change, the unit is 10^8 m^3; η_{p}, η_{E_0}, η_{wc}, η_{gw}, η_{wcp}, η_{coal}, η_{LUCC}, η_{c}, η_{hum} are the contribution rate of rainfall, potential evapotranspiration, surface water consumption, groundwater exploitation, water conservancy projects, coal mining, land use, climate change and human activities to the Yellow River.

2.3.1 Runoff sequence diagnosis method

The hydrological variation diagnostic system takes into account both the trend and the jumping of two variants (Xie et al., 2010). The basic idea is to use a variety of mutation test methods to diagnose the hydrological sequence in detail on the basis of the possibility of preliminary test, and to obtain the efficiency coefficient of the trend and jump diagnosis conclusion, then the degree of fitting of the hydrological sequence to the jumping component or trend component is evaluated according to the efficiency coefficient. According to the actual hydrological investigation and analysis, the variant forms and conclusions were verified, and the final diagnosis results were obtained. The system can solve the problem that the test result of the single method is not reasonable and the results of various methods are inconsistent.

2.3.2 Calculation method of potential evapotranspiration

Based on the 10 weather station data, the basin's potential evapotranspiration from 1956 to 2012 was calculated using the Penman-Monteith formula recommended by the International Food and Agriculture Organization (FAO) (Pereira et al., 2015), as shown in the following equation. At the same time collected and compiled from 1956 to 2012 reservoir, water supply, land use and coal mine data.

$$\text{ET}_0 = \frac{0.408\Delta(R_{\text{n}} - G) + \gamma \frac{900}{T_{\text{mean}}+273} U_2 \cdot (\text{VP}_{\text{s}} - \text{VP})}{\Delta + \gamma(1 + 0.34U_2)} \quad (14)$$

where ET_0 is the potential evaporation, mm d^{-1}; R_{n} is the net radiation, MJ/(m^2 d); G is the soil heat flux MJ/(m^2 d); γ is the wet and dry constant, kPa/°; Δ is the saturation water vapor pressure curve slope, kPa/°; U_2 is 2 m high wind speed, m s^{-1}; VPs are the average saturated vapor pressure, kPa; VP is the actual vapor pressure kPa; T_{mean} is the average temperature, °. Net radiation is the sum of solar shortwave radiation

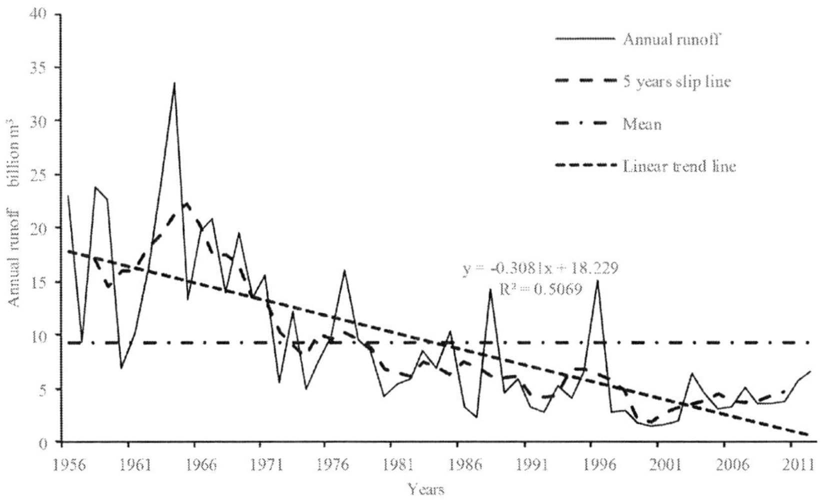

Figure 4. Trend of annual average runoff in Hejin station.

Table 1. The results of runoff diagnosis in Hejin station.

Project	Critical result	Project	Critical result
Hurst coefficient	0.914	Jumping comprehensiveness	4(+)
Overall degree of variation	Strong variation	Trends comprehensiveness	3(+)
Trend of variation	Trend strong variation	Jump efficiency coefficient	52.22
Jumping point	1971	Trend efficiency coefficient	36.72
Jumping comprehensive weight	0.69	Diagnosis conclusion	1971(+) ↓

Notes: 1971(+) ↓ indicates a significant decrease in the sequence in 1971.

and ground longwave radiation, where solar radiation can be estimated as follows:

$$R_s = \left(a_s + b_s \frac{n}{N}\right) R_a \tag{15}$$

where R_s is the solar radiation ($W\,m^{-2}$); R_a is the solar radiation of the top floor of the atmosphere ($W\,m^{-2}$); a_s and b_s are the parameters, where $a_s = 0.25$, $b_s = 0.5$.

2.3.3 Quantitative analysis method of influence of climate change on Runoff

The change of annual rainfall and potential evaporation will lead to the change of water balance, Koster (Koster et al., 1999) and Milly (Milly and Dunne, 2002) give the calculation formula of the change of runoff caused by the change of annual rainfall and potential evaporation in the basin:

$$\Delta W_c = \varepsilon_P \frac{\overline{R}}{\overline{P}} \Delta P + \varepsilon_{E_0} \frac{\overline{R}}{\overline{E_0}} \Delta E_0 \tag{16}$$

$$\Delta W_P = \varepsilon_P \frac{\overline{R}}{\overline{P}} \Delta P \tag{17}$$

$$\Delta W_{E_0} = \varepsilon_{E_0} \frac{\overline{R}}{\overline{E_0}} \Delta E_0 \tag{18}$$

where $\overline{R}$ is the base year multi-year average runoff depth, mm; $\overline{P}$ is the average annual rainfall for the base period, mm; $\overline{E_0}$ is the basis of the average annual number of potential evaporation, mm; ΔP is the difference between the average rainfall of the base period and the change period, mm; ΔE_0 is the difference between the average potential evaporation of the base period and the change period, mm; $\varepsilon_P$is the precipitation elasticity coefficient, ε_{E_0} is the potential evapotranspiration elasticity coefficient. When the climate elasticity coefficient is used to quantitatively calculate the effect of climate change on runoff, the calculation method of the two elastic elastic coefficients has a great influence on the result and requires the underlying surface parameter ω. Sankarasubramanian (Sankarasubramanian and Vogel, 2001) gives a nonparametric method for calculating the elasticity of climate:

$$\varepsilon_X = \frac{\overline{X}}{\overline{R}} \cdot \frac{\left(X_i - \overline{X}\right)\left(R_i - \overline{R}\right)}{\sum \left(X_i - \overline{X}\right)^2} \tag{19}$$

where X is the climatic factor. Based on the results of comparison of various climatic elasticity coefficients, it is found that the nonparametric calculation method of climate elasticity coefficient is reliable.

Table 2. Calculation results of impacts of climate change on inlet runoff in Fenhe River Basin.

Time	$\overline{R}$ (mm)	$\overline{P}$ (mm)	$\overline{E_0}$ (mm)	ε_P	ε_{E_0}	ΔP (mm)	ΔE_0 (mm)	ΔW_P (mm)	ΔW_{E_0} (mm)
1956–1971	45	541	969	1.491	−0.331	−55	−35	−6.89	0.55
1980–2012	13	486	933						

Table 3. The relative contribution of the factors affecting the natural runoff in the Fenhe River Basin from 1980 to 2012.

Influencing factors		Influence quantity (Billion m^3)	Contribution rate (%)
Climate	Rainfall	−2.67	−20.81
	Temperature, wind speed and so on	0.21	1.64
	Subtotal	−2.46	−19.17
Human	Surfacewater exploitation	−2.32	−19.17
	Groundwater exploitation	−3.86	−30.09
	Water Conservancy Project	−0.09	−0.07
	Land use	−1.05	−8.18
	Coal mining	−3.34	−26.03
	Subtotal	−10.66	−83.09

2.3.4 Quantitative Analysis Methods for the Impact of Human Activities on Runoff

The exploitation of surface water affects the runoff of the surface water by the increasing or decreasing of surface water consumption. The groundwater exploitation affects the runoff through the groundwater consumption and the amount of submerged water. The water conservancy project affects the regional runoff and the soil and water conservation measures through the increase or decrease in the regional evaporation Change the land use type area to affect runoff. Therefore, this paper uses the sub-calculation method to calculate the impact of various human activities on the Fen River runoff.

3 Results

3.1 Runoff trends and variation analysis

Showing in Fig. 4 is the Hejin station 1956–2012 runoff trends and 5-year sliding map. It can be seen that the annual runoff from 1956 to 2012 is 994 million m^3, and the runoff is obviously decreasing. Before the 1980s, the runoff was above the mean line, and most of the runoff was below the mean in the 1980s, only in 1988 and 1996, the runoff was greater than the mean.

On the basis of preliminary diagnosis, three flow diagnosis methods of three trend diagnosis methods and 11 jump diagnosis methods were used to determine the flow rate of Hejin station. The first significant $\alpha = 0.05$ and the second significant $\beta = 0.01$ were analyzed. The diagnostic results are shown in Table 1. It can be seen from Table 1 that the runoff mutation period of Hejin station is 1971, and there is a significant downward trend. The trend and variation are significant. The variation efficiency coefficient is larger, indicating that the variation is strong. In conclusion, the runoff of Hejin station showed a significant decrease trend, and the mutation occurred in 1971. Before 1971, the average annual runoff was 1.793 billion m^3. After 1971, the annual runoff was 9.44 billion m^3, and the annual runoff was significantly changed. From the analysis of the hydrological variation diagnostic system, the runoff in the 1956–2012 years of the Hejin Railway Station showed a downward trend and a sudden change in 1971, which was consistent with the results of Zhao et al. (2012).

Figure 5 shows the relationship between rainfall runoff and anomaly in Hejin station. It can be seen from Fig. 5a that the rainfall-runoff relationship in the basin has changed significantly before and after 1971, and the rainfall-runoff curve after 1971 shows that the runoff generated by 1971 Before 1971. It can be seen from Fig. 5b that the change of rainfall sequence is relatively stable, which is basically positive and negative alternately, and the runoff is basically positive anomaly before the 1970s, and it is basically negative anomaly after the 1970s, The At the same time, before and after the mutation, the precipitation is mainly between 306 mm and 652 mm, and the change trend of the precipitation sequence is not obvious. According to the runoff forming mechanism, we judge that the underlying land surface condition in the basin must change. This result is confirmed by some factual data collected in the introduction.

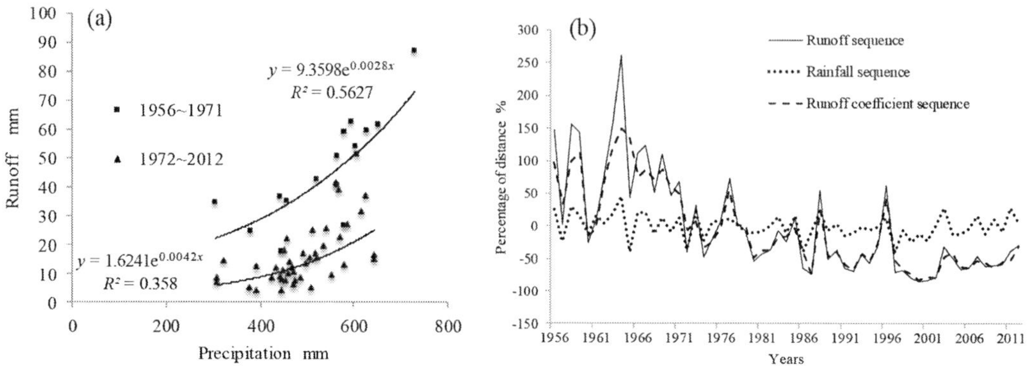

Figure 5. Variation of Precipitation Runoff and Departure from Hejin Station.

3.2 Quantitative Analysis of the Impact of Climate Change on Runoff

Table 2 shows the results of climate change on the impact of Fenhe into the Yellow River runoff calculation, Compared with the base period, the reduction of rainfall in the Fenhe River Basin from 1980 to 2012 resulted in an average annual reduction of 6.89 mm, Which is 267 million m^3. The decrease of the potential evaporation in the Fenhe River Basin in 1980 to 2012 resulted in an average annual increase of 0.55 mm, or 21 million m^3, of the Fenhe River into the Yellow River. The climate change in the Fenhe River Basin in 1980–2012 caused the Fen River into the Yellow River The average annual flow reduction of 246 million m^3. The results of this study are basically consistent with those of Lan et al. (2012).

3.3 Quantitative analysis of the relative contribution of the main influencing factors of runoff reduction in Fenhe River Basin

Based on the quantitative analysis of the main influencing factors of Fenhe runoff sharply decrease, the contribution rate of the main influencing factors of Fenhe River runoff to the sharp drop in runoff is quantitatively described in the attribution segmentation frame of Fenhe River runoff. Table 3 shows the contribution rate of the main influencing factors of the runoff sharply decrease in the Fenhe River Basin from 1980 to 2012. The climate change causes the Fenhe runoff to decrease by an average of 246 million m^3 per year and the contribution rate is 19.17 %. Among them, the reduction of the amount of Fenhe into the yellow runoff is averaged 267 million m^3 per year, and the contribution rate is 20.81 %, The Fen River Basin potential evapotranspiration reduction caused by the temperature, wind speed and other meteorological factors, result in a Fenhe River runoff average annual increase of 0.21 million m^3 and a contribution rate of 1.64 %. Human activities resulted in an average annual decrease of 1.066 billion m^3 and a contribution rate of 83.09 per year. The contribution of groundwater exploitation to the runoff of the Fenhe River flows into the Yellow River was the highest, resulting in an average annual decrease of 386 million m^3 and the contribution of 30.09 %. The coal mining is 334 million m^3 per year, and the contribution rate is 26.03 %. reduction is 233 million m^3, and the contribution rate is 19.17 %; land use is 105 million m^3 and the contribution rate of 8.18 %; After 1980 the new evaporation caused by the new water conservancy project in Fen River Basin result in The Fen River runoff average reduction of 009 million m^3 per year and the contribution rate of 0.07 %. The average annual runoff of the Fenhe River Basin decreased by 1.28 billion m^3 per year from 1980 to 2012. By the calculation of the main influencing factors respectively, the runoff influence by all the factors in 1980–2012 caused an average annual decrease of 1.13 billion m^3, the difference is 29 million m^3. Human activities are the main factors for the sharp drop in Fenhe runoff.

4 Conclusions and discussions

4.1 Conclusions

In this paper, based on the determination of the main influencing factors of the runoff of the Fenhe River flows into the Yellow River, the variation of the runoff series trend of the Hejin River in the Hejin River is analyzed by the hydrological variation diagnosis system. The annual runoff of the main site of the Fenhe River is significantly reduced, The runoff reduce by the rate of 0.158 million $m^3\ a^{-1}$; Fen Fenhe annual runoff has a mutation in 1971. Framework for constructing attribution segmentation of Yellow River runoff reduction in Fenhe River is united. In this framework, different methods is used to calculate the influence rate of climate such as rainfall, potential evapotranspiration and other human activity such as climatic factors and surface water exploitation, groundwater exploitation, land use on the Fen River runoff reduction and the quantity of runoff reduction. The results show that, compared with 1956–1971, the annual runoff of the Fenhe River is reduced by 1.283 billion m^3 per year from 1956 to 1971, and the human activity is the main factor in the Fenhe runoff decrease. The contribution rate of the main human factors to the runoff of the Fenhe River flows into the Yellow River dropped is 83.09 %. From large to small, the contribution factors is: groundwater exploitation > coal mining > surface water exploitation > land use > water conservancy. The contribution rate of the climate change to the runoff of the Fenhe River flows into the Yellow River decrease is 19.17 %, and the contribution rate of rainfall reduction is 20.81 %. The change of meteorological factors such as air temperature and wind speed increases the runoff of Fenhe River, The increase is 21 million m^3 per year.

4.2 Discussions

Based on the principle of water balance, this paper calculates the change of rainfall and potential evaporation and the contribution rate of human surface, groundwater exploitation, land use, water conservancy project and coal mining to Fenhe River runoff decrease, and quantitative evaluate the influence of various factors on runoff from Fenhe River into Yellow River.

However, it is assumed that various factors are independent of each other, But in reality, various factors affect each other, and it is difficult to segment the various factors precisely. Therefore, the influence of the various factors calculated in this paper on the sharp drop in the yellow runoff is only estimated, Moreover, the calculation of the impact of various factors, there must be a repeated calculation part The accurate segmentation and calculation of the influence of various factors on the decrease of the runoff of Fenhe River into the Yellow River need further study.

Data availability. The data are not publicly accessible, because it comes from the the Yellow River Water Conservancy Commission, the state functional organ.

Competing interests. The authors declare that they have no conflict of interest.

Special issue statement. This article is part of the special issue "Innovative water resources management – understanding and balancing interactions between humankind and nature". It is a result of the 8th International Water Resources Management Conference of ICWRS, Beijing, China, 13–15 June 2018.

Acknowledgements. This paper is supported by the key special special project "the mechanism and trend prediction of water and sediment change in the the Yellow River basin" (2016YFC040240203).

Edited by: Bo Pang
Reviewed by: two anonymous referees

References

Awan, U. K. and Ismaeel, A.: A new technique to map groundwater recharge in irrigated areas usinga SWAT model under changing climate, J. Hydrol., 519, 1368–1382, 2014.

Du, X. and Zhang, J.: Implementation of the ecological restoration of the Fenhe River Basin implementation of the central ecological civilization construction grand practice, Shanxi Water Conservancy, 32, p. 4, 2016.

Elfert, S. and Bormann, H.: Simulated impact of past and possible future land use changes on the hydrological response of the Northern German lowland "Hunte" catchment, J. Hydrol., 383, 245–255, 2010.

He, B., Miao, C. Y., and Shi, W.: Trend abruptchange,and periodicity of streamflow in the mainstream of Yellow River, Environ. Monit. Assess., 185, 6187–6199, 2013.

Jiang, S. H., Ren, L. L., Yong, B., Singh, V. P., Yang, X. L., and Yuan, F.: Quantifying the effects of climate variability and human activities on runoff from the Laohahe basin in northern China using three different methods, Hydrol. Process., 25, 2492–2505, 2011.

Jia, X. L., Kang, S. Z., Wei, X. M., et al.: Impacts of climate change and human activities on Yellow River Runoff in the Weihe River Basin, Journal of Northwest Agriculture and Forestry University, 35, 153–158, 2007.

Koster, R. D. and Suarez, M. J.: A simple framework for examining theinterannual variability of land surface moisture fluxes, J. Climate, 12, 1911–1917, 1999.

Ahn, K. H. and Merwade, V.: Quantifying the relative impact of climate and human activities on streamflow, J. Hydrol., 515, 257–266, 2014.

Lan, Y. D., Kang, L. L., Dong, F. F., and Wang, Y. Z.: Effect of climatic change on runoff in Fenhe watershed, Journal of Water Resources & Water Engineering, 23, 70–72+76, 2012.

Li, Z., Liu, W. Z., Zhang, X. C., and Zheng F.: Impacts of land use change and climate variability on hydrology in an agricultural catchment on the Loess Plateau of China, J. Hydrol., 377, 35–42, 2009.

Liang, L. X., Ren, Z. Y., Wang, L. X., and Niu, J. J.: Temporal and spatial characteristics of temperature changes in Fenhe River Basin, Journal of Arid Land Resources and Environment, 24, 52–57, 2010.

Milly, P. C. D. and Dunne, K. A.: Macroscale water fluxes: 2. Water and energy supply control of their interannual variability, Water Res. Res., 38, 1206, https://doi.org/10.1029/2001WR000760, 2002.

Mu, X. M., Chiri, B., Zhang, L., Gao, P., Wang, F., and Zhang, X. P.: Variations of runoff and sediment and their responses to water conservancy and water conservation measures from the Yellow River estuary to Longmen Lake, Sediment Research, 02, 35–41, 2007.

Pereira, L. S., Allen, R. G., Smith, M., and Raes, D.: Crop evapotranspiration estimation with FAO56: Past and future, Agr. Water Manage., 147, 4–20, 2015.

Sankarasubramanian, A. and Vogel, R. M.: Climate elasticity of streamflow in the United States, Water Resour. Res., 37, 1771–1781, 2001.

Wang, G. Q., Zhang, J. Y., Liu, J. F., and He, R. M.: Quantitative analysis of the impacts of climate change and human activities on river runoff, China Water Conservancy, 14, 55–58, 2008.

Wang, W. G., Shao, Q. X., Yang, T., and Luo, Y. F.: Quantitative assessment of the impact of climate variability and human activities on runoff changes: a case study in four catchments of the Haihe River basin, China, Hydrol. Process., 27, 1158–1174, 2013.

Xu, C. F., Dong, W. J., Li, Q., et al.: Analysis of the characteristics and causes of autumn water in the Weihe River Basin in recent 50 years, High Altitude Meteorology, 31, 409–417, 2012.

Xie, P., Chen, G. C., Lei, H. F., and Wu, F.: Diagnostic system of hydrologic variation, Journal of Hydroelectric Power Generation, 29, 85–91, 2010.

Yuan, Z. H., Yan, G., and Liu, Y. F.: Analysis of evolution law and influence factors of water and sediment in Fenhe River since 1950, Advances in Geographical Science, 05, 57–63, 2008.

Zeng, S. D., Xia, J., and Du, H.: Separating the effects of climate change and human activities on runoff over different time scales in the Zhang River basin, Stoch Env. Res. Risk A., 28, 401–413, 2014.

Zhang, G. S., Li, L., Shi, X. H., Xu, W. X., Dong, L. X., and Wang, Q. C.: Climate change in the upper reaches of the Yellow River and its influence on water resources in the Yellow River province, Progress of Water Science, 11, 277–283, 2000.

Zhang, X. C. and Wang, L.: Analysis of variation of natural runoff in the Yellow River, J. Hydrol., 5, 30–33, 2001.

Zhao, F. F., Zhang, L., Xu, Z. X., and Scott, D. F.: Evaluation of methods for estimating the effects of vegetation change and climate variability on streamflow, Water Resour. Res., 46, W03505, https://doi.org/10.1029/2009WR007702, 2010.

Zhao, Y., Hu, C. H., and Hu, S.: Facts Analysis of Significant Reduction in Runoff of Fenhe and Qinhe Basin, Water Resources and Power, 30, 31–34, 2012.

https://doi.org/10.5194/piahs-379-381-2018

Open Access
PIAHS

The application of Mike Urban model in drainage and waterlogging in Lincheng county, China

Qinghua Luan[1,2], **Kun Zhang**[1,3], **Jiahong Liu**[1,3], **Dong Wang**[1,3], **and Jun Ma**[1,2]

[1]School of Water Conservancy and Hydroelectric Power, Hebei University of Engineering, Handan 056021, Hebei, China

[2]Hebei Provincial Research Center of Water Ecological Civilization & Social Governance, Handan 056038, Hebei, China

[3]Institute of Water Resource, China Institute of Water Resources and Hydropower Research, Beijing 100038, China

Correspondence: Qinghua Luan (carol97011202@163.com)

Received: 31 December 2017 – Accepted: 10 February 2018 – Published: 5 June 2018

Abstract. Recently, the water disaster in cities especially in Chinese mountainous cities is more serious, due to the coupling influences of waterlogging and regional floods. It is necessary to study the surface runoff process of mountainous cities and examine the regional drainage pipeline network. In this study, the runoff processes of Lincheng county (located in Hebei province, China) in different scenarios were simulated through Mike Urban model. The results show that all of the runoff process of the old town and the new residential area with larger slope, is significant and full flow of these above zones exists in the part of the drainage pipeline network; and the overflow exists in part of the drainage pipeline network when the return period is ten years or twenty years, which illuminates that the waterlogging risk in this zone of Lincheng is higher. Therefore, remodeling drainage pipeline network in the old town of Lincheng and adding water storage ponds in the new residential areas were suggested. This research provides both technical support and decision-making reference to local storm flood management, also give the experiences for the study on the runoff process of similar cities.

1 Introduction

In the developing countries, the rapid development of economy, industrialization and urbanization have changed the state of climate and underlying surface, and resulted in frequent extreme events such as rainstorms and droughts (Tu, 2014), which is especially significant in China. Furthermore, in China, most of urban drainage pipeline network in the counties have not been constructed, the designed standard of pipeline is low, the standards of municipal are unreasonable and are not uniform with that of water authority (Xie et al., 2016; EWRI of ASCE, 2017), which also resulted in the increase of urban flood disasters, especially in typical mountainous cities. Therefore, it is crucial to study the mechanism of runoff generation and forecast the evolution road of urban waterlogging, especially in the typical developing mountain cities.

Several studies on urban waterlogging have been carried out in China and abroad, and many digital hydrological models have been applied to simulate the process of urban surface runoff and the running situation of drainage pipeline network (Zhou, 2014). Because Mike Urban model has complete urban drainage pipeline system, a friendly version, a good linking with GIS and has been widely used for types of fields, this model was chosen by many researchers. Hlodversdottir et al. (2015) built the surface runoff and drainage network model of the coastal area based on Mike Urban, and evaluated the risk of waterlogging and the net load of drainage pipeline in the central of the Reykjavík. Bouchenafa et al. (2014) built regional Mike Urban model of the urban underground transportation network, simulated the water dynamic processes of urban subway network and improved the scheme of urban flood control.

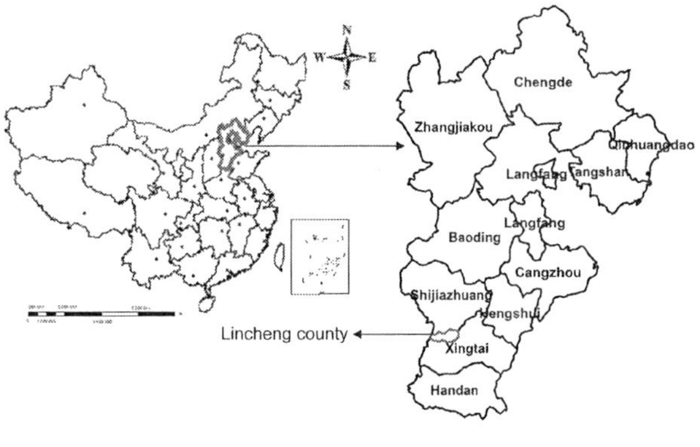

Figure 1. Study area location.

However, the researches on Mike Urban model in the mountain city of semi-arid area is relatively few. So the process study of surface runoff and drainage network in mountainous area is necessary. In this study, the drainage waterlogging situation in Lincheng County in different scenarios would be illustrated through Mike Urban model, which not only can illuminate the regional situation of drainage pipeline networks under different rainwater, but also can provide technology for the urban floods forecast and future waterlogging control. Moreover, the related results also provide decision-making reference for the planning and construction of drainage pipelines network in typical semi-arid mountainous cities.

2 Study area and data collection

2.1 Study area

Lincheng County as a mountainous county is located in Hebei Province, China and in the eastern foot of Taihang Mountain (Fig. 1). There are there geomorphic types: mountainous in the west, hilly in the middle and plain in the east, with the proportions of 35, 50 and 15 % respectively. This study area is the downtown of Lincheng County, covering an area of about 14 km^2, has no hydraulic connection with the surrounding rivers. The south-north water diversion canal through the eastern region and also has no hydraulic connection with the drainage networks. The study area consists of three major parts (NORENDAR International Ltd., 2016): the western part is the old town, the drainage infrastructures are old and the pipeline network is complicated; the central part is a new residential area nearby the mountain; the eastern part as the new urban area is the education land and industrial park under development, the green area of this part is large and fields and ravines are on both sides of the road.

Table 1. The characteristics of designed rainfall processes.

Return period	Three years	Five years	Ten years	Twenty years
Rainfall peak (mm min^{-1})	2.42	2.70	4.00	4.56
Total (mm)	99.60	122.60	158.20	195.60

2.2 Drainage pipeline network

The drainage pipeline networks combining of western part and central part is an independent system and that of eastern part is also independent. The lengths of the drainage pipes and the elevations of the drainage wells can be got through the regional construction planning. Then, through ArcGIS, the drainage pipeline network was successfully digitized and the corresponding properties of the confluence nodes were got. In order to reduce the simulation error, the topologic relation of the pipeline network was checked. The detailed information and data of pipeline nodes and drainage pipeline network were got through the information database of Mike Urban model (Yao, 2012). Eventually, there are 120 confluence nodes and three drainage outlets in the study area. According to the statistic results, there are 121 concrete round pipes in total, of which the total length is about 47 738.6 m, and the main diameter of the pipe is 0.6, 0.8 and 1.2 m. The elevation range of inner base of the pipeline network node is between 77 to 133.8 m (Fig. 2).

2.3 Rainfall data

2.3.1 Observed rainfall data

The observed rainfall processes were based on the gauge monitoring in study area and supplied by Hebei Hydrology and Water Resources Investigation Bureau. Two typical precipitation processes (Fig. 3) were selected for model calibration and validation, which is "7.19" (19 July 2016) and "7.06" (6 July 2017) respectively.

2.3.2 Designed rainfall data

Due to the lack of detailed observed rainfall data, the designed rainfall processes in 24 h of three years, five years, ten years and twenty years were calculated according to the local storm intensity formula in this research (Viglione and Blöschl, 2009; Han, 2015). In order to shorten the rainfall interval, the last processes of different return periods were computed through the coupling of 24 h processes and Chicago rainstorm formula (Cen, 1998). The rainfall peak and total volume of the different scenarios are shown in Table 1.

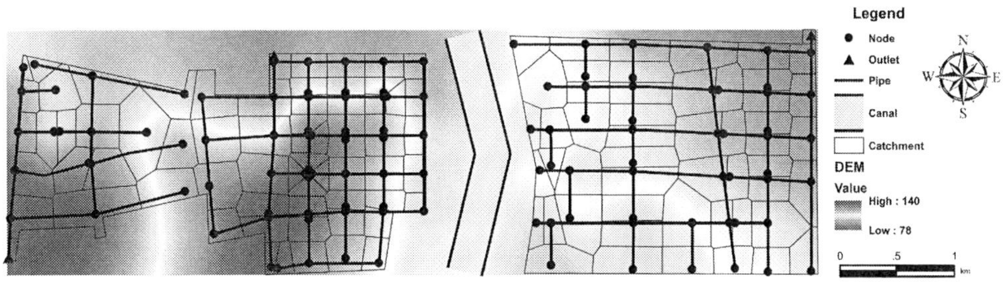

Figure 2. The drainage pipeline networks of the study area.

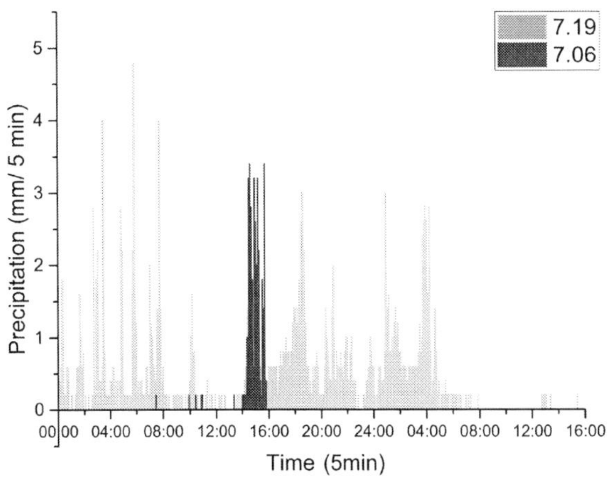

Figure 3. "7.19" and "7.06" typical precipitation processes.

3 Model construction

3.1 Sub-catchments division

The sub-catchment division is based on the digitized results of drainage pipeline network. In Mike urban model system, the confluence node is set as the central of the sub-catchment, and the sub-catchments of study area can be automatically divided according to Theisen polygons (USEPA, 1999; Ahmadi, 2012). Because the elevation of underlying surface in the study area has significant distribution change in the space, the automatic division results were checked and manually adjusted according to the actual confluence situation, in order to improve the simulation accuracy. At last, the study area was divided into 122 sub-catchments (Fig. 2).

3.2 Initial value of model parameter

The parameters of the sub-catchment area were set after the division was completed. Firstly, the underlying surface layers with different land use types were loaded into the model; there are four types of land use in Lincheng: green area, building area, transportation area and integrated land use, of which the corresponding impervious percentages are respectively 20, 90, 80 and 50 % according to the local planning (Kim et al., 2005). Secondly, the corresponding impervious percentages of each layer were set and input to the model. Lastly, the initial impervious percentages of each sub-catchment were automatically calculated through the model. The other model parameters were set according to the recommendation value of the model handbook and the empirical range of actual situation (Son et al., 2016) (Table 4).

Table 2. The empirical parameters of Mike21.

Parameter	Time step (s)	Drying depth (m)	Flooding depth (m)	Manning number
Value	1	0.002	0.003	0.03

3.3 Model calibration and verification

In this paper, the Mike flood model coupled Mike21 and Mike Urban was used to simulate the surface overflowing process and the waterlogging situation of drainage pipeline networks (DHI, 2012). The properties of buildings and roads and DEM were pretreated through ArcGIS and the basic Mike21 files were completed through Mike Zero after the basic files superposition. Through Mike model, the elevation of the buildings and roads were revised and the land values were set in order to ensure the study area is close. The parameters of Mike21 is empirical (Table 2) due to their insensitivity. Mike21 was coupled with Mike Urban through Mike flood platform. Then, the urban pipe drainage networks were linked with two-dimensional surface flow, which is applied for the overflow simulation. The "7.19" and "7.06" rainfall processes were chosen for model calibration and validation, respectively. The parameters were calibrated and validated through comparing with the measured water depth (MWD) and the simulated water depth (SWD) of typical runoff processes. The relative error (RE) between the MWD and the SWD was adopted as the statistical index and the simulated results are shown that all of the REs conform with the related regulation (Rong et al., 2017) (Table 3). The calibrated parameters are listed in Table 4.

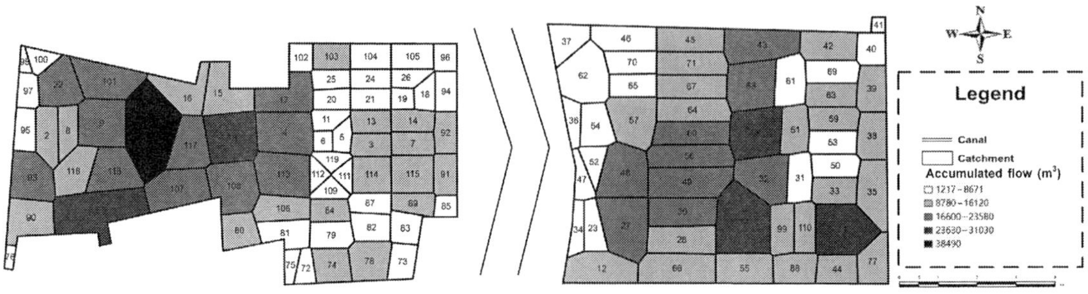

Figure 4. The regional distribution of surface accumulated flow in different sub-catchments.

Table 3. Simulated results of calibration and validation.

Typical process	7.19			7.06		
Node No	MWD (cm)	SWD (cm)	RE (%)	MWD (cm)	SWD (cm)	RE (%)
5	10	9.3	−7	10	9.5	−5
62	13	13.4	3.1	7	7.5	7.1
92	8	8.9	11.3	6	6.4	6.7
107	15	15.6	4	10	10.8	8

Table 4. Calibrated results of Mike urban parameters.

Parameter	Initial value	Calibrated value
Average velocity of surface confluence	0.30	0.20
Hydrological reduction factor	0.9	0.9
Initial water loss	0.0006	0.0015
Manning coefficient of pipeline	0.015	0.013

4 Results and analysis

The designed rainfall processes in different return periods were input and then the corresponding runoff processes were calculated through the calibrated model. In this study, the analysis of the different flood processes contains three aspects: typical waterlogging area, pipeline network load and typical overflow point (SAC, 2008). Detailed analysis in different aspects can indicates the surface runoff processes and the running state of the drainage pipeline network more clearly and be convenient for government to manage urban waterlogging and drainage network construction.

4.1 Typical waterlogging area

According to the simulated results, the peak runoff, accumulated flow and comprehensive runoff coefficient of sub-catchment area increased slowly with the increase of rainfall. The peak of surface runoff in the study area in different return period appeared at 18:30. When the return period is twenty years, the comprehensive runoff coefficient is the largest, the value is 0.658 at 18:30. The waterlogged area includes 6 sub-catchments, and the surface runoff of the No. 99 sub-catchment in the old town area is the largest, and the peak runoff and accumulated flow of runoff during different return periods in this typical sub-catchment are shown in Table 5.

The regional situation of infrastructure construction combining the distribution of surface accumulated flow in twenty-year return periods (Fig. 4) illustrates that the waterlogging frequents of the old town and the joint zone between the old town and the new residential area are higher, because the corresponding elevation is low, thus the confluence speed is fast and the surface flow duration is short. In addition, the lack of drainage capacity in the old town and the improper joint between the new and old urban areas also cause severe waterlogging. Although the surface runoff in some eastern sub-catchments is large (Fig. 4), the related waterlogging situation is better, due to the large water permeability based on the large green area.

4.2 Pipeline network load

The simulated results show that in three-year return period, there are 17 pressured pipelines in the drainage networks, of which 3 and 14 pipelines are under full pressure and partial pressure respectively; while in twenty-year return period, 4 and 20 pipelines are under full pressure and partial pressure, respectively. With the increase of the return period, the number of the pressured pipelines increases evidently, and the pressure and pressured duration also increase in the same

Table 5. Runoff characteristics in different return periods in No. 99 sub-catchment.

Return period	Three years	Five years	Ten years	Twenty years
Peak runoff ($m^3 s^{-1}$)	3.92	4.36	5.20	5.94
Maximum cumulative quantity (m^3)	19 520.9	24 057.1	31 106.9	384 892
Comprehensive runoff coefficient	0.655	0.656	0.657	0.658

Table 6. Total numbers of pressure pipeline in different return periods.

Pressure state (bar)	Three years	Five years	Ten years	Twenty years
Fully	3	4	4	4
Partial	14	16	17	20
Total	17	20	21	24

Table 7. Total overflow nodes numbers of each district in different return periods.

Return period	Three years	Five years	Ten years	Twenty years
Old town	5	5	5	6
New district	4	4	5	6
Industrial zone	3	3	4	5

pressured pipeline. In No. 98 and No. 107 sub-catchment of the old town, their pipelines are pressured at first and under pressure until the runoff ends. The pressured pipelines in the old town account for the main of the total ones, which is consistent with the simulated waterlogging situation and the related analysis. The pipelines nearby the outlet are also easily pressured, due to the runoff confluence. With the duration of rainfall, the pipelines in the other areas are gradually pressured. The number of pressured pipelines in different return periods are shown in Table 6.

4.3 Typical overflow point

The simulated results also show that the overflow firstly occurred at the 96th node in any return period; the overflow appeared earlier (about 01:00), and existed until the rainfall ended in twenty-year return period. Both the water level and the discharge of overflow in the 102th node are largest in any return period. There are 12 overflow nodes in the study area, of which 5 nodes are located in the old town; and the overflow duration of the 96th node is about 20 h, in three-year return period. While, there are 17 overflow nodes, of which 6 overflow nodes are located in the old town, and the overflow duration of the 96th node is about 23 h, in twenty- year return period. The overflow nodes numbers of each district in different return period are listed in Table 7.

With the increase of return period, the overflow nodes number of drainage pipeline network increases, the overflow appears earlier and the overflow duration increases. Additionally, the overflow nodes with high water level are mainly located in the western old town due to the poor drainage capacity and low elevation; while some overflow nodes with low water level are located in the central and eastern urban areas. These results and analysis are both consistent with the simulated results and the corresponding analysis of waterlogging and pipeline network load.

5 Conclusions

In this research, the Mike urban in Lincheng city was successfully constructed. The waterlogging situation of different surface runoff processes and running state of drainage pipeline network in the study area were illuminated. The simulated results show that the water logging pressure in typical mountain part and the old town is greater than that of the plain part and the new area. With the increase of the return period, the accumulated flow, the pressured pipeline number and the overflow nodes number accordingly increase.

The related analysis provides technical support for the early warning of urban waterlogging and construction of drainage infrastructures of study area, and gives references for the simulation of surface runoff process in mountainous cities.

Data availability. The data of this study came from the construction planning of lincheng County. We cooperated with the NORENDAR International Ltd. to complete the program. The content of urban planning and design is open and public in the followed URL: http://www.lincheng.gov.cn/index.aspx?lanmuid=46&sublanmuid=539&id=642. However, the urban drainage network data involved in this study is not publicly available. In addition, Lincheng County is a typical mountainous city with three different types of urban areas, and the urban drainage networks of each type are independent. Therefore, we chose Lincheng County as the research area.

Competing interests. The authors declare that they have no conflict of interest.

Special issue statement. This article is part of the special issue "Innovative water resources management – understanding and balancing interactions between humankind and nature". It is a result of the 8th International Water Resources Management Conference of ICWRS, Beijing, China, 13–15 June 2018.

Acknowledgements. We are grateful to the NSF of Hebei, China (No. E2017402178), the NSFC (No. 51739011) and the NSF of Hebei Education, China (No. ZD2014020).

Edited by: Dingzhi Peng
Reviewed by: Meifang Ren and one anonymous referee

References

Ahmadi, Z.: Investigation of adequacy of Masal city drainage network and suggestions for improvement by using MIKE SWMM model, Iranian Water Res. J., 6, 93–103, 2012.

Bouchenafa, W., Mouhous-Voyneau, N., Sergent, P., and Brochet, J.: Study of flow in a staircase at subway station. Advances in Hydroinformatics, Springer, Singapore, 171–183, 2014.

Cen, G.: Research on rainfall pattern of urban design storm, Adv. in Water Sci., 1, 42–47, 1998.

DHI (Danish Hydraulic Institute): MIKE Urban User Manual, Denmark, 240 pp., 2012.

EWRI of ASCE: Standard guidelines for the drainage of urban stormwater systems, ASCE/EWRI45-05, Reston, Va: American Society of Civil Engineers, 60 pp., 2017.

Han, J. L.: Drainage waterlogging planning of small city based on Mike Urban, M.A. thesis, Zhejiang University of Technology, China, 2015.

Hlodversdottir, A. O., Bjornsson, B., Andradottir, H. O., Eliasson, J., and Crochet, P.: Assessment of flood hazard in a combined sewer system in Reykjavik city centre, Water Sci. Technol., 71, 1471–1477, https://doi.org/10.2166/wst.2015.119, 2015.

Kim, J. S., Ahn, J. H., Oh, T. S., and Moon, Y. I.: Runoff analysis of urban watershed using MIKE SWMM model, J. Korea Water Resour Assoc., 38, 907–916, https://doi.org/10.3741/JKWRA.2005.38.11.907, 2005.

NORENDAR International Ltd.: Control detailed control planning in Lincheng County, http://www.lincheng.gov.cn/index.aspx?lanmuid=46&sublanmuid=539&id=642 (last access: 21 February 2018), 2016.

Rong, M. A., Bai, T., Huang, Q., and Yang, W.: MIKE 21 model and its application on urban waterlogging simulation, J. Nat. Disast., 26, 172–179, https://doi.org/10.13577/j.jnd.2017.0420, 2017.

Son, A. L., Kim, B., and Han, K. Y.: A Simple and robust method for simultaneous consideration of overland and underground space in urban flood modeling, Water, 8, 494, https://doi.org/10.3390/w8110494, 2016.

Standardization administration of PRC (SAC): Standard for hydrological information and hydrological forecasting (GB/T 22482-2008), Standards Press of China, Beijing, China 11 pp., 2008.

Tu, C.: Application of computer simulation of urban stormwater: take Seongbuk of Dayu County as an example, M.A. thesis, Nanchang University, China, 2014.

USEPA (United States Environmental Protection Agency): Preliminary data summary of urban storm water best management practices, EPA, Washington, 214 pp., 1999.

Viglione, A. and Blöschl, G.: On the role of storm duration in the mapping of rainfall to flood return periods, Hydrol. Earth Syst. Sci., 13, 205–216, https://doi.org/10.5194/hess-13-205-2009, 2009.

Xie, J. Q., Liao, Z. L., and Gu, X. Y.: Prediction and evaluation of waterlogging in highly urbanized areas based on Mike Urban: demonstrated on the example of Huoshan-Huimin drainage system in Shanghai, Energy Environ. Prot., 30, 44–49, https://doi.org/10.3969/j.issn.1006-8759.2016.05.010, 2016.

Yao, S. L.: Urban Drainage Systerm Modeling Research Based on Mike Flood, Ph.D. thesis, Beijing University of Technology, China, 2012.

Zhou, Q.: A Review of sustainable urban drainage systems considering the climate change and urbanization impacts, Water, 6, 976–992, https://doi.org/10.3390/w6040976, 2014.

https://doi.org/10.5194/piahs-379-387-2018

Continental and marine surficial water – groundwater interactions: the case of the southern coastland of Venice (Italy)

Luigi Tosi[1], **Cristina Da Lio**[1], **Pietro Teatini**[1,2], **Antonio Menghini**[3], **and Andrea Viezzoli**[3]

[1]Institute of Marine Sciences – National Research Council, Venezia, Italy
[2]Department of Civil, Environmental and Architectural Engineering, University of Padova, Padova, Italy
[3]Aarhus Geophysics Aps, Aarhus, Denmark

Correspondence: Luigi Tosi (luigi.tosi@ismar.cnr.it)

Received: 30 December 2017 – Revised: 1 March 2018 – Accepted: 7 March 2018 – Published: 5 June 2018

Abstract. Understanding the continental-marine surficial water–groundwater exchanges in transitional coastal environments is really challenging at large scale. It requires an image of the saltwater–freshwater relationship which is difficult to be obtained especially in wetlands, lagoons, and marine areas. This study is focused on the coastland of the southern Venice lagoon – northern Po river delta (Italy), a precarious environment subject to both natural changes and anthropogenic pressures. Here, saltwater severely affects farmlands and aquifers. We used an airborne electromagnetics (AEM) survey with the goal of characterizing the continental and marine surficial water–groundwater interactions in such coastal region. The AEM survey allowed depicting a clear image of the fresh water–saltwater occurrence in shallow aquifers along mainland-lagoon – littoral-sea profiles, up to 20 km long. The results reveal that continental groundwater is located in the lagoon subsoil below a 10–20 m thick saline aquifer and extends down to 70 m depth. The whole low-lying farmland located south of the lagoon margin is seriously affected by saltwater contamination, which occurs from a few to about 50 m depth. The integrated analysis of AEM, seismic and borehole data shows that buried morpho-geological structures, such as paleo-channels and over-consolidated clay units control the saline contamination from the lagoon and the sea into the coastal aquifer system.

1 Introduction

Transitional coastal environments have drawn much attention in the last decades because of the high ecological relevance of deltas, estuaries, lagoons, and wetlands with an increasing concern on their vulnerability, especially in view of the expected global changes.

The coexistence of agricultural, urban, and industrial areas with salt- and fresh-water wetlands makes the surface water–groundwater exchanges be impacted by both natural processes and human activities. Whilst the exchange of water among rivers, wetlands, lagoons and sea is fairly well understood at the local scale, the interactions between these waters and the on-shore and off-shore aquifers are often hard to be understood in such complex coastal environments. Therefore, investigating continental and marine surficial water–groundwater interaction, i.e. salt-freshwater exchange, in a unique, consistent, and homogeneous framework is still a challenge. Borehole electrical conductivity measurements, vertical electric soundings, electrical resistivity tomography, and time-domain electromagnetic investigations are usually performed to characterize the saline interface and its evolution in coastlands. Direct measurements such as seepage meters, benthic chambers, and surface water enrichments in natural isotopic tracers are often done to assess submarine groundwater discharge (SGD) in wetlands, lagoons and offshore. Nevertheless, a clear and comprehensive image of the mainland-lagoon – sea hydrogeology in a continuous framework is far from being achieved because inland and offshore surveys are generally carried out separately and by different methodologies. In addition, in-depth and extent investi-

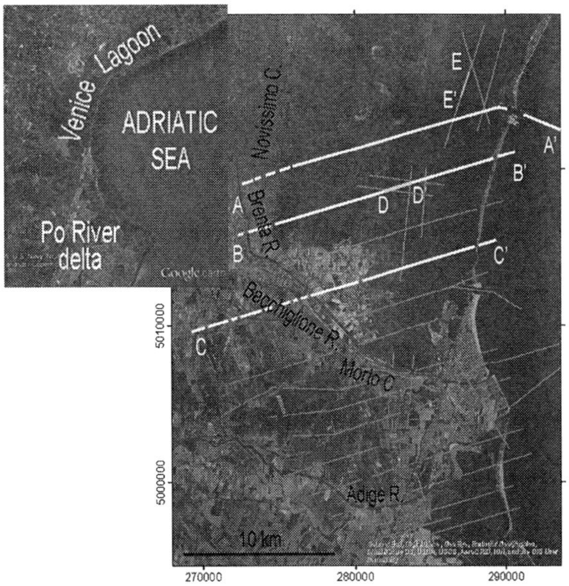

Figure 1. Satellite images of the southern Venice Lagoon – northern Po River Delta coastland. Red and white lines show the AEM acquisition tracks.

gations in lagoon shallows, marshes, rivers and reclamation canals are precluded or at least hardly and locally done.

The Venice coastland (Italy) is an example of heterogeneous transitional coastal zone (Fig. 1), which includes low-laying farmland, lagoon, estuary, delta and littoral environments where water flows are often anthropogenically driven, especially in the low-lying sectors (e.g., Da Lio et al., 2015).

Over the last two decades, recurrent summer droughts and marine water encroachment along the river mouths severely affected the southern Venice coastland, especially the low-lying farmland, with salinization of soils and groundwater. Thus, a number of investigations have been carried out in order to understand the causes and the mechanisms concurring to the aquifer and soil salinization. Although most of the above mentioned methods have been already used (e.g., Carbognin and Tosi, 2003; Mayer et al., 2006; de Franco et al., 2009; Rapaglia, 2005; Tosi et al., 2011) a clear and comprehensive image of surficial water-groundwater interaction in a mainland-lagoon – sea unique, consistent and homogeneous framework, is still far from being achieved.

The aim of this work is to provide a step forward in the delineation of the continental and marine surface water–groundwater interactions and the mechanisms controlling the saltwater intrusion in the southern coastland of Venice. This has been achieved using an airborne electromagnetics (AEM) survey (Viezzoli et al., 2010), consisting of electrical resistivity sections up to 20 km long, acquired from the mainland to the sea, crossing lagoon, estuarine and delta environments. The AEM allows overcoming many intrinsic constrains typical of the ground-based surveys, among all, the "simultaneity" of data acquisition. This method has been successfully tested in the Venice lagoon at local scale (Viezzoli et al., 2010; Teatini et al., 2011) and in this work we explored its potentiality for large scale analysis.

2 Method

About 200 km-long AEM resistivity sections were acquired by the SkyTEM system in 2013 over the Venice coastland. This instrumentation was chosen as its dual moment provides a bandwidth (i.e. a penetration range) suitable for applications where near-surface and deep information is important to refine the hydrogeologic model (Viezzoli et al., 2010; Teatini et al., 2011). The AEM survey was acquired along North-South and West-East profiles, crossing the farmland and watercourses in a continuous and like-simultaneous acquisition (Fig. 1). Data have been processed to eliminate artefacts and assign noise levels at late times, stacked to increase signal to noise ratio while preserving lateral resolution, and inverted using the Spatially Constrained Inversion (SCI) technique (Viezzoli et al., 2010). The accuracy of the resistivity model obtained from AEM data depends on the contrasts and the resistivity range. In highly conductive environment like lagoons and sea it is in the order of ± few tens of %. The interpretation of the AEM outcome has been supported by a number of groundwater and surficial water electrical conductivity measurements carried out simultaneously to the airborne survey (Da Lio et al., 2015) and available chemical analysis, subsoil data, and hydro-stratigraphic information (e.g., Carbognin and Tosi, 2003; Tosi et al., 2007; de Franco et al., 2009; Rizzetto et al., 2003; Tosi et al., 2009; Zecchin et al., 2011; Da Lio et al., 2013). Specifically, in this work, we assumed electrical resistivity <5 and $>10\,\Omega\,\mathrm{m}$, as the upper value for saltwater and the lower value for freshwater, respectively.

3 Results

3.1 Multi-layer vertical resistivity sections

Three vertical resistivity sections have been selected from the whole AEM dataset to illustrate the onshore–offshore relationships between surface water and groundwater.

In most of the examined lagoon area, the salt water infiltrates into the subsoil down to 10–20 m depth, delineating an irregular freshwater–saltwater transition zone (Fig. 2). In the north-western sector, the saltwater intrusion is restricted to the lagoon margin or the surroundings of the Brenta and Bacchiglione rivers, while in the south-western sector the lagoon water contaminates a large part of the farmland. The presence of fresh waters in the lagoon basin has been identified beneath the shallow saline aquifer (Fig. 2). This freshwater flows from the mainland and laterally extends underneath the littoral to the near shore. Although the AEM data

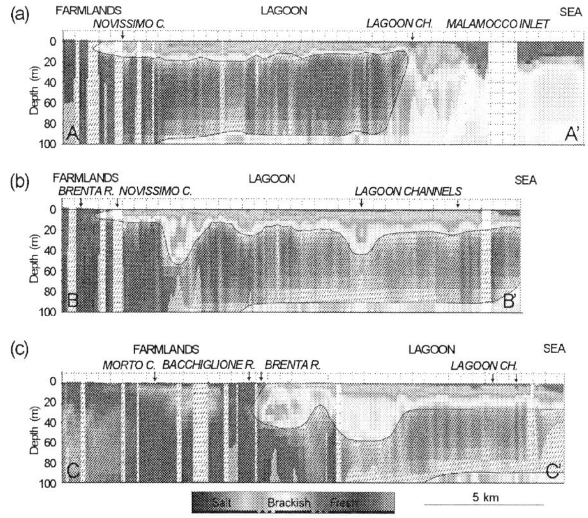

Figure 2. Multi-layer vertical resistivity sections obtained from the AEM survey (see section position in Fig. 1). The freshwaters are highlighted by a black oblique hatching.

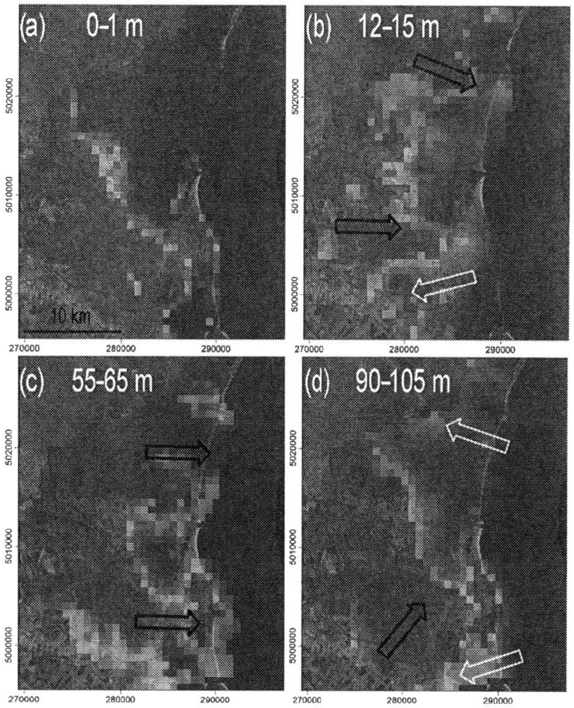

Figure 3. Conceptualization of the freshwater-saltwater relationship at different depths, obtained by integrating the results of the AEM survey with previous available information: **(a)** 0–1 m, **(b)** 12–15 m, **(c)** 55–63 m, **(d)** 91–108 m. Fresh-, brackish- and saltwaters are represented by red, green/yellow and blue colors, respectively. White and black arrows delineate the advancement of the saltwater intrusion and freshwater flow, respectively.

quality rapidly decays with the increase of the lagoon and sea bathymetry, it is reasonably to suppose that continental groundwaters flow up to a few km offshore. In the lagoon sector, groundwater is salty again below 70–80 m depth, probably because of its fossil marine origin, while in the mainland freshwater aquifers occur (Fig. 2). Notice that in the lagoon sector, the AEM survey did not produce reliable data for a rigorous interpretation deeper than 100 m.

3.2 Model of the fresh water–saltwater interaction

The pseudo-3-D electrical resistivity dataset provided by AEM is used to map the presence of saltwater and freshwater at different depths. Electrical resistivity data were interpolated on a 1 km regular grid by the Kriging method and the results classified as freshwater and saltwater for values < 5 and $> 10\ \Omega$ m, respectively. The values between these limits are used to map the brackish transitional zone.

Figure 3 shows the conceptualization of the freshwater-saltwater relationships at four different depths. The first layer (0–1 m depth interval, Fig. 3a) highlights the marine-lagoon and the land areas by blue and red color, respectively, with the exception of some low-lying sectors in the southernmost zone where saltwater contaminates also agricultural soils. Freshwater below the Venice lagoon is unevenly detected between 10 and 20 m depth (Fig. 3b) and more generally occurred in the 60–70 m depth range (Fig. 3c). Southward of the lagoon margin, saltwater intrudes from the lagoon and the sea and extends to the whole mainland, including the northern Po River delta. Saltwater is encountered at depth larger than about 90–100 m in the aquifers below the lagoon while freshwaters are detected in the mainland sectors (Fig. 3d).

4 Discussion

AEM provides a continuous image of the freshwater and saltwater distribution in the transitional coastal aquifers of the southern Venice lagoon and northern Po River Delta. Moreover, the integrated use of additional data, such as very high resolution seismic (VHRS) profiles together with geomorphological and sedimentological analyses, has allowed a clearer understanding of the mechanisms driving surface water and groundwater exchanges.

The analysis highlights that the high conductivity shallow deposits correspond to the Holocene units (Fig. 4), which are characterized by marine and lagoon-back barrier sediments, more permeable and less compacted than the deeper Pleistocene alluvial ones (e.g., Tosi et al., 2007 and references therein). The setting of the saltwater-freshwater transitional zone is very complex and likely related to the location of an aquitard partially confining the Holocene deposits, well known by Venetians as *caranto*. It consists in an overconsolidated stiff layer forming the uppermost part of the Pleistocene units (e.g., Bonardi et al., 2006; Tosi et al., 2007).

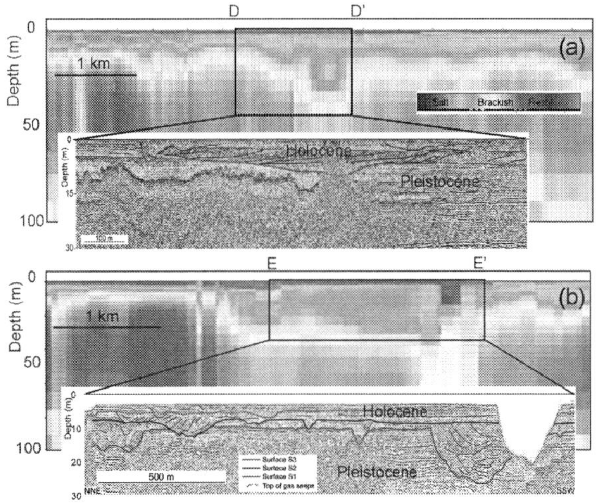

Figure 4. Vertical resistivity and seismic sections along the alignments D–D′ **(a)** and E–E′ **(b)** sketched in Fig. 1. The red line represents the Pleistocene–Holocene boundary.

Localized seepages of saltwater from the lagoon bottom at larger depths involving the Pleistocene deposits (Fig. 4) can be observed. Their location corresponds with the lagoon channels where the *caranto* clay layer misses. Other deeper saltwater infiltrations are related to the presence of buried paleo-channels, which are generally filled with coarse and more permeable sediments than those forming the lagoon bottom.

Similarly to the lagoon basin, also in the inland the morphology and subsoil architecture play an important role in determining the relationship between fresh and salt groundwater (Fig. 3). In the littoral sector, generally characterized by a ground elevation a few meter above the mean sea level, rainfalls favour the formation of rather continuous fresh groundwater lens, about 10 m thick and even more, in the sandy subsoil. Conversely, in the areas lying below the mean sea level (up to 4 m) managed by hydraulic reclamation, saltwater contamination affects the farmland from the subsoil down to 50–60 m depth (Fig. 3). Moreover, the presence of sandy buried paleo-channel systems crossing the farmland, with a main direction from inland to the lagoon boundary, acts as preferential pathways for groundwater flow and solute transport. These structures generally increase the saltwater flow from the lagoon into the low-lying sectors, even though they allow a short-term storage of rainwater in the very shallow subsoil. Conversely, sandy paleo-ridge systems are capable to contain groundwater with lower salinity than that occurring in the paleo-channels, at least in the shallow part. For both paleo-channels and paleo-coastal ridges, the water quality is significantly improved by local rainfalls that rapidly supply freshwater.

5 Conclusions

We have explored the capability of the AEM technique to investigate the continental and marine surficial water–groundwater interactions in a large portion of the coastland encompassing the southern Venice lagoon and the northern Po river delta. The AEM has allowed to overcome the logistical problems typical of the ground-based measurements in coastal transitional environments crossed by rivers, canals and lagoons. AEM has provided a continuous and almost simultaneous image of the relative position of saltwater and freshwater, suggesting the main mechanisms governing the exchange between surface water and groundwater. Significant information on these water interactions has been obtained at "regional" scale and with a good detail even in some local situations. The main results are summarized as follows.

The infiltration of the lagoon and marine waters is less deep than it was supposed in the past.

Fresh and brackish waters have been identified underneath the lagoon basin from only 15–20 m depth, locally even less, and down to about 70–80 m. Below this depth, a saline aquifer probably of fossil origin is detected in the lagoon sector.

Fresh and brackish aquifers extend beyond the littoral to the nearshore and could give rise to local SGD.

The saltwater–fresh water transition zone is very irregularly-shaped and mainly depends on the morphologic setting and the subsoil architecture.

The saline infiltration from the lagoon bottom is controlled by an over-consolidated clay layer and the presence of buried geomorphological structures, generally sandy bodies (e.g., paleo-channels).

In the mainland, sandy buried paleo-channel systems can increase the saltwater flow from the lagoon into the low-lying sectors, while sandy paleo-ridge systems are capable to contain fresh groundwater.

Saltwater intrusion is worryingly spreading in the whole southern sector, mostly in the low-lying farmland, with the exception of the zones around the Brenta river where the contamination is restricted.

Concluding, further quantitative investigations targeted at quantifying the continental and marine surface water–groundwater exchanges, as well as at developing numerical models for the prediction of the future behaviour in light of the expected climate changes, will benefit from the outcomes of this study. Moreover, this work can help detecting the zones at higher risk, where local more detailed investigations must be performed to improve the knowledge of the vulnerability of the farmland to saltwater contamination.

Data availability. The AEM data presented here are part of an ongoing project and, therefore, will be available on completion of the research. They will be available on the website of the Institute of Marine Sciences – National Research Council, Italy (http://www.ismar.cnr.it/index_html-1?set_language=en&cl=en). Geological and hydrogeological data are available through the following websites: Città Metropolitana Venezia (2017) http://webgis.cittametropolitana.ve.it/geologia, http://difesasuolo.provincia.venezia.it/DifesaSuolo/Index?pagina=1&id=banca_dati_idrogeologica; Istituto Superiore per la Protezione e la Ricerca Ambientale (2017) http://www.isprambiente.gov.it/Media/carg/veneto.html; CNR-ISMAR, CORILA, Regione del Veneto, Magistrato alle Acque di Venezia, ARPA Veneto, Consiglio di Bacino Laguna di Venezia, Geoportale Nazionale, IVSLA, Ist. Parco Laguna di Venezia, Ist. Bosco e Grandi Parchi, Museo di Storia Naturale di Venezia, Istituzione Centro Previsioni e Segnalazioni Maree (2017) http://cigno.atlantedellalaguna.it; Regione del Veneto (2017) http://gisgeologia.regione.veneto.it/website/venezia-10k/viewer.htm.

Competing interests. The authors declare that they have no conflict of interest.

Special issue statement. This article is part of the special issue "Innovative water resources management – understanding and balancing interactions between humankind and nature". It is a result of the 8th International Water Resources Management Conference of ICWRS, Beijing, China, 13–15 June 2018.

Acknowledgements. The Flagship Project RITMARE – The Italian Research for the Sea – coordinated by the Italian National Research Council and funded by the Italian Ministry of Education, University and Research within the National Research Program 2011–2013.

Edited by: Depeng Zuo
Reviewed by: two anonymous referees

References

Bonardi, M., Tosi, L., Rizzetto, F., Brancolini, G., and Baradello, L.: Effects of climate changes on the late Pleistocene and Holocene sediments of the Venice Lagoon, Italy, J. Coastal. Res., 39, 279–284, 2006.

Carbognin, L. and Tosi, L.: Il Progetto ISES per l'analisi dei processi di intrusione salina e subsidenza nei territori meridionali delle Province di Padova e Venezia, Grafiche Erredici, Padova (Italy), 95 pp., 2003.

Città Metropolitana Venezia: http://webgis.cittametropolitana.ve.it/geologia, http://difesasuolo.provincia.venezia.it/DifesaSuolo/Index?pagina=1&id=banca_dati_idrogeologica, Banca dati geologica e idrogeologica, last access: 30 December 2017.

CNR-ISMAR, CORILA, Regione del Veneto, Magistrato alle Acque di Venezia, ARPA Veneto, Consiglio di Bacino Laguna di Venezia, Geoportale Nazionale, IVSLA, Ist. Parco Laguna di Venezia, Ist. Bosco e Grandi Parchi, Museo di Storia Naturale di Venezia, Istituzione Centro Previsioni e Segnalazioni Maree, http://cigno.atlantedellalaguna.it, Atlante della Laguna, last access: 30 December 2017.

Da Lio, C., Tosi, L., Zambon, G., Vianello, A., Baldin, G., Lorenzetti, G., Manfe, G., and Teatini, P.: Long-term groundwater dynamics in the coastal confined aquifers of Venice (Italy), Estuar. Coast. Shelf Sci., 135, 248–259, 2013.

Da Lio, C., Carol, E., Kruse, E., Teatini, P., and Tosi, L.: Saltwater contamination in the managed low-lying farmland of the Venice coast, Italy: An assessment of vulnerability, Sci. Total Environ., 533, 356–369, 2015.

de Franco, R., Biella, G., Tosi, L., Teatini, P., Lozej, A., Chiozzotto, B., Giada, M., Rizzetto, F., Claude, C., Mayer, A., Bassan, V., and Gasparetto-Stori, G.: Monitoring the saltwater intrusion by Time Lapse Electrical Resistivity Tomography: The Chioggia test site (Venice Lagoon, Italy), J. Appl. Geophys., 69, 117–130, 2009.

Istituto Superiore per la Protezione e la Ricerca Ambientale: http://www.isprambiente.gov.it/Media/carg/veneto.html, Progetto CARG, last access: 30 December 2017.

Mayer, A., Gattacceca, J., Claude, C., Radakovitch, O., Giada, M., Cucco, A., Tosi, L., and Rizzetto, F.: Radon activity in the southern Lagoon of Venice and the Adriatic Sea, in: Scientific Research and Safeguarding of Venice 2005, edited by: Campostrini, P., CORILA, Research Program 2004–2006, 4, 217–226, Venezia, 2006.

Rapaglia, J.: Submarine groundwater discharge into the Venice lagoon, Italy, Estuaries, 28, 705–713, 2005.

Regione del Veneto: http://gisgeologia.regione.veneto.it/website/venezia-10k/viewer.htm, Progetto CARG, last access: 30 December 2017.

Rizzetto, F., Tosi, L., Carbognin, L., Bonardi, M., and Teatini, P.: Geomorphic setting and related hydrogeological implications of the coastal plain south of the Venice Lagoon, Italy, IAHS-AISH Publ., 278, 463–470, 2003.

Teatini, P., Tosi, L., Viezzoli, A., Baradello, L., Zecchin, M., and Silvestri, S.: Understanding the hydrogeology of the Venice lagoon subsurface with airborne electromagnetics, J. Hydrol., 411, 342–354, 2011.

Tosi, L., Rizzetto, F., Bonardi, M., Donnici, S., Serandrei Barbero, R., and Toffoletto, F.: Note illustrative della Carta Geologica d'Italia alla scala 1: 50.000, Foglio 148–149 Chioggia-Malamocco, APAT, Dipartimento Difesa del Suolo, Servizio Geologico d'Italia, SystemCart, Roma, 164 pp., 2 Maps, 2007.

Tosi, L., Rizzetto, F., Zecchin, M., Brancolini, G., and Baradello, L.: Morphostratigraphic framework of the Venice lagoon (Italy) by very shallow water VHRS surveys: evidence of radical changes triggered by human-induced river diversions, Geophys. Res. Lett., 36, https://doi.org/10.1029/2008GL037136, 2009.

Tosi, L., Baradello, L., Teatini, P., Zecchin, M., Bonardi, M., Shi, P., Tang, C., Li, F., Brancolini, G., Chen, Q., Chiozzotto, B., Frankenfield, J., Giada, M., Liu, D., Nieto, D., Rizzetto, F., Sheng, Y., Xiao, Y., and Zhou, D.: Combined continuous electrical tomography and very high resolution seismic surveys to assess continental and marine groundwater mixing, Boll. Geofis. Teor. Appl., 52, 585–594, 2011.

Viezzoli, A., Tosi, L., Teatini, P., and Silvestri, S.: Surface water-groundwater exchange in transitional coastal environments by airborne electromagnetics: the Venice

Lagoon example, Geophys. Res. Lett., 37, L01402, https://doi.org/10.1029/2009GL041572, 2010.

Zecchin, M., Caffau, M., and Tosi, L.: Relationship between peat bed formation and climate changes during the last glacial in the Venice area, Sediment. Geol., 238, 172–180, 2011.

https://doi.org/10.5194/piahs-379-393-2018

Evaluation of water productivity under climate change in irrigated areas of the arid Northwest China using an assemble statistical downscaling method and an agro-hydrological model

Liu Liu[1,2], **Zezhong Guo**[1,2], **and Guanhua Huang**[1,2]

[1]College of Water Resources & Civil Engineering, China Agricultural University, Beijing 100083, China
[2]Center for Agricultural Water Research in China, China Agricultural University, Beijing 100083, China

Correspondence: Liu Liu (liuliu@cau.edu.cn)

Received: 30 December 2017 – Revised: 27 February 2018 – Accepted: 3 March 2018 – Published: 5 June 2018

Abstract. The Heihe River Basin (HRB) is the second largest inland river basin, located in the arid region of Northwest China with a serious water shortage. Evaluation of water productivity will provide scientific implications for agricultural water-saving in irrigated areas of the arid region under climate change. Based on observed meteorological data, 23 GCMs outputs and the ERA-40 reanalysis data, an assemble statistical downscaling model was developed to generate climate change scenarios under RCP2.6, RCP4.5, RCP8.5 respectively, which were then used to drive the SWAP-EPIC model to simulate crop growth in the irrigated areas of the middle HRB for the future period from 2018 to 2047. Crop yield showed an increasing trend, while crop water consumption decreased gradually in Gaotai and Ganzhou irrigated areas. The water productivity in future 30 years showed an increasing trend in both Gaotai and Ganzhou areas, with the most significant increase under RCP4.5 scenario, which were both larger than 2 kg m^{-3}. Compared with that of the period from 2012 to 2015, the water productivity during 2018–2047 under three RCP scenarios increased by 9.2, 14.3 and 11.8 % in the Gaotai area, and 15.4, 21.6, 19.9 % in the Ganzhou area, respectively.

1 Introduction

The Fifth Assessment Report, issued by the Intergovernmental Panel on Climate Change (IPCC), affirmed the reality of climate warming. Since 1950s, the global climate has changed dramatically and these changes have involved atmospheric warming, shrinking glaciers, sea level rise, and increased levels of greenhouse gases. From 1880 to 2012, the global average temperature increased by 0.85 °C, and it was warmer in the most recent three decades than in any previous decade (IPCC, 2013). A climate model based on greenhouse gas emission is the basis for predicting future climate change. The Fifth Assessment Report of the IPCC presented a new generation of future climate change scenarios called Representative Concentration Pathways (RCPs) (IPCC, 2013). The rate of increase of the ground surface air temperature in China, over the past 100 years, is higher than the world average during this same period. There is a significant temperature increase in the inland areas of Northwest China, with the ground surface air temperature increasing by more than 0.5 °C since the 1980s (Zuo et al., 2004). The mean annual precipitation of China is 649 mm, while that in Northwest China it is only 161 mm. Due to the complex landscape, arid climate, and low vegetation coverage, Northwest China is relatively more vulnerable than other regions to climate change and this is likely to affect the agricultural productivity of the region (Wang, 2016).

The application of downscaling methods in predicting future climate change is an important approach used for studies on regional climate change (Chen et al., 2012; Hu et al., 2013). Downscaling methods can be divided into three types: dynamical downscaling, statistical downscaling, and mixed downscaling that combines both dynamical and statistical methods (Frey-Buness et al., 1995). Dynamical downscaling is time-consuming, requires a large computational load, and relies on boundary conditions provided by the general circulation model (GCM) (Fan et al., 2005). The mixed dynamical and statistical downscaling method is still in the exploratory stage. However, statistical downscaling requires a relatively small computational load. The model is easy to program and implement and operation of the model is less time-consuming. As such, statistical downscaling has been widely adopted (Chu et al., 2010; Huang et al., 2012; Samadi et al., 2013).

The Statistical Downscaling Model (SDSM) is a decision-making support tool developed by Wilby et al. (1999, 2002, 2003), which is used to study the impacts of regional and local climate change. Coupled with multiple regression and a weather generator, the model is easy to operate, and the statistically downscaled scenario could better evaluate the impacts of regional climate change than the scenario directly simulated by the GCM. Fan et al. (2007) found that the SDSM could effectively simulate the temperature in North China where future temperatures indicate a clear upward trend. Dai et al. (2015) demonstrated that the SDSM was highly accurate for monthly temperature simulation, but the simulated value of monthly precipitation exceeded the actual value. Hao et al. (2015) used SDSM to develop a downscaling forecast of the spatial and temporal variations of future temperature and precipitation in the Hexi Corridor. The daily minimum temperature, daily maximum temperature and daily mean temperature were found to be rising, with the increases of the maximum temperature and the mean temperature greater than that of the minimum temperature. The precipitation in the Hexi Corridor had strong regional characteristics, increasing in the western region and decreasing in the central and eastern regions. Xiong and Yan (2013) conducted dynamic downscaling using a high-resolution regional climate model in the Heihe River Basin (HRB). The model was able to simulate the annual and seasonal spatial distribution of precipitation in the HRB and annual variation of precipitation in different regions of this basin. However, the simulated values of oasis areas in the middle reaches of the HRB deviated more from the observed values than in other areas. Cao et al. (2008) conducted a study on corn output in the oasis irrigation areas of the Hexi Corridor indicating that the accumulated temperature in irrigated areas in the Hexi Corridor was significantly increased and associated with global warming. The active accumulated temperature during corn growth is a key factor affecting corn yield. As the climate in the irrigated areas warms, the corn yield in selected local areas has increased. The yield increases, from west to east, were 124, 186, and 301 %, respectively. Zhao et al. (2012) found that temperatures in the HRB increased significantly from 1960 to 2009. The increased temperatures affected the crop growth period, bringing both advantages and problems. There were increased yields of spring wheat, corn, and cotton, but higher pest population levels. Agriculture is significantly affected by climate change.

The arid inland area in Northwest China is an important producer of commodity grains and also a core area of the Silk Road Economic Belt. The HRB is the second largest inland river basin in the arid areas of Northwest China. The variation characteristics of water productivity of oasis in this area related to climate change are significant for the sustainable development of agriculture and the development of the One Belt One Road initiative. We integrated a multi-model adaptability evaluation method and the SDSM model, by using data from observation stations, GCMs, and ERA-40 reanalysis to generate multi-model climate change scenarios, which were used to drive the SWAP-EPIC model to simulate crop growth, and assess the response of water productivity of oasis to climate change.

2 Data and methodology

2.1 Study area

The Heihe River Basin is located in the arid interior area of Northwest China. It is the second largest inland river basin in China and is in the middle of the Hexi Corridor, located between 98–101.5° E and 38–42° N (Wang, 2016). The Heihe River originates in the northern Qilian Mountains and it is adjacent to the Shiyang River Basin in the east and the Shule River Basin in the west. With a total length of 821 km and covering a total area of 142 900 km^2, the Heihe River area has an annual runoff of 1.58 billion m^3 (Zhang et al., 2005; Jiang, 2013). The HRB is divided by the Yingluo Gorge and the Zhengyi Gorge into the upper, middle and lower reaches. Located above the Yingluo Gorge, the upper reaches form the major runoff producing area at an altitude of 1680–5280 m with a cold and humid climate. The middle reaches are located between the Yingshuo Gorge and the Zhengyi Gorge where the terrain is a mix of mountains and plains. With an elevation of 1300–1680 m, the middle reaches have a mean annual precipitation of 140 mm, evaporation of 141 mm, abundant light and heat resources, and well-developed agriculture. The middle reaches are the main water consumption area in the HRB. The lower reaches occur below the Zhengyi Gorge at elevations between 980–1300 m. The lower reaches are characterized by open terrain, dry climate, and sparse vegetation, and this is a main source of sandstorms in North China (Shen, 2006; Guo et al., 2011; Jiang, 2013; Zang and Liu, 2013).

2.2 Data

Data collected by the meteorological stations, the GCM data and the ERA-40 reanalysis data were the basic data used in this study.

1. Data collected by the meteorological stations included the daily precipitation, daily mean temperature, daily maximum temperature and daily minimum temperature. Data were collected by 17 meteorological stations from

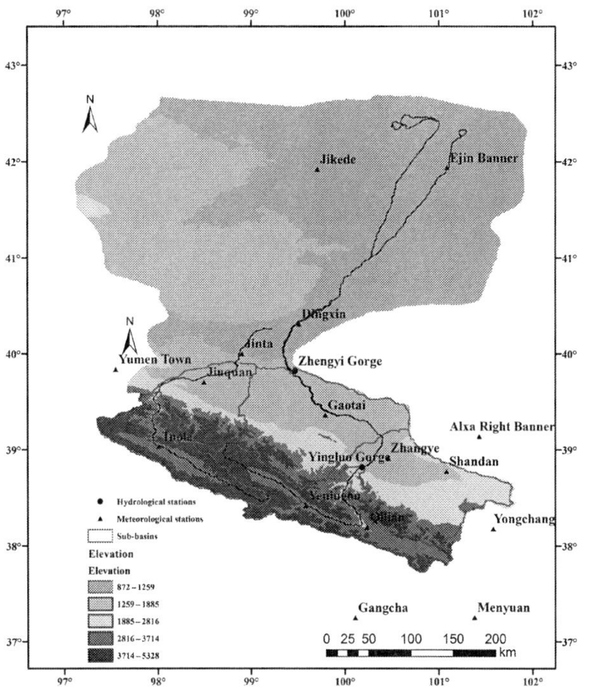

Figure 1. Location of the Heihe River Basin and hydrometeorological stations.

1961 to 2000 in the Heihe River Basin (Fig. 1). The data were extracted from China Meteorological Data Service Center (CMDC) http://data.cma.cn/.

2. Meteorological data of 23 GCMs were selected from CMIP5 in CEDA (Center for Environmental Data Analysis), including monthly and daily data. The monthly data over the time span from 1961 to 2000 were used for the GCM adaptability evaluation and were uniformly interpolated to a 2° × 2° resolution. The daily data were used for statistical downscaling to generate the meteorological data of the baseline period and the future climate scenario data. The baseline period was from 1976 to 2005, and the future climate scenarios corresponded to the time period from 2018 to 2047, including the scenarios of three RCPs (RCP2.6, RCP4.5, and RCP8.5) (Wang et al., 2012). The selected GCMs and their source information refer to Guo (2017).

3. The reanalysis datasets have been widely used in climate change projections, such ERA-40 from the European Center for Medium-Range Weather Forecasts (ECMWF), NCEP (Nation Centers for Environmental Prediction), however the reliability of the datasets at different time-spatial scales shows significant differences. In China, especially in the western part, the reliability of the ERA-40 is higher than the NCEP and other reanalysis datasets (Zhao and Fu, 2006; Chao, 2011; Xie et al., 2011). In this study, ERA-40 reanalysis monthly and daily data from 1961 to 2000 were downloaded from the ECMWF (http://apps.ecmwf.int/datasets/data/era40-daily/levtype=sfc/), with the meteorological elements being the same as the large-scale predictors of the GCM at a 2° × 2° resolution.

2.3 Methodology

2.3.1 Multi-model adaptability evaluation

The GCM is the most feasible current method for predicting large-scale climate changes. However, due to the differences between each GCM in resolution, initial conditions, and mechanisms, the results have significant uncertainty (Cox and Stephenson, 2007). There are large differences between the simulation results of different GCM models, and the accuracy of simulation results is closely related to the simulated area and the simulated climate variables (Jiang et al., 2011; Johnson and Sharma, 2012; Liu et al., 2013). Therefore, before GCM data are used to study regional climate changes, it is necessary to conduct an adaptability evaluation on the simulation ability of each GCM model in the study area and then select the GCM model that is best adapted to the study area.

The rank scoring method is advantageous because it evaluates the GCM with multiple criteria, and rank scoring has been effectively used in the adaptability evaluation of multiple regions (Rosenberg et al., 2010; Liu et al., 2011; Chen et al., 2018). In this study, the rank scoring method was used as the evaluation method. First, the monthly data of 23 GCMs in CMIP5 from 1961 to 2000 were selected, and the GCM output data were uniformly interpolated to a grid of 2° × 2° resolution. The monthly precipitation, monthly mean temperature, monthly maximum temperature and monthly minimum temperature data collected by 17 meteorological stations in the Heihe River Basin were selected as the observation data. A total of 11 statistical eigenvalues of the 23 GCMs and the measured data were compared to evaluate the applicability of the GCM in this area and to select the optimal GCM for each climate variable.

The method was calculated as follows:

$$R_{si} = \frac{X_{\max} - X_i}{X_{\max} - X_{\min}} \times 10, \tag{1}$$

where i is the model number, X_i denotes the relative error between the simulated value and the measured value for a statistical characteristic, and $X_{\min}$ and $X_{\max}$ represent the minimum and maximum relative error, respectively. The smaller the X_i, the larger the R_s.

2.3.2 SDSM

SDSM is a statistical downscaling model coupled with multiple regression and a weather generator. It is easy to operate and is widely applied in the downscaling of regional climate changes (Fan et al., 2007). The main content of the SDSM includes two aspects (Yin et al., 2010): First, build the statistical relationship between the predictand (the climatic elements of the stations) and the predictor (atmospheric circulation factor), and then determine the model. Second, assemble the future sequence data of the climatic elements based on the determined model.

The relationship between a predictand and a predictor can be expressed as:

$$R = F(L) \tag{2}$$

where R is the predictand, L is the predictor, and F denotes a deterministic or random function.

The predictand should be selected according to the study aim, and the selection of a predictor should abide by the following principles (Liu et al., 2008): it must be a factor existing in both the measured data and the GCM data; it must be a factor that can be accurately simulated by the GCM; and it must have a physical connection with the predictand.

After the predictors are screened, the SDSM will establish a statistical relationship according to the selected predictors and predictand and then determine the parameters of the multiple regression equation using the effective dual simplex method. This involves calibration of the model (Yin et al., 2010; Liu et al., 2011). After the model is determined, future climate scenarios output by the GCM can be used to simulate future data of the stations' climatic elements and compare these with the climatic elements in the baseline period. This allows analysis of the future variation trend of climatic elements in the study area (Liu et al., 2008; Yin et al., 2010).

2.3.3 The SWAP-EPIC model

The soil water atmosphere plant (SWAP) model consists of sub-modules such as soil water movement, solute transport, heat transfer, soil evaporation, crop transpiration, and crop growth. Each of the sub-modules influence and interact with each other. The model is mainly used to simulate water and salt transport and crop growth at a farmland scale (Xu, 2006). The environmental policy integrated climate (EPIC) model mainly includes crop growth, hydrology, soil temperature, soil erosion, and nutrient cycling (Fan et al., 2012).

The simple crop module of the SWAP model cannot directly simulate the growth process and yield of crops and so it is necessary to manually input observation data of crop growth, which will then be used in the interactive computations with water and salt transport. The detailed crop growth module WOFOST is complex and requires so many observational data and empirical parameters that the application scope of the model is limited. Therefore, Xu et al. (2013) constructed an SWAP-EPIC model that achieved the improved coupling simulation of soil water and salt dynamic and crop growth at a farmland scale.

Table 1. Changes of precipitation and air temperature during the future 30 years under RCP scenarios.

Change	RCP2.6	RCP4.5	RCP8.5
Precipitation (%)	−4.57	−5.22	−2.40
T_{mean} (°)	0.84	1.14	1.28
T_{max} (°)	1.23	1.35	1.55
T_{min} (°)	1.08	1.18	1.68

3 Results and discussion

3.1 CMIP5 multi-model adaptability evaluation

Figure 2 shows the evaluation results of the 23 GCMs indicating their influences on the precipitation, mean temperature, maximum temperature, and minimum temperature. The GCMs with the optimal overall simulation effects were CNRM-CM5, CCSM4, MPI-ESM-LR, and BCC-CSM1-1-M. There were significant differences between the simulation effects of the 23 GCMs on precipitation. The R_s scores of all the GCMs were above 5.0 in terms of mean temperature simulation, implying good simulation effects. The R_s score of the maximum temperature and the minimum temperature ranged from 3.43–6.33 and 3.32–6.93, respectively. CCSM4 (8.55) had the best simulation effect for mean temperature, however MPM-ESM-MR (8.28) was finally selected as the downscaling GCM due to its more sufficient data series. Therefore, CNRM-CM5, MPI-ESM-MR, MPI-ESM-LR, and BCC-CSM1-1-M were selected as the GCMs to construct the future scenarios of precipitation, mean temperature, maximum temperature, and minimum temperature, respectively, in the Heihe River Basin.

3.2 Construction of multi-model climate change scenarios

Table 1 shows the variation of precipitation compared to the baseline period in the HRB in the three future climate scenarios. According to the table, the precipitation in the future tends to decrease overall with the decrease being especially obvious in the RCP4.5 scenario, reaching 5.22 %.

The mean air temperature in the three future climate scenarios showed an increasing trend compared with that in the baseline period, and the temperature rise was more prominent as the RCP increased. The temperature rise, ranked from high to low, was RCP8.5 > RCP4.5 > RCP2.6. In the RCP4.5 and RCP8.5 scenarios, mean air temperature generally increased above 1 °C, while in the RCP2.6 scenario, the temperature increased within 1 °C. The temperature in RCP2.6, RCP4.5 and RCP8.5 increased by 0.84, 1.14, and 1.28 °C, respectively compared with the baseline period.

From the perspective of the entire basin, Table 1 shows the increase in the mean annual maximum temperature in the entire basin. The maximum temperature in the three RCP sce-

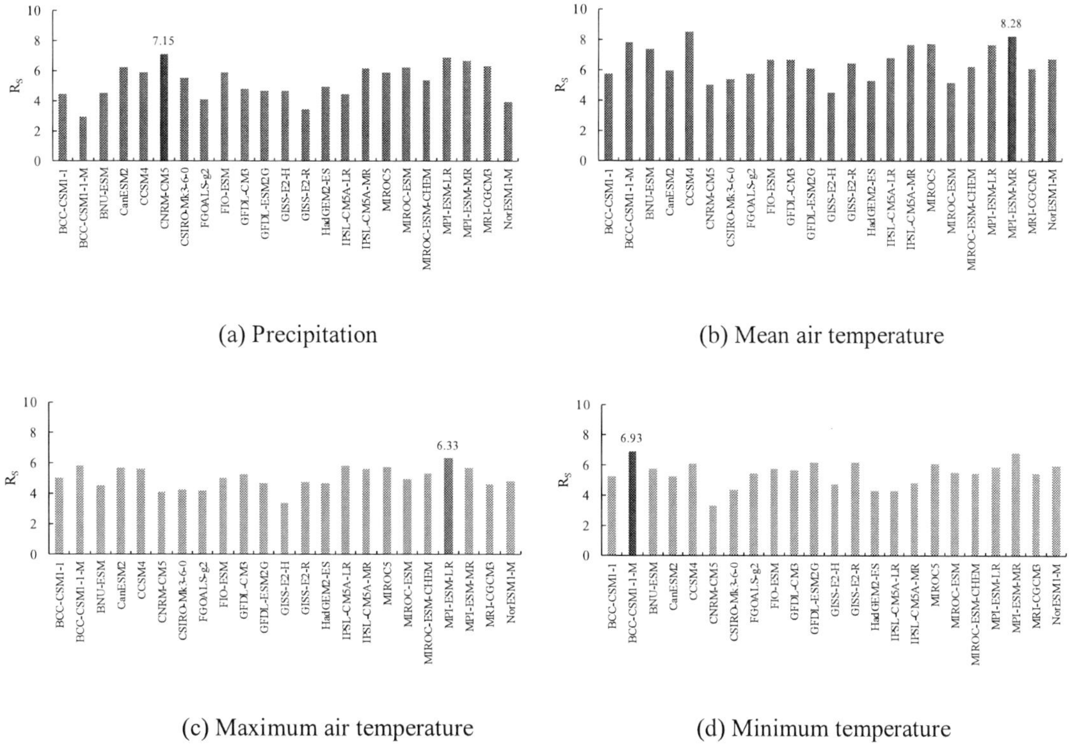

Figure 2. Results of evaluation scores from 23 GCMs by rank scoring method.

narios increased by 1.23, 1.35, and 1.55 °C, compared with the baseline period, which was more prominent than the increase in the mean annual temperature. The minimum temperature in the three RCP scenarios increased by 1.08, 1.18, and 1.68 °C compared with the baseline period.

3.3 Assessment of water productivity under climate change

3.3.1 Crop growth modeling

The SWAP-EPIC model driven by future climate scenario data was used to simulate corn growth in the scenario of RCP2.6, RCP4.5, and RCP8.5, and to obtain the yield and crop water consumption (CWC) ET_a of corn in Gaotai and Ganzhou. The SWAT-EPIC model parameters were based in the calibration and validation results of Jiang (2013) and Jiang et al. (2015) for oasis in the middle reaches of the HRB. The mean annual yield and CWC of corn growing in Gaotai in the three future climate scenarios varied from the 2012 to 2015 data (Fig. 3). The mean annual corn yield in all three future climate scenarios had a decreasing trend, and the amount of decrease ranked from high to low as P2.6 (−5.22 %) > RCP4.5 (−2.44 %) > RCP8.5 (−1.66 %). The mean annual CWC was also had a decreasing trend, and the decrease was above 12 % for all scenarios, with the highest decrease being 14.73 % in RCP4.5.

Figure 4 shows the interannual variation trend of future yield and the CWC of corn in Gaotai. In future climate scenarios, the corn yield in Gaotai featured a wavelike rise, and there were no significant differences between the three scenarios from an interannual perspective. In 2021, there was a projected dramatic increase in the yield and a projected large decrease in 2036. The CWC of corn in the three future climate scenarios did not increase or decrease significantly, showing fluctuating changes.

Figure 5 shows inter-decadal variation and the mean annual yield and CWC of corn growing in Gaotai in the three future climate scenarios compared to 2012 to 2015 data. The mean annual corn yield in the RCP2.6 scenario had a decreasing trend. However, in RCP4.5 and RCP8.5, the corn yield increased by 2.51 and 2.3 %, respectively. The mean annual CWC had a decreasing trend in all three future climate scenarios, and the amount of decrease was similar, ranking, from high to low, as RCP4.5 (−16.27 %) > RCP2.6 (−15.67 %) > RCP8.5 (−15.32 %).

Figure 6 shows the interannual variation trend of the yield and CWC of corn in Ganzhou. The corn yield in Ganzhou had a fluctuating rise in the future; there were no significant differences between the three scenarios from an interannual

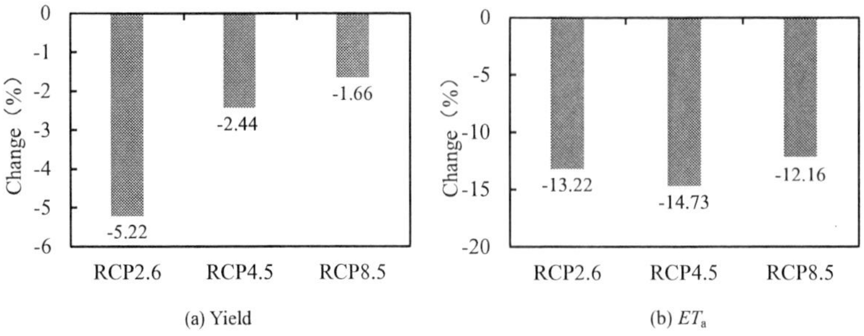

Figure 3. Changes of corn yield and ET_a in the Gaotai area under RCP scenarios.

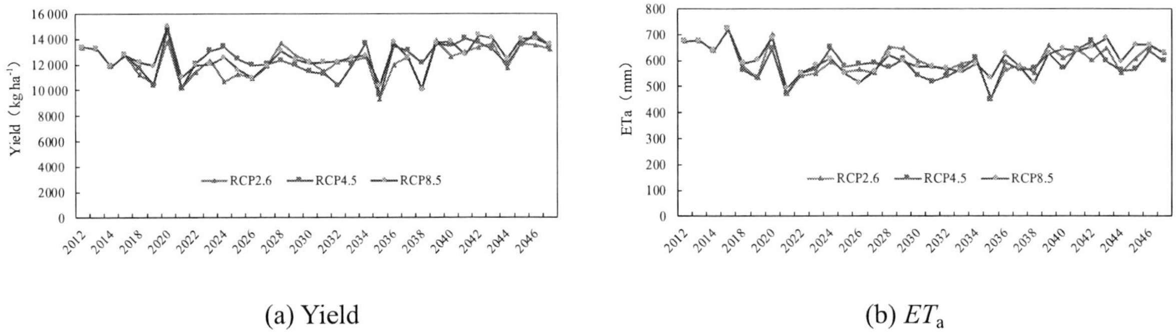

Figure 4. Temporal variations of corn yield and ET_a in the Gaotai area under RCP scenarios.

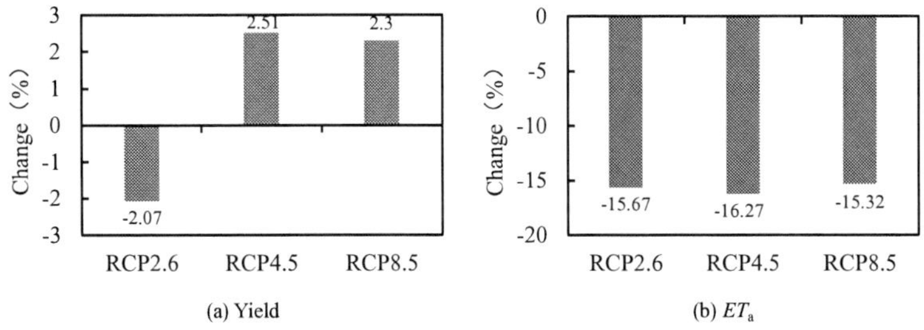

Figure 5. Changes of corn yield and ET_a in the Ganzhou area under RCP scenarios.

perspective, but the overall increase was greater than that of Gaotai. In 2018 and 2021, the yield was predicted to be relatively high, but it would decrease greatly in 2038. The CWC in the three future climate scenarios would generally fluctuate. Compared with the CWC from 2012 to 2015, the CWC in the future would first decrease and gradually show a fluctuating increase from 2022 on. However, the final CWC would remain lower than that from 2012 to 2015.

3.3.2 Water productivity evaluation

Water productivity (WP) refers to the yield or output value obtained from per unit water resources on specific crops and under specific cultivation conditions. With the unit being $\mathrm{kg\,m^{-3}}$, water productivity is mainly used to reflect the water management level or evaluate the farmland water use efficiency in an irrigated area. It can be calculated by the following equation:

$$\mathrm{WP} = \frac{Y}{ET_a}, \tag{3}$$

where Y is crop grain yield ($\mathrm{kg\,ha^{-1}}$), ET_a is the CWC during the crop growth period (crop transpiration + inter-plant evaporation) (mm).

The inter-decadal water productivity of Gaotai and Ganzhou is shown in Fig. 7. As the RCP increased, the mean temperature, maximum temperature, and minimum temperature increased, the wind speed and precipitation decreased,

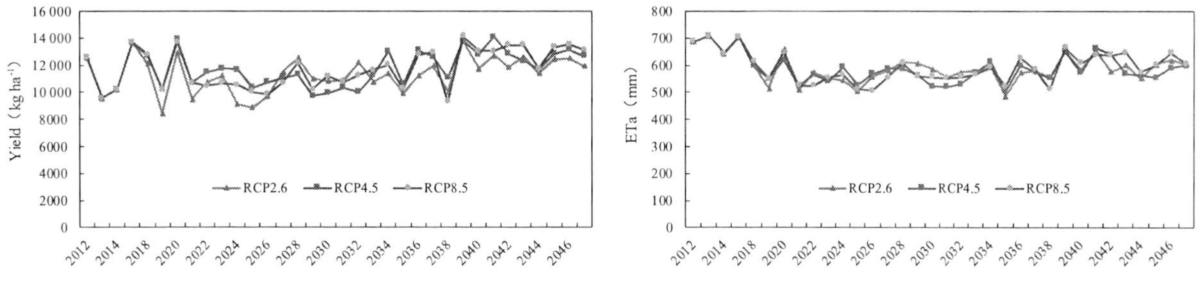

(a) Variation of corn yield

(b) Variation of ET_a

Figure 6. Temporal variations of corn yield and ET_a in the Ganzhou area under RCP scenarios.

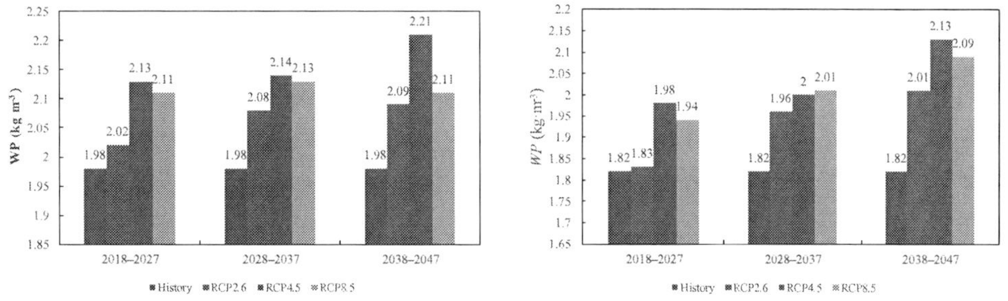

(a) Water productivity in the Gaotai area

(b) Water productivity in the Ganzhou area

Figure 7. Decadal changes of corn water productivity in the middle agricultural region of the HRB under RCP scenarios.

and the water productivity of the two locations increased. The inter-decadal water productivity in all the three future climate scenarios is higher than in the historical simulation series (2012–2015). Among them, the increase of the water productivity in the RCP4.5 scenario was the greatest. Water productivity became higher over time, which was associated with the temperature rise. In addition, increase in the maximum temperature was clearly greater than that of the minimum temperature, which led to an increase in accumulated temperature and promoted crop growth.

Figure 8 shows the variation trend of the annual water productivity in both Gaotai and Ganzhou. In the three future climate scenarios, the water productivity had an upward trend and the increase ranked, high to low, as RCP4.5 > RCP8.5 > RCP2.6. There was a decrease in the water productivity of both Gaotai and Ganzhou in 2038, and as analyzed above, the temperature, yield and CWC decreased in the same year. Therefore, the water productivity also decreased in this year. The overall growth trend in Ganzhou was higher than that in Gaotai, and the two places were similar to each other in the fluctuation trend.

4 Conclusions

The Heihe River Basin was selected as the typical area representing the agricultural oasis of the Hexi Corridor and the Silk Road Economic Belt in this study. Based on observed data at 17 meteorological stations, 23 GCMs and ERA-40 reanalysis data, the SDSM was employed to complete a downscaling projection of climate change during the baseline period and three future climate scenarios, namely RCP2.6, RCP4.5 and RCP8.5. The SWAP-EPIC model was combined to simulate corn growth in the Gaotai and Ganzhou areas located in the middle agricultural region of the HRB to obtain crop yields and CWC data under future climate change scenarios, investigating the responses of water productivity. Conclusions are as follows:

1. The SDSM carried out downscaling on the multi-GCM data and generated data on the three future climate scenarios. Precipitation in the future scenarios had decreasing trends compared to the baseline period, and the precipitation decrease in RCP4.5 was greater than that in RCP2.6 and RCP8.5. The mean, maximum, and minimum temperatures all had upward trends, and as the RCP increased, the temperature rise also increased with

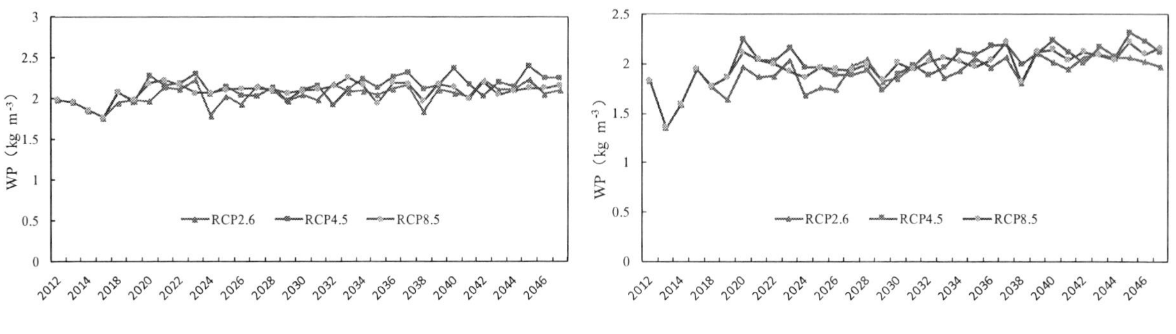

(a) Water productivity in the Gaotai area

(b) Water productivity in the Ganzou area

Figure 8. Temporal variations of corn water productivity in the middle agricultural region of the HRB under RCP scenarios.

RCP8.5 > RCP4.5 > RCP2.6. The increase in the maximum temperature was greater than that in the mean temperature and minimum temperature.

2. The SWAP-EPIC model was used to simulate the corn growth in Gaotai and Ganzhou, and explore the variation patterns of the yield, CWC and water productivity under different climate change scenarios. The water productivity of the two locations showed an upward trend in the next three decades, and the water productivity in both Gaotai and Ganzhou increased the greatest in the RCP4.5, exceeding $2\,kg\,m^{-3}$ in most of the years. Therefore, in Gaotai and Ganzhou, located in the middle oasis of the Heihe River Basin, the water productivity of corn gradually increased as the temperature increased and precipitation and wind speed decreased.

Located in the inland of Northwest China, the Heihe River Basin has an arid climate, so agricultural production mainly depends on timely irrigation. Therefore, the decreased precipitation had little effect on crop growth, while temperature rise caused an increase of accumulated temperature, which in turn promoted the growth of corn and increased corn yield and the increase of water productivity.

Data availability. The meteorological data of observation stations were provided by China Meteorological Data Service Center (http://data.cma.cn/). ERA-40 reanalysis monthly and daily data were downloaded from the ECMWF (http://apps.ecmwf.int/datasets/data/era40-daily/levtype=sfc/). 23 GCMs outputs were released by the World Climate Research Programme's Working Group on Coupled Modelling (http://cmip-pcmdi.llnl.gov/cmip5/index.html). All above data used in this study were reproduced by Guo (2017).

Competing interests. The authors declare that they have no conflict of interest.

Special issue statement. This article is part of the special issue "Innovative water resources management – understanding and balancing interactions between humankind and nature". It is a result of the 8th International Water Resources Management Conference of ICWRS, Beijing, China, 13–15 June 2018.

Acknowledgements. This work was financially supported by the National Key R&D Program of China (2017YFC0403201) and the National Natural Sciences Foundation of China (91425302). The authors are grateful to the Data Center of the Heihe Research Program (http://westdc.westgis.ac.cn/heihe), the World Climate Research Programme's Working Group on Coupled Modelling, which is responsible for CMIP, and the climate modeling groups for producing and making available their model output (http://cmip-pcmdi.llnl.gov/cmip5/index.html). The authors thank the editor and two anonymous reviewers for their constructive comments that have significantly improved this work.

Edited by: Bo Pang
Reviewed by: two anonymous referees

References

Cao, L., Deng, Z. Y., Dou, Y. X., and Yang, X. L.: Influence of climate waring on corn output in Hexi corridor oasis irrigation area and its countermeasure research, Acta Botanica Boreali-Occidentalia Sinica, 28, 1043–1048, 2008.

Chao, Z. H.: The reliability evaluation of three reanalysis air temperature data in West China, Transactions of Atmospheric Sciences, 34, 162–169, 2011.

Chen, H., Xu, C. Y., and Guo, S. L.: Comparison and evaluation of multiple GCMs, statistical downscaling and hydrological models in the study of climate change impacts on runoff, J. Hydrol., 434–435, 36–45, 2012.

Chen, J., Chen, H., and Guo, S. L.: Multi-site precipitation downscaling using a stochastic weather generator, Clim. Dynam., 50, 1975–1992, 2018.

Chu, J. T., Xia, J., Xu, C. Y., and Singh, V. P.: Statistical downscaling of daily mean temperature, pan evaporation and precipitation

for climate change scenarios in Haihe River, China, Theor. Appl. Climatol., 99, 149–161, 2010.
Cox, P. and Stephenson, D.: A Changing Climate for Prediction, Science, 317, 207–208, 2007.
Dai, H. H., Yang, H. B., and Hu, Q. F.: Prediction of climate change over Shule River basin based on a statistical downscaling method, Hydro-Science and Engineering, 5, 46–53, 2015.
Fan, L. J., Fu, C. B., and Chen, D. L.: Review on creating future climate change scenarios by statistical downscaling techniques, Advances in Earth Science, 20, 320–329, 2005.
Fan, L. J., Fu, C. B., and Chen, D. L.: Estimation of local temperature change scenarios in North China using statistical downscaling method, Chinese Journal of Atmospheric Sciences, 31, 887–897, 2007.
Fan, L., Lv, C. H., and Chen, C.: A review of EPIC model and its applications, Prog. Geogr., 31, 584–592, 2012.
Frey-Buness, F., Heimann, D., and Sausen, R.: A statistical-dynamical downscaling procedure for global climate simulations, Theor. Appl. Climatol., 50, 117–131, 1995.
Guo, Q. L., Yang, Y. S., Chang, X. S., and Chen, Z. H.: Annual variation of Heihe River runoff during 1957–2008, Progress in Geography, 30, 550–556, 2011.
Guo, Z. Z.: Evaluation of water productivity in irrigated areas of the middle Heihe River basin under climate change, M.S. thesis, College of Water Resources and Civil Engineering, China Agricultural University, Beijing, China, 2017.
Hao, L. N., Su, X. L., and Wang, N.: Future spatial and temporal changes of temperature and precipitation in Hexi corridor based on SDSM, Journal of Northwest A&F University (Nat. Sci. Ed.), 43, 219–228, 2015.
Hu, Y., Maskey, S., and Uhlenbrook, S.: Downscaling daily precipitation over the Yellow River source region in China: a comparison of three statistical downscaling methods, Theor. Appl. Climatol., 112, 447–460, 2013.
Huang, J., Zhang, J., Zhang, Z., Sun, S., and Yao, J.: Simulation of extreme precipitation indices in the Yangtze River basin by using statistical downscaling method (SDSM), Theor. Appl. Climatol., 108, 325–343, 2012.
IPCC: Working Group I Contribution to the IPCC Fifth Assessment Report, Climate Change 2013: The Physical Science Basis, Cambridge, Cambridge University Press, 2013.
Jiang, X. H., Xu, Z. X., Liu, Z. F., and Liu, L.: Performance of Different General Circulation Models (GCMs) in the Basins of Southeastern China, Resources and Environment in the Yangtze Basin, 31, 51–58, 2011.
Jiang, Y., Xu, X., Huang, Q. Z., Huo, Z. L., and Huang, G. H.: Assessment of irrigation performance and water productivity in irrigated areas of the middle Heihe River basin using a distributed agro-hydrological model, Agr. Water Manage., 147, 67–81, 2015.
Jiang, Y.: Simulation of aro-hydrological processes and assessment of water productivity in irrigated areas of the middle Heihe River basin, M.S. thesis, College of Water Resources and Civil Engineering, China Agricultural University, Beijing, China, 2013.
Johnson, F. and Sharma, A.: A nesting model for bias correction of variability at multiple time scales in general circulation model precipitation simulations, Water Resour. Res., 48, W01504, https://doi.org/10.1029/2011WR010464, 2012.
Liu, W. F., Xu, Z. X., Li, F. P., and Qiu, L. H.: GCM performance on simulating climatological factors in Yarlung Zangbo River basin based on a ranked score method, Journal of Beijing Normal University (Natural Science), 49, 304–311, 2013.
Liu, W. F., Xu, Z. X., Li, F. P., Liu, X. C., and Li, X. P.: Performance of different General Circulation Models (GCMs) in the basins of Southeastern China, Journal of Subtropical Resources and Environment, 6, 13–23, 2011.
Liu, Z. F., Xu, Z. X., Liu, L. L., and Chen, Y. N.: Temporal trends of future maximum and minimum air temperature in the Tarim River Basin, Arid Land Geography, 31, 822–829, 2008.
Rosenberg, E. A., Keys, P. W., Booth, D. B., Harley, D., Burkey, J., Steinemann, A. C., and Lettenmaier, D. P.: Precipitation extremes and the impacts of climate change on stormwater infrastructure in Washington State, Climatic Change, 102, 319–349, 2010.
Samadi, S., Wilson, C. A. M. E., and Moradkhani, H.: Uncertainty analysis of statistical downscaling models using Hadley Centre Coupled Model, Theor. Appl. Climatol., 114, 673–690, 2013.
Shen, Y. Y.: Study of numeric simulative model of groundwater in Heihe River basin and its implementation in water scheduling management, M.S. thesis, College of Environment and Resources, Jilin University, 2006.
Wang, F.: Future Hydrological Variations of RCP-based Scenarios for the Heihe River Basin, M.S. thesis, College of Water Resources and Civil Engineering, China Agricultural University, Beijing, China, 2016.
Wang, S. W., Luo, Y., Wen, X. Y., and Huang, J. B.: New Generation of Scenarios of Greenhouse Gas Emission, Advances in Climate Change Research, 8, 305–307, 2012.
Wilby, R. L., Hay, L. E., and Leavesley, G. H.: A comparison of downscaled and raw GCM output: implications for climate change scenarios in the San Juan River basin, Colorado, J. Hydrol., 225, 67–91, 1999.
Wilby, R. L., Dawson, C. W., and Barrow, E. M.: SDSM – A decision support tool for the assessment of regional climate change impacts, Environ. Modell. Softw., 17, 145–157, 2002.
Wilby, R. L., Tomlinson, O. J., and Dawson, C. W.: Multisite Simulation of Precipitation by Conditional Resampling, Climate Research, 23, 183–194, 2003.
Xie, X., He, J. H., and Qi, L.: A review on applicability evaluation of four reanalysis datasets in China, Journal of Meteorology and Environment, 27, 58–65, 2011.
Xiong, Z. and Yan, X. D.: Building a high-resolution regional climate model for the Heihe River Basin and simulating precipitation over this region, Chinese Sci. Bull., 58, 4670–4678, 2013.
Xu, D.: Review on scaling study in irrigation hydrology, J. Hydraul. Eng., 37, 141–149, 2006.
Xu, X., Huang, G. H., and Huang, Q. Z.: Coupled simulation of soil water flow, solute transport and crop growth processes at field scale and its validation, Transactions of the Chinese Society of Agricultural Engineering, 29, 110–117, 2013.
Yin, Z. Y., Lai, A. W., Gong, Y., Peng, T., and Shen, T. Y.: Research development on downscaling method in meteorology and hydrology coupling, Torrential Rain and Disasters, 29, 89–95, 2010.
Zang, C. F. and Liu, J. G.: Spatial and temporal pattern difference of blue-green water flows in typical reference years in the Heihe River Basin, northwestern China, Journal of Beijing Forestry University, 35, 1–10, 2013.

Zhang, B., Ding, W. H., and Meng, B.: Impacts of land use changes on groundwater resources in arid area – Case study of middle reaches of Heihe River, Arid Land Geography, 28, 764–769, 2005.

Zhao, T. B. and Fu, C. B.: Preliminary Comparison and Analysis between ERA-40, NCEP-2 Reanalysis and Observations over China, Climatic and Environmental Research, 11, 14–32, 2006.

Zhao, Y. F., Zhang, B., Zhang, Y. N., Wang, Q., He, X. Q., and Yin, H. X.: Impacts of climate change on agriculture production in the Heihe river basin in recent 50 years, Journal of Arid Land Resources and Environment, 26, 130–136, 2012.

Zuo, H. C., Lv, S. H., and Hu, Y. J.: Variation trend of yearly mean air temperature and precipitation in China in the last 50 years, Plateau meteorology, 23, 238–244, 2004.

https://doi.org/10.5194/piahs-379-403-2018

Connections between meteorological and hydrological droughts in a semi-arid basin of the middle Yellow River

Binquan Li[1,2,3,4], **Changchang Zhu**[1], **Zhongmin Liang**[1], **Guoqing Wang**[2,3], **and Yu Zhang**[1]

[1]College of Hydrology and Water Resources, Hohai University, Nanjing 210098, China
[2]State Key Laboratory of Hydrology-Water Resources and Hydraulic Engineering, Nanjing Hydraulic Research Institute, Nanjing 210029, China
[3]Research Center for Climate Change, Ministry of Water Resources, Nanjing 210029, China
[4]Institute of Arid Meteorology, China Meteorological Administration, Lanzhou 730020, China

Correspondence: Zhongmin Liang (zmliang@hhu.edu.cn)

Received: 18 December 2017 – Revised: 15 March 2018 – Accepted: 19 March 2018 – Published: 5 June 2018

Abstract. Differences between meteorological and hydrological droughts could reflect the regional water consumption by both natural elements and human water-use. The connections between these two drought types were analyzed using the Standardized Precipitation Evapotranspiration Index (SPEI) and Standardized Streamflow Index (SSI), respectively. In a typical semi-arid basin of the middle Yellow River (Qingjianhe River basin), annual precipitation and air temperature showed significantly downward and upward trends, respectively, with the rates of $-2.37\,\mathrm{mm\,yr^{-1}}$ and $0.03\,^{\circ}\mathrm{C\,yr^{-1}}$ (1961–2007). Under their synthetic effects, water balance variable (represented by SPEI) showed obviously downward (drying) trend at both upstream and whole basin areas. For the spatial variability of precipitation, air temperature and the calculated SPEI, both upstream and downstream areas experienced very similar change characteristics. Results also suggested that the Qingjianhe River basin experienced near normal condition during the study period. As a whole, this semi-arid basin mainly had the meteorological drought episodes in the mid-1960s, late-1990s and the 2000s depicted by 12-month SPEI. The drying trend could also be depicted by the hydrological drought index (12-month SSI) at both upstream and downstream stations (Zichang and Yanchuan), but the decreasing trends were not significant. A correlation analysis showed that hydrological system responds rapidly to the change of meteorological conditions in this semi-arid region. This finding could be an useful implication to drought research for those semi-arid basins with intensive human activities.

1 Introduction

In recent decades, frequently occurred drought disaster has significant impacts on industrial and agricultural production, human's livelihood and ecological environmental protection (Mishra and Singh, 2010). With the social and economic development and population explosion, water resources shortage has been a worsening problem. Thus, drought has attracted more and more attention from all sectors of society. According to the WMO's statistics, the loss caused by meteorological disaster occupied more than 85 % of the total natural disaster's losses; in which, about 50 % of these meteorological disasters were caused by droughts (Obasi, 1994). China is one of the countries that experienced frequent and serious drought disaster (Li et al., 2017, 2015; Zou et al., 2005). For example, a large-scale drought occurred in 2000 spread to 20 provinces of China, resulting in substantial economic losses. More than 12 provinces of China were affected by the 2009 drought event, and especially in some northern provinces, the drought degree reached an extreme level. In another example, a drought with 100-year return period swept across southwest China during summer 2009 to spring 2010, resulting in a large decrease in most river levels (Lu et al., 2011).

Due to the differences of purpose or target in drought research and the complexity of drought formation cause, there are several drought classifications: meteorological drought, hydrological drought and agriculture drought. Meteorological drought describes water shortage phenomenon caused by the imbalance between precipitation and evaporation during a specific period. In general, water shortage degree (calculated with precipitation) would be used as a measure of meteorological drought because of the accessibility of precipitation data. The popular drought indices calculated by precipitation include continuous rainless days, precipitation anomaly percentage, precipitation Z index, and Standardized Precipitation Index (SPI, McKee et al., 1993), etc. The above single precipitation factor based indices lack the consideration of evaporative demand, and thus may be cannot objectively characterize the drought process (Li et al., 2014). To overcome this limitation in drought calculation, many efforts have been conducted to develop new drought indices by introducing extra influencing factors. For example, the Palmer Drought Severity Index (PDSI) uses readily available temperature and precipitation data to estimate relative dryness (Palmer, 1965), and has been widely used in drought research. Vicente-Serrano et al. (2010) proposed a new meteorological index, Standardized Precipitation Evapotranspiration Index (SPEI), combining the sensitivity of PDSI to changes in evaporative demand and the robustness of multi-temporal nature of SPI. Hydrological drought represents the abnormal water shortage caused by the imbalance among precipitation, surface and sub-surface runoff. It measures the anomaly of river streamflow, and its common indices include anomalous percentage of river discharge, anomalous percentage of reservoir storage, and water resource scarcity index, etc. Vicente-Serrano et al. (2011) developed the Standardized Streanflow Index (SSI) for hydrological drought analysis. It evaluates the water shortage degree of a stream or set of streams in multi-temporal scales. Thus, this index could be an appropriate choice for hydrological drought analysis (Li et al., 2014).

The main objective of this paper is to conduct the connection analysis between meteorological and hydrological droughts in a semi-arid region of the middle Yellow River. The SPEI and SSI were used to represent the meteorological and hydrological indices, respectively, and calculated in a typical basin (Qingjianhe River basin) for the period of 1961–2007. Response of hydrological systems to meteorological conditions would be analyzed for better understanding the regional water resources scarcity.

2 Study area and methods

2.1 Study area

Qingjianhe River basin locates between 109°12′–111°24′ E and 36°39′–37°19′ N in the middle Yellow River region (Fig. 1). It covers an area of 4078 km^2, and its elevations increase from 536 m in the east to 1629 m in the west with an average altitude of 1108 m. Annual mean basin-wide precipitation is 451 mm, but is not evenly distributed in space and time. About 65 % precipitation of annual total amount

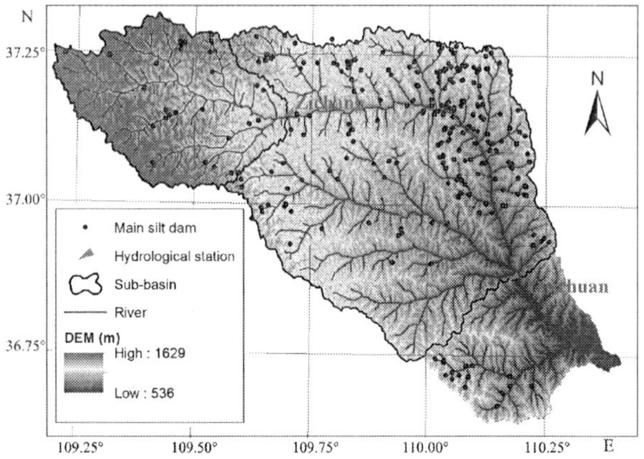

Figure 1. Map of the Qingjianhe River basin.

occurs during the period of July–September. Observed annual streamflow is about 138 million m^3 in this basin. There are two hydrological stations located at the mainstream of Qingjianhe River: Zichang and Yanchuan. The cover areas of these two stations were 913 and 3468 km^2, respectively. Monthly streamflow data (1961–2007) were collected at these two hydrological stations. Daily precipitation data for the corresponding study period were collected from the 0.25° gridded APHRODITE product (Yatagai et al., 2012).

2.2 Drought indices and trend analysis method

Water depict based index, SPEI, could measure the anomaly of the difference between precipitation (P) and potential evapotranspiration (PET) by using monthly precipitation and air temperature (Vicente-Serranto et al., 2010). Thus, the meteorological drought in the study basin was characterized by SPEI that used a three parameter log-logistic probability distribution for P-PET sample fitting. The SPEI values are accumulated to different time scales, and converted to standard deviations with respect to average values. As for the hydrological drought, SSI was used to calculate different time scale streamflow anomaly (Vicente-Serranto et al., 2011). SSI incorporates hydrological processes that determine seasonal loss in streamflow due to the influence of both climate and human use. The calculated SSI value represents an average dry or wet regime in the controlled drainage areas of the hydrological stations. The following drought categories were used for both drought indices (McKee et al., 1993): (a) 2.00 and above: extremely wet; (b) 1.50 to 1.99: very wet; (c) 1.00 to 1.49: moderately wet; (d) −0.99 to 0.99: near normal; (e) −1.00 to −1.49: moderately dry; (f) −1.50 to −1.99: severely dry; and (g) −2.00 and less: extremely dry.

The nonparametric Mann–Kendall (M–K) test (Kendall, 1975; Mann, 1945) was used to detect the trends in time series of meteorological and hydrological factors and drought indices. A positive (negative) value of the M–K test statistic

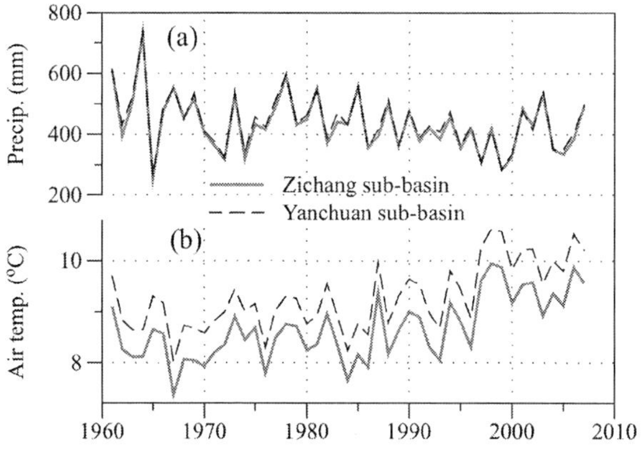

Figure 2. Annual changes of precipitation and air temperature for Zichang and Yanchuan sub-basins during the period of 1961–2007.

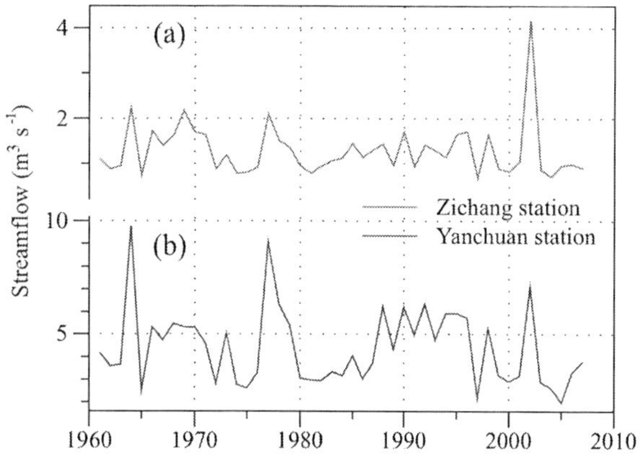

Figure 3. Annual changes of streamflow at Zichang and Yanchuan stations during the period of 1961–2007.

Z signifies an increasing (decreasing) trend. In order to eliminate the effect of significant serial correlation in time series on trend analysis, a trend-free pre-whitening procedure (Yue et al., 2002) was used before applying the M–K test.

3 Results and discussion

3.1 Trend analysis of meteorological factors and streamflow

Trend analyses for precipitation and air temperature were performed for Zichang and Yanchuan sub-basins to understand the evolutions of drought episodes with the effect of climate change. In general, precipitation had a very similar temporal pattern in these two sub-basins, ranging from 254 to 732 mm, during the period of 1961–2007 (Fig. 2a). Annual mean precipitation were 434 and 444 mm for Zichang and Yanchuan sub-basins, respectively. Both sub-basins experienced the decreasing trend of precipitation with the significance at 95 % confidence level. The decreasing rates were -2.11 and $-2.37\,\mathrm{mm\,yr^{-1}}$ for Zichang and Yanchuan sub-basins, respectively. On seasonal scale, only the fall season had the statistically significant decreasing trend in the Qingjianhe River basin. Under the global warming, the basin also showed the remarkable upward trend of annual air temperature in both sub-basins (Fig. 2b). The warming rate was $0.03\,^{\circ}\mathrm{C\,yr^{-1}}$ for the period of 1961–2007. On seasonal scale, all four seasons experienced the statistically significant warming trend with 99 % confidence level, and winter had the largest warming rate of $0.05\,^{\circ}\mathrm{C\,yr^{-1}}$.

According to the M–K test results, downward trends (but not significant) were found at both hydrological stations (Fig. 3). It was noted that streamflow at Yanchuan station (Fig. 3b) included all runoff volume from both sub-basins. In theory, significant decrease in precipitation and increase in air temperature were the indicator of the increase of water deficit of the study basin. However, the final trend of streamflow detected in Fig. 3 reflected both natural and human factors. The human activities are mainly manifested as the large-scale constrictions of soil and water conservation measures in the past decades in the middle Yellow River region, greatly reducing the basin runoff generation and the corresponding streamflow (Li et al., 2018).

3.2 Calculations of drought indices

Figure 4 shows the evolution of two indices over 3 and 12 months intervals for two sub-basins or hydrological stations during the period of 1961–2007. It was obvious that short timescales (3 months) demonstrated a higher temporal frequency of dryness and wetness episodes for both two indices. Correspondingly, long timescales (12 months) produced the lower temporal frequency and longer duration of dryness and wetness episodes. SPEI showed the similar evolution patterns between two sub-basins, indicating of the consistency of two influencing factors (precipitation and air temperature). Generally, the Qingjianhe River basin experienced a near normal condition during the whole study period. This region mainly had the meteorological drought episodes in the mid-1960s, late-1990s and the 2000s depicted by 12-month SPEI.

As for the hydrological drought index, the SSI series calculated at Zichang and Yanchuan stations also had the similar temporal patterns, but showed some differences in dryness or wetness magnitude (Fig. 4). According to the 12-month SSI series, alternant dryness and wetness spells could be found at the Zichang station after the year of 1990. However, most years after 2000 experienced drought events (from moderately to severely and even to extremely dry) except for a wet year of 2002. This temporary change to wet condition could also be reflected in the streamflow series (Fig. 3a).

According to the M–K test of 12-month drought indices, SPEI showed the obviously downward (drying) trend with a significance of 99 % confidence level at both Zichang

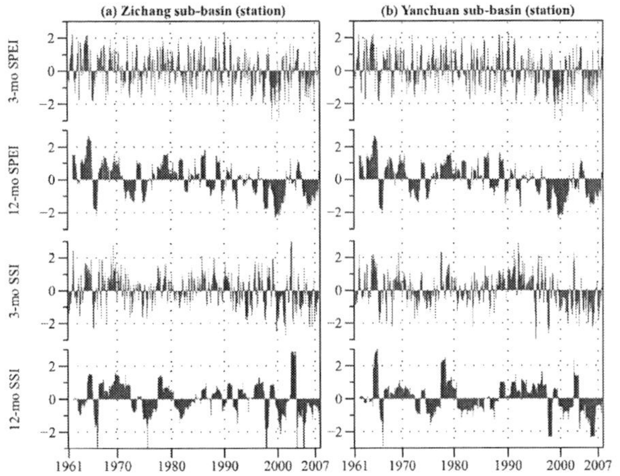

Figure 4. Evolution of 3-month and 12-month SPEI and SSI during the period of 1961–2007.

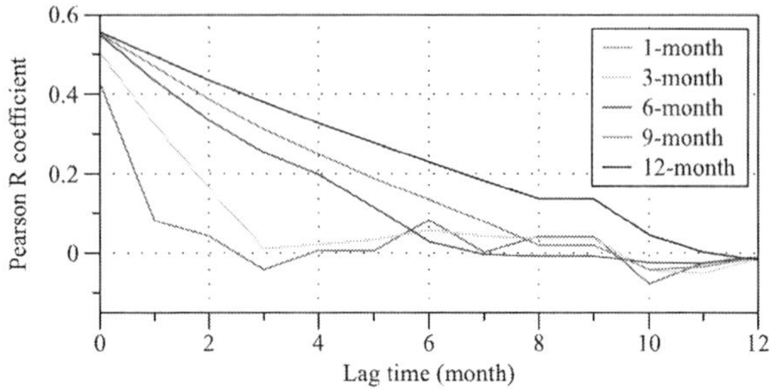

Figure 5. Correlation analysis between SPEI and SSI for different lag times.

and Yanchuan sub-basins. The drying rates were −0.03 and −0.04 for these two sub-basins, respectively. This drying trend could be well explained by the meteorological trends of Qingjianhe River basin (Fig. 2). The drying trend could also be depicted by the M–K test result of hydrological drought index. However, the downward trends of 12-month SSI were not statistically significant at both upstream and downstream stations (Zichang and Yanchuan).

3.3 Connections between meteorological and hydrological droughts

The correlation analysis was carried out between SPEI and SSI at different timescales (from one to 12 months) for intercomparison. The time series of Pearson R correlation coefficient could be calculated between two drought index series. As an example, Fig. 5 shows the Pearson R correlation coefficients between SPEI and SSI with 1, 3, 6, 9, and 12 month time scales and different time lags (from one to 12 months). The time lag represents the time for response of hydrological system to climate variable on a basin scale. Results suggested that the correlation between these two drought indices was relatively low ($R < 0.6$). It indicated that the response relationship of hydrological system to precipitation and evaporation was highly nonlinear, even on annual time scale. The main reason should be the influence of intensive human activities (i.e., the increasing constructions of soil and water conservation measures). For example, the local people have constructed 274 main silt dams from 1955 to 2009, with the total water storage of 251 million m^3 (see their locations in Fig. 1). These main silt dams controlled more than 25 % of total area of the Qingjianhe River basin. This runoff-retaining function should certainly affect the correlation between climatic factors and hydrological response.

The correlation analysis also suggested that there were obvious increasing tendencies of Pearson R correlation with accumulated timescales in SPEI vs. SSI, with the largest correlation values at zero month time lag. This indicated that, in such small size basins, less than one month was enough for hydrological system response to climate factors.

4 Conclusions

In this study, two multi-scalar drought indices, SPEI and SSI, were used to characterize the meteorological and hydrological droughts in the Qingjianhe River basin, respectively. Results showed that the Qingjianhe River basin experienced near normal condition during the study period. As a whole, this semi-arid basin mainly had the meteorological drought episodes in the mid-1960s, late-1990s and the 2000s depicted by 12-month SPEI. The drying trend could also be depicted by the hydrological drought index (12-month SSI) at both upstream and downstream stations (Zichang and Yanchuan), but the decreasing trends were not significant.

The correlation analysis between SPEI and SSI with different time lags suggested that, the response relationship of hydrological system to precipitation and evaporation was highly nonlinear due to the effects of intensive human activities. In addition, results indicated that hydrological system respond rapidly to meteorological conditions in this semi-arid region. The nonlinear response of hydrological drought to meteorological drought was caused by both nature and human factors.

Data availability. The data and code used in this research will be made available upon request; please send request to libinquan@gmail.com (Binquan Li) for data and code.

Competing interests. The authors declare that they have no conflict of interest.

Special issue statement. This article is part of the special issue "Innovative water resources management – understanding and balancing interactions between humankind and nature". It is a result of the 8th International Water Resources Management Conference of ICWRS, Beijing, China, 13–15 June 2018.

Acknowledgements. This work was supported by the National Key R&D Program of China (2016YFC0402706, 2016YFA0601501, 2016YFA0601601), National Natural Science Foundation of China (51509067, 41730750), Special Scientific Research Fund of Public Welfare Industry of Ministry of Water Resources of China (201501004), China Postdoctoral Science Foundation (2017T100388), and Research Funds for Arid Meteorology (IAM201510).

Edited by: Wenchao Sun
Reviewed by: two anonymous referees

References

Kendall, M. G.: Rand correlation methods, Charles Griffin, London, 1975.

Li, B., Liang, Z., Yu, Z., and Acharya, K.: Evaluation of drought and wetness episodes in a cold region (Northeast China) since 1898 with different drought indices, Nat. Hazards, 71, 2063–2085, 2014.

Li, B., Zhou, W., Zhao, Y., Ju, Q., Yu, Z., Liang, Z., and Acharya, K.: Using the SPEI to assess recent climate change in the Yarlung Zangbo River Basin, South Tibet, Water, 7, 5474–5486, 2015.

Li, B., Liang, Z., Zhang, J., and Wang, Q.: A revised drought index based on precipitation and pan evaporation, Int. J. Climatol., 37, 793–801, 2017.

Li, B., Liang, Z., Zhang, J., Wang, Q., Zhao, W., Zhang, H., Wang, J., and Hu, Y.: Attribution analysis of runoff decline in a semiarid region of the Loess Plateau, China, Theor. Appl. Climatol., 131, 845–855, 2018.

Lu, E., Luo, Y., Zhang, R., Wu, Q., and Liu, L.: Regional atmospheric anomalies responsible for the 2009–2010 severe drought in China, J. Geophys. Res., 116, D21114, https://doi.org/10.1029/2011JD015706, 2011.

Mann, H. B.: Nonparametric tests against trend, Econometrica, 13, 245–259, 1945.

McKee, T. B., Doesken, N. J., and Kleist, J.: The relationship of drought frequency and duration to time scales, Proceedings of the 8th Conference on Applied Climatology, American Meteorological Society Boston, MA, 179–183, 1993.

Mishra, A. K. and Singh, V. P.: A review of drought concepts, J. Hydrol., 391, 202–216, 2010.

Obasi, G. O. P.: WMO's role in the international decade for natural disaster reduction, B. Am. Meteorol. Soc., 75, 1655–1661, 1994.

Palmer, W. C.: Meteorologic Drought, US Department of Commerce, Weather Bureau, Research Paper No. 45, p. 58, 1965.

Vicente-Serrano, S. M., Begueria, S., and López-Moreno, J. I.: A multiscalar drought index sensitive to global warming: The Standardized Precipitation Evapotranspiration Index, J. Climate, 23, 1696–1718, https://doi.org/10.1175/2009jcli2909.1, 2010.

Vicente-Serrano, S. M., López-Moreno, J. I., Beguería, S., Lorenzo-Lacruz, J., Azorin-Molina, C., and Morán-Tejeda, E.: Accurate computation of a streamflow drought index, J. Hydrol. Eng., 17, 318–332, 2011.

Yatagai, A., Kamiguchi, K., Arakawa, O., Hamada, A., Yasutomi, N., and Kitoh, A.: APHRODITE: Constructing a long-term daily gridded precipitation dataset for Asia based on a dense network of rain gauges, B. Am. Meteorol. Soc, 93, 1401–1415, 2012.

Yue, S., Pilon, P., Phinney, B., and Cavadias, G.: The influence of autocorrelation on the ability to detect trend in hydrological series, Hydrol. Process., 16, 1807–1829, 2002.

Zou, X., Zhai, P., and Zhang, Q.: Variations in droughts over China: 1951–2003, Geophys. Res. Lett., 32, L04707, https://doi.org/10.1029/2004GL021853, 2005.

https://doi.org/10.5194/piahs-379-409-2018

Guidelines for rainwater harvesting system design and assessment for the city of Johannesburg, South Africa

John Ndiritu, Adesola Ilemobade, and Paulo Kagoda

University of the Witwatersrand, School of Civil and Environmental Engineering, 2050 Johannesburg, South Africa

Correspondence: John Ndiritu (john.ndiritu@wits.ac.za)

Received: 1 January 2018 – Revised: 21 February 2018 – Accepted: 22 February 2018 – Published: 5 June 2018

Abstract. As water demand increases rainwater harvesting (RWH) systems are increasingly being installed for water supply but comprehensive hydrologic design guidelines for RWH do not exist in many parts of the world. The objective of this study was to develop guidelines for the hydrologic design and assessment of rainwater harvesting (RWH) systems in the City of Johannesburg, South Africa. The data for developing the guidelines were mainly obtained from multiple daily simulations of potential RWH systems in the city. The simulations used daily rainfall from 8 stations and demands based on the probable non-potable uses of RWH systems – toilet flushing, air conditioning and irrigation. The guidelines were confined to systems that would typically fill up in the wet season and empty towards the end of the dry season of the same year. Therefore, supply-to-demand ratios ranging from 0.1 to 0.9 were applied. Two generalized design charts of dimensionless relationships were developed. One relates the yield ratio with supply-to-demand ratio and reliability while the other relates the yield ratio with the storage-to-demand ratio and reliability. Reliability was defined as the probability of exceedance of annual yield in order to incorporate the large inter-annual variability of rainfall experienced in the region. The analyses and design of an example RWH system is used to illustrate the application of the design charts.

1 Introduction

Significant population and industrial growth are expected in the City of Johannesburg (CoJ) over the next decade. Therefore, implementing Water Conservation and Demand Management Strategies to enhance water security are current requirements for a sustainable environment. The CoJ considers Rainwater Harvesting (RWH) as a potential source of supply and intends to encourage its use. RWH is generally perceived to be relatively cheaper; simpler to install, operate and maintain; and environmentally more friendly than many other water resource alternatives. It is also essential in Water Sensitive Urban Design as it reduces urban runoff (Armitage et al., 2014) which, could also lead to a reduction in pollutant levels entering receiving waterways (McAlister, 2007). The quality of rainwater is usually superior to surface water and groundwater that may have been contaminated (van der Sterren et al., 2012) although rainwater may not always meet drinking water standards (Evans et al., 2006; Dobrowsky et al., 2014). This paper describes the development of guidelines for the hydrologic design and assessment of RWH systems for the CoJ. This is achieved by developing generalized relationships of the main components of the RWH systems expected to be installed in the CoJ.

2 Materials and methods

The RWH system variables included in the generalization are storage size, catchment area, the rainfall, the yield and its reliability. Relationships amongst these variables were obtained using data from daily time-step simulations of potential RWH systems in the CoJ and the selection of this data applied the concept of hydrologic optimality. A hydrologically optimal RWH system lies on the Pareto front at which storage size is minimized while yield and reliability are maximized (Ndiritu et al., 2017). The RWH system variables required to determine this are the catchment area, the rainfall, the tank size, the demand, the yield and its reliability. The development of the guidelines was based on the numerical data

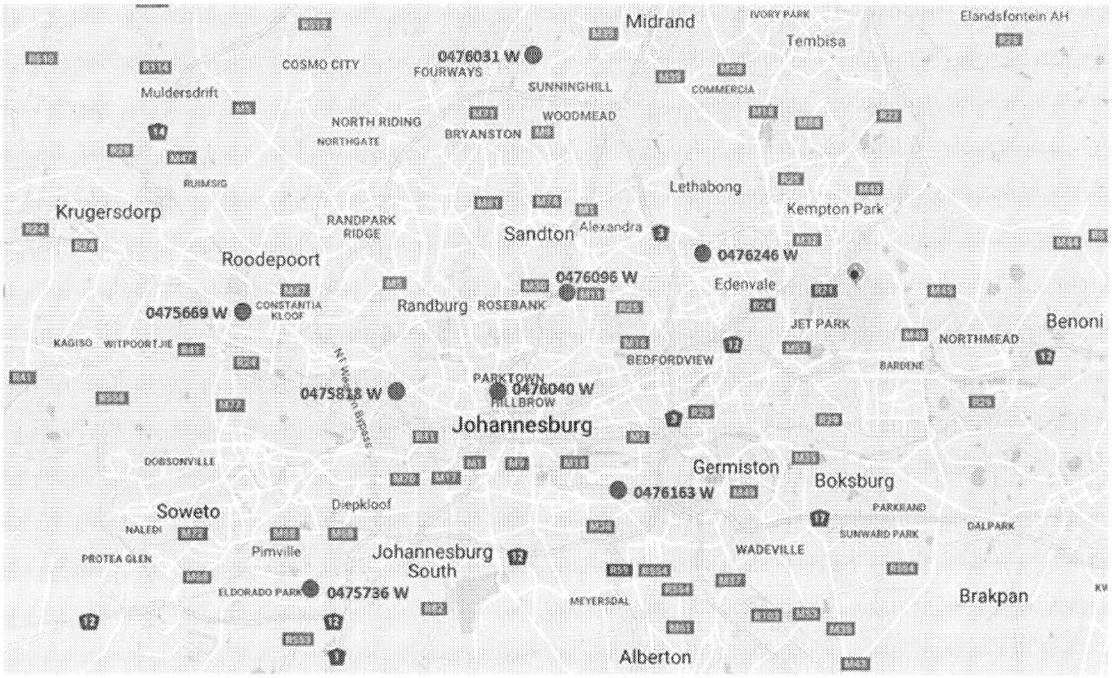

Figure 1. Location of rainfall stations.

of these variables obtained from daily time step simulations of the RWH systems that are likely to be installed in the CoJ. Hanson and Vogel (2014), Campisano and Modica (2012) and Liaw and Chiang (2014) also used simulation to develop generalised methods of RWH analysis. Equations (1) and (2) describe the yield-after-spillage daily water balance simulation that was applied in the study.

$$R_e(t) = \begin{cases} D(t) & \text{if } S(t) \geq D(t) \\ S(t) & \text{if } S(t) < D(t) \end{cases} \tag{1}$$

$$S(t+1) = \min \begin{cases} C - R_e(t) \\ S(t) + \eta R(t) A - R_e(t) \end{cases} \tag{2}$$

where $R_e(t)$ is the volume of water released to meet the demand in period t, $D(t)$ is the demand in period t, $S(t)$ is the volume of water in storage at the start of period t, C is the live storage capacity of the tank, η is the water collection efficiency, $R(t)$ is the rainfall intensity in period t and A is the horizontal projection of the effective roof area.

The quantity and variability of the water available for collection depends on the rainfall available, the roof area and the efficiency of collection. In order to allow for variability of rainfall, 8 raingauge stations spread out across the CoJ (Fig. 1) with long duration and reliable rainfall data were selected for analysis (Table 1). The guidelines needed to be applicable to situations of low, average and high demand. Therefore a wide range of the ratio of available supply to demand was applied by varying the catchment area for a set demand. Various field studies have obtained rainwater collection efficiencies ranging from 80–90 % (Mashau, 2006; Liaw and Tsai, 2004) and a constant rainwater collection efficiency of 80 % was applied. The generalized guidelines were however formulated to be applicable with any collection efficiency. The simulations were also run for a wide range of tank capacities for each combination of roof area and demand in order to ensure that the hydrologically optimal combinations of the variables are obtained. The various RWH demands expected in CoJ have distinct distributions at various time scales and the effects of this variability needed to be incorporated into the guidelines. Domestic household demand is typically constant throughout the year while day school human demand (toilet flushing and handwashing) is

Table 1. Rainfall station information.

Station number	Period of data	% Cont. data	Mean annual rainfall ($mm\,yr^{-1}$)
0475669 W	1886–1993	99	715
0475736 W	1886–2011	91	655
0475818 W	1886–2015	88	719
0476031 W	1886–2017	91	654
0476040 W	1886–2017	89	783
0476096 W	1886–2017	81	615
0476163 W	1886–2005	92	719
0476246 W	1886–2000	93	705

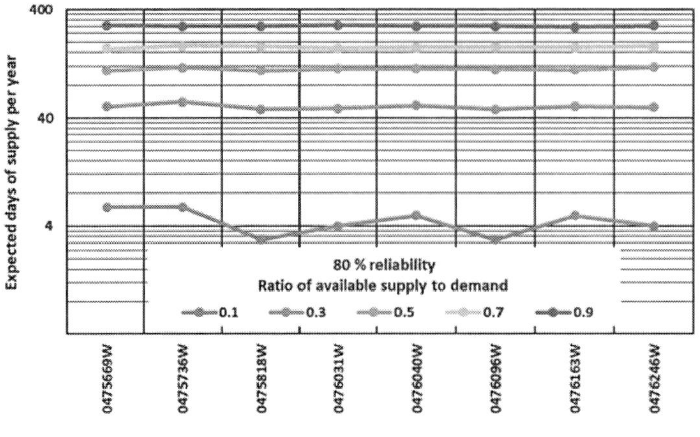

Figure 2. Typical variation of days of supply for different rainfall stations.

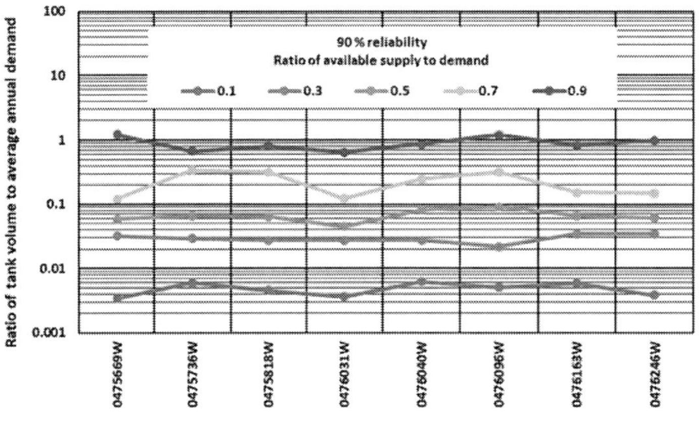

Figure 3. Typical variation of tank volume to demand ratio for different rainfall stations.

higher on weekdays and lower in the non-schooling months of the year. On the contrary, Shopping mall human demand is higher in the weekends and the holiday periods of the year. Irrigation and HVAC (air conditioning) demand vary with seasonal temperature. Irrigation demand additionally varies with antecedent catchment wetness which is rainfall-dependent. Trial simulation runs indicated that increasing the variation of seasonal demand reduced yields (levels of supply). This variation was however found not to impact optimal tank size. Therefore, it was decided to develop the guidelines using the yields obtained from the HVAC/irrigation type demand variation and the largest optimal tank sizes obtained from any distribution of demand. This was considered a way of incorporating safety margins and a reduction of the risk of undersizing system components.

3 The development of design guidelines

Trial runs indicated that for specific ratios of available supply to demand, the yield (number of days the demand is met per year) did not vary appreciably for the different raingauge stations (Fig. 2). This is in agreement with similar observations made for raingauging stations spread over much larger areas of South Africa (Ndiritu et al., 2014, 2017). The variation of the ratios of optimal tank storage to demand varied more appreciably (Fig. 3) but these ratios were found not to correlate with rainfall characteristics ($R^2 < 0.10$) or location within the CoJ. It was therefore decided to develop generalized guidelines that are independent of location within the CoJ by averaging the optimal variables obtained from simulations using the 8 stations. For each rain gauge station data, daily time step simulations were carried out for available supply to demand ratios of 0.1, 0.3, 0.5, 0.7 and 0.9 and the optimal tank volume obtained for reliabilities of 50, 66.7, 80, 90, 95 and 98 %. The respective levels of assurance (probability of exceedance) of days of supply for these reliabilities are 1 in 2 years, 2 in 3 years, 4 in 5 years, 9 in 10 years, 19 in 20 years and 49 in 50 years. Many RWH systems are likely to be complementary (and not the main water supply sources) in the CoJ in which case they therefore would not need to sup-

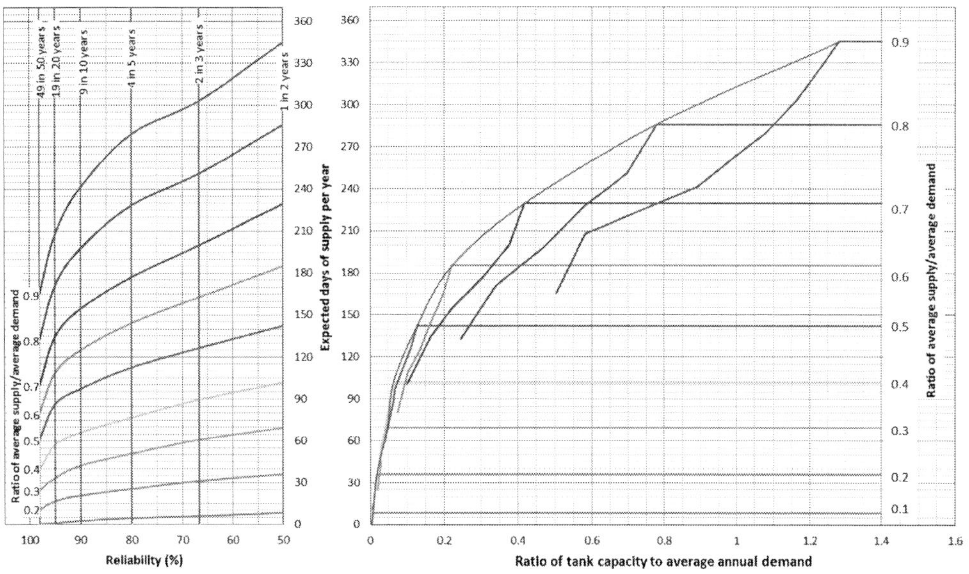

Figure 4. Generalised rainwater harvesting design and assessment charts for Johannesburg.

ply water at the high reliabilities ($\geq 90\,\%$) the overall supply typically requires. Therefore, it was decided to include reliabilities as low as 50 % for the generalization. For probabilistic life cycle or other cost analyses, the expected long-term supply levels correspond to 50 % reliability and generalized guidelines that include this seemingly low reliability would therefore be applicable.

The generalization followed the approach applied by Ndiritu et al. (2017) and was carried out in two steps: (i) finding appropriate non-dimensionless ratios of the optimal hydrologic components obtained from the simulations, and (ii) graphically presenting the relationships of these non-dimensionless ratios and other variables efficiently to obtain design charts. The resulting design charts are presented in Fig. 4. The design charts can be used to size a proposed RWH system or to assess the performance of an existing system and the alternatives for its enhancement. An example illustrating the application of the charts is presented in Sect. 4.

4 Example of application of guidelines

The following hypothetical but realistic example is used to illustrate the application of the RWH design charts. The grass in a 10 000 m^2 recreation park has an annual consumptive use of 870 mm yr^{-1} and 15 % of this is met by rainfall while the rest needs to be provided by irrigation. The City of Johannesburg Parks Department is considering using the stormwater that drains from a nearby housing estate for this. This estate covers an area of 15 850 m^2 and it is estimated that 60 % of the rain falling over this area could be captured as stormwater for irrigation. There is a reservoir of capacity 2590 m^3 available for storing the harvested stormwater. What would be the expected yield of the system? How can the system's yield (expected days of supply per year) be maximized? The MAR for the area is 700 mm yr^{-1}.

The ratio of available supply to demand

$$= \frac{15\,850 \times 700/1000 \times 0.6}{10\,000 \times 870/1000 \times (1-0.15)} = 0.90$$

The ratio of tank capacity to demand

$$= \frac{2590}{10\,000 \times 870/1000 \times (1-0.15)} = 0.35$$

The supply ratio of 0.9 has a minimum ratio of tank capacity to demand of 0.50 (Fig. 5) and the available capacity of 2590 m^3 can therefore not be optimal at any reliability. This means that some of the stormwater supply would not be utilized.

The maximum ratio of available storage to demand that the existing tank with a tank capacity to demand ratio of 0.35 can utilize is 0.67 (Fig. 5).

This would mean that the area of the housing estate that would need to be drained into the tank $= \frac{0.67}{0.90} \times 15\,850 = 11\,799\,\mathrm{m}^2$.

As illustrated on Fig. 5, this would give yields (expected days of supply) of 205, 185, 165, 145, 125 and 95 at assurances of 1 in 2, 2 in 3, 4 in 5, 9 in 10, 19 in 20 and 49 in 50 years respectively.

If the whole area of the housing estate is drained, supply can be maximized by installing additional storage to give a tank volume to demand ratio of 1.275 (Fig. 5). The required extra storage would then be $\left(\frac{1.275-0.35}{0.35}\right) \times 2590 = 6845\,\mathrm{m}^3$.

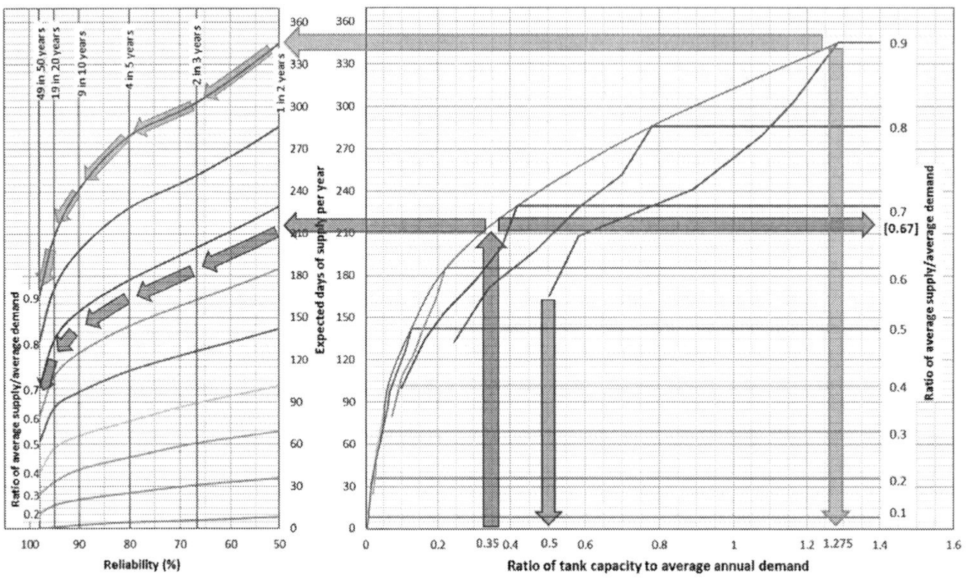

Figure 5. Example of application of rainwater harvesting sizing and assessment chart.

This would provide expected days of supply of 345, 305, 280, 240, 210 and 165 at 1 in 2, 2 in 3, 4 in 5, 9 in 10, 19 in 20 and 49 in 50 levels of assurance respectively.

5 Conclusions

Guidelines for hydrologic rainwater harvesting (RWH) design and assessment for the City of Johannesburg (CoJ) have been developed and their application to a realistic RWH problem has been demonstrated. The guidelines could be enhanced by including more rainfall records in their formulation and by including design and assessment of hydrologically non-optimal RWH systems as demonstrated by Ndiritu et al. (2017). The current guidelines are however considered adequate for the preliminary design and assessment of feasibility of RWH systems in the City of Johannesburg. A numerical version of the design charts can be formulated to enable computer-based design and assessment of RWH systems in the CoJ.

Data availability. The rainfall data for this study was obtained from Lynch (2003).

Competing interests. The authors declare that they have no conflict of interest.

Special issue statement. This article is part of the special issue "Innovative water resources management – understanding and balancing interactions between humankind and nature". It is a result of the 8th International Water Resources Management Conference of ICWRS, Beijing, China, 13–15 June 2018.

Edited by: Zongxue Xu
Reviewed by: two anonymous referees

References

Armitage, N., Fisher-Jeffes, L., Carden, K., Winter, K., Naidoo, V., Spiegel, A., Mauck, B., and Coulson, D.: Water Sensitive Urban Design (WSUD) for South Africa: Framework and Guidelines, WRC Report No. TT 588/14, Water Research Commission, Pretoria, South Africa, 2014.

Campisano, A. and Modica, C.: Optimal sizing of storage tanks for domestic rainwater harvesting in Sicily, Resour. Conserv. Recycl., 63, 9–16, https://doi.org/10.1016/j.resconrec.2012.03.007, 2012.

Dobrowksy, P. H., Mannel, D., De Kwaadsteniet, M., Prozesky, H., Khan, W., and Cloete, T. E.: Quality assessment and primary uses of harvested rainwater in Kleinmond, South Africa, Water SA, 40, 401–406, https://doi.org/10.4314/wsa.v40i3.2, 2014.

Evans, C. A., Coombes, P. J., and Dunstan, R. H.: Wind, rain and bacteria: the effect of weather on the microbial composition of roof-harvested rainwater, Water Res., 40, 37–44, 2006.

Hanson, L. S. and Vogel, R. M.: Generalized storage-reliability-yield relationships for rainwater harvesting systems, Environ. Res. Lett., 9, 1–10, https://doi.org/10.1088/1748-9326/9/7/075007, 2014.

Liaw, C. and Tsai, Y.: Optimum Storage Volume of Rooftop Rain Water Harvesting Systems for Domestic Use, J. Am. Water Resour. Assoc., 40, 901–912, https://doi.org/10.1111/j.1752-1688.2004.tb01054.x, 2004.

Liaw, C. H. and Chiang, Y. C.: Dimensionless analysis for designing domestic rainwater harvesting systems at the regional level in Northern Taiwan, Water, 6, 3913–3933, https://doi.org/10.3390/w6123913, 2014.

Lynch, S.: The Development of a Raster Database of Annual, Monthly and Daily Rainfall for Southern Africa, WRC Report No. 1156/0/1, Water Research Commission, Pretoria, South Africa, 2003.

Mashau, F.: Rainwater Harvesting for Multiple Uses in Siloam Village of Limpopo Province, South Africa; Unpublished Honours Research Dissertation, University of Venda, Thohoyandou, South Africa, 2006.

McAlister, T.: National Guidelines for Evaluating Water Sensitive Urban Design, BMT WBM, Brisbane, Australia, 2007.

Ndiritu, J. G., McCarthy, S., and Tshirangwana, N.: Probabilistic assessment of the rainwater harvesting potential of schools in South Africa, Proc. IAHS, 364, 435–440, https://doi.org/10.5194/piahs-364-435-2014, 2014.

Ndiritu, J., Yashiren, M., and Guliwe, M.: Generalized Storage–Yield–Reliability Relationships for Analysing Shopping Centre Rainwater Harvesting Systems, Water, 9, 1–20, https://doi.org/10.3390/w9100771, 2017.

van der Sterren, M., Rahman, A., and Dennis, G. R.: Rainwater Harvesting Systems in Australia, in: Ecological Water Quality – Water Treatment and Reuse, edited by: Voudouris, K., INTECH, Rijeka, Croatia, 2012.

https://doi.org/10.5194/piahs-379-415-2018

Open Access
PIAHS

Socio-hydrological implications of water management in the dry zone of Sri Lanka

Isurun Upeksha Gamage[1] **and Hetti Arachchige Hemachandra Jayasena**[2]

[1]Post Graduate Institute of Science, University of Peradeniya, Peradeniya 20400, Sri Lanka
[2]Department of Geology, University of Peradeniya, Peradeniya 20400, Sri Lanka

Correspondence: Isurun Upeksha Gamage (isurun.gamage@yahoo.com)

Received: 31 December 2017 – Revised: 2 April 2018 – Accepted: 4 April 2018 – Published: 5 June 2018

Abstract. Water management plays a vital role in the agricultural economy and living conditions of people in Sri Lanka. Though government and non-government organizations have been readily contributing to water management, it is still inefficient, especially in terms of water allocation, consumption and conservation. To identify factors which could be used to implement integrated water resources management (IWRM), a socio-hydrological study was performed in five areas within the dry zone in Sri Lanka. The study covers a comprehensive analysis of how the household income, demography and education level correlating to water usage, purification and disposal methods. The average household income ranges from LKR 2500 to 15 000 per month. The results show that the average daily usage for drinking, cooking, washing, toiletries and bathing are 3, 5, 10, 7, and 85 L per person, respectively. Majority of the families use dug wells and pipe-borne water as the primary source. Correlation coefficients suggest that higher household income or level of education leads to increased water consumption ($R = 0.91, 0.94$). There is no linear relationship between the level of education with the good practices of water purification and disposal. Though these results indicate preliminary assessments based on the dry zone practices, efficient water management could be enhanced by strong socio-hydrological implications through educating people on conservation, usage, disposal practices and health concerns.

1 Introduction

Water management plays a vital role in the agricultural economy and living conditions of people in Sri Lanka (Jayasena and Selker, 2004). Though government and non-governmental organizations have been readily contributing to water management, it is still inefficient, especially in terms of water allocation; consumption and conservation as the water management were based on ad-hoc political decisions (Samad et al., 2017). This paper attempts to examine the causal relationship between the consumption pattern and the trend of water management. The cause is worth studying since a variation of water resources distribution and consumption could be identified based on the educational level and economic status among the families, according to the effects generated due to unequal water supply (UNDP, 2012). There are variations in water usage as shown in principle due to socio-economic and educational diversities. The uneven distribution of rainfall is the basis for agricultural definitions for two seasons "Maha and Yala" (Imbulana and Neupane, 2004). In a previous investigation completed in the same area, it was found that the water supply in this basin is not efficient due to lack of perennial water sources and difficulties in construction and maintenance of water supply schemes (Jayasena and Selker, 2004). In addition, this study focuses on how the relationship between the level of education and the role of economic conditions impact on the water consumption, expecting that the results could be used as a moral support for the management and conservation of water in the domestic level. The variations of distribution and consumption of water are the essential factors to be analyzed as it supports future water management plans in the area. The main objective of this study is to trace the relationships among the rural and peri-urban population with their water consumption, which in-turn will support water management and planning practices in the Deduru Oya basin in Sri Lanka.

Water management involves key stages of assessment, planning, and implementation (Hufschmidt, 1993). The concept of integrated water resources management (IWRM) was developed as an empirical model in 1977 during UNDESA session in Mar del Plata (Hassing et al., 2009) which is used as a reference for this study. In IWRM, economic, socio-cultural and political linkages are interacting on land, water supply and specific water usage. For instance, different perspectives of socio-cultural and political linkages exist between water resources development and people who benefit or adversely affect (Hufschmidt, 1993). Water allocation and usage can be regularized through proper planning of water resource management programs using socio-hydrological data.

2 Methodology

The socio-hydrological survey was carried out using a questionnaire designed to cater field level participation of diverse-income families with an aim to collect basic quantitative and qualitative information. The data were collected from five areas, including Weerakodiyana, Rakogama, Aladeniya, Hedeniya and Chilaw Grama Niladhari (GN) divisions (Fig. 1) from the Deduru Oya basin in Sri Lanka and conducted the questionnaire interview by randomly selecting 50 families from each area. Chilaw is a peri-urban town and the sample was mainly from a fishing community whereas Aladeniya has local merchant community. Weerakodiyana and Rakogama are two different rural agricultural areas with the farming community supplied with reservoir based irrigation water. Hedeniya, an agricultural community from the central highlands supplied with natural springs and groundwater developed through foreign agencies. However, due to local field conditions and considering the diversified family backgrounds, only 30 families were selected from the Aladeniya GN division. The study region is within the Deduru Oya basin and the main focus was given to identify its water management system from several social segments as they have dissimilar economic and social backgrounds.

The main aim of using this sample is to understand their economic, educational, ethnic, and gender backgrounds. The data were collected on many aspects covering several periods from 2003 to 2006. One major component of the study focused on daily water requirement and consumption such as drinking, cooking, washing, toiletries, and bathing. The data were collected by trained enumerators selected from the undergraduates and educated high school students. Interviews were carried out as a parallel data collection method to gather information on livelihood, economic barriers, water-related problems and consumption resolution for this paper.

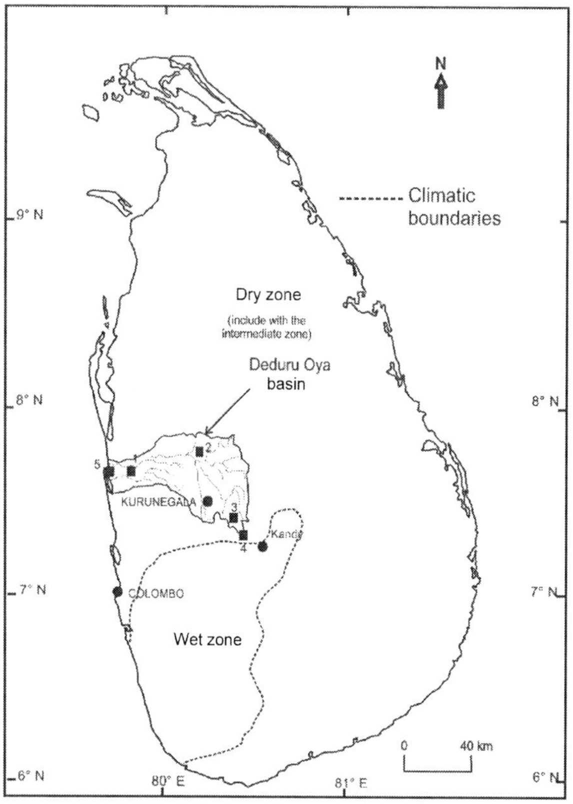

Figure 1. Sri Lanka showing climatic zones (modified after Muththuwatta and Liyanage, 2013), and study GN Divisions. 1. Weerakodiyana, 2. Rakogama, 3. Hedeniya, 4. Aladeniya, 5. Chilaw.

3 Results and discussion

3.1 Water consumption

Per-capita, water consumption has been studied with respect to monthly income and the level of education (Fig. 2) The average water consumption for different income levels has been tested and observed that the water consumption monotonously increases with the income ($R = 0.91$)

This may indicate the freedom of water usage and the easy approach to water availability at the household level that leads to increase in overall consumption. As shown in the village level, public water supply schemes have been incorporated and those who have contributed to the capital and maintenance expenditure got the full share. As given in the previous studies, some communities would prefer to allocate nearly 2–3 % of their monthly emoluments in this respect (Jayasena and Selker, 2004; Jayasena and Dhanapala, 2014). At the same time, food traits or hygiene habits would also change when the buying power is high. As Imbulana and Neupane (2004) reported, about 43–55 % of the families in the Deduru Oya basin have received economic sup-

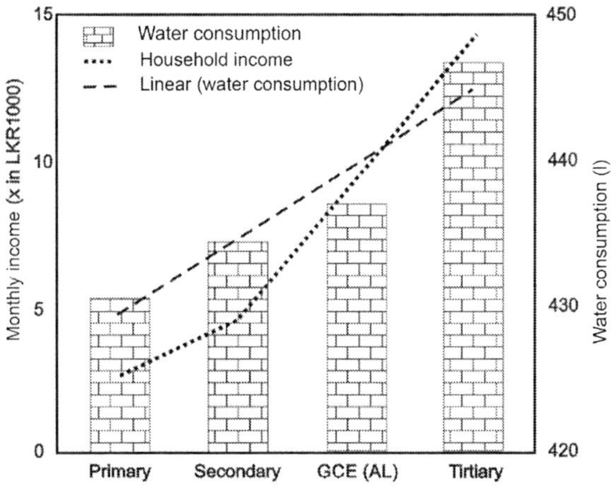

Figure 2. Monthly household income and daily water consumption vs. level of education.

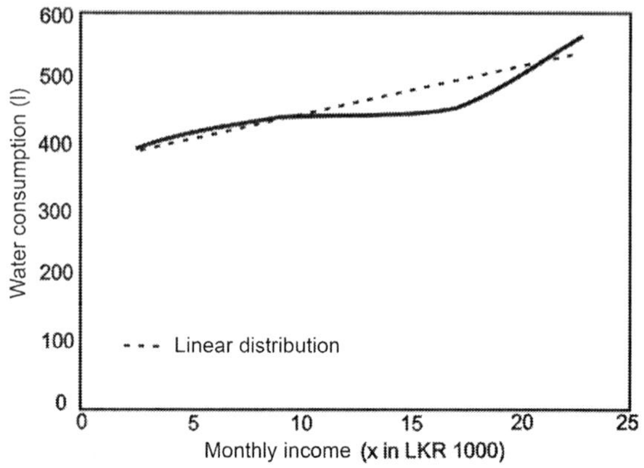

Figure 3. Monthly household income vs. daily water consumption.

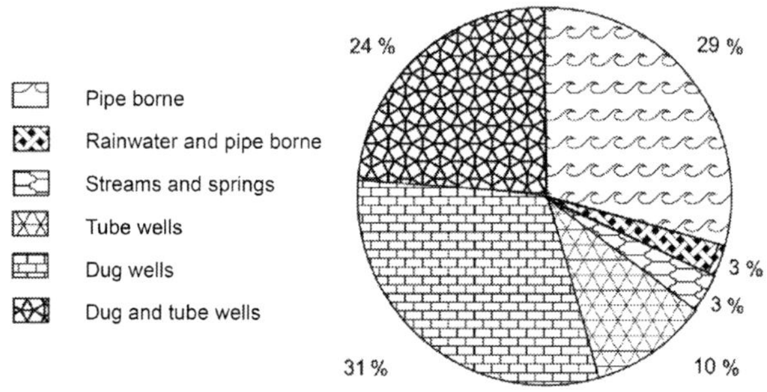

Figure 4. Water sources used by the total sample.

ports during this period. The outcome from the areas of study with low economic development contrast with the developed cities. The positive trend of water and income is still obvious. This could also be the reason behind the increased water consumption among the group with the family income above LKR 5000 (approximately USD 50 in 2004). Based on the outcome, one can argue that the traditional approach to water management or case by case isolated approaches are not appropriate as we have socio-economic factors to consider under a holistic approach to the water management problem in hand (Hufschmidt, 1993). A clear relationship between the two parameters reveals that the water consumption is proportional to the monthly income. According to the "psychological law of consumption" by Keynes (1936), water consumption increases with increasing income, but not to the same extent as the increase in income. Consumption will depend on the level of aggregate income. Thus the monthly water consumption ranges from 96 to 136 L per person per day. A correlation coefficient of 0.94 was obtained when water consumption is related to the level of education (Fig. 3).

Although water consumption increases with the level of education, the rate is smaller than the rate of income rise as specified on the secondary plot (Fig. 3). The average income per person in the low education category is approximately LKR 2600 (approximately USD 26 in 2004) per month. This might be due to low educated consumers with low income attempting to save water by different means. Such trends suggest the importance of general education in making changes to the water-related behavior. Thus the influence of education on water management in the dry zone region will have a positive outcome. The water resources planning and management shall be favored by awareness programs and water conservation campaigns for a better management output.

The average of the total water consumed 110 L per person per day is generally tolerable and taking regular baths is the biggest water usage among the households. As we consider the average daily usage for drinking, cooking, sanitation (water usage for toiletry), washing (particularly clothes) and bathing are 3, 5, 7, 10, and 85 L respectively. With respect to proper management, we cannot expect efficiency by putting up the water price. Sri Lanka has both piped and non-piped water sources so that the influence of price change is insignificant due to the complimentary usage of private well water which is free (Nauges and Van Den Berg, 2009). As findings of Kandy Municipality water conservation suggest that domestic water is neither price-nor income-elastic (Gunatilake et al., 2001). For water management, there should be an estimation of how much water to allocate for different sectors.

Majority households use groundwater though the occurrence is limited to regolith and the fractured network in the basin (Jayasena et al., 1986). As such 31 % of the households use only dug wells, 10 % use tube wells and another 24 % depending on both dug and the tube wells as the primary source of water (Fig. 4). Usually, the Rakogama and Weerakodiyana areas have the maximum usage through domestic dug wells and tube wells. Unfortunately, there is a minimal usage of rainwater, as rainwater harvesting for domestic purposes has not yet been effectively practiced by the villagers (Ariyananda, 1999). From the traditional constructions of tank cascade systems and man-made reservoirs mainly for the purpose of irrigation agriculture, people still consume water; however, there are no significant improvements to the traditional constructions of minor tanks due to the conflict between traditional irrigation systems and mechanized modernization methods. The irrigation system of most dry zone villages is crude in comparison to that existed under the ancient civilization (Panabokke, 2009). On the other hand, short-term political and economic plans and haphazard technical inputs are not yet geared enough for such efficient water management as observed in the history (Gangadhara and Jayasena, 2005). It is imperative to apply suitable socio-technical methods to be used in the modern periods where the outcome of the projects satisfies the needs of society. The water resources in a basin are basically fixed and tend to be relatively scarcer than land resources and they are determined by climatic conditions (Falkenmark, 1995).

The water consumption is mainly dependent on the traditional sources as discussed above. Therefore, the rainwater harvesting systems for a domestic use shall be implemented as a fine project in the dry zone of Sri Lanka. If people can use rainwater in conjunction with the pipe-borne water, the water wasted through infiltrating and soaking into the ground, evaporating into the atmosphere and discharge into the drainage systems can be controlled as such that effective economic management can be placed under IWRM guidelines. According to Jønch-Clausen (2004), the importance of improved water efficiency as part of national water strategies, and thus considered as a part of the IWRM plans. Water resources management shall yield adequate benefits if there is integration between institutions and proper data and scientific analysis of basin hydrology (Hufschmidt, 1993).

Another significant factor as observed in this study is the influence on water consumption due to the unregulated water supply. Availability and access to more water may psychologically tempt consumers to use more water so that the overall usage would be high as quoted by de Miranda Coelho et al. (2016). For instance, having money, big houses and gardens, and a big family promote water waste according to Corral-Verdugo et al. (2003). Overall groundwater usage in the basin is 65 % through dug and tube wells. Only 29 % have access to national pipe-borne water supply (Fig. 4). Accordingly, consumers using pipe-borne water or other natural water resources consume more water compared to dug well or tube well users. The reason for this difference could be twofold. Partly it could be due to low amounts of water produced in the wells and tube wells in the dry zone. On the other hand, one can argue with the natural tendency of humans not to extend their energy to fetch water from areas having difficult access. Psychologically people trying to minimize or share their energy usage, so that they developed various machines to suit their requirements. Corral-Verdugo et al. (2003) study state "using water-saving devices, facing norms demanding a decreased consumption, and living in a place where water is scarce lead to water conservation". Therefore, scarcity of water influences people for lower usage.

Families living in the selected communities use basic methods such as boiling or filtering in order to eliminate suspended particles. Sometimes the villagers do not want to boil or filter as they believe boiling could alter the taste. On the other hand, most dug wells have "kumbuk" (Terminalia Arjuna) trees next to or in the surrounding environment and it was considered as a purifier of water. This has been inherited from their ancestors and common knowledge passing through ayurvedic or Sinhala medicine (Uragoda, 2000). Surprisingly, there is no clear evidence that higher level of education leads to a higher tendency towards water purification. However, it seems that those who have completed a higher level of education are determined at least to boil water which is significantly prominent compared to other education levels. With further data analysis concurrent with similar studies, a formal design to the water management among the families in the dry areas could be achieved.

4 Summary

This paper discusses the data collected from 5 different dry-zone GN divisions in Sri Lanka through questionnaire survey to relate them and identify domestic water demand, water management and how socio-technical parameters viz: economy, inherited knowledge and education act on water management. Water price and income were the most influential parameters of the study. In this paper, attempt to relate collected data with other variables such as occupation level and purification type.

The most popular method of purification was boiling irrespective of the level of education and economic background of the family. Even though there is a direct correlation between economic background and water consumption, it is unhealthy to practice water conservation and management by increasing the price of water since the majority of the families are on the government aid. As a significant factor, the level of awareness and improvement of the IWRM practices in the area for a well-managed water policy need to implement in order to get rid of water consumption imbalances. Thus it is essential to follow the guidelines of IWRM and applying them with the inputs from the socio-technical background of the local regions.

Data availability. The data set that provides information gathered from 50 families each from Weerakodiyana, Rakogama, Hedeniya, and Chilaw, and another 30 families from Aladeniya in the Deduru Oya basin are available at https://www.academia.edu/36404612/Data_Set_2nd_April_2018. These data were collected in the 2003–2006 period with support from the Tharuna Aruna Skilled Development program conducted through Soil Tech (Pvt.) Ltd. These data are comprised of basic socioeconomic, water usage, water purification, and disposal at the household level. This is a subset of the original larger data set which will be available in due course after completion of the detailed analysis.

The Supplement related to this article is available online at https://doi.org/10.5194/piahs-379-415-2018-supplement.

Competing interests. The authors declare that they have no conflict of interest.

Special issue statement. This article is part of the special issue "Innovative water resources management – understanding and balancing interactions between humankind and nature". It is a result of the 8th International Water Resources Management Conference of ICWRS, Beijing, China, 13–15 June 2018.

Acknowledgements. The data collection were supported by enumerators employed through the TharunaAruna Skilled Program of the Government of Sri Lanka and the Soil Tech (Pvt) Ltd, Colombo Sri Lanka. Without their meticulous inputs, the project could not result in such high-quality socio-technical data. The support extended to the senior author during his sabbatical in Canada by the Orbit Engineering is acknowledged. Fellows and friends who supported us in the preparation of this manuscript are acknowledged with many thanks.

Edited by: Zongxue Xu
Reviewed by: two anonymous referees

References

Ariyananda, T.: Rainwater harvesting for domestic use in Sri Lanka, in: Proceedings of the Integrated Development for Water Supply and Sanitation, 25th WEDC Conference, Addis Ababa, Ethiopia, 369–372, 1999.

Corral-Verdugo, V., Bechtel, R. B., and Fraijo-Sing, B.: Environmental beliefs and water conservation: An empirical study, J. Environ. Psychol., 23, 247–257, 2003.

de Miranda Coelho, J. A. P., Gouveia, V. V., de Souza, G. H. S., Milfont, T. L., and Barros, B. N. R.: Emotions toward water consumption: Conservation and Wastage, Revista Latinoamericanade Psicología, 48, 117–126, 2016.

Falkenmark, M.: Land and water integration and river basin management, Proceedings of an FAO informal workshop, Publications Division, Food and Agriculture Organization of the United Nations, Rome, Italy, 31–32, 1995.

Gangadhara, K. R. and Jayasena, H. A. H.: Was rainwater harvesting in the dry zone of Sri Lanka a technically adapted methodology by the ancients?, XII-International Rainwater Catchment Systems Association (IRCSA) Congress, 15–18 November 2005, New Delhi, India, 1–15, 2005.

Gunatilake, H. M., Gopalakrishnan, C., and Chandrasena, I.: The economics of household demand for water: the case of Kandy Municipality, Sri Lanka, Int. J. Water Resour. D., 17, 277–288, 2001.

Hassing, J., Ipsen, N., Clausen, T. J., Larsen, H., and Lindgaard-Jørgensen, P.: Water in a changing world: Integrated Water Resources Management in Action: Side publication series, Dialogue paper, United Nations Education Scientific and Cultural Organization, Paris, France, 22 pp., 2009.

Hufschmidt, M.: Water Resource Management, in: Hydrology and Water Management in the Humid Tropics: Hydrological Research Issues and Strategies for Water Management (International Hydrology Series), edited by: Bonell, M., Hufschmidt, M., and Gladwell, J., Cambridge University Press, Cambridge, 471–495, https://doi.org/10.1017/CBO9780511564468.026, 1993.

Imbulana, K. A. U. S. and Neupane, B.: Integrated water resources management: the relevance of indicators for measuring river basin status and performance, in: Proceedings of Monitoring Tailor-Made IV, International Workshop on Information for Sustainable Water Management, Netherlands, 91–102, September 2004.

Jayasena, H. A. H.: Socio Hydrological data, Data Set, the Academia, available at: https://www.academia.edu/36432381/Data_Set_2nd_April_2018.xlsx (last access: 19 April 2018), uploaded 2 April 2018.

Jayasena, H. A. H. and Selker, J. S.: Thousand Years of Hydraulic Civilization Some Socio-technical Aspects of Water Management, Understanding the Role of Politics in Water Management, in: Proceedings of the Conference: Water and Politics, Understanding the role of Politics in Water Management, World Water Council, Marseilles, France, 225–236, 2004.

Jayasena, H. A. H. and Dhanapala, T. R. W. S.: Analysis of Domestic Waste Water Usage in a Tropical Mountainous Basin in Sri Lanka, in: Proceedings of the World Mountain Forum, Moving mountains towards global sustainability, Cusco, Peru, 23–24 May 2014.

Jayasena, H. A. H., Singh, B. K., and Dissanayake, C. B.: Groundwater occurrences in the hard rock terrain of Sri Lanka – A Case Study, AQUA, 4, 214–219, 1986.

Jønch-Clausen, T.: Integrated water resources management (IWRM) and water efficiency plans by 2005: Why, what and how, Global Water Partnership, Elanders Infologistics Väst AB, Sweden, 5–4, 2004.

Keynes, J. M.: The general theory of employment, interest, and money, 2007 Edn., Macmillan and Co, London, 472 pp., 1936.

Muththuwatta, L. P. and Liyanage, P. K. N. C.: Climate Change Models Shift Boundaries of Agro-Ecological Zones in Sri Lanka IPC climate net, available at: http://climatenet.blogspot.com/2013/09/climate-change-models-shift-boundaries.html, last access: March 2018. 2013

Nauges, C. and Van Den Berg, C.: Demand for piped and non-piped water supply services: Evidence from Southwest Sri Lanka, Environ. Resour. Econ., 42, 535–549, 2009.

Panabokke, C. R.: Small village tank system of Sri Lanka: Their evolution, setting, distribution and essential functions, Hector Kobbekaduwa Agrarian Research and Training Institute, Colombo, Sri Lanka, 84 pp, 2009.

Samad, M., Aheeyar, M., Royo-Olid, J., and Arulingam, I.: The political and institutional context of the water sector in Sri Lanka: An overview, EU, Luxembourg, Europe, 92 pp., 2017.

Uragoda, C. G.: Traditions of Sri Lanka: A Selection with a Scientific Background, Michigan, Vishva Lekha Publishers, 327 pp., 2000.

UNDP: Sri Lanka Human Development Report 2012: Bridging Regional Disparities for Human Development, UNDP, Sri Lanka, 178 pp., 2012.

https://doi.org/10.5194/piahs-379-421-2018

Open Access
PIAHS

Impact of the operation of cascade reservoirs in upper Yangtze River on hydrological variability of the mainstream

Xu Changjiang and Zhang Dongdong

Bureau of Hydrology, Changjiang Water Resources Commission, Wuhan 430010, Hubei, China

Correspondence: Xu Changjiang (xucj@cjh.com.cn)

Received: 30 December 2017 – Revised: 13 February 2018 – Accepted: 14 February 2018 – Published: 5 June 2018

Abstract. As the impacts by climate changes and human activities are intensified, variability may occur in river's annual runoff as well as flood and low water characteristics. In order to understand the characteristics of variability in hydrological series, diagnosis and identification must be conducted specific to the variability of hydrological series, i.e., whether there was variability and where the variability began to occur. In this paper, the mainstream of Yangtze River was taken as the object of study. A model was established to simulate the impounding and operation of upstream cascade reservoirs so as to obtain the runoff of downstream hydrological control stations after the regulation by upstream reservoirs in different level years. The Range of Variability Approach was utilized to analyze the impact of the operation of upstream reservoirs on the variability of downstream. The results indicated that the overall hydrologic alterations of Yichang hydrological station in 2010 level year, 2015 level year and the forward level year were 68.4, 72.5 and 74.3 % respectively, belonging to high alteration in all three level years. The runoff series of mainstream hydrological stations presented variability in different degrees, where the runoff series of the four hydrological stations including Xiangjiaba, Gaochang and Wulong belonged to high alteration in the three level years; and the runoff series of Beibei hydrological station in 2010 level year belonged to medium alteration, and high alteration in 2015 level year and the forward level year. The study on the impact of the operation of cascade reservoirs in Upper Yangtze River on hydrological variability of the mainstream had important practical significance on the sustainable utilization of water resources, disaster prevention and mitigation, safe and efficient operation and management of water conservancy projects and stable development of the economic society.

1 Introduction

Under the impacts by climate changes and human activities, the characteristics of meteorological and hydrological elements in many regions of the world had presented changes in different degrees (Guo et al., 2012). The globally observed evaporation pan evaporation was basically on the decline while the actual evaporation presented in an increasing trend (Kisi, 2015; McVicar et al., 2012). The precipitation changing trend displayed strong regional characteristics (Allan, 2012; Prakash et al., 2015; Wen et al., 2015). Sun and Ao (2013) pointed out that the precipitation changing trend in China in recent 100 years was not obvious; but a slight increasing trend appeared since 1956; and in the future 20–100 years, the annual precipitation in China would appear in an increasing trend. Based on the data observed in about 730 meteorological stations during 1951–2002, the precipitation changes in China displayed strong regional characteristics, displaying a continuous declination in northern China, central China and southern regions in northeastern China and obvious increase in the regions to the south of the Yangtze River basin (Jiang et al., 2014). The studies by Barnett et al. (2008) showed that the water circulation in the western United States had changed dramatically in the second half of the 20th century, where as high as 60 % was due to the environmental changes caused by human activities. The study by Szilagyi (2001) discovered that the mean runoff depth of Republican River in the US during 1977–1996 decreased by around 40 % compared to that in 1949–1968; and this decrease was mainly the result of affections by comprehensive human activities in the basin like reservoir construction, agricultural irrigation, vegetation change and soil and water conservation, etc. The studies by many Chinese scholars showed that the runoff in northern regions of China appeared in a decreasing trend in recent 50 years; for example, the reason of continuous decrease in the actual inflow water amount of the mainstream of Yellow River in recent 50 years was mainly due to the impacts by climate changes and human activities (Liu and Cheng, 2000). The long-series data observed in basins of Yellow River, Hai River, Liao River and Songhua River showed that the river runoff had a tendency to decrease and the runoff during the 1980s and 1990s had decreased by 20.50 % than that during the 1950s and 1960s under the same magnitude of rainfall (Ren et al., 2001). It was found after analyzing the annual runoff series of the Yellow River of nearly 80 years with the detrended fluctuation analysis method that there was an inherent long-range correlation in the annual runoff of the Yellow River of nearly 80 years; and the trend prediction showed that the runoff change of the Yellow River in next ten years and more may appear in a decreasing trend (Li et al., 2008). The data of some basins in southern China regions in recent 50 years showed that the underlying surface changes had resulted in the runoff increase; for example, comparing the runoff in the 1980s and 1990s to that in 1970s, the runoff of the Bei River basin of the Pearl River system increased by 7–10 % (Li et al., 2006) and the runoff of Dong River basin increased by 25.7 % (Lin et al., 2012).

The river ecosystem plays an extremely important role in maintaining regional water circulation, energy balance, climate change and ecological development and is also the most important human life support system, providing water for production, living and ecology. In the 1980s, the "river health management" was on the rise in Europe and North America; and in 2002, the Chinese experts and scholars put forward the idea of "a healthy river ecosystem is bound to become the main target of river management" while introducing the concept of "river health" (Petts, 1979, 1980). The researchers from all over the world had conducted a lot of studies on water requirement of river ecology and environment since the 1970s and had produced many calculation and evaluation methods (Pettit et al., 2001; Poff et al., 1997). In 2003, Tharme (2013) conducted some statistics and found out 207 kinds of ecological flow requirement assessment methods in 44 countries, which could be roughly divided into four categories: hydrologic indicator method (historical flow method), hydraulic method, habitat method (ecological environment) and holistic method.

The river ecosystem is a dynamic flowing water system possessing double attributes in nature and society respectively. Its integrity is largely dependent on the natural dynamic change characteristics of water flow, namely, the river's hydrological regime. The construction of hydraulic structures like reservoir and dam was mainly for the purpose of flood control, navigation and water supply, which also impacted the hydrological regime of downstream river channel (Mathlouthi and Lebdi, 2008). Richter et al. (1996) established a set of indicator system (Indicators of Hydrologic Alteration, IHA) to assess the changing conditions of river hydrology before and after the interference by human activities through a total of 32 indicators in five aspects namely flow, occurrence time, frequency, duration and change rate, and proposed the RVA method (Range of Variability Approach) based on IHA (Richter et al., 1997) for setting the flow of ecological environment as well as the goals and steps of river management, having laid a foundation for analyzing the river's hydrological regime. Richter et al. (1998) applied the RVA method in assessing the hydrological variability before and after the dam construction on the main rivers in the Colorado River basin. The RVA method was regarded as an effective method to assess river's hydrological variability (Chen et al., 2010). Shiau and Wu (2004) used RVA method to investigate the hydrologic conditions before and after the construction of a diversion weir on Chou-Shui Creek, Taiwan, suggesting that restoration of the natural flow is expected to promote the natural stream biota. The RVA has proved to be a practical and effective approach facilitating river restoration planning. However, the RVA method is only used to assess the small-basin rivers in assessing the river's hydrological regime changes, lacking of assessment on the overall alteration conditions of large-basin rivers. To this end, this paper adopted the RVA method, incorporated the runoff regulation and calculation model, and selected the main control hydrological station on the mainstream of Yangtze River to study the impact of the operation of cascade reservoirs in Upper Yangtze River on hydrological variability of the mainstream.

Being rich with hydraulic power potentials, dense river network, abundant rainfall and large runoff volume, the upper Yangtze River basin is the most important rich water area. According to the statistics from "Electric power planning in China (2002)", 34 large reservoirs had been built up or under construction on the upper Yangtze River, and more than 30 large reservoirs had been designed and set up with a total storage capacity of about 200 billion m^3. Since 2000, large reservoirs were constructed mainly in the lower reaches of the Jinsha, Yalong, Dadu and Wujiang Rive basin. In 2020, the layout of major rivers cascade reservoirs will be basically completed in the Yangtze River. Even before, the hydrological cycle was based on precipitation, surface runoff, river and groundwater. But now, the cycle most is also impacted by the strong human activities in four aspects: water taking, water transporting, water usage, drainage and regress (Jia et al., 2006).

To a certain degree, the impacts of human activities on hydrological processes are even more serious than the impacts of climatic change in the upper Yangtze River. Exploring the extent to which human interventions affected the hydrological regimes and related hydrological alterations is crucial

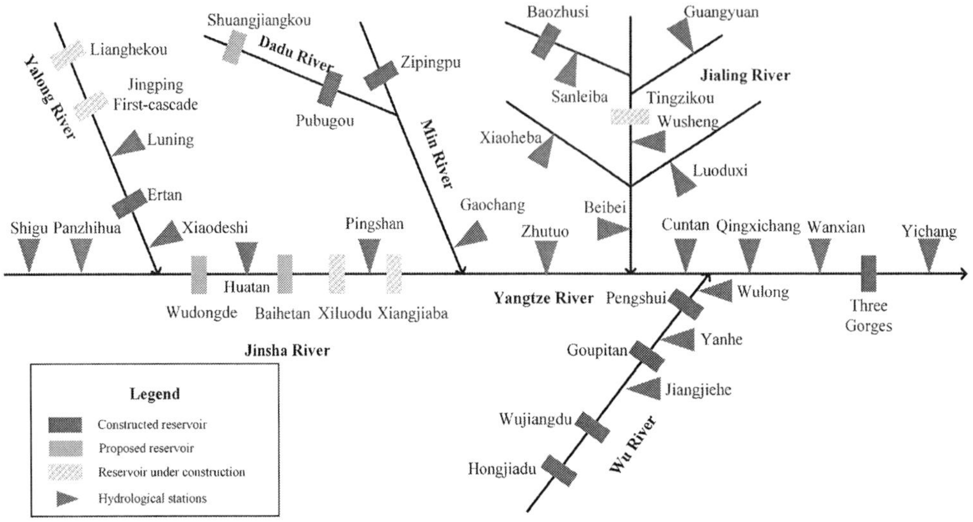

Figure 1. Diagram of Main Reservoirs on the Upper Yangtze River.

for better understanding of human-induced hydrological alterations, and will aid water resources management within the Yangtze River basin. The objectives of this paper therefore are: (1) to present a visually enhanced hydro-alteration assessment method for flow of Yangtze River with data series encompassing pre- and post-alteration periods; (2) to determine the spatial behavior of 33 Indicators of Hydrologic Alteration (IHA) factors which features the hydrologic alteration in Yangtze River; and (3) to evaluate the impact of dams on the hydrological alteration along the Yangtze River of China.

2 Study area and design level year

There are numerous reservoirs on the mainstream and tributaries at the upstream of the Three Gorges. Based on the runoff characteristics of upper Yangtze River, the reservoirs with high regulation capability were selected as the objects of this study. According to the statistics of water resources bulletins of the Yangtze River basin and rivers in the southwest, the construction and impoundment of large and medium-sized reservoirs upstream the Three Gorges had been completed successively during 2003–2013, with total accumulated impoundage of 8.66 billion m^3. The upstream large and medium-sized reservoirs are mainly for the purpose of annual regulation or seasonal regulation, and the impoundment quantity is mainly the dead reservoir capacity. The impoundment of Three Gorges Reservoir starts from 15 September to the end of November. The total impoundment quantity needed for the constructed large reservoirs in the basin upstream the Three Gorges is 10.33 billion m^3; 5.11 billion m^3 for the reservoirs under construction; and 7.08 billion m^3 for the proposed reservoirs; 22.52 billion m^3 in total. As for the selection of upstream reservoirs in this study, in aspect of capacity, the large reservoirs with a single reservoir regulation capacity of more than 500 million m^3 were selected as the objects of study; in aspect of operating time, the reservoirs constructed and put into operation before 2015 were mainly considered. As for the reservoirs constructed after 2015, because of the many variable factors, only the recently developed reservoirs recommended in the comprehensive planning of Yangtze River basin were considered.

According to the aforementioned principles, the reservoirs of Lianghekou, Jinping First-cascade and Ertan on Yalong River, Wudongde, Baihetan, Xiluodu and Xiangjiaba on Jinsha River, Zipingpu on Min River, Pubugou and Shuangjiangkou on Dadu River, Baozhusi on Bailong River, Tingzikou on Jialing River and Hongjiadu, Wujiangdu, Goupitan and Pengshui on Wu River were selected as the objects of study (Fig. 1 and Table 1).

At present, the upstream reservoirs have not been completely constructed; only the constructed reservoirs are under operation at the upstream in current stage; at the same time, the reservoirs under construction will be constructed and put into operation in succession where most of the reservoirs under construction will be constructed around 2015; and the construction completion and production time of the reservoirs planned to be constructed will be even later. Therefore, the constructed reservoirs, reservoirs under construction and proposed reservoirs shall be differentiated appropriately according to their time put into production. In this study, in order to differentiate the production time of the constructed reservoirs, reservoirs under construction and proposed reservoirs, it was divided into three level years according to the reservoir's construction completion status, namely, 2010 level year, 2015 level year and the forward level year.

Table 1. Reservoirs analyzed and calculated in different level years.

Reservoir	Regulating storage volume (10^8 m^3)	Whether involved in analysis and calculation		
		2010 level year	2015 level year	forward level year
Ertan	33.7	√	√	√
Zipingpu	7.74	√	√	√
Pubugou	38.82	√	√	√
Baozhusi	13.4	√	√	√
Hongjiadu	33.61	√	√	√
Wujiangdu	13.6	√	√	√
Tingzikou	17.5		√	√
Goupitan	29.52		√	√
Pengshui	5.18		√	√
Jinping First-cascade	49.1		√	√
Xiluodu	64.6		√	√
Xiangjiaba	9.03		√	√
Lianghekou	65.60			√
Wudongde	26.0			√
Baihetan	104.0			√
Shuangjiangkou	19.0			√

The reservoir construction completion status considered in each level year is as follows.

3 Methods

3.1 Indicators of hydrologic alteration

The Indicators of Hydrologic Alteration (IHA) approach is a method to assess river's ecological hydrologic alterations. Mainly based on such 5 basic characteristics of hydrological regime as flow, occurrence time, frequency, duration and change rate, the IHA approach is divided into 33 indicators in five groups according to the characteristics summarized (as shown in Table 2). The relationship between the parameters of each IHA group and the river ecosystem is as shown in the literature (Richter et al., 1997).

The RVA approach is to determine the river flow status before and after alteration by utilizing the detailed flow data on the basis of IHA, so as to analyze the alteration before and after the river alteration (Richter et al., 1997). The assessment can be conducted in the following steps:

1. Calculate the characteristic values of 33 IHAs by utilizing the daily flow data before alteration;
2. Define the target RVA range of each IHA according to the calculation results of previous step; and the values of the occurrence probabilities of each IHA before alteration of 75 and 25 % were taken as the target RVA range in this paper;
3. Calculate the characteristic values of 33 IHAs by utilizing the daily flow data after alteration;
4. Use the RVA threshold values obtained in Step (2) to judge the alteration of the river's hydrological regime after alteration, determine its impact and represent it with the overall hydrologic alteration. The hydrologic alteration is defined as follows:

$$D_i = \left| \frac{Y_{0i} - Y_{\mathrm{f}}}{Y_{\mathrm{f}}} \right| \times 100\,\% \tag{1}$$

Where: D_i is the hydrologic alteration of number i IHA; Y_{0i} is the number of years of number i IHA in the range of RVA after alteration; Y_{f} is the number of years of IHA in the range of RVA after alteration. Generally, the value of D_i is divided into three sections: 0–33.3 %, low alteration; 33.3–66.7 %, medium alteration; 66.7–100 %, high alteration.

The response degrees of the aforementioned 33 IHAs before and after alteration are not the same and it needs to conduct an overall assessment on the alteration of river's hydrological regime. The overall hydrological characteristic variability is represented with the overall hydrologic alteration D_0 and the actual assessment method is as follows:

It shall be categorized as low alteration when all 33 IHAs belong to low alteration:

$$D_0 = \frac{1}{33} \sum_{i=1}^{33} D_i \tag{2}$$

It shall be categorized as medium alteration when at least one IHA belongs to medium alteration but no IHA belongs

Table 2. Summary of hydrologic parameters used in the IHA, and their features (Richter et al., 1997).

General group	Regime features	Flow parameters used in the RVA
Group 1: Magnitude of monthly water conditions	Magnitude, timing	Mean value for 12 calendar month
Group 2: Magnitude and duration of annual extreme conditions	Magnitude, duration	Annual minimum 1-day means Annual maximum 1-day means Annual minimum 3-day means Annual maximum 3-day means Annual minimum 7-day means Annual maximum 7-day means Annual minimum 30-day means Annual maximum 30-day means Annual minimum 90-day means Annual maximum 90-day means
Group 3: Timing of annual extreme water conditions	Timing	Julian date of each annual 1-day maximum Julian date of each annual 1-day minimum
Group 4: Frequency and duration of high and low pulses	Magnitude, frequency, duration	Number of high pulses each year Number of low pulses each year Mean duration of high pulses within each year Mean duration
Group 5: Rate and frequency of water condition changes	Frequency, rate of change	Means of all positive differences between consecutive daily values Means of all negative differences between consecutive daily values Number of rises Number of falls

to high alteration:

$$D_0 = 33\,\% + \frac{1}{33}\sum_{i=1}^{N_m}(D_i - 33\,\%) \quad (3)$$

Where: N_m is the number of IHA belonging to medium alteration.

It shall be categorized as high alteration when at least one IHA belongs to high alteration:

$$D_0 = 67\,\% + \frac{1}{32}\sum_{i=1}^{N_h}(D_i - 67\,\%) \quad (4)$$

Where: N_h is the number of IHA belonging high alteration.

3.2 Runoff regulation and calculation model

Through establishing a model to simulate the impounding and operation of upstream cascade reservoirs, the runoff of downstream hydrological control stations after the regulation by upstream reservoirs is obtained in different level years, and further the impact of the operation of upstream reservoirs on downstream hydrological variability is analyzed. The runoff regulation is calculated with the "Combined Reservoir Group Water Regulation Model Software 1.0"; the main model structure (Fig. 2) are as follows.

First step, the model first restores the actually measured runoff series affected by the reservoir regulation and impoundment and also obtains the natural runoff series of each hydrological control station and reservoir dam site through the restoration;

Second step, the reservoir at the highest reach of each branch of the upper reach conducts regulation on the natural runoff at dam site, and conducts routine regulation and operation according to the regulation rules, thus obtaining the runoff series after operation; the runoff series after operation is calculated downwards level by level to the reservoirs of the lower levels, and then regulated and operated by the lower-level reservoirs thus obtaining the runoff series of the hydrological control station of each branch after regulation by upstream reservoirs ultimately through the operation level by level.

The model mainly consists of the runoff restoration, upstream reservoir regulation and operation, and runoff calculation modules. The model structure is as shown in Fig. 3.

3.2.1 Runoff restoration module

Since there are some large reservoirs already constructed at the upstream of the Three Gorges and the operation of these reservoirs has changed the natural situations of the runoff of

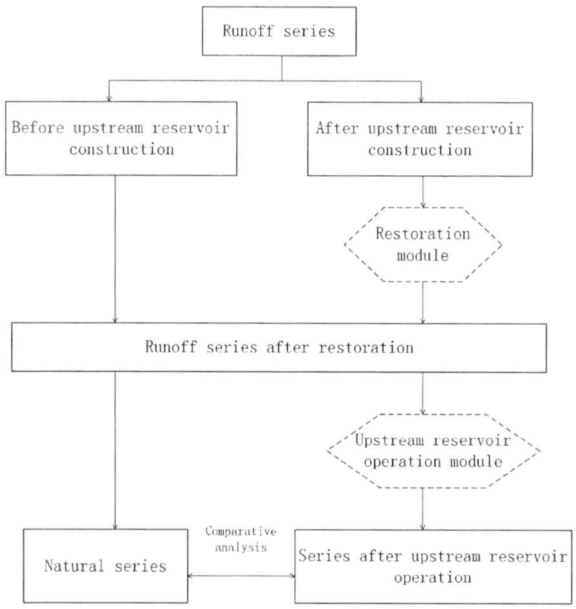

Figure 2. Runoff impact model calculation process.

river channels concerned, in order to achieve consistency in the analyzed and calculated runoff series, the runoff needs to be restored.

The restoration calculation of this module is mainly to calculate reservoir's daily (or pentad) storage variables with the reservoir water balance equation and according to the reservoir's dam water level as well as inflow and outflow. The details are as shown in Eq. (1):

$$Q_{\text{inflow}} = Q_{\text{outflow}} + \frac{\Delta W}{\Delta t} + \frac{\Delta W_{\text{loss}}}{\Delta t} + Q_{\text{led}} \tag{5}$$

Where: Q_{inflow} is the mean reservoir inflow in the time period; Q_{outflow} is the mean reservoir outflow in the time period; Q_{led} is the mean reservoir led-in or led-out flow in the time period; ΔW is the reservoir water storage change value in Δt time period; ΔW_{loss} is the reservoir water loss quantity (including the evaporation and seepage quantity) in Δt time period; Δt is the length of time period.

To simplify the calculation, the reservoir water loss quantity ΔW_{loss} was not considered this time, therefore:

$$\Delta W = V(Z_{t+1}) - V(Z_t) \tag{6}$$

Where: Z_t and Z_{t+1} is the reservoir water level at the beginning and end of t time period respectively; $V(Z_t)$ and $V(Z_{t+1})$ is the reservoir capacity at the beginning and end of t time period respectively.

Take $\Delta Q = \frac{\Delta W}{\Delta t}$ as the mean reservoir water storage flow, then:

$$\Delta Q = \frac{\Delta W}{\Delta t} = Q_{\text{inflow}} - Q_{\text{outflow}} - Q_{\text{led}} \tag{7}$$

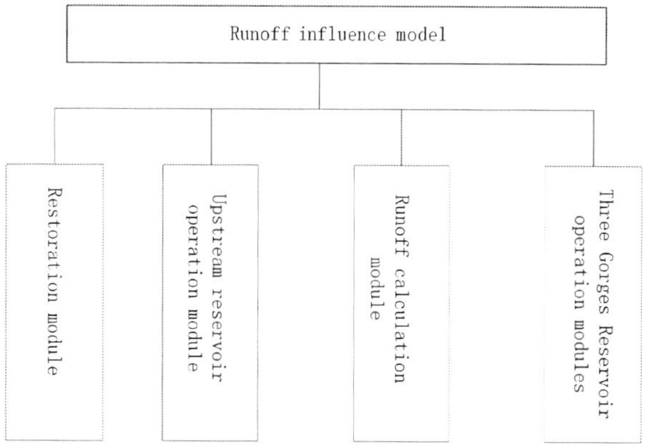

Figure 3. Diagram of Model Structure.

In the case of the reservoir diversion water flow not considered, ΔQ represents the reservoir impoundment when it is positive, and represents that the reservoir is utilizing the regulated reservoir storage capacity to increase the discharge when it is negative.

3.2.2 Upstream reservoir regulation and operation module

The upper Three Gorges reservoir regulation and operation module is to simulate the normal regulation and operation of upstream reservoirs, simulate the reservoir operation according to the operation chart and dam-site runoff of each reservoir, and calculate the reservoir water storage variation in each time period and the alterations of natural runoff after the operation of reservoir. During the reservoir regulation and operation simulation, the operation was conducted in strict accordance with the reservoir operation chart based on the reservoir water level at the beginning of the time period and the reservoir inflow water within the time period. The simulation time period was very long; therefore, the reservoir flood control regulation was not considered in the module. At the same time, in this study, all reservoir operations were based on the regular operation of the operation chart, and the emergency regulation of each reservoir under special circumstances was not considered.

3.2.3 Runoff calculation module

Either in the case of restoring the water storage variable of each reservoir in the calculation or simulating the upstream reservoir runoff after regulation, the calculation needs to be conducted downstream to the reservoirs or hydrological control stations of the lower levels so as to obtain the runoff of downstream hydrological control station or reservoir after the restoration and that after the regulation by upstream reservoirs. The runoff calculation module is mainly used to simulate the calculation of the runoff of upstream reservoirs or

Table 3. Mean flow of Yichang hydrological station from September to November in different level years ($m^3\ s^{-1}$).

Year	Natural	2010 level year	2015 level year	Forward level year	Year	Natural	2010 level year	2015 level year	Forward level year
1955	17 800	17 500	16 300	15 000	1985	19 300	19 100	18 000	16 700
1956	15 400	15 000	13 900	12 600	1986	17 100	16 600	15 600	14 300
1957	13 700	13 500	12 500	11 200	1987	18 500	18 400	17 300	16 000
1958	16 500	16 200	14 800	13 500	1988	20 800	20 400	19 200	18 000
1959	11 500	11 300	10 300	9000	1989	19 900	19 700	18 600	17 300
1960	16 900	16 700	15 600	14 300	1990	18 000	17 700	16 700	15 400
1961	16 900	16 500	14 900	13 700	1991	16 000	15 700	14 700	13 400
1962	17 200	17 000	15 800	14 500	1992	12 900	12 600	11 500	10 200
1963	20 700	20 500	19 300	18 000	1993	20 500	20 200	19 000	17 700
1964	25 600	25 300	24 100	22 900	1994	15 100	14 300	13 000	11 700
1965	22 300	22 100	20 900	19 700	1995	15 900	15 400	14 300	13 000
1966	21 500	21 300	20 300	19 000	1996	14 200	13 900	12 800	11 600
1967	17 400	17 100	15 900	14 600	1997	11 500	11 100	10 100	8900
1968	22 200	22 000	20 900	19 600	1998	16 700	16 400	15 300	14 000
1969	17 600	17 300	16 100	14 800	1999	18 700	18 400	17 200	16 000
1970	16 000	15 800	14 600	13 300	2000	19 700	19 400	18 200	17 000
1971	16 300	16 300	15 100	13 900	2001	20 400	20 100	18 900	17 700
1972	13 300	12 800	11 700	10 400	2002	10 200	10 000	9000	7800
1973	19 400	19 000	17 900	16 600	2003	17 500	17 300	16 200	14 900
1974	22 000	21 700	20 500	19 300	2004	18 000	17 600	16 400	15 100
1975	19 400	19 000	17 700	16 400	2005	16 700	16 400	15 400	14 100
1976	16 100	15 800	14 600	13 300	2006	11 200	10 500	9200	8000
1977	14 900	14 600	13 400	12 200	2007	15 400	15 000	13 700	12 500
1978	14 300	14 000	12 800	11 600	2008	19 500	19 100	17 800	16 500
1979	20 900	20 500	19 400	18 200	2009	12 800	12 600	11 300	10 100
1980	20 400	20 200	19 000	17 800	2010	15 000	14 600	13 400	12 100
1981	17 300	17 100	15 800	14 600	2011	13 200	12 900	11 900	10 600
1982	20 500	20 100	18 900	17 600	2012	17 100	16 700	15 100	14 000
1983	19 100	18 800	17 600	16 300	2013	12 400	12 100	10 600	9200
1984	16 300	16 100	14 900	13 600	Mean	17 200	16 900	15 800	14 500

hydrological control stations to the downstream. The calculation time period of this study was very long and belonged to long-series operation, having not only a very long time span, but also the flood season and the dry season. Currently, the studies on the dry season runoff calculation are relative few and are not mature in terms of methodology compared to the flood season flood evolution. This time, it was disposed with a simplified method where only the propagation time between two hydrological stations was considered, then simple transposition was conducted on the runoff process without taking into consideration the attenuation and deformation, and the upstream flow after transposition plus interval flow was taken as the downstream flow. The propagation time determination was mainly based on the flow and flow velocity in September–October of the water storage period. Since the time period for the reservoir regulation simulation in this time was a long-wait period, the propagation time analyzed was also relatively rough, taking the day as the unit. Upon analysis, the propagation time from upstream reservoirs to the hydrological control station of each branch and Yichang considered in this time of calculation was as follows:

1. Min River: the propagation time was 2d (days) from Pubugou Reservoir to Gaochang, 2d from Zipingpu Reservoir to Gaochang also, and 3d from Gaochang to Three Gorges Reservoir.
2. Jinsha River: the propagation time was 1d from Ertan Reservoir to Wudongde Reservoir, 2d from Wudongde Reservoir to Xiluodu Reservoir, 1d from Xiluodu Reservoir to Xiangjiaba Reservoir and 3d from Xiangjiaba Reservoir to Three Gorges Reservoir.
3. Jialing River: the propagation time was 1d from Baozhusi Reservoir to Tingzikou Reservoir, 3d from Tingzikou to Beibei; and 1d from Beibei hydrological station to Three Gorges Reservoir.
4. Wu River: the propagation time was 1d from Hongjiadu Reservoir to Wujiangdu Reservoir, 1d from Wujiangdu

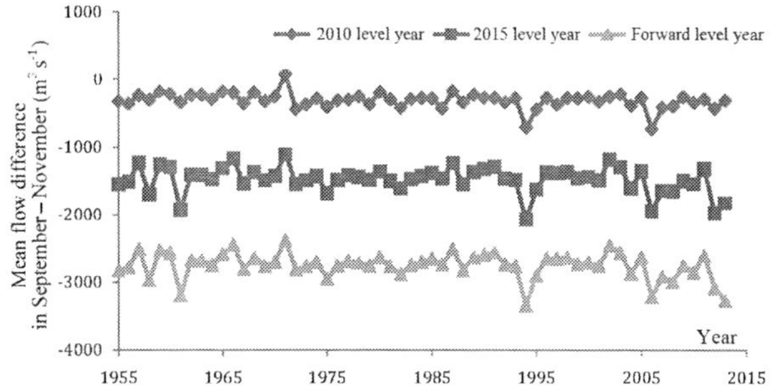

Figure 4. Process of mean flow of Yichang hydrological station from September to November in different level years and the natural condition.

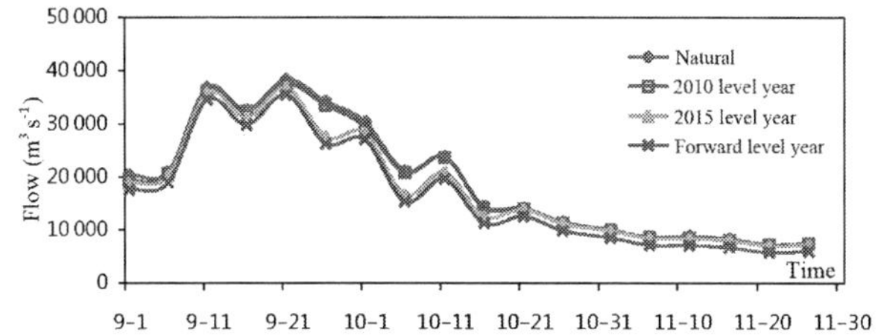

Figure 5. Process of Yichang hydrological station from September to November in slightly rainy year (1973).

Reservoir to Goupitan Reservoir, 1d from Goupitan Reservoir to Pengshui, and 1d from Wulong to Three Gorges Reservoir. The propagation time from Pengshui to Wulong was not considered due to its shortness.

3.2.4 Model calculated series and time period

Since there are many reservoirs on the upper Yangtze River, the design of each hydrological control station and reservoir on each branch is inconsistent due to the different length of series observed in each station, with the shortest starting from 1955 and the observation data missing in some years. The synchronized series of each hydrological station in 1955–2013 was simulated and calculated, and the series of several hydrological stations in the years absent of observation was supplemented with the upstream and downstream runoff or the station's water level. Therefore, all runoff series after the operation of reservoirs mentioned in this study refers to the series in 1955–2013.

In the runoff impact simulation, the calculated time period was controlled by five days. During the operation simulation of each reservoir, the corresponding regulation and operation was all conducted according to the operation chart based on the initial reservoir water level of every five days and the inflow water at the moment.

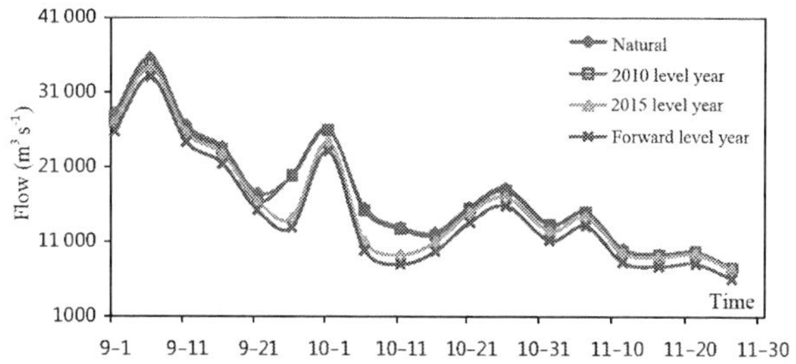

Figure 6. Process of Yichang hydrological station from September to November in normal year (1969).

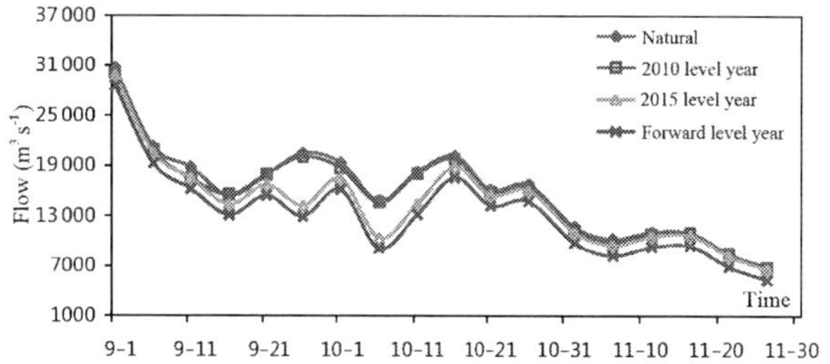

Figure 7. Process of Yichang hydrological station from September to November in slightly dry year (1976).

4 Results

4.1 Hydrological variability analysis of Yichang hydrological station

In order to analyze the impact of large reservoirs at the upstream of Three Gorges on the hydrological variability of Three Gorges Reservoir inflow runoff, the runoff of Yichang hydrological station was taken in representation of the Three Gorges Reservoir inflow runoff. The regulation and impoundment functions of Three Gorges Reservoir were not considered in the three level years.

The mean flow of Yichang hydrological station form September to November in different level years is as shown in Table 3.

The mean flow from September to November after the operation of reservoirs decreased to different degrees; the mean flow from September to November in 2010 level year was $16\,900\,m^3\,s^{-1}$, decreased by $300\,m^3\,s^{-1}$ compared to the natural condition of $17\,200\,m^3\,s^{-1}$ and achieved a decreasing amplitude of 1.8 %; the mean flow from September to November in 2015 level year was $15\,800\,m^3\,s^{-1}$, decreased by $1400\,m^3\,s^{-1}$ compared to the natural condition and achieved a decreasing amplitude of 8.6 %; the mean flow from September to November in the forward level year was $14\,500\,m^3\,s^{-1}$, decreased by $2700\,m^3\,s^{-1}$ compared to the natural condition and achieved a decreasing amplitude of 16 %.

Table 4. RVA calculation results of hydrological variability of Yichang hydrological station in different level years (%).

		IHA factor	2010 level year	2015 level year	Forward level year
Group 1	1	Mean flow in January	16.7	96.7	96.7
	2	Mean flow in February	56.7	96.7	96.7
	3	Mean flow in March	13.3	63.3	83.3
	4	Mean flow in April	3.3	43.3	56.7
	5	Mean flow in May	0.0	0.0	0.0
	6	Mean flow in June	3.3	3.3	10.0
	7	Mean flow in July	6.7	6.7	10.0
	8	Mean flow in August	0.0	6.7	13.3
	9	Mean flow in September	3.3	6.7	13.3
	10	Mean flow in October	3.3	3.3	10.0
	11	Mean flow in November	0.0	6.7	3.3
	12	Mean flow in December	0.0	10.0	80.0
Group 2	13	Annual maximum 1d pentad flow	6.7	3.3	0.0
	14	Annual maximum 3d pentad flow	0.0	6.7	20.0
	15	Annual maximum 7d pentad flow	3.3	10.0	3.3
	16	Annual maximum 30d pentad flow	3.3	13.3	30.0
	17	Annual minimum 1d pentad flow	93.3	93.3	96.7
	18	Annual minimum 3d pentad flow	86.7	96.7	96.7
Group 2	19	Annual minimum 7d pentad flow	70.0	96.7	96.7
	20	Annual minimum 30d pentad flow	23.3	66.7	90.0
	21	Base flow index	71.0	96.8	100.0
Group 3	22	Minimum flow occurrence time	19.0	42.9	66.7
	23	Maximum flow occurrence time	0.0	4.5	4.5
Group 4	24	Frequency of low flow	3.3	13.3	13.3
	25	Duration of low flow	13.3	3.3	0.0
	26	Frequency of high flow	4.3	4.3	17.4
	27	Duration of high flow	20.0	26.7	36.7
Group 5	28	Annual average inter-pentad water fall rate	10.0	23.3	20.0
	29	Annual average inter-pentad water rise rate	0.0	10.0	13.3
	30	Number of inversion	31.8	68.2	63.6
Overall hydrological variability			68.4	72.5	74.3

Figures 4–8 are the comparison of the pentad flow of Yichang hydrological station representing different situations. The operation of reservoirs had a relatively small impact on the runoff in 2010 level year; however, in 2015 level year and the forward level year, the impact increased gradually as the reservoirs of Xiluodu, Xiangjiaba, Jinping First-cascade, Goupitan, Pengshui, Wudongde and Baihetan were put into operation in succession; in the forward level year, the mean flow in September–November after the operation of reservoirs in the slightly rainy year was smaller than that of the natural condition by 14.4 % or so and that in the especially dry year amounted to 28.6 %.

The RVA calculation results of hydrological variability of Yichang hydrological station in different level years are as shown in Table 4. It can be seen that the overall hydrologic alterations in 2010 level year, 2015 level year and the forward level year were 68.4 72.5 and 74.3 % respectively, all belonging to high alteration in the three level years.

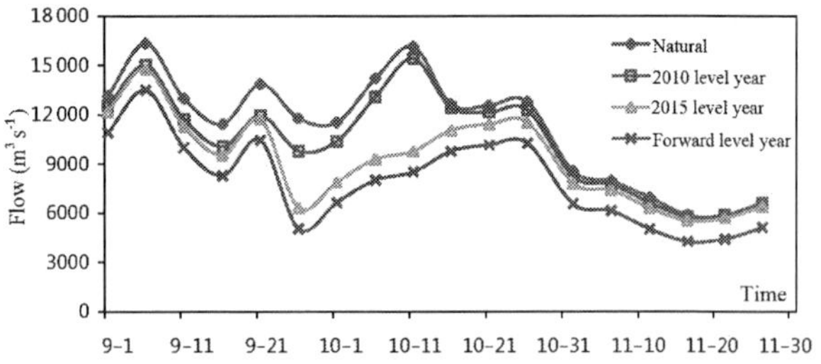

Figure 8. Process of Yichang hydrological station from September to November in especially dry year (2006).

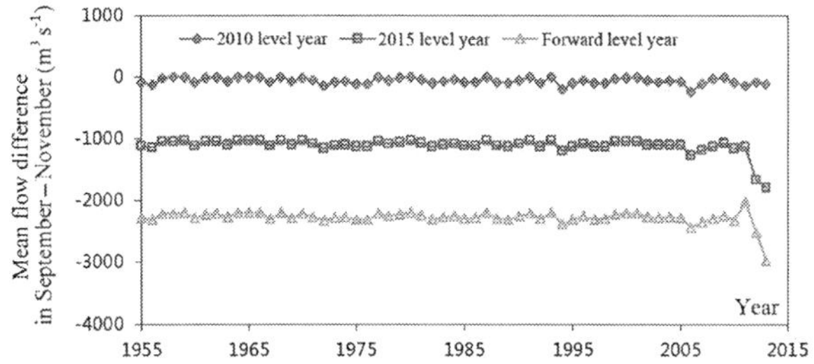

Figure 9. Process of mean flow of Xiangjiaba Hydrological Station from September to November in different level years and the natural condition.

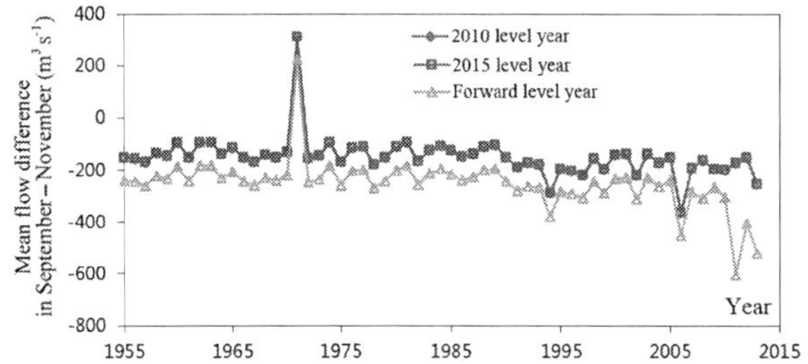

Figure 10. Process of mean flow of Gaochang hydrological station from September to November in different level years and the natural condition.

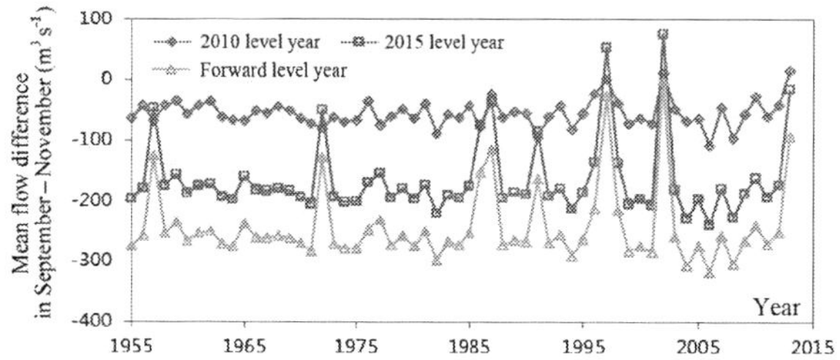

Figure 11. Process of mean flow of Beibei hydrological station from September to November in different level years and the natural condition.

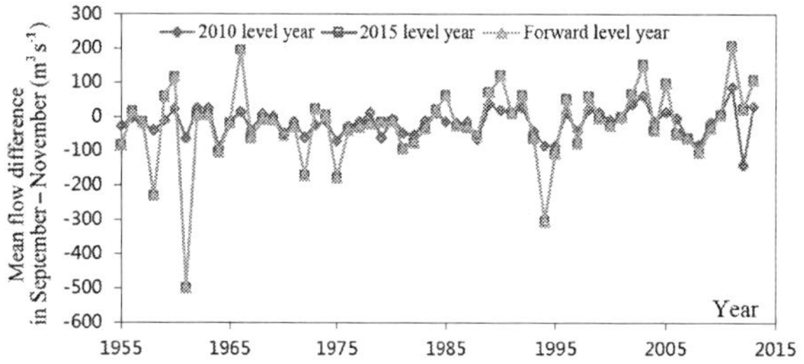

Figure 12. Process of mean flow of Wulong hydrological station from September to November in different level years and the natural condition.

4.2 Hydrological variability of other hydrological stations

The difference process of the mean flow of hydrological control stations on the mainstreams of Yangtze River from September to November in different level years and the natural condition is as shown in Figs. 9–12. The RVA calculation results of hydrological variability of the hydrological stations on each mainstream in different level years are shown in Table 5.

The operation conditions of the reservoirs on Jinsha River in different level years are as follows: only Ertan Reservoir was operated in 2010 level year; the three reservoirs of Xiluodu, Xiangjiaba and Jinping First-cascade will be newly put into operation in 2015 level year; and the Lianghekou, Wudongde and Baihetan Reservoirs will be put into operation in the forward level year. The overall hydrologic alterations in 2010 level year, 2015 level year and the forward level year were 71.1, 80.3 and 82.3 % respectively, all belonging to high alteration.

Gaochang hydrological station is the outlet hydrological control station on Min River. Currently, there are two large reservoirs on Min River, namely, Pubugo Reservoir and Zipingpu Reservoir. In addition, Shuangjiangkou Reservoir on Dadu River is in the planning stage. Therefore, the two reservoirs of Pubugou and Zipingpu were operated at the same time in both 2010 level year and 2015 level year and Shuangjiangkou Reservoir will be newly put into operation in the forward level year. The overall hydrologic alterations in 2010 level year, 2015 level year and the forward level year were 72.5 % for all, all belonging to high alteration.

The outlet hydrological control station of Jialing River is Beibei hydrological station. Only Baozhusi Reservoir was operated on Jialing River in 2010 level year; Tingzikou Reservoir will be put into operation in 2015 level year; and Shuangjiangkou Reservoir will be put into operation in the forward level year. The overall hydrologic alterations in 2010 level year, 2015 level year and the forward level year were 37.1, 68.8 and 68.8 % respectively, belonging to medium alteration in 2010 level year and high alteration in 2015 level year and the forward level year.

Wulong hydrological station is the outlet hydrological control station of Wu River. The upstream reservoirs considered in this calculation are the four reservoirs of Hongjiadu, Wujiangdu, Goupitan and Pengshui. Only Hongjiadu and Wujiangdu Reservoirs were operated in 2010 level year; and the four reservoirs will be operated at the same time in both 2015 level year and the forward level year. The overall hydrologic alterations in 2010 level year, 2015 level year and the forward level year were 68.1, 71.9 and 71.9 % respectively, all belonging to high alteration in the three level years.

Table 5. RVA calculation results of hydrological variability of mainstream hydrological stations in different level years.

River	Hydrological station	Overall hydrological variability		
		2010 level year	2015 level year	forward level year
Jinsha River	Xiangjiaba	71.1	80.3	82.3
Min River	Gaochang	72.5	72.5	72.5
Jialing River	Beibei	37.1	68.8	68.8
Wu River	Wulong	68.1	71.9	71.9

5 Conclusion and discussion

This paper mainly analyzed the reservoir impoundment quantity and evaporation loss and gain quantity of the constructed reservoirs and adopted the IHA approach to study the impact of the operation of cascade reservoirs in Upper Yangtze River on hydrological variability of the downstream. The main conclusions obtained are as follows:

For Yichang hydrological station, the mean flow from September to November after the operation of reservoirs decreased; the mean flow from September to November in 2010 level year was 16 900 $m^3 s^{-1}$, decreased by 300 $m^3 s^{-1}$ compared to the natural condition of 17 200 $m^3 s^{-1}$ and achieved a decreasing amplitude of 1.8 %; the mean flow from September to November in 2015 level year was 15 800 $m^3 s^{-1}$, decreased by 1400 $m^3 s^{-1}$ compared to the natural condition and achieved a decreasing amplitude of 8.6 %; the mean flow from September to November in the forward level year was 14 500 $m^3 s^{-1}$, decreased by 2700 $m^3 s^{-1}$ compared to the natural condition and achieved a decreasing amplitude of 16 %. The overall hydrologic alterations of Yichang hydrological station in 2010 level year, 2015 level year and the forward level year were 68.4, 72.5 and 74.3 % respectively, all belonging to high alteration in the three level years (The impact by the regulation and impoundment of Three Gorges Reservoir was not considered in all three level years). Similar results can be found in Zhang et al. (2012) which showed that reservoir operation in the upper Yangtze River decreases inflow to Three Gorges Reservoir at the end of flood season and increases that in dry season.

Under the impact by the operation of reservoirs and compared to the natural condition, the mean flow of Xiangjiaba hydrological station on Jinsha River from September to November decreased by 0.9 % in 2010 level year, 16.9 % in 2015 level year and 34.8 % in the forward level year; the mean flow of Gaochang hydrological station on Min River from September to November decreased by 4.9 % in both 2010 level year and 2015 level year and 7.9 % in the forward level year; the mean flow of Beibei hydrological station on Jialing River from September to November decreased by 2 % in 2010 level year, 5.9 % in 2015 level year and 8.9 % in the forward level year; the mean flow of Wulong hydrological station on Wu River from September to November decreased by 1.3 % in 2010 level year and 1.6 % in both 2015 level year and the forward level year; and the mean flow of Three Gorges Reservoir from September to November decreased by 1.8 % in 2010 level year, 8.6 % in 2015 level year and 16 % in the forward level year. The overall hydrologic alterations of Xiangjiaba hydrological station on Jinsha River in 2010 level year, 2015 level year and the forward level year were 71.1, 80.3 and 82.3 % respectively, all belonging to high alteration; the overall hydrologic alterations of Gaochang hydrological station on Min River in 2010 level year, 2015 level year and the forward level year were 72.5 % for all, all belonging to high alteration; the overall hydrologic alterations of Beibei hydrological station on Jialing River in 2010 level year, 2015 level year and the forward level year were 37.1, 68.8 and 68.8 % respectively, belonging to medium alteration in 2010 level year and high alteration in 2015 level year and the forward level year; and the overall hydrologic alterations of Wulong hydrological station on Wu River in 2010 level year, 2015 level year and the forward level year were 68.1, 71.9 and 71.9 % respectively, all belonging to high alteration in the three level years.

The construction and operation of reservoirs in the upper Yangtze River aims to reduce the flood disaster and sediment deposition but alters the downstream hydrological regime inevitably. In this paper, the RVA approach was used to assess the hydrological variability resulted before and after the dam construction in the upper Yangtze River. The results of this study will be greatly helpful for the future management of water resources and will be greatly important for further understanding of human impacts (e.g., hydro-dam projects) on hydrological regimes in the Yangtze River. Furthermore, it is significant for sustainable social and economic development in the Yangtze River which is one of the economically developed regions in China.

Data availability. The research data is supplied by Bureau of Hydrology, Changjiang Water Resources Commission which can be obtain by visiting http://www.cjh.com.cn/.

Competing interests. The authors declare that they have no conflict of interest.

Special issue statement. This article is part of the special issue "Innovative water resources management – understanding and balancing interactions between humankind and nature". It is a result of the 8th International Water Resources Management Conference of ICWRS, Beijing, China, 13–15 June 2018.

Acknowledgements. This study is funded by the National Key Research and Development Program of China (grant no. 2017YFC0405302-02). Many thanks are given to the anonymous reviewers for their valuable comments.

Edited by: Dingzhi Peng
Reviewed by: two anonymous referees

References

Allan, R.: Regime dependent changes in global precipitation, Clim. Dynam., 39, 827–840, 2012.

Barnett, T. P., Pierce, D. W., Hidalgo, H. G., Bonfils, C., Santer, B. D., Das, T., Bala, G., Wood, A. W., Nozawa, T., Mirin, A. A., and Cayan, D. R.: Human-Induced Changes in the Hydrology of the Western United States, Science, 5866, 1080–1083, 2008.

Chen, Y. D., Yang, T., Xu, C., Zhang, Q., Chen, X., and Hao, Z.: Hydrologic alteration along the Middle and Upper East River (Dongjiang) basin, South China: a visually enhanced mining on the results of RVA method, Stoch. Environ. Res. Risk Assess., 24, 9–18, 2010.

Guo, L., Xia, Z., Yu, L., Wang, J., and Chen, Q.: Hydrological Variation Characteristics of Rivers in Humid Region: Oujiang River, China, Proc. Eng., 28, 95–98, 2012.

Jia, Y. W., Wang, H., and Yan, D. H.: Distributed model of hydrological cycle system in Heihe River basin: I. Model development and Verification, J. Hydraul. Eng., 5, 003, 2006

Jiang, Z., Shen, Y., Ma, T., Zhai, P., and Fang, S.: Changes of precipitation intensity spectra in different regions of mainland China during 1961–2006, J. Meteorol. Res., 28, 1085–1098, 2014.

Kisi, O.: An innovative method for trend analysis of monthly pan evaporations, J. Hydrol., 527, 1123–1129, 2015.

Li, Y., Chen, X., and Wang, Z.: A Tentative Discussion on the Impact of Human Activities on the Variability of Runoff Serises of the Beijiang River Basin, J. Nat. Res., 21, 12–29, 2006.

Li, Y., Huang, Q., Xu, J., and Zhang, Z.: Long-range Correlations Diagnoses and Trend Prediction of RunoffOver Yellow River-During the Resent 80 Years, J. Arid Land Resour. Environ., 22, 68–72, 2008.

Lin, K., He, Y., and Chen, X.: Identifying the quantitative effect of climate change and human activity on runoff in the Dongjiang River basin, J. Hydraul. Eng., 43, 1312–1321, 2012.

Liu, C. and Cheng, L.: Analysis on Runoff Series with Special Reference to Drying up Courses of Lower Huanghe River, Acta Geogr. Sin., 55, 257–265, 2000.

Mathlouthi, M. and Lebdi, F.: Event in the case of a single reservoir: the Ghèzala dam in Northern Tunisia, Stoch. Environ. Res. Risk Asses., 22, 513–528, 2008.

McVicar, T. R., Roderick, M. L., Donohue, R. J., Li, L. T., Van Niel, T. G., Thomas, A., Grieser, J., Jhajharia, D., Himri, Y., Mahowald, N. M., and Mescherskaya, A. V.: Global review and synthesis of trends in observed terrestrial near-surface wind speeds: Implications for evaporation, J. Hydrol., 416–417, 182–205, 2012.

Pettit, N. E., Froend, R. H., and Davies, P. M.: Identifying the natural flow regime and the relationship with riparian vegetation for two contrasting western Australian rivers, Regul. River., 17, 201–215, 2001.

Petts, G.: Complex response of river channel morphology subsequent to reservoir construction, Prog. Phys. Geogr., 3, 329–362, 1979.

Petts, G.: Long-term consequences of upstream impoundment, Environ. Manage., 7, 325–332, 1980.

Poff, N. L., Allan, J. D., Bain, M. B., Karr, J. R., Prestegaard, K. L., Richter, B. D., Sparks, R. E., and Stromberg, J. C.: The Natural Flow Regime, BioScience, 47, 769–784, 1997.

Prakash, S., Gairola, R. M., and Mitra, A. K.: Comparison of large-scale global land precipitation from multisatellite and reanalysis products with gauge-based GPCC data sets, Theor. Appl. Climatol., 121, 303–317, 2015.

Ren, L., Zhang, W., Li, C., and Wang, M.: Impacts of Human Activities on River Runoff in North China, J. Hohai Univ., 29, 13–18, 2001.

Richter, B. D., Baumgartner, J. V., Powell, J., and Braun, D. P.: A Method for Assessing Hydrologic Alteration within Ecosystems, Conserv. Biol., 10, 1163–1174, 1996.

Richter, B., Baumgartner, J., Wigington, R., and Braun, D.: How much water does a river need?, Freshw. Biol. 37, 231–249, 1997.

Richter, B. D., Baumgartner, J. V., Braun, D. P., and Powell, J.: A spatial assessment of hydrologic alteration within a river network, Reg. Rivers Res. Manage., 14, 329–340, 1998.

Shiau, J. T. and Wu, F. C.: Assessment of hydrologic alterations caused by Chi-Chi diversion weir in Chou-Shui Creek, Taiwan: opportunities for restoring natural flow conditions, Regul. River., 20, 401–412, 2004.

Sun, J. and Ao, J.: Changes in precipitation and extreme precipitation in a warming environment in China, Chinese Sci. Bull., 58, 1395–1401, 2013.

Szilagyi, J.: Modeled areal evaporation trends over the conterminous United States, J. Irrig. Drain. Eng., 4, 196–200, 2001.

Thame, R. A.: Global Perspective on Environmental Flow Assessment: Emerging Trends in the Development and Application of Environmental Flow Methodologies, River Res. Appl., 19, 397–441, 2013.

Wen, G., Huang, G., Tao, W., and Liu, C.: Observed trends in light precipitation events over global land during 1961–2010, Theor. Appl. Climatol., 125, 1–13, 2015.

Zhang, N., He, H. M., Zhang, S. F., Jiang, X. H., Xia, Z. Q., and Huang, F.: Influence of Reservoir Operation in the Upper Reaches of the Yangtze River (China) on the Inflow and Outflow Regime of the TGR-based on the Improved SWAT Model, Water Resour. Manage., 26, 691–705, 2012.

https://doi.org/10.5194/piahs-379-433-2018

Estimating lake-water evaporation from data of large-aperture scintillometer in the Badain Jaran Desert, China, with two comparable methods

Peng-Fei Han, Xu-Sheng Wang, Xiaomei Jin, and Bill X. Hu

Ministry of Education Key Laboratory of Groundwater Circulation and Evolution,
China University of Geosciences, Beijing, 100083, China

Correspondence: Peng-Fei Han (hpf0328@126.com)

Received: 28 December 2017 – Revised: 31 January 2018 – Accepted: 1 February 2018 – Published: 5 June 2018

Abstract. Accurate quantification of evaporation (E_0) from open water is vital in arid regions for water resource management and planning, especially for lakes in the desert. The scintillometers are increasingly recognized by researchers for their ability to determine sensible (H) and latent heat fluxes (LE) accurately over distances of hundreds of meters to several kilometers, though scintillometers are mainly used to monitor the land surface processes. In this paper, it is installed on both sides of the shore over a lake. Compared to the data of evaporationpan, the scintillometer was successfully applied to Sumu Barun Jaran in Badain Jaran Desert using the classical method and the proposed linearized β method. Due to the difficulty in measuring water surface temperature and the easiness to monitor the water temperature at different depths, it is worth thinking that if is feasible to utilize the shallow water temperature instead of the water surface temperature and how much errors it will cause. Water temperature at 10 and 20 cm depths were used to replace the lakewater surface temperature in the two methods to analyze the changes of sensible and latent heat fluxes in hot and cold seasons at halfhour time scales. Based on the classical method, the values of H were almost barely affected, and the average value of LE using water temperature at 20 cm depth is 0.8–9.5 % smaller than that at 10 cm depth in cold seasons. In hot seasons, compared to the results at 10 cm depth, the average value of H increased by 20–30 %, and LE decreased by about 20 % at 20 cm depth. In the proposed linearized β method of scintillometer, only the slope of the saturation pressure curve (Δ) is related to the water surface temperature, which was estimated using available equations of saturated vapor pressure versus temperature of the air. Compared to the values of estimated by the air temperature, while the water surface temperature are replaced by water temperature at 10 and 20 cm depths, in different seasons, the errors of 2–25 % in Δ were caused. Thus was calculated by the original equation in the proposed linearized β method of scintillometer. Interestingly, the water temperature at 10 and 20 cm depths had little effect on H, LE (E_0) in different seasons. The reason is that the drying power of the air (E_A) accounted for about 85 % of the evaporation (i.e. the changes of Δ have only about 3 % impact on evaporation), which indicated that the driving force from unsaturated to saturated vapor pressure at 2 m height (i.e. the aerodynamic portion) has the main role on evaporation. Therefore, the proposed linearized β method of scintillometer is recommended to quantify the H, LE (E_0) over open water, especially when the water surface temperature cannot be accurately measured.

1 Introduction

Accurate quantification of evaporation from water body has always been crucial for water resources use and management reasonably, especially in arid regions (Kustas et al., 2000; Bou-Zeid, 2002; Vercauteren et al., 2009; Mcjannet et al., 2011; Mcgloin et al., 2014). Traditionally, evaporation in open water bodies is directly measured by evaporationpan or indirectly calculated by simple aerodynamic approaches using meteorological measurements. The assumptions made by these methods may usually lead to large errors in evaporation calculation. In the last two decades, better technical methods and equipment such as eddy covariance has been developed to estimate the evaporation of open water bodies directly, whose reliability has also been verified by a large number of researches in all sizes about water body (Blanken et al., 2000; Liu et al., 2009; Nordbo et al., 2011; Mcgloin et al., 2014). Due to the limitation in space in evaporation estimation of eddy covariance method, in recent years, scintillometer technology is developed and popular in accurately quantifying the latent heat flux over a greater spatial extent and accounting for possible spatial variability in evaporation resulting from wind, water temperature and depth (Mcjannet et al., 2011). However, Determination of latent heat flux using scintillometry has mainly been limited to studies over vegetation (Guyot et al., 2009). Until recently, Mcjannet et al. (2011) proposed a new approach for obtaining sensible and latent fluxes from a scintillometer deployed over open water. Compared with the results observed by eddy covariance and alternative scintillometer calculation approaches that are commonly used in the literature, the methodology based on linearization of the Bowen ratio was tested and validated.

In the scintillometry calculation, the most important portion is the definition of Bowen ratio (β). Generally, three approaches are used to define βin most researches: the classical method, the β closure method and the proposed linearized β method. Based on the energy balance in the water body studies, however, the heat flux of water body (G) has the significant contribution in the energy balance (Tanny et al., 2008; Vercauteren et al., 2009; Mcgloin et al., 2014; Mcjannet et al., 2011). The measurement of G in water bodies is difficult and leads to large errors easily (Stannard and Rosenberry, 1991). In order to avoid the measuring uncertainty of water surface temperature and the errors produced by different footprints, the slope of the saturation vapor pressure curve (Δ) was estimated from the air temperature using the polynomial equation derived by Lowe (1977) in the proposed linearized β method, however, this also cause some errors in the calculation. The water surface temperature is also needed in the classical method. Due to the difficulty in measuring the water surface temperature and the easiness to monitor the water temperature at different depths, it is worth thinking that if is feasible to utilize the shallow water temperature instead of the water surface temperature and how much errors it will cause.

In this paper, the scintillometer using the classical method and the proposed linearized β method are applied to Sumu Buran Jaran in Badain Jaran Desert, China, to estimate the evaporation. We also discuss that the effects of water surface temperature replaced by shallow water temperatures at different depths in cold and hot seasons on the scintillometer results.

2 Methods

2.1 Study Area

As shown in Fig. 1, the Badain Jaran Desert (39°20′–41°30′ N, 100°01′–103°10′ E) is located in the western Alxa Plateau in Inner Mongolia, China. It is the second largest desert in China covering 4.9×10^4 km^2 (Wang, 1990). On the north, east and south, the BJD is surrounded by mountains that were mostly formed by granite rocks. The northwest of BJD is Gurinai grassland lying in the downstream area of the Heihe River. Generally, the topography is higher in southeast, lower in northwest. In the desert hinterland, the tall and dense sanddunes and sandhills exist, accounting for as much as 60 % of the desert area, and their relative height are about 200–300 m and up to 500 m. The overall flow of groundwater is from south to north and from east to west, which is consistent with the topography of the desert. Plenty of lakes exist in the BJD, but most of them are smaller than 0.2 km^2 and shallower than 2 m. More than 65 % of the lakes are saline lakes (TDS > 35 g L^{-1}) and the salinities vary from 1–400 g L^{-1}. In this study, the second largest saline lake-Sumu Barun Jaran was chosen, with the area of 1.24 km^2 and the maximum depth of 11.1 m. The elevation of the lake is 1150–1200 m.

2.2 Measurements

A large-aperture scintillometer (LAS-BLS450, Scintec AG, Rottenburg, Germany) was installed on both sides of the Sumu Barun Jaran lake consisting of a transmitter (39°47′44.30″ N, 102°25′19.93″ E) and a receiver (39°46′51.55″ N, 102°25′25.15″ E) with 150 mm aperture, an optical wave signal is transmitted to the receiver over a path length (1801 m). The height of the scintillometer to the lake is about 10 m. The near infrared pulse spectrum emitted by the transmitter is 880 nm, and range of the structure parameter of the refractive index of air (C_n^2) is from 10^{-17}–10^{-10} m$^{-2/3}$ reflecting the atmospheric turbulence intensity. The modulation frequency of the scintillometer is 1750 Hz with an averaging period of 30 min. Combined with temperature, relative humidity, wind speed, net radiation and other meteorological data derived from an automatic weather station, the structure parameter of the refractive index of air can be used to estimate the sensible heat flux using Monin-

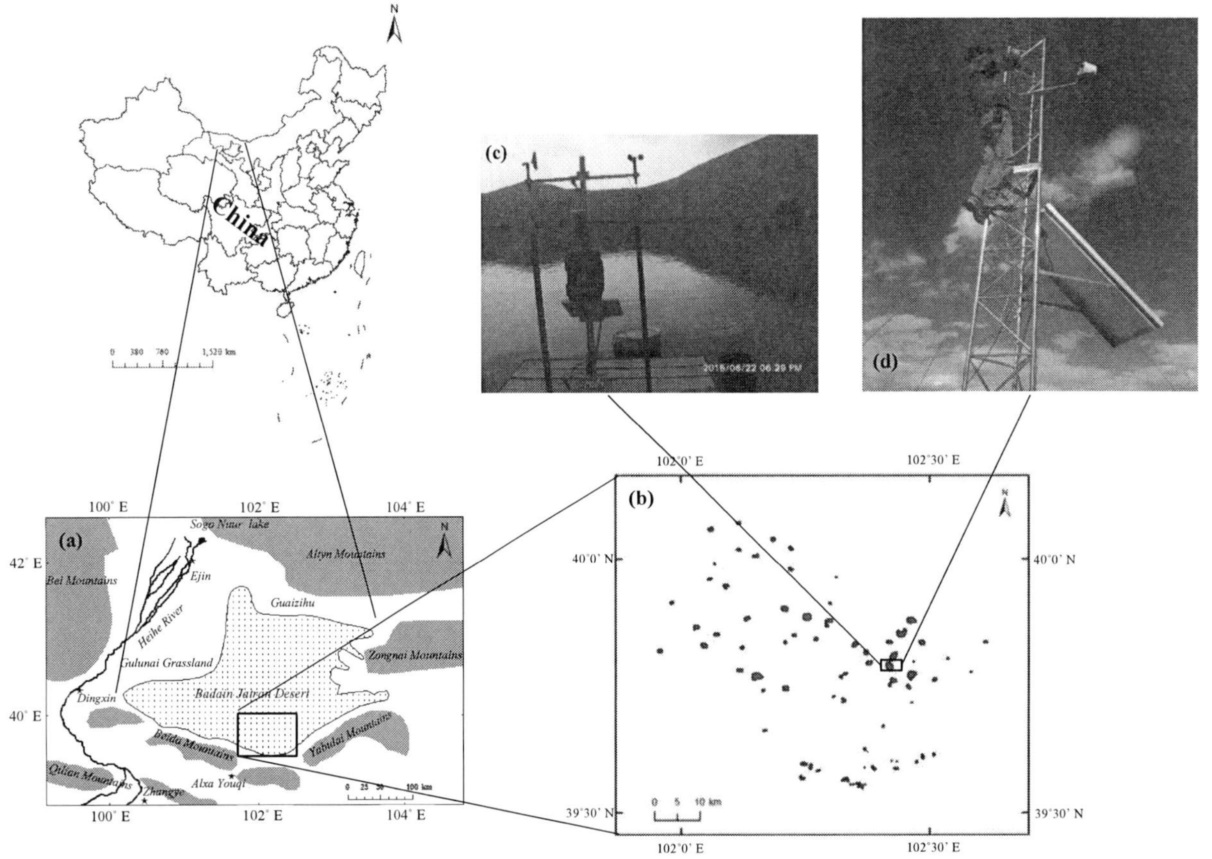

Figure 1. Location of the study area **(a)**, lakes **(b)**, weather station **(c)** and scintillometer **(d)**.

Obukhov similarity theory (Monin and Obukhov, 1954). Once the values of sensible heat flux (H) are obtained, the next step is to calculate the latent heat flux (LE) and evaporation (E_0). The main difference in calculation methods of scintillmeter is the estimation of β. As the water body heat flux (G) is difficult to measure accurately and leads to large errors in results easily, here we do not use the β closure method. However, if we use the shallow water temperature instead of water surface temperature in the β closure method, the heat flux from water surface to the shallow water depth should be estimated when calculating energy balance from net radiation and scintillometer method. The classical method using meteorological data and the proposed linearized β method are studied in this paper. Combined with temperature, relative humidity, wind speed, net radiation and other meteorological data derived from an automatic weather station, the structure parameter of the refractive index of air can be used to estimate the sensible heat flux using Monin-Obukhov similarity theory (Monin and Obukhov, 1954). Once the values of sensible heat flux (H) are obtained, the next step is to calculate the latent heat flux (LE) and evaporation (E_0) on the basis of the β referring to the ratio between the sensible and heat fluxes. Thus, the main difference in calculation methods of scintillmeter is the estimation of β. As the water body heat flux (G) is difficult to measure accurately and leads to large errors in results easily, the classical method using meteorological data and the proposed linearized β method are studied in this paper.

In the classical method, the $\beta$is estimated from the other measurements. Assuming that the turbulent transfer coefficients of heat and water vapor are equal above a water body, then we can calculate βusing the measurements of water surface temperature (T_w) and air temperature (T_a) as follows (Penman, 1948):

$$\beta = \gamma \frac{T_\mathrm{w} - T_\mathrm{a}}{e_\mathrm{w}^* - e_\mathrm{a}} \tag{1}$$

where e_w^* is the saturated vapor pressure at T_w.

In the proposed linearized β method, using the method proposed by Vercauteren et al. (2009), evaporation can be estimated by Penman (1948) equation:

$$E = \frac{\Delta}{\gamma} \frac{H}{L_e} + E_\mathrm{A} \tag{2}$$

where Δ is the slope of the saturation pressure curve. L_e is the latent heat of vaporization and γ is the psychometric constant. E_A is the drying power of the air. The calculation procedure shown in Mcjannet et al. (2011) is an iterative process. Starting values of β and Obukhov length (L) is needed to estimate initial values of H and E. From these values a new βcan be calculated which is then fed back through the iterative process to recalculate the structure parameter of the refractive index of air and H. The procedure is run until the solution is stable.

A detailed description of the calculation processes using the two methods to estimate fluxes and evaporation by scintillometer over open water is introduced by Mcjannet et al. (2011).

There are some differences in the calculation procedures used in this study and presented by Mcjannet et al. (2011). Firstly, z_0 was set to 0.0001 m in Mcjannet et al. (2011), however, considering the fluctuations of the lake, in this study it was fixed at 0.0002 m, which is equal to the experiential value of sea. Secondly, the drying power of the air (E_A) was expressed as (PMU, Monteith and Unsworth, 1990, p. 185):

$$E_A = \frac{\rho c_p}{r_a \gamma^*}[e_s(T) - e] \tag{3}$$

Wherer_a ($s\,m^{-1}$) is aerodynamic resistance and e (Pa) is vapor pressure with subscript s denoting the saturation. More detailed calculation equations were presented in Tanny et al. (2011). Thirdly, as we have net radiation data, the net radiation method is used to determine the atmospheric stability, in which that net radiation > 10 W m^{-2} is considered as atmospheric instability (Kohsiek et al., 2006). Finally, as we have no data of lake water surface temperature, water temperature at 10 cm depth are used to replace in the classical method.

2.3 Experimental data and Data quality control

Meteorological and scintillometer's data during 12:00 LT (UTC+8) on 26 October 2012 to 20:00 LT on 9 March 2013 and 08:00 LT on 10 May 2013 to 20:00 LT on 26 March 2015 with the interval of 30 min in the daytime are collected in this study. Compared to timezone in Beijing, from 08:00 to 20:00 LT is defined as daytime, and the rest is defined to as nighttime.

Based on Wang et al. (2010), the saturation upper limit of the refractive index of air in Sumu Barun Jaran is set to 1.63×10^{-13} $m^{-2/3}$, data beyond which are removed. Negative βvalues tended to occurred later in the afternoon, but we did not remove them directly like Mcjannet et al. (2011), and replaced them using positive values.

3 Results analysis

3.1 Reliability verification of scintillometer with two methods

The comparison periods for the proposed linearized β method and the classical method of scintillometer and E601 evaporationpan measurements covered 62 days between 1 July and 31 August 2014 to test the feasibility of scintillometer (Fig. 2). The daily evaporation estimated by scintillometer using two methods are less than those observed from evaporationpan, while their fluctuations are consistent. The evaporation correlation between evaporationpan and scintillometer using the proposed linearized β method ($r^2 = 0.40$) is better than that using the classical method ($r^2 = 0.01$). The average daily evaporation in July and August measured by pan and estimated by scintillometer using the classical method and the proposed linearized β method is 5.89, 4.90 and 2.36 mm, respectively. Evaporationpan data is 17 and 60 % higher than the corresponding average daily values of scintillometer using the classical method and the proposed linearized β method. That the evaporation calculated by scintillometer using two methods are less than the values observed from evaporationpan has two reasons. One is that the evaporationpan cannot store thermal energy. The other reason is that the data in atmospheric unstable condition in the daytime (i.e. $R_n > 10$ W m^{-2}) are used in the calculation of scintillometer, not all data from 08:00 to 20:00 LT. Moreover, the scintillometer did not record data at some point. These two kinds of data that are not used in the calculation account for 38 % of the total data. Thus, the average daily evaporation in July and August estimated by scintillometer using the classical method and the proposed linearized β method is 6.76 and 3.26 mm, respectively. The average daily evaporation estimated by scintillometer using the classical method and the proposed linearized β method is 15 % higher and 45 % lower than the result (5.89 mm) derived from evaporationpan, respectively. The results derived from the scintillometer using the classical method are closer to the average daily evaporation estimated from E601 pan. The average evaporation estimated by the proposed linearized β method accounts for 48 % of the corresponding average daily value calculated using the classical method. The great difference of evaporation estimated by scintillometer using two methods may be mainly caused by two reasons. One reason is due to the water surface temperature. As the water surface temperature is needed to calculate the β value in the classical method, whereas the water surface temperature here is replaced by the water temperature at 10 cm depth. The other reason is that use of the classical method involves iteratively solving only for L as β is specified from measurements, while both of the L and βin the proposed linearized β method are calculated iteratively until the solution is stable. Although the evaporation estimated by scintillometer and evaporationpan are not exactly the same, they have the similar magnitude

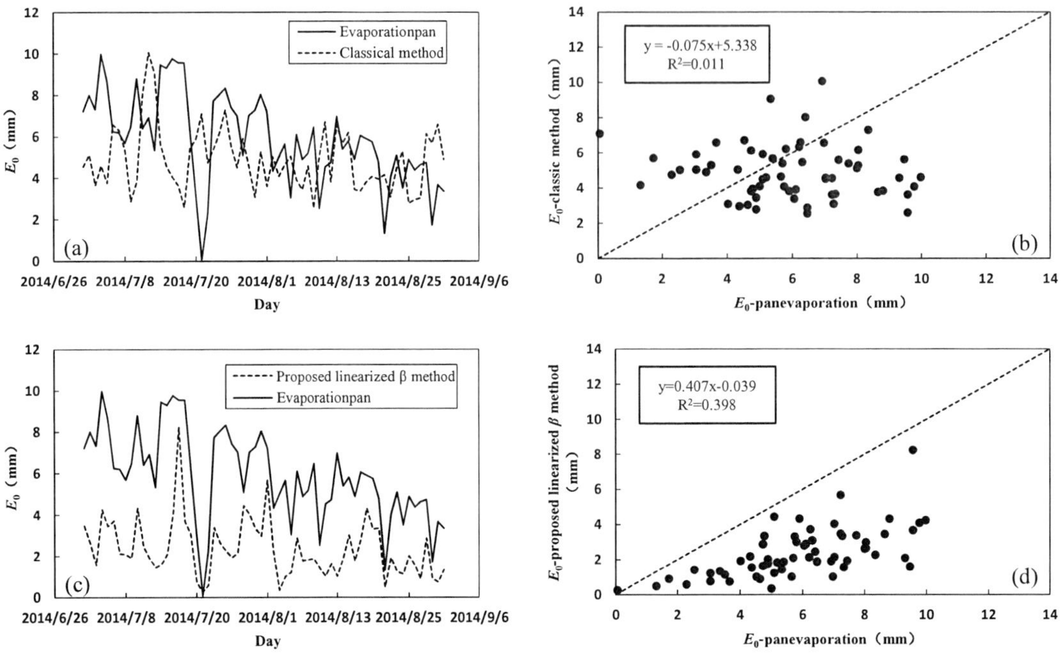

Figure 2. Dynamic variation **(a, c)** and comparison **(b, d)** between the estimated evaporation by scintillometer using the classical method and the proposed linearized β method and observed results by evaporationpan at daily time during 1 July–31 August in 2014.

and fluctuation trend. Therefore, the scintillometer using two methods are considered to be applied to Sumu Barun Jaran successfully.

3.2 Shallow water temperature instead of the water surface temperature

Water surface temperature must be employed to estimate evaporation using the classical method in scintillometer procedure. Measuring water surface temperature accurately is so difficult, while measuring water temperature is easy. Therefore, it is worth thinking that if is feasible to utilize the shallow water temperature instead of the water surface temperature and how much errors it will cause.

The water temperature at 20 cm depth is used to replace the lake water surface temperature, compared to the results replaced by the water temperature at 10 cm depth. The cold seasons and hot seasons are defined from November to February, and May to August, respectively.

3.2.1 The classical method

The results using the classical method at halfhour time scale are shown in Fig. 3. Using water temperature at 10 cm depth, the average value of H and LE are 10.83 and 32.37 W m^{-2} in January and February, 7.00 and 121.17 W m^{-2} in May and June, 6.75 and 119.26 W m^{-2} in July and August, and 14.77 and 30.65 W m^{-2} in November and December, respectively. From cold to hot seasons, the average value of H decreases by 46 %, and the average value of LE increases by 2.8 times. Replaced by water temperature at 20 cm depth, on average, the value of H and LE are 10.90 and 32.12 W m^{-2} in January and February, 8.42 and 95.27 W m^{-2} in May and June, 8.09 and 95.81 W m^{-2} in July and August, and 14.91 and 27.74 W m^{-2} in November and December, respectively. In cold seasons, the average value of H is 36 % lager than that in hot seasons, and LE increases by about 2.2 times. In conclusion, the values of H were almost barely affected, and the average value of LE using water temperature at 20 cm depth is 0.8–9.5 % smaller than that at 10 cm depth in cold seasons. In hot seasons, compared to the results at 10 cm depth, the average value of H increased by 20–30 %, and LE reduced by about 20 % at 20 cm depth. Compared to the results at 10 cm depth, the values of H and LE have greater fluctuation in hot seasons than those in cold seasons.

3.2.2 The proposed linearized β method

The proposed linearized β method in scintillometer procedure to calculate evaporation does not need the water surface temperature as only the slope of the saturation vapor pressure curve (Δ) is related to water surface temperature in this approach, which is assumed that Δ can be estimated from the air temperature using the polynomial equation derived by Lowe (1977). However, the assumption also produces 2–25 % errors of Δ in different seasons.

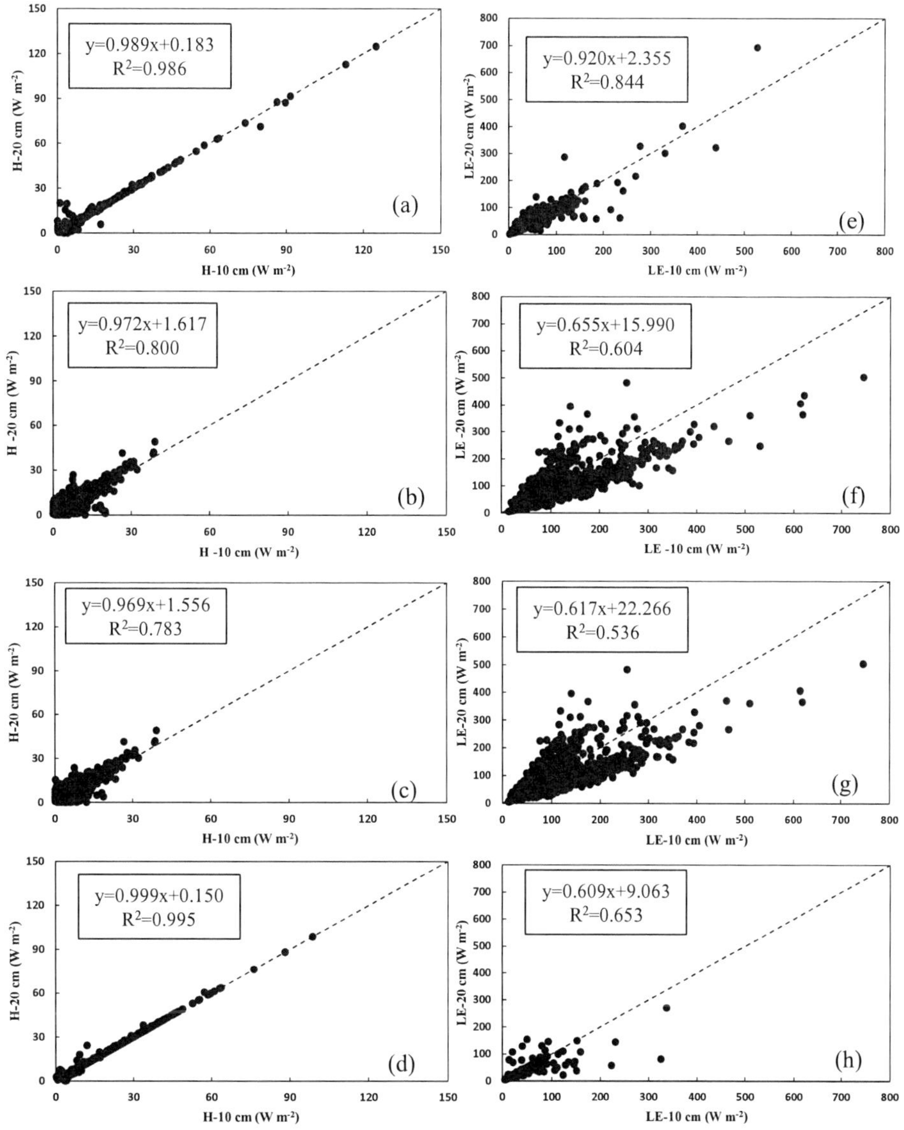

Figure 3. Comparison of the estimated results of H and LE between using water temperature at depths of 10 cm and 20 cm instead of lake water surface temperature in January and February, 2013 **(a, e)**, May and June, 2013 **(b, f)**, July and August, 2014 **(c, g)**, and November and December, 2012 **(d, h)** with the classical method.

In the proposed linearized β method, the results of the Δ estimated by the polynomial equation at air temperature and calculated by the original formula using water temperature at 10 cm depth instead of lake water surface temperature at halfhour time scales are compared (Fig. 4). Using the polynomial equation at air temperature, on average, the value of H and LE are 11.02 and 5.57 W m^{-2} in January and February, 3.01 and 80.64 W m^{-2} in May and June, 2.91 and 66.96 W m^{-2} in July and August, and 14.64 and 7.17 W m^{-2} in November and December, respectively. As the lakewater surface temperature are replaced by water temperature at depths of 10 and 20 cm, in both situations, the results of H, LE (E_0) are hardly affected (Figs. 4, 5).

Compared to the values of Δ estimated by the air temperature, while the water surface temperature are replaced by water temperature at 10 and 20 cm depths, in different seasons, the errors of 2–25 % in Δ are caused. However, changes in Δ have little impact on the values of H and LE. Since evaporation is composed of the drying power of the air (E_A) and the expression of Δ based on the Eq. (1) in the Vercauteren

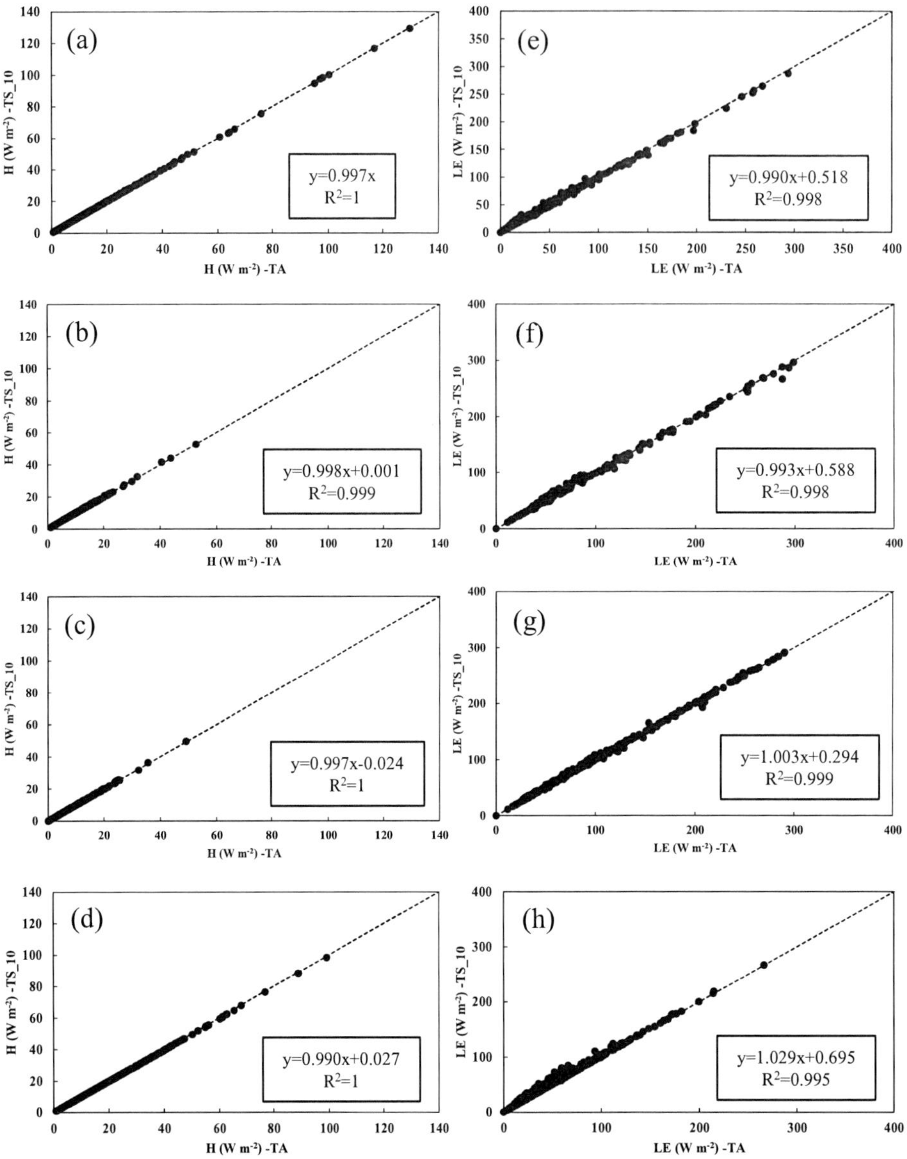

Figure 4. Comparison of the estimated results of H and LE between Δ obtained from the polynomial equation at air temperature and Δ calculated by the original formula at 10 cm depth in January and February, 2013 **(a, e)**, May and June, 2013 **(b, f)**, July and August, 2014 **(c, g)**, and November and December, 2012 **(d, h)**.

et al. (2009), the relationship between E_0 and E_A, shown in Fig. 6. The average value of E_0 and E_A are 0.015 and 0.012 mm in January and February, 0.153 and 0.135 mm in May and June, 0.146 and 0.126 mm in July and August, and 0.024 and 0.018 mm in November and December, respectively. In cold and hot seasons, E_A accounted for about 75 and 87 % of E_0 with good correlation, which indicates that the changes of have an average of 4 and 1 % impact on evaporation, and the driving force from unsaturated to saturated vapor pressure at 2 m height has the main role on evaporation.

4 Conclusions

The scintillometers are increasingly recognized by researchers for their ability to determine sensible (H) and latent heat fluxes (LE) accurately over distances of hundreds of meters to several kilometers, whereas scintillometers are

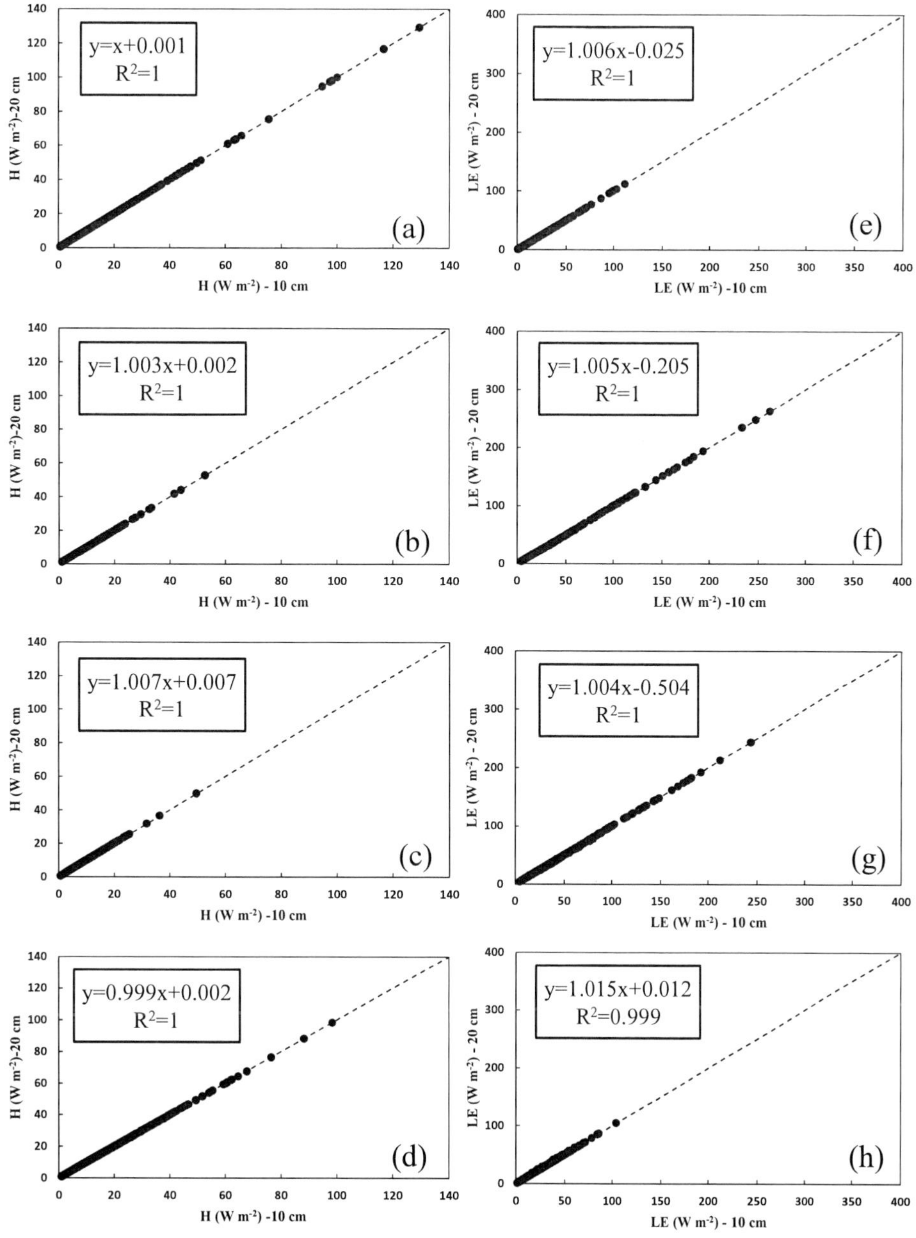

Figure 5. Comparison of the estimated results of H and LE between using water temperature at depth of 10 and 20 cm instead of lake water surface temperature in January and February, 2013 **(a, e)**, May and June, 2013 **(b, f)**, July and August, 2014 **(c, g)**, and November and December, 2012 **(d, h)** with the proposed linearized β method.

rarely used to estimate the evaporative processes over open water. In this study, the scintillometer was installed on both sides of the shore over a lake for deriving sensible and latent heat fluxes to estimate the evaporation. Compared to the evaporationpan, the scintillometer is successfully applied to Sumu Barun Jaran Lake in Badain Jaran Desert, China using the classical method and the proposed linearized β method.

Under the condition of atmospheric unstability during the daytime, using the classical method, the values of H were almost barely affected by the changes of shallow water temperature in cold seasons, whereas in hot seasons, the H change a lot. In different seasons, variations of shallow water temperature have an influence on LE (E_0) with different extent. Using the proposed linearized β method, shallow water temper-

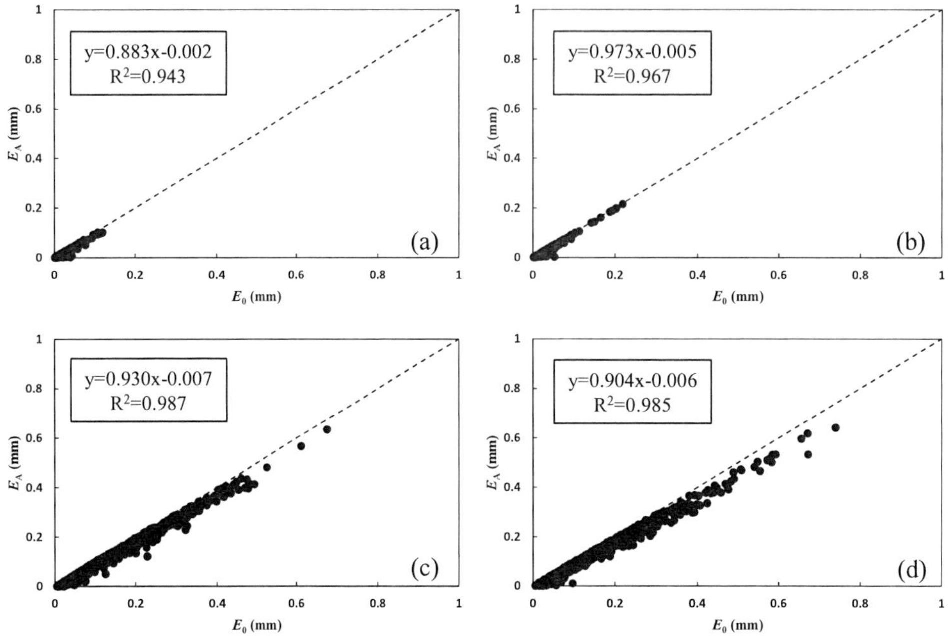

Figure 6. E_A versus E_0 from scintillometer using the proposed linearized β method in January and February, 2013 **(a, e)**, May and June, 2013 **(b, f)**, July and August, 2014 **(c, g)**, and November and December, 2012 **(d, h)**.

ature instead of lakewater surface temperature had little effect on H, LE (E_0) in all seasons. The drying power of the air (E_A) accounts for about 85 % of the evaporation, which indicates that the driving force from unsaturated to saturated vapor pressure at 2 m height (i.e. the aerodynamic portion) has the main role on evaporation. Therefore, the proposed linearized β method of scintillometer is recommended to quantify the H, LE (E_0) over open water, especially when the water surface temperature cannot be accurately measured. Although the proposed linearized β method of scintillometer has more advantages than other approach for estimating the H, LE (E_0) over open water, some iterative convergence problems should be concerned.

Data availability. The Sumu Barun Jaran is located in the hinterland of Badain Jaran Desert. Due to poor environment, data are difficult to obtain. The data in this study are derived from an automatic weather station and a large-aperture scintillometer installed in the Sumu Barun Jaran by China University of Geosciences (Beijing) for understanding the evaporation of the lakes in the desert.

Competing interests. The authors declare that they have no conflict of interest.

Special issue statement. This article is part of the special issue "Innovative water resources management – understanding and balancing interactions between humankind and nature". It is a result of the 8th International Water Resources Management Conference of ICWRS, Beijing, China, 13–15 June 2018.

Acknowledgements. This study was supported by the National Natural Science Foundation of China (91125024). The authors are grateful to two anonymous reviewers for their constructive comments.

Edited by: Zongxue Xu
Reviewed by: two anonymous referees

References

Blanken, P. D., Rouse, W. R., Culf, A. D., Spence, C., Boudreau, L. D., Jasper, J. N., Kochtubajda, B., Schertzer, W. M., Marsh, P., and Verseghy, D.: Eddy covariance measurements of evaporation from Great Slave Lake, Northwest Territories, Canada, Water Resour. Res., 36, 1069–1077, https://doi.org/10.1029/1999WR900338, 2000.

Bou-Zeid, E.: Climate change and water resources in Lebanon and the Middle East, J. Water Res. Pl., 128, 343–355, https://doi.org/10.1061/(ASCE)0733-9496(2002)128:5(343), 2002.

Guyot, A., Cohard, J. M., Anquetin, S., Galle, S., and Lloyd, C. R.: Combined analysis of energy and water balances to estimate latent heat flux of a sudanian small catchment, J. Hydrol., 375, 227–240, https://doi.org/10.1016/j.jhydrol.2008.12.027, 2009.

Kohsiek, W., Meijninger, W. M. L., De Bruin, H. A. R., and Beyrich, F.: Saturation of the large aperture scintillometer, Bound.-Lay. Meteorol., 121, 111–126, 2006.

Kustas, W. P., Prueger, J. H., Hatfield, J. L., Ramalingam, K., and Hipps, L. E.: Variability in soil heat flux from a mesquite dune site, Agr. Forest Meteorol., 103, 249–264, https://doi.org/10.1016/S0168-1923(00)00131-3, 2000.

Liu, H. P., Zhang, Y., Liu, S. H., Jiang, H. M., Sheng, L., and Williams, Q. L.: Eddy covariance measurements of surface energy budget and evaporation in a cool season over southern open water in Mississippi, J. Geophys. Res., 114, D04110, https://doi.org/10.1029/2008jd010891, 2009.

Lowe, P. R.: An approximating polynomial for the computation of saturation vapor pressure, J. Appl. Meteorol., 16, 100–103, https://doi.org/10.1175/1520-0450(1977)016<0100:AAPFTC>2.0.CO;2, 1977.

Mcgloin, R., Mcgowan, H., Mcjannet, D., Cook, F., Sogachev, A., and Burn, S.: Quantification of surface energy fluxes from a small water body using scintillometry and eddy covariance, Water Resour. Res., 50, 494–513, 2014.

Mcjannet, D. L., Cook, F. J., McGloin, R. P., McGowan, H. A., and Burn, S.: Estimation of evaporation and sensible heat flux from open water using a large aperture scintillometer, Water Resour. Res., 47, W05545, https://doi.org/10.1029/2010WR010155, 2011.

Monin, A. and Obukhov, A.: Basic laws of turbulent mixing in the surface layer of the atmosphere, Contrib. Geophys. Inst. Acad. Sci., USSR, 151, 163–187, 1954.

Monteith, J. L. and Unsworth, M. H.: Principles of Environmental Physics, Arnold, London, UK, 1990.

Nordbo, A., Launiainen, S., Mammarella, I., Lepparanta, M., Huotari, J., Ojala, A., and Vesala, T.: Long-term energy flux measurements and energy balance over a small boreal lake using eddy covariance technique, J. Geophys. Res., 116, D02119, https://doi.org/10.1029/2010jd014542, 2011.

Penman, H. L.: Natural evaporation from open water, hare soil and grass, P. R. Soc. London, 193, 120–145, 1948.

Stannard, D. I. and Rosenberry, D. O.: A comparison of short-term measurements of lake evaporation using eddy correlation and energy budget methods, J. Hydrol., 122, 15–22, https://doi.org/10.1016/0022-1694(91)90168-H, 1991.

Tanny, J., Cohen, S., Assouline, S., Lange, F., Grava, A., Berger, D., Teltch, B., and Parlange, M. B.: Evaporation from a small water reservoir: Direct measurements and estimates, J. Hydrol., 351, 218–229, https://doi.org/10.1016/j.jhydrol.2007.12.012, 2008.

Tanny, J., Cohen, S., Berger, D., Teltch, B., Mekhmandarov, Y., Bahar, M., Katul, G. G., and Assouline, S.: Evaporation from a reservoir with fluctuating water level: correcting for limited fetch, J. Hydrol., 404, 146–156, 2011.

Vercauteren, N., Bou-Zeid, E., Huwald, H., Parlange, M. B., and Brutsaert, W.: Estimation of wet surface evaporation from sensible heat flux measurements, Water Resour. Res., 45, W06424, https://doi.org/10.1029/2008WR007544, 2009.

Wang, T.: Formation and evolution of Badain Jaran Sandy Desert, China, Journal of Desert Research, 10, 29–40, 1990 (in Chinese).

Wang, W., Xu, Z., Li, X., Wang, J., and Zhang, Z.: A Study of Applications of Large Aperture Scintillometer in the Heihe River Basin, Advances in Earth Science, 25, 1208–1216, 2010 (in Chinese).

https://doi.org/10.5194/piahs-379-443-2018

Study on the water resources optimal operation based on riverbed wind erosion control in West Liaohe River plain

Sun Wanguang, Li Chengzhen, and Fan Baoshan

China Water Northeastern Investigation, Design and Research Corporation Limited, Changchun 130061, China

Correspondence: Sun Wanguang (sunwanguang@aliyun.com)

Received: 23 December 2017 – Revised: 26 March 2018 – Accepted: 11 April 2018 – Published: 5 June 2018

Abstract. Rivers are drying up most frequently in West Liaohe River plain and the bare river beds present fine sand belts on land. These sand belts, which yield a dust heavily in windy days, stress the local environment deeply as the riverbeds are eroded by wind. The optimal operation of water resources, thus, is one of the most important methods for preventing the wind erosion of riverbeds. In this paper, optimal operation model for water resources based on riverbed wind erosion control has been established, which contains objective function, constraints, and solution method. The objective function considers factors which include water volume diverted into reservoirs, river length and lower threshold of flow rate, etc. On the basis of ensuring the water requirement of each reservoir, the destruction of the vegetation in the riverbed by the frequent river flow is avoided. The multi core parallel solving method for optimal water resources operation in the West Liaohe River Plain is proposed, which the optimal solution is found by DPSA method under the POA framework and the parallel computing program is designed in Fork/Join mode. Based on the optimal operation results, the basic rules of water resources operation in the West Liaohe River Plain are summarized. Calculation results show that, on the basis of meeting the requirement of water volume of every reservoir, the frequency of reach river flow which from Taihekou to Talagan Water Diversion Project in the Xinkai River is reduced effectively. The speedup and parallel efficiency of parallel algorithm are 1.51 and 0.76 respectively, and the computing time is significantly decreased. The research results show in this paper can provide technical support for the prevention and control of riverbed wind erosion in the West Liaohe River plain.

1 Introduction

With the development and utilization of water resources, partial rivers were dried up in arid or semi-arid area. After dried up, bare riverbed becomes the new cradle land of sandstorm. Since 1972, the lower reaches of the Yellow River has began to dry up (Gao, 1998). The reasons for dry up in the watershed of the Yellow River can be concluded into three aspects: (1) natural reasons, the Yellow River basin is located in an arid and semi-arid region, at the same time, the river dry up may be aggravated by global warming; (2) human activities, about 30 billion m^3 runoff were consumed annually (Fu and Chen, 2006), and large-scale soil and water conservation measures decreased the streamflow (Shi et al., 2016). The dry up of river basin has caused many serious problems, such as agricultural and industrial output reduction, environmental degradation and water quality deterioration (Fu and Chen, 2006; Wang and Li, 2007).

Wind erosion caused soil loss and then impoverishes lands (Lal, 2001). The blown dust emission from eroding lands impairs air quality (Lee et al., 2003) and poses a threat to human health (Wilson and Sprengler, 1996). The research of wind erosion prevention and control is mainly focused on the following two aspects: (1) the nonerodible grains are placed on the erodible surface to reduce the degree of wind erosion (Bagnold, 1941); (2) the use of vegetation cover or crop residues to prevent wind erosion (Englehorn et al., 1952; Zhang et al., 2003; Jia et al., 2015). The wind erosion control of bare riverbed or lakebed is quite different from sand land. Due to the technical difficulties, it is hard to implement these measures. Bao et al. (2006) estimated the rational water area and inflow of the Ebinur Lake for controlling wind erosion

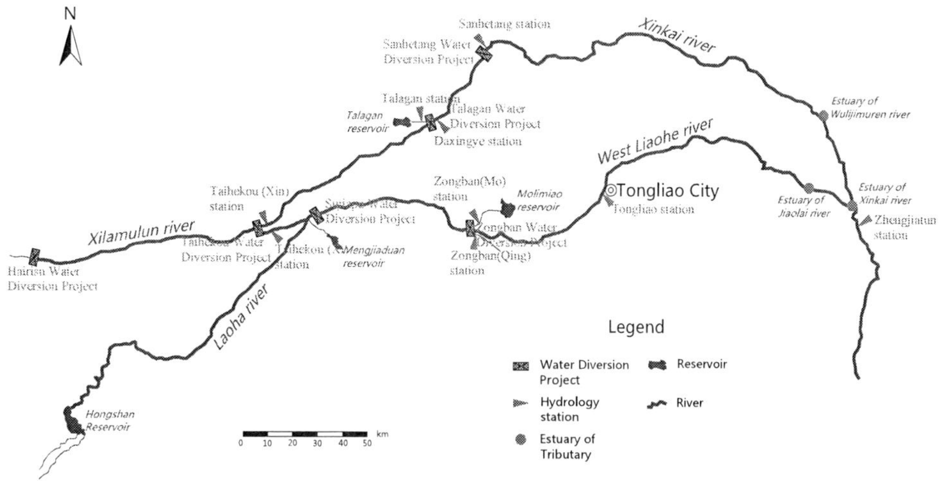

Figure 1. Hydrological station distribution schematic in West Liaohe River Plain.

Table 1. The statistical results about runoff of each hydrological station in the West Liaohe River Plain.

Year	Taihekou (Xi)	Taihekou (Xin)	Talagan	Daxingye	Sanhetang	Zongban (Mo)	Zongban (Qing)	Tongliao	Zhengjiatun
2006	15 016.1	502.9	0.0	0.0	0.0	4598.8	0.0	0.0	646.8
2007	12 013.4	0.0	0.0	0.0	0.0	4078.5	0.0	0.0	296.5
2008	16 450.9	0.0	0.0	0.0	0.0	4770.4	0.0	0.0	10 086.0
2009	10 790.1	0.0	0.0	0.0	0.0	2420.8	0.0	0.0	419.4
2010	195.9	0.0	0.0	0.0	0.0	0.0	0.0	0.0	1398.5
2011	17 994.8	8923.3	1818.3	0.0	0.0	2781.0	903.7	0.0	7632.3
2012	16 771.2	7003.6	1405.9	0.0	0.0	1318.1	1333.9	0.0	8474.8
2013	28 026.2	7729.5	3813.8	0.0	0.0	4795.7	0.0	0.0	10 258.1
Average	14 657.3	3019.9	879.7	0.0	0.0	3095.4	279.7	0.0	4901.5

Units: 10^4 m^3.

in the dried up lake basin (Bao et al., 2006). For intermittent dried up reach, there is no better erosion control measures so far.

In this paper we analyze the hydrological characteristics West Liaohe River plain and establish the optimal operation model for water resources based on river desertification control.

2 Model construction

2.1 Water system of the West Liaohe River plain

The West Liaohe River Plain is formed by alluvial deposits of the West Liaohe River and its main tributaries. West Liaohe River Basin area is 5.2×10^4 km^2. The main rivers of the river system include the lower reach of the Hongshan Reservoir in the Laoha River, the lower reach of the Hairisu Water Diversion Project in the Xilamulun River, the Xinkai River, the main stream of the West Liaohe River, the Jiaolai River and the Wulijimuren River. From late 1990s, the streamway of West Liaohe River Plain frequent drying up. After 2000, due to reduced precipitation, water stored by reservoir, industrial and agricultural water use increased, the trend of streamway dried up aggravated (Wang and Li, 2007).

This paper collected nearly 8 years (2006 ~ 2013) of the West Liaohe River plain river hydrological station (including Taihekou, Daxingye, Talagan, Zongban, Sanhetang, Tongliao, Zhengjiatun) daily flow and sediment data (Fig. 1). The statistical results about runoff of each hydrological station in the West Liaohe River Plain are shown in Table 1.

As shown in Table 1, the annual average runoff of Taihekou (Xi) station is 14657.3×10^4 m^3 and the annual average runoff of Taihekou (Xin) is 3019.9×10^4 m^3. The result indicate that only about 17.1 % of total runoff of the Xilamulun River were diverted into the Xinkai River. The annual average runoff of Talagan station on the Xinkai River is 879.7×10^4 m^3. Because of seepage loss along the streamway, only about 29.1 % of runoff of Taihekou (Xin) can reach Talagan station. The reach below Daxingye station of the Xinkai River were drying up all the time. From 1998,

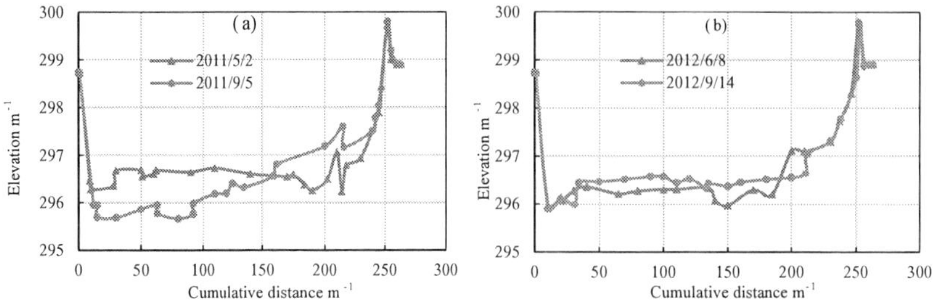

Figure 2. Comparison of large cross-section measured before and after flood season at the Taihekou (Xi) station (**a**: 2011, **b**: 2012).

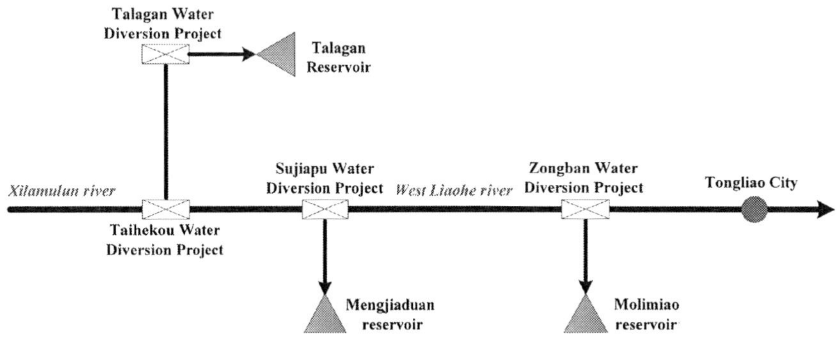

Figure 3. Sketch of water conservancy project in West Liaohe River Plain.

Hongshan Reservoir, which located in the Laoha River has been in a low water level running, so it did not discharge to downstream and the streamway below Hongsha Reservoir has dried up all the time. The runoff of main stream of the West Liaohe River mainly comes from Taihekou (Xi) station. This runoff diverted into Mengjiaduan Reservoir partly, deducting the seepage along the streamway, the remaining runoff were diverted into Molimiao Reservoir and Yinliaojiqing channel. So the reach below Molimiao Water Diversion Project was dried up all the time. The annual average runoff of Zhengjiatun station is $4901.5 \times 10^4\,m^3$ and the runoff is mainly from Wulijimuren River.

According to the analysis results of hydrological characteristics, the reach of West Liaohe River Plain is divided into intermittent river reach and long dry reach. This paper focuses on the prevention and control of riverbed wind erosion in intermittent river reach.

The river of West Liaohe River Plain has a large amount of sediment. According to the Taihekou (Xi) station monitoring data, the average sediment concentration is $9.07\,kg\,m^{-3}$. The sediment concentration increases with the increasing of flow $151\,m^3\,s^{-1}$ at the same cross-section, but the riverbed were deposited seriously. We can come to the conclusion that the longitudinal stability of the riverbed is poor. According to the results of the field investigation, vegetation cover is an effective measure to resist wind erosion in the bare river bed. Due to the longitudinal variation of riverbed, it is difficult to form effective vegetation cover on the riverbed surface. At the drying up time, bare riverbed has become the main source of wind erosion. The optimal water resource operation measures can ensure that the vegetation cover of the river bed is not destroyed frequently by the sediment laden flow.

2.2 Modeling analysis

There are three large beside reservoirs (the total capacity of more than $1 \times 10^8\,m^3$) and four large water diversion project in the West Liaohe River plain (Fig. 3). The runoff from upstream is controlled by the water diversion project and diverted into beside reservoir partly. The water resources stored in beside reservoir are used for local industrial and agricultural water supply. Under the premise of ensuring the local industrial and agricultural water supply, whether the water

Table 2. The parameters of river flow evolution in West Liaohe River Plain.

Reach	Distance (km)	Δt (h)	K (h)	x
Taihekou Water Diversion Project ~ Sujiapu Water Diversion Project	42	3	6	0.38
Taihekou Water Diversion Project ~ Talagan Water Diversion Project	85	3	12	0.38
Sujiapu Water Diversion Project ~ Zongban Water Diversion Project	65.7	3	12	0.32
Zongban Water Diversion Project ~ Tongliao City	67.2	3	12	0.32

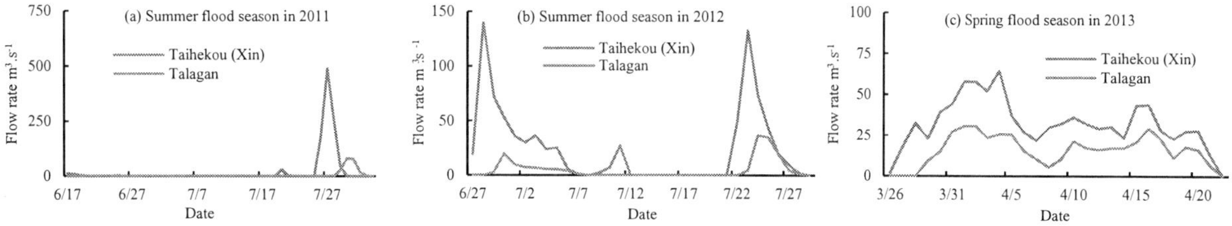

Figure 4. The comparison of daily flow rate of Taihekou (Xin) station and Talagan station.

2.3 Analysis of water flow evolution considering seepage loss

In the intermittent river reach, a thick unsaturated zone formed beneath the riverbed. When the flow of water through the riverbed, seepage loss is very strong. There are two stages of seepage loss in the intermittent river reach, one is initial infiltration stage and the other is stable infiltration stage. Under the initial infiltration stage, the infiltration rate is large, but the decay rate is faster, and the cumulative infiltration of this stage accounted for a large proportion (Cheng et al., 2015).

Due to the difficulty of accurately dividing the seepage stage, the two stages of seepage loss are considered as a whole. From 2006 to 2013, only 3 years (summer flood season in 2011 and 2012, spring flood season in 2013) that the runoff diverted into Xinkai River by Taihekou Water Diversion Project can reached Talagan Hydrological Station. The daily average flow rate of Taihekou (Xin) station and Talagan station is compared (Fig. 4).

By statistics, the runoff of Talagan station accounted for only 20.38 % of the Taihekou (Xin) station in summer flood season in 2011, and the same ratio is 20.07 % in summer flood season in 2012. This indicates that the water seepage loss from the Taihekou (Xin) station to Talagan station reaches about 80 % of the total runoff. In spring flood season in 2013, the runoff of Talagan station accounted for 49.34 % of the Taihekou (Xin) station, and the seepage loss is relatively small, mainly because of the seasonal frozen soil in the riverbed has not been fully melted.

The Muskingum method is used to calculate river water flow evolution in the West Liaohe River plain (Eq. 1). The water flow evolution parameters are the results (Table 2) of the "West Liaohe River flood control planning report" (Northeast investigation and Design Institute of Ministry of water resources, 2005).

$$Q_{\text{down},2} = C_0 Q_{\text{up},2} + C_1 Q_{\text{up},1} + C_2 Q_{\text{down},1},$$
$$C_0 = \frac{0.5\Delta t - Kx}{K - Kx + 0.5\Delta t},$$
$$C_1 = \frac{0.5\Delta t + Kx}{K - Kx + 0.5\Delta t},$$
$$C_2 = \frac{K - Kx - 0.5\Delta t}{K - Kx + 0.5\Delta t}, \tag{1}$$

where $Q_{\text{up},1}$, $Q_{\text{up},2}$, $Q_{\text{down},1}$, $Q_{\text{down},2}$ are the inflow of beginning and end period of the upper section, and the outflow of beginning and end period of the down section respectively, $m^3\,s^{-1}$. C_0, C_1, C_2 are parameters. K is propagation time of the reach under steady flow, hour. x is the flow proportion factor, in addition to reflect on the role of wedge storage flow, also reflect the storage capacity of the river. Δt is the calculation period, hour.

Considering the optimization calculation is too large, and the degree of concern for water quantity is higher than the flow process, simplified flow evolution and seepage loss coupling and used the comprehensive seepage loss coefficient. However, it is necessary to distinguish between spring flood season and summer flood season, expression is as flow:

$$Q_{\text{down},2} = S_{\text{f},r} \cdot \left(C_0 Q_{\text{up},2} + C_1 Q_{\text{up},1} + C_2 Q_{\text{down},1}\right), \tag{2}$$

where $S_{\text{f},r}$ is the comprehensive seepage loss coefficient of reach r. The comprehensive loss coefficient can be approximately considered as linear correlation with the length of the river. Because of the lack of measured hydrological data, the comprehensive loss coefficient of seepage of other reach can be calculated by the correlation between the loss coefficient and the length of the reach which is from Taihekou to Talagan Water Diversion Project:

$$S_{\mathrm{f},r} = \begin{cases} 1 - L_r \cdot 0.8/85 & \text{Summer flood season} \\ 1 - L_r \cdot 0.5/85 & \text{Spring flood season} \end{cases}, \quad (3)$$

where L_r is the length of reach r, km.

To verify the comprehensive river seepage loss coefficient, the measured data of Daxingye station and Sanhetang station on the Xinkai River were selected from 1999 to 2000. The reach length is 36.7 km between the two hydrological stations. The comprehensive loss coefficients of spring and summer flood season were 0.78 and 0.65 respectively using Eq. (3). However, the measured values were 0.72 and 0.69 respectively, so the calculated result is close to the measured value. It is proved that the assumption of linear correlation between the loss coefficient and the length of the reach is reasonable. It is needed to be explained that the comprehensive seepage loss coefficient of reach was obtained under the condition that the river dried up for a long period previously. So the river water conditions will have a certain impact on the coefficient.

2.4 Hydraulic connection

The distribution of river flow is controlled by water diversion project. Through the water diversion project, river flow can be diverted into beside reservoir or lower reach. The hydraulic connections which take water diversion project as a node are described below.

$$Q_{\mathrm{Taihekou\_in}} = Q_{\mathrm{Taihekou\_out}} + Q_{\mathrm{Taihekou\_diversion}}, \quad (4)$$

where $Q_{\mathrm{Taihekou\_in}}$, $Q_{\mathrm{Taihekou\_out}}$, $Q_{\mathrm{Taihekou\_diversion}}$ are inflow, outflow and diversion flow of the Taihekou Water Diversion Project respectively, $\mathrm{m^3\,s^{-1}}$.

$$\begin{aligned} Q_{\mathrm{Talagan\_in}} &= Q_{\mathrm{Talagan\_out}} + Q_{\mathrm{Talagan\_diversion}} \\ Q_{\mathrm{Talagan\_in}} &= Q_{\mathrm{Taihekou\_diversion\_evolution}}, \end{aligned} \quad (5)$$

where $Q_{\mathrm{Talagan\_in}}$, $Q_{\mathrm{Talagan\_out}}$, $Q_{\mathrm{Talagan\_diversion}}$ are inflow, outflow and diversion flow of the Talagan Water Diversion Project respectively, $\mathrm{m^3\,s^{-1}}$. $Q_{\mathrm{Taihekou\_diversion\_evolution}}$ is the diversion flow rate that evolved from Taihekou to Talagan Water Diversion Project, $\mathrm{m^3\,s^{-1}}$.

$$\begin{aligned} Q_{\mathrm{Sujiapu\_in}} &= Q_{\mathrm{Sujiapu\_out}} + Q_{\mathrm{Sujiapu\_diversion}} \\ Q_{\mathrm{Sujiapu\_in}} &= Q_{\mathrm{Taihekou\_out\_evolution}}, \end{aligned} \quad (6)$$

where $Q_{\mathrm{Sujiapu\_in}}$, $Q_{\mathrm{Sujiapu\_out}}$, $Q_{\mathrm{Sujiapu\_diversion}}$ are inflow, out flow and diversion flow of Sujiapu Water Diversion Project respectively, ,$\mathrm{m^3\,s^{-1}}$. $Q_{\mathrm{Taihekou\_out\_evolution}}$ is the outflow rate that evolved from Taihekou to Sujiapu Water Diversion Project, $\mathrm{m^3\,s^{-1}}$.

$$\begin{aligned} Q_{\mathrm{Zongban\_in}} &= Q_{\mathrm{Zongban\_out}} + Q_{\mathrm{Zongban\_diversion}} \\ Q_{\mathrm{Zongban\_in}} &= Q_{\mathrm{Sujiapu\_out\_evolution}}, \end{aligned} \quad (7)$$

where $Q_{\mathrm{Zongban\_in}}$, $Q_{\mathrm{Zongban\_out}}$, $Q_{\mathrm{Zongban\_diversion}}$ are inflow, outflow and diversion flow of Zongban Water Diversion Project, $\mathrm{m^3\,s^{-1}}$. $Q_{\mathrm{Sujiapu\_out\_evolution}}$ is the outflow rate that evolved from Sujiapu to Zongban Water Diversion Project, $\mathrm{m^3\,s^{-1}}$.

According to the analysis results of hydraulic system, the system is a mixed type system and the hydraulic connection between upstream and downstream is closely. The distribution of water resources were controlled by four water diversion projects, so the dimension of the system space is four dimensional. Based on the results of hydrological statistical analysis, the Water Diversion Project of Talagan and Zongban had no outflow in recent 8 years (from 2006 to 2013). So we use regular operation mode for the Water Diversion Projects of Talagan and Zongban, and the two Water Diversion Projects no longer participate in optimal operation. Specific to Zongban Water Diversion Project as an example:

1. When $Q_{\mathrm{Zongban\_in}} > Q_{\mathrm{Zongban\_diversion\_max}}$, then $Q_{\mathrm{Zongban\_diversion}} = Q_{\mathrm{Zongban\_diversion\_max}}$, and then $Q_{\mathrm{Zongban\_out}} = Q_{\mathrm{Zongban\_in}} - Q_{\mathrm{Zongban\_diversion\_max}}$.
2. When $Q_{\mathrm{Zongban\_in}} \le Q_{\mathrm{Zongban\_diversion\_max}}$, then $Q_{\mathrm{Zongban\_diversion}} = Q_{\mathrm{Zongban\_in}}$, and $Q_{\mathrm{Zongban\_out}} = 0$. Where $Q_{\mathrm{Zongban\_diversion\_max}}$ is the maximum capacity of diversion flow of Zongban Water Diversion Project, $\mathrm{m^3\,s^{-1}}$.

After using regular operation mode for the Water Diversion Projects of Talagan and Zongban, the four dimensional optimization problem can be reduced to two dimensional optimization problem, and only the Water Diversion Projects of Taihekou and Sujiapu can be seen as targets of the optimal operation. The outflow and diversion flow are optional state variables for every water diversion project. In order to meet the threshold setting and the realization of the target, the diversion flow is chosen as the state variable in Taihekou Water Diversion Project and the outflow is chosen as state variable in Sujiapu Water Diversion Project.

3 Optimization model

3.1 Objective function

The inflow of Taihekou Water Diversion Project is mainly concentrated in the spring flood season (from 10 March to 20 April) and summer flood season (from 25 June to 20 August). Other times the riverbed is in the drying up state. The maximum inflow were 147 and 1074 $\mathrm{m^3\,s^{-1}}$ of the Taihekou

Water Diversion Project in spring flood season and summer flood season respectively.

Because the water diversion project has no capacity for runoff regulation, the inflow either diverted into beside reservoir or discharged into lower reach. Due to the extreme shortage of local water resources, the water eventually was diverted into beside reservoir regardless of diversion flow or outflow. Therefore, it is a prerequisite to optimize the operation of water resources to meet the requirements of the local industry and agriculture. Based on the prevention and control of wind erosion of riverbed, it is beneficial that the river discharge or dry up for a long time. If the upstream flow is less, the flow is firstly diverted into nearest beside reservoir and reduce interference on the downstream of the river. If the upstream flow is larger, the flow is diverted into long distance beside reservoir preferentially, concentrated flow and reduce the frequency of water through of riverbed. Therefore, as the objective function, it is needed to consider diversion flow or outflow and river length. Specific expressions are as follows:

$$\begin{aligned} &\mathrm{minobj} = \sum_j \mathrm{obj}_j \\ &\mathrm{obj}_j = \begin{cases} L_j \cdot V_{j,\mathrm{cap}}/I_j & \text{When } I_j > 0 \\ P \cdot V_{j,\mathrm{cap}} & \text{When } I_j = 0 \end{cases} \\ &I_j = \sum_t \mathrm{sign} \cdot Q_{j,t} \cdot \Delta t / 10000 \\ &\mathrm{sign} = \begin{cases} 0 & \text{When } Q_{j,\mathrm{s,max}} < Q_{j,\mathrm{thr}} \\ 1 & \text{When } When\ Q_{j,\mathrm{s,max}} \geq Q_{j,\mathrm{thr}} \end{cases}, \end{aligned} \tag{8}$$

where obj is objective function; j is the Number of reservoir; obj_j is objective function value of j reservoir water storage. L_j is the distance from Taihekou Water Diversion Project to j reservoir, km; $V_{j,\mathrm{cap}}$ is utilizable capacity of j reservoir, $10^4\ \mathrm{m}^3$; I_j is the volume of water that had diverted into j reservoir in calculation period, $10^4\ \mathrm{m}^3$; P is penalty coefficient; $Q_{j,t}$ is the flow rate that diverted into j reservoir. $Q_{j,s,\mathrm{max}}$ is the maximum flow rate of flood, $\mathrm{m}^3\ \mathrm{s}^{-1}$; $Q_{j,\mathrm{thr}}$ is the lower threshold of flow rate that diverted into j reservoir, $\mathrm{m}^3\ \mathrm{s}^{-1}$; sign is indicator of status.

The objective function includes two aspects: (1) the objection of water volume that diverted into beside reservoir and (2) the objection of wind erosion prevention and control of riverbed. The objective function is essentially a function of the reach length (L_j) and the amount of water diverted into beside reservoir (I_j). $V_{j,\mathrm{cap}}/I_j$ is a dimensionless variable coefficient, when the amount of water diverted into j reservoir increases, the coefficient becomes smaller. Through the objective function can ensure the amount of water diverted into beside reservoir, and thus guarantee the local industrial and agricultural water supply.

The lower threshold flow is one of the key variables of the objective function. Through this variable, it can avoid the interference of high-frequency of small flow rate to the lower reach. To obtain the lower threshold flow, we can use two methods of optimization and simulation. If the optimization method is used to solve the problem, it is necessary to add two dimension on the basis of the original two dimensional optimization problem; and the use of simulation method to solve, we can learn from the operation experience, given a reasonable lower threshold flow. This paper uses simulation method to obtain the lower threshold of diversion flow of Taihekou Water Diversion Project and outflow of Sujiapu Water Diversion Project respectively. The lower threshold flow setting is directly related to the volume of diversion flow of Talagan and Zongban Water Diversion Project. The optimal value of the lower threshold of the diversion flow is based on the measured water diversion of the Talagan and Zongban Water Diversion Project.

In this paper, the problem of optimal flood control is not studied, and the flood discharge of each river is given in the form of constraint.

3.2 Constraints

3.2.1 Water balance of reservoir

$$V_{j,t+1} = V_{j,t} + I_{j,t} - W_{j,t} - E_{j,t} - A_{j,t}, \tag{9}$$

where $V_{j,t}$ and $V_{j,t+1}$ are water storage at the beginning and end of the period t respectively in reservoir j; $I_{j,t}$ is runoff of reservoir j in period t; $W_{j,t}$ is water supply of reservoir j in period t; $E_{j,t}$ is the loss of evaporation and seepage of j reservoir in period t; $A_{j,t}$ is surplus water of j reservoir in period t.

3.2.2 Constraint of discharge capacity

$$Q_{r,t} \leq Q_{r,\mathrm{max}}, \tag{10}$$

where $Q_{r,t}$ is the flow rate of reach r in period t, $\mathrm{m}^3\ \mathrm{s}^{-1}$; $Q_{r,\mathrm{max}}$ is the discharge capacity of reach r, $\mathrm{m}^3\ \mathrm{s}^{-1}$;

3.2.3 Constraint of upper bound of reservoir water storage

$$V_{j,t} \leq V_{j,\mathrm{nor}}, \tag{11}$$

where V_j is water storage of reservoir j in period t, $10^4\ \mathrm{m}^3$; $V_{j,\mathrm{nor}}$ is upper bound of water storage of reservoir j, $10^4\ \mathrm{m}^3$.

3.3 Search method

3.3.1 Constraint handling method

In the process of water resources optimal operation, the constrained optimization problem is transformed into unconstrained optimization problem by penalty function generally. In this paper, the discharge capacity and the upper bound of reservoir water storage are all mandatory constraint. When the variables violate the constraints, the penalty term is added on the basis of the results of the original objective function. Specific expressions are as follows:

$$\mathrm{obj}' = \mathrm{obj} + F_1\left(Q_{r,t} - Q_{r,\mathrm{max}}\right)^2 + F_2\left(V_{j,t} - V_{j,\mathrm{nor}}\right)^2,$$

$$\text{when } Q_{r,t} > Q_{r,\max} \text{ and } V_{j,t} > Q_{j,\text{nor}}, \tag{12}$$

where F_1 and F_2 are the penalty coefficient of the flow rate and water storage respectively; obj′ is the objective function value that is added by the penalty value.

3.3.2 Solving process

Water diversion project itself does not have the ability to regulate runoff, only transform the inflow into diversion flow or outflow. Importantly there is a time effect of water flow evolution, that is, there is post-efficiency, which is essentially different from reservoirs. Therefore, the solution method of the optimal operation of the water diversion project group is quite different from the reservoir group.

In order to simplify the problem solving, multi-dimensional optimization problems are usually transformed into one-dimensional optimization problems, such as DPSA (Dynamic Programming Successive Approximation) in dynamic planning (Opan, 2010; Yi et al., 2003). Multi-stage optimization can also be translated into a single two-stage optimization problem, such as POA (Progressive Optimality Algorithm) in dynamic plan (Liu et al., 2011; Guo et al., 2011). If DPSA and POA are combined (DPSA-POA), the multi-dimensional and multi-stage optimization problems can be transformed into one-dimensional and two-stage optimization problems (Bai et al., 2015; Zhang et al., 2016), and the difficulty of solving optimization problems reduced greatly, at the same time, the problem of post-efficiency can be solved effectively.

To solve the problem of the water resources optimal operation based on riverbed wind erosion control in West Liaohe River plain, and improve computational efficiency, this paper presents a "multi-core parallel solution algorithm for optimal operation of water diversion project group".

From the above analysis, we can see that the upper and lower reaches are closely linked to each other of the optimal operation system in this paper, and two independent state variables find the optimal combination of states through the DPSA method under the framework of POA to get the optimal operation scheme. According to the relationship between upstream and downstream, Taihekou Water Diversion Project is the first level, Sujiapu Water Diversion Project is the final level. Under the framework of POA, we only change the state variable of the last stage i at a time, and the first state variable is fixed. The full range of two complete series are brought into the calculation method of objective function. According to the calculation method of water flow evolution, the diversion flow and outflow of each water diversion project are calculated, and then the objective function value is calculated. When the traversal of the final level at time i is completed, the state variable of its parent is changed. When all traversal completed at time i, then $i = i + 1$, until all the time were completed. The whole process is repeated n times, until the difference between the two calculation results is less than the set threshold, you can stop the calculation. Calculation process is shown in Fig. 5. In the figure, M and K are the number of diverting states of Tahekou and Sujiapu Water Diversion Project, respectively, and their values can be selected comprehensively according to the state change field and the calculation precision.

From the entire calculation process, the calculation is the largest in the final cycle. It is necessary to calculate the evolution of water flow under fixed stages of two water diversion project. It is necessary to calculate the amount of diversion flow and out flow of each water diversion project and calculate the entire objective function value. In this paper, the last stage is changed into parallel computing mode to balance the load of each thread, so as to avoid the cost of thread synchronization at the end of the cycle. The Fork/Join framework is a framework for performing parallel tasks. In the final cycle, each subtask is independent of each other, so the task can be distributed to multiple threads, and the results of multiple subtasks can be combined into a total calculation result.

The performance of parallel algorithm is evaluated by speedup ratio and parallel efficiency (Peng et al., 2014), and the expression is as follows:

$$S_\text{p} = T_\text{s}/T_\text{p} \quad E_\text{p} = S_\text{p}/P, \tag{13}$$

where S_p is speedup ratio; T_s is serial computing time consuming; min; T_p is parallel computing time consuming; min; E_p is parallel efficiency; P is processor number.

4 Result and discussion

Taking Taihekou Water Diversion Project daily measured flow from 2006 to 2013 as input condition, on the basis of guaranteeing the total water separation of various reservoirs, avoid the destruction of riverbed vegetation by the frequent overflow of the river through the optimal operation of Taihekou and Sujiapu Water Diversion Project, and achieve the goals of riverbed wind erosion control. The combination of the different thresholds of water diversion at the Taihekou and the Sujiapu Water Diversion Project are analyzed. Take Talagan and Zongban Water Diversion Project measured water diversion volume value as the verification. The results are shown in Table 3.

It can be seen from the calculation results in the table that the lower threshold of water diversion for Taihekou Water Diversion Project is set to $150\,\text{m}^3\,\text{s}^{-1}$ and the release threshold of Sujiapu Water Diversion Project terminal set to $15\,\text{m}^3\,\text{s}^{-1}$ is more appropriate (Scheme 2 in Table 3). The simulated diversion water volume of the Talagan Water Diversion Project is $6636 \times 10^4\,\text{m}^3$, and the measured value is $7038 \times 10^4\,\text{m}^3$. The simulated diversion water volume of the Zongban Water Diversion Project is $27\,919 \times 10^4\,\text{m}^3$, and the measured value is $27\,001 \times 10^4\,\text{m}^3$. The simulation results and the measured values are relatively close.

Under scheme 2, Taihekou Water Diversion Project has diverted into Xinkai River for 3 years, summer flood in 2008,

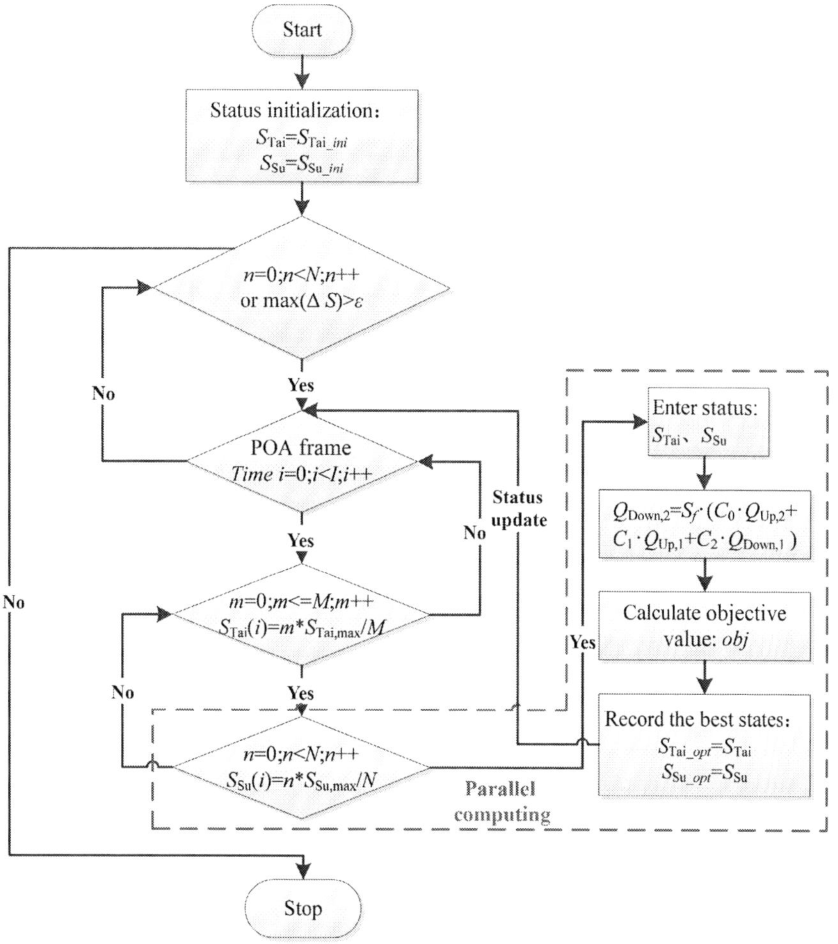

Figure 5. The flow chart of the modeling procedure.

Table 3. Analysis of threshold of diversion flow at Taihekou Water Diversion Project and threshold of outflow at Sujiapu Water Diversion Project. Values in bold indicates that Scheme 2 is the best scheme.

Scheme	Threshold of diversion flow at Taihekou ($m^3\,s^{-1}$)	Threshold of outflow at Sujiapu ($m^3\,s^{-1}$)	Amount of water diverted by Talagan ($10^4\,m^3$)	Amount of water diverted by Zongban ($10^4\,m^3$)
1	140	15	10 215	25 349
2	**150**	**15**	**6636**	**27 919**
3	160	15	4358	29 519
4	150	14	6821	28 876
5	150	16	6066	28 562
Measured	–	–	7038	27 001

summer flood in 2011 and summer flood in 2012 respectively. The number of draining days at Talagan Water Diversion Project is 29 days. In the measured value, Taihekou Water Diversion Project has diverted into Xinkai River for 4 years, spring flood in 2006, summer flood in 2011, summer flood in 2012 and summer flood in 2013 respectively. The number of draining days at Talagan Water Diversion Project is 44 days. The results of the comparison between the calculated and the measured are shown in Fig. 6. In the spring flood season of 2006, the maximum inflow of Taihekou Water Diversion Project was 132 $m^3\,s^{-1}$. In the actual operation, the maximum diversion flow into Xinkai River is

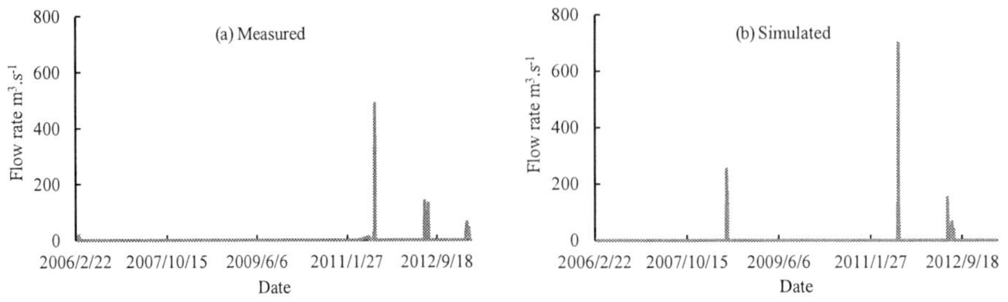

Figure 6. Comparison of simulated and measured diversion flow into Xinkai River at Taihekou Water Diversion Project.

19.4 $m^3 s^{-1}$, and the amount of water diverted into Xinkai River is 503×10^4 m^3 in the whole spring flood season. Leakage along the river consumes the vast majority of water. While not meeting the purpose of diversion water storage, the large amount of sediment carried by the water current also undermined the growth environment of the bed vegetation. After the optimal operation, only when the inflow of Tahekou Water Diversion Project exceeds the set threshold, it will be diverted into Xinkai River. Taking the summer flood season in 2011 as an example, the maximum of inflow of the Taihekou Water Diversion Project is 1074 $m^3 s^{-1}$, and the total flood runoff in this session is $25\,320 \times 10^4$ m^3. After optimization calculation, the maximum flow diverted into Xinkai River is 698 $m^3 s^{-1}$ (the measured value is 489 $m^3 s^{-1}$), and the amount of water diverted into Xinkai River is $11\,900 \times 10^4$ m^3 (measured value is 8655×10^4 m^3). The maximum diversion flow and total amount of diversion water have been greatly increased compared with the measured values.

Under scheme 2, Zongban Water Diversion Project has inflow water for 7 years and the total number of inflow days is 176 days. In the measured values, Zongban Water Diversion Project also has inflow water for 7 years and the total number of inflow days is 200 days.

The results of comparison between the optimized operation and the measured data show that the number of years of water passing through the Zongban Water Diversion Project is the same with the measured value and the number of years that the Taihekou Water Diversion Project diverted into Xinkai River decreased by 1 year after the optimal operation is implemented. For the number of draining days, the number of draining days at Talagan and Zongban Water Diversion Project decreased to some extent contrast with the measured value, but the reduction in the number of draining days at Talagan Water Diversion Project was relatively large.

Through the above analysis we can see that the setting of the lower threshold of diversion flow and the amount of water diverted into the reservoir are conflicting. On the basis of guaranteeing the amount of water diverted into the reservoir, setting a reasonable lower threshold of flow diverted into the new Xinkai River of Taihekou Water Diversion Project and lower threshold of outflow of the Sujiapu Water Diversion Project can make the Xinkai River from Taihekou to Talagan Water Diversion Project (river length of 85 km) and the West Liaohe River from Sujiapu to Zongban Water Diversion Project (length of 65.7 km) through a larger water flow and control the lower flow in the reach. Reduce the frequency of river flow as much as possible, and reduce the frequent damage of water flow to the vegetation cover of the riverbed. The goal of water resources optimization of West Liaohe River Plain based on the prevention and control wind erosion of riverbed has been achieved.

By optimizing the operation, we can summarize the following rules: (1) When the inflow of Tahekou Water Diversion Project is greater than or equal to 150 $m^3 s^{-1}$, the Xinkai River floodgate is opened to preferentially divert the water into Xinkai River, and more water from the same flood can be diverted into the Xinkai River. (2) When the inflow of Tahekou Water Diversion Project is less than 150 $m^3 s^{-1}$, and the flow rate is less than 15 $m^3 s^{-1}$ after it evolves to Sujiapu Water Diversion Project, it should be diverted into Mengjiaduan reservoir as far as possible. When the inflow of Sujiapu Water Diversion Project is greater than or equal to 15 $m^3 s^{-1}$, give priority to the downstream drainage.

The optimized operation solution test environment is a notebook computer. CPU type is CORE i7, dual-core, 4G memory. After calculation, the serial algorithm takes 14.61 min and the parallel algorithm takes 9.66 min. The speed ratio of the parallel computation is 1.51, and the parallel efficiency is 0.76. It can be seen that the parallel computation of multi core shortens the time consuming and improves the computational efficiency.

5 Conclusions

1. According to the measured data and actual operation, the West Liaohe River Plain water conservancy project system is simplified to be operation system that is composed of the four water diversion project and three reservoirs. On the basis of this, through the rules and regulations of Talagan and Zongban Water Diversion Projection, the four-dimensional optimization problem is further reduced to a two-dimensional optimization problem.

2. An optimal water resources operation model of West Liaohe River Plain based on the prevention and control of wind erosion of riverbed was established. The model objective function involves the factors such as the amount of water diverted into the reservoir, the length of the river reach and the lower threshold of the flow rate, and on the basis of ensuring the requirements of water diversion in each reservoir, make the river flow through the reach in a short period of time and avoid the destruction of vegetation above the riverbed by the frequent overflow of the river.

3. Proposed a multi-core parallel solution method of water resources optimal operation in West Liaohe River Plain. Find the optimal combination of states by DPSA method under the framework of POA and adopting the Fork/Join mode design parallel computing algorithm in OpenPM programming model, we proposed the calculation flow.

4. The calculation results show that on the basis of ensuring the demand of water diversion into Talagan and Molimiao reservoirs, the threshold of diversion flow from Taihekou Water Diversion Project to Xinkai River is $150\,m^3\,s^{-1}$, while the threshold of outflow of the Sujiapu Water Diversion Project is $15\,m^3\,s^{-1}$. The optimal operation effectively reduced the water diversion frequency in the Xinkai River reach from Taihekou Water Diversion Project to Talagan Water Diversion Project. Using parallel algorithms, speedup and parallel efficiencies were 1.51 and 0.76, respectively, and computational efficiency improved significantly.

5. The research results of this paper show that it is feasible to achieve the prevention and control of wind erosion of riverbed in the intermittent over-water reach of West Liaohe River Plain through the optimized water resources operation.

Data availability. The hydrological data used in this study all come from the hydrologic Yearbook (Hydrology data of Liaohe River Basin). According to the People's Republic of China hydrological regulations, hydrological data are not allowed to be made public.

Competing interests. The authors declare that they have no conflict of interest.

Special issue statement. This article is part of the special issue "Innovative water resources management – understanding and balancing interactions between humankind and nature". It is a result of the 8th International Water Resources Management Conference of ICWRS, Beijing, China, 13–15 June 2018.

Acknowledgements. The study is financially supported by the National Non-Profit Research Program of China (No.201401015).

Edited by: Depeng Zuo
Reviewed by: two anonymous referee

References

Bagnold, R. A.: The physics of blown sand and desert dunes, Methuen, London, 265 pp., 1941.

Bai, T., Chang, J., Chang, F., Huang, Q., Wang, Y., and Chen, G.: Synergistic gains from the multi-objective optimal operation of cascade reservoirs in the Upper Yellow River basin, J. Hydrol., 523, 758–767, https://doi.org/10.1016/j.jhydrol.2015.02.007, 2015.

Bao, A., Mu, G., Zhang, Y., Feng, X., Chang, C., and Yin, X.: Estimation of the rational water area for controlling wind erosion in the dried-up basin of the Ebinur Lake and its effect detection, Chinese Sci. Bull., 51, 68–74, https://doi.org/10.1007/s11434-006-8209-9, 2006.

Cheng, L., Wang, Z., Hu, S., Wang, Y., Jin, J., and Zhou, Y.: Flood routing model incorporating intensive streambed infiltration, Sci. China Earth Sci., 58, 718–726, https://doi.org/10.1007/s11430-014-5018-x, 2015.

Englehorn, C. L., Zingg, A. W., and Woodruff, N. P.: The effects of plant residue cover and cold structure on soil losses by wind, Soil Sci. Soc. Am. J., 16, 29–33, https://doi.org/10.2136/sssaj1952.03615995001600010010x, 1952.

Fu, G. and Chen, S.: Water crisis in the Yellow River: facts, reasons, impacts, and countermeasures, Water Practice and Technology, 1, 1–8, https://doi.org/10.2166/WPT.2006028, 2006.

Gao, Y.: Analysis of reasons for the Yellow River's dry-up and its eco-environmental impacts, J. Environ. Sci., 10, 357–364, 1998.

Guo, S., Chen, J., Li, Y., Liu, P., and Li, T.: Joint operation of the multi-reservoir system of the Three Gorges and the Qingjiang cascade reservoirs, Energies, 4, 1036–1050, https://doi.org/10.3390/en4071036, 2011.

Jia, H., Wang, G., Guo, L., Zhuang, J., and Tang, L.: Wind erosion control utilizing standing corn residue in Northeast China, Soil Till. Res., 153, 112–119, https://doi.org/10.1016/j.still.2015.05.009, 2015.

Lal, R.: Soil degradation by erosion, Land Degrad. Dev., 12, 519–539, https://doi.org/10.1002/ldr.472, 2001.

Lee, H. N., Tanaka, T., Chiba, M., and Igarashi, Y.: Long range transport of Asian dust from dust storm and its impact on Japan, Water, Air, & Soil Pollution: Focus, 3, 231–243, https://doi.org/10.1023/A:1023254910362, 2003.

Liu, X., Guo, S., Liu, P., Chen, L., and Li, X.: Deriving Optimal Refill Rules for Multi-Purpose Reservoir Operation, Water Resour. Manag., 25, 431–448, https://doi.org/10.1007/s11269-010-9707-8, 2011.

Northeast investigation and Design Institute of Ministry of water resources: West Liaohe River flood control planning report, Changchun, Report No.: 900-CH7-13, 50 pp., 2005.

Opan, M.: Irrigation-energy management using a DPSA-based optimization model in the Ceyhan Basin of Turkey, J. Hydrol., 385, 353–360, 2010.

Peng, A.-B., Peng, Y., and Zhou, H.,: Multi-core parallel computation for deriving joint operating rule curves in multi-reservoir system under the condition of inter-basin water transfer, J. Hydraul. Eng., 45, 1284–1292, https://doi.org/10.13243/j.cnki.slxb.2014.11.003, 2014.

Shi, H., Li, T., Wang, K., Zhang, A., Wang, G., and Fu, X.: Physically based simulation of the streamflow decrease caused by sediment-trapping dams in the middle Yellow River, Hydrol. Process., 30, 783–794, https://doi.org/10.1002/hyp.10649, 2016.

Wang, X. and Li, L.: The drying-up of the West Liaohe River and the countermeasures to the problem, Journal of Arid Land Resources and Environment, 21, 79–83, 2007.

Wilson, R. and Sprengler, J. D.: Particles in our air: concentrations and health effects, Harvard University Press, Cambridge, Mass, 1996.

Yi, J., Labadie, J. W., and Stitt, S.: Dynamic Optimal Unit Commitment and Loading in Hydropower Systems, J. Water Res. Pl., 129, 388–398, https://doi.org/10.1061/(ASCE)0733-9496(2003)129:5(388), 2003.

Zhang, C. L., Zou, X. Y., Dong, G. R., and Lu, Y. Z.: Wind tunnel studies on influences of vegetation on soil wind erosion, J. Soil Water Conserv., 17, 31–33, 2003.

Zhang, W., Liu, P., Chen, X., Wang, L., Ai, X., Feng, M., Liu, D., and Liu, Y.: Optimal operation of multi-reservoir systems considering time-lags of flood Routing, Water Resour. Manag., 30, 523–540, https://doi.org/10.1007/s11269-015-1175-8, 2016.

https://doi.org/10.5194/piahs-379-455-2018

Open Access
PIAHS

Country-level assessment of future risk of water scarcity in Europe

Luis Garrote[1], Ana Iglesias[2], and Alfredo Granados[1]

[1]Department of Civil Engineering, Technical University of Madrid, Madrid, 28040, Spain
[2]Department of Agricultural Economics, Technical University of Madrid, Madrid, 28040, Spain

Correspondence: Luis Garrote (l.garrote@upm.es)

Received: 22 December 2017 – Accepted: 19 February 2018 – Published: 5 June 2018

Abstract. A methodology for regional assessment of current and future water availability in Europe is presented in this study. The methodology is based on a proposed indicator of risk of water scarcity based on the projections of runoff and water availability for European countries. The risk of water scarcity is the combined result of hydrological processes, which determine streamflow in natural conditions, and human intervention, which determines water management using the available hydraulic infrastructure and establishes water supply conditions through operating rules. Model results show that changes in runoff and availability obtained for individual GCM projections can be large and even contradictory. These heterogeneous results are summarized in the water scarcity risk index, a global value that accounts for the results obtained with the ensemble of model results and emission scenarios. The countries at larger risk are (in this order) Spain, Portugal, Macedonia, Greece, Bulgaria, Albania, France and Italy. They are mostly Mediterranean countries already exposed to significant water scarcity problems. There are countries, like Slovakia, Ireland, Belgium, Luxembourg, Croatia and Romania, with mild risk. Northern Arctic countries, like Sweden, Finland, Norway and Russia, show a robust however mild increase in water availability.

1 Introduction

There is a growing concern about the possible impacts of climate change on water resources availability (Arnell, 2004). Regional assessments are an important tool to provide decision makers with a global picture of future situations according to hypothesized future socio-economic pathways (Alcamo et al., 2000). Decision makers require projections several decades ahead, but they are faced with very large uncertainties. There is a multiplicity of future scenarios of physical variables deriving from different emission scenarios and uncertainties inherent to the modelling tools used in the analysis. Other variables are even more uncertain due to the complexity of the socioeconomic system.

Many regional studies have been carried out to obtain projections of water availability. Arnell (1999) presented one of the first regional studies for Europe. The most up to date summary is the last Assessment Report of the Intergovernmental Panel on Climate Change (IPCC) (IPCC, 2014). Most of these studies assimilate global climate model (GCM) results into water resources impact-assessment models, either by downscaling climate model results to a finer scale (Fowler et al., 2007; Fronzek and Carter, 2007) or by developing macroscale hydrological models (Sperna Weiland et al., 2012, van Beek and Bierkens, 2009). Assessment of impacts on water resources requires also analysis of the projected evolution of water demands (Döll, 2002; Alcamo et al., 2007; Wisser et al., 2008) and simulation of water management practices (Gleick, 2003; Alcamo et al., 2003a). In Europe, water resources systems are highly developed and have achieved a profound transformation of the natural characteristics of water flow to adapt to variability and uncertainty. The WaterGAP model (Alcamo et al., 2003b) is a good example of integration of diverse modelling tools to account for the human intervention on the natural system under climate change scenarios.

In this paper we present a methodology to assist decision makers in dealing with uncertainties in projections of water resources scenarios under climate change. The methodology is based on a proposed water scarcity risk index, a global value that accounts for the results obtained with the ensemble of model results and emission scenarios.

The methodology is applied to conduct a country-level assessment of risk of water scarcity in Europe. We first present the methodology to obtain projected changes of runoff and water availability. Projected changes in runoff were inferred from the results of the PCR-GLOBWB hydrologic model and projected changes in water availability were estimated with the Water Availability and Adaptation Policy Analysis (WAAPA) model. We later introduce the methodology to compute the water scarcity risk index and apply it to Europe.

2 Projections of runoff and water availability

In this section we introduce the model developed for the river basins of Europe and the estimation of runoff and water availability projections. A detailed presentation of the model can be found in Garrote et al. (2015a).

2.1 Model configuration

Model topology was taken from the "Hydro1k" data set (EROS, 2008). We identified 1260 subbasins of average size around 5000 km^2. The reservoir storage volume available for regulation in every subbasin was obtained from the ICOLD World Register of Dams (ICOLD, 2004). Dams in the register with more than 5 hm^3 of storage capacity were georeferenced and linked to the corresponding storage capacity and flooded area. It was assumed that all reservoir storage available in a given subbasin was concentrated in a single equivalent reservoir located at the basin outlet. Environmental flows were computed through hydrologic methods. Monthly minimum required environmental flow was defined as the 10 % quantile in the distribution of naturalized monthly flows.

2.2 Climate scenarios

The climate scenarios were provided by five different climate model combinations that were used to force the PCR-GLOBWB model: GFDL-ESM2NM, HadGEM2-ES, IPSL-CM5A-LR, MIROC-ESM-CHEM and NorESM1-M. The models were run from 1960 to 2099 under two Representative Concentration Pathways emission scenarios: "RCP4.5" and "RCP8.5". Three time slices were considered for analysis: "Control (CTL)" (1960–1999), "Short term (ST)" (2020–2059) and "Long term (LT)" (2060–2099).

2.3 Projections of runoff

Naturalized streamflow was obtained from the results of the application of the PCR-GLOBWB model (van Beek and Bierkens, 2009) to the Inter-Sectoral Impact Model Intercomparison Project (Warszawski et al., 2014). The PCR-GLOBWB model was run for the entire globe at 0.5° resolution using forcing from the five GCMs under control conditions and climate change. These results are available for downloading from the CORDEX data portal.

The runoff provided from the PCR-GLOBWB models was interpolated on a finer grid to obtain monthly time series of streamflow in each basin. An example of the average runoff obtained with one of the forcing GCMs (GFDL) for the control period is presented in Fig. 1.

Figure 1 also presents the spatial distribution of relative changes in runoff for the long term period under emission scenario RCP4.5 compared to the control period for the GFDL climate forcing.

Relative changes are calculated following Eq. (1):

$$\Delta R = \frac{R_{\mathrm{RCP}} - R_{\mathrm{CTL}}}{R_{\mathrm{CTL}}}, \tag{1}$$

Where R_{CTL} is the average runoff for the control period and R_{RCP} is the average runoff for the future period.

Average runoff values in the control period were compared to the global composite runoff field of the University of New Hampshire Global Runoff Data Centre (UNH/GRDC), which combines observed river discharges with a water balance model (Fekete et al., 2002). The results are shown in Fig. 2. All climate forcing models produced similar results, with strong positive bias. For this reason, PCR-GLOBWB simulations were corrected for the bias observed in the control period.

2.4 Projections of water availability

Water availability is estimated for every subbasin through the WAAPA model (Garrote et al., 2015b). The objective of the WAAPA model is to simulate the operation of a water resources system to maximize water availability.

In this study WAAPA model was used to estimate maximum potential water availability in the European river network applying gross volume reliability as performance criterion. The simulation consists on estimating the maximum demand that can be supplied at a given point of the river network considering the upstream basin: all the series of streamflow and available reservoir storage in all upstream nodes. All upstream reservoirs are jointly managed to satisfy the total demand located at the basin outlet. The demand is satisfied if it can be met with a given reliability criterion. The criterion used in the analysis was 98 % reliability in volume.

Basic components of WAAPA are inflows, reservoirs and demands. These components are linked to nodes of the river network. WAAPA allows the simulation of reservoir operation and the computation of supply to demands from a system of reservoirs accounting for ecological flows and evaporation losses. From the time series of supply volumes, supply reliability can be computed according to different criteria. Water availability is linked to the point in the river network where it is calculated.

Table 1. Country estimates of changes in average runoff and surface water availability in the long term period (2060–2099) with respect to the control period (1960–1999) under emission scenario RCP4.5. Results show the five GCM 40 forcing models considered and the average value.

	RUNOFF						AVAILABILITY					
	GFDL	HadGEM2	IPSL	MIROC	NorEsH1	AVERAGE	GFDL	HadGEM2	IPSL	MIROC	NorEsH1	AVERAGE
Albania	−0.06	−0.16	−0.12	−0.09	−0.22	−0.13	0.02	0.09	−0.19	−0.22	0.02	−0.06
Austria	−0.02	0.00	−0.13	0.06	0.01	−0.02	0.22	−0.08	0.03	−0.15	0.03	0.01
Belgium	−0.07	0.07	−0.13	−0.04	0.07	−0.02	−0.08	0.00	0.16	−0.08	−0.03	−0.01
Bosnia and Herzegovina	−0.07	−0.12	−0.18	−0.03	−0.08	−0.10	0.34	−0.02	−0.32	−0.24	0.29	0.01
Bulgaria	−0.18	−0.24	−0.18	−0.17	−0.35	−0.22	0.04	0.13	−0.10	−0.38	−0.16	−0.09
Belarus	−0.12	0.19	−0.46	0.16	0.04	−0.04	0.08	0.28	−0.01	0.16	−0.03	0.10
Croatia	−0.09	−0.10	−0.19	0.00	−0.03	−0.08	0.17	−0.10	−0.09	−0.21	0.22	0.00
Czech Republic	−0.12	−0.02	−0.19	0.06	0.02	−0.05	−0.03	0.07	0.02	0.00	0.27	0.06
Denmark	−0.08	0.16	−0.15	0.05	0.21	0.04	0.00	0.00	0.00	0.00	0.00	0.00
Estonia	−0.14	0.21	−0.18	0.12	0.27	0.06	0.00	0.00	0.00	0.00	0.00	0.00
Finland	−0.17	0.19	0.00	0.10	0.22	0.07	0.01	−0.03	0.40	0.22	0.41	0.20
France	−0.06	−0.06	−0.14	−0.05	−0.06	−0.07	0.01	−0.19	0.07	−0.23	−0.08	−0.08
Germany	−0.06	0.07	−0.16	0.02	0.07	−0.01	0.17	0.04	0.07	−0.03	0.22	0.09
Greece	−0.06	−0.22	−0.13	−0.21	−0.32	−0.19	0.03	−0.08	−0.22	−0.16	−0.14	−0.11
Hungary	−0.14	−0.08	−0.21	−0.01	0.07	−0.07	0.49	−0.09	−0.06	−0.43	−0.03	−0.02
Ireland	−0.05	0.10	−0.05	0.00	0.04	0.01	−0.33	0.03	0.15	−0.28	−0.32	−0.15
Italy	−0.04	−0.13	−0.12	−0.04	−0.08	−0.08	0.04	−0.06	−0.18	−0.02	−0.04	−0.05
Latvia	−0.13	0.23	−0.30	0.11	0.37	0.06	0.80	−0.42	−0.31	−0.42	−0.07	−0.08
Lithuania	−0.11	0.22	−0.29	0.11	0.30	0.05	0.38	0.19	−0.13	−0.04	−0.20	0.04
Luxembourg	−0.08	0.07	−0.12	−0.04	0.03	−0.03	0.17	−0.10	0.05	−0.15	0.05	0.00
Macedonia	−0.14	−0.23	−0.21	−0.17	−0.39	−0.23	−0.03	0.07	−0.20	−0.40	−0.14	−0.14
Moldova	−0.14	−0.15	−0.16	0.09	0.00	−0.07	0.28	0.01	0.91	−0.27	0.02	0.19
Montenegro	−0.06	−0.11	−0.08	−0.03	−0.12	−0.08	0.23	0.13	−0.28	−0.17	−0.01	−0.02
Netherlands	−0.05	0.10	−0.11	−0.03	0.13	0.01	−0.07	0.05	−0.12	−0.10	0.02	−0.04
Norway	0.02	0.13	0.04	0.20	0.22	0.12	0.15	−0.01	0.20	0.10	0.14	0.12
Poland	−0.17	0.05	−0.31	0.05	0.14	−0.05	−0.01	−0.07	0.01	0.13	0.05	0.02
Portugal	−0.18	−0.26	−0.08	−0.08	−0.39	−0.20	−0.19	−0.06	−0.15	−0.47	−0.20	−0.21
Romania	−0.13	−0.15	−0.15	−0.03	−0.14	−0.12	0.24	−0.08	0.15	−0.30	0.07	0.01
Russia	0.02	0.23	0.21	0.14	0.23	0.16	0.18	0.20	0.22	0.26	0.16	0.20
Serbia	−0.19	−0.20	−0.24	−0.13	−0.25	−0.20	0.20	0.11	0.09	−0.36	0.19	0.05
Slovakia	−0.12	−0.02	−0.20	0.01	0.03	−0.06	−0.06	−0.09	0.07	−0.25	−0.05	−0.07
Slovenia	−0.01	−0.03	−0.12	0.05	0.04	−0.02	0.15	−0.12	−0.10	−0.26	0.03	−0.06
Spain	−0.07	−0.23	−0.17	−0.12	−0.25	−0.17	−0.26	−0.14	−0.14	−0.35	−0.19	−0.22
Sweden	−0.13	0.15	−0.12	0.05	0.22	0.04	0.04	0.02	0.26	0.04	0.00	0.07
Switzerland	0.00	0.00	−0.13	−0.01	0.01	−0.03	0.23	−0.14	0.02	−0.09	0.09	0.02
Ukraine	−0.05	−0.04	−0.18	0.18	−0.03	−0.02	−0.04	0.13	0.14	0.00	−0.04	0.04
United Kingdom	−0.05	0.10	−0.05	0.02	0.10	0.02	−0.26	0.07	−0.08	0.32	−0.11	−0.01

Water availability values were routed through the river network producing the map shown in Fig. 3. The map shows the projected changes of water availability along the course of the main European rivers for the long term period under emission scenario RCP4.5 compared to the control period for the GFDL climate forcing. The projected changes with respect to current conditions are obtained through the comparison of the control and the climate change situation with an expression similar to Eq. (1). The results are patchier than in the case of runoff, due to the distortion introduced by the irregular distribution of reservoir storage and the changes in flow regime, which changes not only in mean but also in variability. Water availability may be more directly affected by changes in inter-annual or seasonal variability than by changes in mean annual runoff, especially if reservoir storage is small or inexistent.

3 Risk of water scarcity

We estimate the risk of water scarcity from changes in runoff and changes in water availability. The inherent assumption is that in countries under risk of water scarcity there is currently an approximate equilibrium between water availability and water demand, since water availability has been ob-

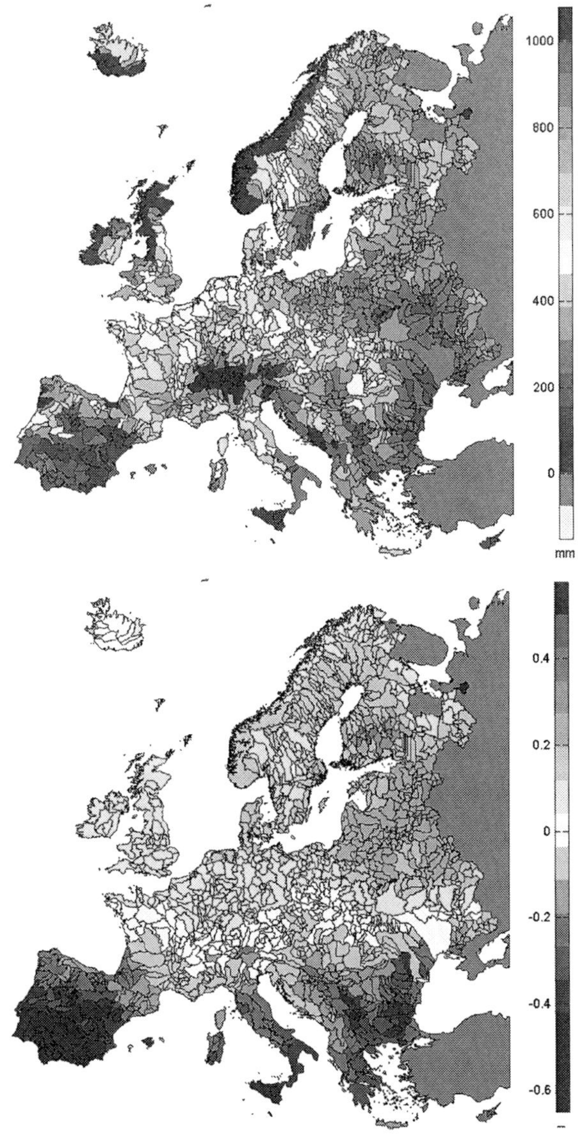

Figure 1. Subbasin average runoff (in mm) in the control period (1960–1999) and relative changes in the long term period (2060–2099) under emission scenario RCP4.5 according to PCR-GLOBWB model forced with GFDL climate data.

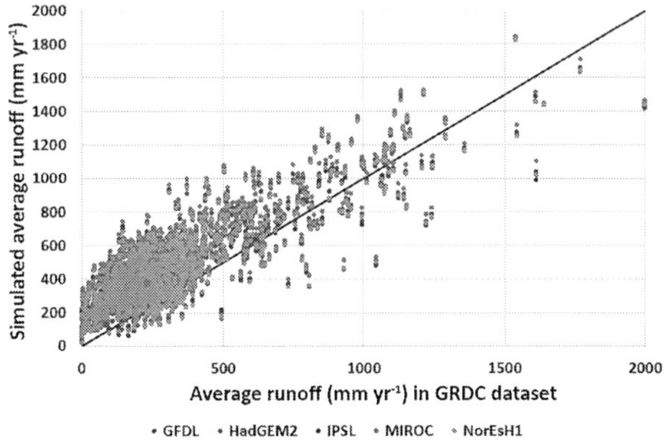

Figure 2. Comparison of GRDC average runoff with simulated average runoff in subbasins according to PCR-GLOBWB model forced in the control period (1960–1999) with the five climate models.

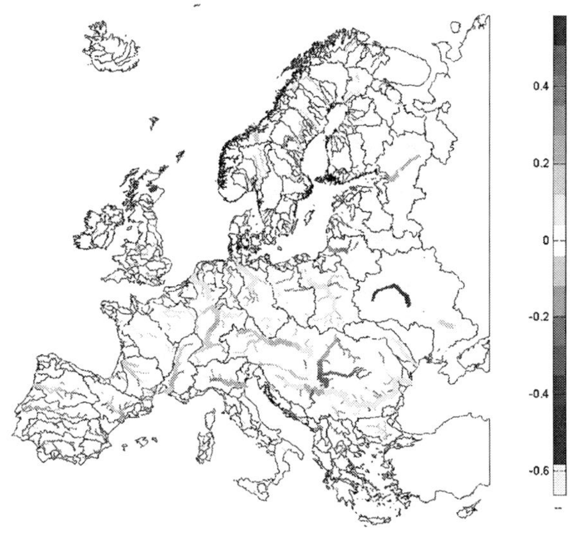

Figure 3. Relative changes of water availability in the long term period (2060–2099) with respect to the control period (1960–1999) under emission scenario RCP4.5 according to PCR-GLOBWB model forced with GFDL climate data.

tained through the progressive development of infrastructure to balance estimated demand. Assuming that we start from an equilibrium condition, we hypothesize that risk of water scarcity is linked to potential reduction of runoff and water availability.

3.1 Rationale for the analysis

The basis for the analysis is the scatter plot of changes in runoff (ΔR) versus changes in water availability (ΔWA) for all basins under analysis. An example is shown in Fig. 4, corresponding to changes in the long term period (2060–2099) with respect to the control period (1960–1999) under emission scenario RCP4.5 with GFDL forcing. This figure shows results for subbasins (lighter colors) and global basins draining to the sea (darker colors).

Four quadrants are identified in Fig. 4. In the first quadrant, both runoff and water availability changes are positive ($\Delta R > 0$ and ΔWa > 0). In this case, there is little risk of water scarcity unless water demands grow more than water availability. In the second quadrant, changes of water availability are positive, but changes in runoff are negative ($\Delta R < 0$ and ΔWA > 0). As discussed above, this can hap-

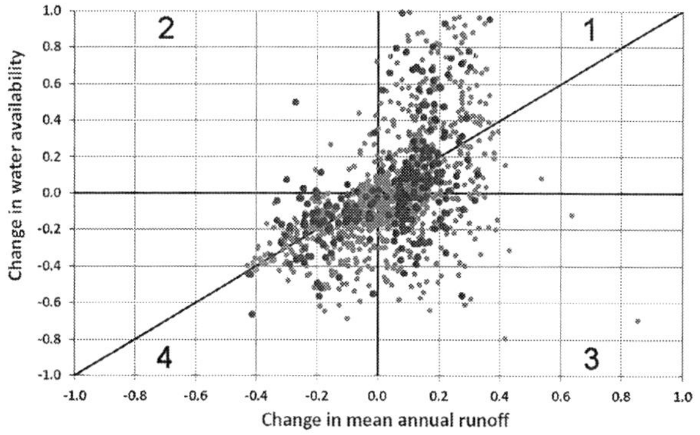

Figure 4. Scatter plot of relative changes in runoff and water availability in the long term period (2060–2099) with respect to the control period (1960–1999) under emission scenario RCP4.5 according to PCR-GLOBWB model forced with GFDL climate data. Light dots correspond to individual subbasins and darker dots correspond to global basins draining to the sea.

pen for certain changes in the flow regime. In this case there is a mild risk of water scarcity because the reduction of runoff might stress the water supply system. In the third quadrant, changes in water availability are negative although changes in runoff are positive ($\Delta R > 0$ and $\Delta\text{WA} < 0$). This corresponds to a significant risk of water scarcity. Finally, in the fourth quadrant, both water availability and runoff experience negative changes ($\Delta R < 0$ and $\Delta\text{WA} < 0$). This is the condition with greatest risk of water scarcity.

The above analysis can be applied to an individual simulation. If there is an ensemble of realizations corresponding to different models, the different results obtained should be summarized in one single index.

3.2 Water scarcity index

We propose a water scarcity risk index based on two factors: the risk corresponding to average conditions I_{AVE} and an uncertainty factor I_{UN}.

The risk corresponding to average conditions I_{AVE} is obtained from the position of the point representing the average of the simulations performed in the scatter plot of changes in runoff and changes in water availability. Its value is the product of two factors, following Eq. (2):

$$I_{\text{AVE}} = I_{\text{QUAD}}\, I_{\text{DIST}}, \qquad (2)$$

Where I_{QUAD} is a risk factor depending on the quadrant and I_{DIST} is a risk factor depending on the distance to the origin of coordinates. I_{QUAD} is given relative values depending on the position of the point representing the average values of the simulations. The basic idea is that a decrease in water availability is a positive risk and an increase a negative risk. If the reduction of availability occurs in conjunction with a reduction of runoff, the risk is considered to be worse than if runoff is increasing. A similar criterion was applied to the increase in water availability. For points in the first quadrant ($\Delta R > 0$ and $\Delta\text{WA} > 0$) I_{QUAD} takes a value of 2, for points in the second quadrant ($\Delta R < 0$ and $\Delta\text{WA} > 0$) it takes a value of 1, for points in the third quadrant ($\Delta R > 0$ and $\Delta\text{WA} < 0$) it takes a value of −1 and for points in the fourth quadrant ($\Delta R < 0$ and $\Delta\text{WA} < 0$) it takes a value of −2.

Table 2. Computation of the water scarcity risk index for emission scenario RCP4.5 in the long term (2060–2099) period.

	Quadrant	Distance	Average Risk	Unvertainty	Final Risk
Albania	−2	0.7	−1	1.3	−2
Austria	1	0.1	0	1.4	0
Belgium	−2	0.1	0	1.6	0
Bosnia and Herzegovina	1	0.5	0	0.7	0
Bulgaria	−2	1.2	−2	1.0	−2
Belarus	1	0.5	1	0.7	0
Croatia	−2	0.4	−1	1.0	−1
Czech Republic	1	0.4	0	1.4	1
Denmark	−1	0.2	0	1.4	0
Estonia	−1	0.3	0	1.0	0
Finland	2	1.1	2	0.8	2
France	−2	0.6	−1	1.6	−2
Germany	1	0.5	0	1.4	1
Greece	−2	1.1	−2	1.6	−3
Hungary	−2	0.4	−1	0.7	−1
Ireland	−1	0.8	−1	0.9	−1
Italy	−2	0.5	−1	2.5	−3
Latvia	−1	0.5	0	0.4	0
Lithuania	2	0.3	1	0.6	0
Luxembourg	1	0.1	0	1.3	0
Macedonia	−2	1.3	−3	1.0	−3
Moldova	1	1.0	1	0.5	1
Montenegro	−2	0.4	−1	1.0	−1
Netherlands	−1	0.2	0	1.6	0
Norway	2	0.8	2	1.6	3
Poland	1	0.3	0	1.0	0
Portugal	−2	1.5	−3	1.1	−3
Romania	1	0.6	1	1.0	1
Russia	2	1.3	3	2.2	6
Serbia	1	1.0	1	1.0	1
Slovakia	−2	0.5	−1	1.5	−1
Slovenia	−2	0.3	−1	1.2	−1
Spain	−2	1.4	−3	1.7	−5
Sweden	2	0.4	1	1.1	1
Switzerland	1	0.2	0	1.3	0
Ukraine	1	0.2	0	1.3	0
United Kingdom	−1	0.1	0	0.9	0
Range	2	1.5	3	2.5	6
Range	−2	0.1	−3	0.4	−5

Table 3. Country estimates of water scarcity risk index for emission scenarios RCP4.5 and RCP8.5 in the short term (ST: 2020–2059) and long term (LT: 2060–2099) periods.

	RCP4.5-ST	RCP4.5-LT	RCP8.5-ST	RCP8.5-LT
Albania	−2	−2	−3	−4
Austria	0	0	0	−1
Belgium	−2	0	−1	−1
Bosnia and Herzegovina	−1	0	1	−3
Bulgaria	−3	−2	−2	−4
Belarus	0	0	1	−1
Croatia	−1	−1	1	−3
Czech Republic	0	1	0	−1
Denmark	0	0	0	0
Estonia	0	0	0	0
Finland	2	2	1	1
France	−2	−2	−2	−5
Germany	0	1	0	0
Greece	−3	−3	−3	−4
Hungary	0	−1	0	−3
Ireland	−1	−1	−2	−1
Italy	−3	−3	2	−5
Latvia	−1	0	0	0
Lithuania	0	0	0	0
Luxembourg	−1	0	−1	−1
Macedonia	−2	−3	−2	−5
Moldova	0	1	1	1
Montenegro	1	−1	1	−2
Netherlands	−1	0	−1	−1
Norway	3	3	3	3
Poland	−1	0	0	−1
Portugal	−4	−3	−2	−5
Romania	−1	1	1	−4
Russia	3	6	2	3
Serbia	−2	1	1	−2
Slovakia	−1	−1	−2	−2
Slovenia	0	−1	1	−2
Spain	−5	−5	−2	−7
Sweden	0	1	0	1
Switzerland	0	0	1	−1
Ukraine	1	0	1	−3
United Kingdom	0	0	0	0
Russia	−1	0	0	−1
Turkey	−2	−2	−2	−3
Range	3	6	3	3
Range	−5	−5	−3	−7

The second factor, I_{DIST}, is obtained through the expression presented on Eq. (3):

$$I_{\mathrm{DIST}} = w\sqrt{\Delta R^2 + \Delta \mathrm{WA}^2}, \qquad (3)$$

Where w is a weighting factor that in our case is set to 5. This value was assigned to obtain a range of I_{AVE} that would allow for a useful classification of countries according to their risk. I_{DIST} represents the square of the distance from the origin of coordinates to the point representing the average of the simulations performed in the scatter plot of changes in runoff and changes in water availability.

The uncertainty factor I_{UN} is obtained from the comparison of all simulations performed in the ensemble. This factor will be high if all realizations provide similar results, reinforcing the average result, and will be low if the realizations disagree, weakening the average result. This factor is obtained from the sum of the distances of the points representing individual realizations to the point representing the average result. Its expression is shown in Eq. (4):

$$I_{\mathrm{DIST}} = \sum_{i=1}^{i=n} w\sqrt{(\Delta R - \Delta R_i)^2 + (\Delta \mathrm{WA} - \Delta \mathrm{WA}_i)^2} \qquad (4)$$

Where n is the number of realizations and ΔR_i and $\Delta \mathrm{WA}_i$ are the changes of runoff and water availability corresponding to realization i.

4 Country-based water scarcity index

The water scarcity index described in the previous section was applied using countries as the spatial units of analysis. Country averaging is presented first, and then the procedure to obtain the water scarcity index is described.

4.1 Country averages

Country averaging introduces some difficulty in the analysis, since several countries may use water that is generated in upstream countries. The solution adopted was to consider locally available water first and add the possibility of using external water (coming from upstream cases) in certain countries.

The averaging was performed in two steps. In a first step, a dataset of country averages was produced for the runoff and incremental water availability variables for each of the hypotheses analyzed (model, emission scenario and time horizon). For example, if we consider RCP4.5 and time horizon 2020–2059, the country averages obtained for the changes of runoff and water availability variables are presented in Table 1. In the second step, significant water flows across boundaries were accounted by increasing water availability of downstream countries accordingly.

The results in terms of water availability do not always follow the pattern of runoff. Water availability is the result not only of mean annual runoff, but also of storage and variability. Climate scenarios imply changes in streamflow variability that affect water availability. Furthermore, the approach is clearly oversimplified in terms of spatial resolution and analytical methodology and has important limitations.

The results presented in Table 1 are also shown graphically in Fig. 5, where the scatter plot of changes in runoff versus

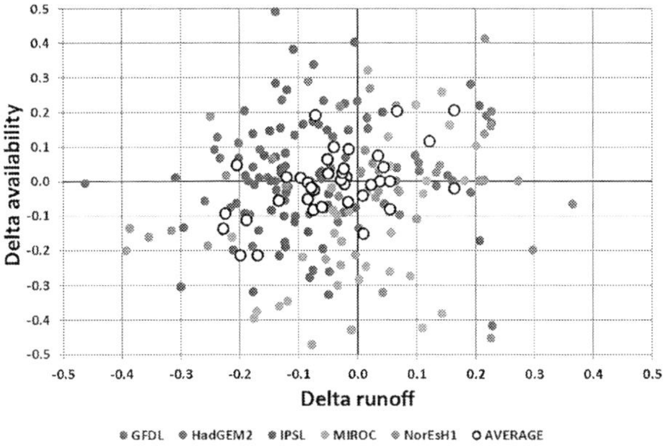

Figure 5. Country estimates of changes in average runoff and surface water availability in the long term period (2060–2099) with respect to the control period (1960–1999) under emission scenario RCP4.5. Results show the five GCM forcing models considered and the average value.

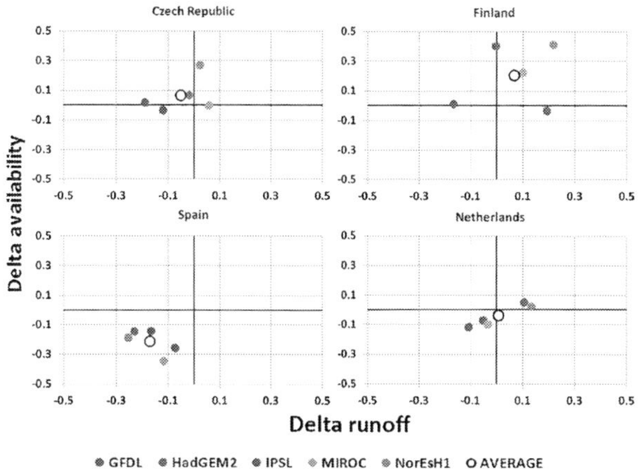

Figure 6. Changes in annual runoff versus water availability for the model ensemble average (circles) and the individual GCMs (coloured dots) for RCP4.5 long term.

changes in water availability is presented at the country level. Individual model simulations are presented as coloured dots, while average values are presented as large white dots. Figure 6 presents examples of four individual countries, one in each of the four quadrants: Finland (first quadrant), Czech Republic (second quadrant), Netherlands (third quadrant) and Spain (fourth quadrant).

4.2 Index computations

The water scarcity indices were computed for all countries following the methodology described above. The detailed computation is shown in Table 2 for the emission scenario RCP4.5 and time horizon 2060–2099.

The final water scarcity risk index ranges from 6 to −5. The country with largest water scarcity risk is Spain (−5), followed by Portugal, Italy, Greece and Macedonia, all with a risk index of −3. These are all Southern European countries that are already exposed to water scarcity to some extent. There are more countries in the fourth quadrant, but with less risk of water scarcity, like France, Albania or Bulgaria. These countries are exposed to a reduction of both runoff and water availability, but the intensity of the reduction is comparatively smaller.

The final results obtained for the two emission scenarios (RCP4.5 and RCP8.5) and time periods (short term and long term) are presented in Table 3. The countries at larger risk are (in this order) Spain, Portugal, Macedonia, Greece, Bulgaria, Albania, France and Italy. All these countries show consistently negative values of the water scarcity index in the four scenarios analysed. They are mostly Mediterranean countries already exposed to significant water scarcity problems. Projections for them are negative, which means that significant adaptation measures are unavoidable.

There are countries, like Slovakia, Belgium, Ireland, Croatia, Luxembourg, Romania, Netherlands, Hungary and Bosnia, with mild risk. Northern Arctic countries, like Sweden, Finland and Norway, show a robust however mild increase in water availability.

5 Conclusions

A regional assessment of water scarcity under climate change has been presented for Europe. The assessment is based on the water scarcity index, which is computed from the simulation results of changes in runoff and changes in water availability. The index summarizes the results obtained with several realizations under a certain emission scenario. It is based on two factors, one accounting for the average results and other accounting for the associated uncertainty.

The application of the water scarcity index to Europe provided a comparative picture of the exposure to risk of water scarcity induced by climate change in the countries of Europe. Countries at larger risk are those located in Southern Europe, particularly in the South East and in the South West.

Data availability. Basic data for this work were obtained from publicly accessible databases. Topographic data were downloaded from the GTOPO HYDRO1k dataset, accessible in USGS Earth Explorer (https://earthexplorer.usgs.gov/), Entity ID: Entity ID:GT30H1KEU. Runoff data from the ISIMIP project were downloaded from the PIK node of the Earth System Grid Federation (ESGF) (https://esg.pik-potsdam.de/search/isimip-ft/), selecting Impact Model "PCR-GLOBWB", Variable "Total Runoff" and Social Forcing "nosoc" (naturalized flows). Reservoir data were obtained from the ICOLD World Register of Dams (http://www.icold-cigb.net/GB/world_register/world_register_of_dams.asp), which is licensed for a fee.

Competing interests. The authors declare that they have no conflict of interest.

Special issue statement. This article is part of the special issue "Innovative water resources management – understanding and balancing interactions between humankind and nature". It is a result of the 8th International Water Resources Management Conference of ICWRS, Beijing, China, 13–15 June 2018.

Acknowledgements. We acknowledge the financial support of the European Commission BASE project (grant agreement no.: ENV-308337) of the 7th Framework Programme and the ADAPTA project, funded by Universidad Politecnica de Madrid.

We acknowledge the World Climate Research Programme's Working Group on Regional Climate, and the Working Group on Coupled Modelling, former coordinating body of CORDEX and responsible panel for CMIP5. We also thank the climate modelling groups of models PCR-GLOBWB, GFDL-ESM2NM, HadGEM2-ES, IPSL-CM5A-LR, MIROC-ESM-CHEM and NorESM1-M for producing and making available their model output.

Edited by: Dingzhi Peng
Reviewed by: Hua Chen and one anonymous referee

References

Alcamo, J., Henrichs, T., and Rosch, T.: World Water in 2025: Global modeling and scenario analysis for the World Commission on Water for the 21st Century, Kassel World Water Series Report No. 2, CESR, Germany: University of Kassel, 1–49, 2000.

Alcamo, J., Döll, P., Henrichs, T., Kaspar, F., Lehner, B., Rösch, T., and Siebert, S.: Global estimates of water withdrawals and availability under current and future "business-as-usual" conditions, Hydrol. Sci., 48, 339–348, 2003a.

Alcamo, J., Döll, P., Henrichs, T., Kaspar, F., Lehner, B., Rösch, T., and Siebert, S.: Development and testing of the WaterGAP 2 global model of water use and availability, Hydrol. Sci., 48, 317–337, 2003b.

Alcamo, J., Floerke, M., and Maerker, M.: Future long-term changes in global water resources driven by socio-economic and climatic changes, Hydrol. Sci., 52, 247–275, 2007.

Arnell, N. W.: The Effect of Climate Change on Hydrological Regimes in Europe, Global Environ. Change, 9, 5–23,1999.

Arnell, N. W.: Climate change and global water resources: SRES emissions and socio-economic scenarios, Global Environ. Change, 14, 31–52, 2004.

Döll, P.: Impact of climate change and variability on irrigation requirements: A global perspective, Clim. Change, 54, 269–293, 2002.

EROS, USGS: HYDRO1k Elevation Derivative Database, Tech. rept., U.S. Geological Survey Center for Earth Resources Observation and Science (EROS), available at: https://lta.cr.usgs.gov/HYDRO1K, last access: 12 March 2018.

Fekete, B. M., Vörösmarty, C. J., and Grabs, W.: High-resolution fields of global runoff combining observed river discharge and simulated water balances, Global Biogeochem. Cy., 16, 1–6, 2002.

Fowler, H. J., Blenkinsop, S., and Tebaldi, C.: Linking climate change modelling to impacts studies: recent advances in downscaling techniques for hydrological modeling, Int. J. Climatol., 27, 1547–1578, 2007.

Fronzek, S. and Carter, T. R.: Assessing uncertainties in climate change impacts on resource potential for Europe based on projections from RCMs and GCMs, Clim. Change, 81, 357–371, 2007.

Garrote, L., Granados, A., and Iglesias, A.: Assessing Water Availability in Europe: a Comparative Study. Proceedings of the EWRA 2015 World Congress – Water Resources Management in a Changing World: Challenges and Opportunities, Istanbul, Turkey, 2015a.

Garrote, L., Iglesias, A., Granados, A., Mediero, L., and Martin-Carrasco, F.: Quantitative Assessment of Climate Change Vulnerability of Irrigation Demands in Mediterranean Europe, Water Resour. Manage., 29, 325–338, 2015b.

Gleick, P.: Global Freshwater Resources: Soft-Path Solutions for the 21st Century, Science, 302, 1524–1528, 2003.

ICOLD: World Register of Dams, International Commission on Large Dams, available at: http://www.icold-cigb.net/GB/world_register/world_register_of_dams.asp (last access: 14 September 2012), 2004.

IPCC: Chapter 3 of Climate Change 2014: Impacts, Adaptation and Vulnerability, Part A: Global and Sectoral Aspects, edited by: Field, C. B. and Barros, V. R., Cambridge University Press, 2014.

Sperna Weiland, F. C., van Beek, L. P. H., Weerts, A. H., and Bierkens, M. F. P.: Extracting information from an ensemble of GCMs to reliably assess future global runoff change, J. Hydrol., 412–413, 66–75, 2012.

van Beek, L. P. H. and Bierkens, M. F. P.: The Global Hydrological Model PCR-GLOBWB: Conceptualization, Parameterization and Verification Utrecht University, Faculty of Earth Sciences, Department of Physical Geography, Utrecht, The Netherlands, 2009.

Warszawski, L., Frieler, K., Huber, V., Piontek, F., Serdeczny, O., and Schewe, J.: The Inter-Sectoral Impact Model Intercomparison Project (ISI–MIP): Project framework, P. Natl. Acad. Sci. USA, 111, 3228–3232, 2014.

Wisser, D., Frolking, S., Douglas, E. M., Fekete, B. M., Vörösmarty, C. J., and Schumann, A. H.: Global irrigation water demand: Variability and uncertainties arising from agricultural and climate data sets, Geophys. Res. Lett., 34, L24408, https://doi.org/10.1029/2008GL035296, 2008.

AUTHOR INDEX

AUTHOR INDEX